PENGUIN BOOKS

THE SILENT DE

'This history of our subma..er
timed ... a wide-ranging s......................................l
written, authoritative and ....
stand the test of time ... its ....ves essential
insight into the nuclear ....David, *Daily Telegraph*

'The particular physical and psychological demands faced by
submariners as they operate in the "silent deep" inform this full and
vivid account of the history of the submarine service ... an account
remarkable for its range and detail, well illustrated and with excellent
maps ... Through all of this they give the submariners a voice,
admire their professionalism and commend their contribution to
national security' Sir Lawrence Freedman, *Financial Times*

'A central theme of this book, a comprehensive study of
British underwater operations since 1945, is that while the men
of the Submarine Service comprise only about a sixth of the
Royal Navy's strength they represent the spearhead. Their trade
involves all manner of perils, even when no enemy is in sight. The
authors' account of submarine accidents is grimly impressive ... The
authors' narrative of some Cold War intelligence patrols makes the
hairs curl ... Although this book is not an official history, it
possesses the authority of one' Max Hastings, *Sunday Times*

'Peppered through the book are confrontations between
British and Russian submarines, some revealed for the first
time ... Hennessy and Jinks raise the submarine service from
its dark lurking place and put it convincingly at the centre of our
modern history and present politics. *The Silent Deep* provides
microscopic analysis of the political storms created by the nuclear
deterrent since the 1950s and it will be a key text as politicians
wrestle with the successor to Trident' Ben Wilson, *The Times*

'Must be the ultimate trainspotter's guide to how Britain's
submarines and submariners work – and have done so since 1945.
They write with the enthusiasm of addicts ... In 1982 the Navy's
submarines were suddenly pitched into a hot war in the Falklands.
This books gives one of the most lucid accounts of the saga ... the
book is radiated by Hennessy's characteristic verve and wit ... there
is so much in this bizarrely entertaining book to make one really
stop and think' Robert Fox, *Evening Standard*

'As you read these words over Christmas, at least one British nuclear submarine is patrolling. Those that say Trident will never be used miss the point: it is deterring every minute of every day. That the UK has been doing so for 47 uninterrupted years is an extraordinary feat. In *The Silent Deep*, Peter Hennessy and James Jinks chart these extraordinary feats, and more, and the high politics in the continuing story of the Royal Navy's submarine service. Most of all, the book charts a labour of love, paying tribute to the fearless professionalism of the submariners upon whose shoulders this mighty undertaking rests'
Admiral Sir George Zambellas, Former First Sea Lord

'A great combination of meticulous historical research, as one would expect of Peter Hennessy, delving into the more secret parts of the government's past decision making. In addition to being about the serious business of the defence of the realm, through Cold-War threats to our security, it also deals with the sheer bravery and professionalism of the most hidden of our armed services and discusses policy issues that matter as much going forward as they did in 1945. An essential read as Labour launches its defence review'
Maria Eagle MP, Shadow Secretary of State for Culture, Media and Sport and former Shadow Secretary of State for Defence

ABOUT THE AUTHORS

Peter Hennessy, one of Britain's best-known historians, is Attlee Professor of History at Queen Mary, Univeristy of London. He is the author of *Never Again: Britain 1945–51* (winner of the NCR and Duff Cooper Prizes), the bestselling *The Prime Minister* and *The Secret State: Preparing For The Worst 1945–2010*. He was made an independent crossbench life peer in 2010.

James Jinks completed his PhD under Peter Hennessy at Queen Mary. His first book was *50 Years of the Polaris Sales Agreement*, commissioned by Her Majesty's Government to mark 50 years of Polaris. He is now at work on *A Very British Bomb*, a history of the British nuclear deterrent.

PETER HENNESSY AND JAMES JINKS

The Silent Deep

The Royal Navy Submarine Service since 1945

PENGUIN BOOKS

PENGUIN BOOKS

UK | USA | Canada | Ireland | Australia
India | New Zealand | South Africa

Penguin Books is part of the Penguin Random House group of companies
whose addresses can be found at global.penguinrandomhouse.com.

First published by Allen Lane 2015
Published in Penguin Books 2016
001

Copyright © Peter Hennessy and James Jinks, 2015

The moral right of the authors has been asserted

Set in 9.18/12.59 pt Sabon LT Std
Printed in Great Britain by Clays Ltd, St Ives plc

A CIP catalogue record for this book is available from the British Library

ISBN: 978-0-241-95948-0

Dedicated to the men and women of the Royal Navy Submarine Service since 1945.

Contents

Illustrations

Endpaper charts © Crown Copyright and/or database rights. Reproduced by permission of the Controller of Her Majesty's Stationery Office and the UK Hydrographic Office (www.ukho.gov.uk).

List of Maps

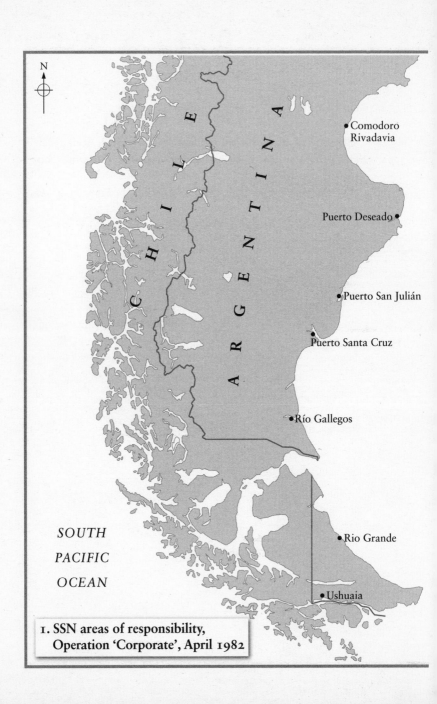

N

SOUTH
PACIFIC
OCEAN

Comodoro
Rivadavia

Puerto Deseado

Puerto San Julián

Puerto Santa Cruz

Río Gallegos

Rio Grande

Ushuaia

C H I L E

A R G E N T I N A

1. SSN areas of responsibility,
Operation 'Corporate', April 1982

SOUTH ATLANTIC

OCEAN

HMS *Spartan*

HMS *Splendid*

Falkland Islands

HMS *Conqueror*

0	50	100	150 miles
0	100	200 km	

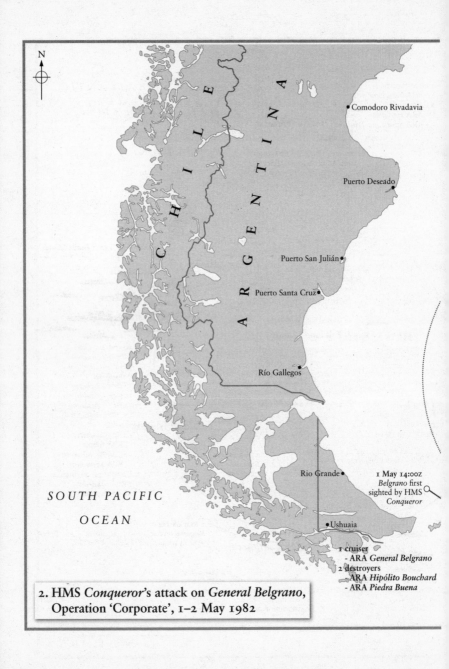

N

Comodoro Rivadavia

Puerto Deseado

CHILE

ARGENTINA

Puerto San Julián

Puerto Santa Cruz

Río Gallegos

SOUTH PACIFIC

OCEAN

Rio Grande

1 May 14:00z
Belgrano first
sighted by HMS
Conqueror

Ushuaia

1 cruiser
- ARA *General Belgrano*
2 destroyers
- ARA *Hipólito Bouchard*
- ARA *Piedra Buena*

2. HMS *Conqueror*'s attack on *General Belgrano*,
Operation 'Corporate', 1–2 May 1982

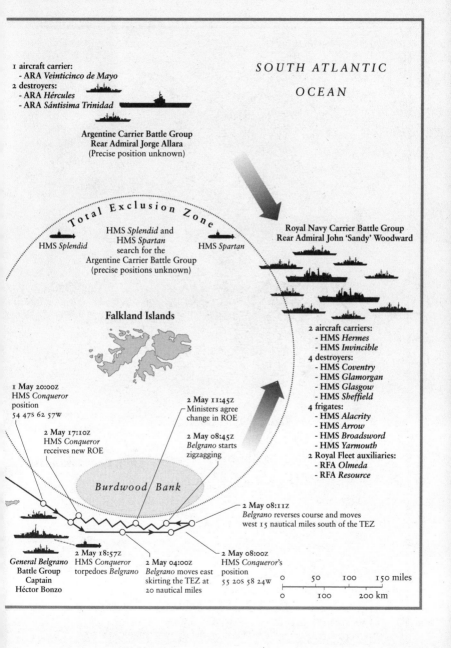

1 aircraft carrier:
- ARA *Veinticinco de Mayo*
2 destroyers:
- ARA *Hércules*
- ARA *Sántisima Trinidad*

**Argentine Carrier Battle Group
Rear Admiral Jorge Allara
(Precise position unknown)**

SOUTH ATLANTIC

OCEAN

Total Exclusion Zone

HMS *Splendid* and
HMS *Spartan*
search for the
Argentine Carrier Battle Group
(precise positions unknown)

HMS *Splendid*

HMS *Spartan*

**Royal Navy Carrier Battle Group
Rear Admiral John 'Sandy' Woodward**

Falkland Islands

2 aircraft carriers:
- HMS *Hermes*
- HMS *Invincible*
4 destroyers:
- HMS *Coventry*
- HMS *Glamorgan*
- HMS *Glasgow*
- HMS *Sheffield*
4 frigates:
- HMS *Alacrity*
- HMS *Arrow*
- HMS *Broadsword*
- HMS *Yarmouth*
2 Royal Fleet auxiliaries:
- RFA *Olmeda*
- RFA *Resource*

1 May 20:00Z
HMS *Conqueror*
position
54 47S 62 57W

2 May 11:45Z
- Ministers agree
change in ROE

2 May 17:10Z
HMS *Conqueror*
receives new ROE

2 May 08:45Z
Belgrano starts
zigzagging

Burdwood Bank

2 May 08:11Z
Belgrano reverses course and moves
west 15 nautical miles south of the TEZ

General Belgrano
Battle Group
Captain
Héctor Bonzo

2 May 18:57Z
HMS *Conqueror*
torpedoes *Belgrano*

2 May 04:00Z
Belgrano moves east
skirting the TEZ at
20 nautical miles

2 May 08:00Z
HMS *Conqueror*'s
position
55 20S 58 24W

0   50   100   150 miles

0   100   200 km

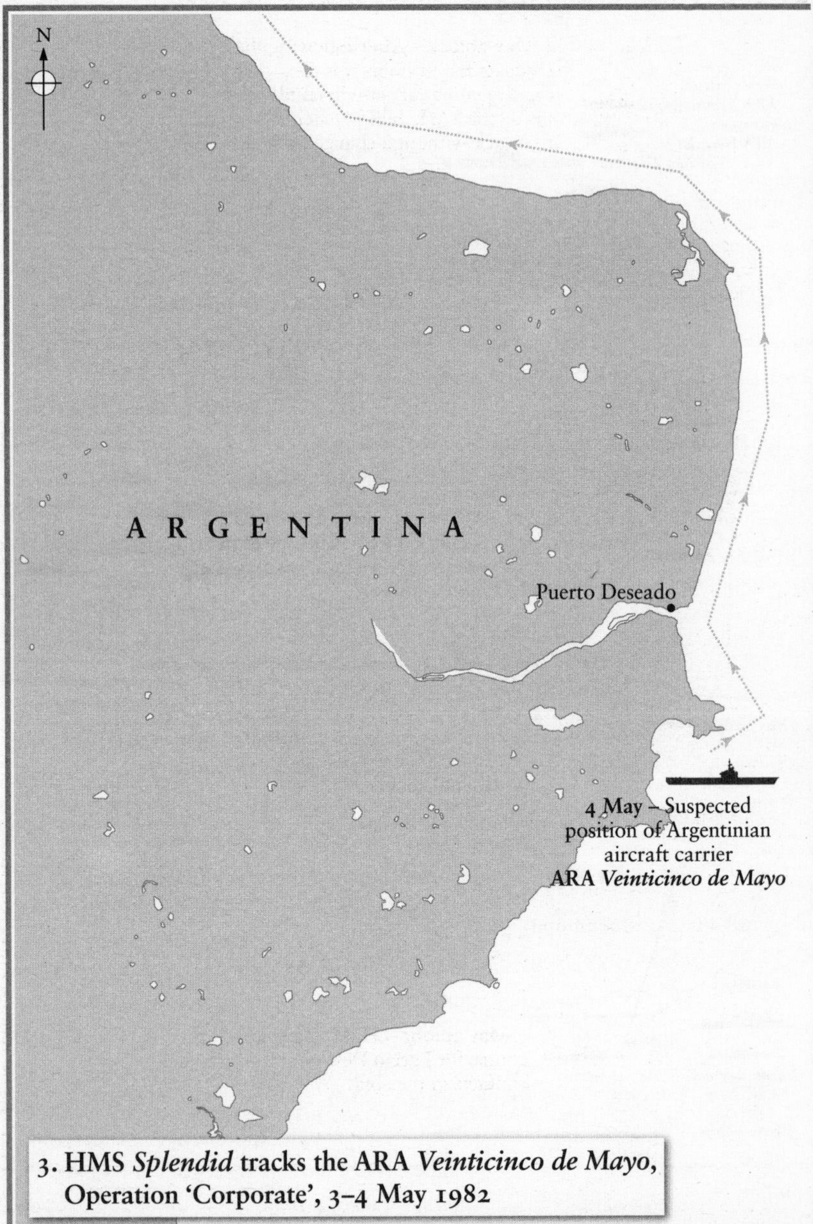

N

ARGENTINA

Puerto Deseado

4 May – Suspected
position of Argentinian
aircraft carrier
ARA *Veinticinco de Mayo*

3. HMS *Splendid* tracks the ARA *Veinticinco de Mayo*,
Operation 'Corporate', 3–4 May 1982

4 May 20:00Z – After following the contacts for 12 hours it is clear that the Argentine carrier will remain well inside the TML. HMS *Splendid* is unable to attack without a change in the ROE.

SOUTH ATLANTIC

OCEAN

4 May 08:00Z – HMS *Splendid* goes to periscope depth but is unable to identify the contacts due to poor visibility. Four contacts are detected heading north-west. HMS *Splendid* follows.

4 May 03:05Z – HMS *Splendid* detects Sonar contacts off Deseado but they are too far away to classify. The contacts, one of which Lane-Nott suspects is the ARA *Veinticinco de Mayo*, appear to be outside the TML. HMS *Splendid* signals Northwood and waits 8 miles outside the TML until sunrise.

3 May 20:00Z – HMS *Splendid* sets course for Puerto Deseado to investigate contacts in the area.

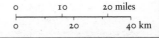

0		10		20 miles
0		20		40 km

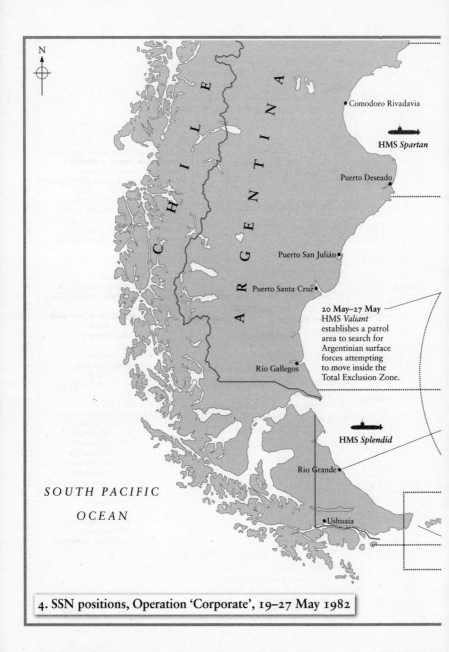

N

Comodoro Rivadavia

HMS *Spartan*

Puerto Deseado

Puerto San Julián

Puerto Santa Cruz

20 May–27 May
HMS *Valiant*
establishes a patrol
area to search for
Argentinian surface
forces attempting
to move inside the
Total Exclusion Zone.

Río Gallegos

HMS *Splendid*

Rio Grande

SOUTH PACIFIC

OCEAN

Ushuaia

C H I L E

A R G E N T I N A

4. SSN positions, Operation 'Corporate', 19–27 May 1982

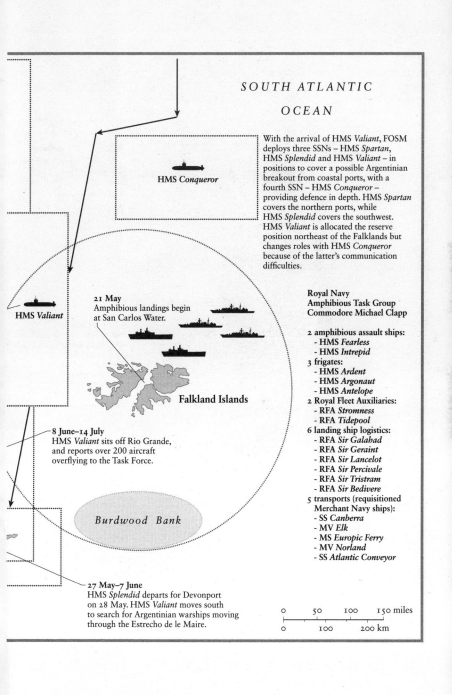

SOUTH ATLANTIC

OCEAN

HMS *Conqueror*

With the arrival of HMS *Valiant*, FOSM
deploys three SSNs – HMS *Spartan*,
HMS *Splendid* and HMS *Valiant* – in
positions to cover a possible Argentinian
breakout from coastal ports, with a
fourth SSN – HMS *Conqueror* –
providing defence in depth. HMS *Spartan*
covers the northern ports, while
HMS *Splendid* covers the southwest.
HMS *Valiant* is allocated the reserve
position northeast of the Falklands but
changes roles with HMS *Conqueror*
because of the latter's communication
difficulties.

HMS *Valiant*

21 May
Amphibious landings begin
at San Carlos Water.

Falkland Islands

8 June–14 July
HMS *Valiant* sits off Rio Grande,
and reports over 200 aircraft
overflying to the Task Force.

Burdwood Bank

27 May–7 June
HMS *Splendid* departs for Devonport
on 28 May. HMS *Valiant* moves south
to search for Argentinian warships moving
through the Estrecho de le Maire.

**Royal Navy
Amphibious Task Group
Commodore Michael Clapp**

2 amphibious assault ships:
- HMS *Fearless*
- HMS *Intrepid*
3 frigates:
- HMS *Ardent*
- HMS *Argonaut*
- HMS *Antelope*
2 Royal Fleet Auxiliaries:
- RFA *Stromness*
- RFA *Tidepool*
6 landing ship logistics:
- RFA *Sir Galahad*
- RFA *Sir Geraint*
- RFA *Sir Lancelot*
- RFA *Sir Percivale*
- RFA *Sir Tristram*
- RFA *Sir Bedivere*
5 transports (requisitioned
Merchant Navy ships):
- SS *Canberra*
- MV *Elk*
- MS *Europic Ferry*
- MV *Norland*
- SS *Atlantic Conveyor*

0	50	100	150 miles
0	100	200 km	

'I can call spirits from the vasty deep'

William Shakespeare,
Owen Glendower in *King Henry IV Part I*

'The Trade'

They bear, in place of classic names,
　Letters and numbers on their skin.
They play their grisly blindfold games
　In little boxes made of tin.
　Sometimes they stalk the Zeppelin,
Sometimes they learn where mines are laid,
　Or where the Baltic ice is thin.
That is the custom of 'The Trade.'

Few prize-courts sit upon their claims.
　They seldom tow their targets in.
They follow certain secret aims
　Down under, far from strife or din.
　When they are ready to begin
No flag is flown, no fuss is made
　More than the shearing of a pin.
That is the custom of 'The Trade.'

The Scout's quadruple funnel flames
　A mark from Sweden to the Swin,
The Cruiser's thund'rous screw proclaims
　Her comings out and goings in:

But only whiffs of paraffin
Or creamy rings that fizz and fade
　　Show where the one-eyed Death has been
That is the custom of 'The Trade.'

Their feats, their fortunes and their fames
　　Are hidden from their nearest kin;
No eager public backs or blames,
　　No journal prints the yarn they spin
　　(The Censor would not let it in!)
When they return from run or raid.
　　Unheard they work, unseen they win.
That is the custom of 'The Trade.'
　　　　　Rudyard Kipling, 'The Trade', *Sea Warfare, 1914–18*

Preface

Submarine Britain

From the very beginning at the start of the twentieth century there has been something special and mysterious about 'Submarine Britain'. In its time, it has stretched from the North Pole to the South Atlantic, the Far East to the Barents Sea. Scarcely a patch of the two thirds of the world's surface that is covered in water has escaped the presence of a Royal Navy submarine at some point over the last century. At home, too, the United Kingdom is girdled with the harbours, facilities, design offices, factories and research laboratories needed to keep the country at the top of the range of the world's submarine powers. Yet for all the mystique and fascination submarine life holds for many people and the political heat generated by the question of whether or not the UK should remain a nuclear-weapons power, very little of this human and physical infrastructure is known to the general public. Both the 'vasty deep' and the land life of the Queen's underwater servants remains very largely a mystery to their fellow countrymen and women.

For 114 years since the first tiny submarine ordered by the Royal Navy was launched at the Vickers Yard, Barrow, on the North Lancashire coast on 2 October 1901, successive British governments have sought to have a presence in the depths of the seas. In this task, the Royal Navy Submarine Service has been their instrument. Yet this famously 'silent service' has never fully gained the place it deserves in the wider historical sun of defence policy, intelligence history or the record of the UK as a nuclear-weapons power. To this day, the Ministry of Defence still responds to all enquiries about submarine operations – past and present – with a simple phrase: 'The Ministry of Defence does not comment on submarine operations.'

In recent years a series of unofficial books have sought to shed a glimmer of light on the 'deep Cold War' between the United States

Navy, the Royal Navy and the Soviet Union. There are two opposing views on revealing these activities, many of which remain some of the most closely held secrets of the British State. Those who favour declassification believe that because the Cold War ended over twenty-five years ago, submariners now deserve more formal recognition and that because there have been so many leaks any such secrecy has long since gone. The contrary view is that submarines of the former Soviet Union, as well as those of other nations, are still at sea, and nothing should be revealed that could jeopardize current and future submarine operations, many of which still involve tactics developed and perfected during the Cold War. The potential threat is still present. The 2013/14 edition of *Jane's Fighting Ships* lists seventeen operational nuclear-powered attack submarines (SSNs) in service with the Russian Northern Fleet, and a further seven more in the Pacific Fleet. The US Atlantic Fleet has twenty-three SSNs, France has four, China five, and India two. All deploy nuclear-powered ballistic missile submarines (SSBNs), armed with nuclear warheads. All have active building programmes.

Today the Royal Navy has just seven SSNs and four SSBNs. On present policies, Britain will remain a submarine nation at least until the 2050s, when the 'Successor' system to the Trident SSBNs as the sole carrier of Britain's nuclear-weapons capability will require replacing, as will the SSNs that protect them. The country will therefore still need small groups of carefully trained young men (and, since 2013, young women) to go silently and deeply into the cold, their lives shaped equally by intimacy and solitariness, in one of the strangest and most singular professions a British citizen can pursue. Their world spans the front line of national defence (surveillance and intelligence gathering) to the last line (nuclear retaliation as the country's near unthinkable 'last resort').

Like all secret services, the submariners have their crown jewels that cannot and should not be given up. Yet has the silent service been silent for too long? Is there a means of highlighting the Submarine Service's contribution to the Cold War, without jeopardizing national security? We believe there is. *The Silent Deep* is not an official history or an authorized history, but it has been prepared with a high level of cooperation from the Royal Navy, for which we are immensely grateful. This cooperation took the form of unprecedented access to documents, personnel and submarines, both operational and decommissioned, all of which has been used to reveal for the first time in significant detail the activities of the Submarine Service since the end of the Second World War.

For all that we have been able to write about, *The Silent Deep* is not a complete history. Many of the files held by the Ministry of Defence, even those covering operations that occurred over fifty years ago, are still too sensitive to release as they show to a high level of detail how certain submarine operations were carried out. Given that these techniques are broadly still used today they would be of great use to any adversary. We hope that this book gives 'Submarine Britain' as much of the visibility as it can safely acquire and that it manages to convey something of the flavour of the lives lived by those who penetrate the silent deep.

Peter Hennessy and James Jinks, June 2015

Abbreviations

The following abbreviations are used in the text:

ABJSM – Admiralty British Joint Service Mission Washington
ABM – Anti-Ballistic Missile
ACCHAN – Allied Command Channel
ACINT – Acoustic Intelligence
ACNS – Assistant Chief of the Naval Staff
AEC – United States Atomic Energy Commission
AERE – Atomic Energy Research Establishment, Harwell
AGI – Soviet Auxiliary General Intelligence
AMP – Assisted Maintenance Period
ANF – Atlantic Nuclear Force
ARS – Auxiliary Rescue Ship
ASDIC – Anti-Submarine Detection Investigation Committee
ASW – Anti-Submarine Warfare
AUTEC – Atlantic Undersea Test and Evaluation Center
AUWE – Admiralty Underwater Weapons Establishment
AWRE – Atomic Weapons Research Establishment
BAE – British Aerospace
BIBS – Built-in Breathing System
BNBMS – British Naval Ballistic Missile System
BNDSG – British Nuclear Deterrent Study Group
BUTEC – British Underwater Test and Evaluation Centre
CAD – Computer-Aided Design
CASD – Continuous at-Sea Deterrence
CDS – Chief of the Defence Staff
CEP – Contact Evaluation Plot
CINCFLEET – Commander-in-Chief Fleet
CINCNELM – Commander-in-Chief US Naval Forces
   Eastern Atlantic and Mediterranean
CINCWF – Commander-in-Chief Western Fleet
CND – Campaign for Nuclear Disarmament
CNO – Chief of Naval Operations

CO – Commanding Officer
COMINT – Communications Intelligence
COMOPS – UK Commander Maritime Operations
COMSUBEASTLANT – Commander Submarine Force,
  Eastern Atlantic
COMSUBLANT – Commander Submarine Force, Atlantic
COMSUBRON – US Navy, Commander, Submarine Squadron
CONMAROPS – Concept of Maritime Operations
CSA – Clear Stern Arcs
COR – Coded Order
CSSE – Chief Strategic Systems Executive
CTF – Commander Task Force
CTG – Commander Task Group
DASO – Demonstration and Shakedown Operation
DCA – Tactical Data Handling System
DIMUS – Digital Multi-beam Sonar
DIS – Defence Intelligence Staff
DOPC – Defence and Overseas Policy Committee
EASTLANT – Eastern Atlantic
EEC – European Economic Community
ELINT – Electronic Intelligence
EOP – Emergency Operating Procedure
ESM – Electronic Support Measures
EST – Eastern Standard Time
FOF1 – Flag Officer First Flotilla
FOSM – Flag Officer Submarines
GCHQ – Government Communications Headquarters
GIUK – Greenland–Iceland–United Kingdom
HE – Hydrophone Effect
HF – High Frequency
HMCS – His/Her Majesty's Canadian Ship
HMS – His/Her Majesty's Ship
HMS/m – His/Her Majesty's Submarine
HMUDE – Her Majesty's Underwater Detection
  Establishment
HTP – Hydrogen Peroxide
HUMINT – Human Intelligence
IBERLANT – Allied Forces Iberian Atlantic Area
ICEX – Ice Exercise
IRBM – Intermediate-Range Ballistic Missile

ISTAR – Intelligence Surveillance Tracking and Reconnaissance

IUSS – Integrated Undersea Surveillance System

JCAE – Joint Committee on Atomic Energy

JIC – Joint Intelligence Committee

JSTG – Joint Steering Task Group

LASS – Launch Area Support Ship

LEM – Leading Electrical Mechanic

LIFEX – Life Extension Programme

LOFAR – Low-Frequency Array

LOP(R) – Long Overhaul Period and Refuel

LRMP – Long-Range Maritime Patrol Aircraft

MAIB – Marine Accident Investigation Board

MDA – 1958 US/UK Mutual Defence Agreement

MEZ – Maritime Exclusion Zone

MIRV – Multiple Independently Targeted Re-Entry Vehicle

MLF – Multilateral Nuclear Force

MOD – UK Ministry of Defence

NATO – North Atlantic Treaty Organization

NIF – US National Ignition Facility

NM – Nautical Mile

NOFORN – Not Releasable to Foreign Nationals

NOPF – Naval Oceanographic Processing Facilities

NOTC – Nuclear Operations Targeting Centre

NOTU – Naval Ordnance Test Unit

OD(SA) – Overseas and Defence (South Atlantic) Committee of the Cabinet

OOW – Officer of the Watch

OPDC – Oversea Policy and Defence Committee of the Cabinet

ORAC – Operation Relentless Assurance Committee

PAC – Penetration Aid Carrier

PINDAR – UK Defence Crisis Management Centre

PM – Prime Minister

PSA – Polaris Sales Agreement

PSPJ – Pre-Swirl Pump Jet

PWR – Pressurized Water Reactor

QT35 – Quenched and Tempered Steel 35

RABA – Rechargeable Air Breathing Apparatus

RAF – Royal Air Force

RAM – Regulus Attack Missile

RAN – Royal Australian Navy

RCNC – Royal Corps of Naval Constructors
ROE – Rules of Engagement
RNSH – Royal Navy Sub Harpoon
SACEUR – Supreme Allied Commander Europe
SACLANT – Supreme Allied Commander Atlantic
SALT – Strategic Arms Limitation Treaty
SAS – Special Air Service
SBS – Special Boat Service
SCOG – Self-Contained Oxygen Generator
SCOSE – Standing Committee on Submarine Escape
SDR – Strategic Defence Review
SDSR – Strategic Defence and Security Review
SHAPE – Supreme Headquarters Allied Powers Europe
SIGINT – Signals Intelligence
SLAM – Submarine-Launched Air-Flight Missile
SLBM – Submarine-Launched Ballistic Missile
SLCM – Submarine-Launched Cruise Missile
SLOCS – Sea Lines of Communication
SMCS – Submarine Command System
SMP – Short Maintenance Period
SNCP – Special Naval Collection Programme
SOA – Speed of Advance
SONAR – Sound Navigation and Ranging
SOSUS – Sound Surveillance System
SP – Special Projects
SPA – SOSUS Probability Area
SPO – US Special Projects Office
SPRN – Special Projects Royal Navy
SSBN – Nuclear-Powered Ballistic Missile Submarine
SSGN – Nuclear-Powered Cruise Missile Submarine
SSGW – Surface-to-Surface Guided Weapon
SSIXS – Submarine Satellite Information Exchange System
SSK – Conventional Submarine
SSN – Nuclear-Powered Attack Submarine
SSPO – Strategic Systems Project Office
STRATCOM – United States Strategic Command
SUBICEX – Submarine Ice Exercise
SUBROC – Submarine Rocket
SURTASS – Surveillance Towed-Array Sensor System
SWFLANT – US Navy Strategic Weapons Facility Atlantic

S3W – Submarine 3rd Generation Westinghouse
S5W – Submarine 5th Generation Westinghouse
SWS – Strategic Weapons System
TEZ – Total Exclusion Zone
TF – Task Force
TLAM – Tomahawk Land Attack Missile
TML – Twelve-Mile Limit
TWCS – Tomahawk Weapons Control System
UHF – Ultra-High Frequency
UK – United Kingdom
USA – United States
USN – United States Navy
USS – United States Ship
USSR – Union of Soviet Socialist Republics
VCNS – Vice Chief of the Naval Staff
VHF – Very High Frequency
VLF – Very Low Frequency
VSEL – Vickers Shipbuilding and Engineering Ltd
WEO – Weapons Engineering Officer
XO – Executive Officer

Introduction

It's an extraordinarily attractive career. You endure hours of boredom waiting for those few minutes where you do something of real value for the Crown. More than anything it's the company you keep. We all come home or none of us come home.

Rear Admiral Simon Lister, Director Submarines,
20 December 2011.[1]

It's the morning of Sunday, 6 November 2011, exactly a week before the Remembrance Day service at the Cenotaph in Whitehall. The place: the Victoria Embankment alongside the River Thames between Waterloo Bridge and Blackfriars Bridge. The occasion: The Submariners' Association Annual Memorial Service at the National Submarine War Memorial, a ceremony that since 1921 has taken place a week before Remembrance Sunday to avoid a clash for those who have to be, or wish to be, at both.

We scribble a diary of the sights, sounds and impressions:

09.43

Grey, chilly November morning, the Thames high on the autumn tide. Current and retired submariners with wives, children and grandchildren walking towards HMS *President* [training ship of First World War vintage, permanently moored] at the Blackfriars Bridge end of the Embankment, some carrying cardboard boxes containing wreaths.

President has bedecked herself in flags. A lone Royal Marine bugler appears on deck. Medals and Dolphins [the golden badge of the Submarine Service] shine against the dark-blue uniforms. Some wear 'Deterrent Pins' [reflecting service on Polaris or Trident patrols]. Sailors' caps emblazoned with HM SUBMARINES.

400 yards upstream from *President*, on the great bend in the river between Westminster Bridge and Blackfriars Bridge with Waterloo Bridge in the middle, is the National Submarine War Memorial. City of London Police have diverted traffic to the east-bound lane, which now becomes two-way.

09.55

Full Band of the Royal Marines, Portsmouth, flows out onto the deck of HMS *President*.

10.00

Everyone forms up in the road. We chat to some of the legendary commanders from Cold War days, Admiral Lord Boyce, Vice Admiral Sir Tim McClement and Admiral Sir James Perowne. The Royal Marine Band strikes up a tune we don't recognize. They look magnificent in their white helmets (which they used to call 'pith pots'). They march impeccably at the head of the procession. A real lump in the throat moment.

We wait for an unannounced officer, who will take the parade, to sweep down the empty Embankment in his pennanted official car. It turns out to be COMOPS [Commander Operations], Rear Admiral Ian Corder (complete with sword). Suddenly, just before 'The Last Post' sounds, the traffic ceases. Very hushed apart from the rustling of leaves. A single seagull cries.

Numbers and names of the submarines lost in the Great War, the interwar years, the Second World War and post-1945 (four of them) are read out. <u>Very</u> moving. Wreaths are laid as the traffic resumes down one carriageway.

18 March 1904	HMS A1	8 June 1913	HMS E5
16 February 1905	HMS A5	10 December 1913	HMS C14
5 June 1905	HMS A8	16 January 1914	HMS A7
16 October 1905	HMS A4	14 September 1914	AE1 (Australian)
14 July 1909	HMS C11	18 October 1914	HMS E3
2 February 1912	HMS A3	3 November 1914	HMS D5
4 October 1912	HMS B2	25 November 1914	HMS D2

4 January 1915	HMS C31	3 April 1918	HMS E1
18 January 1915	HMS E10	3 April 1918	HMS E9
17 April 1915	HMS E15	3 April 1918	HMS E19
30 April 1915	AE2 (Australian)	4 April 1918	HMS E8
4 August 1915	HMS C33	4 April 1918	HMS C26
19 August 1915	HMS E13	5 April 1918	HMS C27
29 August 1915	HMS C29	5 April 1918	HMS C35
4 September 1915	HMS E7	23 April 1918	HMS C3
6 November 1915	HMS E20	28 June 1918	HMS D6
26 December 1915	HMS E6	27 July 1918	HMS E34
6 January 1916	HMS E17	3 October 1918	HMS L10
19 January 1916	HMS H6	6 October 1918	HMS C12
24 March 1916	HMS E24	15 October 1918	HMS J6
2 June 1916	HMS E18	1 November 1918	HMS G7
3 July 1916	HMS E26	22 November 1918	HMS G11
15 July 1916	HMS H3	4 June 1919	HMS L55
9 August 1916	HMS B10	18 October 1919	HMS H41
15 August 1916	HMS E4	20 January 1921	HMS K5
15 August 1916	HMS E41	25 June 1921	HMS K15
22 August 1916	HMS E16	23 March 1922	HMS H42
22 November 1916	HMS E30	18 January 1923	HMS L9
30 November 1916	HMS E37	10 January 1924	HMS L24
19 January 1917	HMS E36	12 November 1925	HMS M1
28 January 1917	HMS K13	9 August 1926	HMS H29
12 March 1917	HMS E49	9 July 1929	HMS H47
17 March 1917	HMS A10	9 June 1931	HMS *Poseidon*
16 April 1917	HMS C16	26 January 1932	HMS M2
17 July 1917	HMS C34	1 June 1939	HMS *Thetis*
20 August 1917	HMS E47	10 September 1939	HMS *Oxley*
16 September 1917	HMS G9	7 January 1940	HMS *Undine*
22 October 1917	HMS C32	7 January 1940	HMS *Seahorse*
18 November 1917	HMS K1	9 January 1940	HMS *Starfish*
14 January 1918	HMS G8	10 April 1940	HMS *Thistle*
18 January 1918	HMS H10	10 April 1940	HMS *Tarpon*
28 January 1918	HMS E14	18 April 1940	HMS *Sterlet*
31 January 1918	HMS K17	29 April 1940	HMS *Unity*
31 January 1918	HMS K4	5 May 1940	HMS *Seal*
1 February 1918	HMS E50	14 June 1940	HMS *Odin*
2 March 1918	HMS H5	16 June 1940	HMS *Grampus*
12 March 1918	HMS D3	16 June 1940	HMS *Orpheus*

6 July 1940	HMS *Shark*	25 November 1942	HMS *Utmost*
9 July 1940	HMS *Salmon*	4 December 1942	HMS *Traveller*
16 July 1940	HMS *Phoenix*	12 December 1942	HMS P222
23 July 1940	HMS *Narwhal*	25 December 1942	HMS P48
1 August 1940	HMS *Oswald*	2 January 1943	HMS P311
2 August 1940	HMS *Spearfish*	24 February 1943	HMS *Vandal*
3 August 1940	HMS *Thames*	27 February 1943	HMS *Tigris*
10 October 1940	HMS *Rainbow*	14 March 1943	HMS
15 October 1940	HMS *Triad*		*Thunderbolt*
18 October 1940	HMS H49	14 March 1943	HMS *Turbulent*
7 November 1940	HMS *Swordfish*	18 April 1943	HMS *Regent*
25 November 1940	HMS *Regulus*	18 April 1943	HMS P615
6 December 1940	HMS *Triton*	21 April 1943	HMS *Splendid*
11 February 1941	HMS *Snapper*	24 April 1943	HMS *Sahib*
26 April 1941	HMS *Usk*	30 May 1943	HMS *Untamed*
12 May 1941	HMS	7 August 1943	HMS *Parthian*
	Undaunted	14 August 1943	HMS *Saracen*
19 July 1941	HMS *Umpire*	16 September 1943	X9
20 July 1941	HMS *Union*	18 September 1943	X8
30 July 1941	HMS *Cachalot*	22 September 1943	X5
18 August 1941	HMS P32	22 September 1943	X6
20 August 1941	HMS P33	22 September 1943	X7
27 October 1941	HMS *Tetrarch*	3 October 1943	X10
6 December 1941	HMS *Perseus*	3 October 1943	HMS *Usurper*
20 December 1941	HMS H31	10 October 1943	HMS *Trooper*
31 December 1941	HMS *Triumph*	19 November 1943	HMS *Simoom*
12 February 1942	HMS *Tempest*	7 February 1944	X22
23 February 1942	HMS P38	16 March 1944	HMS
26 March 1942	HMS P39		*Stonehenge*
1 April 1942	HMS P36	28 March 1944	HMS *Syrtis*
1 April 1942	HMS *Pandora*	16 June 1944	HMS *Sickle*
14 April 1942	HMS *Upholder*	22 November 1944	HMS
29 April 1942	HMS *Urge*		*Stratagem*
8 May 1942	HMS *Olympus*	19 January 1945	HMS *Porpoise*
21 June 1942	HMS P514	6 March 1945	XE11
7 August 1942	HMS *Thorn*	12 January 1950	HMS *Truculent*
17 September 1942	HMS *Talisman*	16 April 1951	HMS *Affray*
10 October 1942	HMS *Unique*	16 June 1955	HMS *Sidon*
4 November 1942	X3	1 July 1971	HMS *Artemis*
11 November 1942	HMS *Unbeaten*		

11.05

Band of the Royal Marines strikes up 'A life on the Ocean Wave' and they all march back to *President*. The Band marches with absolute precision. The Submariners don't (boats are not good places in which to teach drill). We look at the wreaths. The Band up the road is now switching to 'Hearts of Oak'.

You look at the old veterans from the Second World War and wonder what all that time underwater and without natural light did to their bodies. The same applies to Cold War patrols and now. In a sense, when men (and now women) join the Submarine Service they enter into a contract to neglect their own bodies – no real exercise, little healthy food (they run out of fresh supplies after about three weeks, even on the big nuclear boats), poor air, especially in the diesel era (in the nuclear age, the air is cleaner – cleaner than the air we all breathe.) The Submarine Service, in the way that military people often have, possesses a touch of the family about it with everybody knowing each other, but perhaps here it is even more evident: as sustained shared experiences go there can be very few compared to what these men have been through.

The Royal Navy is clad, like blue marble, in its own history, which it polishes and extends anew in each generation. Though the Submarine Service is but a century old, it has its own rich varnishing of lore and legend. In part this is because of a common and sometimes menacing thread that runs through submarine life, from the tiny boats knocked out at Barrow in the first years of the twentieth century to the huge nuclear-missile-carrying Trident boats of the late twentieth and early twenty-first century. Adversaries come and go but one threat, a first order and constant anxiety, is perpetual – the hazard of the deep.

The swiftest and most vivid way to enter this strange and singular world is, fittingly, a total immersion into it – to be invited on board a submarine that will very soon be stretched as close to the limit as it can be, outside a clandestine special operation in peacetime or war itself. Such a boat was HMS *Tireless* in the spring of 2012 as it prepared off the west coast of Scotland for the so-called 'Inshore Weekend' of the course to test the would-be Commanding Officers of the next generation in ways that are themselves transmitters of the history, the tradition and the style of the Submarine Service.

I

The Franchise of the Deep: Perisher

*At one time the course on which we were embarked had been
called the Periscope School; hence the grimly humorous con-
traction 'Perisher'.*

Edward Young, 1952.[1]

*For all the costs, just what does the Perisher course produce?
Arguably the world's finest quality submarine captains. Perisher is the Royal Navy's commitment to making sure that the
men who command their submarines are as good as the boats
themselves.*

Tom Clancy, author of
The Hunt for Red October, 1993.[2]

*What fascinates us [the French Navy] is the feeling that what-
ever you did before is at stake. Nothing counts before. Passing
Perisher gives complete legitimacy. It's a way of redistributing
all the life cards for a very specific thing. Perisher partly reflects
the British spirit – this willingness to have the right man for
the job.*

Commander Rémy Thomas, French Navy,
15 April 2012.[3]

*There is a certain strain in the English, delicate, fastidious, self
despising, which draws some of them to the Arabs, drives them
to adopt their code of chivalry, courtesy, and cruelty, and thus
obtain the franchise of the desert.*

Alan Bennett, 1991.[4]

*The desert is like the sea because it doesn't care tuppence about
you. You can be romantic about it. But the real thing is that it*

doesn't have emotions. And it doesn't think twice before kill-ing you.

> Sir Rodric Braithwaite, former Chairman of the Joint
> Intelligence Committee (1992–3), 30 August 2012.[5]

. . . our own private war over the weekend.

> Commander Hywel ('Griff') Griffiths,
> CO, HMS *Tireless*, 14 April 2012.[6]

A PRIVATE WAR HAS BEEN ARRANGED

To command a Royal Navy submarine is to be given the franchise of the deep – responsibility in modern times for between 120 and 160 lives, with a replacement cost, if the submarine is lost or wrecked beyond repair, of between one and two billion pounds. To acquire this particular commission from the Queen you have to pass the five-month Submarine Command Course, universally known as the Perisher, in your late twenties or early thirties; thereafter a boat will usually (though not invariably) be yours in your late thirties or early forties.[7] It is a task and a duty granted to very few.

Each Perisher course normally consists of a group of four to six aspiring Commanding Officers, usually Lieutenants or Lieutenant Commanders, supervised by an experienced Commanding Officer, known as Teacher. While no two candidates have the same mix of skills and characteristics, when looking back at the Submarine Service's more successful COs it is possible to identify a number of common qualities. 'All accepted responsibility eagerly and were self-confident,' wrote a former Teacher, Martin Macpherson. 'They were strong willed, tenacious and determined; they were brave; and they possessed great physical and mental stamina. They all cared passionately for their ships' companies, had a strong sense of humour and many were surprisingly modest. Their professional experience and training had developed quick calculating brains, the ability to delegate, presence, and "a good periscope eye" ' – the ability to react instinctively and not lose sight of the tactical picture unfolding on the surface.[8]

Since its inception in September 1917, the Perisher course has constantly evolved to take account of developments such as the shift towards greater concentration on Cold War operational roles; the greater emphasis on anti-submarine warfare; the introduction of nuclear submarines with better speed, mobility and endurance; the demise of the Second World War straight-running Mark 8 torpedo; the introduction of new weapon systems such as the Sub Harpoon and land attack Tomahawk Cruise Missile; the reduction in the number of Diesel Submarine Commands available for successful Perisher candidates; the shift to an all-nuclear submarine fleet; and the requirement that all Executive Officers (second-in-command) in nuclear submarines are Perisher qualified. The course has also had to provide newly qualified commanders with a multitude of fresh skills such as doctrinal and tactical knowledge and greater technical understanding of weapon and propulsion disciplines.[9]

While the format of the course has changed over the years and the size of the syllabus has steadily increased, the fundamentals of Perisher remain the same: in order to pass, a candidate must prove that he is able to operate a submarine both safely, effectively and aggressively, in a hostile environment, by completing a variety of tactical scenarios that cover the complete spectrum of submarine activity: everything from attacking a task force or other submarines, gathering intelligence, or landing members of the Royal Marines Special Boat Service (SBS). 'There's a living thread that runs through Perisher,' says Andy Bower, another former Teacher. 'The course is almost the beating heart of the UK Submarine Service. You're a guardian of the flame.' Every submarine officer (aside from engineers) aspires to command his own nuclear submarine. He will spend the early years of his career working towards it. If he fails, he may never set foot on a Royal Navy submarine again. If he passes he has earned the right to join an exclusive club.

Our treatment of Perisher here is in three parts. First, there is Hennessy's diary of the Inshore Weekend of the 2012 course which took place between 13 and 15 April in and around and largely under the Kilbrannan Sound, which separates the Kintyre peninsula, tipped at its southern end by the Mull of Kintyre, and the Isle of Arran. Secondly the last, most intense weekend of the next course, in October 2013, capped by the celebratory 'Perisher Breakfast' in Faslane on 20 October 2013. Finally there is a short history of Perisher as it stood at the end of the Second World War, as those first post-war submariners equipped themselves for an entirely new conflict, the Cold War, between East and West.

The point of entry to Commander Griffiths's 'private war' was Faslane on Friday morning, 13 April 2012. Faslane is the citadel of the Submarine Service, north of Helensburgh towards the end of the Gareloch, a finger off the Firth of Clyde. Today it is bright and shining in the sun. This scene has scarcity value. Faslane excels at cloud and rain, and in the Royal Navy's mental map of its bases Faslane is synonymous with a micro-climate that can go from sun to hail inside half an hour. Those whom Faslane has touched carry its cold and its damp in their bones.

We gather over lunch in HMS *Neptune*, the still new modern officers' so-called 'Super Mess', run for the Navy by Babcock International, which took over a great deal of dockyard and naval support work in 2007. Our group consists of: Captain Paul Halton, Flag Officer Sea Training (FOST in the ubiquitous acronymia of the UK Armed Forces and Ministry of Defence), a former CO of the hunter-killer HMS *Spartan* and soon to be posted to Afghanistan; Commodore Steve Garrett, Commodore of the Faslane Flotilla, former CO of HMS *Turbulent* and most recently the naval Captain seconded to the Cabinet Office to handle its nuclear-weapons responsibilities (including ministerial briefing) and as part of the firing chain that links the Prime Minister to the Trident submarine on patrol in the North Atlantic; and Commander Ryan Ramsey, who works for Halton, recently CO of HMS *Turbulent*, featured in the Channel Five television documentary which followed *Turbulent* from Devonport to the Straits of Hormuz on the Navy's regular Mediterranean/East of Suez deployment. All will be aboard *Tireless* tomorrow and overnight on Saturday/Sunday, in part to put still more pressure on the 'Perishers'. Ramsey will take over the course as Teacher on next year's Perisher. The other observers/guests are Commander Rémy Thomas, the equivalent of Teacher in the French Navy (the French are considering sending one submariner per year to Perisher, which would be quite a step given past difficulties experienced in arriving at *entente nucléaire*), Superintendent Thom McLoughlin from what was soon to become The Police Service of Scotland, who has a special interest in training, and Lieutenant Aidan Riley, a helicopter pilot from the Fleet Air Arm.

Garrett, Halton and Ramsey naturally have taken and passed Perisher in their time. It's a rite of passage no successful Perisher forgets and it is recognized by, and has a cachet well known to, all the other navies of the world which operate submarines. How did they see themselves and the course that had eventually propelled them into positions of command?

Ramsey: 'The pressure is always self-induced to start with. They keep taking over as Captain. It's amazing to participate in, to do and to watch [Ramsey did Perisher in 2000 on HMS *Splendid* and HMS *Triumph*].'

Garrett: 'You have between ten and fifteen years of experience. You'll know tactics and doctrine. It's about learning your strengths and weaknesses ... Commanders see massive potential in them. But it's the technical capacity ... Some come onto the course too early. Now we say "You're not quite there. Go and get some more experience."'

Hennessy: 'What's special about submarine command?'

Ramsey: 'You're on your own.'

Garrett: 'Independence. We give an independence to our Commanders that's quite unique – like an eighteenth-century frigate going round the world – particularly if you're commanding an SSBN [Trident boat]. We can't tell Northwood what's broken or what we're doing tactically because of radio silence.'[10]

Ramsey: 'It's almost the last bastion of mission command.'

Hennessy: 'Does that make it easier or more difficult to command? Don't you have more responsibility than anyone else in their thirties?'

Ramsey: 'Yes. It's how you handle it.'

Garrett: 'It's the environment of the submarine. As a CO you always have in the back of your mind that you're in a hostile environment.'

Halton: 'There is a paranoia that can be used constructively. The good submariner, if he hears the ship transmitting on sonar, thinks "He's got you."'

Garrett: 'It's continuous risk assessment – all the things going on around you. Some guys will get their risk assessment blatantly wrong to achieve the mission. The aim is to live to fight another day.'

Halton: 'Sometimes you have to ask "Why is it too dangerous? Are they being too wet?"'

Garrett: 'You need to know what rests on the mission – the whole war or is it just something that would be nice to know?'

Halton: 'Ultimately it's people and leadership. This weekend, they are still a fairly raw product – just two weeks into their sea phase. We're looking at how they conduct themselves; the standards they set.'

Hennessy: 'Is it a case of you don't know you can do it till you've done it?'

Halton: 'Absolutely ... it's extremely difficult.'

Garrett: 'You add complexity ... to go or no go. They often get wrapped up in that complexity.'

The conversation turns to Perisher itself.

Ramsey: 'I really didn't want to fail that course.'

Halton: 'When I passed it, I thought, "I've made it" . . . it's such a pivotal thing in the Submarine Service; part of the ethos.'

Garrett: 'The course takes place in front of the junior rates. That's the reason why it's irrecoverable. But you need them to help you pass – you can't slave-drive them. If you're shouting at everybody you're not doing it properly.'

Hennessy: 'Is there any psychological testing involved?'

Garrett: 'There's no real psychological profiling before becoming a submariner. But it's how adapted you are. Perisher does a lot of things. Self-education on stress management. They have presentations on that. Teacher did it. I think it should be done a lot earlier. Ideally, it makes people honest with themselves. We do it by teaching people how to operate within the rules – how fast and deep; what you do if a frigate is at hand. It also makes it quite clear that there are the boundaries. In the last phase, Teacher will let them operate outside these rules.'

The lunchtime conversation carries into the 107-mile drive to Campbeltown, towards the Mull of Kintyre, where the party will spend the night before an early-morning rendezvous at sea with a 'Trafalgar' class SSN, HMS *Tireless*. Captain Halton explains that the Submarine Service needs eight Lieutenant Commanders to take Perisher each year and six of them, if possible, to pass. There were four starters (all Brits) on this Perisher and all four are still there. The Dutch and Norwegians are doing their versions of Perisher simultaneously, so three submarines will be prowling around Arran at the same time, but all the Brits will be on board *Tireless*. Ideally the Royal Navy needs to mount two Perishers a year but there won't be enough submarines available to do that until the new 'Astute' class SSNs start coming into service. *Tireless* is the second oldest of the remaining 'Trafalgar' class submarines; her older sister, HMS *Turbulent*, will be decommissioned in July.

It's still sparklingly bright and clear as we travel down the Kintyre peninsula. To the west we see the two Type 23 frigates that will be hunting us tomorrow, HMS *St Albans* and HMS *Monmouth*. *Tireless* will be between them trying to do a periscope reconnaissance of an island off the Mull of Kintyre without the frigates finding her. The beautifully clear weather converts into awful conditions for the submarine, as the bright sun will glint off the periscope as soon as it breaks the surface.

Over dinner in Campbeltown talk ranges widely. Garrett says it's the first time the three navies have done Perisher together. Norway has six diesel submarines, down from fifteen in the Cold War. Ramsey talks about the return of board games on the boats (surprising in the age of DVDs, iPods and iPads). On his *Turbulent* patrol East of Suez they became obsessed with Uckers, the naval version of ludo.

'I, my XO and two key Chiefs started playing again – it became our thing; obsessive. The messes became more talkative. It's spread to other boats – the resurgence of board games in submarine life. There's a special acoustics man who specializes in making Uckers boards.'

Ramsey's *Turbulent* patrol lasted 286 days. On his last night on board he had a steak dinner and later a game of Uckers. 'It really is addictive,' says Halton, 'a spectator sport.'

Saturday, 14 April 2012, SS *Oronsay*, Campbeltown Harbour, 7.30 a.m.

It's bright and clear and the water is gleaming as we eat breakfast at the Craigard House Hotel in Campbeltown. Halton says, 'It's usually dark, wet, windy and early' when they do the boat transfers. 'I'm not sure why we do it to ourselves.'

We can see the two frigates lying outside the mouth of Campbeltown inlet. 'Typical Perisher,' says Halton. 'Frigates are there to make life difficult.' On the way out to rendezvous with *Tireless*, *Oronsay* passes the NATO fuelling jetty, Davarr Island and its lighthouse to the right and the great rock of Ailsa Craig rising out of the sea to the southeast.

Tireless comes into view silhouetted against Arran. She looks a bit battered now but still exudes a certain swagger. The 'Trafalgar' class, after all, was the most sophisticated and highest achievement of the UK's Cold War submarine construction and engineering enterprises – world-beaters in their time. *Tireless* is one of the so-called 'special fit' boats used to carry out the most sensitive of clandestine intelligence missions.

HMS *Tireless*, 8.00 a.m.

As you come alongside, with the backdrop of Arran, the old cliché of the sea monster seems very much not a cliché. We're piped aboard and welcomed on the casing by her Captain, 'Griff' Griffiths, and by Teacher, Commander Andy Bower. We go down out of the chilly morning and into the cramped and womb-like warmth.

First chat in the Wardroom. Griffiths says: 'As a new CO, it's a little counter-intuitive to be hands off. This is my fifth time on Perisher, including my own in 2004.' The boat is, I think, still on the surface, its movement scarcely perceptible. Griffiths says:

'There is a very important relationship between me and Teacher. We've known each other for years. The Inshore Weekend takes place either side of Joint Warrior [NATO Exercise] so we'll have two frigates, a minesweeper and helicopters – our own private war over the weekend. It's a relatively challenging environment operationally. Teacher and I are very well used to the area, which mitigates some of the apparent risk. The risk of collision and grounding if you don't know what you're doing is quite significant.'

Overnight they had been doing surveillance. The Perishers will finish two weeks today, 'after being actively hunted'.

Bower, asked what surprised him about the job of Teacher, replies: 'The realization that for you, unlike the students, hours off to plan and sleep don't happen.'

Griffiths plainly adores *Tireless*, which is due to go out of service in December 2013. (In fact, the boat survived until June 2014, see pp. 683–5.) '*Tireless* is a wonderful old girl, operating extremely well for me, but she does definitely need a refit.'

0730Z

We're now running southwest towards the Mull of Kintyre. We turn our watches to Zulu time, the Coordinated Universal Time used by the military, so it's just after 0730.

0745Z

We're doing a periscope reconnaissance in the vicinity of Sanda Island. Griffiths explains that he has two watch leaders who routinely run the submarine for twenty-four hours a day on six-hour shifts.

Griffiths: 'Above them is the XO. One of us is available twenty-four hours a day. The Perisher students are practising our jobs. I will be on the chart [i.e. navigating]. I'm responsible for everything that happens on board. I do delegate conduct of the boat. I delegate conduct to Teacher. He will take conduct of much of the stressful element. I gave him conduct today just before the boat transfer. We have two parallel navigation operations to maintain safety.'

Halton: 'On our submarines we have two command qualified officers [CO and XO] to enable continuing intensity of operations. Other NATO navies don't and the CO can be worn out. You can sit somewhere hostile for twenty-four hours a day. It's especially important for intelligence gathering.'

Garrett: 'That can be maintained for months at a time.'

Griffiths: 'The strength is in the people. Manpower, equipment, training and stability are the four pillars. But the key thing is the people. It's very much a family.'

Halton: 'Risks and pressures are quite difficult for a submarine. The human condition deals with it in different ways.'

Dan Knight (*Tireless*'s Executive Officer): 'It's a team trying to keep the platform safe.'

Griffiths: 'A hundred and thirty men stuck inside a steel tube have to be massively tolerant of each other.'

Garrett: 'There's rarely a lot of shouting on board or barking of orders.'

Halton: 'Professionalism transcends rank on board. Rank is there and exists but professionalism is all-important.'

Griffiths: 'It's a hugely technical beast. You just inhale this knowledge and it sits in you. It's built over time. The indefinable sixth sense.'

Knight: 'It's hairs on the back of the neck.'

Griffiths: 'There is a feel to the boat. Fifteen minutes ago we looked up at the speedometer as the movement of the boat changed suddenly. It's seat of the pants.'

Halton: 'Subconscious activity is going on all the time.'

Knight: 'You can walk into the Control Room and know something is not quite right.'

Griffiths: 'It's an instinct – a feel about the boat. You can feel when something's changed and it's not right. Particularly when we're in close proximity to shipping. I always look at the eyes of the guy who's on the periscope.'

Garrett: 'You feel the boat change . . . It's almost an internal clock. It might be thirty seconds or several hours or the absence of a piece of information.'

Halton: 'Ninety-five per cent is drills that make you reasonably good. But, in this business, success or failure is in small margins. The hairs on the back of the neck make all the difference.'

Griffiths: 'I learn stuff almost every day when the young men take me through the equipment. This is the best job I'm ever going to have.'

We talk about money. *Tireless* cost £380m [coming into service in 1985]. The figure now for delivering an SSN is about a billion pounds. It costs about £3.5bn a year to be a submarine nation. By 2020 it will be about £5bn.

Halton: 'It's a measure of our kudos as a nation.'

Griffiths: 'We maintain an SSN at very high state of readiness to protect the integrity of the UK at all times. *Tireless* gave it to *Turbulent* four days ago.'

Being at high-level readiness involves a ship, a submarine, a tanker and four Merlin helicopters.

We talk about the capabilities of a Royal Navy SSN. The only thing that limits the duration of a patrol is the amount of food on board – so global reach (ninety days' food – additional dry provisions to get the patrol to 110 days, maximum; fresh water is distilled from sea water). It is capable of carrying out a number of roles, ranging from intelligence gathering and surveillance, land attacks using Tomahawk Cruise Missiles, deployment of Special Forces, force protection of a surface task group, anti-submarine warfare, anti-surface warfare and counter-piracy operations.

Halton: 'An SSN gives tremendous political choice and freedom of manoeuvre. *Astute* is going to be able to use all those capacities together with no external support. It's almost the only platform that can deliver strategic effect without going nuclear.'

0820Z. Control Room

Tireless's vents have just been opened. We're diving. The planesman [responsible for driving the submarine] is at the controls, operating what legend has it is the same steering column that was used on Wellington bombers during the Second World War.

One of the Perishers, Lieutenant Commander Andy Reeves, is in the Captain's seat. We're diving just off the Mull of Kintyre at a 6-degree angle. HMS *Monmouth* and the minesweeper HMS *Brocklesby* are hunting us. We're at periscope depth – 18 metres. We're going down to between 25 and 30 metres and then up to periscope depth. Leak checks made (all fine: no surprise but a mild relief). We dive quickly, but there's no such thing as a crash dive, as in the Second World War films. It takes between seven and eight minutes. That time could be halved during an emergency using the speed of the boat.

Griffiths tells me: 'My job is to maintain the safety of the platform.

That sixth sense and experience. I have in the back of my mind at all times, what I will do if this happens? – the "so what".'

We're in the North Channel between Northern Ireland and Scotland. It's 'a two-lane motorway'.

Tireless's immediate task is to gain intelligence and any indications of activity centred on Sanda Island (a 'perirecce') to the south of the Mull of Kintyre.

0840Z

Periscope up. An anti-submarine Merlin helicopter is sighted. The Control Room is very crowded (crew, Perishers, visitors, Halton, Garrett, Ramsey). The ship's company is a core of 110. There are 130 on board today.

Halton: 'The Perishers will be very aware we're here. They'll think we're watching them more than we are.'

I ask Halton about *Tireless*'s and now *Turbulent*'s role in protecting the Trident-carrying 'Vanguard' class as the submarine at high readiness. 'It's layered defence,' he says. 'Deterrence is the number one priority. It trumps everything else. It's in support of CASD [Continuous at-Sea Deterrence; see p. 218]. It's layered intelligence and support . . . we do feel the loss of the Nimrods [RAF Maritime Patrol Aircraft designed to hunt submarines]. The Merlins don't have the legs.'

We have a brief chat about *Tireless*'s armaments:

The submarine carries torpedoes and missiles. The torpedoes are called Spearfish. They weigh 2 tons and are equivalent to a 550 lb bomb. They explode underneath the target creating a bubble of gas which will break the back of a 50,000 ton ship.

The missiles are Tomahawk Land Attack Missiles (TLAM), each with a range of 1000 nautical miles. Given the near ubiquity of the sea, that means you can strike about 90 per cent of the world. The accuracy is measured in yards. The strike authority is Northwood. A Spearfish and a Tomahawk are about a million pounds each.

We talk about the previous phase of the course when the frigates race at the subs. It was done off Norway this time. A Royal Navy surface ship, a Type 42 destroyer, races in at 30 knots. It's well controlled. The ships spin out at 300 yards.

Inshore is the most perilous – 5000 tons of submarine in 45 metres of water being chased by frigates and helicopters.

There's squeaking on the casing as the helicopter finds us with its powerful dipping sonar, known as 2089, which is lowered into the sea while the helicopter hovers above. You can hear the squeaking above the hum and the hiss of the air conditioning. Just before I'd heard a couple of muffled bangs. Simulated depth charges, I'm told.

Ramsey says, 'We'll be attacked quite a lot today. It's not a nice moment being detected, particularly if they can classify you as a UK SSN.'

The Merlin attacked us as soon as we dived. Depth 21 metres, speed 8 knots.

0950Z

We're on the way to Pladda Island now, off the southern tip of Arran, which carries yet another similar reconnaissance mission. We suddenly descend to 30 metres to get more speed and to make the boat easier to handle. The squeaking stops. I'm told that Commander Sharpe, whom we had met in the Pool of London a few weeks earlier, is up there on HMS *St Albans*. He has apparently said of his and *St Albans'* task, 'We're here to fail the students.' This is one of the toughest days of the Perisher course until the final weekend. Halton and Garrett are adding pressure in the Control Room by making their presence felt.

1000Z

Tireless has escaped her pursuers and we're slowing down again, and Thom McLoughlin and I begin a tour of the boat with Ramsey and Nick Brooks, the Engineer.

Ramsey: 'The hydraulics is the blood of the system.'

Brooks explains what the hydraulics do to maintain the system at pressure:

1. Main hydraulics, bulkheads, valves, raising masts and periscopes, foreplanes
2. External hydraulic system outside pressure hull, masts on bridge, sea water intake
3. Aft hydraulic plant, afterplanes, steering

We then inspect the heads. 'We don't shower very often.' The submarine odour clings to everything, including the towels. Two of Brooks's engineers do the crew's laundry and they get an allowance for it. There are restrictions on water. The boat has two distilling plants. 'We get rid of the water by pumping it overboard. It's noisy so we minimize the use of water for a reason.' It's 1030 and the squeaking has started again.

The Junior Rates Mess is shared by sixty men. There's a tray of apples, oranges and pears on the table.

The Galley is manned by two Chefs during the day. One night Chef. 24-hour cover. Four meals a day for 130 people. John, the Chef, says: 'They base their morale on the food.' Brooks adds: 'Without daylight, I know what time of the day it is from the meals.' 'We're creatures of habit,' says Brooks. 'We have theme nights – Chinese, Italian and curry. Fish on Friday and a pizza night on Sundays.'

We move down the Two Deck passageway, the main thoroughfare of the boat. The furthest you can see is about 20 yards. It doesn't feel claustrophobic to me; but it would to some. Brooks says that when you return to Devonport or Faslane 'your eyes take time to adjust to the long distances. So you leave it for two days before driving [there are no rules, some do, some don't]. There's no sunlight so no Vitamin D so we bring vitamins to sea. There's a calcium problem too.'

In the Escape Compartment, with its oxygen generators and escape tower and the Torpedo Loading Hatch, we're as far forward as we can get. The rescue submersible from Faslane can lock on to the escape hatch. There's an air supply in the tower. Special escape suits in boxes. The other Escape Compartment is at the back in the engine room. It takes 3–4 minutes per person to get out.

As we move back McLoughlin says: 'It's a monastery of the deep.' Above the ladders there's a picture of David Cameron with a rather strange hairdo. It carries a caption:

AN IMPORTANT MESSAGE FROM THE PM
Don't stomp up and down the ladders.

The PM has not been aboard *Tireless*, and he won't know of his photographic role in this particular piece of the Defence of the Realm.

We see the Laundry, consisting of one washing machine and one tumble dryer. We then descend into the Weapons Storage Compartment, or the 'Bomb Shop' to use its informal name. The Spearfish and TLAMs are about 23 feet in length. There are sailors (trainees) asleep next to the torpedoes. There are boxes of baked beans stored here, too. Squeaking again. It sounds particularly eerie down here in the Bomb Shop.

We climb out of the Bomb Shop and enter the Sonar Compartment. It is stacked full of computers. *Tireless*'s towed array [a series of hydrophones towed behind a submarine on a cable], which is about 800 feet long, is fitted just outside the breakwater at Devonport when she sails from Plymouth.

We descend to the lower deck. There are four battery compartments, the Ship's Office, for administration and logistics, and the Garbage Ejector Compartment – tins go outside to the sea but not plastics. Hand weapons are locked in a safe. Air purification space. CO_2 scrubbers. Oxygen stays in; hydrogen goes overboard (compressed as a gas).

1055Z. Control Room

We're at periscope depth quite close to Pladda Island off the southern tip of Arran. The first perirecce of the Pladda lighthouse has been done. We're moving at 4 knots, 30.5 metres above the bottom. We'll do another periphoto later.

In the wider context of UK intelligence gathering this was quite a remarkable event. For in no other branch of the British intelligence trade are outsiders allowed to witness collection techniques in action. Yet here, unfolding bit by bit, was our country's capability for submarine-operated photographic, signals and electronic intelligence against a hostile power well endowed with anti-submarine capacity. It was a striking example of openness and loquacity on the part of Her Majesty's legendary 'Silent Service'.

The frigates and helicopter are still looking for us but the shallow water disrupts the sonar and makes it more difficult for them to pick us up. Teacher gives *Tireless*'s position away to our pursuers by ordering the firing of a white smoke candle. The Perishers can't be put under the stress required if the frigates can't find us. The Perisher has to decide to evade or go for speed. He takes a look through the periscope. Griffiths explains that we have to be careful about the angle of the boat. Thirty metres above the bottom is the limit: 'ordinarily it's forty-five metres. We get special dispensation for this.'

Griffiths says: 'We're just below periscope depth at ten knots. We call it "gulping" – taking a look with the periscope and speeding up. If we put the periscope up we leave a huge feather you can see for miles.' If *Monmouth* sees us she'll approach flat out at 29 knots. The Perisher students have to be able to calculate the distance of the approaching ship and ensure that they have a minute's reaction time, the time it takes to get the submarine to a safe depth.

Monmouth charges and comes about 1500 yards from *Tireless*. There's intense concentration all round in the Control Room.

Ramsey: 'Teacher will fail the student if he thinks he's unsafe. In Command, we're trained to take risky decisions, not dangerous ones.'

Hennessy: 'When did you get most stressed on Perisher?'

Ramsey: 'On the first Inshore Weekend, funnily enough, i.e. this one. I made a mistake. My perioperator got me run over by a vessel. I was duty Captain. The stress you put on yourself.'

Hennessy: 'Did you think you'd blown it?'

Ramsey: 'Yeah.'

Conversation now with fellow weekend rider and Fleet Air Arm helicopter pilot, Lieutenant Aidan Riley. It's his first time witnessing Perisher from inside a submarine. 'A really good insight,' he says. He flies Merlins and we talk of what he would be doing if he was up there hunting *Tireless*:

'There's plenty of opportunity today to see the submarine visually. I would go for visual first and have my radar, which they can detect, turned off. I'd look for the feather-like periscope. It's only a 12-knot wind so not many white horses. I'd tend to hover at about 120 feet to put the sonar in the water. Are we trying to find the sub or stop it taking a picture of Pladda light? I'd hover there. I'd prepare the bottom topography and tidal work. Where is he likely to be? Trying to get inside the mind of the submariner.'

1130Z

Teacher asks for a cup of cold coffee. He's a strong, highly direct and hugely experienced man but no doubt he has his own moments of stress too.

1140Z

Probable picking up of *Monmouth*. It is *Monmouth* at 4700 yards, making 13 knots. This is a world of constant mental arithmetic. We listen to the sound of *Monmouth*'s shafts and blades.

1145Z

One of the Perishers, Lieutenant Commander Andy Reeves, gets on the Attack Periscope. The Trafalgars have two periscopes. The Attack is much smaller and therefore leaves a thinner feather on the surface. But it's not very easy to see through and doesn't have electronic sensors, which means that when you're in close to your hunter, you're more likely to be seen. The Search Periscope has really good vision, electronic sensors and GPS on top. It's big with a huge head. Normally an all-round look that picks up nothing should keep you in the clear for the next twelve minutes.

Once Reeves has finished his stint as perioperator, we snatch a conversation at the back of the Control Room. How long had he been at work today?

'Finished at 1110 last night. Got up at four. Took over at 0515. So six hours. I got about five hours' sleep which is quite a lot. It's part of the stress to try and take the sleep off you. The tiredest people I've ever seen are the Captains who have been on for three days. So it's to make sure that you can still think.'

Reeves explains that this weekend's phase of *Perisher* involves 'a totally different way of thinking':

'When we came on board first we could operate the boat within the parameters set by the book. Now it's pushing navigation, counter-detection and avoiding collision and overall achieving of the aim. It's trying to teach you a balance between those three all the time. Everyone's got their own limits within which they work.'

It's good of him to talk when he's plainly had a seriously stretching morning and I'm struck by his directness, candour and self-awareness. I ask him if on Perisher he's found out new things about himself.

'I've not discovered anything new. But the traits of your personality come out. I'm way too self-critical. I've been told that since I was a child. That part of my psyche that says "bollocks", I almost have to put that on one side just to do it – thinking I'm not going to help the situation by shouting . . .'

Reeves describes for me the earlier, Norwegian phase of Perisher:

'It was just off Bergen, in a fjord. There were three of them – at one point four – charging at you. It was a combination of visual and maths – going back to World War Two. "Eyes Only" we call it. My mental arithmetic was horrendous until last year. Last year I spent thirty minutes a day on mental arithmetic. That sort of stuff you can be taught.

But if you can't take decisions quickly enough, you're never going to do it.'

This last requirement was to be a feature of the afternoon to come – once in a truly eccentric fashion, as we shall see. Decisiveness is perhaps the master-noun of Perisher.[11] Halton had told me earlier that morning that what Commanding Officers really need 'is an ability to see through what's important to cut through the trivia'.

1157Z

In the background we can hear an all-round look being taken on the Attack Periscope. I ask Reeves what have been the most stretching bits of Perisher so far. Without hesitation he replies:

'It's the first time in your career when you can't ask someone else what to do. Quite a strange feeling. A lot of preparation is put into making you not panic. Yesterday we were 1400 yards from the Dutch warship. I was on the periscope at the time. Teacher noticed and the duty Captain noticed the change in our voice. Teacher said: "Don't panic." You assume worst case – that when you put the periscope down he'll come at maximum speed.'

Reeves joined the Navy in 1998 and volunteered for the Submarine Service in 2002. He's served on both 'Vanguard' class SSBNs and 'Trafalgar' class SSNs and was on *Turbulent* during her Libyan operations. So, he says, he's been ten years in preparing for Perisher, 'then all of a sudden, you're making it happen at thirty-two years of age'. Teacher was his Captain on HMS *Vengeance*.

He lists the skills you need to get you to Perisher: mental arithmetic; being able to operate the periscope; dealing with the control; knowing about the TLAM; guiding the weapons; understanding the reactor and propulsion systems. 'My contemporaries are not doing anything near to this in terms of independence and responsibility.' He feels, he says, the 'gravity of it'. He feels, too, 'the lack of appreciation of what we do in the public'. 'Annoying', he calls it. A few months later, the 2012 Olympics threw up a vivid, if absurd, example of this. On 9 August *The Times* led its front page with a story about the future funding for UK sport, 'Britain must build on success of Olympics.'[12] It quoted 'Fuzz Ahmed, who coached Robbie Grabarz to a high jump bronze medal. "It's an irrelevant amount of money compared to a submarine. What would you rather have, Chris Hoy [who had just won a sixth gold medal] or a submarine?" he added.'[13]

1220Z

A Close Quarters Drill. Turns out to be a small lobster-potting boat.

Reeves says the Navy is 'a lot better now' at letting people know what they do. He was, however, 'amazed' by the amount of the deterrent firing chain Richard Knight and I were allowed to record for our December 2008 BBC Radio 4 *Human Button* documentary (so was I). Impressive, honest man, Reeves. In addition to his naval service he has a wife and two daughters and is taking an in-service degree in marine surveying at Portsmouth University. You learn time management in the Submarine Service.

1230Z. Sonar Room

Broadband and narrowband are explained as I gaze, near mesmerized, at the screens. The broadband looks like a formation of Cirrus cloud that's turned green, the narrowband more like grey scrambled egg. Broadband picks up general noise; narrowband will tell you what you're up against. I'm shown where the lobster potter is, the minehunter, *Brocklesby*, and *Monmouth*, which is coming in our direction at 12 knots. If *Tireless* were up against an Akula II submarine of the Russian Navy it would be but a very faint thin green line on the screen. You would not be able to hear it. But you could tell if it was reacting to the presence of another submarine (i.e. you). And the faster a submarine or a surface ship is travelling, naturally the noisier it is. 'It's more an art than a science,' says one of the sonar operators.

1235Z

The Sonar Room picks up *Monmouth*'s Merlin.

1236Z

'She's back,' they say. We're due to take another periscope photograph of Pladda lighthouse in about half an hour.

1240Z

The Merlin is about 9000 yards away.

1245Z

We're slowing down and going shallow. You can feel the boat going up.

1246Z. Control Room

Attack Periscope up.

1253Z

Sound Room says the Merlin has dropped a charge.

Garrett says: 'It's not that close. It's like hitting the sub with a hammer.'

1254Z

The Search Periscope goes up. The camera is on. Pladda light is one mile away.

1254Z +30 Seconds

I take a look through the Search Periscope. The white lighthouse is bright in the spring sunshine and its surrounding buildings are vividly visible.

0101Z

Search Periscope up. Another picture of Pladda light.

0127Z. Wardroom

Teacher, Halton, Griffiths and Ramsey share thoughts on the morning.

Halton: 'It's a little bit casual for me. I'd like it crisper.'

Teacher explains what he is looking for in his quartet of Perishers:

'This is all about them doing it under pressure. I'll occasionally put my hand on the tiller to steer them in the right direction. As we approach the final weekend I won't do that. If I have to intervene then it's a catastrophe for the student. Sometimes there's no right answer; it's shades of grey. No longer are you looking for reasons not to do things. You're looking for reasons to do things.'

Over lunch we talk about the Submarine Service in general. It's now just under a sixth of the entire Naval Service (as of September 2015 there were 4,554 submariners in a Navy of 33,147). Griffiths has all the figures at his fingertips. He's recently returned to operations after a spell in the Ministry of Defence Main Building where he worked on, among other things, the 2010 Strategic Defence and Security Review. He has a taste for policy but returning to operations is 'like coming home'.

Thinking about it later, that first Perisher morning had unfolded like a play – a play punctuated by drop-ins to various working parts of a 'Trafalgar' class submarine at sea plus a recitative of conversations about the submariner's craft, the nature of submarine leadership and the special individual and collective emotional geography that goes with the underwater trade in general and during a Perisher course in particular. As plays go, it's definitely a drama. Why? Because there's never that much depth beneath you and there are large chunks of grey-painted steel above constantly trying to put you off. And the thriller element of the play was to increase as the afternoon deepened.

The afternoon began with another tour of the working insides of *Tireless*, the aft end this time. There are definitely two weather systems in a 'Trafalgar' class submarine. The Control Room is an artificial version of the weather outside, i.e. a nice spring day in Scotland. When you go back over the nuclear reactor (the Trafalgars are powered by a PWR, or pressurized water reactor) to the Manoeuvring Room and the Engine Room, you enter the tropics with temperatures nudging 25–30 degrees. When they are literally east of Suez, the temperature in the Engine Room can reach 60 degrees.

The Health Physics Lab is at this end and handles the radiological and medical life of this underwater village. The team is currently sampling the Control Room, testing oxygen, carbon dioxide, hydrogen and nitrogen levels. SSNs only take a doctor with them if 'we're covert for a length of time'. The Trident boats always carry one given their ninety-day patrols and the requirement not to break the patrol to get a sick sailor off (though in these post-Cold War times, a submariner whose life was in danger would be evacuated by helicopter, with the submarine spending only the briefest time on the surface). The Medic on *Tireless* is a Petty Officer.

We're now in the domain of *Tireless*'s Chief Engineer, Lieutenant Commander Nick Brooks. He's thirty-five and joined the Navy as a direct entry graduate after taking a degree in mechanical engineering at Loughborough. Engineers are like gold dust in the Submarine Service. The boats can't sail without them and there's a growing market for trained nuclear engineers in the outside world as the UK moves towards constructing its next generation of civil nuclear-power stations. The Navy supplements their engineers' pay to increase the chances of retaining them. But it's not the money that keeps Brooks in the Service. 'There was a shortage of engineers in the late 1990s and early 2000s. But I wouldn't be here now, eleven and a half years later if I didn't enjoy it – the professional challenge; the camaraderie.'

Manoeuvring Room

The Manoeuvring Room is like a mini-control room in a nuclear-power station with its bank of dials. There are between five and eight sailors in here (with another three in the engine room), and it's manned twenty-four hours a day in port and at sea. Everything electrical and propulsive in *Tireless* is controlled from here. This is the most baffling bit of the boat for a visiting layman without an engineering training as your gaze travels over the dials that measure pressures, temperatures and flows. In the middle rests the throttle control panel which converts orders from the Control Room into revolutions of *Tireless*'s two main turbines.

Every six months or so *Tireless*'s team train for a 'scram' on a simulator in either Devonport or Faslane. A 'scram' means the reactor has to be shut down to maintain core safety. The word 'scram' goes right back to the very first nuclear reactor that went critical, on 2 December 1942 in the West Stand Squash Court at the University of Chicago's Stagg Field sports ground which housed it. The great Enrico Fermi had a cunning plan to prevent disaster if the core overheated. Cadmium control rods would be dropped down from above. If it happened (which it did not), Fermi would shout 'scram' to a man above the pile holding an axe ready to cut the ropes supporting the cadmium. So 'scram' originally stood for 'safety cut rope axe man'.[14] Nowadays, the control rods drop down automatically; axes are not required.

Overheating can be caused by equipment failure or by too much load on the reactor. On the right of the *Tireless* Manoeuvring Room is the Control Rods panel. If a scram occurs, lights flash red and a bell rings. The main engines might trip. The job is to minimize heat and to switch the submarine to battery power, produced by the diesel generators for a few hours and the entire boat goes onto half-lighting.

Tireless's engineers are reassuringly calm. 'This is the quiet end of the boat,' says Brooks. 'There are normally only eight of us back here.' We move into his workshop with its lathes, drills and grinders. Here, too, is the white smoke signal ejector for use when Teacher wants to give away the boat's position to its pursuers. We inspect the gearbox. Sometimes, in the middle of the night, Brooks's people put potatoes in tinfoil and bake them on the throttles.

0325Z. Control Room

Back in the cool though, things are hotting up. The boat is crossing a shallow stretch of water called 'The Bridge' between Arran and the Kintyre peninsula which is between 40 and 50 metres deep. The next task is intelligence recovery of the approaches to Campbeltown harbour. It's going to be tricky to pull it off because over 'The Bridge' the submerged geography takes *Tireless* into an underwater ravine with only just enough depth beneath the boat. Meanwhile, the hunters on the surface are now all after us – the two frigates, the minesweeper and the helicopters operating off the frigates. It's going to be a testing run to get that photograph and it's made all the harder for the Perishers because they are navigating without GPS (GPS would be turned off if this was for real). Griffiths does have access to GPS so he knows exactly where the boat is.

As the two frigates above go into their fast racetrack manoeuvre (describing an elongated circle to hunt the sub) and the Merlins dip their sonar buoys to ping the boat, the Perisher in the duty Captain's hot seat is Lieutenant Commander Sam Owen. As Owen tries to take evasive action it sounds as if one of the Type 23s is simulating the firing of a torpedo at us.

0415Z

You can feel the concentration and the concern rising in the Control Room. Griffiths and Halton are with the navigators. Griffiths looks up, plainly anxious. Three months later, over dinner in London, I ask him about this. Griffiths says the approach to Davaar is 'an ever-narrowing canyon. He [Owen] was driving himself closer and closer to danger because it's very steep.' 'Were you worried?' 'I was. I vividly remember it.'[15]

In my notebook I record the depth between the bottom of the boat and the seabed becoming a problem as well as the sides of the canyon. A sailor intones the depth at regular intervals in the Control Room like the tolling of a bell. Teacher and Griffiths confer. This is the moment of maximum concentration. Griffiths tells Teacher he's not prepared to carry on with this bit of the mission. Thanks to GPS, Griffiths knows exactly how close we are to the canyon wall. Sam, the Perisher, does not and thinks we are in a safer place than we really are. Davaar light will remain unphotographed. We're turning back north and heading for 'The Bridge'. 'Quite a lot of risk around,' I suggest to Ramsey. 'Yeah. The one who has the greatest risk is the Captain [i.e. Griffiths]. It's very impressive how he's carrying himself through this.' Ramsey recalls that on his Perisher HMS *Triumph* actually hit the bottom. 'Captain Phil Buckley handled that amazingly.' 'Was it you?' 'It wasn't me. I slept through it.'

0425Z

The tension has eased. The pursuers are heading south. We're going north. *Monmouth* is transmitting but she's further away than she was.

0530Z

Owen and I have a chat about the aborted Davaar operation in the Admin Office. 'The world according to Owen', he calls it. It's plainly been a strain but he's recovered pretty well. He drifted a bit after Bangor University, working on building sites, but he's clearly found himself in the Navy and was in HMS *Tireless* in 2004 when she went under the ice to the North Pole. He realizes this patch of Perisher was not his finest hour:

Owen: 'I don't think that run went well. The comfort I'd find in that is they're not designed to run well.'

Hennessy: 'But you kept calm.'

Owen: 'It's the swan approach. It's fairly clear what happened. The unnerving bit is not controlling the target. It was *St Albans* doing the racetrack. The problem was had she turned towards us and charged us. It would have forced us into an emergency action going deep. That would have been rubbish.'

Hennessy: 'Was this the toughest bit of Perisher so far?'

Owen: 'Yes ... Captain Halton is Teacher's boss. It's a bit rubbish when you do a bad run when he's on board ... three and a half hours is an awfully long time not to take a photo. What we basically did was drive ten miles underwater not to take a photo.'

Owen recognizes that he delayed too long in deciding to abort the Davaar run: 'If I was going to self-critique it's that I should have stuck to my guns ten minutes before when I didn't think we could do it. My instinct and gut feeling was that it probably wasn't going to happen. The maths said we could. But I should have baled out. If we had gone in and taken the photo we would have been detected and attacked again.' Though he wonders if Teacher used the white smoke to give his position away, Sam is not downcast by the afternoon's experience. I ask him about Perisher generally.

Owen: 'It's brilliant. You never get to do this again.'

Hennessy: 'Will you be all right?'

Owen: 'I like to think so. I know I can lead men in a submarine.'

The four Perishers know each other very well by this stage and Sam says they are firm friends:

'Whatever happens individually on the course we'll always be friends because of that shared experience. In service there are about sixty to a hundred Perisher-qualified officers. It's a fairly elite club. Ten, maybe twelve, of them are sea-going. The old Russians knew who had been on Perisher.'

Owen thinks the entire Submarine Service is pretty special:

'It's the camaraderie of it. There's an operational element to whatever you do ... We know it's bloody hard to do. It's got some of the largest *esprit de corps*, a lot of band-of-brothers; the lads' professionalism ... And these "T" boats can still compete against the best in the world and it's the people who enable us to do it. *Tireless* is home. We were one of the real sneaky boats. *Turbulent* was the other. Ryan took her on some pretty crunchy patrols.'

As Owen grappled that Saturday afternoon with the depth of the Kilbrannan Sound, the canyon on the approach to Davaar Island and the surface attacks, all of which combined to box him in, not to mention an increasingly anxious Captain Griffiths, Captain Halton, Commodore Garrett and Teacher Bower watching his every move and non-move, I managed a few snatched conversations about submarine life in general with the young men in the Control Room:

'It can be very dull: ninety per cent boredom; ten per cent abject fear.'

'The mates carry you through.'

'The removal from sunlight and society does things to people.'

'Makes you value your time when you're at home!'

'The Bombers [the Trident boats] have a very specific job to do and they do it very well.'

'These submarines are a lot more flexible – practising beach surveys, terrorist training camps.'

'For the Submarine Service, the Cold War was the golden age – North Atlantic, Norwegian Sea, Cold War ASW, protecting the Bombers. We were very good at it.'

I have a chance to talk to Ramsey about the characteristics of each submarine. 'Every boat,' he says, 'has a personality. *Turbulent* was a can-do boat, pretty aggressive. Every CO before me was aggressive. *Tireless* and *Turbulent* are real sister ships, doing the same things but in a different style.'

Ramsey reflects on the way submariners have to learn to live together in an immensely confined space: 'If the world was like submarines there would be no need for submarines.' He thinks we are cutting it fine with our Armed Forces: '180,000 people and two per cent of GDP to protect sixty million people. The Navy could fit into Stamford Bridge and leave room for the away fans.'

Short Visit to the W/T Room

All the communications channels are explained. Very Low Frequency is the primary instrument for receiving traffic. *Tireless* trails a 3000-feet-long buoyant wire aerial to pick up signals. 'T' boats have an email capability that families can use. The 'V' boats don't 'because as soon as they leave the wall, they are on operations'.

Control Room

Another chat with Alex, the Logistics Officer. He confirms something I always notice on submarines. 'ABs [the seamen] are a lot more independently minded on submarines. They are given much more responsibility. They take on things. They're not fazed by anything.' The duty Captain is now Lieutenant Commander Neil Botting, a geology graduate from Imperial who has served in 'Vanguard' class SSBNs and 'Trafalgar' class SSNs (he was in *Triumph* during the Libya campaign). His task at the moment is surveillance of our pursuers and the western coastline of Arran, working towards a periscope reconnaissance of the King's Caves.

A Touch of Protein

Suddenly a rating clambers down the ladder from the Control Room to the Galley carrying a plate on which is resting a piece of raw steak. I ask Ramsey what on earth is going on? It turns out that Botting had been dithering in the Captain's chair. It was too much for Halton, who sent a sailor down to the Galley to get the steak. Halton thrusts it at Lieutenant Commander Botting with the words: 'You need some red meat. Decide! Decide!' Ramsey captured on film a rather sheepish-looking Botting holding the plate. Ramsey thinks this protein moment is a first in the history of Perisher. Later I ask Halton about it:

'He was being wet. He needed some red meat to improve his aggression levels. He was getting a lot of hassle. I tried to lighten it a bit. It's a fascinating, people-watching thing.'

It strikes me as a literal example of the elders of the tribe blooding the young ones. How refreshingly different from the techniques and qualities of most modern HR practice.

A few weeks later, at a Promotions party in the Ministry of Defence Main Building, I recount the 'protein moment' on *Tireless*. 'Ah,' says Captain Andy McKendrick, the former CO of HMS *Vengeance*, 'the theatre of Perisher.'[16] Rear Admiral Ian Corder, Commander Operations at Northwood, is on *Tireless* two weekends later, the last of the course. Later he tells us he had a similar moment to Halton's, reminding the dithering Perisher (he didn't say which one) he was 'driving a sports car not a Lada'.[17]

This attitude towards command is admired all over the world. During a visit to a UK-based US intelligence installation shortly before travelling to Faslane, the subject of Perisher's specialness came up in conversation with a group of ex-submariner Russia-watchers. What's the difference between your submarine commanders and the British? 'Balls, balls' came the unhesitating reply. The explanation? Partly that the US Navy has become more risk averse. But even when it was less so during the Cold War, the Royal Navy's SSNs would take more risks when up against the Soviet Navy probably, our admiring allies thought, because we had so few boats compared to the American Navy and wanted to bring as much intelligence material to the shared table as possible.

Wardroom

Halton, Garrett and Ramsey gather to discuss the Perishers. Owen, for all his tough afternoon, is judged 'safe and effective'. Ramsey tells me Teacher didn't use any smoke on the Davaar section. McLoughlin, who was sitting for a good while between Teacher and the duty Captain, noticed 'that the tension touched everyone'. Later, Ramsey has a session with the four Perishers. He tells me about it early on Sunday morning: 'It was all about making the best of it. Example. Direction. Risk. Communication. Teamwork. They've got two weeks to go now.' And it's plain they still have a lot to do.

This is the first time I've slept overnight on a submarine. I'm put up in the three-man cabin; one of two adjoining the Wardroom. There are three bunks on top of each other with almost no headroom clearance. I'm given the middle one and getting in requires a degree of athleticism I haven't had to find for years. If my things had been put on the top bunk I don't think I would have been able to manage it. Got to bed about eleven. No film in the Wardroom so quite quiet. Slept remarkably well given that I haven't been so confined since spending the night with the Royal Marines in a Norwegian mountain snow hole somewhere north of Narvik in January 1978. The movement of the boat is soothing, sometimes rocking gently from side to side and really pleasant when it rises gradually and smoothly to periscope depth and descends again to its normal 30 metres (surveillance exercises went on through the night).

Sunday, 15 April 2012

Wake up shortly after 0530Z. Wash, but forgo a shave. Cup of tea in the Wardroom then up to the Control Room just after six. *Tireless* is now off the northwestern tip of Arran, and soon to do a 'sensor drop' at the mouth of the small sea loch, Lochranza. For Perisher purposes this is an enemy coastline protected by hostile warships. For 'sensor' read 'mine' in war circumstances. Reeves is duty Captain. *Tireless* has just gone deep to simulate the laying of the sensors. There's a frigate around somewhere to the southwest in Kilbrannan Sound. Ramsey tells me it's the navigational constraints close to a narrow inlet that produce the stress in a Perisher 'sensor drop'.

There's a slight shudder as the sensors are ejected and my ears feel the pressure differential. (The sensors aren't really ejected. It's just water shot through the torpedo tubes.) Another shudder as another simulated sensor goes out. One more to go then a swift exit from the scene. Luckily the frigate is going south and *Tireless* is going north so there's no sonar contact. *Tireless* is taken up to have a look round on the periscope. It's as well to check the sensors have been 'dropped' in the right place. If they're more than 100 metres down they might not work. At 0620Z a fix is taken through the periscope. The sensors have been placed effectively to cover the harbour entrance and the frigates evaded. Andy Reeves has done the job well.

It turns out that overnight *Tireless* suffered a mechanical defect while I was deep in sleep. A valve needed to be fixed in one of the main engines so the Perishers got some sleep.

Captain Griffiths now has a moment to talk to me more widely about *Tireless*'s private weekend war:

'When you're in extremely shallow water, you can't go underneath to avoid a collision. It was concentrating the mind [as we approached Davaar Island yesterday afternoon] because we couldn't duck underneath if we'd been charged. If it does happen you have to aggressively change tack. In training, in the final analysis, we can stick the fin out of the water and talk to the frigate.'

Griffiths's concern and the affection for *Tireless* are apparent once more:

'She's not a machine. She's far more than a machine. She's a home. She has a life of her own – a being with a personality. They're all temperamental. They all have slightly different ways. Little foibles. Each one has to be coaxed to get the best out of her. The boats settle into the water in a different way. My people know the foibles and how to play to the strengths.'

This is the submarine equivalent of anthropomorphism and deserves a Classical name all its own. It reminds me of how engine drivers would talk about the personality of their locomotives in the days of steam. No wonder virtually all the former COs turn up when one of these boats is finally decommissioned.

0650Z

Wardroom. Time for a huge, fried farewell breakfast from the Galley. Nick Brooks, the Chief Engineer, kindly comes in to tell me that West Ham beat Brighton 6–0 yesterday afternoon while we were preoccupied with the Kilbrannan Sound canyon. The internet brings in the football results and the news summaries for the boat.

0712Z

The crew stands by to surface. There is a thump as the air goes into the ballast tanks.

0728Z

I'm on the fin of *Tireless*. Griffiths lights up a cigarette. It's a beautiful, exquisitely bright morning off Arran, the air chilly but delectable. We're moving gently at 3.5 knots in a calm sea and pointing towards the Cumbrae Gap on the Firth of Clyde between the Isle of Bute and the Ayrshire mainland. To the left are Goat Fell and the mountains of Arran in silhouette.

There is an outburst of camaraderie as the weekenders take their leave of Griffiths and *Tireless* before transferring to the *Eva*, a fast craft which sweeps us off towards the Cumbrae Gap, the Gareloch and Faslane at 23 knots (which has to come down to 7 knots before we creep past Helensburgh to avoid swamping the yachts). As *Eva* approaches Faslane, the MOD Police launch comes alongside and puts a man aboard to check our passes.

Faslane is bright in the sun with the huge grey shape of HMS *Bulwark* ready to take part, as is *Tireless*, in the next phase of the NATO exercise 'Joint Warrior' off the northwest of Scotland. Up against the wall is HMS *Turbulent*, which five days ago took over from *Tireless* as the boat at a high state of readiness to protect the deterrent and the UK generally. Ramsey, her former CO, gazes at her like a lover. *Turbulent*, nearly thirty years old, has but three months to go before her decommissioning ceremony, whereupon *Tireless* will become the old lady of the surviving 'Trafalgar' class. HMS *Vanguard* is in Faslane, berthed alongside the huge ship lift, and will relieve HMS *Victorious* when she returns from patrol. The mountains above Arrochar, unlike *Victorious*, look benign today.

Afterwards

Three of the four 2012 Perishers passed: Chris Gill, Sam Owen and Neil Botting. Andy Reeves did not. He left the boat on the penultimate day. I was saddened to hear this, but Reeves has great qualities and will do good things. Who made it and who did not was, I suspect, around the Submarine Service in a flash, probably even before Teacher and his successful students sat down for the legendary Perisher Breakfast when *Tireless* returned to Faslane.

I left the course before witnessing any of the above. However, the following year, in October 2013, Jinks and I boarded another 'Trafalgar' class submarine, HMS *Triumph*, to witness another group of Perisher students, Lieutenant Commander Louis Bull, Lieutenant Commander Ian Ferguson, Lieutenant Commander David Burrill and Lieutenant Commander Ben Haskins, complete the final days of the course under Teacher, Ramsey. They and their predecessor cohort presided over a slightly different course from the one played out on HMS *Tireless* in 2012. After witnessing the 2012 sequence first hand, Ramsey, Captain Halton and Rear Admiral Corder felt that parts of the course needed to better reflect the world that the Submarine Service had been operating within over the past few years and was likely to do over the decades to come. The course continues to evolve, but its core principles remain the same.

HMS *TRIUMPH*, 19 OCTOBER 2013

0725Z

Triumph's CO, Commander Dan Clarke, greets us warmly on the casing. Down into the bright friendly snug of the Wardroom. The steward has hot tea in our hands in thirty seconds. A record in the competitive hospitality stakes. Bacon and sausage butties swiftly follow. The Submarine Service specializes in comfort-food welcomes. It's plain this is going to be quite a final day. The Perishers are about to have the lot thrown at them and they are very, very tired. They have not slept for more than four hours at a stretch for three and a half weeks. However fierce the malign, multiple combinations that are about to assail them, the surviving four will have to convince Admiral Matt Parr, Commander Operations, Northwood; Captain Chris Groves, Captain Flag Officer Sea Training; and Ramsey that they have the poise and judgement under pressure to balance effectiveness of mission with the safety of the boat. It's a question of cope or fail. They can be removed right up to the last minutes. Every Perisher, with just under twelve hours to go, knows that.

The last few days have been stretching. Commander Sarah West on HMS *Portland*, a Type 23 frigate, has been particularly relentless and resourceful and she's up there lurking – a brilliant adversary whose tactics are difficult even for the old sweats on *Triumph* to read. At one point during the previous weekend, *Triumph* came up to periscope depth to find *Portland* close by. Ramsey instantly took over from the Perisher in the Captain's chair and put the boat into a steep dive.

Triumph is about to participate in yet another 'Joint Warrior' NATO exercise involving numerous warships, aircraft, marines and troops. For this particular exercise the UK and Ireland have been turned into something called the 'Wallian Archipelago' and the waters of the west of Scotland into the equivalent of a Middle Eastern flashpoint. The cause of the flashing, as it were, is a country called Pastonia whose borders are largely, though not wholly, coterminous with those of Scotland (running across from south of Ayr in the west to south of Berwick-upon-Tweed in the east). To the south of Pastonia lies Dragonia (whose state line runs from just north of St David's in West Wales to near Grimsby on the North Sea coast). To the south of Dragonia is Avalonia. Ireland, north and south combined, is called Ryania (whether after the airline or the Teacher is unknown).

The NATO countries (continental Western Europe) are deeply worried by the civil war in Pastonia and its consequences for the non-NATO Wallian Archipelago in general. Pastonia is breaking up into two: territory controlled by the Government of Pastonia, the GOP (based in Glasgow), and the far Left Peoples' Republic of Pastonia, the PRP (headquartered in Edinburgh). At one point in the exercise, Edinburgh Castle, the PRP's Ministry of Defence, is destroyed by a cruise missile launched from HMS *Triumph* under the waters between the Isle of Skye and southwest to the south of the Little Minch. The GOP remains the internationally recognized authority in Pastonia. The PRP is actively trying to destabilize Dragonia, its neighbour to the south, through funding and supporting the Free Dragonian Brotherhood, which is a far-left-wing terrorist group active throughout the Wallian Archipelago. At the very centre of the Pastonian civil war are Skye, the other Western Isles and the Isles of Arran and Cumbrae. The Free Dragonian Brotherhood has terrorist groups in the area as well.

0810

The crew are wholly at ease with the 'Wallian Archipelago' scenario and talk about it during the day. 'The shape of the North Channel is very similar to the Straits of Hormuz,' says one of *Triumph*'s officers.

- The tasks will unfold like this: From now until about 1130, *Triumph* will concentrate on the two Cumbrae Islands and Kilchattan Bay on the Isle of Bute, surveilling possible beach landing sites for Special Forces, detecting surface-to-air missile launch sites and taking a look at what intelligence indicates is a 'nest of terrorists' on Little Cumbrae. The frigates and the Merlin helicopters will be giving them a hard time throughout.
- Between noon and 1430, *Triumph* will concentrate on Brodick Bay in the middle of Arran's east coast. Intelligence suggests there may be a dirty bomb factory on the outskirts of Brodick and that a shipment of WMD might be about to leave from Brodick (the real-life ferry from Brodick to Ardrossan, unknown to its serene autumn Saturday afternoon passengers, will simulate this – though, no doubt, they will wonder why two Type 23 frigates are racing around).
- Between 1500 and 1800, concentration will switch to Lamlash Harbour to the south of Brodick, where intelligence suggests there are WMD sites and military raiders which may need to be taken out by Special Forces or cruise missiles.
- Intelligence also suggests that leading Free Dragonian Brotherhood terrorists are active in these installations or close by. Each has a code name: 'Selector 1', 'Selector 2' and so on.

The Perishers know that, in addition to their executing these tasks (of which they have notice and for which they will have prepared), Teacher will throw in reams of the unexpected.

Earlier in the week Ramsey had built in an episode which the crew were still talking about. A young sailor had been chosen for his acting gifts and secretly primed that, at a certain time, he must throw a wobbler and take a hostage on the boat. And one of the Perishers would have to negotiate with him while the rest of the boat carried out their tasks as best they could. The young sailor locked himself into the Wireless Telegraphy Room, taking one of the W/T operators hostage. Then he threatened to wreck the equipment with a wrench. He broadcast music over the submarine's internal speakers. He insisted on ice-cream

all round for the crew. He demanded (and succeeded) in getting a line
through to his mum's phone. Mum was naturally hugely surprised and
not a tad alarmed at hearing her boy was under water and in a bit of
trouble. It took an hour and a half to talk him down and out of the
W/T Room. By general agreement, he deserved an Oscar. His legend
will live on in the accumulated Perisher 'dits'.

There are careers ahead which will see those who succeed on Perisher today through to retirement. These will be the Captains of the
Royal Navy's latest nuclear submarines, the 'Astute' class (they will
hear the boat to which they are to be assigned this evening). And the
pressure is continuing to pile in upon them. The intensity of it can most
vividly be illustrated when Bull was in the Captain's Chair in
mid-afternoon off Brodick. Not only was it stretching for Louis, but
the sequence was a graphic reminder to the non-submariners aboard of
just how many things can go wrong with these boats (and this was but
a fraction of them).

This is what it felt like in *Triumph*'s crowded Control Room.

1440

We're at periscope depth about three miles off Brodick. One of the
Type 23s is on us. We can hear its sonar squeaking. Suddenly the klaxon
sounds 'Emergency. Emergency.' Bulkhead shutdown. *Triumph* effectively stops. Damage control reports come in. 'Emergency. Emergency.
Major steam leak on starboard side.' More squeaking from the frigate's
sonar. 'Casualty. Casualty in the Aft Escape Platform.' More squeaking. 'Major steam leak aft.' A steam leak on this scale means the boat
loses about two thirds of its propulsive power. Working conditions will
be pretty dire aft.

1445

Contact with a hostile submarine: 'Belligerent intent' (i.e. it might be
about to launch a torpedo at *Triumph*). *Triumph* simulates the launch
of a Spearfish torpedo towards the hostile submarine, which is
2500 yards away.

1449

'Emergency stations. Emergency stations. Hydraulic burst in the Bomb Shop.' Bull, as acting Captain, halts the launch of a second Spearfish. We're down to just two torpedo tubes. The others have 'defects'.

1453

'Electrical failure. Electrical failure.' They are trying to work out the cause.

1455

'Fire. Fire in Five Berth.' Everyone puts on their breathing apparatus. You have to do this inside the three minutes it takes for smoke to circulate throughout the submarine. It makes doing your job at least 20 per cent harder. The Control Room resounds to the hissing noise of the oxygen and the crew breathing and talking to each other through the apparatus. Stress is building on stress – five emergencies running simultaneously. The combined concentration in the Control Room is almost off the Richter Scale.

Bull has coped well. He comes down to the Wardroom for a breather. He confirms the five things Teacher, metaphorically speaking, has hurled at him. He got the photos he was tasked to acquire:

- The Brodick ferry (unknowingly masquerading as a carrier of weapons of mass destruction).
- The Brodick jetty, where they had been embarked.
- The grey building on the rim of the town, where the dirty bombs had been put together.

Bull, an enthusiast by nature, is still remarkably chipper. Without self-pity he says: 'I've been nine hours in the Control Room. I've not eaten. I've had quite a lot of caffeine.'

He still doesn't know if he's going to pass but he's truly glad he's on the last stretch with just over three hours to go to Teacher's summons. Bull has a wife and three kids in Glasgow. He and his wife have had a pact that they wouldn't communicate over the last month of the Perisher and they've stuck to it.

Ferguson is back in the acting Captain's seat. Plainly Teacher wants a last look at his capabilities. The final sequence involves identifying

WMD stores and military radar sites behind Lamlash to the south of Brodick and to guide Special Forces towards the location of key 'selectors', i.e. terrorists. You feel for Ferguson. You really want all four to pass. It would be ghastly to stumble as the clock ticks towards seven and the legendary Perisher Sunday morning breakfast is almost within touching distance.

At 1845 Ramsey comes into the Wardroom (now a Perisher-free zone) to tell us all four have passed. Relief and pleasure all round. The room is prepared for the very special rites of passage for the October 2013 Perisher quartet. Just after seven in the evening, one by one they are brought into the Ward Room by Lieutenant Commander 'Bing' Crosby, HMS *Triumph*'s Executive Officer. One of the Wardroom tables is draped in the White Ensign. The other bears the champagne glasses. The Cava (it was an era of austerity, after all) lies chilling in the galley fridge.

Ramsey leaves him in suspense for a few moments more. 'How do you think you did?' he asks. Each exhausted Perisher, face tense, anxiety palpable, manages to get out a few modest words as best he can. Ramsey pauses once more. Then out goes his hand as the magic, confirming words follow. 'Congratulations, Captain.' Applause from the rest of us – then the new 'Captain' is taken out of the Ward Room before the next one arrives. He doesn't know if his fellow Perishers have made it or are about to be put off the boat with their kit and a bottle of Teacher's Whisky for the lonely journey up the Firth of Clyde to Largs, almost certainly never to set foot on a submarine again. Failure brings the cruellest rites of passage.

But they had prevailed after an utterly stretching last day crowded with incident. The handshakes are over and the tension eases. Now a protracted celebration with that Cava in the Wardroom; with beer and Havana cigars on the *Eva*, the boat taking us all back to Faslane. As the *Eva* slices through the dark northwards to a cold and wet and dripping Faslane, Ramsey unburdens. This was his last Perisher. He is due to hand over as Teacher and leave the Navy. 'Tonight it's really mixed emotions. It's as if half my life is done. I'm never going to feel that team cohesion again. I've gone out on a massive high. I've loved the Submarine Service from start to finish.'

There was more late-night drink in the Super Mess when we arrived back at Faslane, followed by a few hours' sleep and then, early in the morning on Sunday, 20 October, we gathered once again for the Perisher Breakfast to complete the rite of passage as all anxiety fled, the relief, the camaraderie and the warmth of the food and the bliss of the booze took hold. There are fourteen of us in the Blue Room, just off the Mess, overlooking the water. The food and drink are heavy duty for the morning:

Starter
Smoked Salmon and Smoked Mackerel served with
bread and a mustard dill sauce

Traditional Perisher Breakfast
Pan Fried Steak
Bacon
Sausage
Grilled Tomatoes
Portobello Mushrooms
Potato Sauce
Scrambled Eggs

Dessert
Stewed Apples blended with natural yoghurt and toasted oats
Coffee
Wines
Port

Admiral Parr distils the significance of passing for the Perishers. Last night in the Wardroom he told them that it did not mean they could now 'walk on water'. This morning he tells them they will be confident as Commanding Officers that they will know what to do when things happen; if they have a feeling something is not quite right with the boat, they must follow that instinct and ensure that things are checked.

The atmosphere is jolly – but not raucous. There is a dash of seriousness. It's a big thing happening to Bull, Ferguson, Burrill and Haskins. Within a few days their names will be inserted on the big board beside the staircase of the Super Mess in Faslane with all the other Perisher-reared submariners who have formed the thin deep-blue line of Commanders since the First World War. All four know where they are going. Haskins will succeed 'Bing' Crosby as *Triumph*'s Executive Officer. Burrill will become XO on *Astute*. Ferguson will become XO on one of *Vanguard*'s two crews. Bull will join *Artful* as XO and see the boat out of Barrow and into its sea trials.

No other club, however exclusive, has a rite of passage to match Perisher's. It's that blending of risk-taking, decisiveness and effectiveness with the crews while still keeping within the boundaries of safety, plus signs that you have that indefinable 'sixth sense', which is so very striking. Commander Rémy Thomas, the French equivalent of Teacher who was a French officer on Perisher 2013, was eloquent about this once we were ashore. 'French COs,' he said, 'are more scientific. They are all engineers. The Royal Navy are more warriors. On board we have discussions between experts – the CO and the Engineer. The Royal Navy is a dialogue between a warrior and a scientist. In France it's between two scientists. We have all this basic culture in nuclear science. And you are paranoid – in a good away – about protecting your secrets.'

It's easy, once you have experienced it, to understand how Perisher has acquired its coating of legend. It is one of the most famous military courses in the world. It's both a training and an initiation rite, all performed in front of an entire crew. Those who pass feel they are a special breed. They never forget their own Perisher and rerun it in their heads like an old movie. It is a theatre whose play, though based on a richly historical plot, grips imaginations anew every year when each set of selected Perishers prepare for the private war that will determine their careers.

2

'The Most Dangerous of All the Services': From World War to Cold War

There is no branch of His Majesty's Forces which in this war has suffered the same proportion of fatal loss as our submarine service. It is the most dangerous of all the Services. That is perhaps the reason why the First Lord tells me that entry into it is keenly sought by officers and men. I feel the House would wish to testify its gratitude and admiration for our submarine crews, for their skill and devotion, which have proved of inestimable value to the life of our country.

Winston Churchill, 1941.[1]

I merely state a fact of life when I say with Kipling, that I received neither 'promotion or pay' from this operation [a Cold War submarine operation]. So, with nothing tangible to show for it, and not even being allowed to mention it over the years, it is almost as if the six months concerned had never happened; leaving something of a hole in my career. When working for the church, I was sometimes addressed by the courtesy title 'Commander'. On one occasion I overheard one secretary say to another: 'Oh, he's not a Real Commander!' I had to 'bite on the bullet', having known a number of 'real' Commanders, particularly – dare I say – non-executive, who have never had the sort of responsibility or experience that I had on those patrols. In those days, it was probably the closest one could get to a real operational command experience, outside an actual 'shooting war'.

Lieutenant Commander Alfred Roake, Royal Navy,
recalling a Cold War operation he conducted
while CO of HMS *Turpin* 1994.[2]

VICTORY

The cost of the Second World War to the Submarine Service in terms of both human life and equipment was vast. Whereas the casualty rate in the surface Navy was 7.6 per cent, in the Submarine Service it was 38 per cent.[3] Very few submariners who witnessed the start of the war lived to see it end. Between 1939 and March 1945, 3508 submariners perished or were captured, over a third of the 9310 that served in the 206 Royal Navy submarines that put to sea. Seventy-four of those submarines never returned from patrol; in the Mediterranean alone, where a single submarine was, in the words of the First Sea Lord, Admiral Sir Andrew Cunningham, 'worth its weight in gold', forty-six were lost between 1939 and 1945.[4] But when the war in Europe ended, the Navy quickly concluded that it possessed far too many submarines: 142 were either in service or under construction.[5]

In May 1945, the head of the Submarine Service, Admiral Submarines, Sir George Creasy, recommended reducing the size of the fleet to 100 by scrapping or expending as anti-submarine targets virtually all of the older 'H', 'L', 'O', 'P', 'R' and 'U' class submarines, while keeping the newer 'S', 'T' and 'A' class for post-war service. But the sudden return to peace, the onset of post-war austerity and the recognition that the United Kingdom had emerged from the war in a seriously weakened position forced Creasy to recommend suspending construction of fourteen of the new 'A' class submarines, leaving just twelve in the programme, and suspending three 'T' class submarines then under construction and scrapping nine old or badly damaged submarines of the class. When the war against Japan ended in August 1945, Creasy recommended additional reductions, from 100 to 85 submarines, consisting of 12 'A' class, 27 'T' class, 37 'S' class and 9 'U' class, with 45 in commission and 40 in reserve.[6] In April 1946, these numbers were once again revised downwards to just 73, with 45 in full commission and 28 in reserve (manned by a one-third crew and refitted every twenty months).[7] Four Second World War era midget submarines, 'XE' class, were also retained.

By the following year all the small 'U' class submarines had been disposed of.[8] Some 66 submarines were still on hand or about to be delivered, 25 'S' class, a similar number of 'T' and 16 'A' class. As the Director of Naval Construction had minuted in 1945, the 'S' class were based on ideas sixteen years old, the 'T' class on ideas ten years old and the 'A' class on ideas three and a half years old.[9] These submarines were essentially surface vessels with the capability to submerge. They were slow, unmanoeuvrable when submerged and could only remain so for limited periods (measured in hours, depending on the speed at which they were travelling), having to surface periodically to run the diesel engines that recharged batteries and replenished air supplies. Submariners, particularly those who operated in the Far East against the Japanese, knew how outdated their submarines were, especially when they were compared with the US Navy's fleet.[10]

Towards the end of 1944, Creasy had spent three months considering 'the problem of the future design of the Submarine after the war'. He was 'swayed by the developments in the U-boat campaign' in the North Atlantic as well as German submarine tactics and technology. At the outbreak of the Second World War the Royal Navy successfully exploited a new underwater detection device, an early form of sonar known as ASDIC (named after the Anti-Submarine Detection Investigation Committee) to detect German U-boats. Although Churchill later claimed that 'The U-boat peril' was the 'only thing that ever really frightened me during the war', the Allies succeeded in exploiting the one major weakness of the U-boat: its need to operate on the surface at frequent intervals to run diesel engines and recharge batteries. Each time a U-boat surfaced it was vulnerable to attack from nearby surface ships and aircraft.[11] Between 1940 and 1942 Allied anti-submarine operations were so successful that Germany's U-boats were forced to adopt a campaign of night surface attacks, with multiple U-boats working as packs, moving ever further afield hoping to stretch Allied forces to breaking point. By 1943, Allied anti-submarine detection and destruction methods were having such a serious impact on the U-boat campaign that in a desperate attempt to reduce their losses the Germans turned to a Dutch device known as the Schnorkel.

The Schnorkel was a piece of equipment consisting of a raised mast through which air could be drawn into a submarine when it was submerged and operating at periscope depth. The new device allowed German U-boats to stay submerged for extended periods of time while running diesel engines. Fitted originally as a defensive measure to reduce the chances of detection by Allied aircraft and surface ships, U-boat crews quickly realized the offensive potential of the Schnorkel and used it to return to waters near the UK mainland in which they had not operated since the early years of the war. Having been driven below the surface by the weight of Allied anti-submarine countermeasures, the

Schnorkel-equipped U-boats started a new offensive, a true submarine campaign. 'As a "submersible" the U-boat has been defeated,' noted Creasy in November 1944, 'but as a "Submarine" it has returned to the attack.'[12]

The Schnorkel had a notable impact in the closing months of the Second World War. In November 1944, the Director of the Anti-U-Boat Division, Captain Clarence Howard-Johnston, concluded that:

> The Schnorkel has had such far reaching results that the whole character of the U-boat war has been altered in the enemy's favour. Frequently he has managed to penetrate to and remain on our convoy routes in focal areas with impunity in spite of intensive air and surface patrols. With more experience in training and with the confidence engendered by his present immunity from air, and often from surface attack, he is likely, in the future, to do us more real harm than he has up to the present.[13]

The war ended before the Germans could exploit the full potential of the new equipment.[14] But the Royal Navy recognized that it was 'evident that the advantages to be gained from the use of the device are very great' and that there may come a point when it was necessary to equip its own 'submarines with "Schnorkel" at short notice for which we must be prepared', but it stopped short of adopting the new device in the final months of the Second World War on anything but a very limited basis. This was because, as Howard-Johnston explained:

> On our side, our Submarines have, happily, never been called on to face any comparable scale of Anti-Submarine Defence, either in weight of numbers or in efficiency of material or personnel. Thus, we have been able to pursue the tactics of diving by day and surfacing by night which have been the basis of Submarine operations since the earliest days of our Submarines. In effect, our Submarines have always operated as 'Submersibles' and continue to do so.

Creasy too argued that the Schnorkel:

> In its present form has several weaknesses . . . e.g. space and weight inside and outside the submarine, inability to operate in rough weather, considerable discomfort, not uncoupled with danger, to personnel. Apart from the raised morale the 'Schnorkel' bestows, by increasing the submarine's chances of survival, it is thought that it will tend to make the submarine captain feel unsafe when it is <u>not</u> being used. This has the direct effect of encouraging submarines to remain submerged, even when it is unnecessary

for them to do so, and thus by decreasing their mobility decrease their offensive spirit and power to inflict damage. Results to date seem to indicate this state of affairs.[15]

Creasy recognized the long-term significance of the Schnorkel. He knew that 'Inevitably the day will come, sooner or later, when the Anti-Submarine efficiency of our enemies reaches the point which we have already achieved.' But he was far more concerned with 'the problem of the more distant future' as he could see 'no development in prospect that would restore to the "Submersible" its surfaced ability'.[16] He recommended that experimental Schnorkels, given a British title, the 'snort', should be fitted in one 'U', one 'S', one 'T' and one 'A' class submarine, 'as soon as it is possible to forecast when they will be required', but that the Royal Navy should wait until it had gained more practical operational experience before equipping its entire submarine fleet.[17]

On 11 January 1945, Creasy assembled the senior Admiralty Staff involved in submarine policy at the wartime headquarters of the Submarine Service at Northways House, near Swiss Cottage, London, to discuss 'The Future Development of the Submarine'. A paper was tabled that outlined three types of future submarine:

(a) A craft with the maximum surface speed, endurance and armament and the best submerged performance and armament that can be achieved, subject to the surface requirements.
(b) A craft with the maximum submerged speed, endurance and armament and the best surface performance and armament that can be achieved, subject to submerged requirements.
(c) A compromise of (a) and (b) with the best alternative combinations of surface and submerged speed, endurance and armament.[18]

The biggest problem was the requirement for maximum submerged speed and endurance. In order to design and build a submarine that met these needs a new type of power plant was required, one that could generate great speeds while allowing a submarine to stay submerged for extended periods. But, as the paper pointed out, 'the production of a propulsive unit around which the future submarine must be designed does not, at the present time, appear to be possible for some years'. This did not mean that other features of a future design could not be developed by using existing means of underwater propulsion, such as diesel engines and batteries, to produce experimental submarines in which to test future designs and equipment. The paper argued that it was 'essential that development of other factors, and in particular the hull form and its attendant problems, be advanced to such a stage that there is a minimum of delay in building the prototype future design, when a suitable type of power plant becomes available'.[19]

The Admiralty had already embarked on some experiments, principally by adapting one of the Royal Navy's 'S' class submarines, HMS *Seraph*, as a fast underwater experimental submarine. Trials with *Seraph* had shown that 'the building of a craft with special hull form, much increased battery capacity and high power motors, increased diving depth and specially designed control gear is a possible and immediate requirement'.[20] Long-term policy became to proceed with experiments and investigations with a view to developing a true submarine, one that could stay submerged for long periods, while in the short term continuing to develop the submersible.

When the war ended the Navy paused its investigations into future submarine designs and started to examine captured German submarine technology. Aside from the Schnorkel, the Navy was interested in two additional German submarine developments.[21] The first was a new class of U-boat, known as the Type XXI, designed to operate submerged for prolonged periods of time, thus evading Allied anti-submarine detection efforts. It was superior to the Royal Navy's submarines with respect to sonar, underwater speed, depth capability and torpedo reload interval, and was fitted with powerful batteries as well as a distinctive streamlined hull which had a rubberized (anechoic) covering to reduce detection from radar on the Schnorkel head and active sonar on the hull.[22] With the ability to approach Allied convoys, attack and evade pursuit at high speed, the Germans hoped that the Type XXI would render obsolete Allied anti-submarine methods which had so effectively countered its existing U-boats.

In 1943 Hitler ensured that the Type XXI construction programme was awarded the highest priority. Albert Speer, the German Minister of Armaments and War Production, brought in Otto Merker as the head of the German Central Board for Ship Construction, who used his experience from the automobile industry to radically reduce the amount of time required to build the submarines by constructing them in eight prefabricated sections which were then fitted together on an assembly line.[23] In late 1943, the head of the German Navy, Admiral Dönitz, placed orders for 170 Type XXIs, and although over 120 were constructed, design faults, manpower and material shortages, short-comings in design and production, as well as a relentless Allied bombing campaign against German shipyards which damaged many of the pre-fabricated sections while they were being fitted together on the slipways meant that only one Type XXI, U-2511, conducted a war patrol. Although this was far too late to have any decisive impact on the out-come of the war the British recognized that the Type XXI 'was a formidable weapon which would have given us a lot of trouble. Nor was it the last word, as submarines of still greater under-water speed were being developed by the Germans.'[24]

Since March 1943 British intelligence had been receiving reports (of varying reliability) about new high-speed submerged U-boats powered by a revolutionary method of propulsion that had the potential to transform submarine warfare. In April 1945 naval intelligence reported on the development of the small Type XVIIB and the larger Type XXVI U-boats, both powered by the Walter gas turbine propulsion unit and capable of producing underwater speeds in excess of 20 knots, fast enough to outrun any surface ship or submarine attempting to pro-tect a convoy. Mindful of the potential threat, the British and the Americans went to extraordinary measures to capture the new U-boats, specifically two of the Type XVIIBs, U-1406 and U-1407.[25]

THE SPOILS OF WAR

On 4 May 1945, the Royal Navy's secret intelligence group, 30 Assault Unit, entered Kiel and proceeded towards an industrial facility known as the Walterwerke, where a German scientist, Helmut Walter, was working on the experimental U-boats. Despite extensive bombing by the Allies, the Walterwerke was around 90 per cent intact. 30 AU quickly apprehended Dr Walter, who was living next to his plant. A 'rather heavy, flabby-cheeked man of 45', Walter had joined the Nazi Party in 1932 and became a leader of industry in Kiel.[26] Initially unwill-ing to divulge the details of his work, Walter only started speaking when Admiral Dönitz, who was by now Hitler's designated successor as Führer, ordered him to disclose everything. 'The effect was instant-aneous, and from that day, May 7th nothing (one hopes) was kept back.'[27] His interrogator wrote that 'It lurks constantly in the writer's mind that he may even yet prove to be one of the sinister crooks of international fame who features so often in filmed fiction.'[28]

Walter's views on submarine design were very advanced. He told his interrogators that he foresaw that the submarine of the future would only operate submerged and that, in war, there was no future for a submarine on the surface operating away from its base.[29] He advocated subordinating all surface qualities in order to obtain optimum submerged capabilities, including the provision of any guns as they made the bridge too big and marred the streamlined shape of the submarine. He had given the bridge design 'ruthless consideration' and argued that it should only be of sufficient size to house the periscope, compass and radar.[30] His views on torpedoes were revolutionary. Stern torpedo tubes, he said, should be unnecessary because of the speed with which a submarine would be able to turn and the additional congestion involved. Walter was obsessed with speed, insisting that any future submarine should be capable of 25 knots in order to outrun escort vessels. He even maintained that submerged speeds of up to 30 knots were achievable.

To achieve these speeds Walter had developed a revolutionary method of propulsion, a closed-cycle system, based on the decomposition of highly concentrated hydrogen peroxide (HTP), which was decomposed to produce steam and oxygen at high temperatures and then combined to ignite diesel oil. Water was then sprayed to cool the temperature and produce steam, which then was used to drive turbines, while the remaining water and residual carbon dioxide were transferred into a condenser before being exhausted.[31] Four Type XVIIA U-boats had been commissioned into the German Navy in late 1943 and early 1944 to prove the technology. While they were successful – one U-boat, U-792, achieved speeds of 25 knots – they suffered from significant mechanical problems and poor reliability. Despite these deficiencies, construction of operational U-boats, known as the Type XVIIB, started in mid-1944. Only three, U-1405, U-1406 and U-1407, had been completed by the time the war ended, and while work on larger XXVI U-boats had also started at the Blohm and Voss shipyard, none were anywhere near complete when it was occupied by the Allies.

The importance of Walter's work was immediately apparent to the Royal Navy. Creasy noted that 'We stand on the threshold of very considerable technical development and the submarine of the future may differ profoundly from the submersibles of the present and past.'[32] The British hoped to capture all three Type XVIIB U-boats, but when they arrived at the Blohm and Voss shipyard they discovered that they had all been scuttled. On 17 July 1945, a meeting chaired by the Third Sea

Lord and Controller of the Navy, Admiral Sir Frederic Wake-Walker, concluded that U-1407 should be raised and brought to the UK and that an experimental high-powered unit that was discovered in one of the occupied factories should be 'bench tested' in Germany and then brought to the UK for further research and development along with the essential German personnel.[33]

Although the UK and the US were still allied with the Soviet Union, the British government was very concerned about Walter's HTP technology falling into Russian hands, even more so when Walter and his fellow scientists made it clear during their interrogations that they were 'indifferent whether they worked for us or for Russia, as long as they were permitted to work somewhere'.[34] There was, Cunningham noted, 'no question that by bringing them over here we prevent the Russians getting at them and also taking advantage of their brains and knowledge'.[35] Despite a concerted effort to prevent the Russians obtaining any of the advanced German technology, Soviet forces had occupied a facility in the town of Blankenburg that had been involved in the design of the Type XVII as well as Walter's hydrogen peroxide drive.[36] They had also captured the plans for the Type XXVI Walter-boats. In July 1946, Cunningham warned that the Walter propulsion system was 'known to other powers and it is essential that we should have, as soon as possible, experience in the construction and use of such submarines'.[37]

MODERNIZING THE WARTIME FLEET

What the British found in Allied occupied Germany had a profound impact on the Royal Navy's future submarine policy. In February 1945, Creasy had concluded that the Royal Navy's future submarine construction should proceed in two directions. First the construction of an experimental submarine to investigate requirements for what was called a 'true submarine'. He envisioned a submarine based on the 'A' class design, but with improved battery capacity to enable short speeds of up to 20 knots to be achieved. Second, he envisaged the construction of three improved submarines in 1945, based on 'A' class, but with a redesigned Control Room to incorporate new radar equipment, a modified snort mast that could be raised and lowered in the same way as a periscope, and a new hull, built from improved steel to allow greater diving depths.[38]

However, in August 1945, Creasy informed the Secretary of the Admiralty that 'as a result of further consideration, including study of German policy and design', his 'view on the development of the submarine of the immediate future' had 'undergone some changes from those expressed' in his February 1945 memorandum. In particular, his initial conclusion that the Royal Navy should construct an improved 'A' class submarine had 'changed profoundly'.[39] 'The decision that has to be made,' Creasy wrote, 'is whether to perpetuate the present policy of high surface qualities and go for improvement on these, which was broadly speaking my original intention, or whether to accept that no future submarine construction would be justified that was not based on the tactical use of the "Snort" and on high submerged performance.'[40]

Creasy concluded that the overriding requirement was for a submarine capable of high submerged performance. Strategically this was 'required in order to enable a submarine to move rapidly from one area to another or to traverse a danger area in the minimum of time'. Tactically, high submerged speeds were required for three reasons. First, to enable an attacking or intercepting position to be reached in the face of air opposition, something that could not be achieved while a submarine was snorting due to the amount of wake caused by the snort tube. Second, for evading surface craft after an attack, primarily in a patrol area close to an enemy's base where a large concentration of anti-submarine operations could be expected. Third, in order to attack surface ships and later, as we shall see, submarines.[41]

Creasy no longer considered it 'desirable to proceed with the design of an improved "A" class' and recommended a 'radical change': a new submarine built around the ' "Snort" at the sacrifice, where necessary, of surface qualities' in order to increase submerged speed. This, in effect, replicated the German policy with the Type XXI U-boat. But Creasy was far more ambitious when he argued that the Submarine Service 'should be prepared to go further than have the Germans in subordinating the attributes required for surface performance to those requisite for submerged performance'. This, Creasy explained, meant 'the surrender of gun power, acceptance of comparatively moderate cruising speed on the surface, and probably of reduced sea-keeping qualities and manoeuvrability on the surface'. Nevertheless, he felt it would be a mistake 'to embark on such a design before we have completed first-of-class trials of the Type XXI submarine and are so able to gain maximum value out of the German mistakes as well as out of the merits of the German design'. He also questioned 'whether it would be justifiable to proceed with the design before we know more about the performance of the Walter engine'. He recommended constructing an experimental 'true submarine' as well as three additional submarines equipped with both the Walter drive and with normal diesel and battery propulsion.[42]

In reaching these conclusions Creasy was well aware that 'requirements change and will continue to change, so that no design can ever be considered more than up to date at the moment of its acceptance'.[43] He believed that the Royal Navy's 'ultimate aim must be to produce the "true submarine" with all surface qualities sacrificed to submerged qualities, with a really high submerged speed, and the ability to remain submerged up to the greatest maximum depth for an indefinite period. This was my view eight months ago and this is my view today.'[44] At the same time, he recognized that such a development depended 'entirely on the production of an engine of the closed-cycle type'. Yet nothing he had learned from German sources convinced him that such an engine was 'yet in sight' and he did not believe that one could be developed 'for a long time yet to come, even if the possibilities of atomic energy are taken into account'.[45] (A few weeks earlier, the United States had dropped the atomic bombs on Japan; the power of atomic energy was on everyone's minds.)

Creasy's mention of atomic energy as a possible means of propulsion was prescient. Individuals within the Admiralty recognized the potential of using atomic energy to propel submarines. 'The atomic reactor is well suited to submarine propulsion, developing full power under all conditions, and quite independent of whether the submarine is on the surface or not,' noted Jack Daniel, a young naval architect from the Royal Corps of Naval Constructors, in a 1947 paper to the Royal Institute of Naval Architects.[46] Between 1946 and 1950 a small team of up to two naval Scientific and two Engineer Officers was incorporated with the Atomic Energy Research Establishment at Harwell to keep under study the possibilities of applying atomic energy to ship propulsion. A few officers from the Naval Staff Divisions and Technical Departments were, under special security restrictions, kept informed of relevant research and development in the atomic field with the aim of guiding and stimulating thought within the Admiralty to the same end. Progress was reviewed at meetings convened from time to time by the Deputy Controller (R&D) but very little was achieved.[47] 'Although such a system appears a possible development, it seems very unlikely that fissile material would be economically employed in a submarine rather than in atomic bombs,' noted one 1947 Admiralty paper on sub-marine development.[48]

Due to lack of fissile material, facilities and trained personnel, it was impossible to institute anything more than the sketchiest preliminary investigations. These identified the problems associated with applying atomic energy to submarines, such as a need for new submarine designs and an extensive testing programme. 'Difficulties associated with the introduction of an entirely new propulsion system would undoubtedly arise,' noted the same 1947 paper. The Navy recognized that 'eventual success is not an unreasonable expectation' but concluded that 'Such a vessel is many years away and so little is known about it that there seems little point in discussing the matter further at this stage except to say that 30 knots for six months does not seem an unreasonable figure for its mobility. (Six times round the world!!)'[49] This early dismissal of atomic energy would plague the Navy and the Submarine Service for years to come. With atomic energy, as well as other types of propulsion systems a distant prospect, Creasy recommended that the Royal Navy 'should proceed with the building of the experimental type "true sub-marine" for the purpose of answering all the problems that will arise in building a craft that will take the engine when it does materialise'.[50]

The Royal Navy's focus in the immediate post-war period developed along two lines: first, the continuation of Walter's work, the development of a Hydrogen Peroxide programme; and, second, the modernization of the Royal Navy's wartime submarine fleet. Both were aimed at devel-oping a submarine with a high submerged speed. Experiments with captured German HTP propulsion units had produced 'satisfactory results'. By 1946 all the important components at the Walterwerke had been transported to Vickers in Barrow, where Walter and seven of his key staff and their families were taken to work. The Royal Navy's Engineer-in-Chief considered that it was 'essential that this research should proceed, as the whole future submarine policy depends on the successful development of these engines; this matter is of outstanding importance and urgency'.[51] As the Navy's knowledge of the technology improved, many in the Admiralty besides Creasy began to believe that HTP was the answer to developing a true submarine. At a February 1946 meeting to review progress, the Admiralty's Deputy Controller (R&D), A. P. Rowe, argued that 'next to the atomic bomb' the hydro-gen peroxide propulsion system and the true high-speed submarine, 'would present the greatest threat to the Commonwealth in war'.[52] In a July 1946 memo to the Cabinet, the First Lord of the Admiralty, George

Hall, added that the 'development was revolutionary and will have far reaching effects on submarine warfare'.[53]

HTP development progressed in two directions. The first involved refitting the raised German U-boat, U-1407, in order to obtain high-speed trials data. The Americans took U-1406 but did not operate her. U-1407's refit was completed at Vickers under Walter's supervision in 1947. In 1948 the submarine was commissioned into the Royal Navy as HMS *Meteorite* and put through rigorous sea trials off the west coast of Scotland. While the trials illustrated that the 'operational possibilities of a very fast moving submarine are obviously enormous', they also revealed a number of disadvantages. The most serious was the high cost of HTP, at least £300 per ton. It was also in short supply and *Meteorite* required a storage ship to carry it as well as a full-time escort, which added further to the cost. The submarine was also very unstable on the surface, unable to stop quickly in an emergency and very noisy. Yet the trials report concluded that:

> It is realized that the disadvantage of expense of an HTP submarine is undoubtedly large. But while it remains the only proven method of very high speed propulsion, it is considered that the disadvantage is outweighed by the speed/time factor. This speed would probably be used mainly for escaping after an attack. With its help big changes of direction and depth could rapidly be made whilst at the same time, large distances are being covered, thus increasing by enormous proportions the difficulties of an escort vessel.

Full-scale trials took place between 17 March and 30 April 1949 and *Meteorite*'s Captain, Oliver Lascelles, concluded that the submarine was 'an outstandingly difficult boat to handle on the surface' but that she was 'outstandingly easy to handle dived'.[54]

Meteorite's trials confirmed what many in the Admiralty already believed, that the Royal Navy's existing submarines 'for all practical purposes' had reached the 'maximum limit of speed and endurance'. 'It is clear,' argued one Admiralty paper at the time, that an HTP-powered submarine:

> would have a very considerable advantage over a battery driven submarine, both in reaching an attacking position and in evading A/S [anti-submarine] hunting craft after an attack . . . In order that our submarine fleet should maintain not only its offensive potential relative to other nations but, equally important, its value in the training of our A/S

forces, it is most necessary that a practical form of such a propulsive system should be developed as soon as possible.[55]

Two experimental submarines, E14, *Explorer*, and E15, *Excalibur*, both equipped with HTP machinery, were included in the 1945/6 and 1947/8 naval building programmes and the Admiralty had tentative plans to build fourteen such vessels to act as fast underwater targets for the Royal Navy's anti-submarine forces.[56] However, concerns about the cost and supply of HTP remained. In wartime Germany adequate supplies of HTP had been produced regardless of the cost. In post-war Britain the circumstances were very different. The capital cost of constructing a plant in the United Kingdom capable of producing 1000 tons of HTP a year was estimated at around a million pounds, while the manufacturing cost of HTP was now around £200 a ton. Once operational, *Excalibur* and *Explorer* were expected to use 30 tons of HTP an hour when operating at full power while submerged. Assuming both submarines ran at full power for 100 hours each year, their combined annual consumption of HTP was expected to be around 6000 tons, a high proportion of the planned total manufacturing capacity of the country, which was expected to reach a maximum of 7700 tons a year in 1952–3. This was insufficient by a substantial margin to support a large HTP-fuelled submarine fleet.

In September 1949, a paper produced for the Ship Design Policy Committee recommended that because of the problems with HTP, work on *Explorer* and *Excalibur* 'should proceed as scheduled, but that no consideration should be given at the present time to proceeding with further HTP-driven submarines unless a sudden emergency demanded reconsideration of this policy'.[57] The Admiralty Board agreed and effectively rejected HTP as a means of propulsion for operational submarines.[58] The twelve additional HTP submarines were cancelled, while work on the two experimental craft, *Explorer* and *Excalibur*, was to proceed as planned as submarines capable of 'speedy evasion or attacks from under a convoy followed by rapid escape were possibilities of such danger to our seaborne communications that it was imperative for us to produce submarines of comparable performance, so that effective counter measures could be developed by our A/S forces'.[59] Due to financial constraints brought on by the 1949 devaluation of sterling and the rearmament programme stimulated by the Korean War, which broke out in June 1950, construction of HMS *Explorer* and HMS *Excalibur* did not begin until July 1951 and February 1952.[60]

The decision to cancel the twelve HTP submarines was the correct one. There were simply too many unknowns. The US Navy, which had been conducting its own experiments with HTP since the end of the Second World War, had also concluded that HTP's economic and logistical problems rendered its use in operational submarines impracticable.[61] Both Navies continued with their HTP research programmes well into the 1950s, but they also explored other means of submerged propulsion which would give comparable results at greatly reduced running costs. Both the Royal Navy and the US Navy concluded that after HTP the most promising means of propulsion was one that used oxygen instead of HTP as the oxidant.[62] The British considered waiting for the results of the American development programme in order to 'adapt to our use the most satisfactory of their designs', but this was dismissed by the Navy's Ship Design Policy Committee because of differences in technical requirements, submarine design and specifications.[63] In November 1949, a UK Working Party on Submarine Propulsion was formed to 'survey the whole of the research and development programme relating to underwater propulsion of submarines by means of stored oxygen, and to direct effort into the most practical channels'.[64]

The United States Navy was also investigating the possibility of using atomic energy to power its future submarines, but up until the late 1940s its efforts consisted of a series of haphazard projects spread across various research laboratories in the United States.[65]

FOSM'S EMPIRE

During the war Admiral (Submarines) and his staff were accommodated in three floors of a modern block of flats in Northways. When the war ended Creasy transferred his headquarters back to the pre-war home of the Submarine Service at Fort Blockhouse in Gosport, next to Portsmouth Harbour. By 1948, the Admiral (Submarines) title had been replaced with the new title of Flag Officer Submarines. In the immediate post-war years the Royal Navy's submarines were organized into five flotillas, which were later known as divisions and eventually renamed squadrons with the adoption of NATO nomenclature in 1952. The flotillas consisted of the 2nd Flotilla, based in Portland and operated from a depot ship, HMS *Maidstone*; the 3rd Flotilla operating from a depot ship, HMS *Forth*, at Rothesay; and the 5th Flotilla in Portsmouth.[66]

Admiral (Submarines) also possessed a far-flung empire of Submarine Flotillas stationed in the Mediterranean and Far East, and later in Australia, Canada and Singapore. The 1st Submarine Flotilla was based in Malta and was serviced by the submarine depot ships HMS *Wolfe* and later HMS *Forth*. Its submarines were painted 'Mediterranean Blue' as the sea was often so clear that an aircraft could visually track submarines operating at depths of up to 100 feet. Submarines assigned to the flotilla spent much of their time conducting anti-submarine exercises with foreign navies such as the Americans, French and Greeks. The 4th Submarine Flotilla, serviced by the depot ship HMS *Adamant*, spent the immediate post-war years moving around the Far East, with periods in Hong Kong and Fremantle, Western Australia, before it finally settled in Sydney in 1949. There it comprised two or three submarines which provided anti-submarine training to the Royal Australian Navy and Royal New Zealand Navy. By 1955, only twenty-six out of the forty available operational submarines were based in the UK.[67]

These flotillas were commanded by seasoned submariners who had served with distinction during the Second World War. From 1945 to 1947 the 4th Submarine Division was commanded by Captain Ben Bryant, a swashbuckling, no-nonsense submariner, tall, with a seadog beard and arrogant eye, with a Distinguished Service Order (DSO) and two bars and a Distinguished Service Cross (DSC) won in the 'S' class submarines HMS *Sealion* off Norway and HMS *Safari* in the Mediterranean.[68] When the 4th Division arrived in Sydney in 1949 it was commanded by yet another distinguished wartime submariner, Commander Ian McGeoch. During the war McGeoch had successfully patrolled the waters of North Africa and the Mediterranean in command of HMS *Splendid*, sinking numerous German warships, for which he was awarded the DSO and DSC. In 1943 a British-built Greek destroyer under German control attacked *Splendid*, scoring a number of hits with depth charges. Though wounded in the eye McGeoch was able to raise the submarine to the surface. For twelve minutes, while the crew abandoned ship, McGeoch remained on board until it sank. He was taken prisoner. Though now blinded in one eye he spent most of his time in captivity trying to escape by various means, including digging a tunnel from an Italian hospital, jumping from a moving train while he was being transferred between prison camps, leaping from a moving car and even attempting to enter the Vatican. He eventually escaped by trekking 400 miles across Italy into Switzerland.[69] Another submarine ace, Captain Anthony Miers, the former CO of HMS *Torbay*, commanded the 1st Submarine Squadron in Malta between 1950 and 1952. Known as 'Crap Miers', in addition to winning the Victoria Cross for a daring and successful raid on Corfu Harbour in 1942 he had been awarded the Distinguished Service Order and Bar for 'courage, skill, enterprise and devotion to duty' in numerous submarine patrols. His ruthlessness, however, had to be reined in after incidents in 1941 when *Torbay* had machine-gunned German soldiers in the water.[70]

A number of distinguished wartime submarine aces also commanded the submarines assigned to the flotillas. Tony Troup, the youngest officer to command a submarine in the Second World War (he had been given command of the training submarine H32 in June 1943 at the age of twenty-one), commanded various submarines before holding a series of senior appointments in the Navy. Arthur Hezlet, the former commander of HMS *Trenchant*, who had destroyed a heavy Japanese cruiser in 1945, for which he was awarded a bar to his DSO and the Legion of Merit, the highest award the United States can bestow on a foreign commander, also remained, as did John Roxburgh, widely regarded as one of the most effective and ebullient submarine commanders to serve in the Royal Navy during the war. (At the age of just twenty-three he had been given command of HMS *United* and over an eleven-month period survived torpedoes, bombs and depth charges and sank some 12,000 tons of enemy shipping. While in command of HMS *Tapir*, he was also responsible for the last British submarine success of the war, destroying Oberleutnant Gerhard Meyer's U-486, which had sunk the troopship *Leopoldville* off Cherbourg with the loss of more than 750 American soldiers.)[71] But with the war over, many submariners were demobilized. Many wanted to leave. They were burnt out and had had enough. 'At the end of the war I was thirty-five and I looked fifty-five,' recalled Commander William King, 'I was a wreck, physically, morally, socially, financially and in every other way. I wanted to get out of the navy, but of course they wouldn't let us go. It took me two years to struggle out by writing letters.'[72]

Those who remained were joined by the first post-war recruits to the Submarine Service. Some, such as John Hervey, who later rose through the ranks to become a Rear Admiral, volunteered after being inspired by an account of the rescue of American submariners from the USS *Squalus*, which sank off the west coast of New Hampshire during a test dive in May 1939. 'I thought those are the sort of people that I'd like to be with and so I was always intent on joining despite a lot of pressure put on us to go into flying because the Fleet Air Arm was the big thing just after the war.'[73] Before the Second World War the Submarine Service had an abundance of volunteers. The service attracted young, ambitious midshipmen eager for responsibility and the opportunity of command at a young age, sometimes as early as twenty-eight. It also appealed to those seeking a stable and healthy work/life balance. Life was predictable, planned around the static or semi-static depot ships that made up the flotillas. It was also relaxed. Many submarines tended to operate three days a week, which allowed the crews to spend time with their families. Money was also an incentive, probably the biggest factor in producing so many volunteers. Submariners were entitled to specialist pay, known as Submarine Pay, which was often enough to make the difference between a young officer getting married or remaining a bachelor. In 1939, a Lieutenant's pay on promotion was 13/6 per day, with an extra 6/- a day Submarine Pay.

The Submarine Service suffered from severe manpower shortages throughout much of the early post-war years as it struggled to recruit adequate numbers of officers and ratings. Prior to the war submariners were nearly all regular servicemen who had volunteered for service. But by September 1942 only 69 per cent of regular officers had volunteered for service; the rest had been drafted. Reservists were also recruited and by 1941 half of all submarine officers were reservists, and their numbers increased steadily, peaking in 1943 at 60 per cent. The service expanded relatively slowly compared to other branches of the Navy, from 2909 regular ratings and 474 reservists at the beginning of the war to a peak of just over 9000 officers and men in September 1944. The high casualty rate meant that over 10,000 ratings were trained between 1939 and 1945, peaking at 3221 in 1943. Many were 'Hostilities Only' men, drafted against their will, an unpopular practice that in 1941 led some men at HMS *Dolphin* to refuse duty. Others deliberately committed offences to have themselves debarred from submarines.[74]

After the war, demobilization and the release of many 'Hostilities Only' ratings led to a severe manpower crisis in the Submarine Service. By April 1946 the position was so serious that Admiral Creasy was warning that 'unless additional personnel arrived within the next 10 days he would not be able to keep 45 submarines in commission, as well as those in reserve'. If the crisis persisted he estimated that by 10 June he would only be able to man 31 submarines, 22 submarines by 29 July and just 18 submarines by mid-September 1946. The Vice Chief of the Naval Staff, Vice Admiral Sir Rhoderick McGrigor, acknowledged that 'drastic steps would have to be taken' and many young officers and ratings were once again drafted or 'volunteered' for service from other branches of the Navy.[75] This unpopular practice continued well into the 1950s as the service continued to fail to attract sufficient volunteers.

In 1949, the submarine fleet numbered 65 submarines, 34 of which were operational. To man these submarines and to accommodate the various shore-based posts, as well as those submarine-qualified COs who were unavailable due to their service in the General Navy, 111 submarine-qualified Commanding Officers were required. To maintain these numbers 36 new junior officers were trained each year, at the rate of about 12 every four months. 16 officers undertook the Perisher course – the maximum that could be accommodated due to the availability of target ships, submarines and first-command appointments.[76] There was also a wartime requirement for 88 'young' Commanding Officers, whose seniority, as Lieutenant Commanders, was less than five years. Past experience had shown that as a general rule older Commanding Officers were 'unlikely to inflict damage on the enemy, and should not be appointed in command of operational submarines in time of war'. These figures varied according to requirements, as did the necessary number of ratings. By the early 1950s, an average of 150 were undergoing training at any one time

The Service struggled to maintain these numbers. By 1952, there were 3564 ratings in the Submarine Service, an insufficient number to allow the manning of all operational and reserve submarines. In September 1952 FOSM was forced to inform the Naval Staff that 'owing to the lack of manpower it will shortly be necessary to place three submarines in reserve'.[77] Part of the problem with manpower was that the Submarine Service no longer appealed to those who wished to find a healthy work/ life balance and see more of their families. With comparatively few submarines, all of which were increasingly required to undertake a significant and ever-changing number of tasks, often in scattered, faraway places, post-war life in the Submarine Service was increasingly characterized by continual and unpredictable disturbances.

Pay was also no longer the incentive it had once been. In 1949, a Lieutenant on promotion received £1.1.6 per day, plus 4/- Submarine Pay, an 18 per cent addition compared to the 43 per cent it had been in the pre-war period. Other branches paid more. For example, those serving in the Fleet Air Arm earned between 9/- and 12/- Flying Pay per day. Income tax, which was nearly double what it had been before the war, also had a significant impact on Submarine Pay. 'In the eyes of the present day young officer the 4/- a day, reduced to little more than 2/- by Income Tax, is likely to appear insufficient compensation for the physical discomfort of life in a Submarine,' noted the Director of Naval Training in November 1953.[78] In 1953, the final intake of Sub-Lieutenants to start their Submarine Training comprised just three volunteers and nine non-volunteers.[79] In November 1953 the Navy was forced to introduce new rates of Submarine Pay in an attempt to attract more volunteers. Commissioned Officers, Senior Commissioned Officers, Acting Sub-Lieutenants and Sub-Lieutenants received 7/- a day extra, while Lieutenants and Lieutenant Commanders received 8/- a day. It had little impact.

One such Royal Navy trainee who received a letter in 1954 notifying him that he had been 'volunteered' for submarine training was a 21-year-old Lieutenant, John Woodward, nicknamed 'Sandy'. If he really hated it, the letter told him, he could apply to leave in eighteen months and would be out in three and a half years. Either way, Woodward was now a pressed man and so began a career in the Submarine Service that would end thirty-one years later:

> all these years later, it emerges as nothing short of an inspired appointment for me, because in a submarine, you are required to become a responsible citizen from Day One. You have to grow up, quickly. The Submarine Service is nothing like being on board surface ships, which by and large tend not to sink, and anyway, if they do, are inclined to do so rather slowly, providing a very sporting chance to its company of surviving the event. In submarines, which are apt to sink rather suddenly, you are expected to understand and to be able to work every bit of equipment on board. I was thus required to become not only a semi-engineer, but also to learn in turn to be the Gunnery Officer, the Navigation Officer, the Communications Officer, the Electrical Officer, the Torpedo Officer, the Sonar Officer, before I could hope for front-line command in about six years' time. Suddenly I was to be permitted, in a position of responsibility, to undertake the very kind of work I had always liked most. It was exactly right for me – though I did not of course know it at the time.[80]

The Royal Navy's submarines were of course much smaller than most surface ships, but they were very complicated and, as Woodward indicated, it was essential that all the crew were trained in the operation of the boat. There was a certain basic submarine knowledge which all officers and men had to possess irrespective of specialization or seniority. The Engineer Officer, for example, was taught the rudiments of attacking technique because he would form part of the attack team when at sea. He also received a small amount of training on wireless telegraphy (W/T) methods, weapons and officer-of-the-watch duties, while at the same time learning about his principal job on the main and auxiliary machinery, systems and maintenance routines.[81]

Classroom instruction was supplemented by days at sea in a submarine, where trainees were allowed to operate machinery, work the hydroplanes, which controlled the depth of the submarine, and get an idea of life on board. 'We were made to understand the maze of pipes, cables, hydraulic systems, air systems, water systems, sewage systems, ship control systems, torpedo firing systems, engines, batteries, electrical systems, motors, pumps, valves, cocks, gauges, masts, periscopes and switches,' recalled Woodward. 'At the end, we were supposed to be able to find any item of equipment quickly and to be able to work quite a few in complete darkness.' When Woodward was examined, his instructor Lieutenant (later Vice Admiral) 'Tubby' Squires ordered him to restore the electrical supply to a submarine. 'Hesitantly I went below,' recalled Woodward:

> A small voice in the back of my mind saying that the answer had to be found in the Motor Room, well aft. In addition I vaguely remembered that a thing called the 'Reducer' was quite likely to solve the problem. Nothing very difficult: find it, make the switches (Navy jargon for turn them on), electrics restored. I was not, however, that confident as I descended into the darkness, and in tentatively 'making' the vital switch I caused it to start arcing and jumped back from it – a dangerous and stupid thing to do. Tubby leant over and blew out the arc quickly before things started to melt. So I tried again, and shoved the switch over, if anything more tentatively, which caused more arcing and another patient but quick puff from Tubby. There was another expectant pause. Finally, he banged the switch home for me and was good enough to say as all the lights came on, 'Well, I'll pass you for knowing which switch to go to, even if you didn't know how to work it when you got there.'[82]

In some respects training was rather primitive. Sam Fry, who joined in June 1951, found himself in the corrugated iron huts that comprised Haslar Gun Boat Yard in Gosport as part of Officers' Training Class 119. He wasn't too impressed:

> We had lots of pictures, some hardware, and not a lot else. There were no handouts, you made your own notes and drawings of pipe systems etc . . .

Some practical things were rather raced over so that we often only had a bit of the overall picture. During my oral exam at the end of the Course I produced a massive spark from the switchboard as I broke a switch without doing the correct thing with the 'field switch', much to my amazement and the horror of the examining officer! Our understanding of the workings of submarine radar was also just about as primitive as the radar itself.[83]

Before any officer or rating could call himself a submariner and draw the extra Submarine Pay to which he was entitled, he had to qualify in submarine escape. This took place in a specially constructed 100-foot submarine escape training tank. The escape method, known as 'Free Ascent', involved exhaling continuously while swimming to the surface, otherwise the expanding air in an individual's lungs would rupture them. 'It was very exciting coming up from 100ft secured to a central wire as leaning back acted as a hydroplane and the tether prevented you hitting the side of the tank so one just circled the central wire on the way up,' recalled Fry. 'If the instructors waiting in the tank at various depths did not see sufficient bubbles you got a hefty belt in the chest.'[84]

Those new recruits who joined the service in the immediate post-war period were fortunate to receive their training from many of the second-generation wartime COs, who had learned their trade from the submariners who had started the war. 'We were very lucky that the people who taught us learned their business in the war and they knew what was required very well,' recalled John Hervey.[85] But the war had left its mark on many. As a young midshipman, Woodward noticed 'there were quite a few men around who were suffering from that very old affliction with the modern name of "stress", otherwise known as "Shell-shock"; "twitch", "lack of moral fibre" were its other, less sympathetic titles'.[86] These problems tended to manifest themselves in a number of ways. Some had serious drinking problems. Others suffered from drastic personality changes. Quiet, studious men became aggressively argumentative, while hell-raisers became introverted. 'Some men never got over it,' remembered Woodward, 'and most never let on that they were anything other than perfectly normal.'[87] Others did. 'There were some who were absolutely shot,' said Peter Herbert, another trainee who entered the service just after the war: 'Several of them were drinking a bottle of whisky and going to sea, taking a bottle of whisky up to their cabin and letting the First Lieutenant do everything. Then coming back drunk because they couldn't take it . . . They were shot. You didn't blame them, they'd had a hell of a war.'[88]

When Hervey was assigned to his first submarine, the 'A' class HMS *Acheron* after completing his initial training, he quickly discovered that the Captain was a 'peculiarly ill-suited officer who had lost his nerve'. Although the 'A' class could dive to around 500 feet, *Acheron*'s CO never went below 90 or 120 feet, because he was scared of his submarine. 'Those sorts of people tend to take it out on their subordinates,' recalled Hervey, 'the fright comes out in terms of anger. The only lesson I learned from him was how not to command.'[89]

After the war, the Service did not have to look far for potential First Lieutenants to turn into Submarine Commanders. They also did not have to look far for 'Teachers' who had the necessary skills and experiences to pass on to future commanders. One of the first post-war Teachers was Commander Hugh 'Rufus' Mackenzie, a wartime submariner whose infectious sense of fun concealed a shrewd and observant mind. When Mackenzie became Teacher in January 1946, in keeping with a return to peacetime conditions, the Perisher course followed an 'easier tempo' than in wartime, but it remained a career breaker with between 20 and 25 per cent of students failing and returning to General Service. Officially, candidates who 'perished' were assured by Flag Officer Submarines that there would be no stain on their record, but this was never really the case. 'By the time I arrived,' continued Woodward, 'not one of them had ever been promoted beyond lieutenant-commander.'[90] One of Mackenzie's pupils, John Coote, a charismatic raconteur notorious for burning the candle at both ends, wrote that if a student failed 'they might as well start reading the Appointments column in the *Daily Telegraph* right away'.[91]

As the first post-war Teacher, Mackenzie ensured that the 'easier tempo' did not 'infiltrate the actual content of the course, which at the same time had to be expanded to incorporate all the accumulated knowledge and experience from the war':

> There was no departure from the strict standards the war had demanded in the conduct of submarine operations. It was a continuing challenge to put all this across in the most effective way I could devise, to a varied intake of pupils over the next two years, mainly British but also including submariners from the Netherlands and Norwegian Navies.[92]

The course typically started with a four-week period at Fort Block-house in Gosport, where students trained in a mock submarine Control

Room, known as the 'Attack Teacher'. The students then switched to a second, more modern Attack Teacher at Rothesay in the Clyde for 'Blind Attack' training, when the attack relied on sonar information only. A mini submarine Control Room revolved around a cyclorama inside the building. Two targets were projected on circular walls by illuminating metal ship models with powerful lights. When viewed through a periscope, students were able to see a rather ghostly image from which they were expected to take ranges and bearings.

Perisher broke many young officers. The most common cause of failure, even for those with years of faultless service from Fourth Hand to First Lieutenant, was for Teacher to realize early on that a student lacked 3D vision through a periscope and was unable to size up a situation in full perspective. 'Towards the end of the Perisher,' wrote Woodward, 'that mental picture, fleeting and ephemeral for some, sharp and clear for others, would not only include the usual fishing boats, ferries, islands, yachts and the like, but would also contain the major confusion factor of five Royal Navy frigates tearing about at full speed, deliberately trying to ruin our day.' Many of those that failed the course lacked the mental agility to cope with a rapidly developing position and were unable to process and hold a mental picture of what was happening on the surface. 'It's not a very good analogy,' explained Woodward:

> But imagine sticking your head out of a manhole in Piccadilly Circus, taking one quick swivelling look round, ducking back down into the sewer and then trying to remember everything that you had seen. The idea is to generate sufficiently accurate recall and timing to avoid a double-decker bus running over your head next time you pop up through the manhole.[93]

Teachers spent a considerable amount of time getting their students to develop an automatic reaction to checking the trim, depth, speed and sonar picture before putting up the periscope to make an observation of a target. Although surface ships generally travel 1000 yards in a minute at 30 knots, many students were tempted to raise the periscope for another quick look in order to reassure themselves that the target was still there. If a student's range estimate was correct then it was possible to determine when it was safe to come to periscope depth after a run towards the target, or when to take a look through the periscope, or even when they might need to go deep to duck under a surface ship.

Near misses were a natural, frequent and deliberate part of the course. Sonar was still in its infancy so 'Blind Attacks' at sea were often over quickly. 'Blind' was a good description as the students often had little or no idea of where the target was. Indeed, Woodward seriously doubted whether anyone would have passed had they been able to see the tiny safety margins allowed – a few feet, and split-seconds only, to separate the hulls of the oncoming frigates and the submarines. Crews liked the course. They would always run a book on who was going to get through and who was not.

Teacher's aim was to push those that possessed a 'periscope eye' to the limits of endurance. 'The real tick in the book that you had to have at the end,' remembered Geoffrey Jaques, another student who took the course in the early post-war years, 'was that for six or eight weeks or however long the sea part of the course went on for you had been put under extreme pressure, you had been put in close contact, you had been made to handle the boat in as many situations as they could devise and they were entirely satisfied that you could do it safely and competently and it was specifically designed for you to be under pressure at all times.'[94]

'At intervals there were consecutive long days and nights intended to stretch the students and see how they reacted when fatigue blurred their judgment,' wrote Coote.[95] It was common to have a party on the Thursday evening to allow the students to let their hair down and meet the officers from the target ships. There was also an ulterior motive. A tired and hungover student on the Friday morning was a very good example of how an exhausted individual would react when under pressure in a fast-moving situation. 'Teacher, if he knew you were the duty Captain the next day, would keep you at the bar until 3 a.m. in the morning, drinking, on the basis that he said: "You may be on a visit somewhere and having to give hospitality and then suddenly you're off to work. I want to see what you're like." '[96] Unsurprisingly Perisher was known as a hard drinkers' course.

Teacher also suffered from exhaustion. Mackenzie found the role both rewarding and exhausting, particularly the long hours at sea closely supervising an endless succession of dummy torpedo attacks, where every latitude had to be given to the embryo CO before possibly having to intervene at the very latest moment and 'pull the plug', to prevent the submarine being endangered. 'It was,' Mackenzie later said, 'nerve-wracking work, concentrated tensely on whichever periscope was not in use by the pupil, and by the end of the day I was very conscious of the strain on my eyes.'[97] Some Teachers, such as Woodward's, Captain Brian Hutchings, were also difficult men to work for:

If you made a minor mistake, he'd tend to rant and rave, which, if you were unused to this sort of thing, could be quite upsetting. However, if you made a really serious mistake, he'd quietly do his absolute upmost to help you out of it. But he was tough-minded and you couldn't pull the wool over his eyes. He generally referred to us collectively as his 'useless officers' and felt entirely free to do so before the ship's company and/or complete strangers. The thing to remember was that he didn't really mean it, did he?[98]

Teachers didn't only have to instil submarine skills; they also had to teach the softer skills of command. John Hervey's teacher, Rud Cairns, another Second World War veteran, 'was absolutely meticulous as a Teacher, very, very conscientious and he didn't just teach you how to do an attack, Rud taught you how to be a CO, including how to write officers' reports and how to get rid of an unsatisfactory chap, right down to what to do on the day you sail from a foreign port and things like that. Some Teachers were very very unsatisfactory in my view and they didn't pay nearly enough attention to the general training of the student. In fact one chap I thought disgracefully didn't even turn up until they got to sea stage, when of course you should be there from the very first day with the student and the Teacher in the base ashore.'[99]

Compared to the hard years of the Second World War, the early post-war years were relaxing affairs with the Royal Navy's submarines conducting numerous peacetime roles. The service returned to its pre-war role of providing the Royal Navy surface fleet and RAF Coastal Command with anti-submarine training (and later, as we have seen, the Royal Australian Navy and the Royal New Zealand Navy). Much of the time of the home flotillas was spent training, what was known as day-running from operational bases such as Portland, the location of an anti-submarine training school, and the Joint Anti-Submarine School, Londonderry. Surface ships would attempt to find submerged submarines participating in the exercises while RAF Shackleton Maritime Patrol Aircraft patrolled the skies above. It was incredibly dull work and many crews, hardened by their experiences in the Second World War, found post-war life at sea very boring. 'It used to be all right during the war; but it gets monotonous now,' complained an Able Seaman from HMS *Taciturn*. 'This clockwork mouse stuff isn't much fun . . . They wind us up at base just like a clockwork mouse and set us running out to sea. We just cruise around and wait for the surface craft to find us. Then back to base. Next day they wind us up again and we do the same thing.'[100]

One of the more exciting events in the calendar was an annual set-piece exercise known as Admiral Submarines Summer War Exercise, later Flag Officer Submarines Summer War Exercise. The first post-war Summer War was held off the Shetlands in late 1946 with up to thirty submarines participating in a series of set-piece encounters spread over a number of days, carrying out dummy torpedo attacks against surface targets, such as ships and convoys. This was the closest submariners came to actual live operations in the immediate post-war years. The service was hardly involved in the only major conflict of the early 1950s, the Korean War.[101] In the Far East, submarines were employed on inconclusive anti-piracy patrols in Mirs Bay and the southern approaches to Hong Kong to counter junks attempting to lift the main telegraph cable from Singapore to cut out long lengths of copper which fetched high prices in the markets in Macao.[102]

The first full-scale mobilization of the Submarine Service was in unexpected circumstances and proved to be a costly and unwelcome distraction. Winter 1947 remains to this day the most severe and protracted spell of bad weather experienced by the United Kingdom in living memory. Towards the end of January 1947 virtually the whole of the British Isles was covered in a blanket of white snow and the sub-zero wind from the east blew for a month without stopping. The Thames froze, Big Ben was silenced (its mechanics were frozen solid) and amid severe snowdrifts most transport ground to a halt. 93 per cent of the nation's energy came from coal, but due to severe mismanagement, the nation's coal supplies were in a dire state even before the bad weather descended on the country. During the first post-war winter in 1946, a mere 6.8 million tons of coal remained in reserve, enough to supply the country's needs for just over a week. In January 1947, coal froze in the pits and there were no trains able to shift it. Power shortages occurred all over the country, and gas too was in short supply, dropping to around a quarter of its normal pressure in most big cities.[103] Thousands of people were laid off work and unemployment shot up from 400,000 to 1.75 million.

As the government muddled through the winter crisis the Submarine Service played a vital role in supporting the electrical needs of the Royal Navy's dockyards. At 2000 on 7 January 1947, a message from the Admiralty in London was brought into FOSM's Operations Room at HMS *Dolphin*, which was lit by a solitary candle, in which had been stuck a replica of the National Coal Board Flag. The signal directed FOSM to supply power to Royal Naval Dockyards by HM

Submarines. By 8 January, the first submarines had fuelled and were moving towards their allotted stations. The 5th Flotilla based at HMS *Dolphin* provided power to Portsmouth, Chatham and Sheerness, the 2nd Flotilla from HMS *Maidstone* to Portland, and the 3rd Flotilla from HMS *Forth* moved south from the Clyde to provide power to Devonport. The operation was demanding, particularly on manpower, with ninety officers and 940 submarine ratings continuously employed on board the submarines, working in considerable discomfort in sub-zero temperatures caused by the strong draught of cold air that was required by the submarines' diesel engines.[104] Despite these difficulties, the volume of work never overwhelmed the submarine crews and the performance of equipment, both mechanical and electrical, was excellent. Most submarines were rotated in and out of operation for maintenance and the periodical discharge of batteries. One 'S' class submarine provided power continuously for twenty-eight days.

The operation, which was codenamed 'Blackcurrant', continued monotonously without major incident until 31 March 1947. Although successful, submariners could not fail to weigh up their contribution against the cost and effort involved. Savings amounted to an estimated 7890 tons of coal, the equivalent of less than ten minutes of normal output from the coalmines, and on the basis of wear and tear the engines of the submarines involved had been run for around 29,000 hours, the equivalent of seven engine refits, each costing around £23,000. Unsurprisingly, in September 1947 Flag Officer Submarines was still complaining that the legacy of Operation 'Blackcurrant' had 'still not yet been written off'.[105]

The service was also involved in trialling new equipment. In April 1946 a 'T' class submarine, HMS *Truant*, became the first Royal Navy submarine to be equipped with an experimental snort mast. *Truant* spent six days testing the new equipment while transiting from Gibraltar to Portsmouth under the watchful eye of an observer from the Royal Naval Physiological Laboratory, who was tasked with studying the impact of snorting on the crew.[106] While the snort enhanced the capabilities of the Royal Navy's submarines by removing the need to surface to charge batteries, reducing the risk of detection, as well as greatly increasing underwater speed thanks to the ability to use diesel engines while submerged, it was uncomfortable for the crew, especially when a submarine was snorting in rough seas. When the head of the snort mast dipped below the water a valve would automatically close to stop water pouring into the submarine. With the diesels still running, they no longer took in air from outside the submarine but from inside. This created a vacuum which built up over time causing men's ears to pop. When the mast eventually cleared the surface, the valve would open again suddenly equalizing the pressure inside the submarine. This was often painful for the crew.

Trials to evaluate how new equipment, submarines and their crews performed in different waters/climates were also conducted in the immediate post-war period. In October 1947, HMS *Alliance* was sent on a thirty-day operation to obtain information about the living conditions on board a submarine during an extended patrol in tropical waters, travelling a distance of 3193 miles over thirty days, completely submerged. Further trials to determine the air quality and the effects of prolonged snorting on a submarine's crew were conducted by HMS *Taciturn* and in 1948 HMS *Ambush* was sent to the Arctic Circle to determine how the snort functioned in northern cold waters. *Ambush* departed from Rothesay on 10 February and found itself having to endure a three-day gale, navigating treacherous icebergs which threatened to collide with the submarine. The storm was so violent that *Ambush*'s men were forced to lash themselves in their bunks to prevent their being thrown out.[107] The pitching of the submarine made it difficult to maintain adequate depth control and waves rising over the snort mast constantly shut off the valve, making life on board very uncomfortable due to the constant pressure changes. *Ambush* was eventually forced to surface where it rode out the storm for three days, pitching and tossing in the violent seas.[108]

Those who joined the Submarine Service in the time immediately after the war experienced what Mackenzie described as 'an immensely exciting period when everyone was consumed with new thoughts, new ideas, on how submarines, their weapons and operations would develop in the future'.[109]

A NEW ROLE

Since the end of the Second World War, the Royal Navy and the Submarine Service had been thinking long and hard about how the Service would fight the next war. During the early Cold War submarines were increasingly seen as one of the few opportunities available to the Royal Navy of taking the fight to the adversary. One 1945 Naval Staff paper concluded that submarines were 'needed to enforce blockade in waters close to the enemy's shores where surface vessels are unable to operate and for special reconnaissance operations'.[110] The Navy recognized that in a future conflict with the Soviet Union involving unrestricted submarine warfare, the submarine was an effective means of intercepting and attacking Soviet submarines deploying from their northern bases into the Atlantic to attack British and Allied merchant shipping. In early 1948, the Admiralty issued the Submarine Service with a new directive:

> In war, the primary operational function of our submarines will be the interception and destruction of enemy submarines in enemy controlled waters. Their other main functions, the importance of which will depend on the circumstances, will be the interception and destruction of enemy warships and merchant shipping, reconnaissance, air/sea rescue and special operations.[111]

Some, such as the Assistant Chief of the Naval Staff, Rear Admiral Geoffrey Oliver, wanted the Submarine Service to assume an offensive role off the well-defended Soviet mainland, attacking Soviet submarines and warships with torpedoes and mining the approaches to their bases. Oliver later said that submarines were an effective means of 'getting to the enemy on his home ground'.[112]

But there was a problem. During the war submarines had been primarily employed against surface ships, attacking enemy warships and convoys, while also conducting other operations such as deploying intelligence agents and Royal Marines. There had been very few submarine versus submarine submerged attacks in the Second World War. The only successful attack occurred on 9 February 1945 when HMS *Venturer*, under the command of Lieutenant Jimmy Launders, sank the German U-boat U-864. Both submarines were submerged. Post-war training was therefore increasingly directed towards meeting this new role and focused on detecting, tracking and attacking submerged submarines. This was not an easy task and required much-improved sonars.

Everything that moves on or under the surface of the sea transmits some form of noise which can be detected at ranges varying from a few yards up to hundreds of miles. Sound is transmitted in water some 4.5 times faster than it is in the air. The lower the frequency the further it goes; the higher the frequency the greater the accuracy with which the bearing of its source can be measured. In the main, noise sources come from a propeller, through cavitation, a form of turbulence; from main or auxiliary machinery noise, or from the disturbance made by the passage of a ship or submarine hull through the water or the passage of water through piping.[113] During and immediately after the war submarines and surface ships were detected by the sounds they radiated in the high audio frequency region (around 10 kc/s) by an operator listening to the 'hydrophone effect' (or HE). In the Royal Navy the equipment used to detect these frequencies was known as ASDIC (named after the Anti-Submarine Detection Investigation Committee), or to use the American terminology, sonar (Sound Navigation and Ranging). Active sonar transmitted acoustic signals that then reflected from an object, giving direction, range and, by plotting successive positions, speed; Passive sonar received sounds generated by a submerged object which, when processed, provided bearing information. During and immediately after the war British efforts were directed towards development of active sonar as anti-submarine warfare was primarily the occupation of surface escorts, which were far too noisy to allow the use of sensitive passive sonar arrays mounted on a ship's hull. For submarines, active sonar had one major disadvantage: it gave away the location of the submarine that was using it. It could also only detect objects a few miles distant.

At the end of the war the sonar equipment installed in Royal Navy submarines was primitive and suffered from a number of limitations. In good conditions its bearing accuracy was plus or minus 2 degrees and it took from five to ten minutes to detect an alteration of course by an enemy at a constant speed. The equipment also only worked on a limited frequency. When noise trials had shown that submarine machinery and propellers generated noise over a wide band of sonic and super-sonic frequencies, lower-frequency noises, which could be detected at greater ranges, could not be heard. There was also no method of obtaining the depth of a target. Unless a periscope or snort mast was seen, the noise of diesel engines heard, or the target restricted by shallow water, there was no means of knowing whether an enemy submarine was operating at a depth of 40 or 400 feet. Self-generated noise was also a considerable problem. Royal Navy submarines attempting to remain silent in order to use passive sonar equipment to detect enemy submarines were required to operate at a low speed of around 2 knots, which prejudiced the chances of reaching a good firing position, or obtaining a firing position at all. Quieter submarines were required.

Another early difficulty was that the Submarine Service's primary torpedo, the Mark 8, was ill suited to anti-submarine warfare. Although the torpedo had been trial-run at depths down to 200 feet, the

equipment which set the depth of the torpedo was incapable of allowing a greater depth than 44 feet once the torpedo had been loaded into
a torpedo tube. The enemy could also often hear the torpedo approaching and move out of its way. Trials had shown that a Mark 8 torpedo
travelling at 45 knots could be detected by an enemy submarine from a
range of as much as 1200 yards. In the 48 seconds it took the torpedo
to cover that distance, an operator on board an enemy vessel had
15 seconds to classify the noise as a torpedo, 15 seconds to warn the
Officer of the Watch and 18 seconds in which to take evasive action, by
increasing speed, altering course and changing depth. Trials revealed
that even with just 18 seconds' warning, an 'A' class submarine, snorting at 6 knots, could manoeuvre in such a way that less than 30 per cent
of it remained in the target area when the torpedo arrived. This reaction time was only likely to improve as more advanced submarines
entered service. The Mark 8 torpedo, FOSM concluded, had 'only a
small chance of success against a reasonably alert enemy fitted with
acoustic equipment similar to our own, unless the firing range is short,
or the depth of water insufficient to permit the target taking evasive
action in depth'.[114] New torpedoes and fire control systems were also
required.

The first serious submarine versus submarine exercises to simulate
attacking Soviet submarines operating near their bases were held in
June and July 1950 in the waters around Skye, with the islands of
Canna, Rum and Eigg acting as the Kola Inlet. HMS *Alcide*, under the
command of Lieutenant Morris 'Roache' O'Connor, played the role of
the Soviet submarine, while the attacking submarine was played by
HMS *Truncheon*, under the command of Lieutenant A. Richardson.
Truncheon was modified by the staff of the depot ship HMS *Montclare*
to reduce the amount of noise that it generated. Noise generated by
water flowing over a submarine's hull, by machinery and cavitation
from propellers served to drastically degrade the submarine's own
sonar performance. It also served to betray its position to a potential
enemy. As with *Tradewind*, *Truncheon*'s 4-inch gun and external bow
torpedo tubes were removed to streamline the hull and reduce the noise
of water flowing around the hull. The submarine was also equipped
with an American hydrophone as well as equipment that allowed the
CO to monitor the amount of noise his submarine was making.

Although these modifications allowed *Truncheon* to manoeuvre
silently close to its target and into an attacking position, *Truncheon*
only claimed eleven successful attacks out of thirty-two. Neither

submarine could remain silent at speeds above 3 knots and *Alcide* succeeded in attacking *Truncheon* on six occasions while the submarine was noisily snorting. The final stage of the exercise consisted of a five-day patrol in the Minches. *Truncheon* was allocated a patrol area of forty square miles and was ordered to hunt HMS *Alcide*, representing an enemy submarine on passage to and from patrol and her base. Over the course of five days and nights, *Alcide* conducted eighteen runs through the area, eleven while snorting, three on the surface, and four dived. *Truncheon*, which had no forewarning, detected twelve of *Alcide*'s runs and carried out ten attacks, four of which were judged to be successful. The remaining six undetected intrusions were explained away by adverse weather, which deteriorated water conditions.[115]

The results of the exercise were disappointing. Further intensive studies led the new Flag Officer Submarines, Rear Admiral Guy Grantham, to conclude that 'a submarine fitted with present equipment stands a good chance of achieving some detections of the enemy, but due to other aspects of the problem, the chances of a successful attack cannot be considered as very high'.[116] Inaccurate sonar equipment imposed 'limitations on the accuracy of the torpedo fire control data' and (as we have seen) there was still 'no method of obtaining the enemy's depth'.[117] Grantham had also concluded that one of the biggest problems was that 'the present general-purpose submarine, with its gun and other projections, all of which reduced its efficiency as a silent listening platform is, for this reason, unlikely to be very successful in its A/S role'.[118] In 1958 one officer on FOSM's staff summed up the state of affairs: 'we are armed with World War II equipment. Our asdic is a short range passive detector; our torpedo is a non-homing straight runner, and we are therefore unable to fire at anything below periscope depth.'[119]

The Service concluded that its future submarines had to be equipped with high-performance sonar, specifically designed for anti-submarine warfare, as well as a torpedo armament capable of homing in on enemy submarines. Prolonged submergence became the priority, especially as submarines were required to carry out their traditional task of patrolling in enemy-controlled waters. Tactically, high, silent speeds were also important to increase the chances of detecting, intercepting and ultimately attacking submerged targets. This represented a significant shift in thinking. The emphasis moved away from the high-speed, low-endurance submarines, such as offered by HTP propulsion, towards moderately low-speed, high-endurance and silent submarines.

To meet these new requirements the Navy embarked on an

ambitious programme to modernize many of its wartime submarines. At the end of the Second World War a number of the 'S' class had also been streamlined and converted to act as fast targets in anti-submarine warfare exercises. These experimental craft, along with the lessons learned from captured German Type XXI submarines, led to major modernizations of the 'T' class. HMS *Taciturn*, the first of the class to undergo such improvements, was fitted with an additional 14 feet of pressure hull to accommodate an extra battery section and an additional pair of motors. Her hull form was streamlined in a manner suitable for high submerged speeds, while the gun and external torpedo tubes were removed from the casing, giving the bows a cleaner line. The fin, which encased seven masts, was also streamlined and the bridge replaced with a modified conning tower. These modifications enabled the 'T' class submarines to remain continuously submerged at periscope depth, dive to greater depths and close a higher percentage of targets to effective torpedo range while submerged with a maximum underwater speed of about 17 knots.

Between 1948 and 1957 eight of the welded 'T' class were modified in a similar way, but with slight differences, progressive modifications that contributed to the boats' nickname of 'T Confusions' but their underwater performance was transformed. They were also known as 'T-Conversions.' It was also decided to streamline five of the riveted 'T' class boats and fit a higher-capacity battery without lengthening them. They were known as 'Streamlined T-boats' or 'Slippery Ts'. The work was carried out in the normal refit cycle. It increased underwater endurance by 30–40 miles and their speed at periscope depth was increased by 1.4 knots and maximum quiet speed by almost a knot. They could go faster than the unmodified boats and make less noise while doing so.

New equipment, as Captain Arthur Hezlet explained in a 1954 presentation on 'The Future of the Submarine', also promised to transform the capabilities of the service and the ability of its submarines to meet their new primary wartime role:

> Firstly we should hear the enemy at double the range and so make many more contacts; secondly we should be able to listen for longer periods and snort for shorter periods and so have a greater 'patrol efficiency'. Thirdly we should be able to follow up the contacts at higher speed without being heard and should have improved weapons and their fire control which will at least double the chances of hitting. Within a few years then, we hope the results already being obtained in exercises will be greatly increased.[120]

This was optimistic. Development of new sonars, fire control equipment and torpedoes was plagued with difficulties. The Submarine Service's standard wartime sonar set, the 129/138 combination was upgraded and renamed the Type 169/168. A sophisticated new sonar, known as the Type 171, or 'Four Square', was intended to provide submarines with a three-dimensional underwater picture. However, it was cancelled in the early 1950s after early trials revealed its performance was unsatisfactory unless used in combination with powerful computers, which could not be accommodated inside contemporary submarines.[121] The Naval Staff struggled to envisage a clear role for the new set and decided that its only useful function was the ability to detect mines.[122] Instead, a new form of sonar known as the Type 187, a medium-range passive directional-listening set that was housed in a distinctive enclosed dome on the front of the submarine, was eventually introduced in the streamlined 'T' class towards the end of the 1950s, alongside a fin-mounted aft-looking hydrophone known as the Type 719.[123] These sets only offered a modest improvement in capability. The ability to detect other submarines remained short, especially when compared to what was to come in the future. The official historian of Britain's sonar programme argued that the many 'false starts' in the immediate post-war period were 'because the Naval Staff found it difficult to formulate requirements'.[124]

New fire control equipment that was capable of handling the uncertainties of bearings and range when attacking submerged submarines was also needed. It was possible to determine a fire control solution using what limited sonar information was available, but this rarely worked in practice. The Navy's submarines were equipped with primitive, rudimentary torpedo fire control systems and attacks depended on the considerable skills of the Commanding Officer, who had to ensure that the submarine was pointing ahead of a target. He would then repeatedly check the target's range and bearing by looking through the periscope while his observations were entered into an electrically driven mechanical fire control computer known as the 'fruit machine'. While effective, the 'fruit machine' was limited and its accuracy was only as good as a CO's visual observations. A more advanced fire control system known as Torpedo Control System (Submarines) Mk 3 (TCS(S)3) was meant to be capable of taking information from sonar, radar and periscope observations and setting the running depth and angle of torpedoes. Although the first sets appeared in 1955 they were criticized for being complicated to operate.[125]

Aside from fire control equipment, new torpedoes were also required. The wartime Mark 8 remained the primary submarine weapon in the immediate post-war years. It was repeatedly modified to increase its chances of striking submerged submarines, principally by allowing it to operate at depths of up to 200 feet. Reducing the time it took for a torpedo to reach its target was also seen as a way of increasing the chances of a successful submerged attack. Development work started on a modified Mark 8 capable of high speeds, fuelled by hydrogen peroxide, known as the Mark 12 'Fancy'. Advanced acoustic torpedoes that were able to home in on submerged targets also offered a possible solution to the problem of inadequate sonar and fire control equipment. There were two types of homing torpedo. Active homing torpedoes transmitted high-frequency acoustic pulses to detect and intercept a target, while passive homing torpedoes used built-in sonars to home in on a target's noise signature. Both had disadvantages. Passive homing torpedoes were limited by their own self-noise, slow, only effective against noisy submarines, and susceptible to decoys and other countermeasures. Active homing torpedoes immediately alerted the target to the incoming torpedo, allowing it time to take evasive action either by escaping or by deploying decoys and countermeasures.[126] Both the Germans and the Americans developed acoustic torpedoes during the war. The American torpedoes were made available to the Royal Navy, which also started to experiment with another modified Mark 8 equipped with active homing capabilities. The torpedo, which was known as 'Trumper', reached the trials stage but was cancelled at the end of the war to save money.[127] Instead the Navy refocused its efforts and started work on a battery-powered passive homing torpedo known as the Mark 20(s) 'Bidder', which it hoped to use against both submarines and surface escorts.

While this new equipment was being developed, submariners compensated for the inadequacies of their existing equipment and torpedoes in a number of ways. They devised complex choreographed actions from which an attacking submarine could arrive at a fire control solution without using active sonar. This depended on the attacking submarine obtaining the bearings of a target at equal time intervals, first with zero own movement, by pointing straight at the target, followed by another set of bearings, with maximum own movement, which was achieved by altering the attacking submarine through 90 degrees. If the enemy remained on a steady course, its course and speed could be devised by overlaying a rule marked off to the right scale at equal time intervals. A vertical perspex plot on which coordinates of time against true bearings were marked was introduced into submarine Control Rooms. On one side stood a rating with a headset from the sonar operator, marking with a pencil each hard bearing reported. On the other side of the perspex plot an officer faired off all the raw data into a smooth line from which he could then read off for the navigator, who would plot the bearings at predetermined exact intervals. Traditional visual attacks, conducted through the periscope, where the target was visible, would be over in a matter of minutes. These new methods required meticulous detective work as well as patience.[128] To compensate for the lack of accurate information about a target's bearing and range and increase the chances of carrying out a successful attack, submariners fired salvos of Mark 8 torpedoes. Finally, to bring about 'steady improvement' in submarine versus submarine operations a submarine versus submarine phase was introduced into FOSM's annual Summer War Exercise and in 1951 and 1952 the whole of the Summer War was devoted to submarine versus submarine operations.

The service was also hampered by the fact that the rest of the Royal Navy tended to think that the purpose of the submarine fleet was to provide anti-submarine training to its surface and air forces. 'I think you will agree that the submarine must become perfect in its operational functions of attacking both U-boats and surface targets,' pleaded Admiral George Simpson, the Flag Officer Submarines from 1952 until 1954.[129] He urged the Navy as a whole to 'think of us as the greatest spearhead – off enemy-controlled ports in enemy-controlled waters – for sinking U-boats in parts of the ocean through which they have got to pass to get to their bases'.[130] Exactly when the Submarine Service might be called on to carry out its new role was an uncertain question, but by the early 1950s who the enemy would be was alarmingly plain.

THE EARLY COLD WAR

Immediately after the Second World War the possibility of conflict with the Soviet Union seemed distant and improbable. However, by 1947, East–West relations had deteriorated markedly. In March, President Truman, responding to Soviet-backed communists in the Greek Civil War as well as growing Soviet influence in Turkey, told the US Congress that 'It must be the policy of the United States to support free peoples who are resisting attempted subjugation by armed minorities or outside pressure.' Two months later, in June 1947, the European Recovery Programme, the so-called Marshall Plan – financial aid to help rebuild European countries – was rejected by the Soviet Union and its Eastern European allies. In March 1948, a communist coup in Czechoslovakia and ongoing disagreements about the future of Germany led to the first major crisis in the Cold War, when the Soviet Union decided to blockade the land and water routes from the western zones of Germany into the West's enclaves in Berlin. On 9 July 1948, British ministers and the Chiefs of Staff considered the question: 'Should we be prepared to go to war over Berlin?'[131]

Meanwhile, British naval planners, contemplating a future war, operated under a number of assumptions. First, 'In the foreseeable future there is no maritime power other than Russia with whom we are likely to go to war.' Second, 'We will not go to war with Russia unless America is on our side from the outset.' Third, 'The size of the fleet which this country will be able to support in the future will be severely restricted by economic factors.' Fourth, 'The main task for our Navy in war will be to support what the Chiefs of Staff have stated to be the main pillars of our strategy. From a naval point of view these are the defence of sea communications, the defence of the U.K., and the defence of the Middle East.'[132] Prior to the formation of NATO in 1949, the United States Navy and the Royal Navy identified the Barents Sea, the White Sea approaches, and the Danish Straits as key Soviet submarine operating areas that would be defended by Royal Navy and United States Navy submarines. During 1949, the US Joint Chiefs of Staff prepared a plan called 'Dropshot' that, although never formally adopted, revealed the outlook of British and American strategists. In the event of war with the Soviet Union, Dropshot foresaw strikes by four US and two Royal Navy aircraft carriers operating in the Barents and Norwegian Seas against enemy submarines and airbases in the Kola Peninsula.[133]

Despite the limits of the Submarine Service's anti-submarine capabilities, many submariners felt that if war should break out, they would still retain a considerable advantage over the Soviet opposition. Two prevailing thoughts about the capabilities of the Russian Navy and the Russian submarine fleet dominated the immediate post-war period. The first was that anything the Russians had done well with submarines since the war was the result of German know-how acquired after VE Day in 1945. The second was that the Russians were not natural seamen and that because of this the submarine threat was, in fact, minimal. While a spring 1946 Joint Intelligence Committee report noted that Russia possessed about 210 submarines, including ten German ones; that 'She takes a great interest in submarine warfare and in this particular arm of the Naval Service she has shown herself to be more proficient than in any other'; and that 'German assistance and methods particularly in connection with prefabricated submarines, would enable her to construct a formidable Submarine Force in a comparatively short time', the JIC also noted that Russian submariners 'were still inexperienced in attack tactics'.[134] An October 1946 Naval Intelligence assessment also noted that:

> The Russians are far from being a nation of seamen, and this weakness is reflected in the operation of their submarines, however technically good these boats may be. Their attack technique is amateurish to a degree . . . The submarines themselves are probably capable of carrying heavy armament a long way with reliability, but are by no means certain of hitting the target when they get there.[135]

In the early 1950s, the chief advocate of this view was the Flag Officer Submarines, Rear Admiral George Fawkes, one of the few surviving British naval officers who had served in Russian submarines during the Second World War. In 1954 Fawkes warned that there was a 'great danger that we can over-estimate the Russian submarine menace' and explained that:

> The Russian submarine threat as it was in 1941 was practically non-existent. They hadn't a clue how to submarine . . . I would say that there were very few Russian naval officers who are seamen and even fewer who are submarine officers. The Russian submarines in the last war achieved absolutely nothing, and they had every opportunity to achieve a lot, both in the Black Sea and off the Norwegian coast. Now I am quite certain that they are not too content to let things remain like that, and I would be the last person in the world to say that you can dismiss the Russian submarine menace as meaning nothing. Our country has been nearly brought to its knees by the submarine in two world wars and therefore we must have a very healthy respect, as of course I have, for the submarine. [But] I should be very surprised if the Russian submarine service has got to anything like the stage of what we are apt to be led to believe. I would put it roughly like this – I am open to correction – the Russians could deploy in the event of war, by the end of this year,

something like thirty submarines and maintain that number ... How they would employ their submarines, I do not know, and I rather doubt that they know; they have no experience and they have only recently, in the last year or so, developed powerful radar stations by means of which they could control them ... I do feel very strongly that we are apt to over-rate the Russian submarine menace – at the same time we must not under-estimate it.[136]

Not everyone agreed with Fawkes's assessment, which was heavily influenced by his own wartime experiences. A Naval Intelligence officer, Lieutenant Commander McMillan, argued that:

The fact is that they [the Russians] have some three hundred and fifty submarines now; they built forty new ones last year; they are going to build over fifty this year, and over seventy next year. So just going on the numbers alone, the question of whether they are going to provide able submarine commanders, looms as an almost secondary one, because with those numbers they will soon gain experience ... they are taking their peacetime very seriously, and they are determined to catch up on the lack of experience they had had in the past. I don't think the Russians have ever been more earnest in anything in the Naval field than this. They are determined to catch up with the other great navies of the world, particularly in submarine matters.[137]

By 1953, the US and UK had drawn up an Emergency Joint US/UK Submarine War Plan that carved up the North Atlantic into British and American patrol zones. Initially, the areas north of latitude 67°N were confined to US submarines, while areas south of latitude 67°N, extending along the west coasts of Norway and Denmark as far as the north coast of Holland, including the Skagerrak and Kattegat, were allocated to the Royal Navy.[138] Rear Admiral Sidney Raw, FOSM 1950–51, was not entirely thrilled at this prospect, complaining that:

It is evident that until the Russians use advanced bases outside their own territory this area of European waters South of 67°N gives no opportunity for maintaining patrols outside enemy submarine bases. Our own submarines will therefore be limited to transit area patrols where even the chances of Russian submarines passing through are limited.[139]

Whether the development of the war plan showed that transit-area patrols were more profitable than patrols off enemy bases was a matter of conjecture, but Raw argued that it was 'clear that at the outset of the war a very big dividend may be reaped by having submarines on patrol outside the enemy submarine bases' and that 'to achieve the best A/S results, the earliest establishment of submarine patrols outside the enemy submarine base is essential. This can best be achieved by the employment of British-based submarines <u>North</u> of Latitude 67°N.' Raw recognized that if enemy bases in the Kiel Canal, the Great Belt (the strait between the main Danish islands of Zealand and Funen, effectively dividing Denmark in two) and the Sound (the strait that separates Zealand from the southern Swedish province of Scania) were disregarded, 'the only possible European enemy submarine bases at the outset of the war are to East of North Cape'.[140]

While it would have taken seventeen days for US submarines from the Atlantic coast to reach the North Cape of Norway, it only took nine days for Royal Navy submarines based on the Clyde to reach the same areas. Raw argued that:

> whatever the length of warning time available before hostilities break out there is therefore a better chance of covering these bases by submarines starting from the United Kingdom rather than from the U.S.A., particularly if there is a long period of strained relations when it would be most difficult to maintain such a patrol with submarines based in the U.S.A.

The Americans accepted these arguments, and, after detailed discussions with the US Submarine Command, the Submarine War Plan was amended to allow Royal Navy submarines to patrol outside enemy submarine bases north of latitude 67°N at the earliest possible date on the outbreak of war, until the Americans arrived in greater numbers.[141] How effective this would have been is questionable. Raw himself 'admitted that even if our submarines can reach North Cape before hostilities break out the Russian submarines are likely to have left for patrol, but there is still a good possibility of picking up the second flight or even the first flight returning from patrol'.[142]

If war had occurred, the 2nd, 3rd and 5th Submarine Flotillas would have concentrated at Rothesay in the west of Scotland at anchorages two miles apart in anticipation of a nuclear attack, while the arriving US submarines would have operated from a base further up the Clyde, at Rosneath. The early Submarine War Plan envisioned the 3rd Submarine Flotilla, led by the depot ship HMS *Montclare*, and the 2nd Submarine Squadron with its depot ship, HMS *Maidstone*, arriving in Rothesay between ten and twenty days after the start of hostilities. The submarines comprising the 5th Submarine Squadron would have been split between the 2nd and 3rd Flotillas. If the war lasted more than three months, a further depot ship, HMS *Adamant*, would also have berthed in the Clyde and formed a new 6th Submarine Squadron, primarily concerned with training. A second war plan existed, which would have seen Royal Navy submarines conducting an offensive mine-laying campaign in the Baltic. But the Admiralty expected Denmark to be 'overrun in the early days of a war in the immediate future', and 'the Allied minesweeping effort and the air cover necessary to enable the minesweepers to operate is unlikely to be available to provide reasonable protection for the passage of our submarines through the Baltic entrances'.

In 1954, the Submarine Service's war plans were integrated with a NATO submarine war plan, known as the Eastern Atlantic Submarine War Plan. It consisted of some 25 British, 23 US and 2 Dutch submarines declared to the Supreme Allied Commander Atlantic (SACLANT).[143] The Flag Officer Submarines was appointed as the NATO Commander Submarine Force, Eastern Atlantic (COMSUB-EASTLANT) and worked closely with his American counterpart, the Commander Submarine Force, Atlantic (COMSUBLANT).[144] The Submarine Service also participated in NATO exercises. In 1952 the Submarine Service took part in Exercise 'Mainbrace', a major NATO exercise involving 203 ships from nine navies participating in operations off Norway and Jutland. Seventeen submarines took part in the exercise, playing the part of Russian submarines attacking surface ships in the North Sea. In some of the first reported encounters of the exercise, HMS *Taciturn*, under the command of Lieutenant Commander J. Mitchell, acting as a submerged Russian submarine, claimed to have torpedoed the aircraft carriers HMS *Eagle*, HMS *Illustrious*, USS *Midway* and USS *Franklin D. Roosevelt* between the Isle of Arran and Ailsa Craig while they were deploying from the Clyde.[145] The Blue fleet disputed this outcome and argued that an RAF aircraft had attacked and crippled *Taciturn* before it was able to carry out its attack.

While the Submarine Service was finding its way and preparing for its new Cold War role, the early 1950s were characterized by a series of disasters that reminded submariners of the dangers of the sea.

'THE ADMIRALTY REGRETS . . .'

On 12 January 1950, the 'T' class submarine HMS *Truculent* was return-ing to Sheerness after completing a series of sea trials following a refit at Chatham dockyard. At 1900, as the submarine moved through the Thames Estuary on the surface, the Swedish oil tanker *Divina* was sighted in the channel. *Truculent*'s Officer of the Watch mistakenly decided that the *Divina* was stationary and as *Truculent* was unable to pass to the starboard side without running aground, the order was given to turn to port, across the channel and the *Divina*'s bows. This was a costly mistake. According to sound seamanship and the 'Rules of the Road' *Truculent* should either have held its course, turned to starboard or stopped. The two vessels eventually collided and remained locked together for a few seconds. As a result of the collision a large split was torn in *Truculent*'s pressure hull. Immediately before the collision the Captain had ordered all those on the bridge to go to their stations below, but only the lookout was able to obey the order before the remaining officers and ratings on the bridge were swept into the water. The Captain had ordered all water-tight doors closed, but there was insufficient time to establish a watertight boundary throughout the entire submarine before the collision occurred. As a result the whole forward part of the submarine and the Control Room were flooded and *Truculent* sank rapidly, bow down. In the Con-trol Room, the First Lieutenant, realizing what was happening ordered the crew to move aft. As 50–60 men crammed into the Engine Room and after-ends of the submarine, water started pouring into the Engine Room through the snort induction system. The men were trapped.[146]

In April 1946, the Admiralty had initiated a new review of escape policy and equipment. Under the Chairmanship of Rear Admiral P. A. Ruck-Keene, the Committee studied every known escape from sub-marines of all nations in peace and war and made a number of recommendations, including the production of a special immersion suit for each man fitted with a pressure-tight light that would activate auto-matically when it came into contact with sea water; the development of an indicator buoy with sufficient buoyancy to carry a radio transmitter and reflectors; a means of operating air purification apparatus in the event of a failure of electrical power; and improved training through the construction of a special 100-foot escape tank. By the beginning of 1951, although significant progress had been made, not all of the Com-mittee's extensive recommendations had been implemented.

As the most senior officer trapped inside the submarine, *Truculent*'s First Lieutenant decided to escape immediately. Propeller noises were heard overhead and with 50–60 men concentrated in the after compartments, air conditions were likely to deteriorate quickly. Preparations for an immediate escape were made in both compartments. In the engine room, it took 40–45 minutes to flood the compartment and ten sets of the Davis Submerged Escape Apparatus were issued to the weak or non-swimmers. *Truculent* carried one set of Escape Apparatus for each person on board with a small margin of spares, which were unavailable on account of the flooding of the forward compartments. Opening the escape hatch when the pressure inside the submarine is being equalized with that outside is a difficult task. In the after compartment, the First Lieutenant, in attempting to speed up the slow flooding of the compartment, went up into the escape trunk on two occasions to vent it. On the third occasion, anticipating that he might be blown out in the air bubble when he opened the hatch, he ordered a rating to hold his legs. In spite of this, on opening the hatch, the First Lieutenant was blown to the surface. The remaining men in the compartment waited until the rush of air from the escape tower had ceased, and the water had risen sufficiently to form a seal around the escape trunk. Although no one took charge after the First Lieutenant was ejected there was no panic and everyone left in an orderly manner.

The Master of the *Divina* did not immediately realize that his ship had sunk a submarine. The first signal to the shore was made forty-five minutes after the collision by a Dutch vessel, *Almdyk*, but it only said that the *Almdyk* was rescuing men, not that a submarine had sunk. The *Divina* picked up ten men; all the remaining survivors who succeeded in reaching the surface alive were swept out to sea by the strong ebb tide. Areas to which survivors might have been carried were searched throughout the night by ships and at daylight by naval aircraft. (To make matters worse, the crew of an Avro Lancaster were killed while taking off from RAF Kinloss en route to RAF Leuchars to pick up a party of Royal Navy divers who had been instructed to assist in the rescue operation.) Of the 79 men on board *Truculent*, 64 were lost. Of the 15 survivors, 5 had been swept off the bridge before *Truculent* disappeared beneath the waves; the remaining 10 escaped from the submarine. Post mortem examinations on recovered bodies revealed that many of the men had died from decompression injury and drowning while escaping from the submarine. As a result of the disaster a special navigational light for use on all surfaced submarines, known as the

Truculent light, was introduced on all submarines, and to speed up the implementation of the recommendations of the Ruck-Keene Committee, a Standing Committee on Submarine Escape (SCOSE) was set up to monitor progress.[147] It held its first meeting in October 1951, a few months after news of another tragedy spread throughout not only the Submarine Service, but also the country, and the world.

On 16 April 1951, HMS *Affray*, one of the Royal Navy's 'A' class submarines, departed for a week-long practice war patrol, codenamed Exercise 'Training Spring', in the English Channel. *Affray*, under the command of the 28-year-old Lieutenant John Blackburn, was carrying seventy-five officers and ratings, as well as four Royal Marines from the Royal Marine amphibious school at Eastney. At 2100 that evening, Blackburn sent a signal back to HMS *Dolphin* in Gosport: 'Diving at 2115 in position 5010N, 0145W for Exercise Training Spring.' When *Affray* failed to make radio contact at 1000 the next morning, Flag Officer Submarines, Rear Admiral Sidney Raw, issued the emergency submarine incident alert 'Subsmash One', calling on all available ships and aircraft to begin searching for a missing submarine. By noon, the search had failed to locate *Affray* and Raw upgraded the alert to 'Subsmash Two'. Every Royal Navy ship and submarine available at the time put to sea, alongside US and French warships, and scoured the English Channel, looking and listening for the missing submarine, 'a dismal and depressing task', recalled Joel Blamey, one of the men involved in the search, 'the weather was poor, and once darkness set in there was little hope of sighting anything. Dangerous . . . as there were so many ships milling around close by.'[148]

At one point hopes were raised when a tapping sound, thought to be the crew banging on *Affray*'s hull, was detected in an area where the submarine was thought to have been operating. A rescue flotilla closed on the position and dropped explosive signals into the water to let *Affray*'s crew know that it was safe to exit the submarine using the onboard Escape Apparatus. But, as they waited, no one came to the surface. The Admiralty later concluded that the tapping sound was actually the noise of other ships searching for *Affray*. On Thursday, 19 April, sixty-nine hours after *Affray* had dived, the Admiralty cancelled the Subsmash order and called off the search for survivors. Given the limited underwater endurance of the 'A' class submarines, no one could have survived beyond this point as the air inside the submarine would have run out. A scaled-down search continued for another two months until June, when the submarine was discovered in 300 feet of water, thirty-seven miles from its reported diving position. When divers examined the wreck they discovered that *Affray*'s snort mast had broken free and was lying on the seabed, leaving a 14-inch-diameter hole above the engine room exposed to the sea. The mast was later raised and sent to one of the Navy's metallurgical laboratories for tests. The results revealed that metal fatigue had caused it to break free of *Affray*. But how this had happened was the subject of intense debate. Some suspected the mast had been knocked off by a passing ship, or by the rescue ship *Reclaim* during subsequent investigations. Others could not understand why the snort snapped when weather conditions were ideal, and not when the mast was subject to more severe stresses such as when the submarine was rolling heavily during previous exercises.

A Board of Inquiry, one of the most controversial in British naval history, was convened to find answers. It adopted a policy of trying 'to eliminate unfruitful lines of investigation' and removing 'at least . . . a certain amount of inevitable conjecture'. Possible causes that were considered ranged from a battery explosion, metal failure, human error, hitting a mine and collision with another vessel. All were ruled out and the Board concluded that *Affray* 'was lost due to the snapping of her snort mast and that this was due to a failure of the material of which that mast was made'. The crack occurred above the point where the bending movement of the snort was the greatest and where a high number of welds were present. As the break occurred the mast fell over to port, the remaining portion twisting so that only a small piece of metal kept the snort from separating from the submarine as it lay down over the port side of the submarine. The circular tube was welded and its material, though of satisfactory composition, was found to have been grossly overheated and untreated after being welded at a very high temperature. This accounted for its extreme brittleness at normal air and sea temperatures. The Board's description of what they believed happened on board *Affray* is vivid:

As a result a 14in hole was suddenly open direct to the sea, flooding the Engine Room. A delay of three seconds in closing the valve could allow enough water to enter to produce a stern-down angle of 16 degrees if not quickly rectified. If she had been snorting through the night, she would have reached her actual position somewhere between 0500–0700. It would have been quite reasonable and in accordance with submarine practice for the majority of officers and ship's company to be turned in and the submarine to be at watch snorting stations. The Commanding Officer would very likely have turned out at dawn (about 0530) and having satisfied himself that all was well, have turned in again.

In consequence, those inside the submarine started at a disadvantage. That something serious had happened would be evident, but precious seconds would be likely to have been lost in diagnosis . . . Our evidence shows that unless remedial action was taken within 15 – and possibly 10 – seconds, the situation would start to get out of hand . . . There would have been little chance of escape or even of releasing the indicator buoys or firing smoke candles due to poisonous fumes, pain in the ears, exhausting effects of pressure and lack of concentration and coherent thinking. There would also have been a short-circuiting of electrical equipment causing loss of power and lighting throughout the submarine at an early stage. There would have been electrical fires and fumes.

The chances of escape at a depth of over 250 ft have to be regarded as virtually negligible. Perhaps the odd exceptional man might reach the surface and survive, only to die from the effects of air embolism (the bends). From 200–250 ft the escape rate is down to 10 percent and at over 250 ft the survival rate using the 'twill trunk' method is negligible.

All on board would have died within a very short time. If any personnel had managed to shut themselves into the foremost watertight compartment, they in turn would very rapidly have lost their lives.[149]

In reaching this conclusion the Board had been strongly influenced by a theory put forward by the Flag Officer Submarines known as 'FOSM's hypothesis', which postulated that the snort mast snapped while the submarine was snorting near the surface, water rushed in before anyone could do anything about it and the submarine sank stern down, striking the bottom at an angle of 65 degrees. However, if this was correct the submarine would have hit the bottom with a large stern-down angle, causing considerable damage to the rudder. According to the evidence there was no damage to the stern and many in the Admiralty, such as the Director of Torpedo, Anti-Submarine and Mine Warfare and the Director of the Operations Division, noted that it was 'difficult to reconcile the absence of any visual damage to the submarine's stern with the wholesale acceptance of FOSM's hypothesis'.[150] One of the harshest critics was the Director of Naval Construction, Sir Victor Shepherd, who argued that:

The conclusions of the Board of Enquiry as to the cause of the accident are not concurred in. Inspection of the wreck has revealed nothing amiss except the broken snort mast, which may have broken off after the submarine had come to rest on the bottom – in fact, the position in which the mast was found and examination of the fracture suggests this was so.

To conclude that the loss of the ship was due to the fracture of the snort mast merely because no evidence has been obtained pointing to any other cause, ignores the fact that many possible sources of flooding could never be revealed by external inspection. It is also considered that the evidence against the Board of Enquiry's conclusions has not been given the weight it deserves.

There is no evidence as to the state of the valves in the snort induction system, but the Board's conclusion implies that neither of them was closed. To explain this, the Board have accepted FOSM's hypothesis 'how it happened'. If the snort mast snapped off as is premised in the hypothesis, it is not disputed that a large angle of stern trim would be taken up and that unless one of the valves was closed within a short time the situation would get out of control. There is, however, no satisfactory explanation why the snort mast should snap in perfect weather and at a speed which produced a maximum tensile strength of only one ton/sq.in. in the tube, even allowing for its material defect. Whether this can be explained or not the hypothesis is considered disproved by the absence of damage to the rudder and by the position and state in which the snort mast was found. If the hypothesis does not fit the facts, it follows that grave doubt must be cast on the premise.[151]

Shepherd's own theory was that a battery explosion that 'might have fractured the battery ventilation outboard trunking and valves, so admitting water to the ship and at the same time starting a fire, could explain the ship having sunk with the snort mast up and with little hope of survival of the crew'.[152] In June 1950, HMS *Trenchant* had suffered extensive damage from a battery explosion, caused by a build-up of gas during the charging process. However, the Board of Inquiry concluded that 'Batteries in *Affray* were likely to have been in very good condition and that if the ventilation system was in good order and correctly operated, a battery explosion is very unlikely to have occurred, even with batteries in poor condition. As the evidence suggests that both the batteries and ventilating system were in good order, the likelihood of a battery explosion is discounted.'

The Head of Naval Law came away from the Inquiry with the 'impression that the Board of Inquiry by adopting the policy of trying "to eliminate unfruitful lines of investigation" and to "remove at least a certain amount of inevitable conjecture" has perhaps closed its eyes to too much.'[153] But further diving work on the wreck to try and establish what had happened was abandoned in early November 1951 after

Affray began to list, making diving operations increasingly dangerous. During the search the Admiralty identified that, compared with other British submarines, the 'A' class, because of their narrow underwater lines, had 'a tendency, if heavily flooded, to bottom with a large list'.[154] Thereafter the Government judged that any attempt to salvage the submarine was dangerous, expensive and not at all certain to be successful. In the absence of any other positive evidence, FOSM's hypothesis was deemed to be the most likely cause of the accident. A flap valve, known as the '*Affray* valve' was introduced, designed to slam shut in the event of water flow. Locking pins were also installed to ensure snort masts remained upright. On 14 November 1951, the First Lord of the Admiralty, James Thomas, informed the House of Commons that 'with the high risk of total failure, there is no justification for this substantial diversion of our resources. There will therefore, be no further operations in connection with *Affray*.'[155]

Publicly that was the end of the investigation. However, behind the scenes in the Admiralty, their Lordships were seeking to apportion blame. In December, one of the Admiralty's naval lawyers stated that 'there is no doubt that the officer responsible ... for sailing *Affray* in such circumstances is deserving of censure'. That officer was Captain Hugh Browne, the Captain of the 5th Submarine Flotilla. The Admiralty's lawyer argued that as it was Browne who had issued the orders for *Affray*, and although a copy was received by Flag Officer Submarines, Rear Admiral Raw, and by the Admiralty, 'the responsibility for ensuring that the submarine was in a fit state for sea and her crew properly worked up for the operation ordered rests with Captain Browne'. Browne was later informed that 'Although there is no definite evidence as to the cause of the loss, My Lords consider that you made an error of judgment in sailing *Affray* with a training crew and folboat party, a team of Royal Marines with special canoes, embarked on a training patrol, before she had been given opportunity for the working up which was clearly desirable after her protracted refit and the many changes in her crew.'[156] No further action was taken against Browne. There was no court martial and no public announcement. No one outside of the Admiralty knew that the Captain of the 5th Submarine Flotilla had been censured.

For the families and relatives of *Affray*'s crew, the search for answers continued throughout the twentieth and into the twenty-first century, fuelled by the lack of an explanation and the failure to raise the submarine. In 2007, the then Armed Forces Minister, Bob Ainsworth, rejected a request on behalf of some of the relatives to reopen the inquiry after a number of allegations were made about a cover-up. That year the

Royal Navy's Naval Historical Branch carried out a study of claims that the true cause of the loss was known and suppressed in order to spare the embarrassment of senior naval officers. It found no evidence of a cover-up. Despite this, calls for a new inquiry resurfaced in January 2012, but the then Armed Forces Minister, Nick Harvey, told the House of Commons that:

> Submarines are complex ships, operating in an environment that is extremely dangerous, even in peacetime. Submariners operate at the limits of human ingenuity, and that is to their credit. They are among the bravest men in the Royal Navy, and soon to be the bravest women too. The loss of *Affray* and the men who served on her was a national tragedy, as well, of course, as a personal tragedy for many. We all understand only too clearly why those who were personally affected want definitive answers, but nothing can bring the fallen back, and after more than 50 years, there seems to me to be no realistic likelihood that we can ever provide the answers that, for understandable reasons, they crave. I cannot see that any new evidence is available to us now that was not available to the original Board of Inquiry. The passage of so much time seems to me to make the prospect of discovering anything new infinitesimal.[157]

Today, HMS *Affray* and her crew remain entombed on the seabed on the northern edge of Hurd Deep in a site protected by the 1983 Protection of Military Remains Act. The site is strictly off limits to divers, unless permission has been obtained from the Ministry of Defence.

The loss of *Affray* was crippling to the officer structure of the Submarine Service and only worsened the manpower problem. Not only did the Navy lose an entire ship's company, an entire officer's training class was also on board. After the incident the number of officers volunteering for submarines dropped dramatically. Some reflected on the future of the service. In 1952, Captain Bertram Taylor, from the Staff of Flag Officer Submarines, told a conference on anti-submarine warfare that:

> The fact is that every single submarine we possess today is either in operational service or under refit or conversion. What is more, there is no prospect of improvement until new construction comes to sea and the additional manpower is allocated. Unhappily the immediate prospects are somewhat gloomy; we may be obliged to pay off 3 submarines into reserve (at least temporarily) for lack of manpower.[158]

But the Submarine Service was about to be called upon to undertake another important, dangerous task, one that would take submariners

into the icy northern waters of the Barents Sea and to the front line of
the Cold War.

THE RISE OF UNDERWATER
INTELLIGENCE GATHERING AND
RECONNAISSANCE

By the mid-1950s the Royal Navy's submarines were increasingly
involved in gathering intelligence and conducting reconnaissance operations in the ice-free areas of the Barents Sea near Soviet waters. The
Russians started to introduce new ships, submarines and aircraft,
together with new missile, gun, radar and communication systems.[159]
The Royal Navy and the US Navy recognized that the submarine was
ideally suited to collecting intelligence about these new weapons in the
areas in which they were tested, the supposedly secure waters adjacent
to the Soviet mainland. 'A submarine collects the high-quality data
derived from monitoring opposition activity – *when they think no one
else is around!*' explained John Hervey. 'If you send a surface ship
to do the job, a marker is put on it and all the interesting activity it
might have witnessed is transferred to another area. If a reconnaissance aircraft is sent, all emissions worth recording will be switched off
long before it gets close enough to intercept them. The same thing happens during the passage of an electronic orbiting satellite.'[160] The
existence of these operations, as the Director of Naval Intelligence,
Vice Admiral Sir Norman Denning, told the Joint Intelligence Committee in September 1960, 'had always been kept a very closely guarded
secret'.[161]

The United States Navy conducted the first post-war intelligence-
gathering operations in its own recently modernized submarines. The
first operation appears to have taken place in the summer of 1949 when
the USS *Tusk* and USS *Cochino* sailed to the Arctic Circle to gather
intelligence on Soviet nuclear-weapon tests.[162] The operation ended in
disaster when a battery fire on board *Cochino* resulted in the loss of the
submarine north of Norway and the deaths of a number of crewmen
from *Tusk*. It would be many years before the Americans went back.
The first post-war Royal Navy intelligence-gathering operation in
northern waters took place in 1952.[163] The Submarine Service had spent
the immediate post-war years ensuring that its submarines designed for
patrols in the tropics operated in the Arctic, particularly the Barents

Sea, areas of which relatively few Commanders had any experience. What little information there was came from submariners such as Edward Young, who had operated in the Barents during the Second World War. The conditions were challenging:

> In our patrols between Kola inlet and North Cape we encountered no floating ice, but during the brief daylight visibility was hampered by the spray which froze on the periscope and by frequent snow storms which blotted out the land for hours on end. And on the surface in the long hours of darkness we faced the beastliness of spray which turned to ice even before it struck our faces. It froze on the gun, on the periscope standards, in the voice-pipe, and all over the bridge. Icicles hung from the jumping wire from one end of the submarine to the other, and sometimes formed so much top weight that the Captain became concerned about our stability.[164]

In February 1948, HMS *Ambush* conducted an operation between Jan Mayen and Bear Island in order to obtain information about how the Royal Navy's new 'A' class submarines operated under Arctic conditions. Three years later, in April 1951, the Admiralty informed the Foreign Office that it intended to send a submarine further into the Barents Sea:

> Our war plans involve submarine patrols in the Arctic Ocean and Barents Sea north of latitude 67 degrees N., bounded on the west by longitude 80 degrees East. The success of submarine operations in this zone will depend entirely upon the efficiency and reliability of the communications systems, particularly on the reliability of submerged W/T reception by submarines. We have no experience of the conditions in the area in this respect and we therefore propose that in July–August 1951, one of our submarines should carry out a cruise in the Barents Sea approximately between longitude 30 degrees E and longitude 80 degrees E with the object of obtaining as much information as possible on the efficiency of W/T communications in the area . . .
>
> We appreciate how important it is that we should do nothing at the present time which might be looked upon by Russia as a provocative act, but we attach considerable importance to this trial of communications, and as our proposal involves only a cruise in the open sea and there is no question of the submarine entering, or going anywhere within miles of Russian waters, we do not consider that even if the submarine is seen by the Russians they can take much exception to the operation.[165]

The Foreign Office agreed on the condition that the Admiralty stick 'religiously to their undertaking that the submarine will not go anywhere within miles of Soviet waters (as the Soviet authorities conceive them, not merely what we think they should be)'.[166] As one official explained:

> The Soviet Government like their Tsarist predecessors have invariably claimed territorial waters extending for a distance of 12 miles from the coast of the Soviet Union. H. M.G. have, of course, never recognised that territorial waters extend more than 3 miles and have reserved their rights as regards Soviet regulations claiming 12 miles.[167]

British fishing trawlers regularly strayed inside the twelve-mile limit. Although the Russians generally tolerated these intrusions, in 1950 Soviet authorities detained a Hull-based trawler, the *Etruria*. However, the Foreign Office was supportive: 'I don't think we have anything to fear even if the Russians spot it and make some propaganda insinuations,' noted one official. 'After all, their submarines, I believe, do the same kind of exercise in the North Sea and Atlantic.'[168] HMS *Andrew* was ordered north but the patrol was abandoned because the Admiralty was 'not satisfied with the performance of the submarine's Snort apparatus'.[169]

The Admiralty tried again a year later. In April 1952 the Foreign Office approved another operation on the condition that the CO of the submarine observe certain safeguards. He was instructed 'to keep well clear of territorial waters claimed by the Soviet Union and, as far as possible, of Soviet shipping'.[170] The Naval Attaché in Moscow was also informed in case something should go wrong.[171] On 30 August 1952, HMS *Alcide* sailed from Rothesay for the Barents Sea under the command of HMS *Venturer*'s former CO, Commander Jimmy Launders. The patrol was codenamed Operation 'Adamaston' and *Alcide* was ordered to proceed north of a large island off the Russian mainland known as Novaya Zemlya (Russian for 'New Land'), which was used as a testing facility for Soviet weapon systems. *Alcide*'s CO was ordered to remain unseen and undetected, to carry out communication reception trials and obtain data on water conditions, anti-submarine operations, navigational data, ship-to-shore communications, and physiological data as well as testing material and equipment. Launders was also ordered 'to sight and subsequently report on ships, both Naval and Mercantile, to the East of NOVAYA ZEMLYA'.[172]

The risk of detection was considered to be very small. A 1952 assessment by the Director of Naval Intelligence concluded that:

Such Soviet A/S equipment as is known to exist is elementary in design by modern standards. Some asdic-type sets are fitted but multi-unit hydrophones are thought to be in use by a number of surface vessels. A limited number of 'Hedgehog' type ahead throwing weapons [an anti-submarine weapon which fired a small number of motor bombs which exploded on contact] may be in service, but the main A/S weapon is still the depth charge. A/S tactics and operating skill are likely to be poor, as may also be cooperation of ships in a hunt. Attacks are likely to be more ferocious than accurate, large numbers of depth charges being used.[173]

The Royal Navy's first operation in northern waters was a failure. Launders was forced to abandon any attempt to pass north of Novaya Zemlya after a number of *Alcide*'s fuel tanks were damaged during the operation and he had forgotten to fill the submarine's emergency fuel tanks prior to departure. The Flag Officer Submarines, Rear Admiral Simpson, was so angry that he considered taking 'disciplinary action with regard to the Commanding Officer and Engineer Officer ... for the lack of seamanlike judgment and professional knowledge they displayed in this matter'.[174] Launders incurred Simpson's 'severe displeasure for ... the curtailment of this important operation'.[175] The Director of Naval Intelligence, Rear Admiral Sir Anthony Buzzard, was also disappointed and wrote of 'the need for further operations of this nature since our knowledge of this area is not otherwise covered'.[176]

In February 1953, the Admiralty tried again, this time with an operation specifically aimed at collecting intelligence about a Soviet Fleet exercise. But the Prime Minister, Winston Churchill, refused to authorize the operation after taking scientific advice that a submarine stood little or no chance of getting anywhere near the exercise undetected.[177] The Chief of Staff to Flag Officer Submarines, Captain Arthur Hezlet, disagreed and ordered a submarine to conduct surveillance against the Royal Navy's Combined and Home Fleets based in Gibraltar in order to convince the Prime Minister that 'there was little risk in allowing a submarine to undertake a similar operation in the Barents Sea during Russian fleet exercises'.[178]

HMS *TOTEM* – 'A SLIPPERY CUSTOMER'

Hezlet selected the wartime submariner and CO of the recently converted 'T' class submarine HMS *Totem* to conduct the operation, now codenamed Operation 'Cravat'. By the mid-1950s the modified 'T' class were proving far superior to the 'A' class such as Jimmy Launders' HMS *Alcide* and Coote had made a number of unauthorized modifications to *Totem* that greatly increased its capabilities. The most valuable was a prototype S-band receiver antenna that Coote had obtained while visiting a scientist friend who worked at the Admiralty Signal and Radar Establishment in Portsdown Hill.[179] After a good lunch in a nearby pub, Coote persuaded the scientist to part with the prototype and later drove out of the establishment with it in his raincoat pocket. Once it was fitted to *Totem*'s radar periscope, the intercept equipment granted *Totem* immunity against specific airborne and shipboard S-band search radars.[180] It also provided *Totem* with a considerable range advantage. 'We certainly ran rings around various destroyer flotilla captains who dominated night exercises by ordering X-band silence, then easily detecting and putting down snorkeling submarines by using their S-band radars,' remembered Coote. 'In this manner we were never caught napping by surface or air forces, and my "luck" became a byword.'[181]

Totem was also one of the fastest 'T' conversions in the Royal Navy, if not the fastest. This was thanks to yet another unauthorized modification. On most of the 'T' class the streamlined fin was broken by an inverted tub-shaped protrusion at its top, known as the 'bird bath', which was meant to house an antenna from another new radar. 'As soon as I reached the depot ship away from the prying eyes of the Admiralty I had it cut away and replaced with a perfectly faired plating,' explained Coote.[182] The result was an extra 0.75 knots on *Totem*'s maximum dived speed. 'It was like money for old rope,' he said, 'we'd just pop under . . . wind on 18 knots and we would clear a mile in four minutes. Follow that with ten minutes' silent running . . . at 12 knots and we would be 3 miles from the escort.'[183]

Coote was a highly competent submariner and during the 1953 Summer War Exercise he had earned a reputation for unconventional thinking by coming up with an innovative means of intercepting 'enemy' submarines entering or leaving a bay near the Outer Skerries lighthouse just off the Shetland Islands. The lighthouse had an excellent view of the approaches to an area that had been designated as an enemy base. Seeing an opportunity, Coote launched a folboat with two officers armed with a portable voice radio, several bottles of Scotch and a haversack full of 1.25 lb charges. He told the men to paddle ashore to the lighthouse, capture it and persuade the keepers not to report their presence by bribing them with Scotch and cigarettes. For two days the officers kept a lookout over the surrounding area while *Totem* patrolled nearby. Whenever an enemy submarine approached the bay the officers would toss a signal charge into the water. This was the signal *Totem* needed to raise her VHF antenna and listen on a prearranged frequency to one of the officers: 'Customer for you, outward bound in position Alpha Lima Four Seven. Course 135 degrees at 15 knots.' *Totem* would then move in for the kill and Coote was able to maintain a watertight blockade of the enemy base throughout the rest of the exercise. 'I did not get the roasting I expected and probably deserved,' he later wrote.[184] Flag Officer Submarines was forced to accept that the ploy was a realistic possibility.

Totem also had a first-rate crew. Coote's second-in-command was the future Chief of Defence Staff, Lieutenant John Fieldhouse. Known as 'Snorkers' after the Australian First Lieutenant of *Compass Rose* in *The Cruel Sea* with whom he shared a love of sausages, Fieldhouse was regarded by Coote as 'without equal' on personal and professional grounds.[185] *Totem* was also one of the first submarines in the Royal Navy to put to sea with an Electrical Officer. This was Lieutenant Peter Lucy, who arrived on his first morning to find a distinctly unimpressed Coote nursing a terrible hangover and in no mood to pass the time of day with anyone. 'We've managed in submarines for half a century without specialist Electrical Officers,' remarked Coote. 'The First Lieutenant has always been responsible. When something goes wrong he just gives it a kick and it re-starts. What do you have to offer that's so unique?' 'Ah yes, sir,' Lucy replied, 'but I went to University for three years to learn *where* to kick it.'[186]

Coote's brief for Operation 'Cravat' was to act independently, as if *Totem* were a Soviet submarine, and obtain as much information as possible about Royal Navy activities in the vicinity of Gibraltar. The operation was so secret that only the Commanders-in-Chief, Home Fleet and Mediterranean, and their Chiefs of Staff knew that it was taking place. A small party from the Radio Warfare Section of HMS *Mercury*, the Royal Navy's Signals School near Petersfield in Hampshire, embarked on board *Totem*.[187] These men were radio communications operators, specially trained in signals intelligence, and they were responsible for the 'special fit' communication and intercept equipment that was installed in the submarine.[188] To make the operation realistic, *Totem* was only allowed to act on information that was available in publicly accessible sources, such as the press. The

Hampshire Telegraph provided Coote with an outline of the Home Fleet's Spring Cruise programme, as well as the dates and venues of the various Combined Fleet sporting fixtures, with finals to be contested in Gibraltar. The *Times of Malta* was also a useful resource as it listed the departure dates of many of the heavy units in the Mediterranean Fleet. To maintain operational secrecy, Coote even went as far as chartering HMS *Dolphin*'s yacht and sailing off for an overnight stop at Bosham Harbour just outside Chichester in West Sussex, after too many shore staff at the base started asking questions. On 4 March 1954, the yacht quietly slipped back into Haslar Creek in Gosport.

Totem sailed an hour later and ran south, diving by day but operating on the surface at night. At Cape St Vincent she turned east and started the 200-mile voyage into the Western Mediterranean. Between 10 and 20 March, *Totem* patrolled undetected between Gibraltar and Algiers, during which considerable radio, tactical and equipment intelligence was obtained from observing and monitoring ships, submarines and aircraft of the Royal Navy's Home and Mediterranean Fleets. In total fifty-two recordings of radars and ultra-high-frequency (UHF) and very-high-frequency (VHF) circuits were taken and 497 high-frequency communication messages intercepted. Although Coote was ordered to remain undetected south of the 45th parallel, on a number of occasions he was forced to take *Totem* into 'lethal proximity' of surface ships conducting anti-submarine exercises in order to collect intelligence. Although *Totem* enjoyed a considerable advantage – Coote possessed detailed knowledge of the capabilities of air- and shipborne anti-submarine equipment which would have normally been denied to a foreign submarine – the prevailing natural conditions of unlimited visibility, anomalous radar propagation and calm, isothermal seas more than redressed the balance in favour of *Totem*'s targets.

At 0730 on 11 March, *Totem* advanced towards what intercepted radio traffic indicated was a Submarine Exercise Area to observe an exercise between two submarines, the 'A' class HMS *Artful* and the 'S' class HMS *Scorcher*, and the units of the Royal Navy's 6th Destroyer Squadron and 6th Frigate Squadron. At 0730 *Totem* was caught between two destroyers, HMS *Battleaxe* and HMS *Crossbow*, while they were engaged in an anti-submarine exercise, hunting HMS *Artful*. 'In flat calm and alarmingly good asdic conditions, detection seemed certain,' wrote Coote. 'The two ships passed either side of *Totem*, which was at 175 feet trying to keep end on simultaneously to the two ships both apparently in contact at 2500 yards range. Perhaps the fact that the echo was outside the exercise area and inside Spanish waters saved us; at all events, on regaining periscope depth at 0840, both squadrons were busily engaged hunting their official targets. A series of excellent recordings were taken.'[189]

Coote avoided detection (later analysis revealed that the two destroyers were in contact with *Totem* but they failed to classify the submarine correctly) and after recording and photographing enough of the 6th Destroyer Squadron's exercise with *Artful* he moved *Totem* southwest to monitor the 6th Frigate Squadron's exercise with HMS *Scorcher*. After recording the exercise Coote returned once again to observe the 6th Destroyer Squadron:

> The afternoon started with some useful 'X' and 'S' band radar recordings from *Apollo* and *Artful* whilst they waited for 'Go' time. They were photographed. From then till 1745 a most interesting recording was made of a complete [exercise] yielding all manner of tactical intelligence from the moment of the aircraft's first contact till he handed over to the ships, their search and final kill.[190]

The next morning, 12 March, Coote moved *Totem* to the southwest, near to Gibraltar airfield, in time to catch the Royal Navy fleet as it returned to harbour after completing weapons training. There, he gained yet more intelligence, especially about the Royal Navy's latest operational fleet carrier, HMS *Eagle*:

> At 1337 EAGLE accompanied by the 4th Destroyer Squadron was sighted bearing 122° as she steamed towards EUROPA Point [the tip of Gibraltar]. At this stage TOTEM was a mile offshore to the east of La Linea where an apparently incessant stream of Neptunes, Shackletons, Skyraiders, Fireflies and S.51 Helicopters either orbiting whilst waiting to land or actually approaching the runway discouraged prolonged exposure of the masts needed to monitor EAGLE's traffic. The sea was, as always, calm and translucent, however photographs were taken, together with a recording of the carrier's 'S' band radar and a check of her rev-not ration [revs per knot].
>
> The three hours spent there were by no means wasted. The Rock's air warning mattress arrays were plotted and noted to stop operating after the tenth Shackleton had landed soon after 1500. A very powerful 'S' band radar ... was thought to be part of the fortress's defence; it was heard out to 50 miles that night ... 6 different frigates or destroyers were recorded ... talking on UHF circuits as they entered harbour.

Coote withdrew *Totem* to the northeast and hugged the Spanish coast until the next morning. He had intended on moving southeast to intercept a convoy; however, he became suspicious of a Neptune search aircraft which was sighted intermittently between 1200 and 1325. 'The very fact that an aircraft had been put out on an intensive patrol on a Saturday afternoon during a stand-off period was clear indication of the serious measures being taken to discourage uninvited spectators,' wrote Coote. He decided to abandon his plan to intercept the convoy and instead withdrew *Totem* to the North African coast and to wait in the Gulf of Arzeu for the night.

For the next few days Coote attempted to intercept the main exercises between the Royal Navy's Home and Mediterranean Fleets, codenamed 'Touchline One', 'Touchline Two', 'Touchline Three' and 'Touchline Four', between Oran and Algiers. But with surface ships and aircraft participating in the exercises expecting to come under attack from submarines that were also taking part, Coote had to be careful. *Totem* was almost detected on 16 March while attempting to intercept a convoy escorted by the Royal Navy's last operational battleship, HMS *Vanguard*. At 0914, two 'Weapon' class destroyers, escorting *Vanguard*, passed astern of *Totem* and one of them, HMS *Venus*, came perilously close to the submarine. 'We went deep fine on her port bow, moving across and out, keeping stern on,' wrote Coote. 'At 0931 she pinged all over us, but moved on. There remained but one more source of transmissions to ping over the submarine at 0940, before I was able to regain periscope depth and see that the port wing destroyer, which must have passed nearly overhead, was one of the Daring Class.' Coote continued to observe *Vanguard* from four miles away, advancing into the waters astern of the battleship's starboard bow escorts, taking photographs of the battleship and two training carriers, and recording the accompanying destroyers' radar and communication transmissions. He remained in the area until 1020, when a Shackleton Maritime Patrol Aircraft started patrolling astern of the convoy.

For the next two days, 17 and 18 March, *Totem* attempted to intercept HMS *Eagle*, which was due to follow a convoy from a distance of twenty-five miles. However, the carrier proved elusive and the heavy presence of frigates and aircraft fully alerted to the presence of submarines participating in the exercise restricted *Totem*'s freedom of movement. Worse, *Totem* was suffering from a number of mechanical problems and at 1128 Coote was forced to withdraw his submarine after its remaining ballast pump became defective. As *Totem* withdrew, a sharp explosion in the Control Room, 'like a baby volcano erupting from the dark recesses', put the hydroplanes out of action, damaging the main vents and *Totem*'s snort air intake valve, which could only be shut by hand. Coote had little choice but to surface *Totem* as the main hydraulic system was spewing liquid onto the Control Room deck, something that only became apparent when 'people started floundering like novice skaters all over the control-room' floor as the room was in dimmed red lighting at the time. By the time the defect was repaired over 70 gallons of oil had escaped from the system.

Despite these difficulties *Totem* was able to dive again and continue the operation. As night descended on 19 March, Coote moved his submarine towards the North African coast to recharge batteries. The next morning he dived *Totem* again and set course for Gibraltar Harbour to carry out close range photographic reconnaissance of the many ships that made up the Combined Fleet. *Totem* entered Algeciras Bay close to Carnero Point lighthouse, unobserved by either the powerful Gibraltar Rock Radar or by the lighthouse keeper, who as Coote recorded 'chose that moment to answer nature's call on the foreshore'. *Totem* spent the next hour and a half eavesdropping, collecting call signs and frequencies from the various ships in the harbour. At 1146 the submarine swept down past the entrance to Algeciras Harbour and photographed HMS *Indefatigable*, two cruisers, HMS *Implacable*, HMS *Eagle* and HMS *Vanguard* 'in such a manner as to show the maximum detail of their antennas. It took 9 shots to get all *Eagle*'s aerials,' recorded Coote.[191] He then took *Totem* so close to the Combined Fleet that on one occasion he looked out of the raised periscope and found himself looking straight at the planking of an Admiralty 14-foot sailing dinghy. 'No one sighted us, or if they did, they did not believe their eyes,' he recalled.[192] 'I wondered whether any of the boats racing in the bay at the time would mistake the periscope for a mark on the course, particularly RN 688's dinghy, which was nearly skewered by it.'[193]

When Coote returned to the UK he presented his report of the operation, along with some conclusions about the intelligence he had been able to obtain, to Arthur Hezlet. 'In assessing the validity of this intelligence,' he wrote:

> consideration has been given as to how much of this information could have been picked up by a suitably placed and equipped Russian submarine which did not possess *Totem*'s prior knowledge of events and frequencies. The surprising answer emerges from a necessarily brief

analysis that every single item . . . could easily be in Russian hands now. Sometimes the most insignificant interception may yield a clue of great importance.[194]

The Admirals on the Admiralty Board were so shocked at how much information Coote had been able to obtain, especially from VHF and UHF voice circuits, that they forwarded a summary of his patrol report to the Commanders-in-Chief of the Royal Navy's various fleets:

> It will be observed that it was possible for a submarine to remain un-detected in the vicinity of the fleet for many days and to derive valuable intelligence, especially from VHF/UHF voice circuits. Some of the information gleaned, particularly the transmission characteristics of the different radar sets fitted to ships and aircraft, could have been obtained by other means. Nevertheless, Their Lordships view with serious concern the indiscriminate use of plain language communications and of radar and wish to stress that steps must be taken to minimize the opportunities provided for a potential enemy to obtain vital tactical intelligence.[195]

For Hezlet, the man responsible for masterminding the operation, Coote and *Totem* had proven that 'a corresponding probe among Soviet forces was a justifiable risk, unlikely to create an embarrassing international incident'.[196]

This view was reinforced two months later when HMS *Trenchant* and HMS *Sentinel*, two submarines with the 1st Submarine Squadron in Malta, conducted a successful intelligence-gathering operation against a Soviet 'Sverdlov' class cruiser and two 'Ognevoi' class destroyers.[197] Sailing from Malta on 2 June, the two submarines were able to obtain propeller revolution and speed counts, asdic transmission frequencies and periscope photographs of the Soviet vessels as they returned to the Black Sea. 'In view of the calm sea I had decided to keep about four thousand yards off track, this being adequate to get a good revolution count and positive identification, and yet being out of sonar and periscope sighting range,' wrote *Trenchant*'s CO. 'The force was formed with one destroyer about 1500 yards ahead of the cruiser with the second destroyer in close order astern, all steaming a steady course.'[198] The Director of Naval Intelligence concluded that *Trenchant* and *Sentinel* had obtained 'valuable' intelligence, while the Director of Undersurface Warfare noted that 'the valuable information derived of rev. counts and asdic frequency fully justified the time spent by the submarines on this duty'.[199]

At some point towards the end of 1954 Churchill was informed that *Totem* 'was successful in escaping detection during the whole of the 10 days of Operation "Cravat" and was able to use her special equipment with full success'.[200] That autumn, the Prime Minister changed his mind and approved the Royal Navy's first submarine intelligence-gathering operation in northern waters, codenamed Operation 'Defiant'.[201] The following year, once again under Coote's command, *Totem* rounded the North Cape of Norway, entered the Barents Sea, remained totally submerged for about six weeks and 'battled against serious opposition with some unexpected results'.[202] She experienced considerable difficulty with the specially fitted intelligence-gathering equipment coupled to the snort mast and after periscope to 'hoover up' electronic intelligence. When the equipment started to malfunction, Coote suspected that the connection on the periscope was loose, as when *Totem* was stationary it appeared to work as intended. With faulty intelligence-gathering equipment Coote was faced with the very real possibility of curtailing his operation and returning home. But he decided to try and fix the equipment. He withdrew *Totem* to the western limits of the patrol area in order to surface and physically check all the connections on the periscope.

After withdrawing *Totem* surfaced. Coote ordered Peter Lucy, the Electrical Officer, to climb up the conning tower and carry out repairs on the equipment. He told him that if *Totem* was detected he would immediately close all hatches and crash-dive, leaving Lucy to fend for himself in the freezing waters of the Barents Sea. 'Lucy, who did the job alone,' recalled Coote, 'was under no illusions that if we were jumped by unfriendly forces, I might have to dive the boat under him.'[203] One of the Communications specialists on board, Tony Beasley, recalled what happened next:

> Waiting over, the boat was brought up to a level which made the Fin protrude completely out of the water. A quick all round view with the periscope ascertained no contacts visible, it was all systems go. Fortunately the weather had abated to a force two or three . . . [Lucy], armed with a bag of tools and a tin of white Vaseline [the medium used to protect connections from salt water], waited for the First Lieutenant (number one) to open the hatch and sample the freezing salty air of the Barents, before he started to ascend the iron ladder to the Fin. Everyone in the Control Room wished him luck as he crept ever upwards. Someone even offered him a week's tot. The First Lieutenant [Fieldhouse] went after

him, remaining on the ladder to pass on progress made. He was also responsible for closing the hatch in the event . . .[204]

Fieldhouse opened the hatch and inhaled what for submariners was a precious commodity, 'fresh air, lovely fresh air'. As it poured into the submarine, everyone in the Control Room took advantage of the 'gift from heaven'. On the fin, Lucy carried out 'a prolonged and tricky operation' to repair the equipment while *Totem* was dangerously exposed and wide open to detection.[205] 'The whole crew were on tenter hooks,' remembered Beasley. 'It seemed hours before the number one and . . . [Lucy] descended the ladder. Clips in place, we immediately dived back to periscope depth. Straddled around the periscope, the Captain made a couple of 360 degree turns, lowered the scope and asked . . . [Lucy] what the problem was. "Loose connection sir, it was crossed [*sic*] threaded."[206]

With *Totem*'s intelligence-gathering equipment repaired Coote continued with the operation. Eight days later and with only two days remaining, *Totem* encountered yet more difficulties after detecting an unknown HE contact. 'This was good news,' recalled Beasley:

> The ole man decided to venture further with the hope of obtaining a visual contact of this HE contact and establish the source. At this stage the boat was heading dangerously close to shore, way closer than our brief allowed. Known 'S' band Radar intercepts were being received but were not near enough to cause any undue concern. Completely out of the blue an 'X' band intercept, very near, coincided with a 'near' HE contact dead ahead. We crash dived. The periscope was lowering as we went down to 120 feet.[207]

As *Totem* dived the noise of propellers could be heard as a surface ship passed overhead. Then *Totem*'s crew heard the unmistakable sound that every submariner dreads: splashes. *Totem* was under attack. Beasley described how:

> The first depth charge exploded way under our depth of 120 feet, followed by others, from different directions. A rather loud 'clunk' on our forward casing was followed by an enormous explosion which shook the boat, followed by others at a greater depth. Another depth charge exploded close above us rocking the boat much as before . . . Depth charging continued for longer than I care to remember.[208]

In order to avoid the depth charges Coote took *Totem* deeper and deeper, eventually levelling off at around 280 feet. But *Totem* had now

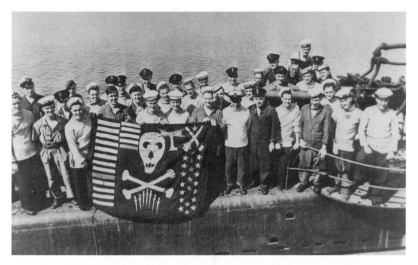

1. 'The most dangerous of all the services' (Winston Churchill). The crew of HMS *Safari*, one of the most successful British submarines in the Second World War, September 1943. The Jolly Roger includes white bars for each ship sunk; daggers for operations involving the delivery or recovery of shore parties from enemy territory; crossed sabres for the boarding of another vessel; and the lighthouse for the submarine's use as a navigational marker for an invasion force.

2. Admiral of the Fleet Sir George E. Creasy, Admiral Submarines, September 1944–November 1946, who steered the Submarine Service from World War to the post-war world.

3. Rear Admiral Guy Grantham, Flag Officer Submarines (FOSM) August 1948–January 1950.

4. Rear Admiral Sydney M. Raw, FOSM, January 1950–January 1952.

5. The German Type XVII 'Walter-boat' being transferred to the captured Walterwerke, Kiel, shortly after being salvaged, July–August 1945. The U-boat was later commissioned into the Royal Navy as HMS *Meteorite*.

6. Operation 'Blackcurrant'. The diesel engines of submarines supplied many Royal Dockyards with electricity during the 1947 winter fuel crisis.

7. HMS *Affray*. The Wartime 'A' class Submarines were the only new class of submarine built by the Royal Navy during the Second World War and entered service just as the war was nearing its conclusion.

8. The 'A' class also underwent conversion. Here HMS *Alliance* leaves Portsmouth in the early 1950s. Note the streamlined hull, larger sail and forward sonar dome.

9. HMS *Turpin* in its wartime configuration. The 'T' class also underwent modification, ranging from simple streamlining to more extensive conversion.

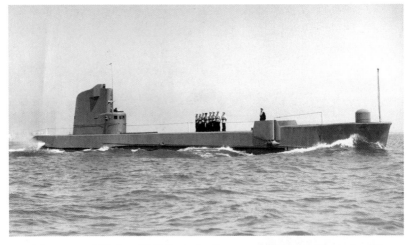

10. The converted 'T' class HMS *Totem* ('a slippery customer', according to her CO, John Coote), pictured here in 1953, shortly before Operation 'Cravat', the pioneering intelligence-gathering operation off Gibraltar.

11. Commander John Coote, CO HMS *Totem*, on his promotion to Commander, 1952.

12. HMS *Totem*, showing the eponymous totem pole with John Coote (left) and *Totem*'s First Lieutenant, John Fieldhouse (right), the future First Sea Lord and Chief of the Defence Staff.

13. (*top left*) Photograph taken through HMS *Totem*'s periscope on the approach to Gibraltar during Operation 'Cravat', 20 March 1954.

14. (*top right*) HMS *Vanguard* at Gibraltar, photographed through HMS *Totem*'s periscope during Operation 'Cravat', 20 March 1954.

15. (*bottom left*) HMS *Eagle* photographed through HMS *Totem*'s periscope during Operation 'Cravat', 20 March 1954.

16. (*bottom right*) Skyraiders Nos. 313 and 308 on HMS *Eagle* photographed through HMS *Totem*'s periscope during Operation 'Cravat', 20 March 1954.

17. With the creation of NATO in 1949, the Submarine Service prepared for war against the Soviet Union, in alliance with other submarine nations. Pictured here in October 1960 is the shore installation and home of the Submarine Service, HMS *Dolphin* at Gosport, during a NATO exercise. From left to right: HMS *Trenchant*, USS *Sailfish*, HMS *Thermopylae*, USS *Dogfish*, HMS *Talent*; (second row) HMS *Tireless*, HNLMS *Zeeleeuw*, USS *Tirante*, USS *Halfbeak*; (back row) unnamed vessel, USS *Threadfin*, USS *Chopper*, USS *Picuda*.

18. The submarine depot ship HMS *Adamant* in the late 1950s with an impressive brood of submarines. By the mid-1950s many Second World War submarines were nearing the end of their lives.

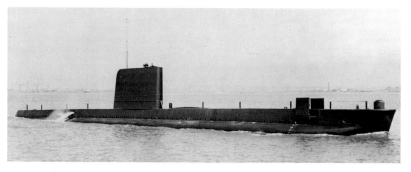

19. The first post-war submarine design to enter series production was the diesel electric 'Porpoise' class, which incorporated many of the features of the German Type XXI submarines. Pictured here is the first of class, HMS *Porpoise*, in April 1958.

20. The culmination of the Royal Navy's experiments with hydrogen peroxide propulsion were the experimental submarines HMS *Explorer* and HMS *Excalibur* (pictured). Hydrogen peroxide was very unstable: *Explorer* was nicknamed HMS *Exploder* and *Excalibur* nicknamed HMS *Excruciator*.

21. HMS *Sidon* salvaged in Portland Harbour after an explosion caused by a 21-inch Mark 12 hydrogen peroxide torpedo, codenamed 'Fancy', 16 June 1955. Twelve men died in the explosion. The accident marked the end of the Royal Navy's experiments with hydrogen peroxide.

22. Admiral Lord Louis Mountbatten, First Sea Lord, 1955–9, operating the prototype nuclear reactor of the USS *Triton* while Admiral Hyman G. Rickover, director of the US Naval Reactors Branch and 'the father of the US nuclear navy', looks on, 20 October 1958. Mountbatten's relationship with Rickover was crucial in securing American assistance in the foundation of the UK nuclear fleet.

23. The 1958 Mutual Defence Agreement allowed the US Navy to assist the Royal Navy with its own nuclear-propulsion programme, specifically permitting the purchase of an American-made S5W Pressurized Water Reactor for the Royal Navy's first nuclear submarine, HMS *Dreadnought*. *Dreadnought* was designed with a distinctive teardrop-shaped hull and moved the Royal Navy into the nuclear age.

drifted towards an area that was listed on the charts as a possible Soviet minefield. Beasley's account continues:

> Depth Charges continued to be picked up quite a distance from our position until they eased. The deathly silence was itself as frightening as the bombardment. Our HE operator reported that his Sonar was 'bent' (not working). The boat was on silent routine, 'no movement throughout the boat'. Within the control room, we looked at each other with an air of bewilderment, wondering what was next on the agenda. The silence was again broken by a rasping noise clearly audible down our starboard side. This was followed by another some moments later ending with a muffled 'twang', as if something was 'caught' then managed to free itself. These noises were firstly attributed by some as being pieces of ice, however reality soon came to the fore that the noises were hawsers rubbing on our outer casing attached to mines floating above.[209]

Checks to *Totem* revealed no internal structural damage, but there were slight traces of water running down the periscope and snort mast. Coote was worried. Until it was safe to return to periscope depth it would be difficult to assess the extent of the damage to the snort and periscope masts. Fortunately he was able to use *Totem*'s undamaged gyro compass and charts in the Control Room to move *Totem* away. With the periscope inoperable, the batteries dangerously low on power and foul air gradually making the crew ill, Coote had little choice but to surface blind somewhere off Kirkenes, Norway.

Once on the surface, *Totem*'s crew inspected the damage. According to Beasley:

> Our periscopes and snort were bent at an angle, un-usable. Guardrails, aerials, asdic dome, everything on the upper casing had been blown away, including most of the Fin. Later we found the forward hatch was warped and could not be opened. Not knowing what further damage the boat may have sustained, we remained on the surface, like a sausage in the water, (a phrase made by one of our destroyers who found us).[210]

When *Totem* returned to HMS *Dolphin*, Coote was called up to London for a debriefing with the First Sea Lord, Admiral of the Fleet Sir Rhoderick McGrigor. Coote argued that as well as a wealth of new raw intelligence data there was some hard evidence that the UK's naval policy should exploit the Soviets' 'demonstrable weakness in anti-submarine warfare, with priorities adjusted accordingly'. Although Flag Officer Submarines 'purred' with pride, this did not go down well with the rest of the Admirals. 'There was a silence among the ranks of the Naval Staff,' recalled Coote. It was only broken by an 'incredulous top planner, addressing his remarks to the Chief of the Naval Staff: "You will appreciate, Sir, that the logical conclusion of this presentation could have far-reaching adverse effects on our Strike Carrier programme and the development of the Buccaneer aircraft." '[211] This was not what some Admirals wanted to hear given their commitment to the carrier programme for which McGrigor and the Naval Staff were fighting hard and successfully against both the Prime Minister and Duncan Sandys, Minister of Supply.[212]

In the years that followed Operations 'Cravat' and 'Defiant', British 'T' class submarines engaged in similar intelligence-gathering operations in northern waters, pioneering the collection of vital intelligence on the capabilities of Soviet warships and submarines as well as their equipment, tactics and weapon systems.[213] With each patrol the Submarine Service refined its intelligence-gathering craft. After Operation 'Defiant', for example, the Admiralty Signals Division undertook a security survey of *Totem*'s radio transmissions and warned the Director of Naval Intelligence that the KGB's electronic listeners might be able to pick up 'unusual very secret traffic on a home station submarine broadcast' over a number of weeks and might also notice that *Totem* was absent from normal exercise areas. They suggested implementing a suitable cover plan with 'dummy communications'.

These highly secret operations were the post-1945 equivalent of wartime patrols. Yet while many wartime patrols often went without incident, Cold War operations almost always produced incidents, some of which went on for days and weeks. (Some historians have questioned whether British ministers knew about these operations and the incidents they often produced.)[214] Each operation required political authorization and, as we have seen, in the case of Operation 'Defiant', it came directly from the Prime Minister. In April 1956, Churchill's successor, Sir Anthony Eden, cancelled all intelligence-gathering operations after the notorious 'Buster' Crabb incident when a Royal Navy frogman was killed while attempting to carry out underwater surveillance in Portsmouth Harbour on another 'Sverdlov' class cruiser, the *Ordzhonikidze*, which had carried Nikita Khrushchev and other Soviet leaders on a visit to Britain.[215] The incident, which caused a major diplomatic row with the Soviet Union, led a furious Anthony Eden to suspend British intelligence-gathering operations while Sir Edward Bridges, the Head of the Civil Service, completed a report into their oversight and approval. This had an impact on the Navy's use of submarines to gather intelligence. At the time of the Crabb incident the Submarine Service had plans to send another submarine into the Barents Sea but because, to use the words of the Director of Naval Intelligence, Rear Admiral John Inglis, 'political approval was not forthcoming', the operation, which was codenamed Operation 'Pontiac', was cancelled.[216] This had wider repercussions.

In April 1955, after leaving HMS *Totem*, Coote was appointed as Staff Officer to the Admiralty's British Joint Services Mission in Washington (ABJSM). Headed by Rear Admiral Robert Elkins, the BJSM was responsible for maintaining relations with the United States Navy (USN). Coote was assigned as an Assistant Attaché with liaison duties with the USN submarine force. As part of this assignment he was invited to take part in a US Navy submarine intelligence-gathering operation. In return, the Royal Navy agreed to allow a USN officer to take part in one of its own submarine operations. Coote was assigned to the USS *Stickleback*, a 'Guppy' conversion submarine, similar in capability to the Royal Navy's converted 'T' class.[217] The *Stickleback* spent thirty-four days off Petropavlovsk intercepting the annual reinforcements to the Soviet East Asiatic Fleet as it emerged from the Bering Sea. 'It was a very long trip, including five weeks consecutively dived, all for not more than three days' action,' wrote Coote. 'Time passed with endless cups of coffee and two movies a day.'[218] But it was a productive patrol. The *Stickleback* intercepted two 'Sverdlov' class cruisers, four 'Riga' class frigates and twelve 'Kronstadt' class patrol craft and regularly observed Soviet 'L' and 'W' class submarines exercising west of Cape Shipunski.[219] The *Stickleback* recorded considerable VHF voice, IFF and radar traffic and it was clear that the Soviet Navy possessed a low anti-submarine capability. 'Local training and defences emphasise Soviet readiness to deal with unfriendly air intrusion but no difficulties placed in way of submarine visitors,' noted Coote.[220]

The Americans were impressed with the results of *Stickleback*'s patrol. Coote was later hustled into a debriefing with the US Commander Submarine Force Pacific:

> I made some cautious comments suggesting how we might have got more out of the trip if our objective had called for more sophisticated electronic intercept and recording gear, so as to gather raw tactical data which only a submarine could do by playing it pretty close to the horns. This seemed to strike a sympathetic chord, for I was paraded the next day to repeat my

views in front of the Commander-in-Chief Pacific himself . . . I said my piece without interruption from the tight-lipped senior aviators on his staff who clearly resented this upstart Limey making a direct pitch on behalf of the USN submariners.[221]

Coote noted that 'CINCPACFLT's intention is to press for early relaxation of 12 mile limit restriction and to increase scope and scale of submarine intelligence activity within his command.'[222]

With the Prime Minister refusing to authorize any submarine intelligence-gathering operations, not only was the Royal Navy unable to reciprocate, but the US Navy was having to fill the intelligence gap in the Barents Sea by sending its own submarines into northern waters to observe the Soviets. This created problems for other operations the Royal Navy was conducting against the Soviet Union, especially in the aftermath of the 1956 Suez Crisis.

OPERATION 'NIGHTJAR'

Five Royal Navy submarines participated in Operation 'Musketeer', the Anglo-French invasion of Egypt, in November 1956: HMS *Totem*, HMS *Tudor*, HMS *Trenchant*, HMS *Sentinel* and HMS *Sea Devil*. HMS *Tudor* was assigned to air–sea rescue duties alongside a French submarine, *La Créole*. The operation was uneventful apart from one episode when *La Créole*, which was dived, detected an unidentified submarine operating close by. Concerned that the unidentified boat was Russian in origin the French CO manoeuvred *La Créole* into a firing position and in accordance with his rules of engagement signalled the Task Force Commander, who instructed that *La Créole* could only fire in self-defence. The unidentified submarine turned out to be HMS *Tudor*, which had altered course due to a navigational error.[223]

As the Suez Crisis neared its culmination in early November 1956, the Soviet Union invaded Hungary in response to a national uprising. The US, already deeply concerned about Soviet influence and possible intervention in Egypt, moved its forces to high alert. A number of US Navy submarines, including the *USS Quilback*, took up stations between Greenland and Iceland, where they waited to intercept any Soviet warships and submarines that attempted to transit through the Denmark strait.[224] The Royal Navy was informed of these operations and complied with a US Navy request to withdraw two submarines,

HMS *Acheron* and HMS *Trespasser*, which were conducting equipment trials in the same area.[225]

At some point in early November 1956, the First Sea Lord, Lord Mountbatten, and the Secretary of the Navy, Sir John Lang, met with the First Lord of the Admiralty, Lord Hailsham, to discuss whether the Royal Navy should conduct a similar operation and send submarines into the Barents Sea and to positions off the Norwegian coast to undertake reconnaissance with the aim of detecting any abnormal and significant submarine or surface activity which might indicate that the Soviets intended to start a war. Hailsham approved the operation, but only on the understanding that the 'submarines did not approach the Russian coast nearer than between 50 and 100 miles' and he 'decided that as the operations were in no way connected with other intelligence gathering operations about which we were bound to consult the Foreign Secretary and Prime Minister, there was no need to do so in this instance'.[226]

Orders for the operation, codenamed Operation 'Nightjar', were produced by FOSM on 19 November and two submarines, HMS *Tabard*, under the command of Lieutenant Commander Peter Samborne, and HMS *Artful*, under the command of Lieutenant J. T. Mitchelmore, departed Rothesay on 21 and 23 November, and took up positions off the Norwegian coast by the North Cape, where, in some of the worst weather imaginable, they remained until late November.[227] Both submarines were under strict orders to remain undetected. They were not to commit hostile acts towards Soviet Forces unless authorized to do so and if they were detected they were to do their utmost to break contact. If they failed and were attacked with persistence by Soviet forces they were to surface and inform the Admiralty that they were being attacked. The submarines were also required to maintain complete radio silence throughout the operation. They were only permitted to break it to report:

(a) A large scale movement of ships, submarines and depot ships e.g. in the region of 50% order of battle.

(b) Five or more Submarines transiting to the West in a period of 24 hours.

(c) A total of 10 or more Submarines transiting to the West in a period of 72 hours.

(d) Three or more cruisers proceeding West of 31 degrees East independently.

(e) A cruiser task force (e.g. 2 Cruisers and 8 destroyers proceeding west of 31 degrees East).

(f) If attacked . . .

(g) If assistance is needed in the event of a serious breakdown.

(h) If forced to leave patrol prematurely.[228]

Although HMS *Tabard* detected sixty-nine contacts, only three were considered to be of Soviet origin, the remainder attributed to neutral, or possibly Soviet, merchant ships, trawlers or fishing boats.[229] HMS *Artful* encountered twenty-four merchant ships and fishing vessels and one Royal Navy warship, HMS *Hound*, which was on passage near Bear Island. Mitchelmore suspected that *Artful* had detected two possible diesel submarines moving to the southwest, but he was unable to identify them.[230] Although uneventful, both COs found the patrols valuable from the point of view of training. Samborne treated the operation as a war patrol, which he said 'was of great benefit to myself and the ship's company; morale remained high throughout – the feeling that the job is worth while being an excellent stimulus'. But he was frustrated by the restrictions imposed on *Tabard*'s movement. 'Whilst it is gratifying that the "raison d'être" of this operation was not fulfilled, it was a great disappointment that the opportunity could not be taken to spend a period in an area only twelve snorting hours further east.'[231]

Two additional submarines, HMS *Totem* and HMS *Andrew*, were to sail from the UK on 5 December to relieve *Artful* and *Tabard*. However, on 30 November the Admiralty decided to hold them at seventy-two hours' notice due to improvements in the international situation.[232] On 19 December the requirement was reviewed once again and Inglis concluded that it was no longer required. 'I consider that there is no indication of global war and that tension has diminished considerably since Operation Nightjar was instituted.' Operation 'Nightjar' was suspended and *Totem* and *Andrew* were stood down. *Artful* arrived back in the UK on 16 December and *Tabard* on 19 December. With the last of the British and French troops due to withdraw from Egypt on 22 December, the judgement was that the international circumstances no longer warranted a continuous submarine patrol in northern waters. But Inglis suggested that the 'Nightjar orders should be retained for use at short notice in times of tension. The operation could then be remounted with a minimum of delay.'[233]

Inglis attempted to use the operation to prompt Eden to reconsider his position on intelligence-gathering operations. Reports from HMS *Tabard* and *Artful* were sent to the First Lord, Lord Hailsham, who found them 'most interesting' and insisted the COs of both submarines come to London to brief him personally. But Hailsham felt there was little point in approaching Eden, who was in poor health and politically weakened:

> It is not much good trying to get the P.M. to alter his post-Crabb policy unless a powerful case can be deployed based on either (i) change of circumstances (ii) substantial advantages to Anglo American cooperation or (iii) prospects of really impressive yield. However I favour vigorous patrolling as the best possible training and a valuable source of information and will help all I can.[234]

It was the second of these cases, the consequences for Anglo-American relations, that led to the resumption of intelligence-gathering operations.

Inglis had hoped to deepen relations with the US Navy by inviting American observers to participate in Operation 'Nightjar', but he was unable to deliver a formal invitation to the USN before the operation was cancelled.[235] Another opportunity arose a month later, in December 1956, when COMSUBLANT privately informed the Flag Officer Submarines, Rear Admiral Woods, that the US Navy intended to send the USS *Tirante* into northern waters in February 1957.[236] A worried Elkins informed the First Sea Lord, Lord Mountbatten that:

> It is apparent the U.S.N. see this as the first of a series of patrols to provide thorough coverage of the Northern Fleet by submarine reconnaissance. It is also plain that the decision to institute these patrols was influenced by two factors: firstly, the U.S.N. feel that, since the cancellation of PONTIAC, we are no longer providing sufficient cover in an area where we have hitherto been a reliable and productive source; secondly, having recently been given access to the reports of our own submarine operations off the Murmansk coast, the U.S.N. have been able to persuade the State Department of the feasibility and value of such operations, and that the risks of detection are negligible . . . It appears that our prestige will suffer in the operational and intelligence fields (where it stands high at present), unless we resume these activities ourselves, though, of course, I appreciate it was a political decision to discontinue them.[237]

Elkins urged not only restoration but also a 'bigger and better operation'.[238] He told Mountbatten that 'we have here an opportunity to co-ordinate our efforts with the U.S.N. to our mutual benefit'.[239] However, any such operation would require Prime Ministerial authorization and the Prime Minister, Anthony Eden, was consumed with the aftermath of the Suez Crisis. However, on 9 January 1957, Eden resigned and was succeeded by Harold Macmillan. At some point between Macmillan assuming office on 10 January and 4 February 1957, the new Prime Minister approved the Royal Navy's 'submarine operation in Northern Waters this spring', codenamed Operation 'Offspring'.[240]

With the Americans now operating in northern waters, Inglis informed Mountbatten that it was 'necessary to co-ordinate our operations with those of the U.S.N.' There were two ways of achieving this. The first was to ask the US Navy to cease its operations in northern waters. The second was to allocate areas in which US Navy and Royal Navy submarines could operate. Woods strongly favoured the second option and felt that because 'his relations with ComSublant are so good' he could easily arrange it.[241] Mountbatten agreed and on 4 February he wrote to his opposite number in the US Navy, the Chief of Naval Operations, Admiral Arleigh Burke, and recommended coordinating US Navy and Royal Navy submarine operations in northern waters and repeated the offer to host US observers on Royal Navy submarines.[242] Burke replied on 3 March:

I was pleased to read that we agree as to the necessity for closely coordinating submarine surveillance operations in areas of mutual interest. Various means of implementing an effective liaison procedure have been reviewed and I believe the following system would be advantageous for both of us. It is proposed that we utilize national command structures rather than the NATO organization. Thus, through the offices of Admiral Boone, we could ensure more intimate coordination with over-all U.S. planning and operational data. This more direct access to the facilities and resources of CINCNELM [Commander-in-Chief US Naval Forces Eastern Atlantic and Mediterranean] would facilitate the essential exchange of information concerning these sensitive operations.

If you concur, I will direct the Commander in Chief, U.S. Atlantic Fleet to establish direct liaison between the Commander Submarine Force, U.S. Atlantic Fleet and Flag Officer Submarines utilizing the offices of Admiral Boone. There are several excellent officers on Admiral Boone's staff who are well qualified to serve as a strong link between the Commander Submarine Force U.S. Atlantic Fleet and your Flag Officer Submarines.

It is our sincere hope that we will be ready to take advantage of your kind invitation to send an observer to one of your submarines this spring.

Implementation of such close coordination and teamwork in peacetime will materially increase the over-all readiness of our submarine forces.[243]

Operation 'Offspring' was the first intelligence-gathering operation coordinated with the United States Navy and was conducted by HMS *Tabard* under the command of Peter Samborne.[244] 'We were patrolling and nothing much was happening,' recalls Richard Heaslip, then a junior navigator on board *Tabard*. 'Then the entire Russian fleet came out. The natural thought is "They've got us", but that proved not to be the case. They were all coming out for an exercise. But they were all around us, the whole lot: planes, helicopters, destroyers. We were at action stations for about four hours in the Control Room. Samborne loved that. The tighter it got, the more he enjoyed it.'[245]

This is where the accessible records end. However, there were clearly other secret submarine operations. The catalogue at The National Archives in Kew lists a series of files whose titles include submarine patrol reports such as Operation 'Sanjak' (HMS/M *Turpin* Submarine Patrol, 1954–1955); Operation 'Tartan' (HMS/M *Turpin* Submarine Patrol, 1954–1955); Operation 'Offspring' (HMS/M *Tabard* Submarine Patrol, 1957); Operation 'Tripper' (1957–9); Operation 'Orion' (1958); and Operation ADAMIS (1958–1959).[246] All are withheld under section 3.4 of the Public Records Act 1958 and will remain so for the foreseeable future.

PIN-PRICKING A COLOSSUS

What was it like on board one of the Royal Navy's 'T' class submarines during a Cold War intelligence-gathering operation? A 22-year-old Leading Electrical Mechanic (LEM) named Michael Hurley kept a diary during two patrols conducted by HMS *Taciturn*, under the command of Lieutenant Commander Roache O'Connor, in September 1957 and April 1958. According to Hurley, O'Connor, 'a patient, polite and understanding' CO, was one of the few individuals on board *Taciturn* who knew exactly where the submarine was going and what it was going to do.[247] However, the arrival of extra personnel on board alerted *Taciturn*'s crew to the nature of the operation. One addition was a Lieutenant Commander named Lucas, described by Hurley as one of the 'special team' and as 'fat, foreign looking with a slight accent but seems good natured'. Rumours quickly spread that he was 'an ex-Russian who can speak 14 languages':

Everyone has already got their own ideas about this trip but we will know a great deal more when the Captain tells us later what its purpose is . . . A report in yesterday's paper seems significant as it reports that the Red Fleet is exercising in the Arctic from Sept 10 to Oct 15 and warns all ships to stay out of the area. In this morning's paper the warning is repeated plus statements saying live ammo and charges will be used. We will see if this affects us. Morale at present is high.

When *Taciturn* sailed from the Clyde, O'Connor briefed his crew:

We are going to snoop on the Russian Fleet exercises and the Captain made no bones about what would happen if we were caught: 'It will be very unpleasant and most dangerous'. The charts will be screened and we are not to look at them. We shall retain a complete Radio silence apart from emergency. He also warned us of the complete necessity for keeping our mouths shut when we return. For some time we will be watched and so will our families. We must not say a word even to wives and mothers. Aboard a different routine will be worked, lights out as much as possible to conserve the batteries. All the time we will be in Defence State 2 (ready for deep diving). Finally the Captain repeated his earlier warning by saying 'This will be a wartime patrol, we must get good results'. Most people had a fair idea before but now we know for certain it is different although everyone is so far being humorous.

Before sailing north, *Taciturn* dived and exercised against a number of British warships. While conducting 'listening watches' on the British ships the submarine suddenly 'jumped a couple of times and rocked a little':

My first thought was of 'depth charges' (shows how dramatic we've become!). The Control Room asked for damage reports if any and told us we had 'grazed' the bottom. No internal damage was discovered and we were told we had hit a rock pinnacle and would remain at periscope depth until time to surface. At 2030 we surfaced and went into Passage Routine.

After completing the exercises *Taciturn* set course for HMS *Adamant*, the submarine depot ship stationed off Greenock. O'Connor informed his crew that when they arrived no one was allowed to leave the submarine as the depot ship had been quarantined due to an outbreak of Asian Flu. 'No one at first believed this and thought it just an excuse to prevent us breaching security.' Once *Taciturn* had been loaded with stores, the submarine slipped away from *Adamant* and 'opened up for diving'. The next day the crew were struggling to keep the submarine stable while snorting due to bad weather – 'it was a job to hold our depth properly' – and conditions deteriorated to such an extent that even at 50 or 60 feet below the surface life was uncomfortable for the crew. The weather was also taking its toll on *Taciturn*:

> During this spell our anchor became loose and was continually thudding against the bows. The weather of course was too rough for anyone to go on the casing so we continued below at 150 ft, snorting when necessary. The weather was the same if not worse on Saturday with snags coming up on Husk due to oil and salt water sloping over it. On Sunday the weather calmed down enough for 'Scratch' and Silver to secure the anchor. We then carried on as before snorting until the forenoon when we surfaced to complete operations on the anchor as now the weather was quite calm. We had to dive in a hurry due to an aircraft and remained below all day until 2145 when we stopped snorting and went deep.

Two weeks into the operation the 'buzz' on board was that the submarine was somewhere in the Arctic Circle. It was very cold and, when *Taciturn* snorted, ice would form around the Engine Room bulkhead door. The submarine was also making a lot of noise while submerged: 'we might as well tow a bell for the amount of noise we are making. It's a serious problem as it is imperative all possible noise is eradicated to avoid detection,' wrote Hurley. O'Connor repeatedly surfaced *Taciturn* at night and sent teams outside to try and find the source of the noise. After two days Hurley wrote that:

> Unfortunately this mysterious noise has continued and caused us a great deal of trouble. Last night we surfaced and dived no less than three times. It is a fact that the Skipper seriously considered turning back. This may sound silly but unless this noise was located it was pointless to go on as if located we would either be a grave embarrassment to HM Government or dead! However at 1730 this evening we surfaced once more and a real attack on the casing was made and I think the trouble has finally been removed although we will not be sure for a few hours.

Taciturn's crew was now working under a very arduous regime. O'Connor repeatedly emphasized the need for silence. An 'ultra-silent' state was introduced whereby all machinery was stopped and all unnecessary movement was discouraged:

He told us that although in many ways we had cut down on noise we would in future have to be really quiet all the time. No crashing of stores, shouting, hammering, or dropping of hatch covers. He then told us what was happening. Although he was unable to say where we are he said that we were in the area and that starting from yesterday afternoon we would be passing through a heavily patrolled U-boat area for about 48 hours then after a further two days we would enter or I should say cross another line which would have both Submarines and surface vessels patrolling. On the other side of the line would be an area which would be 'most interesting' and busy. It appears we have not got one particular stretch to cover but a roving commission in the area.

'T' class submarines operating in northern waters would stay well outside the twelve-mile limit off the Soviet coast recognized internationally as territorial waters. The submarines would remain dived during the day, hoovering up communications and electronic signals from special aerials stuck on top of the periscope, augmented with visual sightings on the rare occasions anything was observed. When operating outside the twelve-mile limit in the harsh Russian winter the coastline in bad weather was barely visible and thick snow tended to cover everything. The only time the submarines would surface was at night to establish a star fix. Access to the Control Room was strictly limited and only the officers were allowed to look through the periscope or consult the charts which were used to keep track of the submarine's location. The majority of the crew had little idea of where exactly they were operating. As a Leading Electrical Mechanic, Hurley was assigned to the Motor Room in the aft sections, where he and the rest of the crew would speculate about *Taciturn*'s exact location. 'There seems to be no doubt,' he wrote, 'that at the moment we are going round the "top" as we have gradually been changing course from n/e and are now going due East.'

As the weeks passed, living conditions on board *Taciturn* deteriorated. Although no two 'T' class submarines ever had the same complement, from an accommodation point of view the post-war modernized 'T' boats were no different from their wartime counterparts. During the war the 'T' class had carried an average of 53 men. Following conversion and modernization, the addition of various specialists and intelligence operators known as riders increased the number to 68. To accommodate these extra personnel there were two sittings for Wardroom meals and the torpedo stowage compartment became a mess deck complete with bunks. The growing size and complexity of electronic equipment also had a material bearing on habitability, further reducing the space available for accommodation of the personnel whose duty it was to use and maintain it. Officers and ship's company lived in conditions that were best likened to caravanning with much reduced headroom, and complete exclusion of daylight. Wooden boxes containing tinned food would be stowed as a false deck in the gangway throughout the submarine, forcing the crew to move about in a more crouched position than ever.[248] Instruments and controls of all sorts obtruded into what were primarily living spaces. To open a locker or cupboard usually entailed disturbing messmates who were using it as a seat or bunk. Coote wrote that conditions in *Totem* were 'pretty squalid, there was no doubt about it. I would hardly pretend it was comfortable for anybody.'[249]

Crews also struggled to maintain personal cleanliness. Water was in such short supply that *Taciturn*'s crew had to do everything possible to conserve it. At best there was a small mug of water for brushing teeth and the men rarely shaved. Many gave up the struggle to stay clean completely, which meant working and sleeping in the same clothes for weeks on end. Supplies of food were also very short:

> We bake our own bread which soon goes, tea, milk, sugar and water must be watched carefully and often dinner is very small as its cold though normally supper is a good large meal. It's not that we are starved so much as the long gaps between good meals (which are really good) and the fact that if one is hungry there is no bread to fill up on as is normal. But before we are through things will be a lot worse. The 1st Lieut informed us this morning that we had used 300 galls of water a day while the distiller had been running but it is stopped now and we must not exceed 250 galls, if we do he will shut all the water off the next day to make up for it and have a waterless day, no washing, tea, etc. At our present rate of consumption we have 19 days ration left and 29 days to go. As for gash [waste such as empty tins etc.], rags and boxes have now to be kept on board in the trenches [stowage space in the fore-ends] until we return. As one can see things are becoming tighter and living harder.

As conditions inside the submarine deteriorated many of the crew would feel rather 'crabby'. The temperatures decreased substantially and condensation became a real problem. Hurley complained that 'one is always getting large drops of icy water on us'. Worse, in order to avoid unnecessary noise, the crew was not allowed to blow the toilets (heads) for several hours. This did 'not help the general smell'. Tempers would grow short and most people were irritable. 'It's surprising how easily things happen, probably due to boredom, lack of regular food, cold and headaches, (which most people seem to have) and Rum which is I think the main cause!' wrote Hurley.

When on station off the Soviet coast in the winter the crews lived in near constant red lighting. The Arctic sun was only up for a few hours each day and got progressively less as *Taciturn* moved north. The periscope had an inbuilt light loss of around 40 per cent so good night vision was of paramount importance for the safety of the submarine. There was no alcohol for the CO or the periscope watch keepers. As well as being very wearisome to the eyes and the senses, the near constant red lighting tended to make food less appetizing. Italian tinned tomatoes, a regular on the menu, used to disappear.

As the weeks passed *Taciturn*'s crew fell into a daily routine:

We do a short burst of snorting in the morning about 1100 to clear the air and start snorting proper at about 1800 until the charge is completed around 0100 in the meantime stopping occasionally for an all round listen. During the rest of the time we just snoop around at about 5 knots. The weather outside is now so cold that we have to raise and lower the Snort Mast every so often when snorting to keep the valve clear and also shut and open the flap valve. This keeps the ice clear from the top of the mast. The weather seems to be calm although there is a swell. Sea temperature is 41f.[250]

During the operation *Taciturn* was involved in 'one of those unfortunate submarine incidents which if a little worse could be fatal':

We suddenly started to have a stern down angle which got quite bad within two minutes. I was in the Motor Room on watch when suddenly Bellis, the Afterends watchkeeper, came running out and rushed past us towards the Control Room. Such was my state I fully expected to see him followed by a column of water! Even as he was passing we had to break the charge as we were at 65ft. Then the Captain took charge, blew fours [four mail ballast] and increased speed to stop us slipping back. And for about 20 minutes he sorted things out. It appears what had happened was that some officer of the watch, believed to be 'Engines' as he was on, had opened Z Kingston with the Inboard vent open, consequently we took water into Z tank at an alarming rate so the watchkeeper shut the vent, when he opened it again the air was compressed and came out like an H.P. jet. So owing to the noise he decided against using the phone or Tannoy and went and told them himself what was happening. Once again the Captain was the one who averted anything more serious by his calmness and quickness.

– calmness and quickness: two of the many essential prerequisites for a good submarine Captain. O'Connor successfully averted a possible disaster. A good CO had to know exactly what was going on in his submarine at every moment of every day.

After a week conducting 'deep field' listening and watching, *Taciturn*'s crew had eaten their way through all the bread. 'It has lasted well and we did not even have to throw any away although it has been green on the outside for a few days now,' wrote Hurley, and while the Chef had started to bake a new batch, Hurley complained that it was 'a mistake to wait for the bread to run out before baking as everyone is so hungry when the new bread is baked it will go in no time'. While on patrol the crew celebrated Easter, with a small service for the Catholics on board, eggs for breakfast and chicken for dinner. 'It's peculiar but on days of celebration one is more inclined to think of home,' reflected Hurley.

As the operation entered its most dangerous phase the need for silence was so great that the *Taciturn*'s crew were told to stay in their bunks unless they were on watch:

All lights and heaters were off and no meals were cooked all day, so we had sandwiches and tin fruit for dinner and ship's biscuits and ham and soup for supper. We stooged around all day apart from a couple of quick Snorting runs and remained deep, we were going slow on one Motor most of the time and at one point moved 500yds in an hour! The trouble with being deep is that one cannot blow the 'heads' so everything has to go in buckets and sugar tins and after a time that becomes unpleasant. The boat of course becomes pretty cold as well with no heating on so it was quite a relief when we were able to get Snorting properly again in the evening and we completed the Charge one side and nearly did so the other. It had not only been Rackets [contacts detected on radar] which prevented us from charging earlier but the fact that the sea was so glassy our masts left a wake for hours.

After almost forty days dived, rumours started to circulate that *Taciturn* was due to head home. 'We have not been told whether we will or not,' wrote Hurley. 'Some say we are now staying until Saturday, others that things are as originally planned which means today. One thing is obvious to me and that is that we are going somewhere.' *Taciturn* continued north in search of the Soviet fleet, but, apart from a trawler and a suspected submarine contact, found nothing. The next few days were spent dived, occasionally snorting, until *Taciturn* surfaced and returned to HMS *Adamant* to find a very relieved Captain Submarines standing on the dockside.

These early Cold War patrols were often stressful experiences. 'Short of actually being in a "shooting" war,' wrote Basil Watson, 'these were the most testing conditions imaginable for a CO and for all the team around him.'[251] Many found that the requirements of collecting intelligence were not always compatible with good CO-manship as taught on the Perisher, especially when it came to using the periscope. COs had to strike a balance between remaining undetected at all times and the need to bring home the best possible intelligence. Too much caution would lead to an unproductive patrol, but too little could result in an embarrassing international incident such as surfacing in enemy territorial waters or, even worse, sinking.[252] 'I had to consider my evasion tactics if counterdetected,' wrote another CO who conducted intelligence-gathering operations in 'T' class submarines, Lieutenant Commander Alfred Roake. 'Balanced against the C.O.'s first duty of saving the lives of the Ship's Company, was the question "how do you defend yourself to avoid becoming another *Pueblo*?"'[253]

As more and more 'T' boats conducted intelligence-gathering operations, a new breed of CO – officers who had refined their craft in icy northern waters – began to assume command of their own submarines. By the summer of 1958 John Coote's former First Lieutenant, John Fieldhouse, now the CO of HMS *Tiptoe*, was preparing for his own operation. *Tiptoe* departed Faslane on 26 September 1958 as heavily laden as it was possible to be with fourteen specialist intelligence staff and eventually returned 'with results that far exceeded expectations'.[254]

Cold War patrols also started to have an impact on the ageing 'T' class. In March 1956 the Controller of the Navy admitted that 'these submarines, particularly the T Class, are getting very old and have been run very hard indeed all their lives'.[255] When Lieutenant Commander Alfred Roake assumed command of HMS *Turpin* in 1959, he found a submarine that was described as the Royal Navy's 'latest and best' but was actually sorely lacking. *Turpin*'s engines were ex-HMS *Tradewind* and had already completed over 12,000 hours' service. This resulted in cracks in the main engine frames due to metal fatigue and Roake's first patrol ended in a humiliating thirty-day, 5200-mile tow back to Gosport from Kingston, Jamaica. He was 'very conscious of the limitations of the submarine'. *Turpin* was 'nowhere near, for example, up to American standards as an operational vehicle; nor indeed as regards either accommodation or comfort'. Although *Turpin* was 'a wee bit primitive for the job it was being asked to do', Roake did admit that 'we kept our end up'.

In order to prepare for his first intelligence-gathering operation Roake read other submarine patrol reports, picked the brains of other Commanding Officers and underwent a formal briefing at the Old Admiralty Building in London, and at GCHQ. Roake then put *Turpin* and her crew through their paces on a number of practice runs:

> The boat had been fitted with anti-cavitation very quiet screws, and we carried out comprehensive noise trials in an attempt to eliminate every possible source of noise. As part of the Work-Up we needed to calibrate the tubes in Loch Long, for which we went up to Arrochar . . . We were fitted with sonar and radar intercept equipment; a useful stubby whip aerial on the periscope; and painted out our identification pennant number. We also embarked periscope cameras, both still and cine. These we gave a trial round the Scilly Isles, producing a panoramic set of pictures from as close inshore as possible.[256]

Turpin set off on patrol on 21 October 1959 and almost immediately Roake discovered that as CO he had to be prepared for anything, even when sleeping:

> I had left the intercom speaker switched on, as I invariably did, to pick up what was happening throughout the boat, and keep in touch with the 'tune'. In the background, I could hear a muttering: 'I can't trim this b. . . . . boat – the f. . . . . bubble is all over the place!' Old Patrol reports always used to warn about the problem of 'leaky blows' to the Main Ballast Tanks; so I had given strict instructions to leave the Main Vents open, in order that there would be no build up of air pressure in them. I heard the Engineer Officer on watch say 'let's try opening main vents', which in fact had been shut, contrary to my orders. This was followed by 'Open one main vent!' I shot into the Control Room, and gave a stream of orders: 'shut main vents – blow all main ballast – full ahead together – planes hard – arise'. That was the voice of my Guardian Angel. We plummeted down, and pulled out at somewhere past 350 feet, – our supposed 'safe

diving depth'. Meanwhile we caught a 'main ballast trim', until we had sorted things out. With thousands of fathoms under the keel, we could have been crushed like an egg shell. Had my instructions been followed, we would never have got into that situation – but we had to learn the hard way.

Turpin was also fitted with a new design of filter for special oxygen generators intended to prolong the amount of time a submarine could remain underwater without snorting and replenishing the atmosphere inside the submarine. This consisted of a cylinder, with a breach at the bottom, into which was inserted a 'candle' whose chemical compound produced oxygen when a hotwire element was switched on. But the new filter was faulty. While *Turpin* was in northern waters an 'oxygen assisted' fire erupted:

> In the confined space of a submarine, this was a very unpleasant experience indeed, particularly with three potentially hostile destroyers in the vicinity. It necessitated switching off, isolating it, tackling the fire, and clearing the foul smoke – eventually through the snort mast. Any fire in a submarine is nasty, and in this situation it was particularly so; but the same thing was to occur again on our second patrol. Even so, when we got back, we had a job to convince the pundits that it was not our own fault.[257]

In addition Roake found himself having to deal with an Electrical Artificer who had developed a very high temperature. There was no doctor on board, only a 'book of words' that contained medical advice, together with a wallet of 'spanners' (a medical kit). Roake was dogged with 'much heart searching and worry' as all of the indicators seemed to point to polio. Worse, this occurred just as *Turpin* was in the thick of the action:

> We were in a very vulnerable situation and unable to break radio silence. With no one to ask you have to make your own decision and decide your own priorities – rather as in Nelson's day. Supposing he had some contagious disease, or died? Do you head for home? Bury him at sea – in 'peace time'? These are the sorts of questions that go through your mind; although no different from those which face any other C.O. who may find himself in this sort of operational situation.

Fortunately the man recovered.

Summarizing his patrol years later, Roake wrote that:

We photographed those warships and submarines which came our way, all of which are now very dated. There were a number of unidentifiable bangs and bumps: we heard in the distance 'close encounters of the third kind' and once thought we heard a torpedo. We took avoiding action by going deep and silent, and combed the possible tracks – just in case. Submarine activity included a snorting Submarine, which passed close enough for me to see his periscope and snort mast. One of the more interesting occasions occurred on a bright sunny morning, flat calm, when we sighted a submarine, which at the time was of some interest, but now long since overtaken by other developments. I photographed this from his beam, approaching on a 90 track and went to dip under him at the last minute. To my horror he pulled the plug and started diving right on top of us. I went very deep and fast to get out of his way – it was an exciting moment.

Roake took *Turpin* 'as far north as the ice line' and 'proceeded very cautiously' as he did not want to damage the masts, which 'were absolutely vital to both our safety and the operation'. This was 'not much fun' as it is difficult to spot low ice at night and poor visibility: 'It has no navigation lights.'

On another occasion, a group of five or six Soviet destroyers managed to get a 'sniff' of *Turpin*:

Already in strict 'silent routine', we again went deep, and took every evasive action in the book, but had great difficulty shaking them off. In 'silent routine' we couldn't speed up too much, nor run pumps, as this would have given them a confirmed contact: so we got heavier and heavier, deeper and deeper, with the stern sagging, and the shafts grinding, as the stern down angle put them in very deep water. We finished up at something like 425 feet. When we got back, the Constructor on the Admiral's Staff told me we would probably have collapsed [the hull collapsed due to pressure] at about 470 feet – but we had successfully shaken off the opposition. Altogether it was a lucky escape: and I know how it feels to be hunted for real. Good training – but not very nice!

Turpin returned safely, to a reception that gave no hint of her true exploits:

We flew no 'Jolly Roger' listing our achievements, and had no special welcoming party ... we had left and entered harbour like a 'thief in the night'. I called on the new Flag Officer Submarines, who had been changed during our second patrol, and the Captain (S/M), neither of whom officially knew where we had been, nor what we had achieved – although they must have had a very good idea. I then went to London for de-briefing. We had no feed-back as to how we had done, and a verbal enquiry only elicited a non-committal reply, so I didn't pursue it. Meanwhile we were all ordered not to breathe a word about our adventures, so I deliberately expunged it from my consciousness.

These submarine intelligence-gathering operations remain one of the British State's most closely guarded secrets. The crews that took part in them knew little of what they were doing, and they were forbidden from ever speaking about their exploits. Despite these restrictions, many came away with a profound sense that they had been involved in an exciting and unique activity. Hurley later wrote that:

no one really can know what life in a boat is like, not even the General Service ratings until they do a trip. Some of it is unbelievable, the condensation which is like rain at times, the fog (literally) when we surface quickly, the varying pressure when snorting on one's ear drums, the damp and cold and absolute cramped style, (35 bodies in a space smaller than our kitchen at home), lack of water, fresh air, daylight, sleeping in one's clothes for weeks. No one is a hero because of this and no one really grumbles, but anyone who says submariners have an easy life and don't deserve the extra pay ought to see for themselves.[258]

Roake, who unlike Hurley did know where the submarine was and what it was doing, later wrote that he felt like:

David against Goliath, in my small diminutive 'T' boat of some 1,320 tons carrying out a tiny pin prick of an operation against a colossus. We were on our own with the nearest support and succour thousands of miles away. While this is normally nothing unusual; for a submarine in our situation – endurance becomes of paramount importance, coupled with the need for anything detectable, such as the snort mast being exposed for the minimum of time.[259]

He later wrote that the crews of the 'T' boats that conducted intelligence-gathering operations in northern waters in the mid- to late 1950s 'really were "Cold War Warriors" '.[260] But, as these accounts make clear, the Royal Navy's new underwater warriors desperately needed new chariots to replace their Second World War-era submarines.

3
'A New Epoch':
Towards the Nuclear Age

The present day submarine is the cheapest warship per ton to build, it provides the finest form of naval training and opportunities of command in peace and the present man power of the submarine branch is less than 3% of the Navy. Apart from blowing the submarine trumpet, doesn't a large proportion of the Navy's future lie in this direction?

Rear Admiral George B. H. Fawkes, Flag Officer Submarines, 1954–5.[1]

We are calling this ship Dreadnought *because it is opening a new epoch just as was the old* Dreadnought, *built fifty years ago.*

The Earl of Selkirk, First Lord of the Admiralty, July 1959.[2]

The arrangements made with the United States to help with the building of Dreadnought *have proved an absolute godsend and it is a fact that not one single piece of United States equipment arrived a day late at Barrow [the Vickers shipyard]. This has enabled us to concentrate on learning the complex lessons of how to build a nuclear submarine.*

Admiralty presentation to Peter Thorneycroft, Minister of Defence, May 1962.[3]

THE COLD WAR FLEET

While Royal Navy 'T' class submarines were conducting intelligence-gathering operations against the Soviet Union in the Barents Sea, discussions were taking place inside the Admiralty about what to replace them with. By the early 1950s, the post-war conception of very fast underwater submarines, such as HMS *Explorer* and HMS *Excalibur*, had given way to a requirement for submarines with very large submerged endurance and high, quiet (non-cavitating) speeds. Throughout the 1950s, the Admiralty embarked on an ambitious construction programme, building twenty-three new submarines and modernizing a further eighteen with the overall aim of converting its entire fleet of snort-fitted submersibles into submarines that were fully effective in the submarine versus submarine role, giving the Royal Navy a powerful addition to its anti-submarine forces.

Alongside the 'T' conversions and streamlined 'T' class submarines, fourteen of the sixteen 'A' class were taken into the dockyards and streamlined between 1955 and 1960, a process that involved the removal of the external torpedo tubes and the fitting of a new lightweight, aluminium superstructure. The Bridge was also raised and the bow modified, giving the submarines a flatter profile. The batteries were enlarged, although the original motors remained. As more and more submarines of both classes were taken into the dockyards for modernization, the Submarine Service struggled to maintain a fleet of sufficient size to meet requirements. Towards the end of 1951, the service had concluded that, with just fifty-three submarines in the fleet, it could no longer meet its peacetime training commitments and that it did not possess enough submarines to fight a future war. A large submarine fleet was required because at the very beginning of any conflict there would be very few US submarines operating in European waters (see above, p. 84).[4]

The Second World War had also shown that submarines were seldom mechanically satisfactory for operations against an enemy after three years' service, and with refit periods lasting between eighteen and twenty months, combined with expected losses, the Admiralty concluded that it would struggle to maintain adequate numbers in any prolonged conflict. Wartime experience had also shown that the submarine was an effective weapon not because it was invulnerable, but because it was readily replaceable. Although the Royal Navy had started the Second World War with just fifty-four submarines and in the course of six years lost over seventy, the total number at the end of the war was over double what it had been at the beginning. This was largely because the 'T' and 'U' class designs were sufficiently advanced when the war started for them to be easily built in quantity.[5] Post-war studies carried out by FOSM indicated that on the outbreak of a future war the Navy's operational submarine strength would initially rise as reserve submarines were reactivated but that in a prolonged conflict the number of operational submarines would fall steadily due to losses.

In June 1953, FOSM submitted a paper to the Admiralty on the 'Future Composition of the Submarine Fleet', in which he argued that:

> The present fleet of 53 boats normally contains at least 14 refitting, leaving 39 in commission for operations, submarine and A/S training and trials. The present submarine Priority list requires 22 submarines to be continuously employed for submarine and A/S training and trials, and even so not all commitments can be met. We have already declared 21 submarines to N.A.T.O. for operations in home waters, and 12 in the Mediterranean, leaving only 6 for trials and training. The 21 submarines for operations is a very small force only capable of maintaining 7 submarines or so on patrol at a time. Our fleet is therefore insufficient even to meet our present N.A.T.O. commitments. Casualties would cause the situation to deteriorate still further, and if these are as high as those of the last war, the total submarine fleet will fall from 53 to 29 before new construction becomes available. This would cause a corresponding reduction in the number of submarines available both for operations and training.[6]

FOSM wanted to increase the overall size of the submarine fleet by retaining older submarines in service after new construction became available after 1954. In order to maintain an adequate force, a number of old 'S' class submarines were commissioned back into the Royal Navy. In 1949 there were just three 'S' class submarines in service. By 1953 there were thirteen, and by 1955 fifteen.[7] The Navy was even forced to commission some of the smaller pre-Second World War 'U' class submarines, two of which, HMS *Upstart* and HMS *Untiring*, were returned from Greece in 1952 (they had been loaned to the Greek Navy) and operated in the Royal Navy until 1956. In 1955–6 the total number of active Royal Navy submarines came to a post-war peak of forty-three. Fourteen more were in reserve or under refit.[8]

1956 also saw the launching of the initial two members of the first class of post-war conventional submarines designed and built for the Royal Navy; they were known as the 'Porpoise' class. Described by FOSM as 'virtually improved "T" Conversions', they and their incrementally improved 'Oberon' successors eventually became the backbone of the Royal Navy's conventional submarine fleet throughout much of the Cold War. Their design incorporated all the experience gained from wartime operations, trials with surrendered U-boats, and the British 'T' conversions. They were a little bigger and slightly shorter than the 'T' conversions and were designed from the outset to be deep-diving and capable of a submerged endurance of fifty-five hours at 4 knots, nearly three times that of previous British submarines. Great attention was given to habitability, with the incorporation of air conditioning, and the six bow torpedo tubes had rapid-reloading gear so that a second salvo of torpedoes could be fired very soon after the first.

The 'Porpoise' class also incorporated a number of innovative design features such as a new diesel electric-drive system and a series of acoustic-isolation measures, such as resilient mountings, which reduced vibrations transmitted to the hull. This all resulted in a class of submarine that was 'significantly quieter at a given speed either on electric motors or main engines than any known class in Western or Soviet Navies', including the US Navy 'Guppy' class.[9] The Porpoises were so quiet that their radiated noise was reduced to just 3 per cent of the previous norm.[10] They were also equipped with greatly improved sonar: the Type 187, a medium-range passive directional-listening set that was housed in a distinctive enclosed dome on the front of the submarine, as well as the Type 186, an offshoot of a complex series of programmes into fixed shore-based passive sonar undertaken by the Admiralty in the early 1950s under the codename Project Corsair.[11] The Type 186 was fitted in the saddle tanks on each side of the submarine's hull and was capable of detecting snorting submarines from sixty to eighty miles away.

Although the 'Porpoise' design was conceived in 1946, it was only completed in 1950, and though the first submarine, HMS *Porpoise*, was ordered in 1951, construction at Vickers-Armstrongs, Ltd, Shipyard at Barrow-in-Furness did not begin until June 1954 due to concerns within the Admiralty about the complexity of the design. Admiralty studies indicated that in the event of war the industrial capacity existed to lay down eighteen 'Porpoise' class submarines each year. But due to the complexity of the design it would take around two years to complete them, meaning replacements would only enter service in sufficient numbers to keep the strength of the submarine fleet at around thirty submarines. Many argued that if the Navy built a class of smaller, simpler submarine, of around 1000 tons, compared to Porpoise's 2000 tons, construction would only take a year, meaning the fleet could return to full strength in two years, with numbers increasing steadily thereafter.[12] The Director of Torpedo, Anti-Submarine and Mine Warfare argued for 'simplicity' and 'something much smaller':

> Our early submarines were simple in the extreme and have only become complicated as the result of a long, slow process of evolution. Although there was always some very sound and logical reason for each new complication, alteration or addition which applied at the time, the same good reason may not still apply under present or future operation conditions.[13]

Arguments in favour of something smaller and simpler were fuelled by the change in the envisioned wartime role of the Royal Navy's submarines. As we have seen, by 1952 the Navy believed that 'Our submarines are most unlikely to be called upon to fight a war against large convoys in the Atlantic ... We have, on the other hand, various duties for them to fulfil [anti-submarine warfare, anti-surface warfare, clandestine patrols] which might well be undertaken by a submarine of small tonnage and we have limited financial and shipbuilding facilities which will not support a large building programme of large submarines.'[14] On 1 January 1952, the Director of Plans suggested the Royal Navy 'should explore the minimum characteristics of a submarine that would be capable of carrying out our requirements with a view to replacing some or all of the PORPOISE class by a larger number of small submarines in programmes after 1953/54 and particularly in an emergency war programme'.[15]

In December 1952, the Submarine Service issued a staff requirement for a class of simple, less capable and general-purpose submarines. Initially known as the 1953 design, the 'Boreas' class were intended to be submarines of about 1200 tons, with a submerged 'high quiet speed' and a high rate of battery charging, rather than high surface speed. The torpedo armament would be small, with four tubes forward and two aft, and the position of the auxiliary machinery was planned with self-noise very much in mind. Machinery was grouped away from sonar sets and the Sound Room and new hovering gear was included in order to eliminate propeller noise. There was also a serious debate about whether to use one or two propeller shafts. One was much quieter, but two provided an important backup should one of the two shafts fail while the submarine was at sea. By late 1954, the Admiralty planned to build six twin-screw 'Boreas' class submarines, followed by possibly nine single-screw versions, some of which would be powered by HTP machinery.[16]

However, the 'Boreas' class was terminated in 1955 after the Navy concluded that such a small submarine would be too reliant on frequent support from depot ships, many of which would be vulnerable to attack in war, especially one that involved the use of nuclear weapons.[17] Thereafter the Submarine Service reverted back to the original 'Porpoise' design, construction of which had been authorized on 15 June 1954. However, the first of class, HMS *Porpoise*, did not emerge from Vickers until April 1958 due to construction of the second experimental HTP submarine, HMS *Excalibur*, which was also being built at Barrow.[18]

THE UK NUCLEAR PROGRAMME

Meanwhile, the United States Navy had decided to go in an entirely different direction pregnant with possibilities for the future. In July 1951, the US Congress authorized the construction of the US Navy's first submarine powered by a nuclear source, USS *Nautilus*. In May 1950, when the British learned that the US intended to build an atomic submarine, the Prime Minister, Clement Attlee, enquired about the status of the British atomic-submarine programme. The First Lord of the Admiralty, Viscount Hall, informed Attlee that 'We have hitherto preserved the strictest secrecy about our preliminary work on the atomic submarine in order to avoid the slightest chance of advancing in any way its similar construction in Russia.'[19] By 1950, the small team of naval officers and Royal Naval Scientific Service staff attached to the Atomic Energy Research Establishment (AERE) at Harwell, known as 'the T party', had concluded that an atomic reactor suitable for submarine propulsion was practicable. In October 1950 the Defence Research Policy Committee concluded that the development of such a submarine was a research project of the first importance.[20] In December 1949, a three-stage Research and Development programme was formulated, with Stage I consisting of Admiralty and AERE design studies into a submarine, nuclear reactor and associated power plant. It also involved the construction of a prototype for running tests to prove design features, endurance and performance and to determine shielding requirements. Providing Stage I was successful, Stage II would consist of the design of an experimental operational submarine, for construction within 5–6 years, with sea trials in eight years. Stage III involved construction of a fleet of nuclear submarines from 1959 onwards.[21]

Crucially, by 1950, an important obstacle to any UK nuclear-submarine programme had been removed. By June 1950, AERE was reporting that by 1953 production of fissile material was expected to exceed the immediate needs of the Ministry of Supply. It was therefore in a position to allocate a small amount of fissile material, enough to power one nuclear submarine, with the possibility of further supplies for twelve additional submarines.[22]

Royal Navy support for a nuclear-submarine programme increased throughout the early 1950s and senior officers were rightly worried about being left behind in an important new technological field. From a strategic point of view, the Navy recognized that nuclear power had the potential to increase the effectiveness of the submarine by allowing it to operate at far greater distances from bases, which could be situated further away from possible sources of attack. From the tactical point of view the Navy was also attracted to the potentially high submerged speed and endurance of nuclear submarines, which promised to significantly increase their attacking capability, making it much easier for submarines, once detected, to both force their way through opposition, carry out attacks and escape from any counter-attacks, particularly in heavy weather when surface ships' speed would be reduced. It was also argued that nuclear submarines had the potential to reduce the value of anti-submarine aircraft, as well as making submarines difficult targets for other submarines.[23]

The Admiralty recognized that in any future war its existing conventional submarines would need support in enemy waters for considerable periods against air and surface oppositions. The greatly increased mobility and virtual avoidance of having to operate on the surface afforded by a nuclear submarine would, the Admiralty recognized, make operations far less hazardous than they had been before. But equally, after some analysis highlighted the threat of atomic submarines, the Navy concluded that:

> The atomic submarine will be a much more effective weapon, and much more difficult to counter by surface forces, aircraft or our own submarines, than any submarine so far produced. Our experience in the last two wars shows that, given a large enough effort, our enemies could, by the use of submarines, reduce this country to a condition in which she would cease to be an effective participant in war. It follows that the possession of atomic submarines by Russia would materially reduce the effort required to achieve this result, and that their advent therefore constitutes a grave disadvantage to us.[24]

This particular Admiralty paper concluded that 'while this country could be forced out of a war by the use of submarines, it is unfortunately a fact that our own submarines ... could never, by themselves force Russia out of a war. It follows, therefore, that the development of nuclear propulsion for submarines is likely to do this country more harm than good.'[25] So the sense of transformational possibility aroused by the prospect of nuclear-powered true submarines was tinged with anxiety.

By the 1950s, the Navy suspected that the Soviet Union was also attempting to construct its own nuclear submarines. Intelligence reports indicated that the Russians had made little progress with developing the German HTP machinery, which assessments concluded 'was allowed to proceed in a rather haphazard fashion'.[26] Intelligence assessments estimated that the Russians were unlikely to be more than two years behind the United States in developing a nuclear submarine. In fact, the Soviets had secretly started developing their first nuclear-powered submarine in late 1952.[27] The prospect of the Soviets operating nuclear submarines worried the Admiralty. As one brief pointed out:

> It must be assumed that Russia will eventually develop atomic submarines. It is vital we should be able to master such a dangerous threat to our sea communications, but until we have ourselves produced atomic submarines, the counter-measures to them will remain largely conjectural.[28]

On 31 January 1950, the Controller of the Navy, Admiral Sir Michael Denny, and the Assistant Chief of the Naval Staff, Rear Admiral Ralph Edwards, wrote to the First Sea Lord, Admiral of the Fleet Lord Fraser of North Cape, and the First Lord of the Admiralty, Viscount Hall, and recommended 'that the Board should offer its support to the development of a nuclear powered submarine' as it would give 'the Navy a submarine of performance transcending that of any other type, at remarkably small additional cost to the country'.[29] The Board approved and 'recommended that every effort should be made to produce the nuclear submarine as soon as possible'.[30] Cost was an important consideration. Not only did the nuclear submarine offer the only known means of obtaining prolonged submerged endurance, up to 100 days at 25 knots, but it was significantly cheaper when compared with HTP.[31]

Denny set up a Special Propulsion Sub-committee of the Ship Design Policy Committee, under the Director of Naval Construction to coordinate activities of Admiralty departments and Harwell, where the small naval/civil team was assessing the various reactor systems likely to be suitable for submarine propulsion. But little was achieved. The Admiralty suspected that Harwell was not very keen on developing a reactor for a submarine, and at the end of 1951 William Penney, the architect of the UK Atomic Weapons programme, had written that he could not think why Britain wanted a nuclear submarine. 'What we have done,' he wrote, 'is to contract to build a power reactor in the most cramped of all ships, such that we lose the ship, the reactor, and the crew if the reactor fails when the ship is underwater as it must be most of its normal cruising.'[32]

This lack of enthusiasm can be explained by Penney's overriding priority for nuclear weapons, for which fissile material was in short supply, as well as Harwell's preoccupation with the UK civil nuclear programme. From the outset any research into nuclear-submarine propulsion had to tie in with Harwell's civil nuclear research.[33] At the time, Harwell was developing advanced gas-cooled reactors for the UK civil nuclear programme. One consequence of this work was that many of Harwell's scientists and engineers showed little interest in conducting research into the highly enriched uranium fuel pressurized water reactors (PWR) that the United States Navy was intending to use in its first nuclear submarine, USS *Nautilus*, as they saw no future for PWR reactors in the UK civil power programme. Although Harwell could forecast a development programme for a PWR 'with the least uncertainty of any of the new proposals' and, given enough priority, a prototype reactor could be built by 1955–6, Harwell's scientists and engineers concluded that the PWR was, at best, an 'interim solution' and was unsuitable for use in a submarine as it suffered from 'the disadvantages of low temperature ... and a development programme quite different from the remainder of the Ministry of Supply reactor programme'.[34] Harwell concluded that a PWR-powered nuclear submarine was not a priority, 'especially in view of our limited capacity to carry out special hull and machinery development as to justify a recommendation to the Ministry of Supply to take extraordinary steps to develop what is considered to be an interim solution to the submarine propulsion problem'.[35]

Instead Harwell concentrated its initial research efforts on a graphite-moderated, gas-cooled reactor that used helium for heat exchange. But by January 1952 detailed studies had shown that a submarine powered by such a reactor would have been far too large to be of any value to the Royal Navy.[36] The research ceased and instead Harwell identified two liquid-metal-cooled reactors that were more acceptable in terms of size and technical superiority than the graphite-moderated, gas-cooled reactor and the PWR, and so would warrant further study and development. A revised research programme initially focused on assessing the characteristics of both types of liquid-cooled reactor was submitted to the Admiralty and approved by the Controller of the Navy, Vice-Admiral Sir Michael Denny, on 19 January 1952.[37] However, resources were only sufficient to allow research into one of the liquid-metal-cooled reactors alongside the PWR.[38]

Harwell's research programme was only possible because the Ministry of Supply had announced that it intended to build a plant capable of producing highly enriched fissile material that could be used to fuel the reactors. Although the government had allocated a small amount of fissile material to the nuclear-submarine programme, the vast majority of supplies were still allocated to the UK nuclear-weapons programme and the civil nuclear programme. Within a year it became apparent that a nuclear plant with the characteristics required would need a large quantity of highly enriched fuel. This could not be spared from the nuclear weapons programme and towards the end of 1952 the Chiefs of Staff effectively suspended the research programme.[40]

The Navy now had a choice: it could wait and hope that enough fissile material would become available in the near future to allow it to restart its nuclear-submarine project, or it could attempt to produce a submarine of similar capabilities using an alternative means of propulsion. Further research into HTP had shown that it was possible to produce it 'at a price of about one fifth of the current price with a fair prospect of still further reduction'. Work on the HTP machinery sets for HMS *Explorer* and *Excalibur* was also nearly complete and the scientific and engineering research team at Barrow was facing the prospect of dispersal due to lack of work.[41] With a stalled nuclear programme the Navy concluded that it was too much of a risk to lose the HTP research team. It once again reversed course and in September 1952 the Naval Staff asked the Flag Officer Submarines, Rear Admiral George Simpson, to prepare a proposal for an operational class of submarine using low-powered HTP machinery.[42]

Simpson's paper, which was completed quickly in September 1952, argued that HTP machinery could deliver considerable underwater performance and provide the Royal Navy with a powerful platform, capable of fulfilling the Submarine Service's primary operational role.[43] He envisioned a submarine of around 1400 tons which was capable of staying submerged for extended periods, powered by specially silenced HTP machinery to allow long-range interceptions. Simpson was:

> firmly of the opinion that Great Britain leads the world in the development of HTP machinery for submarines, and that there is in prospect in the near future an operational submarine of exceptional capabilities. Moreover I consider that we may have in our possession a design of a nearly 'true-submarine' many years before the Nuclear type can be completed.[44]

Meanwhile, the United States Navy, which had access to adequate supplies of fissile material, was continuing to construct USS *Nautilus*, which was powered by a pressurized water reactor, the same type of reactor that the scientists at Harwell had dismissed.

By 1953 the rapid advancement of the US nuclear programme, as well as the prospect of some highly enriched fuel being made available, persuaded the Admiralty to commit more resources to the UK nuclear programme. The case for a Royal Navy nuclear submarine was restated in a paper considered by the Defence Requirements Policy Committee in June of that year and the project was revived. By November, Harwell had abandoned research into the liquid-metal-cooled reactor after its investigations revealed that it required a large submarine of around 3000 tonnes and that there was little prospect of reducing its size unless it was redesigned in a way that used more highly enriched uranium, supplies of which were now not expected to be available until 1956. It was also 'very vulnerable to underwater explosions', which made it far too unsuitable for use in a submarine.[45] This left the Naval Section at Harwell with the type of reactor it had dismissed three years earlier. It had independently reached the same conclusion as the United States Navy, that a pressurized water reactor would best meet the needs of a submarine propulsion plant, offering high power density and control stability, and above all the facility to be engineered into a small submarine hull.[46]

Around the same time the Americans sent a clear signal that they saw their future as nuclear. In 1954 the US Navy abandoned its HTP research programme and terminated all funding into non-nuclear submarine propulsion.[47] Conscious that the Royal Navy was falling behind, the Government awarded the nuclear programme more resources. In 1954 the Naval Section at Harwell was expanded under the leadership of Captain Harrison Smith, with officers from the Admiralty, the Royal Naval Scientific Service, the Royal Corps of Naval Constructors and the Yarrow Admiralty Research Department, as well as engineers from

Vickers-Armstrongs, Ltd, the main contractor for the prototype set of nuclear-submarine machinery, Rolls-Royce Ltd, the main subcontractor responsible for the design and production of the reactor and associated equipment, and Foster Wheeler Ltd, responsible for the design and manufacture of the reactor pressure vessel, primary circuit and steam generators.

The Royal Navy finally had a nuclear-submarine programme, but it had taken four years to get to where the US Navy had been in 1951. The Admiralty had been quick to appreciate the potential of nuclear submarines but from 1946 until 1954 it struggled to complete anything but tentative research. While there was plenty of support for a nuclear-submarine programme its development was never afforded the same priority, and supplies of fissile material, as the British nuclear-weapons programme and the civil nuclear programme. The work into nuclear-submarine propulsion that was completed before 1954 was hampered by the requirement that research had to correspond with the civil reactor programme. Scientists and engineers at Harwell wasted vital years investigating different reactor designs, all of which were eventually dismissed as unsuitable for use in a submarine. Perhaps Harwell's biggest mistake was the early dismissal of the pressurized water reactor, the type the US Navy had selected to power *Nautilus*. Collectively these decisions cost the Royal Navy precious time, something that is immediately apparent if the British programme between 1950 and 1954 is compared with the US Navy's over the same period.

One other important feature that the early British nuclear-submarine programme lacked was an individual who was prepared to champion and drive through its development. This changed in 1955 with the appointment of the new First Sea Lord, Admiral of the Fleet, Lord Mountbatten of Burma. The younger son of Prince Louis of Battenberg, and great-grandson of Queen Victoria, Mountbatten had begun his long and distinguished naval career in 1913, joining the Royal Naval College Osborne as a cadet. After a series of naval appointments, and wartime service, with spells as Chief of Combined Operations and Supreme Allied Commander, South East Asia, in 1947 Mountbatten was appointed Viceroy of India, with the task of transferring sovereignty from the British crown to independent rule. Following independence Mountbatten resumed his naval career and returned to service at sea. He climbed quickly through the ranks, with appointments as Fourth Sea Lord in 1950, responsible for supplies and transport, and CinC of the Mediterranean Fleet between 1952 and 1955, followed in 1955 by the post of First Sea Lord, the pinnacle of the naval profession. Mountbatten was determined to move the Navy into the nuclear age. He championed the nuclear programme, ensured it was given 'decisive new impetus' and sought to enlist the assistance of the United States Navy.[48]

SECURING AMERICAN HELP

The United States Navy and the Royal Navy had always enjoyed a particularly close relationship, and it was at its tightest between the two submarine services. The 1950s represented a high point in this relationship. Both services were cooperating in the highly secret surveillance and intelligence-gathering operations directed against the Soviet Union and on the complex problems of anti-submarine warfare too. In 1955, this culminated in the appointment of a Royal Navy (Submarine) Exchange Officer, Lieutenant Commander A. M. B. Buxton, to the United States Navy's Submarine Development Group Two (SubDevGruTWO) at the Submarine Base in New London, Groton, Connecticut. Working as an operational analyst alongside USN submarines, SubDevGruTWO was given, by the USN Chief of Naval Operations, 'the sole task of solving the problem of using submarines to detect and destroy enemy submarines'. All other operations of any nature, even training, were subordinated to this mission. A Royal Navy officer has been a part of the US organization and its successors since that date.

Despite the close collaboration in the Manhattan Project and the wartime agreements at Quebec in August 1943 and Hyde Park in September 1944 that promised continued post-war cooperation in the atomic energy field, when the Second World War ended, the United States Congress passed the McMahon Act which prohibited the passage of classified atomic energy information to all foreign countries, including Britain.[49] Thereafter, ironclad atomic energy legislation ensured that US officials and companies were severely limited as to what they could show to and discuss with their British counterparts. Throughout the late 1940s and early 1950s successive British governments pushed the United States for renewed nuclear collaboration by demonstrating the success of the independent UK nuclear-weapons programme.[50] Following the Soviet nuclear test in August 1953, President Eisenhower told Congress that the McMahon Act was a 'terrible piece of legislation'[51] that undermined the United States' relationship with its NATO partners. He set about a lengthy process that would eventually result in the restoration of nuclear collaboration between the United Kingdom and the United States.

The Admiralty attempted to enlist American help with its nuclear programme in early 1954.[52] Several high-level approaches to the US Navy were nevertheless unsuccessful, not because the US Navy was unwilling to assist, but because they could not 'go ahead without the approval of the Atomic Energy Commission'.[53] On 20 June 1955, Congress ratified a US/UK Military Atomic Cooperation Agreement that stated, inter alia, that 'the USA may exchange with the UK such atomic information as the USA considers necessary for the development of the UK's defence plans . . .' However, in the wake of the exposure of Soviet spies Klaus Fuchs and Donald Maclean, Congressional opposition and US interagency disputes made the negotiation of further agreements on nuclear propulsion difficult to achieve. The US Sub-committee on Agreements for Co-operation of the Joint Committee on Atomic Energy regarded nuclear reactors as a new 'secret', and during hearings on the two Anglo-American agreements, Congressional members insisted that the US preserve its technological lead in the field. Behind the scenes, the US Defense Department asked the US Attorney General for a confidential opinion on whether the AEC could legally transfer information on submarine reactors to other states. British officials remained optimistic that 'an inter-Service approach to the U.S. Navy would bear fruit'.[54] But without approval from the US Atomic Energy Commission the US Navy was unable to talk.

At that time, anyone interested in the US nuclear-submarine programme knew of Admiral Hyman G. Rickover, the head of the US Naval Reactor Division. Known as the 'father of the nuclear navy', due to his personal role in championing nuclear power in the United States Navy and driving through the design and construction of USS *Nautilus*, Rickover was unique in that although he was a naval officer who held a commission in the US Bureau of Ships, he also held an appointment with the civilian US Atomic Energy Commission. A complicated, difficult, sensitive and aggressive character, he was bitterly unpopular with many naval officers because of the dictatorial methods he used to get things done. But he enjoyed strong support from Congress, which effectively enabled him to do what he liked. For three decades Rickover personally interviewed and approved every Commanding Officer in the US nuclear-submarine programme. He was so powerful that when a young Ensign from Georgia who once worked in Rickover's office became President of the United States and Rickover's Commander-in-Chief, Jimmy Carter admitted that he 'never really felt like his boss'.[55] Carter and four other Presidents would intervene to keep Rickover on active duty twenty years beyond the standard retirement date in the US Navy.

Rickover had spent two years of the Second World War in the United Kingdom as naval attaché, studying battle-damaged Royal Navy warships.[56] He knew the British and the Royal Navy well and was willing to help the Royal Navy develop a nuclear-submarine programme. In early June 1955, he told a member of the British Joint Services Mission in Washington that 'It would be a pity if we traversed the same ground, it would be good if we could have an agreed joint program and finally that if we had a firm submarine project of our own he was certain that we would be given the information on Nautilus provided the approach was made through Service channels and not through the A.E.C.'[57] When Rear Admiral Fawkes, the Flag Officer Submarines, bumped into Rickover during a visit to Pearl Harbor, he again stressed how much he could save the UK in both time and money with nuclear 'know how' and repeated that any request for information should be made through military, rather than commercial, channels.[58] The naval team at Harwell welcomed the prospect of collaboration with the United States but, in June 1955, Captain Harrison Smith warned that 'Unless detailed information on Nautilus is obtained it is quite clear that we shall have to go through all the difficulties and set-backs which the Americans encountered during development of the submarine project.'[59]

In late 1955, in order to obtain the necessary information the Admiralty attempted to bring about a meeting between Rickover and Mountbatten. John Coote, fresh from his pioneering intelligence-gathering operation in HMS *Totem*, was ordered to prepare a brief about Rickover for Mountbatten. Coote attempted to meet Rickover and got as far as having coffee in the Admiral's outer office, where the young Lieutenant Carter quickly shuffled top-secret papers out of Coote's sight. When he was eventually introduced to Rickover the Admiral refused to even look up from his desk. Coote's second encounter with Rickover occurred at Honolulu Airport, while both men waited for a small US Navy aircraft which had been sent to bring Rickover back from the first holiday he had taken in seven years. 'Well,' Rickover said, 'after all this is the second busiest airport in the world after O'Hare in Chicago.' Mentioning Heathrow as a possible contender for

the dubious honour of busiest airport, Coote 'promptly got both bar-
rels at full caliber, complete with statistics and all their sources'. The
ensuing ride to San Francisco was in total silence.[60]

The brief Coote produced for Mountbatten captured Rickover's
achievements as well as his complex character:

> The story of the building of *Nautilus* reveals a restless, lonely and ruth-
> less man, working around the clock to get the boat to sea, regardless of
> opposition. It is said that he stoops to quite unethical methods if the end
> justifies the means – as they always have in his book. For example, he
> would play off one firm or department against another by attributing
> quite untrue statements to each other.
>
> It is difficult to assess Admiral Rickover's true importance to today.
> One must presume that, by virtue of continuing to hold the chief respon-
> sibility for the development of naval reactors both in the Bureau of Ships
> and the AEC, he is now engaged in implementing the USN's recently
> declared policy to put nuclear propulsion into all ships which will be
> employed in offensive roles, including CVAs [aircraft carriers] and their
> escorts.
>
> But there are many who are eager to point out that his mission was
> accomplished with the launching of *Nautilus*; and that he will now be
> discarded as not being whole-heartedly behind nuclear power for surface
> ships. In any case, they say that he is not a physicist so much as a
> hard-driving coordinator of a specific engineering project, who has made
> too many enemies in the process for his survival in the Navy.
>
> Much of this is wishful thinking by his detractors. The USN Submar-
> ine Force freely admits that, but for him, *Nautilus* would not be at sea
> today. They regard him as being more influential than any other serving
> officer, except perhaps the CNO [Chief of Naval Operations] himself.
> They accept as a painful necessity his dictatorial methods. Some even
> explain his boorish manners and insulting conversational gambits simply
> as devices to make his listeners remember what he has to say.
>
> One thing is certain: Rickover will pursue his self-appointed course
> regardless of the opinions of friend or foe, particularly friend. Even his
> bitterest enemies cannot deny his single-minded devotion to duty. Nor
> have they ever attributed to him ambitions for commercial or financial
> rewards – now or in the future. Nor can he be dismissed outright as a
> megalomaniac. Whilst he shuns personal publicity of any sort, he has
> carefully built himself a solid political lobby and the support of the
> most influential voices in the media and in the highest corridors of power

to maintain the priorities he needs for the continuation of his nuclear programme.

In this respect he bears an uncanny resemblance to the Royal Navy's Jacky Fisher, who became First Sea Lord in 1905 [1904] at the age of sixty-four and proceeded to ram though his revolutionary *Dreadnought* programme, which gave Britain the lead in high-speed, hard-hitting capital ships long before the Great War broke out. Nothing was allowed to stand in his path, even those amongst the highest in the land who mistrusted his methods and feared for sacred naval traditions being dismantled.

To further his aims, Fisher used the Press and his friends in Parliament to a degree hitherto unknown. He wrote anonymous articles in *The Observer.* Friendly journalists got special briefings from him. Soon the public took up his cry: 'We want eight, and we won't wait.' The resemblance does not end in the two Admirals' drive and ruthlessness, as their portraits show. But to find out what makes Hyman Rickover tick, it is necessary to start by taking account of a lifetime spent in antipathetic surroundings. In the end one has to settle for the fact that he is motivated by a love of the Service which he joined as an expedient and which has repeatedly reminded him how unwelcome he is, rather than by hope of honours, self-aggrandisement or financial reward.[61]

Mountbatten was due to visit the United States in October 1955 to meet his US counterpart, the Chief of Naval Operations (CNO), Admiral Arleigh Burke. Burke and the US COMSUBLANT, Rear Admiral Fred T. Watkins, thought it would be fitting if Mountbatten were to become the first foreign officer to go to sea in *Nautilus.* However, when Mountbatten arrived on 27 October 1955 Rickover refused to meet him and ensured that Mountbatten was not allowed anywhere near *Nautilus.* Coote later wrote that an irreverent headline writer could have called the non-meeting 'Rickie Snubs Dickie'.[62]

Rickover's behaviour must be set in the context of the complex Atomic Energy Legislation and his bitter battles with the US Navy and Atomic Energy Commission. With a resolution on the legality of passing information about the US nuclear-submarine programme to the British from the Attorney General still pending, Rickover was unwilling to show Mountbatten anything. Allowing him on board *Nautilus* was in Rickover's mind an unnecessary risk that could have provided valuable ammunition to his many enemies. At the time, he was at the centre of a battle between the Joint Committee on Atomic Energy (JCAE) and the US Navy over his possible promotion to Vice Admiral.

When the promotion was refused by the US Navy promotion board, Rickover responded by waging a long and bitter struggle with the US Navy that would continue until he retired in 1982. He disregarded the normal promotions system and used his connections with the AEC and Congress, and a well-run publicity campaign (which culminated in an appearance on the cover of *Time* magazine), to become one of the best known officers in the Navy.

To Mountbatten, hoping to enlist American assistance with the UK nuclear programme, it appeared as if 'the uranium curtain had been well and truly run down'.[63] Admiral Burke was also deeply 'embarrassed' and instead offered Mountbatten a visit to the second most significant submarine in the US Navy, the USS *Albacore*. The USS *Nautilus* was essentially a conventional twin-screw submarine, with a conventional hull form, fitted with an experimental nuclear reactor. The USS *Albacore* was a hydrodynamic test vehicle with a single shaft and highly streamlined and distinctive teardrop hull, powered by electric batteries and capable of speeds up to 30 knots.[64] According to Admiral Ignatius Galantin, who was on board during Mountbatten's visit, 'Mountbatten never forgot the enthusiasm and conviction with which *Albacore*'s young skipper described his ship and put her through her paces, the ship heeling and turning, climbing and diving steeply in a way no other ship could.'[65] The visit had a profound impact on Mountbatten's thinking about the Royal Navy's future submarine policy.

FUTURE SUBMARINE POLICY

Prior to leaving the United States, Mountbatten told the Chairman of the United States Atomic Energy Commission, Admiral Lewis Strauss, that he was 'afraid of the consequences on the relations between the two navies if, in this process, we developed some new and possibly more efficient method and would then be unable to talk to the Americans on grounds of lack of reciprocity. He did not wish to see our paths diverge in this manner.'[66] Strauss, an advocate of full cooperation, regarded the American position 'as contrary to good sense and logic and therefore as untenable in the long run'.[67] But like his colleagues he was hamstrung by legislation. Even President Eisenhower wrote to Mountbatten and advised 'for the moment we must await favourable legislative action before anything further can be done'.[68]

When Mountbatten returned to the United Kingdom he accelerated the development of the British nuclear-submarine programme, in part to speed up its development, but also to show the Americans that the Royal Navy was determined to press on, with or without US assistance. Mountbatten also secured the necessary funding from the Treasury, a complex process that involved commissioning a 20-inch model of a nuclear submarine which opened up to display the inner workings for the Chancellor of the Exchequer, Derick Heathcoat Amory, who came from an old naval family. Mountbatten placed the model in front of the Chancellor's seat at a meeting of the Cabinet's Defence Committee where Heathcoat Amory played with it throughout the discussion of earlier items on the agenda. He was fascinated and when the question of nuclear propulsion came up, he merely looked across the table and asked, 'How much?'[69] The Treasury sanctioned the development of a land-based prototype reactor as well as the hull and machinery for one submarine.

Mountbatten also ensured that the first submarine was given a highly symbolic name. The Admiralty originally intended to name its nuclear-powered submarines after former battle cruisers, starting with 'In-' as, like battle cruisers, nuclear submarines were regarded as 'ship[s] of the future' that would 'undoubtedly control sea communications'. Names such as *Invincible* and *Inflexible* were briefly considered, but rejected, in the case of the former because of its unfortunate sinking at the Battle of Jutland, and the latter because 'the epithet of inflexibility is the last one that we should wish to apply to the Navy and to these new submarines in particular'.[70] *Indefatigable* was then considered appropriate to convey 'the essential quality of these nuclear boats'. However the Ship's Names Committee, the body responsible for naming all Royal Navy vessels, agreed that the most suitable name for the submarine would be *Dreadnought*, a name that had 'a unique attraction in that it already represents a land-mark in Naval history, associated as it is with most revolutionary war-ship design ... it should be used for the first "Jet-Age" submarine.'[71]

The acceleration of the nuclear programme also had implications for the Royal Navy's existing submarine policy. The constantly changing requirements, false starts and misdirection were deeply frustrating to the senior Naval Staff, such as the Controller of the Navy, Ralph Edwards, who wrote to Mountbatten in September 1955 that 'The more I see of the way submarine matters are conducted, the more disquieted I become. Year after year we see the opinion of the submarine commanders switch 100 per cent and for reasons which I do not personally believe are valid.'[72]

In February 1956, as part of Mountbatten's 'Way Ahead' studies into the future size, shape and role of the Royal Navy, a comprehensive review of submarine policy took place. On 2 February, senior Naval Staff assembled for a conference to define the size and shape of the Royal Navy's submarine fleet for much of the Cold War. Mountbatten opened the conference with a warning: 'By 1960 all submarines in the present Fleet will have become unfit for service, regardless of any modernization measures which may be taken in the meantime.'[73] He explained that:

> it had become clear that our present building programme [eight Porpoises building and four more to commence in 1956/57] was no longer in step with requirements. The priorities accorded to the Atomic Energy Commission had been for power stations and the bomb, but they were now pressing forward with ideas to produce our first atomic powered submarine. We must, therefore, now make up our minds on submarine requirements for the future, e.g. as to whether we need more than one type, and, if so, how many.[74]

Aside from nuclear propulsion, in 1955 the US Navy had deployed a submarine capable of launching cruise missiles, known as the Regulus Attack Missile (RAM). Fitted with a nuclear warhead and capable of hitting targets in the Soviet Union, Regulus was designed to complement the US Navy's carrier-based strike capability. After tests in USS *Cusk*, the missiles, which featured folding wings and tail fin, subsonic jet propulsion and booster rockets that detached after launch, were housed in hangars for two missiles, which were retrofitted abaft the fin of two existing submarines, USS *Tunny* and USS *Barbero*. Larger hangars to take four missiles were fitted forward in two purpose-built submarines commissioned in 1958, USS *Greyback* and USS *Growler*. Finally, in 1960 a nuclear-powered Regulus submarine was commissioned, USS *Halibut*, with a forward hangar for five missiles. Launching the missiles required the submarine to surface, its crew to extract the missile from the hangar manually, place it on a launching ramp and extend its wings and tail fin before firing it at a target.[75]

Mountbatten wanted to know whether the Royal Navy should follow US Navy policy and build cruise missile submarines, what he called a 'shore-strike' submarine capable of launching medium-range airborne missiles against enemy bases. If the answer was yes, he wanted to know if the Royal Navy should standardize its weapons with the US Navy and seek to purchase Regulus to save time and development costs, and take advantage of US logistic support. He also sought answers to wider questions: could a submarine take part in the strategic offensive in war? Could the Royal Navy justify building very large and costly vessels? Would they be suitable for any other form of submarine warfare? If not, should they be built solely for such a role? There was little consensus on any of these questions and the conference simply agreed to revisit them in two to three years.[76]

Discussion then shifted to the role of submarines in 'Peace', 'Global War' and 'Limited' War. These were listed as follows:

Peace and Limited Wars:
1) To prepare for war by training and trials
2) Reconnaissance patrols off potential enemy bases in times of international tension and off hostile bases in war
3) In a limited war, reconnaissance and attack on warships in the war area

Global War:
1) Anti-Submarine Operations
2) Attacking enemy bases in order to destroy enemy submarines and their supplies
3) Attack against surface warships, in particular the Sverdlovs
4) Attack against merchant ships and supply lines.

Although the conference agreed that the primary role of the Submarine Service continued to be anti-submarine operations, FOSM argued that 'such distinctions were dangerous' and that 'what was needed was not a good anti-U boat submarine or a good anti-ship submarine but a good all-rounder'.[77] The Navy, he continued, should not 'build a special type of submarine for each role'; nor should it 'try to incorporate too many special features in one submarine to suit various roles'. Any submarine that was constructed with the primary aim of anti-submarine warfare 'should be versatile only to the extent of being able to attack different types of ships: i.e. submarines, surface warships and merchant ships'.[78]

Mountbatten's views about the design of such submarines were heavily influenced by his earlier visit to USS *Albacore* and he wanted to cancel the 'Porpoise' construction programme and switch the design teams to work on an 'Albacore' design. He was convinced that 'the Russians would in all probability adopt such a type in due course and felt that the R.N. could not afford to be out of the picture'. Why, he asked, had the Royal Navy 'not already designed such a craft?' The Director of Naval Construction responded that British Naval Architects 'had known for some 60 years that the ALBACORE shape performed best under water' and that the Admiralty design departments were still 'experimenting with the true submarine shape on a small scale'. Mountbatten argued that given that the *Albacore* already existed, the design was 'sufficiently advanced' for the Royal Navy 'to profit by any mistakes made'. He recommended that the Navy 'should go straight to the operational design, and save both time and money in cutting out the experimental stages'.[79] He wanted to know if it was 'possible, and within what time, either to design ourselves or alternatively to produce on U.S.N. designs, an Albacore'. Mountbatten regarded 'this as the most urgent and important decision we have to take'.[80]

There was also a detailed discussion about the merits of continuing with the HTP programme. It was impossible to proceed with the HTP and nuclear programmes at the same time due to limited resources and Mountbatten warned that 'the HTP idea would have to be dropped if it was likely to retard development of the nuclear form of propulsion'. However, FOSM warned that:

> It might be fifteen years before we could get any real contribution to the fleet from the nuclear side. For the next five or six years we should have to rely on the PORPOISE, and we could not afford to wait ten more years for something to follow. We were already lagging behind both the Americans and the Russians. H.T.P. was the only form of propulsion likely to close the gap and we should think hard before allowing effort on the nuclear form to stop it.

The conference reached a number of conclusions. First, it agreed with Mountbatten's proposal to abandon the 'Porpoise' programme and design two or three conventional (i.e. non-nuclear) 'Albacore' submarines as soon as design teams became available. These would then be followed by an improved 'Albacore' design powered by recycled diesel propulsion running on HTP, 'subject to logistic and financial implications of HTP supply being acceptable and also to effort on the nuclear powered submarine design not being jeopardized'. Thirdly, to revisit the Strike submarine proposal in two or three years to 'see what the Americans are doing, and decide on what was feasible' for the Royal Navy.[81]

COLLABORATION RESTORED?

Towards the end of January 1956, the US Attorney General, Herbert Brownell, ruled that under the 1954 US Atomic Energy Act the Eisenhower administration could legally negotiate agreements to transmit nuclear submarine propulsion to the British.[82] A new agreement between Britain and the US was signed on 13 June, and on 20 June the Foreign Secretary, Selwyn Lloyd, told Parliament that it would 'permit a broader exchange of materials in the Atomic Energy programmes of the two countries' and 'provide for the exchange of information concerning military package power reactors and other military reactors for the propulsion of naval vessels, aircraft and land vehicles'.[83] But British hopes that the new agreement would lead to the transfer of information about nuclear submarine propulsion were quickly dashed.

On 25 June, Senator Thomas E. Murray, a member of the US Atomic Energy Commission, notified the Committee that he disapproved of the agreement because of inadequacies in British security procedures. Although the agreement took effect on 16 July, certain members of Congress protested that they had been kept in the dark by the Eisenhower administration and argued that the 1954 Atomic Energy Act did not, in fact, permit the exchange of information about nuclear propulsion with the British. The Committee insisted that President Eisenhower suspend the new agreement for up to sixty days, which Eisenhower, a minority president without a majority in either Senate or House, facing an imminent presidential election campaign, duly did.[84] Congress then adjourned until 3 January 1957, leaving the prospect of any immediate resolution looking very thin. 'Domestic politics, I am afraid, have once again put a brake on any action for the present on the Agreement to Exchange Information on Submarine Reactors,' wrote the Admiral, British Joint Services Mission in Washington, Vice-Admiral Sir Robert Elkins, to Mountbatten.[85]

But all was not lost. Between 20 and 31 August 1956, Admiral Rickover conducted his first visit to the United Kingdom in order 'to discuss exclusively' with Mountbatten 'the obstacles with which he was confronted and the methods he had to use to surmount them in order to get Nautilus built'.[86] Rickover was 'most insistent' that, due to the problems with Congress, 'discussions could only be arranged by personal invitation from you and that any suggestion of such discussions reaching the press might do incalculable harm'.[87] Mountbatten invited Rickover and his wife to Broadlands, his country estate in Hampshire, with the Controller of the Navy, Admiral Sir Ralph Edwards, the Engineer-in-Chief of the Fleet, Admiral Mason, and the head of the Atomic Energy Authority, Sir John Cockcroft. Their wives were also invited in order to make the meeting a social occasion. Over drinks and dinner, Rickover, the 'introvert iconoclast from the Ukraine [sic] . . . fell under the spell and aura of Queen Victoria's great-grandson'.[88] Rickover was amused to learn that he and Mountbatten were born in the same year and that while Rickover was growing up in Makow in tsarist Poland, Mountbatten and his family were occasional guests of Nicholas II at St Petersburg.[89]

Throughout the discussions Rickover restricted his talking points to organizational matters associated with the development of nuclear propulsion. Due to the legal restrictions surrounding the Atomic Energy Act and the difficulties with Congress, he was unwilling to talk about technical aspects. When he was asked how fast *Nautilus* could go, he went silent, leaving his wife to give 'away the fact that by hearsay nothing could catch the NAUTILUS'.[90] Rickover was emphatic that the Royal Navy should award its development work to a private company, what he called a firm, and that in no way should the Royal Navy rely on a Government department to run the programme. He explained that 'some may believe that projects such as Nautilus emerge from routine organisations, systems and established procedures. Our experience with the Nautilus shows that none of these lead to really significant results. A project like the Nautilus calls for dedicated people who do not permit obstacles to stop them.'[91] When Mountbatten asked what exactly the Royal Navy should do Rickover told him to disregard the normal Admiralty machinery and appoint an independent man with direct access to the top to lead a small staff of about a dozen individuals, all with engineering experience, and that they should stay in the job for a long time, possibly up to seven years. In the words of the minute taker, Rickover:

> proposed that we should buy a reactor from Westinghouse in order to save time and the initial large development cost. Thereafter he said we could go off on our own. He was only speaking of the reactor and said that the rest of the machinery could be designed and made in the UK. He

added that the development will cost a lot of money and its scope should not be under-estimated. He mentioned 20 million dollars as the probable cost of developing the first machinery installation. He suggested that we use Rolls Royce both as our reactor designers and because they have a link with Westinghouse. We should buy a reactor from the latter and while it was building send naval and civilian personnel to the firm to learn the technique. This would save time and money and he appeared to be genuinely concerned that we should be able to save both. He made it clear that he was all for giving us their latest information and was working towards that end.[92]

(The Rolls-Royce link with Westinghouse took the form of a 1953 agreement on the exchange of classified information about gas turbines.)

Mountbatten recognized that if he accepted Rickover's offer he 'would have to have someone of high calibre to work with him and be his connection with this country'.[93] Edwards was concerned about 'how on earth we are going to find the right chap to work with him and so make use of all that would accrue from Rickover and his organisation?'[94] Mountbatten, Edwards and Mason were, Rickover said, 'quiet and contended cows [sic]' and unsuitable, no Engineer had the necessary qualities and any retired Admiral 'was not worth a damn'.[95] Rickover was adamant that what the British programme needed was a man with 'fire in his belly'. 'In other words we have got to find another Rickover, maybe he has a brother' was Edwards's rather glum conclusion.[96] Although Rickover implied on many occasions that no Englishman qualified for the job he was very impressed with Edwards, who, after the meeting, wrote to Mountbatten and explained how Rickover 'talked at great length – or rather he talked and I listened' and:

> Ended up by embarrassing me horribly by telling me that I had better take the job on and then went and spent twenty minutes trying to persuade my poor wife that being a Commander-in-Chief was as nothing to being the Head of this new development organisation. He finished the evening by saying he wouldn't take no for an answer and was going to see Mountbatten. I explained to him that in our country no matter whether we were Admirals or midshipmen we did what we were told and could not go choosing our way.[97]

Mountbatten was 'left with the impression that, provided Admiral Rickover was handled with care and that we were prepared to take at least some of his advice, he would be an ardent helper'. Rickover told Mountbatten in the margins at Broadlands: 'If you have allies, and good ones at that, you would be foolish not to take them into your confidence over one of the most important developments of war today.'[98] Mountbatten later wrote of the 'extremely friendly and helpful attitude which Rickover has shown during his visit here ... he could not possibly have been more friendly to us and I am hoping for great things from our contact'.[99]

But Mountbatten was not, at that stage, prepared to accept Rickover's main proposal – purchasing a US-made reactor from Westinghouse. He told Sandys that while, eighteen months before, American help would have been valuable, 'we have had to develop our own ideas' and the need was largely past.[100] Rickover's recommendations about reforming the Admiralty organization responsible for supervising the Navy's nuclear programme also had little immediate impact. When the Flag Officer Submarines, Rear Admiral Woods, argued that 'a far more closely knit design team than is possible under the existing Admiralty organization is required. Ideally this team should be responsible to one man with the authority to take decisions necessary to ensure proper co-ordination of effort,' Admiral Edwards disagreed, stating that 'the arrangements which have been made to handle the nuclear-submarine project should ensure that it is dealt with as expeditiously as possible. Unless and until this is proved otherwise I think it is unnecessary to consider the exceptional measures postulated by F.O.S.M., which would in any event involve duplication of effort and increase in staff.'[101, 102]

A VISION OF THE FUTURE

The nature and punch of the Soviet adversary changed gradually but significantly in the mid-1950s. The post-Stalin leadership in Moscow was determined to narrow the very considerable gap between its naval capabilities and those of NATO, the United States especially. When the new First Secretary of the Communist Party of the Soviet Union, Nikita Khrushchev, came to power in the mid-1950s, the Soviet Northern Fleet, with its headquarters at Severomorsk on the Kola Peninsula had only a few surface ships and around thirty submarines – less than 10 per cent of the underwater capability of the Soviet Navy at that time.[103] In 1956 Khrushchev appointed as Commander-in-Chief of the Soviet Navy an individual who would have a profound influence on

Soviet naval policy for the next thirty years, Admiral Sergey Geor-
gyevich Gorshkov, who immediately set about transforming the Soviet
Navy from a primarily coastal defence force to a modern long-range
offensive fleet. He scrapped the Soviet Navy's First World War battle-
ships, the cruisers built in the 1920s and 1930s, and most of the German
and Italian war prizes worn out after extensive post-war trials and
operations. In his memoirs, Khrushchev recalled how he and Gorshkov
'decided to place the construction of submarines literally on an
assembly-line basis, to create a powerful submarine fleet that could
threaten the enemy on all oceans and seas'.[104]

Khrushchev made no secret of what he was trying to do. In April
1956, during a visit to the United Kingdom, he told a largely British and
naval audience that had gathered at the Royal Naval College in Greenwich that:

> Your country 'rules the waves,' but that is a thing of the past. We have
> to look at things realistically today. Everything has changed ... In a
> future war the chief military questions will not be decided by cruisers,
> not even by bombers. They too are outdated ... Today the submarine
> fleet has come to forefront as the chief naval weapon, and the chief aerial
> weapon is the missile, which can hit targets at great distances, and in the
> future the distances will be unlimited.[105]

Given the Soviet Union's limited number of land-based intercontinental
strike aircraft and missiles, submarines offered a potentially attractive
means of launching nuclear weapons against the United States. In
1954 the Soviet Government had ordered the development of ballistic-
missile-armed submarines and later that year tests took place of the
short-range (ninety miles with a nuclear warhead) R-11 missile (better
known as 'Scud') on a swaying platform simulating a rolling submarine, This quickly led to mating the R-11FM with a 'Zulu' class
submarine B-67, with two missile tubes. The missile had to be hoisted
out of the fin using a 'horn and hoof' arrangement to keep the missile
perched on the top. The first launch from the Kola test range on 16 September 1955 was the first-ever ballistic missile launch from a submarine.
In 1956 production began of five new Project 611AB missile submarines known in the West as 'Zulu Vs'. These also carried two missile
tubes and four were in service with the Northern Fleet by the end of
1957; a fifth was added to the Pacific Fleet in 1959.

At the same time as it had ordered the Zulu conversions the Soviet
Government also called for a new missile submarine, Project 625,

known as the 'Golf' in the West. These were to carry a new missile in three tubes in a lengthened fin, the 360-mile-range D-2. However, due to delays in the D-2 development programme, the first three submarines commissioned in 1959–60 carried the R-11FMs with non-nuclear warheads, while the rest carried the D-2 equipped with a high yield (1.45 megaton) thermonuclear warhead.

In parallel with these ballistic-missile developments there were cruise missile projects, the most promising of which was the P-5, a jet-powered rocket-boosted weapon fired from the launching tube. It was tested on board a 'Whisky' class submarine at the end of 1957. This led to the conversion of six 'Whisky' class submarines with two launch cylinders as Project 644 ('Whisky Twin Cylinder'). This was a not entirely successful conversion, one boat capsizing in 1961.[106] Improved types, the Project 665 'Whisky Long Bin' and the specially designed Project 651, known as the 'Juliett' class, each with four tubes, were deployed in the early 1960s. The role of the cruise missile changed from land attack to anti-ship warfare when an improved missile, the P-6, was deployed. Even the ballistic missile firers were temporarily re-tasked for naval purposes after Khrushchev gave the land-based Strategic Missile Forces a monopoly of the main strategic land attack role.

Nevertheless the land attack potential of the Soviet submarine-based missile force soon made its mark. In a 1956 paper 'NATO Control of Shipping Exercise, codenamed LIFELINE', the reaction to the introduction of only three submarines equipped with missiles threatening the American seaboard was so violent that all other maritime air operations were brought to a halt. The Directing Staff eventually had to dispose of the submarines in order to avoid ruining the overall exercise.[107] The same year the US Chief of Naval Operations, Admiral Arleigh Burke, told a US reporter that 'the Soviet submarine threat is tremendous right now. They are still building new submarines at an unprecedented rate. Some will probably have guided missiles. We have to be able to meet the threat if war comes, or we will be in big trouble, world-wide. This is a serious problem.'[108]

In light of these developments, the Flag Officer Submarines, Rear Admiral Woods, was 'concerned that such decisions as have been reached on new Submarine design and construction have been based on the fulfilment of present roles and priorities with the effort now available, rather than on a reassessment of the place of the submarine in the Fleet, after full consideration of its greatly increased potentialities'. Woods produced a paper that included 'a reassessment in the light of the opportunity the submarine now offers to increase the offensive power of the Navy at comparatively small cost'. He argued that since 1948 'three revolutionary advances have been made which combine to alter completely previous concepts of submarine warfare':

Those advances and their effects can be summarized as follows:-

(a) <u>Nuclear Power.</u> Allows the submarine virtually unlimited endurance at a speed comparable to, and probably exceeding that of the fastest surface ship. A relatively high quiet speed is already possible, and further development on existing lines will inevitably lead to very high quiet speeds indeed. The Submarine has not only regained the advantage which it had over the surface ship before the advent of asdics and anti-submarine weapons; but has become a flexible weapon of decision as opposed to one of chance encounter. At the same time, the difficulties of detecting and attacking it from the air or surface have become truly formidable.

(b) <u>Detection Equipment.</u> With the advent of VLF asdics, submarines are capable of detecting surface ships, or other submarines proceeding above their silent speeds or inadequately silenced, at very many times the range previously possible. Thus, they need no longer be confined to operating near enemy ports, or at other focal points established by careful intelligence of enemy movements. There is no prospect of similar detection ranges being achieved by surface ships; and, though the actual ranges may be reduced by enemy silencing developments the submarine will always remain the best A/S detection platform (even if forced to rely upon active asdics), and will probably remain the most deadly counterweapon if properly armed.

(c) <u>Guided Missiles.</u> Two American Submarines now in operational service are capable of launching a missile (Regulus I), which has a range of 500 miles with a megaton warhead, and which can be controlled by the launching submarine for 180 miles with a terminal accuracy of 1000 yards. By 1964, the Americans expect to be able to throw from one submarine three (TRITON) missiles with a smaller warhead and terminal accuracy, for 1500 miles (at Mach. 3.5). To launch those weapons, the submarine has to surface for about six to ten minutes. Since, however, even Regulus I has a total range of 500 miles, and its terminal guidance can be provided by a second submarine stationed inshore, there is no need for it to be launched from close to the coast, or for the launching submarine to be navigated with pin-point accuracy. The submarine is therefore not only able to attack ships at sea, or in relatively unprotected harbours, but can destroy tactical or strategic targets far inland.[109]

Woods argued that:

> By contrast with the immensely increased flexibility and destructive power of the Submarine ... Land based aircraft and missiles remain totally dependent on static and vulnerable logistic support. Surface ships (and hence carrier borne aircraft and missiles) must devote an increasing proportion of their armament to defence against aircraft and submarines, and must be surrounded by powerful screening forces, if they are to stand even a reasonable chance of survival. The submarine, on the other hand, needs no defensive support and very little defensive armament. It is, in fact ... an outstandingly economical and effective weapon of offence.[110]

Woods believed that because the Royal Navy could not afford to develop submarines, aircraft carriers and other ships simultaneously at the same speed as the Americans, it was important to review the roles which the various arms of the Navy would perform in ten years' time with the submarine in the forefront.

When Woods produced his paper, the Submarine Service was absorbing 5.9 per cent of the total naval manpower and only a small percentage of the Naval Estimates, exactly the same as it had in 1938. As far as Woods was concerned, this was unacceptable. But he had to tread carefully lest he upset the other, more traditional and more influential branches of the Royal Navy:

> It is not my purpose to suggest that the submarine can or should replace the carrier, the cruiser or the surface escort. It is to show not only that the submarine can undertake the tasks at present allotted to it more effectively than hitherto, but that it can now take its place as a complement to the carrier borne and shore based aircraft or missile in mounting the deterrent in global war. In this role it has the unique advantages of invisibility coupled with mobility, and hence the ability to remain on station unknown to the enemy (or potential enemy) for long periods. In addition the submarine will always be needed in considerable numbers in peace or war for training our A/S forces; it forms a potent threat in cold war and is a powerful offensive weapon in limited wars. Since all these assets can be realised with very considerable relative economy in manpower and cost, I am convinced that the time has come to devote a higher proportion

of the available money, manpower and effort to the early development of a modern Submarine Fleet of effective size.[111]

Woods prepared a list of functions which he believed the submarine could carry out 'effectively and economically':

  (a) <u>The Deterrent</u> – it can supplement the carrier borne and shore based deterrent forces.

  (b) <u>Global and Limited War</u> –
    (i) Attack on naval bases and communications.
    (ii) Attack on strategic and static tactical targets.
    (iii) Attack on enemy submarines.
    (iv) Attack on enemy surface forces.
    (v) Deterrent to and prevention of seaborne invasion.
    (vi) Provision of intelligence.

  (c) <u>Cold War and Peace</u> –
    (i) Imposition of expensive A/S defences on potential enemies.
    (ii) Training (Pro submarine and A/S) for war.
    (iii) Flag Showing – in conjunction with other ships displaying offensive naval power.[112]

This was a bold vision of the future. As a first step, Woods promoted the submarine over what he regarded as the expensive carrier striking force. He argued that there was a strong case for a 'strike submarine', armed with long-range guided missiles such as the American Regulus II and Triton, operating from friendly waters, capable of striking tactical and strategic targets 1500 miles away.[113] The 'strike submarine' could, he argued, remain on station in times of tension and its relative invulnerability to surprise attacks was a considerable asset compared to fixed installations, bomber bases and carrier task forces which were vulnerable to pre-emptive surprise attacks. Woods also drafted a set of standing orders to supersede the Admiralty's 1948 directive to hunt and kill enemy submarines. The proposal assigned a whole series of new tasks to the Submarine Service:

Their Lordships have reviewed the functions of our submarine forces in peace and war in the light of the advances in submarine capabilities made possible by nuclear propulsion, their ability to launch and control guided missiles with nuclear warheads and the development of long range detection equipment. They have decided that the staff requirements, development and training of submarines should be framed to achieve the following main objectives:

(a) To form part of the forces providing the nuclear deterrent to limited and global war.
(b) To perform their operational functions in war.
(c) To provide realistic targets for the development of anti submarine material and tactics.

In war, the primary operational functions of our submarines will be:-

(a) The delivery of nuclear weapons to strategical and tactical shore targets.
(b) The interception and destruction of enemy submarines in enemy controlled waters and in transit areas.

The relative importance of these two roles will depend on circumstances.

The other main functions of our submarines, the importance of which will also depend upon circumstances, will be the interception and destruction of enemy warships and merchant shipping, reconnaissance, air/sea rescue and special operations.[114]

Woods's vision of the future would take many years to become a reality. Although the Navy's Director of Plans, Captain Duncan Lewin (a distinguished naval aviator), agreed that 'a final decision on the future composition of the submarine fleet is a matter of urgency', he disagreed with Woods's proposals on the grounds that if they were implemented they would radically alter the Navy's future submarine policy. The Director of Plans argued that the nuclear deterrent could best be provided by 'Strategic Bomber Forces' and that the submarine should have no part to play in providing or delivering nuclear weapons against strategic targets.[115] He also argued that there was 'no case for the employment of submarines as guided missile carriers against tactical targets such as permanent or temporary submarine bases because should the destruction of these bases become essential then it could be done more economically and more accurately by strategic bomber forces'.[116] Instead, Lewin presented a more modest set of requirements for the Submarine Service in global, limited and cold war:

In global war, setting aside the nuclear exchange, the Russian submarine fleet, which is still expanding in size and potential, poses the biggest single maritime threat to our survival. The submarine can play an important part in the A/S operations required to meet this threat because of its ability to operate unsupported close to enemy bases and on enemy transit routes. In addition, the submarine must provide our surface and air forces with realistic A/S training facilities.

In limited and cold war our naval forces require mobility and flexibility and this dictates that, for some years, these forces will be centred round the aircraft carrier, which despite its cost and vulnerability, is the only type of vessel that can undertake the multitude of tasks required. D. of P. considers that the submarine's functions in limited and cold war are offensive operations, reconnaissance, covert operations (in periods of strained relations) and the provision of A/S training facilities.

D. of P. concludes therefore that the role of the submarine should continue to be: -

(a) In peace
  (i) To train to perform its operational functions in war.
  (ii) To provide realistic targets for A/S training and for the development of A/S material and tactics.
(b) In war
  (i) The interception and destruction of enemy submarines in enemy controlled waters and transit areas.
  (ii) The interception and destruction of enemy war and merchant shipping, reconnaissance, air sea rescue and special operations.[117]

In terms of guided missiles launched from submarines Lewin recommended that:

R. N. activity should be restricted to keeping in close touch with U.S. developments. A time may come when considerations of vulnerability or expense may force us to replace the aircraft carrier with the strike submarine ... in the foreseeable future however, our meagre financial resources will not be able to stand a venture, even on a meagre scale, into the field of either building or conversion to a guided missile submarine.[118]

Undeterred by this opposition, Woods turned his attention to the UK nuclear-propulsion programme. Due to its 'unexpectedly swift' progress and the prospect of cooperation with the Americans, he urged Mountbatten to reassemble the February 1956 submarine conference. When it reconvened in October 1956, Woods recommended it reverse previous policy decisions and instead 'continue the development of nuclear propulsion at maximum intensity, making the best possible use of American experience'. He proposed abandoning any attempt to design an entirely new class of diesel electric submarine, including one based on the US Navy's 'Albacore' hull and argued that the existing 'Porpoise' design should be used as the basis for an improved diesel electric submarine, which could be designed cheaply and built quickly to keep the submarine fleet up to strength and bridge the gap until sufficient numbers of nuclear submarines entered service.[119]

Mountbatten agreed. The 'Albacore' project was effectively abandoned, although he insisted that it 'should continue to be borne in mind in case surplus design capacity or American help became available'.[120] The Admiralty's design teams started work on an improved version of the 'Porpoise' design, what would later become the 'Oberon' class.[121] The meeting also made two other important decisions concerning the strike submarine and the experimental HTP programme. Mountbatten continued to insist that 'the strike submarine was a desirable weapon', but the meeting concluded that it 'should not be allowed to interfere with the development of the nuclear submarine and the improved PORPOISE and that an ALBACORE with good A/S qualities had a greater claim on any surplus design effort'.[122]

The meeting made one other important decision. It terminated all research into alternative forms of submarine propulsion. Design work on recycled diesel engines was discontinued (although an existing recycled diesel design, was retained for possible future use) and the long-running HTP programme, including the proposed operational fleet conceived in 1952, was also brought to a close. The Submarine Service had already learned the hard way the dangers of attempting to harness the volatile hydrogen peroxide in its submarines. At 0325 on 16 June 1955, the 'S' class submarine HMS *Sidon* was alongside a Submarine Depot Ship, HMS *Maidstone*, in Portland Harbour, preparing to sail on Stage II trials for the new Mark 12 'Fancy' HTP torpedo. While *Sidon* was alongside HMS *Maidstone*, members of her crew were in the process of re-loading a torpedo in to one of the torpedo tubes. The stop valve on the torpedo was opened and the starting level accidentally triggered. The torpedo exploded and blew open the submarine's bow caps, as well as the rear door of the torpedo tube. Some of the torpedo was immediately ejected into the sea through the tube, but other components, including carbon dioxide, were ejected into the submarine. Two officers and ten men were killed instantly, six by injuries from the explosion and six from asphyxiation by carbon monoxide or carbon dioxide.

Eyewitnesses on board HMS *Maidstone* described watching a sheet of flame shoot up through *Sidon*'s conning tower, followed by pieces of equipment, furniture and items of clothing such as hats and coats, all of which were flung into the air. As the injured, dazed and partly asphyxiated crew evacuated the submarine through the conning tower hatch (as *Sidon* was at Harbour stations all other hatches were closed), rescue parties armed with breathing equipment prepared to descend into the submarine. *Sidon* was full of smoke and debris littered the Control Room, preventing the boarding party from going forward or aft. The explosion caused severe damage to the submarine's watertight doors and bulkheads, and as water poured into the torpedo compartment the rescue teams realized that *Sidon* was sinking. They too were forced to evacuate, escaping through the conning tower and after hatch. One of those who had boarded the submarine to help treat the wounded was Surgeon-Lieutenant Charles Rhodes, a doctor from HMS *Maidstone* who was unfamiliar with the layout of the submarine, as well as how to use the specialist Davis breathing equipment. He died from asphyxiation after helping three casualties escape. *Sidon* sank at 0850, with a 25-degree list to starboard in 39 feet of water.

The Board of Inquiry convened to investigate the explosion concluded that the activation of the starting lever caused a pressure build-up, which led a pipe-line to burst, spraying HTP fuel and lubricating oil onto the torpedo. When the volatile HTP came into contact with the torpedo's metal components, it decomposed into oxygen, gas and steam, which then exploded. The Board concluded that 'it was humanly impossible for the Ship's Company . . . to have prevented the submarine from sinking' and *Sidon*'s CO, Lieutenant Commander Hugh Verry, was cleared of any responsibility. The Submarine Service had been very lucky. The Flag Officer Submarines, Rear Admiral Fawkes, wrote in October 1955 that it was 'widely known that had the explosion taken place at sea, the loss of life would have been much greater and the cause of the explosion or indeed of the accident would probably never have been determined'.[123]

The Board of Inquiry recommended that work on the development of the Mark 12 torpedo should be suspended until a number of modifications could be incorporated into its design. However, not long after the accident another Mark 12 exploded at the Arrochar torpedo range in Loch Long, Scotland. 'The safety problems with HTP are so great that it is an uncomfortable shipmate in a Submarine or, for that matter, in a depot ship,' complained one naval officer from FOSM's staff in

1958.[124] The torpedo was cancelled, yet another post-war disappointment for the Submarine Service. Work on a replacement known as the Mark 23 'Grog' started in 1955. Based on a wire-guided torpedo known as 'Mackle' which had been cancelled in 1956 due to its complexity, the Mark 23 was an anti-submarine, passive homing torpedo, essentially a modified Mark 20 with a wire guidance system inserted between the battery and the after body which allowed an operator on board the firing submarine to listen to hydrophones inside the torpedo, manually steering it towards a target and away from any decoys and countermeasures. The Mark 23 entered service in 1966.

The two experimental HTP submarines, HMS *Explorer* and HMS *Excalibur*, also suffered from considerable problems once they were commissioned into the Royal Navy in 1956 and 1958. Both submarines were designated primarily as anti-submarine targets, but they were rarely used due to the high cost of HTP. HMS *Explorer* only managed twenty-two hours of exercises during its first commission, while HMS *Excalibur* only achieved 100 hours.[125] They were also unsurprisingly regarded as unsafe. In *Explorer*, the hydrogen peroxide was fed into a catalyst chamber where oxygen became disassociated from water with a great release of heat. The resulting steam and oxygen were then passed into a combustion chamber where sulphur-free fuel was injected which burnt and considerably raised the temperature. Water was then injected to cool the gas, producing yet more steam, which was then used to drive a turbine. The steam was subsequently condensed in a condenser where carbonic acid was removed and then injected back into the combustion chamber again while the carbonic acid was pumped into the sea. The whole process of starting and running the HTP machinery in *Explorer* was known as 'fizzing' and to the unwary bystander 'fizzing' in harbour was 'like a preview of doomsday'. The sight of exhaust gases, emerging at speed, towered above the submarine in great plumes of grey smoke, and was accompanied by a roar which shook windows a hundred yards away. When *Explorer* first 'fizzed' after joining the 3rd Submarine Squadron at Faslane, HMS *Adamant*'s officer of the watch was so convinced that the submarine was about to explode that he called out the fire and emergency party and summoned the local fire brigade.

The volatile HTP could only be stored in containers and passed through pipes made of 'compatible' materials such as glass, porcelain, PVC, some forms of rubber, certain types of stainless steel, and, for a limited exposure time, aluminium. It reacted vigorously with incompatible materials, such as mild steel, brass, wood, clothing or human tissue, instantly producing both heat and oxygen – two of the three essentials to establish combustion. On both *Explorer* and *Excalibur* the HTP was carried in fifty-four special bags, outside the submarine's pressure hull. Filling these bags with HTP was a dangerous operation in itself as the bags had a worrying tendency to explode. During sea trials in February 1957, one of *Explorer*'s HTP bags burst, exploded and flooded much of the HTP system with sea water.[126] 'Any small leak in any of the plastic fuel bags needed a docking to change the whole lot,' recalled Michael Wilson, one of HMS *Explorer*'s COs. 'It was VERY frustrating.'[127] *Explorer* was eventually banished to a small timber jetty a few hundred yards from *Adamant* and awarded the nickname 'Exploder'.

At sea, however, HMS *Explorer*'s performance was both impressive and complicated. The HTP propulsion machinery gave short periods of very high underwater speeds. The same weight of hydrogen peroxide provided thirty-five times the energy that could be stored in an electric battery. But the HTP machinery suffered from repeated breakdowns and was notoriously unreliable. Those in charge of operating and maintaining it resorted to unusual practices to carefully nurture the equipment. 'If I, as Engineer Officer, failed to do my usual rounds and make my daily obeisances, the turbines would not perform' remembered John Pratt (hereafter referred to under his pen name, John Winton), one of the specially trained and highly attuned engineers who served on board HMS *Explorer*. 'They would not, in any case, perform on Sundays or holy days; break-downs on those days happened too often to be coincidence. Once, after we had slogged for 36 hours into a raging Atlantic gale, neither turbine would start. Later, I checked and found it was Yom Kippur.' It sometimes took weeks for *Explorer* to accept a new operator and superstition was widespread. Some members of the ship's company were forbidden to move aft of the Control Room bulkhead while *Explorer* was 'fizzing' because of the so-called 'evil eye' effect.[128] Despite the dangers inherent in operating both *Excalibur* and *Explorer*, their crews grew very fond of the two submarines. 'We did not look upon her as being dangerous. The crew took the bangs and fires as a matter of course,' recalled another of *Explorer*'s COs, Commander Christopher Russell.[129]

One of the most dangerous incidents on board HMS *Explorer* occurred on 5 October 1961, off the Mull of Kintyre. *Explorer* was fizzing on the surface, acting as a target for another, dived submarine. 'It was the first "fizz" of the day, indeed the first for many days, after lengthy and exhausting repairs,' wrote Winton. 'There was much jubilation on the turbine platform when both turbines got under way and settled down to the required r.p.m. with only the minimum of bangs and alarms. It seemed that for once we were going to have a good day.'[130] After fifteen minutes' fizzing, the watch keepers in the Control Room became concerned about the volume of smoke pouring down the conning tower. The First Lieutenant, a new arrival, was standing at the foot of the tower ladder with painful eyes, struggling to catch his breath. He mistakenly assumed that the smoke was a normal occurrence in *Explorer*, having heard that anything was possible as far as HTP was concerned.

As carbon dioxide poured into the submarine the equipment designed to measure the gas content showed such unprecedented results that all three indicators were reported as defective. 'Looking back now, it does seem that I was extraordinarily slow to take the point which was being hammered in on me from all sides,' reflected Winton. But he, along with the rest of the crew inside the submarine, was suffering from the effects of carbon dioxide poisoning: headaches, dizziness and nausea. Their judgement and reasoning were also impaired. As he recalled:

> I myself felt perfectly fit, although one or two men around me were screwing up their eyes in concentration and complaining of slight headaches. But there seemed no reason to stop the turbines. It cost our department so many back-breaking man hours to maintain them, and we had to overcome so much 'bad joss' to start them, that subconsciously we must all have resisted the idea of stopping the turbines unnecessarily or prematurely.[131]

Explorer's CO was with the Navigating Officer on the bridge, where there was no sign or smell of gas. When he descended into the submarine he found a Control Room that was full of smoke and a number of crew members asking to be relieved. The CO immediately ordered a full stop and evacuated the submarine, and as the crew clambered onto the casing some were very sick. Others lay face down on the casing, their foreheads pressed into their fists. A few just sat, looking bewildered. HMS *Explorer* was decommissioned in June 1963, followed by HMS *Excalibur* in May 1964.

Why did the Royal Navy pursue HTP as a means of propelling its future submarines for so long, and not atomic energy, which we know with hindsight was to prove as revolutionary as the transition from sail to steam? In ideal circumstances it would have followed the US Navy and adequately funded research and experiments into a number of different forms of submarine propulsion. But this was not an option in the immediate post-war period. The country was exhausted from war and financial and material resources were in short supply. Given the choice between atomic energy and HTP propulsion, the choice at the time was clear. The UK's civil nuclear programme and the Attlee Government's decision to embark on a largely secret programme to build a British atomic bomb were the immediate post-war priorities. With fissile material in short supply and the United States unable to exchange atomic information without a change in legislation there was only one option open to the Navy. The German HTP programme was already well advanced and the experimental U-boats that had been produced by the end of the war as well as much of the technology, research and technical expertise in the form of German scientists and engineers were there for the taking. Even after the operational problems with HTP became apparent, the Navy rightly continued with *Explorer* and *Excalibur*, recognizing that HTP propulsion offered a means of quickly obtaining a submarine that could be used by the surface Navy to develop countermeasures against Soviet submarines of similar capability. As a 1949 Admiralty memorandum noted:

> The policy to equip the target submarines E14 [*Explorer*] and E15 [*Excalibur*] with H.T.P. engines was adopted because of the comparatively advanced state of the technique in using this oxidant, the plans made for the supply of their H.T.P. and of the need to get these submarines to sea at the earliest practicable date, to prove British-made H.T.P. machinery, and to study the handling of fast submarines and the anti-submarines involved.[132]

The Navy had decided its future was nuclear. However, events in the Middle East overshadowed any further steps towards resolving the legislative roadblock that prevented increased collaboration with the United States as the Suez Crisis caused the Anglo-American political relationship to deteriorate to new lows. Following Anthony Eden's resignation in the aftermath of the crisis the new Prime Minister, Harold Macmillan, was determined to use his close wartime relationship with President Eisenhower to rebuild the Anglo-American alliance. Eisenhower, too, was anxious to make amends for in effect pulling the plug on the Anglo-French assault on the Suez Canal. He was conscious of the need to repair the Anglo-American relationship in order to counter the growing threat perceived from the Soviet Union. In January 1957, he invited Macmillan to a conference in Bermuda where Eisenhower demonstrated that he was determined to play a more active part in achieving closer Anglo-American nuclear cooperation. On 5 February, the

President, no longer constrained by an impending election, ordered the AEC, Department of Defense and the State Department to implement the June 1956 agreement concerning nuclear propulsion with Britain.

The Royal Navy's Submarine Service had reached the rim of a new nuclear age – of the true submarine with immense endurance and great agility – perhaps the most transformational technological moment since the development of the first submersibles. It was also a very special moment in the history of US–UK relations.

THE SPECIAL RELATIONSHIP

With the legal obstacle to collaboration removed, the British finally had access to the US nuclear-propulsion programme. 'We opened up to them more of our information and facilities than we had ever opened up to any other group,' said Rickover.[133] But relations between Rickover and the British quickly deteriorated as he became concerned that collaboration would distract his own team from their main task of advancing the US Navy's own nuclear programme. He explained:

> We would give them blueprints, complete design information for something, where they could have gone and manufactured the item, if they wished, or to buy it from the United States, but that is not what they wanted. They wanted to come over here and have us come over there, and they would say, 'Why did you make the gasket this dimension?' or 'Why did you make this bolt of such and such a dimension?' So the net results would be, if we acceded to any extent to requests of that kind, we would be tying up our people and we would not be able to do our own work.[134]

Rickover was so angry that during a meeting with British officials on 8 April 1957 he 'made an extraordinary intervention expressing anger at the detailed questions, and at the great interference with U.S. work involved in answering them' and accused the British of 'fouling up his helpful intentions'.[135]

In May 1957, when Rickover visited the UK to inspect the British nuclear-design team at Rolls-Royce, based in the nuclear-research laboratory at Derby Old Hall, he was, according to the Head of Advanced Research at Rolls-Royce, Sir Alex Smith, 'at his obnoxious worst, Corinthian in his execrations of standards in the engineering industry, and all but spat upon the design efforts of the Old Hall team, so critical was he'.[136] When Rickover was shown some of the British reactor designs, his withering comment was 'That's not bad for a high school design.'[137] It took a lot of calming interventions from Rickover's Chief Technologist, Dr Harry Mandil, to prevent the meeting from collapsing in disarray. After one particular morning of strained meetings the Chief Engineer at Rolls-Royce, Adrian Lombard, introduced Rickover to the company Chairman, Lord Hives. 'A lord eh? Chairman eh? And what are you then, a banker or a lawyer or what?' said Rickover. 'What, me? No, no, no, no, no! Me, I'm just a mechanic, just a bloody plumber,' said Hives, with a big broad smile on his face. Rickover was thrown, he had not expected that kind of answer. When Hives explained that he had known Henry Royce himself, had worked for him, and had absorbed from him his passion for engineering excellence Rickover's mood and attitude started to improve. Hives spent the lunch telling a captivated Rickover about Royce's obsession with achieving perfection. He then took Rickover on a personal tour of the works, not in a Rolls-Royce or a Bentley, but in a very modest Hillman.

This seemed to impress the volatile American Admiral. Hives was the type of Chairman who knew where the shop floor was, who could show a guest around, and who would be recognized by and interact with the workforce. According to Smith, when Hives returned after showing Rickover around the Rolls-Royce works the Admiral was 'a changed man, subdued, pleasant, cooperative, uncritical, and no further put-downs or denigrations of British engineering escaped his lips'. Smith felt that he had witnessed something important:

> It is my view that that hour of dialogue between Rickover and Hives, followed by the short tour of the works, changed everything in the British nuclear submarine programme, even though nuclear submarines were never mentioned. Rickover began to realise the great strength of Rolls-Royce, in which he undoubtedly saw a company with a superb tradition of engineering excellence, a company which could reach the very high standards which he saw as essential in nuclear submarine engineering. He made up his mind on that day that, if there were to be established a form of cooperation between his organisation in America and Britain, then Rolls-Royce was the company to which it should be entrusted.[138]

When Rickover returned to the United States, he cancelled arrangements for a second UK technical mission and replanned the visit to suit what he thought would be to UK advantage.

When a combined team of Admiralty, UK Atomic Energy Authority and contractors' representatives, led by Rear Admiral Wilson, arrived in the US in June 1957, Rickover ran the team 'into the ground by arranging a series of visits that covered about 30,000 miles, with a typical visit beginning at 10pm, and finishing at 3.30am to enable [the British team] to get 2 hours sleep before going off on another trip, continuing in that vein for some 3 weeks'. 'We stood the strain quite well,' recalled the senior Admiralty scientist on the tour, Professor Jack Edwards. Rickover quickly developed an 'innate and unfair dislike' of Rear Admiral Wilson – a man who was easily offended – as well as other officials on the British team, whom he repeatedly undermined. Eventually it became obvious that Rickover was attempting to 'expose any weaknesses in our nuclear programme and personnel, and determine whether or not the UK could be trusted to observe both secrecy and advance the state of the art of nuclear propulsion in the UK'.[139] During tours of various facilities associated with the US nuclear programme, US officials and companies continued to protect their interests zealously. When a naval architect from the Admiralty's Royal Corps of Naval Constructors, Jack Daniel, attended a meeting at the Electric Boat Company, 'one of the firm's people giving a brief talk strayed into forbidden material and was literally pulled off the stage by two other men while in mid-sentence!'[140]

Despite these underlying difficulties, the UK team acquired some valuable information. When it returned to the UK, Mountbatten wrote in his monthly First Sea Lord's newsletter, which was distributed throughout the fleet, that:

> Rickover was as good as his word and laid on an extremely good series of presentations at all these places. From what I hear, no questions were barred. The general opinion is that the visit was of great value in corroborating that the lines on which we have been working in the design of our plant have been basically sound and the extent to which our calculations line up with there is very reassuring. Nevertheless, we have learned a great deal from their experience, particularly in installational design and we now need time to collect our thoughts and to take a number of decisions on possible changes which could do much to improve the final ship at the expense of some delay.[141]

By the end of 1957 the British programme had advanced considerably. A site on the northern tip of the Scottish mainland at Dounreay had been selected for the prototype and a full-scale wooden mock-up of the plant was nearing completion at Vickers-Armstrongs Limited Works at Southampton. At Harwell, 160 professional staff were directly employed on R&D, and *Neptune*, a zero energy reactor used to check the design calculations for the Dounreay prototype, was taken critical at Harwell on 7 November 1957.[142]

Meanwhile, the Russians too were demonstrating their appreciation that the future was nuclear. On 9 August 1957, as the Admiralty had long feared, the Soviet Navy launched its own nuclear-powered submarine, the Project 627 'November' class, K-3. Commissioned in July 1958, the K-3 was the first in a series of Soviet nuclear-powered submarines produced beginning in December 1959 and running through to 1964. In 1958, the Soviet government had taken the decision to mass-produce nuclear-propelled submarines and during a 1958 meeting in Moscow, dedicated to the future development of the Soviet Navy, Khrushchev 'spoke in favour of creating about 70 nuclear submarines with ballistic missiles, 60 with anti-ship cruise missiles, and 50 with torpedoes'.[143] Using the 'November' design, the Soviet Navy soon commissioned 'Hotel' class nuclear-powered submarines, equipped with submarine-launched ballistic missiles, as well as 'Echo' class nuclear submarines carrying anti-ship cruise missiles.[144] Together, these first-generation, twin-reactor, double-hulled nuclear submarines were known in the West as the 'HENs' and between 1958 and 1968, just as the Royal Navy was constructing its first nuclear submarines, the Soviet Navy deployed a total of 13 Novembers, 8 Hotels and 34 Echos.[145] These came to be known to the Royal Navy as 'Type I' Soviet nuclear submarines.

Two days before the launch of the first Soviet 'November' class submarine, the Prime Minister, Harold Macmillan, asked the First Lord of the Admiralty, Lord Selkirk, for a short paper on the history of the British nuclear-submarine programme.[146] In a memo Selkirk urged Macmillan (wrongly, as it turned out) to think of the nuclear submarine not as 'an isolated project' but as 'the beginning of as revolutionary a development for ships as was the transition from sail to steam. Within 25 years the conventional warships of to-day will have largely given way to nuclear-propelled warships of much greater speed and endurance. An entirely new type of Navy lies ahead of us.'[147] While Selkirk acknowledged the 'need to avoid extra calls on the Defence Budget, especially in the years ahead', he urged Macmillan to 'remember, however, that the development is, in effect, an investment in the interest of the country as a whole':

> I am convinced that the capital expenditure involved, which will be spread over six years, is a small price to pay for what is offered in return:

the building of the first nuclear-propelled ship to go to sea; the first ex-perience for the shipbuilding industry of the problems of applying nuclear techniques to the field of marine engineering; the achievement of the first stage towards the gradual introduction of nuclear propulsion into the fleet as a whole.[148]

However, by October a 'major controversy' was raging behind the scenes between the proponents of the nuclear submarine and those who supported the development of a nuclear-propelled surface ship. The lat-ter argued that the Admiralty should redirect its limited resources away from the nuclear submarine and concentrate instead on surface propul-sion.[149] One of the biggest sceptics of the nuclear-submarine programme was the Chancellor of the Exchequer, Peter Thorneycroft. On 14 Octo-ber he wrote to Macmillan:

> I think we shall be bound to consider very carefully whether we ought to go on with this project. One important role envisaged for our submarines is, I understand, anti-submarine warfare. Even if we accept that we must have submarines for this purpose in addition to other anti-submarine measures, the nuclear submarine seems to me a very doubtful propos-ition. The cost is prodigious – £14½m. for research and development and £12½m. for production of the fleet submarine. How many of these are we likely to be able to afford? How soon can we get them? When they arrive, will they be already obsolescent as compared with American or Russian versions? We need satisfactory answers to these questions before we can decide that this is a sensible way of deploying our scarce technical and financial resources.[150]

The Admiralty Board, now deeply committed to the *Dreadnought* pro-gramme, was alarmed that it was in danger of cancellation. A worrying number of articles appeared in the press with headlines such as 'Atomic Sub Plan May be Scrapped', 'Check to British A-Ship Plans' and 'Navy to Lose I-Sub'.[151] What was it to do?

OPERATIONS 'RUM TUB' AND 'STRIKEBACK'

In October 1957, the USS *Nautilus* arrived in the UK to take part in Operation 'Rum Tub', an exercise that would mark the first time that the Royal Navy's anti-submarine ships, submarines and aircraft, fitted with the latest equipment, were matched against a nuclear submarine. Mindful of the controversy over the UK nuclear programme, Mountbatten arranged a tour for Duncan Sandys, now Minister of Defence, to impress on him the extraordinary capabilities of the American vessel. Sandys was indeed impressed. 'I believe it has tipped the scale in the minds of our Government for the need to press on with *Dreadnought*,' wrote Mountbatten after the visit. 'The main result is that we now appreciate that we are in the presence of a revolution in Naval warfare; in some ways more far reaching than the transition from sail to steam.'[152] But Mountbatten was also deeply worried that Sandys had been far too impressed with *Nautilus*. He later told the Commander-in-Chief of the Home Fleet that 'there is a very real danger that he [Sandys] may decide that the nuclear-propelled submarine has made our present Navy completely obsolete'.[153]

'Rum Tub' revealed that the Royal Navy's air and surface anti-submarine defences were incapable of protecting a surface force against a high-speed nuclear submarine. *Nautilus*'s mobility and speed were terrifying, typified by one occasion when a helicopter jumped to a position within 500 yards of a green grenade fired by *Nautilus*, which was then already 3500 yards away. In all respects *Nautilus* was able to operate with immunity from any form of enemy action. The CO of *Nautilus* later said that the very considerable Navy and RAF air opposition 'simply never entered his calculations, and as far as he was concerned they might just as well have stayed on the ground'.[154] *Nautilus* also proved to be an effective anti-submarine platform, successfully attacking the US Navy submarine USS *Quilback* and HMS *Auriga*.[155]

On board *Nautilus* throughout the exercise was John Coote, the first non-American to spend more than a few hours at sea in a US nuclear submarine. He witnessed *Nautilus* carry out an attack on the Royal Navy aircraft carrier HMS *Bulwark*, with devastating results:

She then went about 50 miles to the eastward to await the final run of this carrier force back to Londonderry. She heard them coming, positioned herself head on, and she made what you might call the only classical approach of the 24 she carried out in the exercises . . . She turned in and got in two very good undetected attacks on BULWARK and then started following her sitting underneath. One hour later BULWARK – again BULWARK – gained contact and started to shake her off. This time BULWARK used her speed but did not in fact succeed in shaking her off. She knew she was sitting there, and we have a signal sent by BULWARK to everybody saying – 'Will somebody for God's sake come and take this bloody submarine away from under me'. The reaction to that was that at one time she had five escorts within two cables [a nautical unit equal to one tenth of a nautical mile, approximately 185.32 metres] of BULWARK at night trying to pick it out from underneath – without success. The exercise closed with NAUTILUS having the last say with a final salvo at BULWARK.[156]

In order to salvage at least some of the Royal Navy's pride Coote asked the medical team on board HMS *Sea Eagle* to 'shoot some X rays' at the small dosimeter he was made to wear while on board *Nautilus* which measured any unusual exposure to radiation from the submarine's reactor.[157] Coote returned the dosimeter, which indicated that he had been exposed to dangerously high levels of radiation, to *Nautilus*'s sickbay. The surgeon had the last laugh, when a priority signal was read out at the end of a presentation to several hundred of those who had been involved in 'Rum Tub'. It read, 'Analysis of dosimeter worn by Commander Coote reveals unacceptable levels of radiation exposure. Please arrange one gallon sample of his urine to be shipped to USS *Nautilus* for further checks.'[158]

Nautilus's performance during Operation 'Strikeback' only confirmed the Navy's worst fears. With its unlimited power and endurance *Nautilus* conducted simulated attacks against sixteen different ships: two aircraft carriers, one heavy cruiser, two oilers, two auxiliary cargo ships and nine destroyers. On one occasion she detected a carrier group steaming almost directly away from her at 20 knots. To carry out an attack on the group *Nautilus* travelled 219 miles in 10¼ hours at an average speed of 21.5 knots. Sixteen hours later she carried out another attack against a lone destroyer 240 miles away from the previous attack. From the start of the exercise until the conclusion *Nautilus* remained submerged, steaming 3384 miles over a ten-day period at an average speed of 14.4 knots.[159]

The Royal Navy came away from Operations 'Rum Tub' and 'Strikeback' 'deeply impressed and depressed by the realization that we had no counter to NAUTILUS'.[160] Woods later said that he would have got on better without any other submarines in the Norwegian Sea – just the *Nautilus* alone against the Strike Fleet.[161] 'It was a devastating demonstration of her potential,' wrote Coote, 'which changed our thinking forever.'[162] The Commander-in-Chief Home Fleet, Admiral Sir John Eccles, was so disillusioned that unusually he did not ask his staff to draft a report on the operation but wrote his own and sent it to the higher echelons of the Admiralty. In it, Eccles argued that *Nautilus* 'confirmed our worst fears of the threat to our sea communications posed by a nuclear submarine, even when it is armed only with existing torpedoes'.[163] Its performance highlighted the following facts:

(a) Not only has the nuclear submarine complete freedom of action in three dimensions: its ability to manoeuvre at high speed and make rapid changes of direction and depth far exceeds that of conventional submarines. Furthermore its endurance is such that it can continue to carry out this type of manoeuvre virtually indefinitely.

(b) A nuclear submarine can disregard the threat from the air due to the fact that she need not for days on end put anything on the surface. The anti-submarine aircraft's supreme asset of speed therefore remains of no avail until a major research break through arms it with the means of detecting and killing a deep nuclear propelled submarine.

(c) With modern equipment she has as good a picture of what is going on around her on the surface, and an even better picture below the surface, as a surface ship, and due to her unlimited manoeuvrability she can make the best possible use of this information.

(d) In her ability to attack and destroy submarines (conventional or nuclear) and surface ships she is vastly superior to surface ships and conventional submarines in the attack role. During Exercise STRIKE-BACK, NAUTILUS constituted a greater threat to the opposite forces than did all the other 21 Snort fitted submarines put together. Exercise RUM TUB demonstrated that she can command the freedom of the seas wherever she chooses to take the initiative.[164]

Eccles could only conclude that *Nautilus* showed that 'unless science comes to the aid of ships and aircraft with some unexpected technical development, only another nuclear submarine, acting as convoy escort or on barrier patrol, can locate, track and destroy a nuclear-propelled opponent of similar characteristics'.[165] 'It is my firm belief that time is against us' he wrote:

We are now at least six years behind the United States Navy and, possibly, the Red Fleet. Accordingly, we must make fundamental and immediate revision of our priorities, drawing freely from recent transatlantic experience. In particular we should note the energy with which the United States Navy tackled the problem once they appreciated its true significance.

Although this clearly is a field for closer co-operation with the United States Navy (including the establishment of many more exchange and liaison appointments at all levels), we must now emulate the following steps and decisions they have already taken, or are clearly moving towards.[166]

In order to counter the threat, Eccles argued:

(a) That extraordinary methods and personalities with over-riding powers are needed to get a nuclear submarine into the Fleet.

(b) That the highest priority be given to developing a range-determining adjunct to VLF passive sonars and to adapting a long range active sonar to submarine use.

(c) That compatible weapons, with nuclear and conventional explosives, must be developed for use at the longer ranges at which submarines will fight one another.

(d) That a new overall command of all anti-submarine defence forces is needed.

(e) That submarine opinion and experience must in future be adequately represented in the Navy's highest counsels.

(f) That there is no place in the submarine new-construction programme for battery-powered submersibles, even though existing conversions will be invaluable for years to come as missile launchers or as mobile sonobuoys.[167]

The Commander-in-Chief Portsmouth, Admiral Sir Guy Grantham, later told an anti-submarine conference that the 'advent of the nuclear submarine ... with its fast underwater speed and almost unlimited endurance and the characteristics of the true submarine has clearly introduced an entirely new problem in the Anti-Submarine World'.[168] His remarks were fully substantiated by the post-exercise analysis of 'Rum Tub' conducted by the Joint Anti-Submarine School at Londonderry, the first sentence of which read: 'The overall impression gained from the Exercise was that our existing air and surface A/S defences are incapable of protecting any surface force against the high speed, deep, true submarine.'[169]

Woods felt vindicated. 'For the past twelve years,' he wrote in a letter to the Secretary of the Admiralty:

Flag Officer Submarines has consistently advocated by letter and on Admiralty Dockets that the true Submarine, capable of high submerged speed, is the logical successor to the 'submersible' of World Wars I and II and that, to counter the threat posed by its possession by a potential enemy it is essential that we ourselves should develop such a submarine ... Now that practical experience has been obtained, in such a potentially devastating manner, of the offensive power and elusiveness of a nuclear propelled true Submarine, it has become more than ever important to speed up the building of HMS *Dreadnought*, for I am convinced that as the Commander-in-Chief, Home Fleet remarks in his paragraph 13, time is no longer on our side. A radical change in sea warfare is presaged by the exploits of U.S.S. NAUTILUS in Exercises STRIKEBACK and RUMTUB, which cannot be ignored or lightly pushed aside if the future safety of this country is to be preserved.[170]

No longer was it a matter of whether the Royal Navy could afford nuclear attack submarines. The question now was whether it could afford to be without them. The Admiralty Board was clear. 'The advent of such naval vessels must change the strategy and tactics of naval warfare,' it concluded. 'If the Royal Navy did not acquire these submarines it would cease to count as a naval force in world affairs.'[171] Woods urged the Admiralty to think of *Dreadnought* as the first 'and not a lone venture into the atomic field; and that steps should be taken to lay down at least four more of her class as soon as her design is approved'.[172]

While *Nautilus* was in the United Kingdom, the Soviet Union launched Sputnik, the first satellite, into orbit, sending shockwaves throughout the Western world. Within three weeks Macmillan was in Washington, where the Prime Minister and President issued a 'Declaration of Common Purpose' that promised increased interdependence in the defence field.[173] Conscious of the increased Soviet threat, President Eisenhower also announced that he intended to ask Congress to amend the Atomic Energy Act, restoring nuclear cooperation with the British. In the President's State of the Union Address on 9 January 1958, he stressed that it was 'wasteful ... for friendly allies to consume talent and money in solving problems that their friends have already solved'.[174]

THE AGREEMENT

At the end of January 1958, Rickover visited the United Kingdom for his third and most decisive visit. In the final few months of 1957 relations between the British and American nuclear-propulsion programmes had deteriorated to such an extent that 'the British naval attaché delivered an unsigned, undated letter to the Navy Department, stating that they were not getting adequate information from' Rickover and the US Navy.[175] In early January, Elkins informed the Admiralty that when Rickover arrived he was 'likely to suggest to us (a) that we buy a reactor on very favourable terms, which he will arrange & (b) that we go on with our own development causing as little interference as possible with his'.[176] Mountbatten called, but did not attend, a meeting of what he called the Admiralty's 'special nuclear committee' to discuss what to do 'if the UK were pressed to acquire a nuclear submarine or parts of it with Rickover's help'. Those who attended the meeting felt 'bruised by the past few months' fruitless attempts to conduct information exchange in the USA' and they had concluded that 'Rickover was ad-amantly opposed to firm-to-firm business.' They 'completely misjudged Rickover's opinions and his intentions' and concluded that any offer of American assistance should be refused for three main reasons.[177] First, lack of money, principally dollars; second, the belief that American assistance would not accelerate the *Dreadnought* programme; third, the belief that, by accepting the American reactor, the UK would lose all the advantages of having to work out its own indigenous reactor design.

Rickover arrived on 24 January and met with Mountbatten in his office at 10.30 a.m. 'I told him that I understood that this powerful nuclear submarine propulsion committee was going to advise strongly against accepting the steam propulsion plant,' wrote Mountbatten:

> We spent the next hour going through this in very great detail. He abso-lutely convinced me that this would be a mistake. He told me that whereas the nuclear reactor was a reasonably straightforward job as they knew what they were doing, the problem of the heat converter and the steam turbine was one that gave them the most trouble. They had had endless little failures and unexpected difficulties and now at last they had got a homogenous whole in the Skipjack. He now had the authority to offer us the whole propulsion plant and he thought we would be absolutely crazy to cut out the steam propulsion unit.[178]

At 11.30 a.m. Mountbatten left Rickover and met the committee. He listened to their arguments. 'I then said I had an hour with Admiral Rickover and had heard his arguments.'[179] He was convinced that Rick-over was right and the committee was wrong. He warned them that he intended to recommend that the First Lord back the proposals to acquire the entire American propulsion plant. According to Denning Pearson of Rolls-Royce, one of the attendees at the meeting, this announcement 'produced a deathly hush of disapproval'.[180] 'I must say that this caused consternation,' remembered Mountbatten. 'I do not think I have ever seen top class people quite so horrified and so hostile at the attitude I took up.'[181] Mountbatten told the committee that he was 'quite determined to persist in this attitude'. He then called in Rickover and explained the committee's views, that the British should not take the entire propulsion plant, but should instead purchase the nuclear reactor. Rickover remained 'very calm and lucid'. He told the committee that:

> he was out to help as much as possible and if eventually the United Kingdom decided to go ahead with the DREADNOUGHT project on present lines he would assist as much as he could, including firm-to-firm contacts. He would, however, most strongly advise that the easiest and cheapest way in which the Royal Navy could achieve its aim of having a nuclear submarine was for the Admiralty to designate Rolls-Royce Ltd. as their representative with intent that the firm should place a contract with Westinghouse of America for a complete machinery propulsion plant for a submarine. What he had in mind was that the United Kingdom should acquire everything in the way of drawings, spares, training and so on including such facilities from other contractors like Electric Boat Company and also facilities to have representatives of the contractors watching the manufacturing processes of the machinery including the nuclear cores ... He emphasized that every part of the nuclear machinery was tied up with every other part and that you could not change one thing in a nuclear machinery plant without taking account of its effect on others. He suggested that the United Kingdom should decide on the type of submarine to aim for and thought that the best choice would be the U.S.S. 'Skate'. This was a proven type and was a ship of high performance with a speed of over 20 knots – admittedly less than 'Nautilus' and also smaller than 'Nautilus', but nevertheless a very satisfactory ship. He estimated that about 18 months to two years would be taken in building the machinery unit, but the United Kingdom could go on with the hull in the meantime.[182]

On 5 February, Selkirk informed the Cabinet's Defence Committee that it was 'desirable that we should respond to this overture as rapidly as possible, since it should ensure that the Royal Navy would be provided with a nuclear submarine at a considerably earlier date than could be envisaged if we had to rely solely on our own efforts'. The Committee approved the purchase as it would 'provide valuable evidence of the practical application of the policy of inter-dependence which the Prime Minister and President Eisenhower had recently endorsed'.[183]

During Rickover's discussions with the Admiralty he rightly 'emphasised that he was not making a proposal from America to England, he was speaking for himself and it would be for the English Government to put forward a request to the United States Government'. Any sale also required further amendments to the Atomic Energy Act. Rickover promised to 'do his best to ensure that the answer of the American Government was favourable to such a request' and Bills incorporating the proposed amendments were introduced to both the House and Senate on 27 January 1958.[184] A month later, Rickover testified before a sub-committee of the US Joint Committee on Atomic Energy. He explained that:

> About a year ago agreement was reached between the United Kingdom and the United States committing exchange of submarine nuclear propulsion information between the two Governments; that agreement has been consummated. The British had people over here studying what we are doing, we gave them the necessary design information. They took this information and went back to England, but what they then started doing was not fully conducive to the conservation of scientific and engineering talent of the two countries. They were starting to go into minute detail and learn how to design on their own, every nut, every gasket, every bit of equipment in order to make a complete British design out of the American design. To supply them with all the necessary details would be an endless, time consuming proposition and would require very large expenditure of effort on our part and would seriously interfere with our ability to do our work. You know how it is if one Company makes a machine and another wants to make one exactly like it but use their own standards and techniques. It is an almost impossible situation.

Rickover then explained why he was standing before the Committee:

> In a meeting I had with British Admiralty officials it was agreed that to conserve their money, it would be best that the British designate a commercial company to act for their Government. The British company would enter into a commercial relationship with an American Company to buy a submarine nuclear propulsion plant just as they would buy

anything else in this country. In this way they would keep completely out of the transaction; it would be a completely commercial transaction.

He then elaborated on the advantages of a commercial as opposed to government agreement:

If our Government were to furnish the British a reactor and something went wrong it would invariably end up as our fault, but if they buy the reactor on a commercial basis from us, they have their own inspectors, they can decide whether or not to accept the various items but once they accept them through their own inspection, it is their responsibility and we, as a Government, have no further responsibility.

It also ensured that the British kept out of Rickover's nuclear programme:

The gist of it is they would like to station people all over – in my office, in every laboratory – and be right there all the time getting every question answered. That means a great deal of extra work for our people, and we cannot do that and keep on with our own job. This was why I recommended to them and they agreed with the idea of procuring by commercial arrangement a complete plant, on the condition they keep out of our laboratories, and let the American firm . . . handle the matter for them.[185]

Rickover argued that it was 'better for the strength of both our countries that the United States naval nuclear programme should not be interfered with'. He also argued that 'There are probably no two countries in the world that have as similar a cultural and legal and political background and are the same kind of people. I think they are probably the most reliable outpost in Europe as far as we are concerned, and we are in a sense helping ourselves, too.'[186] But Rickover was equally clear that the term 'exchange', which was often used to describe the proposal, was inaccurate. 'They do not have anything to give us,' he said, 'if I confine myself to nuclear power, there is at present no contribution they have that is evident at this time to our program.'[187]

But there was one area where the British did have something worthwhile to exchange, the British civil nuclear programme, in particular the gas-cooled plant at Calder Hall, the first nuclear reactor in the world to deliver commercial quantities of electricity. Some members of the Joint Committee were very interested in this British natural-uranium-fuelled gas-cooled reactor technology. In January 1957, Congress had requested information from the British and although a limited amount was provided, a formal proposal to exchange information on US submarine reactors for British data on gas-cooled reactors was rejected by Macmillan's Cabinet, which feared that the United States would pass the information on to private companies who would then use the design and compete with the UK commercially in the gas-cooled reactor field.[188] This angered the Joint Committee. 'I think it's wrong that when we give them the secrets of the *Nautilus* they hold back from us,' declared the Democratic Senator for New Mexico, Senator Clinton P. Anderson.[189] Hard bargaining followed. Rickover told Elkins that 'Congress was sore on this point with the British' and that if 'the U.K. would clear information to the U.S.A. on gas cooled reactors, then the U.S.A. would give the U.K. practically any information they wanted'. The response of the head of the British Joint Services Mission in Washington, Vice-Admiral Sir Robert Elkins, was simple. 'I told him the British were sore with the U.S.A. on the question of the McMahon Act where they had sucked our brains and then closed down.'[190]

The Eisenhower administration eventually abandoned attempts to link the two. With the prospect of an agreement tantalizingly close, the British continued to do all they could to ensure the legislation passed successfully. Mountbatten even claimed that he stopped the demolition of a house in London in which Eisenhower had stayed in the pre-'Overlord' days in 1944. 'Ike was furious when I told him,' said Mountbatten, 'but I prevented it and so we got the propulsion plant.'[191] Privately some officials were worried that Rickover's 'extreme unpopularity throughout the U.S. Navy' would jeopardize any agreement. 'I have to be careful that none of it sticks to me,' wrote Elkins.[192] 'The fact is that Rickover has many enemies and, while they cannot get at him over the U.S. programme, they could quite easily stir into activity the latent resentment which exists in Congress against the British.'[193] For his part, Rickover continued to do all that he could to ensure that Congress passed the legislation. When it looked as if the negotiations were going to take longer than expected he put pressure on Congress. He even suggested 'that in order to obtain a nuclear submarine for anti submarine training at the earliest possible moment [the UK] might be able to borrow one the year after next, when more would be available'.[194]

On 30 June 1958, Congress finally passed the amendments to the Atomic Energy Act of 1954. The day after the act became law on 2 July, the US and UK signed a new bilateral agreement for 'Cooperation on the Uses of Atomic Energy for Mutual Defense Purposes'. The new agreement provided for the unprecedented exchange of a wide range of nuclear secrets and established a framework for an Anglo-American nuclear partnership that continues to this day.[195] Under article 2a of the agreement, the UK and US were permitted to exchange information for 'the development of defense plans, the training of personnel in the employment of, and defense against, atomic weapons and other military applications of atomic energy, the evaluation of the capabilities of potential enemies in the employment of atomic weapons and other military applications of atomic energy, the development of delivery systems compatible with the atomic weapons which they carry, and research, development, and design of military reactors to the extent and by such means as may be agreed'. Article 13 permitted Westinghouse to sell to the British government a complete nuclear-submarine propulsion plant and spare parts, together with 'information relating to the safety features and information for the design, manufacture and operation of the reactor'.[196]

THE *DREADNOUGHT* PROGRAMME

Advanced design work on *Dreadnought* could now start. During the meeting on 24 January, as we have seen, Rickover suggested 'that the United Kingdom should decide on the type of submarine to aim for and thought that the best choice would be the U.S.S. "Skate" '.[197] Commissioned in December 1957, the USS *Skate* was smaller and slower than *Nautilus* and was powered by an S3W (Submarine, 3rd Generation, Westinghouse) nuclear reactor. The existing *Dreadnought* design was bigger and would have required significant modifications in order to accommodate the twin-screw *Skate* machinery. It would also have led to a relatively small submarine compared to the original *Dreadnought* and if the Admiralty accepted it, 'nothing done by us up to now would be directly applicable to the resulting submarine, and the resulting submarine would not be directly applicable to our future programme as now envisaged'.[198] The alternative was the reactor and machinery used in the US Navy's then newest submarine, USS *Skipjack*, the S5W (Submarine, 5th Generation, Westinghouse). The *Skipjack* was bigger and faster, it had considerably greater endurance at sea under operational conditions, and it was large enough – which the *Skate* was not – to fit the new advanced sonar then under development. The US Navy also intended to use it to power its new Polaris-missile-carrying submarines (of which we shall see more in the next chapter).

Due to the quantity of information which had been received from the Americans before the decision to purchase a plant from the US Navy, the 'Dreadnought' design was very similar to the *Skipjack* design and it was possible to instal the machinery in the 'Dreadnought' hull with the minimum of reworking, resulting in a submarine with high speed, better manoeuvrability, and three times the operational endurance of the *Skate*.[199] On 13 February 1958, the Admiralty Board settled on the 'Skipjack' class machinery.[200] There was some discussion about whether it was appropriate to name the submarine *Dreadnought* now that it embodied an American propulsion unit but the Board decided that the name should not be transferred to a later vessel, 'if only because of the criticism which would arise on account of the inordinate time which would then apparently elapse between the inception and completion of the submarine bearing the name "Dreadnought." '[201]

Rolls-Royce was named as the British agent and Westinghouse as the US agent. Vickers-Armstrongs (Shipbuilders) was named as the shipbuilders and Vickers-Armstrongs (Engineers) as the machinery installation subcontractors. The shipbuilder responsible for building many of the US Navy's nuclear submarines, Electric Boat Company, was named as the subcontractors to Westinghouse and in January 1959 the Naval Section at Harwell was disbanded and Rolls-Royce Ltd, Vickers-Armstrongs Ltd, and Foster Wheeler Ltd joined together to form a new private company, Rolls-Royce and Associates Ltd. The final commercial agreement between Westinghouse and Rolls-Royce was in two parts – first, a supply contract covering the sale of machinery, the training of British personnel and certain maintenance services; and, second, a licensing agreement which permitted Rolls-Royce to manufacture the reactor equipment and fuel elements for nuclear reactors in the United Kingdom.

The British hoped to continue exchanging design and manufacturing information with the Americans for at least ten years but the US Navy refused and insisted that future cooperation and information about reactors could only take place if a new agreement was negotiated.[202] 'It is very disappointing,' Macmillan wrote to Selkirk.[203] Selkirk agreed, but argued that 'enough remains to make the agreement worth while'.[204] Later attempts to work around these restrictions caused further problems with the endlessly volatile Rickover, who continued to insist 'that in future there could be no informal discussion on British needs and that the British could not expect him to help them in breaking the "law" '.[205]

Sourcing fissile material was also a problem because of the continuing demands of the UK nuclear-weapons programme. In November 1958, the Chairman of the UK Atomic Energy Authority, Sir Edwin Plowden, informed the Cabinet's Defence Committee that 'information obtained from the United States about the design of nuclear weapons would greatly extend our capacity to manufacture nuclear warheads from our existing supplies of fissile material'. *Dreadnought* required around 50 kilograms of U-235, which would cost £1,000,000 if purchased from UK sources, compared to around £300,000 if purchased from US sources.[206] In November 1958, the Defence Committee authorized the Atomic Energy Authority to reopen discussions with the US AEC about whether there was any possibility that, as an alternative to the construction of an expensive new gaseous diffusion plant in the UK to produce additional fissile material, the US would be prepared to supply the UK with adequate quantities of U-235 on acceptable terms. The Americans agreed. In March 1959, Plowden led a delegation to Washington in order to remove the remaining difficulties and an amendment to the 1958 Mutual Defence Agreement was signed in May, and came into force in July, which significantly widened the range of cooperation to include the sale of non-nuclear components and the exchange of British plutonium for US uranium-235, tritium and lithium.[207]

One of the many questions the Admiralty had to answer following the American purchase was whether the Royal Navy's post-*Dreadnought* nuclear submarines should depend on US technology and machinery, or should the UK follow its own development path, taking on board information gleaned from the American plant. When the possibility of purchasing a US reactor was first mooted in 1956, officials in the Admiralty insisted that 'although any U.S. information which we obtain may well result in a slight reduction of our own work, it cannot replace it'.[208] One of the reasons Rickover had suggested selling the British a complete plant was because it would provide the Royal Navy with, as Mountbatten explained, 'practical knowledge which would greatly facilitate our own research work in this field so that when we came to build the second and subsequent submarines we could stand on our own feet'.[209] Rickover 'did not want to see us abandon our own project: indeed he hoped that his organisation might in due course benefit from us'.[210]

Independent work at Harwell and construction of the first British submarine machinery and the Dounreay containment hull at Barrow were effectively halted following the American purchase. The decision to continue was finally taken in early April 1958, following a review that concluded that any attempt to Anglicize the 'Skipjack' class plant would involve a significant amount of redesign work. Continuing also avoided breaking up the teams which might be required if the agreement with the Americans subsequently failed to materialize.[211] Dounreay was also needed to proof test the first all-British plant, subsequent development and training. Further development proceeded on the basis of the original *Dreadnought* machinery, except for the reactor core and control mechanisms and any other items that could be substituted for more advanced American designs.[212]

The Admiralty accepted that some substantial organizational changes were required to coordinate its efforts with the Atomic Energy Authority and the many contractors involved in the *Dreadnought* programme. The organization inside the Ships Department at Bath which had been responsible for the programme was altered to enable the 'Dreadnought Project' – to which the Admiralty Board had now accorded high priority – to be driven through using a compact and dedicated in-house team. Authority was concentrated (perhaps following Rickover's exhortation) in the hands of a single authority known as the Technical Chief Executive, Dreadnought Project. A brilliant naval constructor, Rowland Baker, was appointed as Technical Chief Executive and given responsibility for the design and construction as well as controlling the cost of *Dreadnought*, a task that left Baker with mixed feelings that 'alternated between elation at the prospect and terror':

> Terror because just when they were about to sign a government to government agreement, I realised that not one of those who would have to be on my staff . . . approved of me in any way, or of the scheme. They had pottered about for several years, and now had not only a solution, but a chieftain imposed on them. Of course they all hated it . . . My terror derived in part from the conviction that even if we had a Rickover reactor, all and sundry would want to 'improve' it and feed in their national ideas throughout the ship.[213]

Baker quickly established his authority and brought together the many departments and contractors involved in the programme. He concentrated all nuclear-submarine activity at the Admiralty's design departments located in Foxhill, Bath, a change that greatly enhanced the efficiency of the Dreadnought Project Team.

The *Dreadnought* design was the responsibility of the Admiralty's in-house naval architects, the Royal Corps of Naval Constructors

(RCNC). Before the decision to purchase the S5W plant the arrangement and structural design of *Dreadnought* was largely complete. After the purchase, the design was altered to match the hull particulars for American plant. The Admiralty Board was initially critical of the RCNC's first revised design concept as it was noticeably heavier than the USS *Skipjack*. The naval architects had attempted to match the new design to the original *Dreadnought* staff requirements. However, the Board was 'not satisfied that all the superior requirements which had been incorporated justified the extra weight involved' and directed that *Dreadnought*'s staff requirements should be reviewed. Revised requirements, which compromised a number of *Dreadnought*'s original features, were issued in September 1959. Maximum speed was reduced from 25 knots to 23 knots. The use of resilient mountings for the main turbines and gearing was scaled back because it was impossible to guarantee any silent submerged speed above 4½ knots with the US machinery. The target diving depth of 750 feet was also reduced to 700 feet, the depth at which the 'Skipjack' machinery was designed to operate. Endurance was also reduced as the 'Skipjack' air conditioning, refrigeration and stores arrangements only allowed for a maximum endurance of 75 days, not the original Royal Navy requirement of 90 days. The weapons fit was also reduced from 36 torpedo reloads (i.e. in addition to those in the tubes) to just 31.[214]

In all previous submarine designs there had always been a conflict between the different characteristics required for operating on the surface and when submerged. The *Dreadnought* design was optimized for submerged operations. The hull was based on the streamlined teardrop design of USS *Albacore*. *Dreadnought*'s stern was almost identical to that of a US 'Skipjack' class, while the bow was designed by the naval architect Louis Rydill and incorporated the best of the traditional RCNC design procedures, tried and tested in previous ships and submarines.[215] The bow included many new hull systems and equipment for ballasting, diving control, ventilation and atmosphere content, as well as a new water ram torpedo discharge system. The bridge fin was further aft than in US submarines, partly to take account of the layout of the front end, but also to reduce the roll induced when the submarine was manoeuvring at speed. *Dreadnought*'s designers also positioned the forward hydroplanes near the bow rather than on the fin as per US Navy practice, as this improved handling at low speeds, particularly when the submarine was operating at periscope depth, but at the expense of interference with sonar performance.[216]

The US Navy advised the Royal Navy to aim for a three-year build-ing programme, the time it had taken to construct USS *Skipjack*. The Admiralty felt this was unrealistic due to the more complicated circum-stances of *Dreadnought*'s design, a view that was quickly reinforced by delays in drawing up and signing the contracts between the several US and UK companies involved in the programme. The Admiralty recog-nized early on that having solved the major technical difficulty on the nuclear side, construction was now likely to cause significant prob-lems. As one Admiralty paper explained:

> Submarine construction itself, overall, now presents a very likely chance of failure. For this submarine is, as a submarine, a very much more for-midable project than any earlier vessel and because we are now getting all the machinery from the U.S., not only are all the design parameters no longer all in our hands but when we come to the actual installation in the ship we shall not have the advantage, that we had promised ourselves formerly, of merely copying what we had at Dounreay. So therefore up to the time of the Rickover deal it was wise to stress mostly the nuclear end, now that this is solved it is essential to stress the shipbuilding end.[217]

The shipbuilder, Vickers-Armstrongs, had an illustrious record of submarine construction, as well as a long tradition, dating from the turn of the century, of working with the American shipyard responsible for building the US Navy's nuclear submarines, the Electric Boat Div-ision of General Dynamics Corporation, located in New London, Connecticut.

Fabrication of what was then the largest pressure vessel con-structed in the UK started in early 1959 and involved an estimated 635,000 man-hours. *Dreadnought*'s hull was constructed out of special steel capable of withstanding enormous pressures in the depths of the oceans, known as QT35 (QT standing for 'quenched and tempered'). Vickers managed to construct *Dreadnought* with QT35 'without too many problems' although, as we shall see later, there were worrying problems with cracking.[218] Advice and assistance from Electric Boat were invaluable, especially when problems arose.[219] 'So many new materials and new methods of construction were introduced for the first time that technique development was a major problem,' wrote Gregg Mott, at the time the head planner of the *Dreadnought* build at Vickers. 'Most of the difficulties arose through failure to meet stand-ards of acceptance which were much higher than those acceptable for merchant or surface naval ships or even for previous submarines.'[220]

Aside from construction problems the *Dreadnought* programme also suffered from a serious security breach involving the new sonar designed to detect and track Soviet submarines. In April 1957, Her Majesty's Underwater Detection Establishment (HMUDE) at Portland started to design a new sonar system for *Dreadnought*. Known as the Type 2001 in recognition of its complexity compared to the Type 187 fitted to the 'Porpoise' class, the new sonar was designed against the background of improved understanding about the acoustics of the world's oceans and used advanced digital multi-beam sonar (DIMUS) techniques.[221] It consisted of a 40-foot fixed sonar array arranged like a horseshoe around the upper part of the bow of the submarine, 6 feet high, tilted backwards 20 degrees from the vertical to fit in with the streamlining of the hull, and was designed to give continuous sonar coverage over an arc of 240 degrees, both active and passive. When operating in active mode, and depending on *Dreadnought*'s speed and aspect, it was capable of detecting targets at a maximum range of 25 miles at 5 knots to 5 miles at 20 knots, the decrease in range with increase in speed caused by self-noise from the submarine. When operating in the passive mode, the Type 2001 could detect a snorting submarine moving at 8 knots at distances of up to 30 nautical miles, reduced to 17 nm and 6 nm at speeds of 10 knots and 20 knots respectively.[222]

Given the advanced nature of the new sonar, HMUDE, at Portland, was one of the Soviet Union's prime intelligence targets. In 1952, an ex-Royal Navy Master at Arms, Harry Houghton, offered his services to the Polish intelligence service while serving as an Admiralty civil servant at the British embassy in Warsaw. Houghton passed copies of numerous top-secret naval documents to the Poles, who in turn shared them with the KGB. When Houghton was posted back to Britain he started working at HMUDE as a clerk in the personnel department, before being transferred to the unit responsible for the maintenance of the small number of vessels assigned to the establishment, where he eventually became responsible for all the papers and correspondence that passed through the unit. Houghton was finally handed over to 'Gordon Lonsdale', an illegal (i.e. not acting under diplomatic cover) KGB agent named Konon Trofimovich Molody, who pressed Houghton for more detailed technical information. At the time, Houghton's marriage had disintegrated and he had entered into a relationship with a clerk at HMUDE, Ethel Gee, who worked in the stores section until 1955, when she was posted to the HMUDE Drawing Records Office. 'Lonsdale' and Houghton were able to convince Gee to pass them classified information and she eventually became a conduit for channelling information from HMUDE to the KGB.

In April 1960, the CIA alerted MI5 that it had learned from a Polish intelligence officer, Michal Goleniewski, that at some point in 1951 Polish intelligence had recruited an agent in the British naval attaché's office in Warsaw. MI5 quickly identified Houghton as the prime suspect and placed him under surveillance. By July 1960, he had led MI5 to 'Lonsdale', who in turn led MI5 to an antiquarian bookseller and his wife, Peter and Helen Kroger, American KGB illegal agents acting as Lonsdale's radio operators and support team and responsible for transmitting the information to Moscow.[223] The Portland Spy Ring, as the group was later known, was convicted in March 1961. Houghton and Gee received 15 years each, the Krogers 20 years and Lonsdale 25.[224] The Krogers were eventually exchanged for the British teacher Gerald Brooke, who had been arrested in the Soviet Union for smuggling anti-Soviet leaflets, in 1969, while 'Lonsdale' was exchanged in 1966 for Greville Wynne, an Englishman accused of spying in Russia. Lonsdale came to be regarded as a hero in the Soviet Union.

It is still difficult to gain a clear insight into the damage caused by the Portland Spy Ring. A 1962 brief on the projected Soviet submarine threat in the 1970s, prepared two years after the ring was closed down, for the Director of Under Surface Weapons, concluded that 'The Soviets are behind the West in the development of long-range active and passive sonars but it is probable that by 1970 they will have sets of similar performance in use today by the West such as the 2001 and 186.'[225] Embarrassingly, the British were forced to explain to the Americans that the secrets of DIMUS had been given to the Soviets before the Royal Navy had even put the new technology to sea in its new nuclear submarine.[226]

By 1960 the Admiralty was beginning to understand how to work effectively with Rickover. They recognized that he had little time for social chitchat and that he would, as a brief for Duncan Sandys explained, 'only cooperate with people that he likes . . . if he takes a dislike to a particular individual, it is a complete waste of time to use that individual to deal with him'.[227] Rickover would only unwind, slightly, when in the company of certain people: the Queen, Harold Macmillan, Lord Mountbatten and later Solly Zuckerman, the Chief Scientific Advisor at the Ministry of Defence. With everyone else he was both brusque and demanding. Jack Daniel encountered him once: 'he grunted to me to get out of the way when he and I met in a doorway'.[228] But, as the brief to Sandys also explained, 'in spite of his very peculiar and temperamental behaviour and in spite of the fact that he is undoubtedly a megalomaniac who considers himself sent by the Deity into this world to see that America remains supreme, Rickover is very anxious indeed to help the British and can be relied upon absolutely to carry out any firm promise that he makes'. But, the brief also warned: 'it is clear that his desire to help the British does not extend to tolerating any interference with his own programme'.[229]

Rickover was still unimpressed with how the Admiralty was managing the 'Dreadnought Project'. While showing Mountbatten around USS *Skipjack* in October 1958 – a 'fantastic peep into the future', Mountbatten later said – Rickover was brutally honest: 'Admiral, I think your British set-up is lousy . . . What you want to run a show like this is a real son-of-a-bitch.' Mountbatten's reply delighted Rickover: 'That is where you Americans have the edge on us, you have the only real son-of-a-bitch in the business!'[230] Rickover refused to acknowledge Rowland Baker, the new Technical Chief Executive of the Dreadnought Project Team. '[H]e would not see me,' recalled Baker. 'The second time we met (in the Admiralty Board Room) he again ignored me.' Whenever Rickover visited the UK to inspect progress Baker took to recording detailed accounts of what took place. 'I next saw the Admiral late on Wednesday evening at Duffield Bank House,' he wrote of a February 1959 visit:

> He was supposed to arrive for dinner but was late (on purpose?) There were present beside me Messrs. Pearson, Rubbra, Barman, Fawn, Bellings, Pepper and Marsh of Rolls Royce and Redshaw and Storey of Vickers. The Admiral felt the party was too big and was a bit disgruntled. He tried to ring Washington and could not get Lascora, the man he wanted. This made him worse. He did not approve of the billiards. Mandil did try one shot.[231]

On another occasion, Baker was with Rickover during a visit to Derby and Barrow. On boarding the train in London both men entered a compartment in which two window seats had been reserved for them. Rickover entered first and sat down. Baker went and sat opposite him. Rickover moved to the corridor, as far as it was possible to be from Baker, all in total silence. Shortly before the train departed a man entered the compartment and sat opposite Rickover. After some time he caught Rickover's eye and said: 'Pardon me, sir, but aren't you Admiral Hyman G. Rickover, the father of the nuclear submarine?' Rickover asked how it was that the man recognized him, to which he replied that he had seen his picture on the cover of *Time* magazine and had never forgotten the story of Rickover's achievement. According to Baker, Rickover was clearly pleased and almost polite for the rest of the day, even nearly saying sorry when he trod on Baker's foot getting out of the train at Derby.[232] But this was a rare display of civility.

By 1961, relations between Rickover and the *Dreadnought* team had once again become strained and the Chief of the British Naval Staff in Washington was 'worried . . . particularly about his very bad relationship at present with Admiral Rickover'.[233] On another of his visits to the UK, Rickover and Baker stopped off at the then Royal Navy Air Station Lossiemouth while en route to Dounreay, where the Captain of the Air Station, Michael (later Vice Admiral Sir Michael) Fell and his wife, Joan, decided to host a dinner party. A young Gunnery Officer, Peter Kimm, found himself sitting next to Rickover at the dinner table. At first the conversation was low key, with Rickover contributing little or nothing to it. Then, about halfway through the first course of shrimp cocktail Rickover suddenly came alive and said to Kimm: 'Why are you here? Why aren't you home studying?' The table went quiet. 'I've been invited to meet you, sir,' replied Kimm. Rickover took another spoonful of shrimp cocktail, wiped his mouth, looked again at Kimm and said, deliberately and slowly: 'I've got you figured out. You are an effete young English lord. You are interested in nothing but your fitness reports. I wouldn't have you in my program if you were the last man alive.' Everyone around the table was speechless as Rickover returned to his cocktail. Captain Fell eventually broke the silence and from the other end of the table quietly told Rickover that 'you really have got Lieutenant Commander Kimm wrong. He is in fact a very good officer.' Rickover was unimpressed. 'If you hadn't got your best here tonight, you'd be a damn fool,' he said. 'I stick by what I said. He is everything I dislike in the English. He is a socialite. He is an effete young English lord. He is interested in nothing but his fitness reports.'[234]

Another silence descended on the table. Then Rickover once again turned to Kimm and said: 'Alright, if I'm wrong, prove it. I require all officers in my program to work forty hours a week formal technical studies in addition to their normal duties. Would you do that?' No, replied Kimm. 'Ah!' said Rickover with great satisfaction. Kimm argued that he was not ashamed of his work, that he already led a full and balanced life, that he had a wife and children and that a man needed to re-create himself. 'You can divorce your wife any day,' said

Rickover before tearing apart every part of what Kimm had tried to say. 'In short,' Rickover said, 'you're no more than a traitor.' That was too much for Kimm, who stood up, white as a sheet, trembling with anger and said: 'Admiral, *nobody* says that to me.' 'Aw don't be so corny,' replied Rickover. Another silence descended on that table. 'Admiral,' Baker said gently, 'you really ought not to say that sort of thing to a Brit.' Rickover backed down, 'Alright. I take back "traitor".' The rest of the dinner went on, with Rickover repeatedly returning to press Kimm. The following morning Kimm called on Michael Fell to apologize for his outburst. 'Peter,' Fell said; 'If you had struck him, I would have defended you at your Court Martial; and if he'd been a British admiral, and you hadn't hit him, I would have fired you!'[235] Such was working with Rickover.

When Mountbatten left the Admiralty to become Chief of Defence Staff in 1959, he ensured that his successor as First Sea Lord, Admiral Sir Charles Lambe, was aware of just how important Rickover was:

> One very important point is that stormy petrel of the American Navy, Vice-Admiral Hyman G Rickover, arrives next Sunday on a visit to our nuclear propulsion activities. As we virtually owe him the ability to complete DREADNOUGHT two or three years ahead of time with a saving of millions of pounds on R&D we must all keep in with him. In his unbelievable, egotistical way, he has always regarded the First Sea Lord as his opposite number, and went so far as to tell Geoffrey Thistleton-Smith [Vice-Admiral in the British Joint Services Mission, Washington, 1958–1960] that he had begun to doubt whether the British Government were taking nuclear propulsion sufficiently seriously, since they were allowing me to leave the Admiralty before the project was properly through![236]

New appointments to the Admiralty, such as Lord Carrington, as First Lord in 1961, were simply told that 'Rickover mattered'.

As *Dreadnought* neared completion there was a considerable amount of interest among submariners. When Sandy Woodward passed his Perisher in late 1960, Teacher politely informed him that 'you've done quite well. So I'm giving you first choice of the available appointments in Command. Where would you like to go?' Woodward told him, deadpan: 'I'd rather like to drive *Dreadnought*.'[237] He was disappointed and would have to wait a few more years for his own nuclear submarine. The Admiralty had already selected *Dreadnought*'s Commanding Officer. Born in September 1924, Peter Samborne was one of the early pioneers who had taken HMS *Tabard* into the Barents Sea in 1956 on one of the Royal Navy's early intelligence-gathering operations. Samborne's appointment as *Dreadnought*'s captain was the source of some controversy with Rickover. When the senior officers of *Dreadnought* were selected in 1959 they were all summoned to meet Rickover, along with Mountbatten. But the meeting was cancelled, twice, before being reinstated because Rickover insisted on having the final say over the appointments, just as he had done with every US Navy officer selected to serve on board nuclear submarines. 'On the whole it made the prospect of being caught red handed by the KGB with a roll of film of the Kola Inlet in one's briefcase seem greatly to be preferred,' noted Coote.[238] Mountbatten eventually persuaded Rickover to change his mind. But when the officers were finally brought into his presence 'He met us with little grace,' recalled Peter Hammersley, *Dreadnought*'s first Senior Mechanical Engineer.[239]

Alongside the nuclear plant, the US also provided the UK with practical experience and training. While the exchange of complex technical information was difficult enough, the training of personnel – shipyard personnel, shore-based personnel, shipboard personnel – was, according to Admiral Burke, the US Chief of Naval Operations, 'nearly impossible to solve ... they needed the experience which the United States people had accumulated without going through the expense and the false starts which we experienced. On our part the training of additional people put a severe burden not only on our school system and our shipboard training facilities, but also on our normal operating procedures.' Samborne spent nine months on board USS *Skipjack* learning about the S5W and several of *Dreadnought*'s officers and ratings were given about six months' sea training in US nuclear submarines.[240] The US Navy offered to train a total of around fifty officers and men in sea-going nuclear submarines and at dockyards. Officers and ratings were sent to the United States for practical instruction and experience in handling a nuclear-power plant and its associated machinery. Officers then completed theoretical instruction at Greenwich, where a Chair of Nuclear Science and Technology was created.[241] Ratings were trained at HMS *Collingwood*, the Naval Electrical School, near Fareham, and at HMS *Sultan*, the Mechanical Training Establishment at Gosport.

HMS *Dreadnought* was launched on Trafalgar Day, 21 October 1960, by the Queen to considerable pomp and ceremony. At the post-launch lunch laid on for the Royal Party, Lord Carrington explained that 'This is a very special occasion for the Royal Navy, which I have the honour to represent, for today we have seen the birth not just of a new ship but of a new era . . .'[242] The Queen also praised *Dreadnought* as 'a fine achievement on the part of designers and ship-builders and a great step in the maritime history of our island' and paid tribute to the US for providing the nuclear reactor and machinery that 'had been ably pioneered under Admiral Rickover'.[243] Rickover, who was also present, understood that the British were rightfully proud of their submarine. 'We should be careful not to look as if we were trying to grab any glory from them. She's their ship, let them have full credit,' he said.[244]

Although most of *Dreadnought*'s internal structure had been completed by the time the submarine was launched, relatively little installation of machinery and equipment had been carried out. Fitting out was completed in a floating dock, where the submarine's hull was opened up and the remaining items were installed. The fitting-out process, compressed into the two years between launch and start of tests and trials, was a considerable challenge as 'in every compartment of the submarine there were major piping systems and main cable runs to be installed in what were, despite the large size of *Dreadnought*, often very congested conditions, with the different trades of the workforce competing for access'.[245] Vickers worked around the clock to complete *Dreadnought* on time. 'It seemed that whenever I visited, morning, noon or night, weekday or weekend, George Standen, the manager responsible for the reactor and machinery, would be there. No matter when, it seemed, George would emerge from a hatch looking pale and drawn but always in complete control of the situation,' recalled Jack Daniel.[246] Under Standen's supervision *Dreadnought*'s nuclear core was installed on 8 July 1962. It was taken critical in November 1962, the power gradually increased, step by step, as a special Reactor Test Group carried out safety checks, including radiation surveys to measure the radioactivity levels inside and outside the submarine.

Sea trials took place in mid-December 1962 in the very stormy Irish Sea. Although designed and optimized for submerged propulsion no one knew for certain how fast *Dreadnought* would go on the surface. 'She went like a racehorse,' wrote Daniel, 'through her resistance hump and away, swept along on a great wave at about 20 knots with her bow casing dry and her stern well down. I was on the bridge when it happened and could hardly believe my eyes when suddenly the bow came up and the forward deck was dry.'[247]

Dreadnought's crew embraced the Anglo-American nature of their submarine, placing a sign on the bulkhead between the fore and aft sections of the submarine that read 'Checkpoint Charlie – You are now entering the American Sector'. A chief stoker, a veteran of seventeen years' service in conventional submarines, said proudly: 'Wherever you go on this boat there's a feeling of power. But it's the spaciousness that means so much ... when my wife came to look round she said she'd never seen anything like it for comfort: told me I was living in the lap of luxury – and I had to agree.' Samborne was more realistic. 'It isn't cramped, but it isn't a hotel – we are a warship after all.'[248] For the first time a water-distilling plant provided virtually unlimited supplies of fresh water. There were washing machines, improved toilet facilities, air conditioning, a galley equipped with the most modern equipment, including bread makers and pressure cookers. There were three messes – the officers' wardroom and separate cafeteria-style dining halls for senior and junior ratings. The standard of accommodation was of a quality far removed from any previous British submarine.

THE 'VALIANT' CLASS

Thus the Royal Navy had its first nuclear submarine, an entirely new weapon of war. But it was still years behind the United States and the Soviet Union. In August 1960, the Royal Navy placed an order for its second nuclear submarine with Vickers. Its nuclear plant, known as the Pressurized Water Reactor 1, PWR1, or the 'Son of S5W' as it was sometimes called, was supplied by Rolls-Royce and Associates Ltd.[249] Originally named HMS *Invincible*, the name was withdrawn after objections were raised on account of the disaster to the former battle cruiser HMS *Invincible* at the Battle of Jutland (it blew up with almost total loss of life, due to poor ammunition handling).[250] The Committee then returned to the earlier suggestion of *Inflexible* (see p. 145), with *Implacable* reserved for the third submarine. But, again, an objection to *Inflexible* was raised on the grounds that the name did not truly indicate the essential quality of nuclear submarines. The Committee, determined to select names with the prefix 'Im-' or 'In-', settled on *Implacable* as the first-of-class name. However, a year later, in 1961, the name was dropped and *Valiant* adopted instead.

In design terms *Valiant* was what *Dreadnought* would have been had the British refused Rickover's offer.[251] The basic design concept was to use the forward end of *Dreadnought* together with an aft end consisting of the British nuclear machinery based on the Dounreay prototype but with a few minor alterations, such as the use of different materials. Improvements on *Dreadnought* included more explicit reactor safety requirements; a 75 per cent increase in diesel fuel; a 50 per cent increase in communications fit and the repositioning of the forward hydroplanes away from the main bow sonar array to reduce self-noise. The diving depth was increased, back to the original UK specification of 750 feet; a secondary means of propulsion was fitted independent of the main propulsion system, which could be retracted into the submarine's streamlined hull when not in use; sound isolation against sonar noise and reduced machinery noise was also introduced as well as raft-mounted machinery, a considerable achievement given the size and weight of much of it. At normal operating speeds, the raft essentially floated free of the hull, giving the submarine a noise advantage over both American and Soviet submarines. However, at high speeds the raft was locked into position and *Valiant* made a great deal of noise.[252]

According to the naval architect David Brown, 'Valiant was the finest British post-war design of its day, surface or submarine.'[253] However, it had a number of deficiencies, especially with machinery operation and maintainability. Insufficient attention had been given to the propulsion and auxiliary machinery sited outside the reactor compartment.[254] The secondary plant, as it was known, was regarded as established technology because of its similarity to steam propulsion plants found in Royal Navy surface ships. The designers at Dounreay mistakenly concluded that the systems and equipments unique to a diving submarine, which had successfully evolved over the preceding forty years, particularly for the HTP programme, would only require extrapolation to meet the requirements of the nuclear programme. The design was also very complicated. A full-scale mock-up constructed of wood and cardboard was built in Barrow to aid those responsible for fitting out the submarine. When the layout in an area was confirmed and accurately represented in wood, the pipe shop would bend a wire to conform to the shape of the mock-up, and go back and fabricate the real article in the appropriate material. 'Though it was effective it was ridiculously expensive,' recalled Patrick Middleton, who worked on the mock-up when building a later 'Valiant' class submarine. 'The lower reaches of the structure were like a forest floor, strewn with broken bits.' (It was also covered in 'suggestive' graffiti aimed at the female cleaners who cleaned the mock-up in the small hours when no one else was around.)[255]

Why did Rickover help the British? His respect and admiration for Mountbatten certainly played a part. In September 1957 he wrote to the Commanding Officer of USS *Nautilus*, Commander William R. Anderson, and explained why he had gone 'all out' to help the British. 'I did this,' wrote Rickover, 'because of my feeling of urgency about the international situation, my admiration for the British, and particularly my great liking for Admiral Mountbatten.'[256] But there were clearly other, calculated reasons. As Vice Admiral Sir Robert Hill has concluded 'it seems clear that his intention from the outset, having personally established that the UK had the ability and determination to have a nuclear-submarine force, was to give a single, time-limited boost to the UK programme. He knew that we were poor, and would become poorer; and that given half a chance would scrounge on the US rather than apply our own thought and our resources to solving the problems that would inevitably arise.'[257]

Whatever his motives, Rickover's offer had a dramatic impact on the Royal Navy's nuclear-submarine programme. Two of its key architects later admitted that 'The UK's debt to the US Navy, and to Admiral Rickover in particular, is incalculable.'[258] According to Baker, 'in so far as our nuclear submarine programme has been a success it is mainly due, first to RICKOVER selling us his bit and second to me for insisting that this S5W plant be used by us in an environment similar to Skipjack, and that we should buy from America a complete machinery installation'. Designing and building *Dreadnought* provided the Admiralty's naval architects and engineers, Vickers and Rolls-Royce as well as associated parts of industry, with a sound basis from which to progress to the all-British nuclear submarines from then on to the present day. Since *Dreadnought* put to sea submarine reactor development in the UK has steadily increased reactor core life until the reactor cores currently being manufactured by Rolls-Royce are designed to power a submarine throughout its lifetime.

Reflecting in January 1982 on the achievement of the earlier pioneers of the nuclear programme, Vice Admiral Sir Ted Horlick concluded that:

Today DSMP [Dounreay Submarine Prototype] is seen as one of the corner stones of our successful nuclear propulsion programme. Not only has it provided a test bed for new technology and the lead plant for sea going submarine propulsion systems, but it also provided initial practical training for the operators of nuclear submarines. Furthermore it afforded a test bed for the solution of technical problems which could be expected to arise in the submarine plants.[259]

There is still a healthy debate about just how much impact Rickover's offer had on the Royal Navy. According to Rickover's biographer, Francis Duncan, 'Events had made Rickover the "stepfather" of the British nuclear Navy. Its officers and officials found him hard to deal with, hard to predict, and often abrasive, but without him they would never have gotten their first nuclear submarine to sea.'[260] This view is shared by Vice Admiral Horlick: 'There is little doubt that the Admiralty's objective in having the Royal Navy's first nuclear-powered submarine at sea by 1963 would not have been achieved without the purchase of the S5W.'[261]

Professor Jack Edwards had a different view:

> I suppose some would regard him as some sort of foster father of the British nuclear fleet. Personally I am still convinced we would have built our nuclear submarine entirely on our own efforts – it would not have been as good as *Skipjack*, and it would have taken us some 2 years longer to get to sea. But it would have been entirely our own design and would not have made us so dependent on the whim of the US Congress on the passage of further information to us.[262]

The truth probably lies somewhere in between. Peter Hammersley, HMS *Dreadnought*'s first Mechanical Engineer:

> Without belittling the achievements of those who had been working on the British design, I believe that two or three years is an understatement. I also believe, and have always believed, that we would have lost most of that advantage had we bought the American reactor and primary plant only to match to our propulsion machinery as was proposed by some in the MOD but opposed by Mountbatten.[263]

This fits with William Crowe's view that 'The minimum estimate was six months given by a scientist in the British program. The maximum was five years given by a high ranking officer who was on the Admiralty Board at the time and instrumental in driving through the exchange.'[264] Rickover estimated the time saving as a minimum of three years and the cost saving as between $50m and $75m.[265] Wherever the truth lies, Sir Solly Zuckerman, who later became the Government's Chief Scientific Advisor, was surely right when he said that 'Without Rickover's cooperation, we would never have had the Skipjack reactor. While we may have devised corresponding equipment on our own, we were certainly enormously helped by the American Government.'[266] This was not lost on the Navy. 'The American Agreement not only advanced it [the UK nuclear programme] but it also enabled the UK to establish the necessary planning, quality and management standards,' Hammersley recalled:

> We in *Dreadnought* were American trained. The key people in the ship-builders and Rolls-Royce had also learned from the Americans. Between us, we transferred Rickover's standards to our programme. We were lucky. We had a proven plant with an experienced crew. *Valiant* and her successors had a new British plant, which was more complicated, and a crew who had only such experience as we could give them, which wasn't very much. They had a much more difficult time and deserved great credit for what they achieved. I don't think that was ever fully recognised.[267]

The term 'special relationship' is often overused, but can rightly be applied here. 'We knew from the very beginning that we were closer politically, culturally, and in many other ways with the British than we were with other countries, and that is why we treated them differently from other countries,' Rickover later said.[268] When parts of the Eisenhower administration raised the possibility of sharing US reactor technology with France, Rickover was horrified. 'The head of the French Atomic Energy Commission is a proselytizing Communist,' Rickover is said to have complained. 'He makes frequent trips to Moscow for instructions. He has five hundred known Communists in his atomic program. It's nuts!' In a special meeting of the AEC Rickover sat in silence as each commissioner outlined why he thought the proposal was a good idea. 'Does anyone here doubt how the American people would vote if they were given a chance to express themselves on this issue?' he asked. 'Knowing that, it is *immoral* for you to proceed otherwise.'[269] The US never did sell its nuclear-propulsion technology to France, or any other nation.

With HMS *Dreadnought* on trials, HMS *Valiant* under construction and the Dounreay prototype beginning to show significant progress by the end of 1962, a small pattern of nuclear-submarine building had begun in the UK. Important questions remained. 'The problem now facing us, in a nutshell, is this,' said Lord Carrington in a presentation to the Admiralty in 1962. 'On the assumption that the Royal Navy is to get into the nuclear submarine business, what sort of building programme is required, what is the best way to use our shipbuilding capacity for it, and what first steps ought we take now against the background of a sensible and feasible future programme? And by "feasible" I mean financially as well as technically.'[270]

Although Anglo-American naval relations were at something of a high point, one of the most 'potentially destructive crises in Anglo-American relations' was about to thrust the Royal Navy and the Submarine Service into an entirely unexpected and unfamiliar role: as custodians of the United Kingdom's independent nuclear deterrent.[271]

4

'Move Deterrents out to Sea': The Bomb Goes Underwater

Indestructible retaliation. Indestructible retaliation. That is the secret. Never forget that.

> Winston Churchill to Lord Hailsham,
> First Lord of the Admiralty, 1955.[1]

Move deterrents out to sea, where the real estate is free and where they are far away from me.

> Admiral Arleigh Burke, Chief of Naval Operations,
> US Navy, to Admiral Lord Mountbatten,
> First Sea Lord, 1959.[2]

We planned in 1963 to fire our first missile at 1115 EST (Eastern Standard Time) on 15 February 1968; we failed by 15 milliseconds. We were told in 1963 that there must be a continuous deterrent from July 1968[9]; this was achieved.

> Rear Admiral Charles Shepherd,
> Deputy Controller Polaris.[3]

Give us the will, but never the wish, to obey the order to fire. Oh God, if it is thy will, grant that that order may never need be given, Amen.

> Captain Michael Henry, CO HMS *Resolution*.[4]

FIRST CONTACT

Throughout the 1950s, responsibility for carrying the United Kingdom's independent nuclear deterrent lay with the Royal Air Force. In order to maintain its effectiveness and credibility into the 1970s, the Ministry of Aviation began to develop Blue Streak, a fixed-base, liquid-fuelled medium-range ballistic missile (MRBM) to replace the Royal Air Force's ageing and increasingly vulnerable nuclear-armed Vulcan, Victor and Valiant bombers.[5] Although the 1957 Defence White Paper reaffirmed this policy of airborne deterrence at the expense of conventional forces (ominously for the Royal Navy stating that 'the role of naval forces in total war is somewhat uncertain'), by 1970 the Royal Navy and the Submarine Service were on the front line of British defence policy, as the custodians of the Polaris weapon system.[6]

Polaris was 'possibly the most revolutionary development in weapons technology in the twentieth century'.[7] For a deterrent system to be truly effective it must be capable, reliable, available and invulnerable and the political will to use it must exist or be thought to exist. With three quarters of the Earth's surface covered by the sea, the nuclear-powered Polaris submarine combined the advantages of land-based ballistic missiles with the flexibility of air-launched missiles but with the disadvantages of neither. Land-based ballistic missiles and the aircraft that carried air-launched missiles were vulnerable to pre-emptive attacks. Polaris, as one of its key US architects explained, provided 'assurance of retaliation' and fulfilled 'the new function of military force – that of preventing war – by being so attuned and adjusted to grand strategy requirements that battles do not occur'.[8]

The Royal Navy's association with Polaris began in November 1955 when Mountbatten learned of the United States Navy's programme to develop a solid-fuel ballistic missile that could be fired from a submarine. He was so impressed with the determination of the Chief of Naval Operations, Admiral Arleigh Burke, to push ahead with development that he asked if the Royal Navy could be associated with the programme. Burke agreed. Mountbatten was amenable to anything that would give the Royal Navy a new role.[9] Polaris was one way of achieving this. Although he remained 'sceptical about the reality of Britain's independent deterrent . . . [he was] determined that it should be maintained in the most efficient manner'.[10] During a discussion on submarine policy in February 1956, Mountbatten explained that if 'the Americans got through their teething troubles and were successful in developing such a missile, and were willing to give us the design, then . . . we might consider such a project'.[11] A year and a half later, he believed that it was a question of 'when, and not whether, the Admiralty should seek the resources to introduce' a submarine-based ballistic missile into the Royal Navy.[12]

When the Naval Staff first heard of Polaris they 'found it a little difficult to believe', as did most of the Admiralty technical departments in Bath, which 'regarded it akin to Science Fiction'.[13] Throughout 1956 and 1957 its development in the USA was watched with great interest. In October 1957, the Admiralty Board concluded 'that it would be inexpedient for the present to urge any claim to equip Royal Navy submarines with ballistic missiles', but the Naval Staff did decide to discreetly promote it to ministers, including the Prime Minister.[14] When the First Lord of the Admiralty, Lord Selkirk, visited the United States at the end of 1957 he received a personal briefing from the Director of the US Special Projects Office, Admiral William 'Red' Raborn, outlining the technical, financial and managerial aspects of the US Polaris programme.[15] Selkirk was deeply impressed and on his return to London wrote to his predecessor, Lord Hailsham:

> I am very anxious to percolate slowly but gradually into the minds of our colleagues the future possibilities of submarines equipped with the IRBM [Intermediate-Range Ballistic Missile] Polaris which is being developed in the United States and is likely, I believe, to be operational about 1961 . . . I am sure that by 1967 or so missile sites will be out of this island and at sea in submarines.[16]

'Out of this island and at sea' was an attractive prospect, given the public and political uproar that had arisen over the recent deployment in the UK of the US land-based IRBMs, known as Thor.[17] Selkirk was particularly anxious to avoid a position whereby 'we are completely committed to a policy of IRBM sites in this country before the full implications of the guided missile submarines are appreciated'. But he had to 'get the idea across slowly without frightening or offending or promising that it will be available'.[18] With the UK also developing its own land-based MRBM, Blue Streak, Selkirk knew that it was highly unlikely ministers would agree to stop work on the missile, which had strong supporters, including the Minister of Defence, Duncan Sandys, who had initiated its development, and the MOD's Chief Scientist, Sir Frederick Brundrett, who was 'wedded to Blue Streak'.[19]

By then the Admiralty had already decided to increase its links with the US Polaris programme. The Admiralty British Joint Service Mission (ABJSM) in Washington developed particularly good contacts with the Special Projects Office, whose staff operated 'an open door' policy for the Royal Navy. Relations were so close that when Polaris was publicly unveiled at a Navy League Symposium in Washington in April 1957, ABJSM officers were granted unprecedented access to a test firing of a Polaris missile. A slightly perplexed ABJSM reported back to the Admiralty that:

> Our Staff Ordnance Engineer Officer attended this firing and was taken very much into the confidence of those present – even to the extent of being offered a skin diving trip to view the business under water! Whilst his attendance was officially cleared, in fact he saw and heard rather more than I fancy was really intended by the authorities and I therefore feel that it would be wise not to publicise that we managed to get in so very well 'on the ground floor.'[20]

But the United States Navy was unwilling to make Polaris available until it had completed its own programme. Despite this, in April 1958 Admiral Burke expressed his hope 'that the Royal Navy will join with us in the Polaris weapon system at some time in the future', suggesting participation in about one or two years, once the US Polaris fleet had achieved operational capability.[21] Burke's advice to wait and see was welcomed by many in the Admiralty.[22] In May 1958, Mountbatten confided to Burke that:

> what we are aiming to do at the moment is to keep the Polaris pot boiling over here so that the manifest advantages of this weapon system shall not be overlooked. We can only do this if we can show that the USN, backed by your government, is willing to give us every possible assistance to get such a weapon into service or at all events to be prepared to hold discussions on this subject. But whether or not this happens, I hope you would agree to release to us on a strictly, Navy-to-Navy net as much information as your Acts will allow us to have in advance of your getting the weapon into service. It is only by this means that we shall be in a position to take advantage of your suggestion to join in with you, without great loss of time.[23]

Although the development of Polaris was under the highest US security 'NOFORN' classification (Not Releasable to Foreign Nationals), Burke provided Mountbatten with a great deal of assistance and advice through letters carried between Washington and London by a senior Commander on Admiral Raborn's staff. (For years the SPO's copies of the correspondence were kept in what was affectionately known as the 'Dear Dickie File'.)[24] By May 1958, Mountbatten was requesting so much information about Polaris that Burke had to tell him: 'for God's sake Dickie stop pestering me, put one of your men in our Special Projects Office and he can tell you all you need to know'.[25] In order to

establish a more efficient means of channelling Polaris-related information, Burke suggested a direct liaison between the Admiralty and the
SPO would facilitate the flow of general and technical information
between the two navies.[26]

Whenever doubts were expressed about Polaris, Mountbatten sought
out information to rebuff the claims.[27] He also secured from Burke a
commitment to supply the Navy with 'Polaris missiles (less warhead) at
production cost' with all research and development costs paid for by
the USN.[28] However, both men were discussing something they had
little power to deliver as ultimately only the President of the United
States could permit the sale of Polaris to the UK. Burke did not expect
there to be any 'serious difficulty' at the political level, but he could
only give 'the U.S. Navy position in this matter, with no guarantee that
it will finally work out that way'.[29] In these circumstances, the British
Government unsurprisingly remained committed to developing Blue
Streak. Mountbatten advised that 'the arguments for deploying the UK
contribution to the deterrent in submarines remain as strong as ever'
and continued to walk the line between discretion and advocacy.[30]

When Mountbatten stood down as First Sea Lord and became Chief
of the Defence Staff (CDS) in May 1959, Polaris lost one of its most
important advocates. Mountbatten's successor as First Sea Lord, Sir
Charles Lambe, did not share his enthusiasm for the new weapon and
immediately adopted a more sceptical approach. When asked by Selkirk for direction on Polaris, Lambe advised that it was 'unwise to go
into final battle'.[31] He also sought to reduce the temperature of the
'Polaris pot' by tempering the plans Mountbatten and Burke had set in
train. When Lambe accidentally opened a letter from Burke addressed
to Mountbatten, he wrote back in a highly critical tone:

> As this whole subject of the British contribution to the Deterrent is a
> pretty hot one politically, I am rather glad that your letter was inter-
> cepted in this way, as it could be unfortunate if anyone else in the Ministry
> of Defence became aware of how closely you and Dickie are working on
> this matter.[32]

Instead, Lambe counselled caution, writing to his senior Admiralty
colleagues:

> We in the Admiralty need a much clearer picture than we have at present
> of the probable repercussions of the Polaris programme on the rest of the
> Navy before we start any official pro-Polaris propaganda. Indeed, I doubt

whether it is right for the Navy to undertake any such propaganda at all. I believe we would be in a far stronger position if we were (at any rate, apparently) pushed into the Polaris project rather than to push it ourselves. My advice for the present would thus be to let sleeping dogs lie, and in the meantime continue our own internal Admiralty examination of the Polaris project so that we are on the top line to answer any queries when they come – as, in my view, they undoubtedly will.[33]

When Blue Streak was cancelled in early 1960, due to its high cost and technical and strategic obsolescence (it was vulnerable to a pre-emptive Soviet strike), the government decided to purchase an American air-launched ballistic missile – Skybolt – after the RAF enthusiastically promoted it, backed by the United States Air Force and the British aircraft industry.[34] Although the Cabinet's Defence Committee discussed Polaris, moving the deterrent underwater was deeply unattractive politically, as over £500m had already been invested in the V-bombers which would carry Skybolt.[35]

Lambe believed that Mountbatten would use his new position as CDS to do 'everything in his power to see that the merits of Polaris are brought to the notice of HM Government'.[36] But Mountbatten wanted to avoid inter-service rivalry and any impression that the Admiralty was attempting to 'take over' the deterrent from the RAF, and, for once, 'maintained the impartiality expected of a CDS', supporting the Air Staff's bid for Skybolt.[37] But he knew that Skybolt was only a short-term solution to the UK's deterrent problems. He was aware that there was 'no absolute certainty' that it would be successful, and that the US 'would not develop it for our use alone'.[38] Burke informed him in April 1960 that Skybolt was 'a very expensive and vulnerable system'.[39] Mountbatten could afford to 'sit back and let the Americans establish that the air-to-surface missile was technically and economically impossible' while continuing to subtly prepare the case for Polaris.[40]

When the US Navy approached the Admiralty about establishing a naval base in Scotland for US Polaris submarines, the government saw an opportunity to secure Polaris for the Royal Navy in exchange for the facilities and as an 'insurance against the failure of Skybolt'.[41] The Admiralty was eager to grant the US request: allowing US SSBNs access to British waters would further improve relations between the two navies and expand the Royal Navy's 'knowledge in the general field of nuclear submarine operating'.[42] The Government also hoped to obtain two complete US Polaris submarines in exchange for the facil-ities.[43] However, in August 1960, the Eisenhower administration made it clear that it was 'not prepared at the present time to leave the impres-sion that they were negotiating a purely bilateral arrangement for supplying' the UK with Polaris.[44] In September, the Cabinet agreed to allow the US to establish a Polaris base at Holy Loch in Scotland. Min-isters also agreed to drop the request for Polaris submarines, noting that 'Any requirement there might be would be more likely to be met through the close relations between the Royal Navy and the United States Navy and the willingness of the United States Government fur-ther to develop technical co-operation.'[45] Carrington as First Lord was instructed to 'develop as far as possible even closer technical cooper-ation with the US Navy, with a view to making it possible in due course for the UK to build its own missile submarines if it were decided to do so'.[46] The Government was pushing the Admiralty towards Polaris.

Admiral Sir Peter Reid, the Controller, who felt that the transfer of responsibility for the strategic deterrent to the Navy was inevitable, had already commissioned an investigation by the former Director General Weapons and future First Sea Lord, Michael Le Fanu, into the possible organization that could manage the procurement of Polaris should the Government decide it wanted it.[47] When Le Fanu presented his report to the Admiralty Board in July 1960 only Flag Officer Sub-marines, Rear Admiral Sir Arthur Hezlet, was supportive.[48] The Naval Secretary, Sir John Lang, who was bitterly opposed to Polaris, was scathing. He did not believe 'that the proposals had been fully thought out in detail' and he outlined 'the fundamental weakness' of Le Fanu's recommendations:

> he seeks to put Polaris into a privileged position as against the rest of the
> Navy programme and yet the Board, however whole-heartedly they may
> put their backs into Polaris, cannot divorce themselves of their parallel
> responsibilities for the rest of the Navy. There is no escaping the fact that
> there will be frequent clashes, in all fields, between the interests of Polaris
> and the rest of the Navy, irrespective of whether money is no object or
> not.[49]

The Controller and the First Sea Lord agreed 'broadly' with Lang, while the rest of the Board were concerned about 'wounding the fleet'.[50] Le Fanu's report was pigeonholed, filed away in case the need for it should ever arise.[51] 'It fell so flat,' he later recalled, 'that I wondered for a moment if I had been wrong.'[52]

Towards the end of 1960 a small Admiralty working party was established under the chairmanship of the Director General Ships, Sir Alfred Sims, to keep in touch with the US Polaris programme in anticipation of a future UK programme.[53] There was some debate about whether, if the Government decided to purchase Polaris, the Royal Navy should purchase complete Polaris submarines from the US Navy or build them in the United Kingdom. After exploring the options the working party advised that if the Government wanted Polaris it should either purchase the submarines completed from America, or buy all the American components and assemble them in British shipyards. In November 1960, Burke offered what he felt 'might well be a deciding factor in acquiring the British Polaris submarine', the use of support facilities in both Holy Loch and the United States.[54] He also allowed a small Admiralty team to visit the US to study in greater detail the implications of a Royal Navy Polaris programme and 'how strong the case is . . . for buying or building Polaris submarines'.[55] A small technical team under the Deputy Director of Naval Construction, Sidney Palmer, visited the US from 26 February to 11 March 1961 to 'obtain sufficient facts to enable the Admiralty to advise the Government, should the occasion arise, of the cost in manpower, materiel and money of building Polaris submarines in the U.K., and of supporting them'.[56] Palmer's team concluded that there was only one logical answer: 'To be sure of success from the very outset and to save time the only sensible thing to do is to buy the whole of the Polaris system from the U.S. as a "do it yourself" kit and fit it straight into British built submarines.'[57]

Palmer's team identified the best means of constructing a British Polaris submarine, by following US practice and grafting the missile section onto an existing SSN design.[58] When the US Navy started its Polaris programme it took an existing SSN, the USS *Scorpion*, cut it in half, lengthened the hull by 130 feet, and installed two rows of eight missile tubes in the middle to create the US Navy's first SSBN, the USS *George Washington*.[59] Palmer's team 'became convinced' that the 'British Polaris submarine should be basically to the SSN02 (*Valiant*) design with a Polaris midship section' of sixteen missiles.[60] However, *Valiant* was one foot wider in diameter than the *George Washington*. *Valiant*'s main machinery, shape, and spacing in the frames and the reactor compartment were also completely different from those in US submarines, and the hull was made out of a different type of steel. Moreover, to reduce vulnerability, US SSBNs were designed to operate at much greater depths than British submarines. If the operating depth of the

British SSBN was to be changed, the *Valiant*'s hull, machinery and hull valves would all have to be redesigned at considerable time, cost and effort. Nevertheless, Palmer's team believed they could work around the installation problems, and the Americans did not think the reduced vulnerability afforded by greater diving depths was sufficient to justify the extra cost.

Examinations of the Palmer Report continued throughout 1961 and into 1962. In July 1961, Sir Arthur Hezlet, the Flag Officer Submarines and a keen Polaris supporter, produced a report on the feasibility of constructing an underground base in Scotland to shelter a future Royal Navy Polaris fleet, either in Loch Glencoul in Sutherland, just south of Cape Wrath; in Loch Nevis, near Mallaig opposite the southern end of Skye; or in Loch Striven, in the Clyde Approaches.[61] The plan envisioned submarines entering the base via a submerged entrance, being drawn inside by winches after settling on a smooth slipway.[62] The facility would house a dry dock, engineering, electrical and periscope workshops, armament stores for the Polaris missiles, and torpedoes, spare-part stores and communication facilities.

Hezlet's proposal found little support. 'It would be absurd at the present stage to incur any expenditure at all with Polaris submarines in mind, since we have no idea whether there will even be any successor of any kind to the V-Bombers with Skybolt,' noted one official.[63] From 1960 onwards the Admiralty's Polaris policy had aimed at 'the limited object of advising the Ministry of Defence on the scale of effort and resources required'.[64] By the end of 1961, with the detailed technical information from the Palmer Report, the Admiralty felt it had gone as far as it could. It was unprepared to fund any further studies without a firmer guarantee from the Government that Polaris would succeed the V-bombers. This whole question was debated in a special committee known as the British Nuclear Deterrent Study Group (BNDSG) or the 'Benders', under the Chairmanship of the MOD's Permanent Secretary, Sir Robert Scott.[65]

However, by the end of 1961, after sitting for almost a year, the group had failed to reach any definite conclusions. Discussion had become polarized between the supporters of a submarine force and the advocates of a continued airborne deterrent. In an attempt to resolve the stalemate, Solly Zuckerman set up an independent group of experts which it was hoped would avoid the inter-service politics that dominated the discussions of the full BNDSG. However, the discussions of the technical subgroup also quickly deteriorated as Zuckerman attempted to steer its deliberations in favour of Polaris. After bitter debates with the supporters of the airborne option the group produced a report acknowledging the need for a mobile delivery system, but left open the decision on whether it should be delivered by a submarine or an aircraft. Zuckerman held up its circulation and then modified its conclusion, tilting it towards Polaris. This so angered the supporters of the airborne option that they produced a minority report which forced Zuckerman to back down and reinstate the original balanced conclusion.[66] His manoeuvrings offer a possible explanation for a widely held belief in RAF circles that Zuckerman and Mountbatten conspired to wrest the nuclear deterrent from Bomber Command in order to restore the 'image of a rapidly declining Navy'.[67]

In November 1961, frustrated at 'seeing the merits of sea power undersold' by the Air Ministry, and mindful that most of the BNDSG's members 'thought the Navy had the stronger case' for the successor system, the Admiralty Board produced a paper for the Minister of Defence. It highlighted 'the military advantages of an underwater vehicle for the next generation of the deterrent', outlining a dual-purpose submarine capable of operating both as an SSN and as an SSBN, 'as a satisfactory alternative to the single-purpose Polaris vessel'.[68] The paper explained that 'Admiralty studies had shown that the "hybrid" submarine itself was technically and operationally feasible and the proposed programme of construction realisable.'[69] The Minister of Defence, Harold Watkinson, immediately expanded the work of the BNDSG to cover the hybrid, and the Admiralty set about preparing a more detailed analysis, which it intended to present to the BNDSG by October 1962.[70]

By October 1962, Whitehall was awash with reports that Skybolt was in trouble. After reviewing its development, the US Secretary of Defense, Robert McNamara, and his group of Pentagon statisticians and management experts recommended the cancellation of the missile. If Skybolt were cancelled, Britain would be left without a new means of delivering its nuclear deterrent.[71] The Admiralty was first asked to present its alternatives to Skybolt on 14 November 1962 at a meeting of the BNDSG. The Deputy Chief of the Naval Staff told the group that 'if we are to continue to deploy an independent strategic deterrent, the only sensible possibility remaining would be a Polaris type submarine'. Over objections from the Air Ministry, the Admiralty was asked to consider how it could deliver a 'crash' Polaris programme.[72]

When the Admiralty reported under a month later on 4 December, three options were outlined: (1) purchasing four complete Polaris sub-marines from the US on 1 January 1966, 1967, 1968 and 1969; (2) hiring two complete Polaris submarines from the US on 1 January 1966 and 1967 and building a fleet of seven dual-purpose submarines in British shipyards, completing the first submarine by 1 April 1970; (3) hiring three submarines from the US on 1 January 1966, 1967 and 1968, followed by a fleet of seven UK-built dual-purpose submarines.[73] These dual-purpose submarines met one of the Navy's main concerns: that a Polaris programme must not deprive the fleet of conventional and nuclear attack submarines. It was also attractive because it would place the Admiralty and the government 'in a much more flexible pos-ition in that if at any time in the next thirty years it should be decided to discontinue the UK deterrent, these submarines could without con-version be redeployed in a wholly anti-submarine role'.[74] Significantly, it was not possible for the Admiralty to deliver UK-built submarines by 1965, the projected in-service date for Skybolt and the estimated date beyond which it was estimated that the V-bombers would be unable to penetrate Russian airspace and target defences successfully. Unless swift measures were taken, there would be a gap in the deterrent.

Macmillan was initially reluctant to abandon Skybolt, but by early December, as news reached Whitehall that the Americans indeed intended to cancel the missile, he concluded that all efforts should be focused on securing Polaris and borrowing American Polaris submar-ines as an interim measure, while the Royal Navy built its own fleet of submarines.[75] 'It seems that we may find ourselves in the position in which the Government may say "you must operate a POLARIS deter-rent for us" – always provided the Americans let us have it!' wrote Admiral Sir Caspar John, who had succeeded Lambe as First Sea Lord in May 1960.[76] But would Washington make Polaris available? Although there were strong elements in the Kennedy administration (which had come into office in January 1961), such as the US Under-Secretary of State, George Ball, who wanted to see an end to Britain's independent nuclear deterrent, Kennedy, backed by the Defense Department, agreed in principle to make Polaris available to the British on the condition that when the submarines had been completed they were committed to 'a multilateral or multinational force in NATO'.[77]

The complex negotiations between Harold Macmillan and Presi-dent Kennedy that ultimately led to the purchase of Polaris took place at Nassau in the Bahamas in December 1962. The conference was one of the most significant and explosive encounters in the history of

Anglo-American relations. Macmillan and the British delegation arrived in the Bahamas on 18 December deeply angry and suspicious, amid embarrassing press reports that the United States was attempting to force the United Kingdom out of the nuclear business. Macmillan desperately needed a replacement for Skybolt. 'We would have been in a very, very nasty position politically,' recalled the then Foreign Secretary, Lord Home. 'I think that the government would probably have been beaten. It might well have been a case for an election'.[78]

Amid frank and vigorously contested discussions, the Prime Minister delivered an emotional performance, speaking with a memory that 'perhaps went back further than anyone else in the room', as a veteran of two world wars, recalling the United Kingdom's long struggle for freedom and the wartime partnership that had led to the development of the first atomic bomb.[79] 'There was not a dry eye in the house,' remembered Sir Philip de Zulueta, Macmillan's Foreign Affairs Private Secretary.[80] Kennedy wanted to help, but many of his advisors were reluctant to make Polaris available, fearing repercussions in Europe and the collapse of the earlier policy that aimed to abolish independent nuclear capabilities. As a concession the President offered to continue with Skybolt, splitting its development costs equally between the two countries. Macmillan refused. 'I observed that while the proposed marriage with Skybolt was not exactly a shot-gun wedding the virginity of the lady must now be regarded as doubtful,' wrote Macmillan.[81] If Skybolt was not suitable for the United States, it was not good enough for the United Kingdom.

Washington's unwillingness to make Polaris available bilaterally, as a straightforward replacement for Skybolt, was the most difficult aspect of the Nassau negotiations. Kennedy reportedly told Macmillan that 'We did not mind too much about giving you Skybolt because we thought it might not work and in any case it would have been obsolete by 1970. But Polaris is an entirely different matter as it will last much longer.'[82] As Polaris would last into the 1980s, the Americans felt they were perfectly entitled to press harder. Kennedy eventually accepted that Macmillan could have Polaris, but there was one condition: the United Kingdom would have to assign its new Polaris force to NATO. There would no longer be an independent British nuclear capability. Macmillan was unable to agree to such terms. He had to ensure that the independent control of the deterrent was clearly and unambiguously expressed, not only for himself, but for future Prime Ministers. He offered to assign Polaris to NATO 'provided the Queen had the ultimate power and right to draw back in the case of a dire emergency similar to that in 1940'.[83]

With time running out, the Prime Minister threatened to walk away. He told Kennedy, bluntly, that 'if an agreement was impossible the British government would have to make a reappraisal of their defence policies throughout the world'.[84] The threat of a deep rift in Anglo-American relations, a rupture in the special relationship, worked. On the final day of the negotiations the two sides agreed on a formula, a set of formal words that preserved the independence of the United Kingdom's nuclear capability. Macmillan conceded that Polaris was 'a different kind of animal' and paid the 'political price which the Americans demanded' by agreeing to assign the UK's deterrent to NATO.[85] But the British government reserved the right to use its nuclear forces independently in circumstances *where Her Majesty's government may decide that supreme national interests are at stake*'.[86] Macmillan got what he wanted and, contrary to what many of Kennedy's advisors thought best, the young President had given it to him. 'The Americans pushed us very hard,' Macmillan wrote in his diary. 'The discussions were protracted and fiercely contested. They turned almost entirely on "independence" in national need. I had to pull out all the stops – adjourn, reconsider, refuse one draft and demand another etc, etc.'[87]

Yet what would constitute a 'supreme national crisis'? In what circumstances would a British Prime Minister invoke the 'supreme national interests clause' and use the United Kingdom's nuclear capability independently of NATO and the United States? Kennedy admitted that it was 'difficult to conceive of such a situation' but he acknowledged that if the British 'were threatened with the bombardment of their island, they might feel they wanted to have the capacity to respond, or at least say they had the capacity, and if there was an attack, to respond'.[88] Kennedy later cited to the French Ambassador both the Suez Crisis in 1956 and British action to support Kuwait against territorial claims by Iraq in 1961 as examples of how the 'supreme national interests' formula might be invoked.[89] Some, such as Zuckerman, regarded any discussion of independent use of the United Kingdom's nuclear capability as of no consequence. 'If our Polaris force were ever used, we would never even know whether our missiles had struck,' said Zuckerman. 'There would be no newspapers to tell us, no television to show what had happened, and maybe no "us", just the crews of those Polaris boats that had been at sea.'[90]

Not everyone was convinced that the independent deterrent had been preserved. Sir Robert Scott, the MOD Permanent Secretary, protested about the 'very serious risks' of the Polaris deal.[91] He persuaded Zuckerman, who 'had looked very happy with the Polaris Agreement', to co-sign a document stating that the claim that the UK deterrent remained 'independent' could be made only with difficulty – particularly as a cardinal feature of Kennedy's policy was to deny his allies independent nuclear forces.[92] The UK would be spending vast sums of money contributing to a multilateral force, the object of which was political rather than military. They also pointed out that by committing to Polaris the UK would lose some of its freedom of action in deciding how best to support NATO militarily, both in the nuclear and in the non-nuclear field. The note was delivered by hand to Thorneycroft, their departmental minister, who seems to have ignored it. The only minister who ever referred to it was Julian Amery, the Minister of Aviation. 'One day,' he said to Zuckerman, 'you will be proud of that Minute; not that it has made any difference.'[93]

George Ball described the Nassau conference as 'one of the worst prepared . . . in modern times'.[94] The product, the Nassau Agreement, was 'intolerably vague', written in 'the worst drafted language anyone had ever seen'.[95] It has been described as a 'monument of contrived ambiguity' and the attempt to reconcile interdependence with independence remained a source of continuing difficulties over the next two years as the two countries disagreed over what exactly had been agreed.[96] In the long term, this unpreparedness worked to Britain's advantage. Zuckerman, for example, with remarkable foresight, insisted that the phrase 'on a continuing basis' be added to the section of the agreement governing the transfer of Polaris, thus making it 'legally straightforward' to transfer future US Fleet ballistic systems, such as Trident.[97] David Ormsby Gore, Macmillan's Ambassador in Washington, and a close personal friend of Kennedy's, concluded that Nassau was a 'compromise that no other ally could have achieved'.[98] It was important not just because it saved a quasi-independent British deterrent, but because it served as 'evidence of the continued closeness of Anglo-American relations'.[99]

However, there was a price. Failure to clarify or even raise certain issues created problems for both governments and their navies. At the geopolitical level, in 1963 the French President, Charles de Gaulle, vetoed Britain's first application to join the European Economic Community, citing Nassau and the reassertion of the special Anglo-American nuclear relationship as one of the many reasons why the United

Kingdom was not yet committed to the European project.[100] At the technical level, because there were no representatives from the US Navy or its Special Projects Office present at the conference, 'professional/technical subjects were hardly discussed at all'.[101] Little, if any, attention was given to the implications of the agreement for the Royal Navy. Le Fanu was 'certain that the Prime Minister and the . . . Ministers at Nassau [Alec Douglas-Home, Foreign Secretary, and Peter Thorneycroft, Defence Secretary] do not give a damn about anything except getting an "independent deterrent" as quickly and as cheaply as possible'.[102] Macmillan had succeeded in securing Polaris, but the 'detail had not been worked out . . . it had been postponed to a later date'.[103]

Le Fanu met McNamara briefly after the signing of the agreement. McNamara 'was keen to force on as hard as he could go' and wanted an Admiralty mission to travel to Washington immediately to begin technical talks and resolve 'the many grey areas of the Nassau agreement'.[104] Le Fanu reported that McNamara was 'unimpressed' with the Admiralty's proposal for a dual-purpose submarine: 'tinkering with the 16 missile module meant trouble' and operationally, he argued, rightly, that it was:

> contrary to the whole US philosophy, by which submarines emerged from their base, are lost in the ocean without making any transmission on [wireless telegraphy] for 60 days, when they return to base. He is frightened that the Hybrid Concept would pinpoint the submarine's position when taking part in its secondary function and allow shadowing and destruction by an enemy.[105]

Macmillan had also dealt an unintentional blow to the dual-purpose submarine by failing to ask if the Navy could hire US Polaris submarines as an interim measure. Even if the US had agreed to loan submarines to the Royal Navy, whether this would have worked on the practical level was highly questionable. Macmillan's brief for Nassau pointed out that the President would most likely 'encounter difficulties with the Atomic Energy Committee (and especially Admiral Rickover) and perhaps with some sections of Congress'.[106] It would also have been difficult to sell to the public and Parliament. As the Deputy Leader of the Labour Party, George Brown, eloquently pointed out, 'You can rent a telly, you can rent a car; but, good God, you cannot rent a nuclear independent deterrent.'[107] Politically, it was now imperative to get Polaris operational as quickly as possible while also finding an alternative means of filling the 'gap' in the deterrent capability.[108]

The Naval Staff had been given 'broad assurances' that its existing naval programmes 'would not be swayed' by going for Polaris. But as there was no money in the Navy's ten-year long-term budget and 'whatever immediate safeguards we may make, it might become a long term drain on our money and resources', distorting the balance of the service, just as the Naval Staff had always feared.[109] Polaris was successful in the US Navy because it was 'devised and built by true believers'.[110] It was imperative that the senior Naval Staff replicated this enthusiasm and set an example to the rest of the Fleet. Le Fanu had warned in 1960 that:

> neither the organization I have suggested, nor any other, will work unless the whole Navy, in and out of uniform, is brought to realize that although a successful POLARIS submarine programme may wound the fleet, an unsuccessful one will kill it, and that once given POLARIS we must unite to make it a success. A Polaris programme will be a challenge to our resourcefulness, our brains and our energy but above all it will be a challenge to our leadership.[111]

Caspar John knew that uniting the Navy was going to be difficult. The Nassau Agreement brought to the surface a distinct 'undercurrent of opposition to Polaris'.[112] Vice Admiral Sir Aubrey Mansergh took to the pages of the *Naval Review* to complain about:

> the unpalatable task that the Navy seems shortly to be stuck with . . . the Admiralty (if such an institution is to survive the changes) is destined to wage a constant, and probably losing battle against the Treasury to prevent the 'deterrent force' swamping and distorting the 'balanced fleet' . . . If this is a good bargain the writer will eat the editorial hat.[113]

For Polaris to succeed the Board had to quickly alleviate such concerns. John wrote to the senior Admirals in the Royal Navy that, though 'the die is now cast and it is up to us to go full speed ahead, and to make the greatest possible success of this vital new task that we have been given', he was determined to preserve the Navy's existing plans, secure the necessary money and manpower, and 'get across that the Polaris programme is quite outside our normal naval task, and that the latter must go on as before'.[114] But John, the first naval aviator to be First Sea Lord, was in reality deeply unenthusiastic. While professionally he had advised Macmillan to purchase Polaris, privately, as his biographer has pointed out, 'Caspar was not in the least enamoured by the prospect of having to accept the responsibility to deploy Britain's nuclear deterrent.'[115] Shortly after the Nassau conference he made the following entry in his diary:

> A filthy week . . . this millstone of Polaris hung round our necks. I've been shying off the damned things for 5½ years. They are potential wreckers of the real Navy and my final months are going to be a battle to preserve some sort of balance in our affairs.[116]

The Admiralty was about to embark on what was to become the largest programme it had delivered since the Second World War, an undertaking of immense magnitude and complexity. Implementing it heralded vast unknown problems, all of which would have to be identified and then solved over the coming months. Did the Royal Navy have the capacity and indeed the talent to meet the challenge it had been set by the Government? One thing was certain, as Le Fanu noted, 'We are now "IN" Polaris in a big way. We are in the big time.'[117]

PLANNING FOR POLARIS

The Nassau Agreement 'sparked off an explosion of activity within the Admiralty', where many saw Polaris as a threat, believing that 'a strategic nuclear force would require a sharp downgrading of other valued parts of the service'.[118] Carrington and John both sought to reassure the Royal Navy that the budgetary consequences of the programme would not be culled from the surface fleet. The January edition of the Admiralty's monthly news bulletin pointed out that while the addition of Polaris was 'a great day for the Navy' it 'must not cause us to lose sight of the fact that this is a defence burden that is to be accepted as an addition to present tasks' and that 'whatever may be decided about the composition and size of this force, one thing is certain: the traditional role of the Navy remains unchanged'.[119] But both Carrington and John must have known that Polaris was going to change the Navy: it would upset traditional procedures and habits, and would force the Naval Staff whose minds were concentrated on the surface fleet, the future carrier and Mountbatten's proposals to reorganize the MOD (by the amalgamation of the three service ministries into the Ministry of Defence) to think in an altogether different way, and about an entirely unfamiliar role. The Navy's reputation would rest on its ability to complete a programme more demanding than anything it had undertaken since the end of the Second World War. The challenge was to ensure that all the elements – the submarines, the missiles, the base, support facilities, the training of crews, logistics – everything came together at the right time, an undertaking that necessitated what would now be called project management of the highest order.

Admiral Le Fanu's 1960 report was dug out, examined in detail and found to provide firm foundations on which further action could be based.[120] The Admiralty recommended the creation of an organization along the lines outlined in Le Fanu's 1960 report, 'something in the nature of a full time Project Team, under a leader, with senior representation from the Professional Departments and from the Secretariat'.[121] The Flag Officer Submarines, Rear Admiral Sir Hugh Mackenzie, was appointed to run the programme and was told that he would have to spend 'at least five years in the saddle', running it through to completion.[122] 'If you say "no",' John told him, 'I'll twist your arm until you bloody well scream'.[123] He had little choice but to accept. When he first arrived in Whitehall, on New Year's Day, 1963, to head the Polaris Project Team, known as the Polaris Executive, he was assigned an empty room in the North Block of the Old Admiralty Building, 'empty, that is, except for a chair and a large desk, on which sat a telephone not yet connected, no staff, no paper-work . . . an unusual and perplexing situation'.[124] He spent most of January assembling an 'initial nucleus' of officers and civil servants in London and Bath.[125] New arrivals often spent their first weeks 'owning sometimes a desk, sometimes a chair and sometimes a telephone but never all three together'.[126] The 'initial nucleus' spent the first few weeks operating out of two rooms and a closet but a common sense of purpose and comradeship quickly grew out of the 'bustling chaos'.[127] As Mackenzie put it, the Polaris Executive was:

> the cornerstone on which the whole project ultimately depended: for supervision and direction, for ensuring financial support and, when or where necessary, for fighting battles with higher authority and gaining political approval. In time of major difficulty or crisis – which were many in the early years and never finally disappeared – it was where 'the buck stopped'.[128]

Having an identifiable individual, a central authority, initially located in the Admiralty, followed by the MOD Main Building, with an overall view of the programme and the right of access to members of the Admiralty Board, became extremely valuable in obtaining prompt solutions and maintaining the drive of the programme.[129] Mackenzie quickly secured the backing of the Admiralty Secretariat, the Admiralty's highly efficient civil servants. The new Admiralty Secretary, Sir Clifford Jarrett, appointed in January 1961, and his civil servants were 'enthusiastic that their Department should not be found wanting on a national task'.[130] Jarrett recognized that 'if the Polaris programme was to be completed on time, very radical measures were needed in order to bring this about'.[131] The Admiralty's technical departments in Bath were reorganized, changes that were disruptive but necessary and 'fundamental to the success of the entire project'.[132] Polaris was so complex that Mackenzie and the Admiralty Board had to repeatedly stress to the Treasury, the MOD, even the surface fleet, Submarine Service and Fleet Air Arm, that it was going to be the 'biggest' project the Royal Navy had ever undertaken.[133] When a paper entitled 'What's So Special about the Polaris Programme?' was circulated to government departments in early March 1963, it stressed that:

> the size of the programme, physical and financial – the inter-connection of the ship and system, design and production, procurement and installation – the elements of bricks and mortar; assembly and stowage; selection and training, have all to be combined by January, 1968, to produce the first of a class of submarines which achieve consistently standards of reliability higher than we have ever demanded before.[134]

Past experience of procuring smaller and less advanced weapons systems had been 'far from simple and subject to delays and difficulties over which we inevitably have no control'.[135] The political directive to get the Polaris fleet to sea as quickly as possible meant that the programme could not afford to be delayed in any way by 'teething problems' common in other first-of-class ships.[136]

Although Macmillan had agreed to assign the Polaris fleet to NATO as a contribution to the Western Strategic Deterrent, the 'supreme national interests' clause in the Nassau Agreement meant that it could also be used independently. The level of deterrent strength for national purposes was therefore an important factor in determining the size of the Royal Navy's Polaris fleet. In a paper for the Cabinet's Defence Committee, Thorneycroft defined the requirement as 'the continuous ability at any moment to achieve 50 per cent destruction of between 15 and 20 Russian cities including Moscow and Leningrad', which could be achieved by having between 24 and 32 Polaris missiles on station at any one time.[137]

Before Nassau, the Admiralty envisaged using seven smaller,

dual-purpose submarines, each armed with eight Polaris missiles, to meet this criterion.[138] However, Carrington advised Thorneycroft that 'the position may have to be reassessed' after learning of McNamara's concerns about the technical delays and vulnerability of the dual-purpose concept.[139] A Polaris fleet consisting of four non-dual-purpose submarines, each carrying sixteen missiles, was now the favourite.[140] One Polaris submarine, carrying sixteen missiles, constituted 'a 7 or 8 city deterrent', while two submarines carrying a combined total of thirty-two missiles constituted a twenty-city deterrent.[141] A sixteen-missile fleet could be constructed and deployed two years earlier than the hybrid, in 1969 as opposed to 1971, a date that was now seen as 'unacceptable and would open the door to major expenditure on gap filling'.[142] A fleet of sixteen-missile submarines was also more cost effective. Although estimates indicated that each submarine would cost £38m compared to £32m for a hybrid, the combined cost of four sixteen-missile submarines in money and manpower was considerably lower per missile on station, £220m as opposed to £290m for a hybrid fleet.[143]

The size of the Polaris fleet was also linked to the second major decision that was taken: what to do about the Navy's nuclear-submarine programme? The Admiralty had briefly considered converting HMS *Valiant* into a Polaris submarine, just as the Americans had done with the USS *George Washington*. However, this was discarded because *Valiant* was practically complete and Vickers would have to slow down work considerably, just at a time when they had built up a 'good tempo' on nuclear-submarine construction. It was also considered far better to overcome the problems of building an all-British submarine before 'superimposing it on the difficult task of a Polaris programme'.[144] *Valiant* was important because it was powered by the unproven British nuclear-propulsion plant that would also propel the Polaris submarines.[145] Completing *Valiant* and the third nuclear submarine, HMS *Warspite*, would allow the Navy to 'meet both building problems and early running problems on them rather than on the Polaris boats'.[146] The Navy therefore decided to keep both nuclear submarines, 'not least', wrote Carrington, 'because it seemed to me to offer an economical means of increasing the Navy's nuclear hunter-killer capability at the same time as we built up the deterrent force'.[147] Operationally, having 'three SSNs available by the time the Polaris fleet was complete to sweep out exit routes' (a process that submariners quickly dubbed 'delousing') when the Polaris submarines sailed from their operating base would also be of considerable advantage.[148]

When the Cabinet's Oversea Policy and Defence Committee met on 23 January 1963, Thorneycroft argued for a fleet of five sixteen-missile-carrying submarines, whereas the Chancellor, Reggie Maudling, argued for either four sixteen-missile submarines or a fleet of six dual-purpose submarines. Thorneycroft countered that 'the arguments in favour of the larger class of boat seemed overwhelming' but in order to appease the Treasury agreed, along with the rest of the committee, to settle for four submarines, the minimum required 'to maintain an adequate deterrent capability at all times', though it also said it would revisit the question of a possible fifth submarine at the end of 1963.[149] It accepted that construction of HMS *Warspite* should continue, but to avoid additional expenditure and straining the resources of the shipbuilding industry agreed that 'no further Killer submarines should be taken in hand during the POLARIS submarine programme'.[150] Although the Polaris Executive had only been in existence for a little over a month on 6 February 1963, the first 'Longcast' for the whole programme had been produced, showing all major critical dates, realistic targets for the construction, fitting out and trials of the submarines and base support facilities. The date for the operational deployment of the first submarine was given as June 1968, with the rest following at six-monthly intervals.

Having settled on the minimum size of the Polaris fleet, the Navy turned to the Polaris missile. By 1963, the SPO had developed and deployed two versions of Polaris, the A1 with a range of 1200 nautical miles and the A2 with a range of 1500 nautical miles. It was also in the process of developing a third, the Polaris A3, with a range of 2500 nautical miles, capable of carrying more advanced re-entry systems and warheads. The Admiralty wanted to 'merge into the American development programme and pick up those items which the Americans had developed, i.e. they would not go back and accept the A1 missile since they would then be two years astern of the US; nor would they insist on A3 if this was insufficiently developed'.[151] The British were therefore initially attracted to the A2, which was available 'off the shelf'. However, the SPO had already announced that it intended to phase out the A2 once it had completed the A3 development programme: if the Navy selected the A2, the SPO would have to keep its production lines open solely to meet British requirements. This was technically difficult and potentially very expensive. More importantly, the Royal Navy would have found itself in a position where it was the sole user of the A2 System and would have lost commonality with the US Navy.[152]

By January 1963, five out of six A3 test flights had failed, with the sixth described as a 'partial' success.[153] Although Galantin told Zuckerman that it 'would undoubtedly be successful and would be coming into service with the USN in August 1964', relying on yet another undeveloped American missile so soon after the cancellation of Skybolt was difficult, and politically very dangerous.[154] 'If we go for the A3 missile and they do a Skybolt,' warned one official, 'we shall look foolish if we have paid over large amounts of dollars with nothing to show for them.'[155] Carrington was 'quite sure that the two cases are not comparable'.[156] But he realized 'that the great argument against it, especially after the Skybolt failure, is the element of risk in going for another weapon still under development'. But in reality the differences were really considerable. Skybolt had never been tested as a whole, nor its principles proven, whereas the American incentive to complete the A3 was much greater than Skybolt. Investment in the US Polaris programme was 'immense'.[157]

There was one major problem with the Polaris A3. Should the British decide to purchase the A1 or A2, the question of development charges would not arise, since they had been incurred by the United States before the signing of the Nassau Agreement. If the British government purchased the A3 and any future missiles, the US wanted the British to make a contribution towards research and development costs. This was a reasonable request. McNamara told Ormsby-Gore that as the British were benefiting from 'the 2.4 billion already spent on development and quite probably billions more in the future' they were getting a 'fantastic bargain'.[158] But politically it was very embarrassing for the Prime Minister, who, in playing up the benefits of the Nassau Agreement, had told reporters that 'in Polaris Britain was getting the benefit of £800m worth of expenditure to which the British taxpayer had contributed nothing'.[159] Now the Americans were asking for an additional payment estimated at anything from £11m to £33m up until 1968.[160] 'They will tear me limb from limb in the House if this ever comes out' was Thorneycroft's initial reaction.[161] Macmillan described it as 'a great shock'.[162] Fearing a 'very important and perhaps decisive' two-day Defence debate in the House of Commons on the Nassau Agreement on 30 and 31 January 1963, Macmillan was concerned that if he had to tell the House that the supply of Polaris was provisional, dependent 'on a satisfactory settlement of the financial arrangement', it 'may result in the defeat of the Government with all the consequences which would follow'.[163]

Hard bargaining followed.[164] Macmillan 'refused to make an open-ended contribution to an unknown bill for Research and Development'.[165] He felt that he had already paid for Polaris by accepting American demands over 'assignment, allocation [to NATO] – call it what you will . . . a very difficult thing for me to defend here with the press and the Party'.[166] To have to pay 'a quite different form of commercial price to that which was included in the original Skybolt agreement – which has only broken down because of the failure of the American Government to go on with it' was, in his view, unacceptable.[167] He instructed Ormsby-Gore to go directly to Kennedy and offer just 5 per cent on top of the production cost of each missile.[168] The President was 'determined there should be no Anglo-American row' over the issue and he instructed McNamara to accept the offer.[169] 'If,' as a very satisfied Macmillan later recorded in his diary, 'we bought fifty million pounds of missiles, we would pay fifty two and a half million pounds . . . Not a bad bargain.'[170] The final settlement saw Britain spend just over £90m on missiles and spare equipment between 1963 and 1971/72.[171]

In fact, the settlement represented a fantastic bargain. The 'on a continuing basis' clause in the Nassau Agreement meant that the 5 per cent formula applied to Polaris and 'any later Marks of missile which we may want to buy'.[172] The importance of this clause would become apparent almost twenty years later when Margaret Thatcher's Government entered negotiations with the United States to purchase a replacement for Polaris known as Trident. The deal was so good that Ormsby-Gore had to warn Macmillan that it 'was such a poor one from the American point of view that they may suggest that you do not spell out the effect of the agreement quite so precisely'.[173] He urged the Prime Minister, when announcing the deal in public, 'to explain that ours is a unique case' as there was 'likely to be strong criticism here of a deal which is so favourable to us'.[174] McNamara was furious. He had told Zuckerman 'that if ever a Polaris submarine was fired he hoped that the opening shot would be the then MOD Public Relations Officer, who had done a grand job in painting the picture of how the US had let us down over Skybolt; and no-one would have known that he, and not a missile, was in the tube!'[175]

There was only one shipyard in the United Kingdom capable of constructing a Polaris submarine within the very tight timetable – Vickers-Armstrongs at Barrow-in-Furness. As a paper to the Cabinet's Economic Policy Committee put it, 'they alone of British shipbuilders have experience of nuclear submarines, they already possess at Barrow much of the special equipment that is required for nuclear submarines [the government had already invested £1.5m in capital equipment for the construction of HMS *Dreadnought*] and they have very good working relations with the Electric Boat Company of America whose advice and technical assistance will be essential'.[176] Vickers provided the Admiralty with assurances that it had the capacity to complete the submarines in the time required, without any detriment to the completion dates and cost of *Valiant* and *Warspite*.[177] However, Vickers' assessment was seen as 'over optimistic' and the Admiralty concluded that the shipyard had insufficient capacity to build more than two Polaris submarines at the same time. A second shipyard was required. Two shipyards with recent experience of submarine construction, Cammell Laird at Birkenhead and Scotts at Greenock, were considered. Cammell Laird had the strongest case. It was the second-largest shipbuilding firm in the UK and had recent experience constructing conventional submarines, as well as a modernized shipyard and excellent docking facilities.[178] However, as they had never built a nuclear submarine before, Baker warned quite presciently that 'we must not be surprised if Cammell Laird fail to honour their delivery promises'.[179] Vickers would construct the first and third Polaris submarines while acting as 'lead yard', guiding Cammell Laird with the second and fourth.

An equally urgent question was where the Polaris submarines, their missiles and warheads would be based. From the outset, and largely derived from a recommendation in Palmer's 1961 report, Devonport was 'generally thought to be the best choice'.[180] A number of documents prepared in early February 1963 all assumed that Devonport would be the operating base.[181] But Mackenzie's temporary replacement as FOSM, Commodore Ernest Turner, was 'struck by the assumption that Devonport' was 'the best choice for a forward operating base' and objected. He thought Devonport was in 'danger of being accepted without a proper study of the requirement having been made'.[182] A working party was established to examine, evaluate and make comparative costings of various locations across the UK.[183] In total, the working party conducted eleven detailed site studies: Portland; Devonport; Falmouth; Milford Haven; Gareloch; Clyde Area (other than Gareloch); Fort William; Kyle of Lochalsh; Loch Ewe; Invergordon; and Rosyth.[184] The three most important factors were operational, safety and cost. Safety was overriding. As one Admiralty brief concluded, 'should the worst happen, Ministers must be able to stand up to searching fundamental questions about the reason for deciding on the acceptability of risks in a particular location in the first place and about safety measures taken on the spot. One question is bound to be "Was there no safer location? If so why was it not used?"'[185]

The working party concluded that on safety grounds Devonport was not suitable due to the dockyard's proximity to the population of Plymouth. On operational grounds the dockyard only had a limited number of berths for unrestricted regular access by nuclear submarines and it was too far away from sound-ranging facilities, which could not be provided locally. There was also only one possible site in the area for a new armament depot and, as the Admiralty did not own it, planning permission for development would be required. It was 'unrealistic to ignore the probability of strong objections on the part of the present occupiers' and the public when attempting to acquire the new land.[186] 'In view of its serious safety objections and operational limitations,' concluded the working party's final report, 'we cannot recommend Devonport.'[187]

The working party concluded that Rosyth and Faslane on Gareloch were the most suitable locations. On safety grounds there was very little between them, but on operational grounds Faslane was judged to have a number of advantages. The Captain of the Third Submarine Squadron, Ian McGeoch, had transferred the Squadron and its Depot ship, HMS *Adamant*, to the area from Rothesay in 1957.[188] The location was close to the Royal Navy's northern exercise areas, a sound range in Loch Fyne and the Clyde degaussing range, a means of erasing the magnetic field from a submarine to camouflage it against magnetic detection. This was important because of the rigid requirements of the operating cycles. In order to ensure that two submarines were kept on station as much as possible it was important to use the lay-off periods as productively as possible rather than wasting time moving the submarines to exercise areas (Rosyth was over 500 miles from the nearest sheltered sound range).[189] Faslane was also ideal because the large amount of shipping and numbers of submarines moving in and out of the Clyde area would help disrupt enemy surveillance of the Polaris fleet. The approaches to Gareloch offered very unfavourable conditions for hostile anti-submarine surveillance, whereas the approaches at Rosyth were much more open. But Faslane did have some disadvantages. It was estimated that constructing the base would cost £20m–£25m, whereas Rosyth would cost around £2.5m less. The proposed site was also very close to a Ministry of Power emergency oil storage installation which might have to be moved as the whole base would become a target for a

Soviet nuclear strike. In addition, Metal Industries Limited had just secured a lease for jetty space after years of protracted negotiations with the Ministry of Transport. The prospect of renegotiating the contract, at considerable time and expense, was not attractive to the Admiralty.

Faslane's proximity to two suitable sites for a Royal Naval Armament Depot to assemble, test, store, load and unload the Polaris missiles and warheads overcame these disadvantages.[190] A site at Glen Douglas was ruled out almost immediately because to load and unload the missiles from the submarines, it would have been necessary to moor within a few yards of the busy Glasgow to Oban road. The Glen Douglas site was also adjacent to a NATO depot and the working party concluded that there would be strong political objections to putting the national strategic nuclear deterrent in such close proximity to a NATO facility. The second site at Coulport was far more suitable. Although configuration of the land meant that development was likely to be expensive, the fact that the Admiralty already owned an existing Hydro Ballistic Research Establishment there was of considerable advantage, as was its deep-water jetty and proximity to Faslane.[191] From start to finish it took the working party a fortnight to recommend Faslane.[192] The Admiralty Board agreed that 'the operational advantages and other conveniences of Faslane made it the proper selection, notwithstanding the additional cost'.[193]

The Admiralty desperately needed more information about Polaris from the Americans. But in order to access it a new agreement was required that would turn what had been agreed at Nassau into a practical working arrangement. In mid-February 1963 a technical and political team was dispatched to Washington to conduct a series of negotiations that would lead to one of the most important and enduring US/UK post-war agreements, the 1963 Polaris Sales Agreement (PSA). Once in Washington the technical team split into various sub-committees, each responsible for studying a particular area such as the re-entry body, warhead, shipyard services, security, patents, warranties, indemnities and financial matters.[194] Both sides agreed early on that it was impossible to produce an agreement that included the full specifications of items and equipment required to meet British needs. The aim rather was to produce a 'permissive and not binding' agreement that ensured the British had 'the option to buy, if we wished, any equipment which the Americans produced as part of the Polaris weapons system'.[195] They also agreed to set out the detail at a later date in a series of technical arrangements, basic ground rules to implement the programme at the working level. Article II of the agreement authorized each country to enter into 'such *technical arrangements* consistent with this Agreement, as may be necessary'. By 1970, once all four British Polaris submarines had put to sea, a total of forty-eight Technical Arrangements were in operation.[196]

Negotiating terms that governed the supply of the missiles and related equipment were relatively simple, although talks on aspects of the warhead and re-entry body were 'made all the heavier because of the bevy of lawyers on the US side whose main aim is to protect themselves by reflecting all conceivable legislative requirements into the agreement'.[197] However, in general, there were 'no insuperable problems'.[198] An agreement was quickly assembled consisting of fifteen articles and a number of confidential secret annexes setting out the terms, conditions and arrangements for the supply of Polaris missiles (minus the warhead) and equipment, and measures for governing development work carried out by the US on behalf of the UK. It granted the UK use of the Atlantic Missile Range at Cape Kennedy, now Cape Canaveral, for test launches, and set out a financial procedure for payments consisting of twenty-four equal instalments of $729,167 payable quarterly into a special Polaris Trust Fund.[199] It also outlined the institutional links between the Polaris Executive and the SPO, principally a commitment to establish a Joint Steering Task Group (JSTG) to meet four times each year in London and Washington to monitor the programme. Significantly, the agreement made no provision for termination.[200]

During the negotiations the Navy argued that in order to build Polaris submarines it required information about the latest US nuclear reactors. But the final agreement did not 'authorize the sale of, or transmittal of information concerning the nuclear propulsion plants of United States submarines'. The problem was Admiral Rickover. As we have seen, having fulfilled his obligations with respect to HMS *Dreadnought*, Rickover was unwilling to provide any further assistance to the British.[201] Rickover also had a difficult relationship with the SPO. He had repeatedly tried and failed to expand his role in the US Polaris programme, but he had been deliberately shut out, confined to supplying the reactors for the US Navy's Polaris submarines.[202] The immediate impact of this restriction on the UK programme was limited. Although US technical information would have been useful, Mackenzie had already concluded that the British nuclear reactor based on the Dounreay prototype should power the Royal Navy's Polaris submarines. Although there were important legal and technical reasons for doing so, Galantin had also advised him to 'go for our own design of nuclear reactor in order to insure that Admiral Rickover had no grounds for interfering in our own Polaris programme'.[203]

After some minor amendments, the Polaris Sales Agreement was signed in Washington on 6 April 1963 and placed before Parliament on 9 April 1963. It was the ultimate piece of that 'great prize' Macmillan had spoken about in 1958 when Anglo-American nuclear collaboration had been restored.[204] Two hours after the agreement was signed, the Polaris Executive Special Projects Royal Navy (SPRN) team landed in Washington and left the Americans 'impressed by the nice timing'.[205] Serious work could now begin. But the Sales Agreement was a political document; how it would work at the operating level was now for the Polaris Executive and the SPO to determine. As Captain Peter La Niece, Mackenzie's representative in Washington, put it, it 'was really only like signing adoption papers. What this lusty new obligation really meant took a lot longer to work out.'[206]

CONSTRUCTING POLARIS

Following the signing of the Polaris Sales Agreement, the Polaris Executive began to bring together all the various components of what was known throughout Whitehall as the British Naval Ballistic Missile System (BNBMS): the submarines, warheads, re-entry bodies, communication facilities, navigation facilities, maintenance support and logistics systems, all of which needed to come together at the right time to deliver a credible deterrent. The timescale was so short that all major activities had to start concurrently, requiring the systematic breakdown of the project into systems and sub-systems. By this time a total of 275 naval officers and civilian staff were now working full time on the programme, fifteen in Washington, twenty-six in London and the remainder in Bath. The shipbuilders, Vickers and Cammell Laird, had started to tackle what was to become one of the most highly organized efforts of planning ever undertaken by the British shipbuilding industry, defining precisely what was needed in terms of works, drawings, orders, programmes, manpower and costs, and upgrading facilities, recruiting staff and coordinating the hundreds of British and American firms involved in constructing the submarines and the many complex components that went inside them.[207] In Scotland, the Admiralty embarked on a 'gigantic effort . . . at breakneck speed' to settle the requirements for the support facilities, and began to construct what would likewise become one of the largest works projects ever under-taken by the Admiralty.[208] In the Ministry of Aviation and the Atomic Weapons Establishment at Aldermaston, clarity on the type of missile allowed work on the 'one – and vital bit of the POLARIS weapon sys-tem which is British made', the nuclear warhead, to proceed with all dispatch.[209]

Normally, a new submarine design evolves out of a series of studies into requirements and desired capabilities, which are then traded off against costs.[210] With the Polaris submarines, the staff requirements (the main features, functions and concept of operations of the submar-ine), were completed using what was described as an 'abbreviated procedure' in order to have them completed by the end of February 1963.[211] These requirements specified a submarine that could patrol undetected for fifty-six days and fire its missiles within one hour's notice while operating under dual NATO – national control. The sub-marine had to be able to defend itself, mount a counter should it come under attack, maintain simultaneous, continuous and instantaneous communications with shore installations, and possess the ability to change the target allocation of the missiles within five minutes, all while achieving a reliability factor of 99 per cent.

By 1962, the Admiralty's naval architects were ready to exploit the knowledge they had gained from designing and building *Dreadnought* and *Valiant*. The fact that the *Valiant* design was so far advanced was crucial to the success of the programme.[212] By using it the architects were able to produce one for a Polaris submarine in a remarkably short space of time and avoid 'the evil of everything new at once'.[213] But it would be a mistake to think of the Polaris submarine as a simple cut and shunt job. This was not just a case of cutting an existing British submarine design in two and inserting a self-contained missile com-partment in the middle, which would give 'a false impression of the magnitude of the task'.[214] The design 'was entirely new' and involved 'the installation of a highly sophisticated and closely integrated weap-ons system of US design and manufacture in a British built nuclear submarine of advanced design' that at that stage was still unproven.[215] This was a 'stupendous, mammoth task', but it was completed on time and submitted to the Admiralty Board for approval on 6 February 1964, 'a quite remarkable achievement'.[216]

With a submerged displacement of around 8550 tonnes (surfaced 7750 tonnes), the submarine was 43 metres longer than 'Valiant', only half of which was directly attributable to the missile compartment; the rest was required for other associated changes.[217] In total, around 50–60 per cent of the hull construction, propulsion machinery and hull services were the same as in 'Valiant'.[218] The design also differed considerably from US Polaris submarines. The most obvious superficial alteration was the relocation of the hydroplanes (horizontal rudders) from the conning tower (as in US submarines) to the upper part of the forward hull. The challenges involved in grafting the Polaris weapon system onto the 'Valiant' design were considerable. In nearly every boundary between the US equipment and the UK submarine, piping, wiring and power supplies had to be integrated.[219] To facilitate the fitting of US equipment, the designers decided that the frames, decks and bulkheads in the missile compartment should follow those of the 'Lafayette' class submarine. But it was not an exact 'Chinese copy'. The hull diameter of the missile compartment had to be reduced by 3 inches in order to make it fit with the 'Valiant' hull and there were some minor changes at the top and bottom of each missile tube.

Mackenzie knew that 'If our Polaris submarines are to mean anything at all they must be as near 100% perfect as possible.'[220] This placed a great responsibility on the shipbuilders. If they failed to construct the submarines on time, the government would be unable to maintain its deterrent policy. Mackenzie was initially unimpressed with the way both shipyards, in particular Vickers, approached the complex task of managing the construction programme. As far as he was concerned both shipbuilders 'had no conception of what they were taking on in building, fitting out, and testing and tuning of an SSBN.'[221] Vickers, he said, was 'virtually still working in the nineteenth century' and had yet 'to move into the twentieth century in their methods of organizing themselves to address such a complicated task'.[222] It took 'months and months of argument and persuasion' before both shipyards 'committed the resources necessary for detailed planning, which was absolutely essential if the SSBNs were to be built on time'.[223]

Before physical construction of the submarines could start 'both shipyards embarked on a task which involved an expansion of facilities, an increase in staff, a change in the structure of staffing, and a profound impact on the standards of work and the nature of management operations'.[224] In 1963, the workforce at Vickers numbered 3100 men. By 1967, this had risen by 45 per cent to 4500.[225] The most dramatic change was in the numbers of support staff. In 1963 there were 800. By 1967 there were 2400 involved in expanded planning, testing and quality control organizations.[226] The ratio of qualified to unqualified staff increased from 1:5 to 2.5:4.[227] At Cammell Laird, employment also reached new highs, averaging 8280 between 1960 and 1964 and reaching a peak of 11,400 by 1969. Polaris dragged 'the ship-builders . . . into the modern age of submarine construction'.[228]

Towards the end of 1963, the new First Lord of the Admiralty, George Jellicoe, steered the Navy through what the Flag Officer Sub-marines, Admiral Sir Horace Law, later called 'one of the most painful episodes' in the entire Polaris programme.[229] The US S5W reactor first fitted in USS *Skipjack* and used by the Royal Navy in *Dreadnought* contained valves, piping and fittings that were constructed out of stain-less steel. Those responsible for designing the British prototype plant at Dounreay concluded that stainless steel was very liable to stress corro-sion and they did not want to use it in situations where it might come into contact with sea water. Despite warnings from Rickover the designers decided to use a different material, a chrome-moly steel and a material known as Inconel.[230] This was a costly mistake. In Novem-ber, as Dounreay was nearing completion, engineers discovered serious flaws in some of the welds in certain pipes and valves of the British prototype nuclear plant. The successful installation and proving of the plant in *Valiant* was 'seen to be absolutely critical' as it was going to power the Polaris submarines.[231] In total there were 1658 affected welds and installation work had already started in both *Valiant* and *Warspite*. None of them could 'be regarded as trustworthy' and by December Rolls-Royce had ceased all work.[232] Engineers were unable to find a 'scientific or engineering explanation' and 'no remedy' was 'available or in sight'.[233] The only solution to this serious, 'not calam-itous but . . . serious enough' problem was to replace the defective parts in Dounreay, *Valiant* and *Warspite* with a stainless steel alternative.[234]

The technical capability to design and produce such components in the UK did exist, but it was impossible for UK industry to produce what was needed without causing significant delays to both the *Valiant* and *Warspite* construction programme. This had important implica-tions for the Polaris programme as there would be, Le Fanu argued, 'less and less time in which to prove at sea . . . the operation of the nuclear propulsion plant on which the SSBNs are utterly dependent'.[235] Using a UK-sourced alternative in *Valiant* and *Warspite* was out of the question as 'the probability of delay to SSBN.01' was judged to 'be very strong'.[236] 'The only feasible alternative,' argued Mackenzie, was 'to acquire the stainless steel pipes and valves from the only other source, the United States.'[237] But that meant going back to Rickover.

The Chief Scientific Advisor at the Ministry of Defence, Solly Zuckerman, was sent to the United States to negotiate with Rickover. The two had first met in 1962, after Rickover had read an article Zuckerman had written on 'Judgement and Control in Modern Warfare' for the journal *Foreign Affairs*.[238] Thereafter the scientist in effect replaced Mountbatten as Rickover's preferred channel of communication with the Royal Navy and the British Government.[239] Rickover, who in Zuckerman's view had become 'more paranoid as the years passed' since the 1958 Agreement, agreed 'somewhat reluctantly' to supply the steel, by opening up US naval stores and allowing the British to purchase what they needed.[240] 'Unfortunately,' as Zuckerman told Thorneycroft, 'he did so very largely on the basis of the personal relationship he has now established with me.'[241] And there was a price. Rickover saw an opportunity to use the fact that the British required his help to secure concessions in the 'lengthy and difficult' negotiations that had been taking place since 1960 over a memorandum of understanding permitting the use of Holy Loch in Scotland as a base for US Polaris submarines.[242] Rickover informed Zuckerman that 'The US helping us over our reactor troubles would be conditional on our coming to an agreement over the Memorandum of Understanding.'[243]

By the middle of January 1964 negotiations had come to 'a complete halt' due to Rickover's refusal to allow British inspectors access to US submarines using Holy Loch so that they could conduct safety inspections and monitor radiation levels.[244] Rickover had become very 'bitter' about the number of Foreign Office redrafts of the sections of the Holy Loch memorandum dealing with safety and inspections and he was 'determined to scrutinize carefully all further changes and suggestions'.[245] But it was apparent that there were other factors driving Rickover's behaviour. One of the fundamental reasons he had agreed to the sale of the S5W plant for *Dreadnought* 'was to prevent the British having continuous and direct access to the American project, because he believed that responding to their queries would disrupt American progress and deflect him from his singleminded determination to develop nuclear plants for the USN'.[246] Yet here were the British asking for yet more help. 'They ask for too much,' complained Ted Rockwell, one of Rickover's officials.[247] On 22 January, Mackenzie warned Le Fanu that:

> the most serious aspect of further delay in receipt of the stainless steel is that each week shortens by the same time the now marginal period for proving *Valiant*'s reactor plant, the pilot U.K. seagoing system, and

incorporating the almost inevitable modifications which will prove necessary before reactor plant installation in SSBN.01 must start ... The risk
that further delays in receiving stainless steel from the US will be reflected
in delays to SSBN.01 is therefore present from now on.[248]

Far more worrying was the fact that 'both *Valiant* and *Warspite* in
their final months at Barrow will clash with the initial stages of
SSBN.01 testing and tuning'.[249]

Rickover still refused to provide what was needed.[250] 'If the safety
clause in the Holy Loch Agreement is satisfactorily settled,' Zuckerman
repeated, 'supplies of stainless steel will be forthcoming quite quickly.'[251]
Eventually, the Foreign Office conceded that Nassau and the Polaris
programme had put the government 'very much at the Americans'
mercy' when negotiating the Holy Loch memorandum.[252] When the
Cabinet Committee responsible for the Holy Loch negotiations, GEN
836, eventually approved the memorandum on 24 January 1964, without the right to access certain areas of US submarines, it did so because
ministers feared that Washington would 'withhold certain vital equipment'.[253] Rickover never wavered from the pursuit of US interests and
the British government had little choice but to accept his demands to
keep the Polaris programme on schedule.

The fallout was considerable. Although Rickover had got what he
wanted, according to Vice Admiral Louis Le Bailly, the Washington
Naval Attaché, Rickover was furious with how the whole affair had
been handled and he 'severed relations with the British embassy including my office'. Thereafter, Le Bailly wrote, Rickover's 'gall was well
reserved for the UK end of the Polaris affair'.[254] Aside from conceding
over Holy Loch, the British also accepted tighter security restrictions,
including an assurance that 'No information on this equipment will be
published; nor will it be provided to foreign nationals.'[255] This was
accepted, but it put the government in an awkward position. It had
always maintained that the Dounreay prototype was British designed
and built. In reality, it owed a 'very great deal to American information' and was sometimes referred to by the Navy as the 'Son of S5W'.[256]
Putting US-purchased auxiliary systems in both *Valiant* and *Warspite*
meant that it was harder to sustain the argument that the end product
was more British than the sum of its parts, as one official, Philip Newell, explained:

The fact remains that the plant is of British design, for which the Americans bear no responsibility whatever. Equally, however, it would be

churlish and unrealistic not to recognise that it is based on American information and experience. We cannot spell all this out in too much detail without infringing security as agreed with the Americans; in particular we <u>cannot</u> say we are procuring any items from America.[257]

Newell was 'tempted to make use of the analogy that e.g. the fact that Sir Winston Churchill had an American mother does not make it difficult for us to regard him as very typically English', but admitted that 'the analogy could break down'.[258]

THE FIFTH SUBMARINE

When, in January 1963, the Cabinet's Overseas Policy and Defence Committee decided to order four submarines and postponed a decision on a fifth, it '<u>expected</u> to have two boats permanently at sea from a force of four' and a deterrent capable of inflicting unacceptable damage on 15–20 Russian cities.[259] The operational availability of the Polaris force depended on factors such as reactor core life, inter-patrol maintenance periods, the time taken to refit the submarines by overhauling and replacing machinery and equipment, crew training and weapons trial requirements. The frequency and length of time to refit the submarines were two of the most important factors. When the original four-submarine decision was taken the Royal Navy had no experience of refitting nuclear submarines. It had to rely on information from the SPO, which at that time 'lacked experience' of refitting Polaris submarines.[260] Thus information and analysis it provided to the Royal Navy were 'hypothetical'.[261] By the end of 1963, the initial 'availability figures' had been 'disproved' and the SPO had 'revised (sharply upwards)' its estimates of likely refit times.[262]

Armed with this revised American information the Polaris Executive embarked on 'exhaustive studies' which revealed that with a fleet of only four submarines, two could be kept on station for around 250 days. For the remaining sixteen weeks, split into periods of one month or more, it would only be possible to keep one submarine on station – the Royal Navy could not meet the government's deterrent criteria as one submarine was only capable of providing a seven- or eight-city deterrent. And, obviously, should something happen to that submarine on patrol, the UK would be left without a deterrent. Only with a fleet of five submarines could two submarines (carrying a total of thirty-two missiles, providing a twenty-city deterrent, comparable to that of the V-bomber force) be kept on station at all times.[263] The larger force would also provide a margin against unforeseen events such as an accident or an act of sabotage. Mackenzie was concerned about the 'inevitable pressures' a four-boat fleet would impose on those tasked with operating the submarines and was determined to reduce the pressure 'to tolerable levels'.[264] A fleet of four submarines, he later said, 'was basically unsound', as it 'would impose, in peace time, an unnecessarily high degree of stress and strain on sea-going crews and base staff alike'.[265] He 'maintained that the Polaris force should consist of five . . . instead of four' as 'only in this way could the constant credibility of the deterrent be totally guaranteed'.[266]

The Admiralty Board supported these arguments. Although constructing an additional submarine would delay the resumption of the hunter-killer programme by 6–9 months, increase the financial pressure on the Admiralty and 'exacerbate' an already severe manpower problem (a fifth boat required an extra 45 officers and 360 ratings), the First Sea Lord, Sir David Luce, accepted that the Navy's 'inability with four submarines to keep more than one submarine on station at certain times of the year would not be a very seamanlike method of operation'.[267] The Vice Chief of Naval Staff concluded that 'in spite of the formidable problems arising from the manpower requirements, the sounder <u>military</u> solution is to have five boats'.[268] But the Board recognized that any decision went beyond mere operational considerations.[269] The debate over the size of the Polaris force went to the very heart of the arguments for Britain's nuclear deterrent, as Jellicoe explained to Thorneycroft on 2 December 1963:

a proper judgement of this difficult issue largely turns on whether one regards the British POLARIS fleet primarily as a contribution to the Western deterrent as a whole or whether one views it primarily, albeit in the last resort, as an independent national deterrent. If one takes the first view, then I really do not think it makes a lot of difference if there are four or five British boats. The difference between 45 and 46 SSBNs in the Western POLARIS deterrent force is obviously quite marginal. But if one takes the second view, then the difference is of a quite different order of magnitude. Here something may turn on whether we need a 15 or 20 city independent national deterrent or whether we can do with less. All in all, however, I do not myself think that anything less than two POLARIS submarines 'on station' all the year round constitutes a credible independent deterrent. The logic of this is that if we regard the British POLARIS force primarily as an independent deterrent, then we should plump for the five rather than the four boat concept.[270]

The Chiefs of Staff certainly recognized the military case for a fifth submarine but they continued to harbour considerable doubts about its expense, deciding that 'no more money should be spent on the Polaris Force than was necessary to achieve a credible deterrent'.[271] They advised Thorneycroft that they were unwilling to 'exacerbate the man-power problem' in the Royal Navy and were 'unwilling to accept the inevitable reductions elsewhere which would result from the expend-iture on a fifth boat'.[272] In a democracy, they wrote, 'the final decision on whether to have a fifth submarine was, and must remain, a political one'.[273]

At the ministerial level there was considerable disagreement. Ini-tially, Thorneycroft agreed 'with some reluctance' with the Chiefs' advice and concluded that 'we should not add the fifth boat to our programme'.[274] The Chancellor, Reginald Maudling, and the Chief Secretary to the Treasury, John Boyd Carpenter, were also 'both strongly opposed to a fifth Polaris submarine on financial grounds'.[275] However, the Prime Minister, now Sir Alec Douglas-Home, a firm believer in the deterrent 'and Britain's ability to use it independently', took a different view.[276] During a visit to Washington in January 1964, he dis-covered that the US intended to propose an international 'freeze' on nuclear delivery vehicles.[277] Although he was assured that the Royal Navy's Polaris fleet would 'escape being caught' by the US proposal, there were wider implications.[278] If the freeze were implemented, the NATO Alliance would still enjoy nuclear superiority over the Soviet Government. However, if NATO collapsed and the United States with-drew from Europe the Soviets would enjoy permanent superiority over the UK and French nuclear arsenals.[279]

When Douglas-Home returned to the United Kingdom he told the Cabinet that the proposed freeze made 'it even more essential that we should be seen to retain an independent capacity to inflict an unacceptable degree of nuclear damage upon an enemy'.[280] In order to reinforce the case for the fifth submarine and to convince those ministers who continued to oppose it, he deployed an entirely new argument that aroused what Michael Quinlan, a former Permanent Secretary at the Ministry of Defence, called the 'national gut feeling'.[281] France had started to develop its own indigenous submarine-based deterrent, which was going to consist of five submarines. Douglas-Home argued that if the freeze became a reality, the French could end up with a superior nuclear force. 'We should not wish to be inferior to the French,' he said.[282] The perception of inferiority to the French had and still has the ability to 'twitch an awful lot of very fundamental historical nerves'.[283] Despite concerns about cost and opposition from the Treasury, the Cabinet approved the fifth submarine on 25 February 1964.[284]

The day after the Cabinet approved the fifth submarine, the Director General Ships, Sir Alfred Sims, presided over the laying down of the first 100-ton prefabricated circular keel of the lead submarine at the Vickers shipyard. This small but important milestone, Mackenzie later recalled, 'was a momentous, even historic moment achieved on time'. It was a 'significant indication of progress' and 'an encouraging augury for the future' which showed the country and indeed the world that the programme was on schedule.[285]

But just as physical construction of the submarines was beginning, the Polaris programme was becoming a controversial political issue.

THE 1964 GENERAL ELECTION

Harold Wilson had taken over leadership of the Labour party a month after the Nassau Agreement was signed. In the twenty months that followed, Labour, as Wilson later wrote, 'opposed the decision, and opposed still more the pretence that Britain had an "independent" nuclear weapon'.[286] The party certainly spent much of the almost two-year period between the signing of the Nassau Agreement in December 1962 and the general election in October 1964 criticizing the Polaris programme and the government's obsession with the independent nuclear deterrent, much to the annoyance of the Conservative government. When Douglas-Home became Prime Minister on Macmillan's resignation in October 1963, he had pledged, in his first speech to the House of Commons, to put the whole question of Polaris and the future of the nuclear deterrent before the electorate at the next election.[287] When the two parties released their manifestos shortly before the start of campaigning in September 1964, the Conservatives declared that 'Britain must in the ultimate resort have independently controlled nuclear power to deter an aggressor' and that 'Only under a

Conservative Government will we possess it in the future.'[288] Labour appeared to adopt the opposite position:

> The Nassau agreement to buy Polaris know-how and Polaris missiles from the USA will add nothing to the deterrent strength of the Western Alliance, and it will mean utter dependence on the US for their supply . . . [Polaris] will not be independent and it will not be British and it will not deter . . . We are not prepared any longer to waste the country's resources on endless duplication of strategic nuclear weapons. We shall propose the re-negotiation of the Nassau agreement.[289]

The government had started to prepare for a possible election confrontation over Polaris immediately after Douglas-Home became Prime Minister. When Lord Jellicoe succeeded Lord Carrington as First Lord of the Admiralty in October 1963, he was surprised at how much of the controversy surrounding Polaris was 'based on a widespread ignorance of the facts'.[290] This ignorance was, Jellicoe said, 'dangerous, especially in the present political situation' and he informed Thorneycroft that he intended 'to bring the facts about the British Polaris submarine force home to a wider audience – in Whitehall and in the Navy as a whole'.[291] He attempted to lessen the controversy surrounding the programme. A number of Conservative politicians were concerned that the name of the first submarine, HMS *Revenge*, was inappropriate. 'These strike me as most excellent names,' Jellicoe explained to Le Fanu:

> But I have some doubt about the order. I suppose that *Revenge* is about the best name for a ship which one could possibly have. Nevertheless, I am inclined to wonder whether we would be really wise to Christen SSBN 01 *Revenge*. The essence of the British Polaris fleet is deterrence and the concept of *Revenge* seems to me to imply that deterrence would have failed. I am a little bit inclined to feel that those who are opposed to the concept of a British deterrent would find this grist to their mill.[292]

The Royal Navy promptly renamed the first submarine *Resolution*.[293] *Revenge* was consigned to the last submarine in the hope that by the time it was launched in 1969, much of the controversy surrounding the Polaris programme would have passed.

The prospect of a new government cancelling the programme was very attractive to those in the Royal Navy who continued to harbour reservations about it. Zuckerman noted in his memoirs that 'some of the Navy Chiefs had cautiously assumed that the new Government would scrap Polaris', especially those who believed that Polaris was having a very real impact on the 'proper', conventional Navy.[294] When Peter Nailor, a civil servant who served in the Polaris Executive, was conducting research for his analysis of the Polaris programme, *The Nassau Connection*, eight out of eleven 'very senior people' he interviewed, who had not been directly involved in the programme, used precisely the same words when asked about its impact: 'a frightful chore'.[295] He could only conclude that there was a 'strong current of feeling' among senior officers that the programme was having a detrimental effect on the naval service.[296]

The government put enormous 'political pressure' on Mackenzie to press on with the programme so that a Labour Government would face 'quite a large bill should they decide to cancel it'.[297] He was told to redouble 'efforts to keep the programme forging ahead' in the hope that 'if enough progress could be achieved and sufficient money firmly committed by the time of the election, the future would be more assured'.[298] The technical departments in Bath were told to spend as much money as possible. 'Apart from being a good management technique,' Jack Daniel later recalled, 'it was, in part, to make it more costly and difficult for a Labour government to cancel the Polaris programme.'[299] The Atomic Weapons Establishment at Aldermaston also came under considerable pressure to conduct the test of the Polaris warhead before the election, 'in September or as soon as possible thereafter'.[300] Eventually, Labour's negative campaigning came to have an 'immobilizing effect' on the programme and the Polaris Executive was 'hard put to it to maintain the momentum which the programme required'.[301] Mackenzie later referred to a 'never-ending fight against anything that could cause the programme to falter'.[302] There was 'no doubt', he later wrote, 'that the wide publicity given to Labour's views in the run-up to the election gave rise to doubts and fears amongst many, in industry and elsewhere, whose wholehearted co-operation was critical to the Polaris programme'.[303]

This was especially apparent when it came to the two shipbuilders, Vickers and Cammell Laird. Both were certainly expecting cancellation and each sought assurances from the Navy about the operation of the Shipbuilding Break Clause that was written into the contracts.[304] Recruitment became a problem. Both shipyards struggled to hold on to and recruit skilled workers 'as a result of the political uncertainty'.[305] In Barrow, the Vickers workforce realized that because of the importance of the programme they could effectively blackmail the management by striking.[306] Polaris also became an important local-election issue in Barrow. Conservative Head Office bombarded the town with vast amounts of campaign material: 'No Polaris, no Barrow. Do not take a gamble. Polaris means peace for Britain, prosperity for Barrow'; 'Hands off Polaris. Britain has 50 million lives in danger. Barrow has 4,000 jobs in danger'; 'Wilson's anti-Polaris means a defenceless Britain and a destitute Barrow'; 'Polaris is Britain's only sure defence in the 1970s. Barrow is building this defence.'[307] The local Labour MP, Walter Monslow, a seasoned campaigner, who had been involved in no fewer than seven general elections, remarked that he had 'never known such scare tactics as those indulged in' by the Conservatives.[308] They 'had a tone of jingoism at its worst'.[309]

In Whitehall, the Chiefs of Staff were certainly expecting cancellation.[310] The CDS, Lord Mountbatten, was 'dismayed' at the position Wilson had adopted.[311] On 29 September, he assembled the Chiefs and told them that he wanted a paper that he had commissioned in 1962 'on the importance of the retention of the United Kingdom independent nuclear deterrent' updated for use 'in the event of a new Government proposing to abolish this force'.[312] In his view, the Chiefs 'were under a moral obligation to put the military aspects of the problem without delay to any Government which might consider abolishing our independent deterrent'.[313] Mountbatten firmly believed that Wilson would cancel the programme if he became Prime Minister:

> On the subject of the independent nuclear deterrent however the Leader of the party was known to be in favour of abolition; the majority of the party were in agreement and abolition was included in their election manifesto. There was nevertheless a body of opinion in the party who were uncertain on this point and it was possible that, if a strong case could be made out for retention, Labour policy on the subject could be reversed, providing that a face-saving formula could be found.[314]

Mountbatten thought that 'the best method of achieving this lay in convincing an incoming Labour Government that the Polaris submarine project was so far advanced that very large nugatory expenditure and impracticable conversion problems would arise if any attempt was made to convert the Polaris submarines (SSBN) into Hunter Killer submarines'.[315] He proposed that the Chiefs should prepare and personally sign a paper which stated that:

> it was the traditional responsibility of the Chiefs of Staff of the Services to defend these Islands against all forms of attack, and recall that this

responsibility had originally been discharged by the Royal Navy with the
Army in support, and then by the Royal Air Force up until the advent of the
thermo-nuclear weapon. The paper should go on to say that now and in the
foreseeable future the Chiefs of Staff saw no way of continuing to discharge
this defence responsibility except by the possession of a nationally controlled nuclear deterrent force, of such a capability that it could inflict upon
any aggressor such a degree of damage as to outweigh any possible benefit
which he might obtain from his aggression. The paper should conclude by
asking that if the Government wished to do away with our deterrent force,
they should formally absolve the Chiefs of Staff from further responsibility
for the defence of the United Kingdom against attack.[316]

Such an approach would have almost certainly set the Chiefs on a collision course with a new government.

There were two problems with Mountbatten's 'face-saving formula'.
First, the Polaris programme was not that far advanced. Cancellation
was entirely feasible and the 'beyond the point of no return' argument,
as Zuckerman notes in his memoirs, 'varies arbitrarily case by case [in
the] defence world'.[317] The second problem was that the submarines
could still be converted into hunter-killers. In July 1963, Sir Alfred
Sims had told the Admiralty Board that it would be possible 'technically to complete this ship in the hunter/killer role', but that the
submarine 'would be longer and less manoeuvrable than the *Valiant*'.[318]
In March 1964, Treasury officials were also told, by the naval architects in Bath, that 'If necessary, it would be possible to remove this [the
missile] section or construct these boats without it, although it would
be by no means a simple job and there would be a certain amount of
waste space since there would no longer be any need to house various
computers and firing equipment.'[319] There were even plans for conversion. To satisfy his curiosity, Jack Daniel had removed the missile
system from the submarine and rejigged what was left of the design to
produce a balanced SSN with the minimum waste of ordered materials
and equipment. He had also drawn a profile and plans of a submarine
that was about 30 feet longer than *Valiant* and named the HMS *Harold Wilson*.[320] 'In the current political climate we thought it to be a
good joke,' Daniel later wrote. 'It was shown to a few friends in the
Naval Staff and put away.'[321]

Perhaps this was why the Chiefs put a stop to Mountbatten's plans.
When they next met on 6 October, Mountbatten was away overseas in
West Africa and although the Acting CDS, General Sir Richard Hull,

firmly believed 'that the Chiefs of Staff should resist any attempt to abolish the British Strategic Nuclear Deterrent', he thought that they 'should reconsider their ideas on how best to present their views on the matter'.[322] After further discussion, the Chiefs decided to hold back the paper with Mountbatten's 'face saving formula' on converting the SSBNs to hunter-killers 'for use if required', as:

> it would deploy their complete argument, and this might be premature since the attitude of a Labour administration could not be known for certain until it had gained access to the true facts of the matter. Furthermore, it would be tactless to present an in-coming Labour administration with a bald statement flatly opposing what had been a major plank in their Election platform.[323]

Instead the Chiefs came up with a new three-stage approach. First, they prepared a factual brief on British nuclear forces for the new Secretary of State. Second, they produced a paper that set out the case for retaining the nuclear deterrent that they would send to the Secretary of State in the event that the new administration advocated abolishing the deterrent. Third, they agreed that if the new Secretary of State refused to forward the second paper on to the Oversea Policy and Defence Committee, or if the Committee rejected it, the Chiefs would 'exercise their prerogative and forward to the Prime Minister a third paper putting forward all the facts, and deploying all the arguments, some of which would best be kept in reserve for this purpose'.[324] Great care was taken when preparing these papers. Paragraphs that were deemed 'liable to infuriate a Socialist Minister and thereby spoil any good effect there may be from the main arguments' were removed.[325]

On 16 October 1964, Labour won the general election with an overall majority of just four seats. Harold Wilson became Prime Minister, Denis Healey the Secretary of State for Defence, and Patrick Gordon Walker the Foreign Secretary. The future of the Polaris programme seemed very uncertain.

'GO' OR 'NO GO': DECIDING TO CONTINUE[326]

Wilson's government was forced to confront serious economic problems during its first weeks in office, including an £800m balance-of-payments deficit, double what had been expected.[327] Despite all the political grandstanding of the election campaign, the new government quickly decided to carry on with the Polaris programme. Shortly after the election Healey was informed that it was indeed possible to convert all five Polaris submarines to hunter-killers provided that a decision was taken during the course of the next few months and that 'there would be no technical difficulties in finishing four boats as at present planned as SSBNs, with the fifth boat becoming an SSKN[SSN]'.[328] He was also made aware – almost certainly by the anti-Polaris elements in the Navy – that 'most of the senior admirals were reluctant to take on the Polaris force within their existing budget at the expense of other ships, and were uncertain whether they could find the additional skilled personnel to operate and service Polaris'.[329]

According to Healey, all of this was 'unexpected news'. When he went to deliver it to Wilson and the new Foreign Secretary, Patrick Gordon Walker, they both told him 'not to let other members of the Cabinet know' as they 'wanted to justify continuing the Polaris programme on the grounds that it was "past the point of no return"'.[330] There appears to be a parallel here between Wilson's 'past the point of no return' justification and Mountbatten's 'face-saving formula'.[331] Mountbatten told the Chiefs on 5 November of his efforts to encourage the government 'to retain, in some form, our independent nuclear capability'.[332] He had written to the new Prime Minister three days after the election, setting out the case for maintaining the deterrent.[333] He also claims that he sought 'an opportunity for private discussion of the matter' and that 'after some three hours of closely reasoned argument, Mr. Wilson declared that he was fully convinced of the necessity for Britain to maintain her independent strategic nuclear deterrent'.[334]

Did Mountbatten continue to privately pursue his 'face-saving formula' as he so often did in many other areas of foreign and defence policy, much to the annoyance of many of his colleagues? To do so would not have been out of character. As Zuckerman has noted, 'When Dickie wanted something to happen, big or small, he would use all his wits, his guile, in whatever way seemed appropriate and get it done'.[335] Healey told Wilson and Gordon Walker that when the time came, he would go along with the deception and not reveal the true position.[336] Given the uncertainties, he later said, citing the Cuban Missile Crisis and the first Chinese nuclear test on 16 October 1964, 'we felt, on the whole, it was wise to continue with [Polaris]'.[337] He strongly believed that 'once the United Kingdom had become a nuclear power, it could not turn its back on nuclear power'.[338] Wilson always maintained that he 'kept the Polaris submarine programme because production was beyond the point of no return'.[339]

Why did Labour keep Polaris specifically? 'The basic reason was that the deal which Macmillan had got out of Kennedy was a very good one,' explained Healey. 'It was a very cheap system for the capability it offered.'[340] But cost was not the only factor. 'The real question was whether it was worth continuing with a programme whose real value lay in the ability to have a handle on the Americans.'[341] This mistrust of the Americans and a determination to maintain a measure of British influence in Washington was also driving Wilson's thinking, as he admitted in 1985:

> I didn't want to be in the position of having to subordinate ourselves to the Americans when they, at a certain point, would say 'we're going to use it,' or something of that kind . . . We might need to restrain the Americans, if we learnt about new things that could happen of a devastating character.[342]

There was also the problem of the NATO Multilateral Nuclear Force, or MLF. Labour and the Conservatives had both strongly opposed the MLF since it was first proposed in the late 1950s. Healey regarded it 'as militarily unnecessary, economically wasteful and politically dangerous'.[343] As the Americans showed every sign of being determined to press ahead with nuclear-sharing schemes after the October 1964 election, keeping Polaris gave the government a bargaining tool that it could use to influence discussions about the MLF proposal; it would also help maintain influence in the Alliance and a veto over any possible use of the new force. In the days following the election, the British advanced as an alternative to the MLF the Atlantic Nuclear Force (ANF). The ANF would consist of the British V-bombers (excluding the aircraft that were needed for commitments outside the NATO area), the British Polaris submarines, at least an equal number of United States Polaris submarines (these forces would be 'nationally manned and not mixed manned') and some kind of smaller mixed-manned, jointly owned element, in which the non-nuclear powers could participate. France could also sign up if it wished to do so.

Abandoning Polaris would therefore have created more problems than it solved. Wilson knew, as he later admitted, that the deterrent 'had an emotional appeal to the man in the pub'.[344] His real problem was with the Cabinet, which included a number of ministers who were vehemently opposed to nuclear weapons. When the question of continuing was put before the full Cabinet on 26 November, it was so bundled together with the ANF proposals that the majority of ministers endorsed the decision to continue without any objections.[345] The 'face-saving formula', that construction of two of the submarines 'was already sufficiently advanced to make it unrealistic to cancel the orders', worked.[346] Even left-wing members of the Cabinet such as Frank Cousins, the Minister of Technology, Anthony Greenwood, the Secretary of State for the Colonies and a founding member of CND, and Barbara Castle, the Minister for Overseas Development, all of whom 'were noted for their fervent campaigning in past years for unilateral nuclear disarmament', endorsed the decision.[347] The only consistent opposition to keeping Polaris came from George Wigg, the Paymaster General, and Alun Chalfont, the Minister for Disarmament.[348] Others kept their doubts private. Richard Crossman, the Minister of Housing and Local Government, was struck by Wilson and Healey failing to see how retaining Polaris was 'incompatible with our election pledges because they would claim that our Government was consciously giving up the attempt to have an independent deterrent'.[349]

To secure the Cabinet's approval, Wilson proposed reducing the number of Polaris submarines 'to make it clear that we no longer contemplated the maintenance of an independent nuclear force'.[350] He had already convened a special Cabinet committee (MISC 16) on 11 November 1964 – the smallest ever group of Cabinet ministers to discuss nuclear-weapons questions. There, he along with Gordon Walker and Healey all agreed that:

> three submarines would represent the minimum force which would be acceptable to us in the event of the dissolution of the NATO Alliance . . . The provision of three submarines alone would not make it possible to guarantee that there would always be one United Kingdom submarine on station, but since it would be a part of the agreement that an equivalent number of United States submarines would be committed to the force, sufficient coverage would be provided.[351]

But this went against the advice of the Chiefs of Staff, who argued that while three submarines were militarily sufficient as a contribution to the ANF because they would end up operating alongside an equal contribution of US submarines, such a small fleet would not constitute a viable national force as an insurance against the break-up of NATO, due to the inability to keep one submarine continuously at sea.

The Chief of the Naval Staff, Admiral Sir David Luce, advised

Healey that it was 'nearly, but not quite' possible to keep one submarine on patrol at all times if the size of the fleet was reduced from five submarines to just three. A three-submarine Polaris fleet would have left 'no margin for unforeseen contingencies' and put the Navy 'in a straight-jacket without any days, let alone weeks, to spare'. 'It will not just be accidents which will break the cycle,' explained Luce, 'it will be wearing out or dislocation to the machinery from time to time, however good the workmanship and the maintenance.' He was also concerned about the crews, 'however good the officers, it will be the very devil to keep up the morale of the men who are doing these very long patrols, unless they are convinced that the job is effective and convincing'.[352]

Not everyone shared this view. The Assistant Chief of the Naval Staff (ACNS), Sir Michael Pollock, a gunnery officer who would later become Flag Officer Submarines, argued that 'whether we like it or not' the Navy was probably going to end up 'with a three boat Polaris force committed to an ANF'. He did not believe 'that a three boat force could not provide <u>an independent national deterrent</u>' and felt that now the Navy was 'faced with this probability . . . the proposition deserves re-examination'.[353] In his view:

> We have allowed ourselves to become mesmerised with the American method of operating these boats in their present function of 'bastion of the West': as an independent British force operating nationally after the breakdown of the North Atlantic Alliance there would be no question of them fulfilling this function: all they would be required to do would be to make it clear that the UK could not be trampled on without a price to be paid.[354]

Pollock argued that it was 'quite unnecessary to maintain two or even one boat constantly on patrol' as it was 'inconceivable that the type of circumstances in which we would require to use this deterrent independently of the Americans could arise without a preliminary period of greatly heightened tension between this country and the USSR'. This, he argued, allowed the UK to 'disregard the possibility of a pre-emptive strike from a "clear sky" – which is the only justification for the present pattern of operating'.[355] He recommended operating a three-boat force 'on more naval principles', which would allow 'at least two' to be on station within forty-eight hours, 'which could then maintain a patrol over a period of tension lasting two months – surely a sufficiently long period in which to resolve a crisis of the nature which might have caused them to be sailed initially'.[356]

This was very much a minority view. The Navy would have found it very difficult to maintain a credible and continuous second-strike capability with only three submarines. Establishing sensible and efficient operating cycles with such a small force would have been very complicated. The efficiency and rapid response required of the Polaris system depended on it being maintained at a state of high readiness on a regular cycle by efficient, highly trained crews. Alternative operating arrangements, based on deploying a submarine or submarines in times of tension, were not attractive and in some cases could have escalated crisis as a potential aggressor might interpret the sailing of a submarine during a crisis as an act of aggression.[357]

At a major defence meeting on 21 and 22 November 1964 at Chequers, attended by senior ministers concerned with defence issues together with their Permanent Secretaries and the Chiefs, ministers agreed that two submarines were inadequate to achieve ANF objectives and that five 'was unnecessarily high'.[358] There was considerable support for a force of three submarines, as it represented the 'contribution best calculated to achieve the political objectives in negotiation' and 'the maximum saving for the United Kingdom economy'.[359] George Brown, First Secretary of State, who was 'at first inclined to cancel the Polaris programme altogether', felt that because three submarines:[360]

> would not represent a credible independent deterrent, it would make it
> apparent that we had abandoned any idea of regaining independent
> national control at any time in the future and had committed ourselves
> irrevocably to an international force. While this admittedly did not pro-
> vide full national insurance should NATO break up, not only did the
> latter seem most improbable, but even if it were to happen we could not
> hope to maintain our national security alone, but should be bound to
> seek the negotiation of alternative alliances.[361]

Healey disagreed. He favoured 'the retention of our capability to recover the force to national control, as an insurance against the break-up of the NATO Alliance'.[362] He argued that 'national importance was attached to the maintenance of one submarine always on station' and that a 'force of three submarines would not enable us to achieve this and the sense of national purpose would therefore suffer'.[363] But he was unable to carry all his colleagues.

By December, it was becoming increasingly difficult to postpone a decision on the size of the Polaris fleet, as the fifth submarine was due to be ordered in January 1965. Vickers was complaining that that 'uncertainty' about the government's intentions towards Polaris was 'seriously impeding their efforts to recruit and hold labour'.[364] Healey pressed Wilson to make a decision.[365] On 29 January 1965, the OPDC decided to cancel the fifth submarine. Debate then centred on the fourth. Ministers agreed that the 'strong financial arguments for reducing the force to three submarines were reinforced by political considerations' – namely that it would 'more clearly demonstrate the Government's policy of not retaining an independent nuclear deterrent' – and came to the 'general view that these arguments were outweighed by other political and defence considerations'. Healey argued, successfully this time, that it was by no means certain that the proposals for the ANF would succeed and that the country's negotiating position would be far stronger with four submarines; a smaller force would also jeopardize a plan to deploy the submarines in the Far East. He also used some of Mackenzie's earlier operational arguments such as the 'strain on personnel concerned' and how 'seeking to maintain one submarine always on station from a force of three' would be 'harmful to morale'. It was the possibility of an incident or accident disabling one of the submarines that had the most impact on the discussion. In summing up, Wilson noted that 'there was general agreement that the balance of argument was in favour of a force of four submarines, having regard particularly to the weakness of our position if one of a force of three were involved in an accident'.[366]

The Navy accepted the decision.[367] However, Mackenzie was very disappointed. 'In my heart,' he later wrote:

> I continued to believe that keeping a fifth submarine in the programme would have been a wiser decision: the additional cost would have been small, and the advantages to all who had to operate and maintain the force would have been immeasurable. I believe the true reason for its cancellation was political, not financial: a sop to the Left Wing of the Labour Party. All that their clamour achieved was to lay an almost intolerable burden on the men, and women, responsible for the efficiency of the deterrent.[368]

Had the fifth submarine been completed, it would have commissioned into the Royal Navy as HMS *Royal Sovereign*. For years, there was a long-running joke within the Ministry of Defence that the Navy intended to call the fifth submarine HMS *Reconsideration* because of the 'long and torturous' discussions that surrounded its fate.[369]

COMPLETING THE PROGRAMME

By the summer of 1965 the Navy had 'broken the back of the procurement of equipment'.[370] Up until the end of 1964 construction of all four submarines remained on schedule. But by the time the keel of the third submarine, HMS *Repulse*, was laid down in August 1965, the programme at Vickers was around ten months behind schedule.[371] Vickers had struggled to manage the workload induced by running both the SSN and SSBN construction programmes in parallel and, by December 1965, Mackenzie and the Controller of the Navy were involved in a 'constant battle' with Vickers to get the shipyard to speed up progress.[372] As the submarines and their weapons systems were to constitute the principal element of Britain's nuclear deterrent, the operational availability and reliability of each was of paramount importance. This demanded unparalleled attention to the construction of the hulls and the installation and performance of equipment and systems. For the first time in history on a shipbuilding contract there were formal standards of quality control and the standard of equipment reliability was degrees higher than anything which the MOD had asked for in the past.[373] The shipyards had to control quality 'to an unprecedented degree' and independent quality control organisations, reporting directly to top management, had to be set up to assure the MOD.[374]

The first submarine, HMS *Resolution*, was launched on 15 September 1966. By May 1967, testing and tuning had been carried out in less time than the Americans had taken on their own submarines and on 22 June 1967, only two months later than originally planned, *Resolution* left Barrow for months of 'intense, trying and tiring' sea trials off the west coast of Scotland, testing equipment, conducting exercises, carrying out dummy fire drills and action stations.[375] The Royal Navy followed US Navy practice and assigned two crews to each submarine to ensure they remained operational for as long as possible. The Flag Officer Submarines, Vice Admiral Sir Horace Law, discarded the terminology the US Navy used to identify its own Polaris crews, 'Gold crew' and 'Blue crew', and instead adopted a more traditional British means of identifying them: 'Port' and 'Starboard'.[376] The question of who should command HMS *Resolution*, the first Royal Navy vessel where command was equally shared between two COs, had been solved in October 1965, with the appointment of two men of equal rank, the extremely ambitious and confident Commander Michael

Henry, thirty-seven, who assumed command of the Port crew, and the calm and collected Commander Kenneth Frewer, thirty-five, who assumed command of the Starboard crew.[377]

The Submarine Service took its new responsibility very seriously, naming the Polaris squadron after one of its most decorated and battle hardened Second World War submarine squadrons, the 10th Submarine Flotilla, which had operated from Malta and Maddalena between 1940 and 1944.[378] But the rest of the Navy continued to struggle with the impact of Polaris. There was a general feeling that the Polaris programme was taking valuable resources away from the conventional fleet. Antipathy was still present when *Resolution* returned to Barrow in August 1967 after successful Contractors' Sea Trials. There was, according to Henry, 'a deafening silence'. No one said anything, good or bad. Only the Polaris Staff Officer to CINCFLEET congratulated the crew on their achievements.[379] When the Commander-in-Chief Western Fleet, Admiral Sir John Bush, visited *Resolution* shortly before the submarine departed on patrol he told the entire crew that as far as he was concerned *Resolution* was just another ship in the Fleet, 'no different to one of the minesweepers'.[380] 'Perhaps he meant well,' Henry later recalled, 'perhaps consciously or unconsciously he was expressing the Fleet's view.'[381] Listening to Bush, the mood of the men who had dedicated themselves to getting *Resolution* ready for sea was, understandably, 'mutinous'.[382]

It was also to a degree a reflection of the government's determination to avoid publicizing Polaris in any way. Privately, Labour ministers were happy to give 'their support wholeheartedly, and even enthusiastically', to the Polaris programme.[383] But in public they 'remained sensitive to anything to do with' it 'and were reluctant to encourage much in the way of publicity for . . . what it was achieving'.[384] Healey even attempted to delay the Royal Navy's first test firing of a Polaris missile in order to avoid a clash with the publication of the 1968 Defence White Paper, but wisely abandoned the proposal when he was told it would have wrecked the tight timetable.[385] When *Resolution* was commissioned into the Navy on 2 October 1967, not a single representative of the Government attended the commissioning ceremony as it coincided with the week of the Labour Party Conference.[386]

As the programme neared completion it suffered some embarrassing setbacks. When HMS *Repulse* was launched at Barrow on 4 November 1967, the submarine grounded in the Walney Channel, where it remained for twelve hours, before it was floated off on the next tide.

Fortunately for Vickers and the Navy, a thorough inspection of *Repulse* revealed no damage. 'Her paint has hardly been scraped,' said a Vickers official.[387] In January 1968, a severe storm hit Central and Western Scotland, badly damaging 250,000 houses, leaving 2000 people homeless and twenty people dead, and damaging the construction works at Faslane and Coulport.[388] Although there was some 'structural damage to some of the buildings' at Faslane this did not affect the opening of the base in March 1968.[389] However, Coulport was badly damaged. The Armament Depot was meant to provide limited support from mid-1967 and to have been fully operational by March 1968. By June 1968, Coulport was running 12–15 months behind schedule and had to work up in 'highly unsatisfactory conditions'. The Armament Depot was even unable to receive certification from US Authorities in time to conduct the tactical missile out-load for *Resolution*'s first patrol and was only able to load the missiles with extensive US oversight.[390]

Another unfortunate incident occurred during HMS *Resolution*'s Demonstration and Shakedown Operation (DASO) (where both the Port and Starboard crews carried out the first test firings of a Polaris missile), which took place off Cape Kennedy on 15 February 1968. The first test firing by the Port crew was a complete success. 'I gave permission to fire and up popped the first Polaris missile to have Royal Navy painted on the side, arching away into the clear blue sky, to deliver the dummy warheads "in the pickle barrel" a thousand or more miles down the range,' remembered Henry.[391] However, the second missile firing by the Starboard crew, under Commander Frewer, nearly ended in disaster. Prevailing conditions were terrible, with short choppy seas and wind force of 6–7. While *Resolution* was dived at a safe operational depth, the accompanying US escort, the destroyer USS *Fred T Berry*, lost visual and radar contact with the submarine's 100-foot red telemetry mast, which was specially fitted to the submarine for the DASO and raised above the ocean to track the missile. The *Fred T. Berry* passed down *Resolution*'s starboard side and then over the top of the submarine and knocked off the telemetry mast.[392] Fortunately for the Navy, the incident occurred before thirty-five representatives of the world's media arrived on board the destroyer to observe the launch.[393] A replacement mast was fitted and the subsequent testing completed in time to enable *Resolution* to recover the original programme for the second UK test firing on 4 March.

When the first 'Longcast' outline programme had been issued in February 1963, it had given February 1968 as the first date for the 'First

24–26. Many of those who led the Submarine Service into the nuclear age were former Second World War submarine aces. *Left:* Rear Admiral Wilfred J. Woods, FOSM, December 1955–November 1957; *middle:* Bertram W. Taylor, FOSM, November 1957–November 1959; *right:* Rear Admiral Arthur R. Hezlet, FOSM, November 1959–September 1961.

27. The first all-British SSN, HMS *Valiant*, sails up the Clyde, May 1967.

28. (*top left*) Rear Admiral Hugh S. Mackenzie, FOSM, September 1961–May 1963. Mackenzie went on to head the Royal Navy's Polaris programme.

29. (*top right*) The architects of the US and UK Polaris programmes: Admiral Ignatius J. Galantin, USN; British Weapons Officer; Rear Admiral Hugh S. Mackenzie; Captain Charles Shepherd; Rear Admiral Levering Smith USN, February 1968.

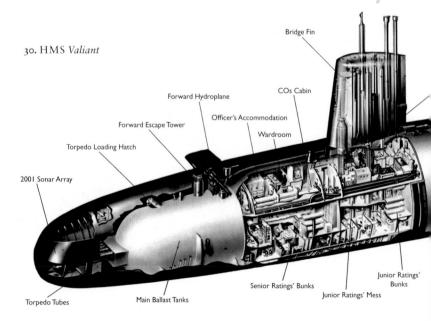

30. HMS *Valiant*

31. (*top left*) The bow section of HMS *Renown* under construction at Cammell Laird, Birkenhead, February 1966.

32. (*top right*) HMS *Repulse* under construction at Vickers, Barrow-in-Furness, November 1966.

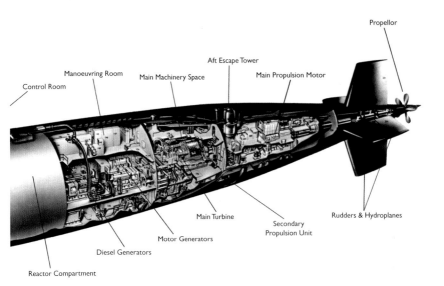

33. HMS *Resolution* and HMS *Repulse* under construction at Vickers, Barrow-in-Furness, 1966.

34. HMS *Resolution*

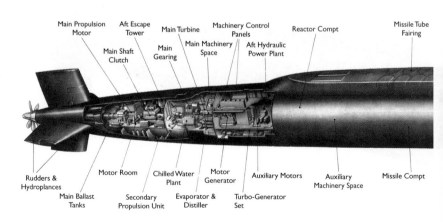

35. HMS *Resolution* launched at Vickers, Barrow-in-Furness, 4 November 1967.

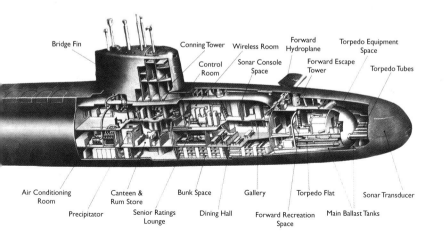

Bridge Fin · Conning Tower · Wireless Room · Forward Hydroplane · Torpedo Equipment Space · Control Room · Sonar Console Space · Forward Escape Tower · Torpedo Tubes

Air Conditioning Room · Canteen & Rum Store · Bunk Space · Gallery · Torpedo Flat · Sonar Transducer · Precipitator · Senior Ratings Lounge · Dining Hall · Forward Recreation Space · Main Ballast Tanks

36. (*top left*) The Control Room on board HMS *Resolution*.

37. (*top right*) Commander Ken Frewer (CO of *Resolution*'s Starboard crew) inserts his missile-firing key into the panel in *Resolution*'s Control Room.

38. (*middle left*) The Missile Compartment on board HMS *Resolution*.

39. (*bottom left*) The Missile Control Centre on board HMS *Resolution*. The Weapons Engineering Officer (front) is holding the launch trigger.

40. (*bottom right*) A Polaris A3 missile test-fired from HMS *Resolution*, 15 February 1968.

41. Polaris required extensive new infrastructure. The Royal Navy settled on a site at Faslane to build the Polaris operating base. In 1962, before construction work started, the ship scrapyard is to the north and the submarine depot ship with a small brood of submarines to the south.

42. The Polaris operating base at Faslane in the early 1980s, complete with administration and stores facilities, jetties for submarines and floating dock for dry-docking submarines.

43. An extensive armament facility for storing the Polaris missiles and mating them to the nuclear warheads was also constructed across the water from Faslane at Coulport, in the late 1960s. Note the missile storage bunkers to the north and the Explosives-Handling Jetty to the west.

44. The Explosives-Handling Jetty at Coulport with a 'Resolution' class submarine alongside embarking missiles before departing on patrol. Note the re-entry bodies containing the warheads covered with protective tarpaulin.

Training Firing' by SSBN 01. The First Sea Lord, Admiral Sir Varyl Begg, rightly informed Healey that 'by completing this firing on 15th February 1968, we could not have got much nearer to the target set five years ago'. It was a 'very considerable achievement'.[394] However, the delivery of the other three submarines, in particular HMS *Renown* and HMS *Revenge*, did not go according to plan. The closing years of the Polaris programme, especially between 1968 and 1970, were notable more for their setbacks and failures than for their successes. By December 1966, progress on the two submarines at Cammell Laird had slipped considerably and Mackenzie was forced to impress on the workforce just how important it was to 'complete the programme ON TIME'. The Minister of Defence (Equipment), Roy Mason, was 'shocked' to discover that the programme was running five months late and 'to hear that it might slip even further was most depressing'.[395] The Cammell Laird workforce was largely responsible for this poor performance and the submarines were routinely referred to as 'gravy boats', in reference to the easy money that they represented. The Government issued repeated warnings to the company: 'Work yourselves into a job and not out of it.'[396] The MOD was so unimpressed with Cammell Laird's performance that officials considered towing both *Renown* and *Revenge* to Barrow so that Vickers could finish the job.[397] A consensus formed inside the MOD that there would be 'no more orders for SSNs, if Cammell Lairds keep on letting us down over *Renown* and *Revenge*'.[398]

One of the most 'awkward issues' surrounding the delays to the two Cammell Laird submarines was that it was impossible to conduct Contractors' Sea Trials on two submarines concurrently. When Cammell Laird delayed the in-service date of *Renown* by six months in October 1967, the entire programme, whereby each submarine joined the Fleet at six-monthly intervals, became disjointed. Unless Vickers' second submarine, HMS *Repulse*, was delayed well beyond June 1968, she would hold up Cammell Laird's first submarine, HMS *Renown*. 'This would severely damage the whole programme,' noted Mason.[399] One official noted that the delay 'would have to be reported to S of S [Secretary of State, Healey], the PM, NATO and Lord knows who else'.[400] However, as Mackenzie later recalled, 'As the year advanced it became clear that progress on *Repulse* at Barrow (the third SSBN in the programme) was now so good due to completion of *Resolution* and *Valiant* that she would be finished before *Renown* at Birkenhead.'[401] Progress at Cammell Laird eventually slipped even further, resulting in *Repulse* leaving Barrow 6–10 weeks ahead of *Renown*. Although this altered the original programme, the Navy was confident that there was 'as yet no reason to doubt' it could meet the promised overall deployment date of one boat operational in April 1968 (*Resolution*) and another in June 1969, with the whole force operational by 1970.[402]

Cammell Laird was also struggling to build its two submarines, not just because of its sluggish workforce. Late in 1966 it emerged that the length between the bulkheads of the torpedo stowage compartment in *Renown* differed by one inch from that of *Resolution*. 'Consternation erupted all round,' remembered Mackenzie. 'Laxity in adherence to, or in interpretation of documents and drawings, somewhere along the line between the lead yard and follow-up yard, was deemed the cause. It was a horrifying discovery at the time' that fortunately 'gave little more than a ripple of disturbance to the overall programme'.[403] Then, in November, a 'tinkling' noise was reported in *Renown*'s reactor compartment.[404] Eleven pieces of broken metal from *Renown*'s thermal sleeve were later found in the primary circuits.[405] The Navy attempted to keep the 'most disturbing situation' concealed from the press and after intensive investigation nine pieces were recovered, but a further two remained undetected.[406] Fortunately the remaining two pieces were soon located and removed. Had the engineers failed to do so Cammell Laird faced the possibility of having to remove the core, which would have meant at least a six-month delay and costs in the order of £500,000.[407] Even so, the defect caused an additional two-month delay.

This was only the beginning of problems with *Renown*. In February 1969, when leaving Cammell Laird prior to Acceptance Trials, the submarine's port bow struck the southern side of the entrance to the No. 7 dock at Cammell Laird, causing some internal bulkheads to buckle and damage to a number of torpedo tube bow shutters.[408] It took four weeks to repair the damage, which was fortunately absorbed prior to the Starboard crew leaving for the US for the DASO in July and August 1969.[409] The submarine finally became operational in November 1969, but not before colliding with a merchant ship, the MV *Moyle*, on 13 October 1969, while operating in the Irish Sea.[410] *Renown*'s CO, Commander Kenneth Mills, was court-martialled, found guilty of hazarding his submarine and relieved of command.[411] By this time *Renown* had earned a reputation as a troublesome submarine, a reputation it lived up to in 1974 when it suffered 'intensive structural damage' after hitting the bottom while on sea trials off Scotland. One of her COs recalled that *Renown* 'gave us quite a few problems and wrecked the careers of a number of COs. Simply keeping her operational was difficult enough, never mind avoiding detection by the Soviets. A patrol felt like a technical endurance test from start to finish.'[412]

As the construction of the submarines neared completion the Government was forced to answer a series of questions about the future of Polaris. The first concerned whether or not to improve the system in response to advances in Russian Anti-Ballistic Missile (ABM) defences, of which we shall see more in Chapter 8.[413] By the end of 1967, these questions became bound up in a wider, inter-departmental review of the Polaris programme and nuclear policy that aimed to draw together 'the considerable amount of work' which had already been done in various fields of nuclear-weapons policy into a 'single comprehensive memorandum' for the new standing Ministerial Committee on Nuclear Policy (PN), which Harold Wilson created in the early autumn of 1966 in order to coordinate the government's decision making on nuclear matters.[414] The review allowed the small group of ministers involved in nuclear policy, especially Healey, to 'ask some provocative questions about the value of the POLARIS system, in order to crystallize the policy case for the present programme'.[415]

The review, which was part of a wider defence review, was carried out over a six-month period by the Defence Review Working Party, and asked a number of questions, including whether or not the Polaris system should be improved and (even at this stage) whether or not the Government should continue with the programme or abandon it altogether. The result, a paper setting out the arguments for and against retention, was then scrutinized by the Defence and Overseas Policy (Official) Committee and next placed before Wilson's Nuclear PN Committee. When it met on 5 December, there was only 'tacit acceptance' that Polaris should be retained.[416, 417] When it met again on 5 January 1968, the nature of the debate had changed and the case for keeping Polaris was overshadowed by wider economic considerations. In November 1967, the pound had been devalued and the Cabinet had been asked to agree to a package of cuts and an expenditure review, 'the most formidable task I had attempted in over three years of government,' Wilson later wrote.[418] Healey was under enormous pressure to reduce defence expenditure and was forced by the new Chancellor, Roy Jenkins, to choose between Polaris and the F-111, an American strike aircraft that was to be purchased for the RAF as a replacement for the cancelled TSR2.[419] When the Cabinet met on 12 January, the balance of opinion was in favour of cancelling the F-111 and 'of retaining our nuclear capacity'.[420] Polaris survived.

COMMAND AND CONTROL
ARRANGEMENTS

While the Government debated the future of Polaris the Navy started to implement the procedures to operate, command and control the Polaris force. Although the government still maintained that it was committed to internationalizing the Polaris force, by 1967 any way of doing so by means of an ANF, in a way that would have nullified the national 'escape' clause, had failed to develop in any concrete form.[421] During the nuclear review in late 1967 ministers admitted that 'there was no sign' that it would be possible to internationalize the Polaris force 'at present'.[422] The command and control arrangements were therefore 'planned on the basis that in the absence of any political directive on assignment ... the force could operate under National control, at the same time, they can be tailored to meet the requirements of any assignment of the force to the Western Alliance'.[423] But in both roles the Polaris force was 'retained under ultimate UK control by virtue of the firing arrangements'.[424]

The command authority for the Polaris force was the headquarters of the Commander-in-Chief Western Fleet (CINCWF) at Northwood in Middlesex.[425] CINCWF was designated Commander Task Force 345 (CTF 345), and a special underground Polaris headquarters was constructed beneath Northwood incorporating a USN liaison cell and direct communication links with national and NATO authorities, all of which continue to this day. CINCWF was responsible for ensuring that when the force was operational, one submarine was continuously on deterrent patrol and that maximum use was made of the availability of a second submarine. He was also responsible for ensuring that each submarine on patrol was prepared to receive an order to fire at any time and could react to such an order within fifteen minutes (except for two non-consecutive periods of thirty-five minutes' response time every day to allow maintenance). He was also required to keep the remaining submarines in harbour at maximum notice of twenty-four hours to fire thirteen missiles.[426]

Throughout the late 1950s and early 1960s, procedures for nuclear retaliation had been created, revised and then re-revised by the Macmillan and Douglas-Home governments.[427] In 1967, Healey commissioned a review of the machinery of Government in War and a Cabinet Committee on Nuclear Retaliation Procedures (NRP(67)1) was tasked with identifying arrangements and procedures for the transitional period when the changeover from the V-bombers to Polaris was taking place and for the Polaris period thereafter.[428] The NATO Commander, SACEUR, could, 'in an emergency, obtain the approval of the North Atlantic Council to release nuclear weapons' but, in order to protect the UK's 'supreme national interests', authority for ordering a missile launch in both the NATO and national roles rested with the Prime Minister.[429] Wilson wrote to Healey that it was 'of vital importance that no submarine Commander shall be authorized to fire the Polaris weapons without my specific authority'.[430] CINCWF was told that he was 'responsible for ensuring that your command arrangements meet this requirement of the Prime Minister without any possibility of failure'.[431] If the North Atlantic Council authorized SACEUR to release nuclear weapons, the order would have been transmitted by means of an 'R' hour message (SACEUR's release message for general nuclear war, broadcast in the clear over all available NATO communication systems) direct to Northwood. Once received, Northwood would have immediately re-transmitted the message to the Polaris submarine(s) on patrol where the Commanders were under strict instructions not to obey the message until they had received a further order to fire from CINCWF. CINCWF would only have done so after receiving authorization from the Prime Minister.[432]

The Navy Department considered it 'essential that the POLARIS control officer at Northwood should be able to see the Prime Minister as he gives the order to release and that, similarly, the Prime Minister should see the Control Officer'.[433] A two-way closed-circuit television system was installed between No. 10 and CINCWF at Northwood and the secondary Polaris headquarters in the Old Admiralty building. A camera was fixed to the wall in the office of Harold Wilson's Principal Private Secretary, Michael Halls, and permanently focused on a table in Halls's office on which there was a blue telephone connected to Northwood. The act of lifting the phone and pressing three buttons activated the link to the Polaris Duty Officer in Northwood, who would appear on a small monitor alongside the table. The system, the existence of which was highly secret and known only to a limited number of people, was tested daily by the Duty Clerks in Downing Street, as and when the pressure of work in the Prime Minister's Private Office allowed, usually first thing in the morning, at lunchtime or during a quiet time in the evening. Strict procedures were developed to ensure that the individuals testing the system 'could in no way imitate the Prime Minister' and 'that if the tester went haywire or indeed had a mental aberration someone else was present to deal with the situation'.[434]

From March 1968 onwards, Duty Clerks in No. 10 followed a strict set of procedures to ensure that the system was working correctly. It went like this:

(a) remove the lens cap from the camera;
(b) sit at the table with the blue telephone instrument and –
  (i)   press the brown button and slide the top to the left
  (ii)   lift the handset,
  (iii) press the button marked 'PRIM'Y' ('SEC'Y' in the case of tests with Whitehall Wireless)
  (iv)   press the button marked 'CALL';
(c) when the Duty Officer at POLARIS HQ can be seen on the monitor screen, say –
  'This is the Duty Clerk testing the television link. Report how you see and hear me.'
In reply the Duty Officer will say –
  'This is the Duty Officer. Your test message received. Vision is clear and sound is good.'
To this, the reply from No.10 will be –
  'Test completed.'
(d) Replace the telephone instrument and slide the top of the brown button to the right.
(e) Replace the lens cap on the camera.[435]

In the event of technical problem with the CCTV system there was a 'fall back of a direct telephone connection using a simple authentication system'.[436] Two pairs of pads containing authentication tables were housed in two separate combination lock boxes located in the inner or outer Private Office in No. 10. Only two individuals knew the combination settings to each box (four individuals in total) and neither pair were permitted to know the other combination.[437]

What if the UK was attacked in a so-called 'bolt from the blue' scenario? The recently declassified Appendix Z from the 1969 Government War Book indicates that in the event of such an attack CINCWF was to seek instructions from the Prime Minister or the Nuclear Deputies, ministers selected by the Prime Minister with the authority to authorize the release of nuclear weapons in the event that, at the critical moment, the Prime Minister was not available.[438] If CINCWF was unable to contact the Prime Minister or the First Nuclear Deputy, he was to seek authority from the Second Nuclear Deputy, who, the instructions assumed, would have moved to Northwood.[439] There were no instructions about what to do in the event that the Second Nuclear Deputy failed to reach Northwood or in the event that CINCWF and the Second Nuclear Deputy were wiped out before being able to issue a firing directive.[440] According to the files, arrangements existed 'to enable authority for the firing of POLARIS weapons to be transmitted direct by the surviving Prime Minister to the submarines'.[441] While it is not clear what these were, it presumably involved the 'secondary firing HQ for Polaris' in London.[442]

What if the Prime Minister, the Deputies and CINCWF were wiped out before being able to issue a firing directive? When the Royal Navy assumed responsibility for the strategic nuclear deterrent there was no 'in place' system for authorizing the launch of Polaris missiles in the event of a complete breakdown of communication with the submarines. Authority to fire was not delegated to CINCWF, as was the case with the Royal Air Force, nor to individual Commanding Officers on board the submarines.[443] There was no sealed order, locked in the inner safe in the control room on board the submarine, with the Prime Minister's 'response from the grave', as exists today.[444] In the event of the communication system failing, Polaris Commanding Officers were given certain actions to carry out, including trying to receive BBC radio. In extremis, they were to try and place themselves under US command.

It was only in 1972 that concern grew about the degree of uncertainty in the actions to be taken in the event of a surprise bolt-from-the-blue attack resulting in destruction of large swathes of the United Kingdom and the Command and Communication arrangements for the release of the weapons. The solution, which had first been devised in the late 1960s, was the issue of the so-called 'letter of last resort', containing written orders from the Prime Minister which relieved the CO of any personal responsibility for initiating a nuclear response. From 1972 onwards Polaris submarine commanders were issued with instructions that if they encountered indications of a nuclear attack on the United Kingdom or if all Polaris and other naval broadcasts transmitted from the United Kingdom went silent for four hours they were to open a sealed envelope held on board each submarine. Each envelope contained further instructions from the Commander-in-Chief Fleet, which laid down the conditions under which another sealed envelope containing specific instructions from the Prime Minister should be opened.[445]

The CO of a Polaris submarine was incapable of physically launching the missiles – part of the failsafe lay in the fact that he merely gave permission to launch. A combination of careful drills and tight discipline ensured that a submarine was unable to launch without proper authority or be misled by spurious instructions.[446] On receipt of a properly formated firing message the Polaris Systems Officer and the First Lieutenant would have been summoned to the Control Room to open a two-man safe in the Navigation Centre which contained material to properly authenticate the message. That signal was in two parts. The CO or the First Lieutenant had access to decode one part of the signal, while the Polaris Systems Officer had access to decode the other. Once decoded and authenticated the first CO would give permission to fire by moving the submarine to 1SQ, ready to launch missiles, and inserting a key into the Attack Centre Indicator Panel, located in the Control Room, which activated the firing circuits. The Polaris Systems Officer was responsible for pressing the trigger and launching the missiles. If any of the procedures were followed incorrectly, personnel in the Navigation Centre, who were aware of the correct procedures, would have physically restrained the CO from using his key. This was just one of a number of physical breaks in the firing chain that prevented any one person on board the submarine, including the CO, from launching the missiles without authorization.[447]

What about the possibility of a mad or even rogue crew launching the missiles? For Rear Admiral Whetstone, the first CO of HMS *Repulse*, a scenario in which he and the Polaris Systems Officer were 'in cahoots' was unlikely. 'We both had to have a brainstorm if we wanted to amuse ourselves before lunch by starting World War Three. The chance was, had the other people around the bazaars seen us behaving in this bizarre manner without a signal having arrived or gone through the procedure somebody would have wrapped us up, locked us in our cabins and sent for the Doctor!'[448] Ken Frewer, CO of HMS *Resolution*, agreed: 'If you can envisage 143 people, all going mad at once, I suppose it is just conceivable,' he said, before adding: 'In my opinion it is not conceivable because the attitude we take to this job is a professional attitude. We are not either morally or philosophically involved. We are in the Royal Navy, this is our job and we do as we're told. We also have as an added adjuncture a doctor on board to keep an eye on us all.'[449] Indeed, all personnel, whether serving on board the submarines or working at Faslane, Coulport and the Operating Headquarters had their medical records checked in order to safeguard the weapon system against sabotage, inadvertent arming, launching or firing. COs were also given strict instructions to keep 'constant watch on all personnel under their command' for any officers or ratings who 'showed signs of mental or emotional instability'.[450]

When the Polaris force first went to sea in 1968 the missiles on board were targeted at the Soviet Union. We also know that the primary target plan was allocated under the NATO nuclear-response plan.[451] Targets were allocated by SACEUR to CTF 345, who in turn issued target plans to individual SSBNs and a forecast of the operational availability.[452] In order to fire the missiles each submarine carried target tapes which fed target information into the missile system. The provision of such tapes was 'extremely complicated' and involved the 'selection of targets in broad terms, choice of aiming points, establishment of aiming points to the required standards of accuracy, translation of geodetic data into a form suitable for incorporation into a tape, and the manufacture of the tapes, independent check of their correctness and, finally, their timely distribution'.[453] The Chiefs directed that the tapes had to be completed by March 1968, when *Resolution* was able to deploy in an emergency capacity and fire its missiles. Preliminary discussions with NATO's targeting authority started in May/June 1966, target planning in October 1966, the firm and detailed target list was issued in February 1967 and production of tapes began in April 1967.[454]

The submarines also carried alternate national target plans, details of which were 'known only to a small circle in the UK and they have never been communicated to either the US or NATO authorities'.[455] A December 1963 Chiefs of Staff paper stated that 'The 32 missile load of two POLARIS submarines constitutes a 20 city deterrent; and of one submarine, a 7 or 8 city deterrent. These figures allow for the probabilities of inaccuracies and for the two biggest targets to be hit more than once.'[456] The figures also took into account the reliability of the weapons system (calculated at 99 per cent), the probability that each weapon would hit its target (calculated at 70 per cent), failures in the fire control system, misfires due to the influence of weather conditions and, finally, in-flight failures. Overall, it was estimated that if a Polaris submarine did launch its missiles, and assuming none were intercepted while in flight, eleven out of the sixteen missiles on board would hit their targets. Eleven missiles on target represented a seven-city deterrent, as Moscow required four hits and Leningrad two.[457] CINCWF was also required to keep submarines alongside at Faslane at maximum notice of twenty-four hours to fire thirteen missiles in order to increase the number of missiles available. But as the submarines were alongside, they were at risk of a 'bolt from the blue' attack or sabotage so the credibility of the threat could be questioned. There was also a school of thought in the MOD that with adequate 'political warning . . . it could well be that the whole Polaris force could be on patrol'. However, given the complex maintenance and refit arrangements of the Polaris force this would have been very difficult to achieve and sustain.

The capability to set targets manually also existed on board the submarines. This was done using an item of equipment called the Target Indication Panel and Target Acquisition Panel (known as the Tip Tap Panel) which was used to feed information into the missile system. Whetstone recalled how 'One or two bright sparks, said: "How about Paris for a start?"'[458]

ON BOARD A 'RESOLUTION' CLASS SUBMARINE

On 14 June 1968, HMS *Resolution* slipped out of Faslane and departed on her first eight-week patrol. For most of the patrol *Resolution*'s carbon dioxide absorption plants – 'scrubbers' – did not work and Henry was forced to bring *Resolution* up to periscope depth each night to 'ventilate', renewing the atmosphere in the submarine through the snort inlet and exhaust masts. 'It was glassy calm for most of the patrol with clear skies, and I sat in the Control Room for the hour or so that it took each night with both periscopes manned, glumly contemplating our chances of detection,' remembered Henry.[459] *Resolution*'s primary method of navigation, the Loran C Chain (a radio navigation system that allowed a submarine to determine its position by listening to low-frequency radio signals) also failed to function throughout the entire patrol, meaning the submarine had to rely on 'bottom contouring' – reading the map of the ocean bed for navigational updates. At the end of the patrol news of the suppression of the Prague Spring by Soviet forces in Czechoslovakia came through and Henry prepared for an extended patrol. 'We had emergency provisions on board, and could certainly have extended from 56 to 90 days [far short of later record patrols], but happily it was not to be so,' explained Henry.[460]

With the commissioning of HMS *Repulse* on 28 September 1968 and HMS *Renown* on 15 November 1968, the Navy was ready to assume its new role. On 14 June 1969, Commander Henry Ellis, a Royal Navy Commander in the Navy's Plans Division, conveyed by hand to his opposite number in RAF Plans a signal stating that the Navy was ready to assume responsibility for the UK strategic deterrent. 'I knew, and he certainly suspected,' said Ellis, 'that it had been a close run thing.'[461] Despite the problems and the setbacks the Royal Navy had achieved its overall aim. From that moment, the RAF's V-bomber force ceased to maintain the extremely high state of readiness at which it had been kept, although it retained its nuclear capability and weapons.

The first Polaris patrol for which the Royal Navy was responsible for the strategic nuclear deterrent was conducted by HMS *Repulse*, under the command of Tony Whetstone.[462] He quickly discovered that commanding an SSBN was entirely different to commanding any other submarine. Traditionally, a submarine's role was to detect and identify a target, and then close and attack it. In an SSBN, this concept was completely reversed: if a target was detected, no matter how tempting, the SSBN's duty was to evade. Only when the missiles had been successfully launched would it take up the hunter-killer role for which it was fitted with long-range sonar equipment and six torpedo tubes. This took a while to get used to, as Whetstone explains:

'Whereas one's previous submarine experience had all been directed towards going on patrol and looking for people and trying to get close to them, we now had the problem of trying to make sure that we never got close to anybody for six weeks. There were various different opinions as to whether it was safest to stay on a single course very slowly, or whether in order to make sure you had detected any potential people you had to alter course, and would altering course too frequently upset your navigation system? We found out that it didn't as a matter of fact and after a while we found that the freedom of operation you're allowed was much greater than perhaps we'd feared when we first set out . . . we were still careful, but we were careful within sensible limits.'[463]

Living and working conditions were vastly different to other submarines and it took some time for the crews to adjust to the unique manning and operational arrangements. Publicly, the Navy wanted to give the impression that the Polaris crews were not fazed by their new responsibilities. When asked what it felt like to be in Britain's first Polaris submarine the standard response from officers and ratings was 'I am a submariner. This is a submarine. It is a job like any other.'[464] But with each crew conducting two patrols per year, on roughly a three-month cycle, service in Polaris was unlike any other job in the

Navy. Begg later pointed out to Healey that 'service in SSBNs in time of peace poses special morale problems which have no counterpart elsewhere or in the other Services'.[465] This was apparent to Mackenzie when a rating complained that because of the length and regularity of the patrol cycle the crews would never be at home for the birth of their children. Mackenzie could only say: 'Sorry about that, but I promise that I will have you there for the conception!'[466]

Adapting to these new operating methods was not easy and there was some disharmony between the first Port and Starboard crews of HMS *Resolution*. Much of this was due to Henry, who had 'thrown himself wholeheartedly into his important mission, but possibly with a higher degree of egotism than some of his superiors felt to be appropriate'.[467] His insistence and contrivance that the Port crew did everything first engendered an aura of superiority.[468] Lessons were learned and the COs of *Repulse*, *Renown* and *Revenge* worked carefully on inter-crew relationships and ensured that the Port crew were first to sea and conducted Contractors' Sea Trials, while the Starboard crew conducted the workup and deployed on deterrent patrol.

For submariners who had served in the Navy during and after the war, living and working conditions on a Polaris submarine were unrecognizable. In 1972, the former submariner, turned author and *Daily Telegraph* naval obituarist, John Winton, wrote about his visit to HMS *Renown*, an experience that made him feel 'not just old but positively antediluvian':

> Antediluvian is right. In the decade or so since I left . . . there has been a flood of changes in the submarine world. It is as though every aspect of the submarine life and technology I used to know has now been raised to a much higher power.
>
> For instance, the Navigating Officer used simply to point at the coffee stain on the chart and tell the Captain, 'We're somewhere about there, sir.' Not any more: *Renown* has a miraculous table gyro known as the Ship's Inertial Navigation System (SINS) which can fix her position on the earth's surface to within a few yards. Ship's course, speed, tides, ocean currents, magnetic variations, make no difference to SINS.
>
> The First Lieutenant used to trim the boat by the seat of his pants and a few simple calculations. Every dive was likely to be something of a venture into the unknown. The First Lieutenant in *Renown* still works out the trim, but the depth keeping, steering and planing can all be done automatically, untouched by human hand.

In the old days, the Coxswain was the boat's medicine man, and dispensed from the PO's mess his own empirical brand of diagnosis. Above the waist – aspirin. Below the waist – the Number Nine depth charge, an explosive laxative capable of moving the bowels of the earth. Now they have a qualified doctor on board, and a properly equipped sickbay.

Submarine food always used to have a certain spectacular unpredictability; quality and quantity depended upon the progress of the Chef's sex life and the accuracy of the Coxswain's arithmetic. We seemed to subsist on a staple diet of 'bangers, beans and babies' heads' (babies' heads: an especially glutinous variety of steak and kidney pudding) often with a rib-sticking Cabinet pudding, known as 'figgy duff' or 'zizz pud' for afters. And there was always the favourite brand of tinned, skinned tomatoes known as 'train smash.'

Actually they do still have 'train smash' in *Renown*, but it looked much less macerated than of old, and it was just one item from a very good menu. The food in *Renown* was excellent, with a choice of main courses and frequent salads, served cafeteria-style in a dining-hall – a vast compartment by old submarine standards. Polaris submarines carry Supply Officers, the first time officers of this branch have gone to sea in submarines – yet another innovation.

That *smell* has gone, that characteristically pervasive submarine attar of diesel oil, rubber boots and boiled veg., underlaid with something more sinister, as though there had been a recent human sacrifice somewhere down in the bilges. Compared with that, *Renown* smells like a rather superior clinic.

They don't even wear the same clothes at sea any more. Where are all those exotic 'steaming rigs,' striped football shirts, leather jackets, Davy Crockett hats and woolly caps? All gone, apparently forever. They wear uniform now, blue shirts and trousers, with smart gilt lapel badges.

No more 'hot bunking' – where a man climbed into every bunk vacated by his relief. Every man has his own bunk, with ventilation louvre and reading light. Never more the great glad cry of 'One all round!'; smoking is virtually unrestricted. They've never heard of the old ritual of Ditching Gash, when bins of rubbish were hauled up the conning tower and ditched over the side from the bridge. *Renown*'s garbage is chopped up, packed in special weighted containers and fired overboard through a garbage ejector – a fitting like a miniature torpedo tube.

No need to do your dhobeying [a sailor's term for clean laundry] in a bucket. They have a laundry. No more rationing of fresh water on a long patrol. They have hot showers and distilling capacity to spare. Above all, no more jealous hoarding of every amp in the main battery. The reactor has enough power to supply a small town.

But *surely*, I thought, looking around me, it can't all be changed?[469]

Winton took comfort in the fact that 'at first sight', the 'blokes', the crew, appeared to be 'typical submarine company: typically cheerful, and cynical, and competent – the very best people in the world to serve with'. Although *Renown* and her sisters were big compared to other Royal Navy submarines, they still had that unmistakably 'small ship atmosphere'. There were still some tattooed forearms, the apt nicknames and phrases, and the special submarine brand of black humour was still very much intact. On board *Resolution* 'friendly rivalry grew between the different branches, and Polaris people earned the nickname "polaroids", while the stokers were known as the "propulsion gang"'.[470] New types of crewman, technicians, were also introduced in far larger numbers than on any other submarines. There were seven science degrees in the wardroom of HMS *Resolution* and many of the junior ratings had specialist technical qualifications.[471] Winton was forced to conclude that 'They have changed.'

Winton observed that Frewer 'has awesome responsibilities which hardly bear thinking about'. The COs of Polaris submarines were responsible for leading their crews through the long tedium of patrol, while at the same time keeping them on their toes. 'In this sense, he is like a coach training a team to run a steady successful marathon and yet be ready to sprint a hundred yards in even time at any moment in the race,' noted Winton. Those first Polaris COs took their responsibilities very seriously. Each came to terms with the destructive power under his command in different ways. For Michael Henry this involved squaring his new responsibilities with his religious beliefs. He composed a 'Prayer for Polaris', which he included in the Christian service he conducted while on patrol:

Lord thou command us saying 'thou shall not kill'. Thou knowest that we prepare ourselves constantly to kill, not one but thousands, and that by this preparation we believe we help to preserve peace among nations. Do thou, who gave man the knowledge to fashion this terrible weapon, give him also the sense of responsibility to control its use; so that fear for the consequences may indeed maintain peace until that day when love, not fear, shall control all men's actions. Give us the will, but never the wish, to obey the order to fire. Oh God, if it is thy will, grant that that order may never need be given, Amen.[472]

For Tony Whetstone, the first CO to conduct an operational patrol providing the UK with its nuclear deterrent: 'Having rationalized my views on this, some people would say sensibly, some people might argue incorrectly, but having rationalized them, I can't say that I lost any sleep while on patrol by thinking about the destructive power of these missiles. I think one had to sort it out well before you sailed for patrol and if you couldn't sort it out I don't think you've got any right to be there. In the same way you can't go through your life continually dreading falling under a bus.'[473]

This was a view that was shared by many subsequent COs of the Polaris force. 'I don't think any of the other COs ever had any hangups about what on earth they would do if we ever got a firing signal,' said Geoffrey Jaques, HMS *Revenge*'s First Lieutenant on commissioning and later its CO. 'It certainly didn't worry me. One always felt that in the scenario in which you might have to press this button there would be so much chaos and mayhem going on and that everything you'd ever stood and lived for was probably in smithereens anyway that you would probably be more than delighted to hit the button. There's no delight in letting loose these megatons . . . what I really mean is that the scenario would have been so full of doom and gloom that I don't think you would have had a problem if you'd still got a signal to go, to get on with it. Also, I always had the feeling that it was a deterrent and it was going to work and because we'd got it and because it was so horrible we were most unlikely to ever be faced with a problem. The system itself seemed to be so good, and so reliable, it was so confidence building, you did have a system that was going to work and I think it was recognized worldwide that it was a system that was going to work. If you're going to have an effective deterrent a) its got to be something which is going to deliver something that the other side does not like, b) they've got to know that it is going to work, c) they've got to know that in extremis you'll certainly use it.'[474]

But not everyone selected to command a Polaris submarine felt this way. Some officers doubted their resolve. Toby Elliott, a former CO of HMS *Resolution*: 'I knew several of my colleagues who went through the commanding officers' course and who were then selected to command Polaris submarines who said they couldn't do it,' Elliott explained, 'it was because they turned down the opportunity, or the invitation, to command a Polaris submarine because they had doubts about their ability to carry out the ultimate act . . . They were very brave to do so. In [some] cases they lost their sea-going appointment and effectively ended their Naval careers.'[475]

As for the crews, Winton discovered that they had been asked the question 'ten thousand times before':

A patient, long suffering look comes over a Polaris sailor's face when it is asked yet again. Actually there is no clear cut answer. Broad generalizations about a Polaris crew's feelings can be dangerously misleading. 'Drop-outs' on moral grounds are very rare, but that does not mean that the great majority are insensitive. There are, of course, some doubters who suppress their misgivings and say 'anyway, it'll never happen.' There are those who are 'all for it.' And there are some who clearly have never thought about it at all: to them it is, in a quite literal sense, unthinkable. But most have given the subject a great deal of thought and now console themselves with the absolute certainty that under no conceivable circumstances would Britain ever start a nuclear war. Polaris is therefore a 'second strike' weapon, used as morally justified retaliation. With that certainty to support them, the ship's company are free to get on with the job they are paid to do, to address themselves to its undoubted professional challenges, and to achieve a high degree of personal involvement.[476]

'Professional' is the key word. All the Polaris weapons specialists in *Resolution* agreed to give up their daily ration of alcohol, known as the 'tot', because of the immense responsibility that now lay in their hands.[477] That responsibility was embodied in the sixteen Polaris missiles, each with a range of over 3000 statute miles, stowed in colossal vertical tubes, 33 feet long and 7 feet in diameter, which reached down through the three decks that comprised the missile compartment in the centre of the submarine like great white columns.

Launching the missiles was routinely exercised, both while on and off patrol. 'To keep us on the ball they used to send us a signal which in every respect resembled a firing signal,' recalled Whetstone. 'It had to be decoded in exactly the same way and verified in exactly the same way. But instead of an order to fire, it said WSRT, which stands for Weapons System Readiness Tests. On receipt of a WSRT you had to go through all the procedure except that the missile firing circuits were not activated and you didn't actually fire a missile. And the whole of the actions taking place during a WSRT were recorded by a black box and analysed to check that your response to a firing signal was satisfactory.

We used to have about six every patrol, at odd times, you could never predict it.'[478]

Practice launch drills only really amounted to the briefest flurry of excitement in a patrol. After only a few days at sea, the submarines' pulse would then slow right down and the hours would pass in what John Winton described as 'a suspended state rather like a controlled hibernation':

> All the normal parameters of submarine existence – speed, depth, course, even time itself – no longer have the same meaning or relevance. The routine eats time and a patrol passes strangely quickly, in a way which the crew almost resent. 'It's two months out of your life, cut out just like that. It's gone. When you get back and meet somebody, you forget it's weeks since you last saw them. It hasn't been weeks to you.' Although time of day means nothing, the watches change, meals are served, the lights are dimmed and raised again, as though obeying some atavistic memory of a solar day. As one wag said 'We're like battery hens.'[479]

The standard eight-week patrol tended to pass through various stages, as Arthur Escreet, a member of HMS *Resolution*'s first Starboard crew, explains:

> For the first two weeks the crew settled down, having just left loved ones at home and not at this stage missing them. The second two weeks became a matter of going on watch, coming off watch and perhaps watching one of the 56 movies which we carried on board. The fifth and sixth weeks saw some of the crew becoming a bit bored and petty niggling took place. Midway through the patrol a 'Sod's Opera' was put on. Items of ladies underwear were produced, making one wonder what sort of people you were at sea with. However, it was an enlightening opera and relieved some of the boredom. By the seventh week morale began to improve as we realized that home was not that far away. The eighth week saw the onset of 'Channel fever' and thoughts of an evening in female company. And then back alongside the wall and 'homers'.[480]

Although it was common knowledge that Polaris submarines remained dived on patrol for extended periods Her Majesty's Customs still insisted on boarding and inspecting each submarine as it came back into Faslane because the Ministry of Defence refused to disclose exactly where each submarine had been while it was on patrol.[481]

The Navy paid a great deal of attention to the physical and mental well-being of the Polaris crews while on patrol. A few men studied for

GCE 'O' levels or took courses, but many spent their leisure time reading, sleeping, playing cards or Uckers. For the energetic there were weight-lifting, rowing, cycling machines and even table tennis. Ship's contests, quizzes, bridge tournaments, darts competitions were all organized and a film was shown every night. Each man was issued with a daily ration of canned beer and a keg of beer was often on tap in their mess. To foster a sense of community spirit a ship's newspaper was published, in *Renown*'s case the *Hi-Ho Journal*, which included news, contributions from ship's departments and *Andy Capp* cartoons, supplied *before* national publication, courtesy of the *Daily Mirror*. Personality clashes on board were inevitable, but surprisingly rare – the boats were big enough for crew members to get out of each other's way.

While the Polaris force was finding its way in those final months of 1969, work on HMS *Revenge* continued. Although the submarine had been launched on 15 March 1968, work had been repeatedly pushed back due to delays with HMS *Renown*. But since *Renown* was launched progress had been 'particularly unsatisfactory. Low productivity, weakness in shipyard management and emphasis on commercial work' all continued to 'impede her [*Revenge*'s] construction programme'.[482] The submarine finally commissioned into the Navy in December 1969, and following intensive sea trials sailed to Cape Kennedy (formerly, and latterly, known as Cape Canaveral) for the final test firing in June 1970. 'All on board were fully aware that to date the UK Polaris firing programme had been highly successful – one more ballistic flight down the Test Range and we would achieve a 100% record,' recalled Captain R. W. Garson, at that time a member of the Royal Navy staff in Washington.[483] As the countdown began the 'atmosphere in all compartments was confident and outwardly calm but with an understandable feeling of excitement and tension'. When it reached zero:

> the boat shuddered responding to the initial vertical movement of the missile as it lifted off, exited the launching tube, and almost immediately the first stage motor ignited – the enormous thrust, while the missile was still below the surface, blasting it upwards on its sub orbital path. As it broke surface all onboard the accompanying Destroyer experienced a thrill one had to be present to appreciate. The flight monitoring reports were eagerly awaited and received with no-one in the submarine wanting to break the silence which had followed the launch. Then it finally came, 'On Track. All OK', when an enormous cheer echoed throughout the submarine.[484]

The task was done. On return to harbour – as was the custom after each SSBN completed its missile firing – a formal dinner had been arranged in the Patrick Air Force Base Officers' Club: 'this one was more special than all the others being the end of programme,' noted Garson.[485]

18 June 1970 was also the date of a general election in the UK. As the crew sat down to eat, the election results started to come in. The Polaris programme ended on the same day that the Conservative Party was re-elected to government. Edward Heath's administration would have to take the difficult decisions about whether or not to improve Polaris that Labour had been very effective at avoiding.[486]

5

Mixing It with the Opposition:
The Cold War in the 1960s

'Our generation was incredibly lucky. We joined an illegit-
imate piratical fringe and nuclear power came up behind us.
And having come up behind us we were swept along on the
bow wave. There's no other way of putting it.'
Captain Richard Sharpe, CO, HMS Courageous, 1974–7.[1]

To deter war we must make any potential enemy realise that we
have the strength to overcome him in the area where he wants
to dominate and force his policy on us. To have a fair chance of
achieving this we must have, and be seen to have, weapons systems which can easily be got to scene of action and which,
when there, are capable of winning. As for winning a war at
sea, were it forced on us, the balance of advantage has swung
between the anti-submarine forces on the one hand and the
submarines on the other. With the advent of nuclear propulsion
the balance is tilting in favour of the submarine: only the addition of SSNs to the anti-submarine forces can prevent the battle from becoming one-sided. Ten years from now our most
possible opponents are likely to have nuclear submarines. As
we cannot just reach up and take this weapon system off the
shelf, we must build them ourselves now, and this is accepted
policy.

Admiral Sir John Frewen,
Vice Chief of the Naval Staff, 1964.[2]

From the middle 1970's the main striking power of the Navy,
apart from the Polaris submarines, will be provided by the
growing force of fleet submarines.
July 1967 Defence White Paper.[3]

The threat is REAL and very much more real to those who have mixed it with the opposition than to most of us who tend to regard it as an academic exercise or an opportunity to score off our contemporaries in fleet exercises.

Commander John (Sandy) Woodward, 1970.[4]

THE COLD WAR AT SEA

While the Royal Navy was occupied with constructing its first SSNs and SSBNs, the Submarine Service was conducting its own deep Cold War, reacting to the growing Soviet naval and strategic threat, mounting offensive and defensive anti-submarine and anti-ship operations; forward surveillance; Special Forces operations; training surface and air forces; weapons development; and showing the flag around the world. Since the mid-1950s, Soviet submarine development had moved forward in three simultaneous trial and construction programmes. First, the development of ballistic-missile submarines, similar to US Navy and Royal Navy Polaris submarines, to contribute to strategic deterrence; second, the development of cruise-missile-firing submarines, which gave the Soviets an offensive capability against surface task forces, principally NATO Strike Fleets; and third, attack submarines, both diesel and nuclear, for traditional submarine tasks and anti-submarine warfare.

Intelligence received in 1959 confirmed the JIC estimate that the expansion of the Soviet Navy was complete. Obsolete vessels had been scrapped and replaced with new classes of ship and submarines fitted with advanced weapon systems, which considerably improved both the offensive and defensive capability of the Soviet Union. Although the Soviets possessed a powerful Navy, its surface fleet was limited to operations within the range of shore-based fighter cover, due to its lack of aircraft carriers. But in 1960 the Soviet Navy began to alter its strategy away from the traditional notion of a purely defensive navy towards one that was capable of taking the battle to the enemy. Plans were developed for a large-scale interdiction campaign against sea lines of communication in which Soviet submarines would transit through the Greenland–Iceland–United Kingdom (GIUK) Gap to attack NATO (primarily US) Carrier Battle Groups and convoys in the North Atlantic tasked with resupplying Europe. In February 1960, the JIC produced an assessment of 'The Employment of the Soviet Navy and Soviet Air

Forces in the Maritime Role at the Outbreak of Global War – 1960–64', which outlined the five main tasks of the Soviet Navy as the JIC saw them:

(a) The defence of the USSR against attack by aircraft and surface ship/submarine launched missiles.
(b) The defence of Soviet coastline, ports and shipping.
(c) Missile launching strike submarine operations against the North American continent.
(d) Attacks on Allied sea communications, particularly those between North America and Europe, including mine laying operations.
(e) Flank support of land operations and amphibious assaults.[5]

The JIC assessed that at the start of a global war the Soviets would employ both submarines and shore-based naval aircraft to strike at NATO Carrier Battle Groups, the small number of US Polaris strategic missile submarines in service by the early 1960s, and the all important sea lines of communication, without which Europe would not have been able to survive. The JIC concluded that 'the main weight of the attack on shipping will develop in the North Atlantic and western approaches', particularly in the Eastern Atlantic, because of the still limited operating radius of the majority of Soviet submarines.

This new strategy was reflected in the deployment of Soviet forces, especially submarines, which ventured increasingly further afield in the 1960s. This process began in 1958, when for the first time Soviet submarines operated in the Atlantic, supported by a tanker, the *Vilyuisk*. In late 1959, *Vilyuisk* and another vessel, the *Mikhail Kalinin*, and two submarines made a five-month voyage from Murmansk via the Atlantic, Indian and Pacific oceans, to Soviet Fleet bases in the Far East. One British Naval Intelligence assessment concluded that this 'dramatically illustrates the natural development and progress in Soviet Naval ocean navigation, research and operations that can be expected from a growing Navy "spreading its wings" to gain experience, particularly with regard to the strategic operation and control of submarine forces'.[6]

By the beginning of the 1960s, the Soviet submarine fleet comprised a total of 427 submarines. The JIC estimated that on 1 January 1960 'the Northern Fleet included 5 [nuclear] missile-firing and 109 long-range conventional submarines'. There was also evidence of up to three Soviet nuclear-powered submarines, a figure which the JIC expected to increase to twenty-two, with eight armed with nuclear missiles by 1965.[7] The deployment of US Navy Polaris submarines, armed with ballistic missiles aimed at the Soviet mainland, had also caused the Soviet Union to rethink the composition of its Navy. As a 1961 JIC assessment of Soviet defence policy in the period up to 1970 pointed out:

> The arguments which convinced the Soviet leaders after the last war of their need for a powerful navy have been given new point by the development of the Polaris submarine. Thus, we believe that the Soviet Union still needs a navy, but of somewhat different form and with a much greater emphasis on anti-submarine warfare, including the use of helicopters, aircraft and anti-submarine submarines ... Submarines will remain the most important arm of the service. As well as maintaining its very powerful anti-shipping capability, made more effective by the introduction of nuclear-powered attack submarines, the Soviet navy has acquired for the first time, by the development of its own ballistic missile-firing submarines, a strategic attack capability which will play an increasing part in the Soviet deterrent.[8]

To meet these new threats, not only did Harold Macmillan's Conservative Government approve construction of the Royal Navy's first nuclear attack submarines and nuclear ballistic-missile-carrying submarines, it was also responsible for taking crucial decisions about the size and shape of the Navy's conventional submarine fleet. In line with the conclusion reached at the reconvened October 1956 submarine conference, the 'Porpoise' design, of which eight were built between 1956 and 1961 – HMS *Porpoise*, HMS *Rorqual*, HMS *Narwhal*, HMS *Grampus*, HMS *Finwhale*, HMS *Cachalot*, HMS *Sealion* and HMS *Walrus* – was used as the basis for an improved class of conventional submarine known as the 'Oberon' class.

Very similar to the 'Porpoise' class, the 'Oberon' class was fitted with improved detection equipment, was quieter and possessed greater diving depth because of the use of improved steel. They were also the first class of submarine to use plastic and glass fibre laminate to construct the casing. Between 1960 and 1967, thirteen of the class – HMS *Oberon*, HMS *Orpheus*, HMS *Odin*, HMS *Olympus*, HMS *Osiris*, HMS *Onslaught*, HMS *Otter*, HMS *Oracle*, HMS *Ocelot*, HMS *Otus*, HMS *Opossum*, HMS *Onyx* and HMS *Opportune* – were constructed for the Royal Navy in various shipyards, including Vickers, Chatham and Scotts. In May 1961, the Government also decided to stop providing submarines to the Royal Canadian Navy, Royal

Australian Navy and Royal New Zealand Navy for anti-submarine training.[9] The Royal Australian Navy eventually purchased six British-built 'Oberon' class submarines – HMAS *Oxley*, HMAS *Otway*, HMAS *Ovens*, HMAS *Onslow*, HMAS *Orion* and HMAS *Otama* – while the Royal Canadian Navy purchased three: HMCS *Onondaga*, HMCS *Okanagan* and the already completed HMS *Onyx*, which was transferred to the RCN and renamed HMCS *Ojibwa*. A replacement for HMS *Onyx* was later built for the Royal Navy and given the same name.

In 1961, the Navy put forward proposals that envisioned maintaining thirty-one submarines in operational service, which required a total fleet of forty-three to allow for modernization and refits. During a meeting of the Cabinet's Defence Committee in May 1961, ministers agreed that the 'main global war requirements for submarines were to deter the Soviet fleet from operating freely against allied fleets and to hunt and destroy Soviet missile-carrying submarines'. But they also recognized that the 'requirement for limited war was perhaps more important' as foreign navies such as those of Egypt, China and Indonesia were being supplied with Soviet submarines 'which could cause serious interference with limited war operations if they were not neutralised'.[10] Although ministers agreed that nuclear-powered submarines were 'far more effective than the conventional submarine' and that 'strong arguments could be adduced for converting to this type as rapidly as possible and building no more [conventional] submarines', they acknowledged that conventional submarines were still 'a very efficient instrument of war which could not be considered out of date'. Cost was also important. Whereas a nuclear-powered submarine cost around £18m, a conventional submarine only cost £3m. In summing up the discussion, Macmillan concluded that 'strong arguments had been advanced to show that there would in any event be some further requirement for conventional submarines'.[11] However, the Macmillan Government eventually decided to cease construction of conventional submarines after the twelfth 'Oberon' class was completed and 'turn over to an all nuclear building programme as fast as our finances allow'.[12]

The Royal Navy's new 'Porpoise' and 'Oberon' class submarines spent much of their working lives conducting reconnaissance patrols and collecting intelligence about Soviet forces. In 1962, the Soviet Navy undertook the first in a series of annual summer naval exercises in the Norwegian Sea. These exercises allowed the Royal Navy to form some sort of judgement on the current state of Soviet equipment and tactics as well as a unique opportunity to assess the capabilities of Soviet ships, submarines and aircraft under operational conditions in what would have been a vital operating area in any future conflict. Surveillance of earlier small-scale Soviet exercises had been largely carried out by aircraft, but by 1962 the Flag Officer Scotland and Northern Ireland, Vice Admiral Sir Arthur Hezlet, complained to the Vice Chief of the Naval Staff, Vice Admiral Sir Varyl Begg, that:

> we have missed a very great deal by not having sent our own submarines on surveillance in this exercise. If we had done so, we might have learnt a great deal about how the Russian groups of submarines communicate and operate; about co-operation between the conventional and nuclear submarines, and how they fit in the operation of their aircraft and Krupnys [Soviet guided-missile destroyers] in the same area. I think we might also have found out whether this operation was directed against Polaris submarines as well as against the Strike Fleet.[13]

Hezlet believed that compared to surface ships and aircraft, submarines were far more suited to conducting surveillance on Soviet exercises as they were 'better for monitoring ship or submarine/air communications as the Russians should not know they were there and order a radio restriction policy as they would probably do if a frigate was used'. Begg agreed. He was concerned that in the days leading up to the exercise 'the Soviets achieved such a large deployment of their submarines without our getting wind of it'.[14] Begg, and many others, always 'thought that we should get a clear indication if a submarine movement on the scale of that apparently involved here in fact took place'.[15] The movement of Soviet submarines was one of a large number of possible indicators about Soviet intentions, especially in the run-up to war. The Navy had to be able to detect it.

THE CUBAN MISSILE CRISIS

In October 1962, when the Soviet Union deployed medium- and intermediate-range ballistic missiles in Cuba, the United States established a sub-air barrier consisting of ten diesel electric submarines, covering the 600 miles of ocean between Newfoundland and an area 300 miles northwest of the Azores, to try to detect Soviet submarines deploying to the Caribbean to support Soviet naval operations.

Although the Soviets deployed five diesel electric submarines to Cuba during the Cuban Missile Crisis, they were actually only detected when they encountered American quarantine forces blockading Cuba.[16]

Royal Navy submarines from the 6th Submarine Squadron, based in Halifax, Canada, also played a role, largely unacknowledged, in the Cuban Missile Crisis. From 1954 through to 1965, the Royal Navy's 6th Submarine Squadron was tasked with assisting the Royal Canadian Navy with anti-submarine training. Submarines assigned to the squadron were manned by a large number of Royal Canadian naval personnel, receiving training prior to the arrival of the 'Oberon' class submarines from 1965 onwards. Ratings were drawn from each Navy. The Navigating Officer was typically from the Royal Canadian Navy while the Commanding Officer was always a Royal Navy officer. When the Cuban Missile Crisis began, the US Chief of Naval Operations, George Anderson, formally asked the Royal Navy and the Royal Canadian Navy to assist the US Navy track Soviet ships and submarines that had sailed from their Northern Fleet bases and were proceeding by the quickest route to Cuba.[17] Although the Royal Navy submarines assigned to the 6th Submarine Squadron were under Canadian operational control, the Canadian Maritime Commander Atlantic, Rear Admiral Ken Dyer, agreed with the Flag Officer Submarines, Admiral Hugh Mackenzie, that two Royal Navy submarines from the squadron, HMS *Astute* and HMS *Alderney*, should be stored for war as a contingency, and if the crisis escalated automatically revert back to FOSM and thus Royal Navy operational control.

On the morning of 23 October, HMS *Astute*, under the command of Lieutenant Commander J. Ringrose-Voase, sailed from Halifax after a night of provisioning stores and ammunition and took up a forward surveillance patrol to the northeast of the Grand Banks to cover the Strait of Belle Isle (the water separating Newfoundland from Labrador) to provide early warning of Soviet submarines entering North American waters en route to Cuba.[18] HMS *Alderney*, under the Command of Lieutenant Commander E. Cudworth, was also recalled to port to prepare for war under a cloak of secrecy. 'Because we were at sea and missed President John F. Kennedy's speech to the world on the evening of 22 October, few of us knew exactly what was happening or where we were going,' explained Peter Haydon, HMS *Alderney*'s Canadian Navigating Officer.[19] *Alderney* then sailed to a patrol area to the northeast of the Grand Banks. Both submarines, like their US counterparts, failed to detect any Soviet submarines. 'We were in extraordinarily difficult sonar conditions where the Gulf Stream and the Labrador Current meet,' said Richard Sharpe, *Alderney*'s Torpedo Officer. 'You get the cold Labrador Current coming in under the Gulf Stream. I mean if you go from one to the other you are undetectable. We wouldn't really, I don't think, have detected anything anyway with the sonars that we had, which were pretty basic.'[20] Both *Astute* and *Alderney* remained in barrier positions northeast of Newfoundland conducting surveillance patrols until the end of the crisis on 28 October 1962.

SURVEILLANCE

In the aftermath of the crisis, the possibility of severe political repercussions dictated caution, particularly when high-level discussions about the Nuclear Test Ban Treaty were taking place. In early 1963, the Navy had plans for a far more ambitious surveillance operation against what now seemed to be an annual Soviet Fleet exercise, involving submarines, frigates and destroyers. However, the new Minister of Defence, Peter Thorneycroft, wrote to the First Lord of the Admiralty, Lord Carrington, and questioned 'whether results that we obtain from these various operations are worth the political and other risks involved'.[21] Thorneycroft was so concerned that he considered 'reporting these misgivings to the Prime Minister', but before doing so he asked Carrington if he was 'fully satisfied that the results are worth the risk and that the "productivity" of these operations is fully analysed within your Department?'[22] Nearly a month later one of the Royal Navy's 'Porpoise' class submarines, HMS *Sealion*, was forced to surface after being detected by Soviet surface vessels. The Prime Minister was provided with regular updates as *Sealion* was escorted out of the Soviet exercise area by Soviet surface forces, and a press statement was drafted with a cover story which stated that 'at the relevant time HMS *Sealion* was semi-submerged making trials of reception of wireless signals'.[23]

Despite Thorneycroft's concerns, this incident appears to have had little impact on other operations. When the Soviets deployed for yet another annual Fleet exercise in the summer of 1963, the Royal Navy initiated Operation 'Bargold', sending three submarines – HMS *Onslaught*, HMS *Olympus* and HMS *Rorqual* – into waters codenamed Piccadilly and Leicester Square, to detect the initial sailing of Soviet forces. However, 'after a week of unproductive surveillance, it became clear that the Soviets had not, in fact, started to deploy for their exercise'. 'The patrol was uneventful,' wrote the CO of HMS *Onslaught*. 'No Soviet forces were encountered and no intelligence gained.'[24] The Director of Naval Intelligence was, once again, 'most concerned at the lack of knowledge and measures for detecting large scale movements of submarines from Northern Fleet bases to the Atlantic'.[25] Although submarines were ideally suited for this task their ability to obtain such information had 'not been proved because insufficient numbers of British submarines have been deployed in past operations'.[26] Hezlet was so convinced of 'The importance of obtaining warning of Soviet submarine deployment and the great training value obtained' that he urged the Secretary of the Admiralty to ' "divert" far more submarines from their routine functions for this purpose' because 'The intelligence gained will be out of all proportion to the effort expended and much excellent training obtained as a by-product.'[27]

When the Soviets finally deployed for their annual Fleet exercise in early August 1963, about a month later than initially expected, the Royal Navy obtained a great deal of intelligence, particularly about how the Soviets intended to use the new 'Krupny' class destroyers with their surface-to-surface missiles against a NATO Strike Fleet. The Director of Naval Intelligence concluded that 'Bargold' 'was a valuable operation from the naval intelligence point of view whereby confirmation was obtained of many aspects of Soviet Naval tactics and fresh evidence produced of intended tactics and practice in the area concerned'.[28] Overall, it was clear that the Soviets were attempting to push their defences outwards into the Atlantic in order to keep NATO Strike Fleets at greater ranges.

In June 1964, the Ministry of Defence sought political approval for Operation 'Clash', a 'special maritime surveillance operation' against the Soviet Northern Fleet exercise in the North Atlantic. Similar to Operation 'Bargold', Operation 'Clash' placed rather less emphasis on aerial surveillance during the exercise itself and more on preliminary watch by submarines and aircraft. The first phase of the operation saw submarines and aircraft attempting to detect the initial deployment of Soviet forces, especially submarines, before the exercise began. A considerable number of UK forces were earmarked for the operation, including a frigate, four conventionally powered submarines, two of which were available on patrol at any one time, as well as five squadrons of maritime Shackleton aircraft, operating from Scotland and Norway. The United States, Canada and Norway also committed forces.

The second phase of the operation involved following the Soviet exercise when it was under way. Air surveillance was scaled back considerably and concentrated mainly on detecting the position of the Soviet submarine barriers. Royal Navy surface forces, including frigates, were ordered to shadow Soviet surface forces, while Royal Navy submarines were tasked with shadowing Soviet submarines. Royal Navy forces were instructed to observe a number of general limitations while carrying out these activities, such as maintaining a distance of three miles by day, or five miles by night. Surface ships were also forbidden from passing between Soviet vessels or interfering with their Fleet exercises in any way. If the Soviets asked them to leave the exercise area, then they were to do so immediately. Submarines had significantly fewer restrictions placed on them. They were instructed not to approach closer than 2000 yards from a Soviet submarine. As the Defence Secretary, Peter Thorneycroft, told the Prime Minister, Alec Douglas-Home, in June 1964:

> The operation would, of course, take place entirely on the high seas and our forces would remain well clear of foreign and territorial seas. As our submarines would not go east of 24°E there would be no risk of their infringing Soviet territorial limits. However, they would at all times give overriding priority to avoiding detection and, if discovered, they would withdraw clear of the Soviet forces before resuming their task.[29]

Royal Navy submarines were also increasingly involved in counter-intelligence operations against Soviet surface vessels, designed to protect the British and Allied activities against intelligence-gathering by Soviet ships. In March 1964, the Cabinet Secretary, Sir Burke Trend, told Douglas-Home that:

> This new requirement arises from the fact that since 1958 more than 30 trawler-type vessels, laden with electronic equipment, catching no fish and unashamedly designed to pick up communications and electronic intelligence (COMINT/ELINT), have been identified. Some of the vessels may be capable of underwater – as well as radio – intelligence gathering. Their activities in the U.K. Communications Security Area and in the vicinity of U.K./NATO naval exercises have caused us growing concern during the past two years. If the security of important exercises and trials is to be properly protected, we must be in a position to know the whereabouts of any of these vessels and sometimes, e.g. when a COMINT/ELINT vessel is near the limit of the range of U.K./NATO emissions, to keep a regular and continuous watch on them.[30]

At the end of March 1964, the First Lord of the Admiralty informed the Prime Minister that 'we have mounted a considerable number of counter-intelligence operations with special ad hoc approval, to protect selected naval exercises and trials during the past 12–18 months'.[31]

In October 1964, three Soviet submarines, a submarine rescue tug, a destroyer and a minesweeper all left the Baltic and headed around the north of Scotland towards the Londonderry exercise area, where a Joint Maritime Exercise was due to take place, with UK, US and Canadian forces, including a series of trials involving HMS *Dreadnought*, a US Navy nuclear submarine, four conventional submarines and nine surface escorts. Five Soviet electronic intelligence-gathering trawlers (ELINT trawlers) were also in the southwest approaches to the UK, waiting for the exercise to begin.[32] The Soviets would even intrude on the Perisher course, as Sam Fry, a future Teacher, recalled:

> One of the periods when exercising the submarine v submarine attack, a Soviet 'Whisky' class submarine was spotted by the Officer of the Watch in HMS *Finwhale*, then the submarine allocated to the COs' course. As the Russian interloper snorted by we formed up with *Finwhale* close astern and our proper target, an A class submarine on the Russian's quarter. Safety was not a problem as we could see the wake of our fellow trailer quite well. We kept this up for a few hours until we ran out of our allocated area and let the Russian blithely carry on![33]

The Navy also continued to mount specific intelligence-gathering operations in northern waters. In September 1964, the Navy was preparing 'a possible submarine operation in the Barents Sea to observe Soviet fleet movements' codenamed Operation 'Catgut'.[34] This caused a problem in that the best period to mount submarine operations in the Barents Sea, during October and November, when the darkness/daylight conditions were optimal, coincided with the next general election in the United Kingdom. The Prime Minister, Alec Douglas-Home, was well aware of the political repercussions should a submarine be discovered and trigger an embarrassing international incident which could affect the outcome of the election. He agreed that the 'exercise should go on' but urged that it should take place 'in late October or November rather than earlier. We don't want any complications at election time.'[35]

During the planning of Operation 'Catgut', a heated debate took place between the Navy and the Foreign Office about how close Royal Navy submarines could approach the Soviet coast. The Foreign Office had approved previous submarine intelligence operations on the proviso that the Submarine Service took 'special care ... to avoid giving any ground for an accusation that Soviet territorial waters had been infringed, or for any action based on such an excuse'.[36] For Operation 'Catgut', the Navy wanted to allow the submarine carrying out the operation to patrol right up to the twelve-mile limit. However, the Foreign Office 'thought this was too close'. The Ministry of Defence weighed in, suggesting a compromise of thirteen miles. It is not clear what the outcome was, but Royal Navy submariners almost certainly patrolled close to Soviet territorial waters during special intelligence-gathering operations. Commanding Officers differed in their approach to their orders. Some were cautious and remained well clear of Soviet territorial waters, while others were more daring and tended to patrol as close as they could get. One naval officer, John Coward, then an intelligence watch keeper on board HMS *Oracle* under the command of Robin Morris, recalled one particular patrol off Murmansk: 'Morris didn't have any respect at all for the Russians. He'd spent his life in intelligence and he just wanted to patrol as close to the coast as possible preferably with the radio aerial up so he could listen to the test match. That was quite hair raising ... If the Russians detected you and they thought you were not one of theirs they fired. I've often seen rockets going over the periscope and things like that. But Morris didn't care. He had the utmost contempt for them and that was that. He didn't bother to pull the mast down.'[37]

Following Labour's election victory in October 1964, the new Labour Government continued to authorize special submarine intelligence-gathering operations, in January 1965, approving 'the special submarine operation Findon'. Like their Conservative predecessors, Labour ministers were always concerned about the political repercussions should anything go wrong during an operation. In January 1965, George Thomson, Minister of State in the Foreign Office, was particularly concerned about the timing of Operation 'Findon' as 'it could coincide with visits here by Gromyko, and possibly Kosygin', the Soviet Minister of Foreign Affairs and the Chairman of the Soviet Council of Ministers. 'We must do what we can to avoid the risk of incidents during these visits,' noted Thomson.[38]

Harold Wilson's Labour government continued to mount surveillance operations against Soviet naval exercises. In June 1965, Labour was given a stark reminder of how a submarine surveillance operation could go wrong. On 21 June, one of the Royal Navy's 'Oberon' class diesel electric submarines, HMS *Opportune*, was on patrol in the

Norwegian Sea when it received an Intelligence Summary from the Flag Officer Scotland and Northern Ireland, giving the position of a Soviet force conducting exercises to the northeast. *Opportune*'s CO moved his submarine into range and at 0424 on 22 June came into contact with two Soviet 'Skory' class destroyers, transmitting on sonars. As *Opportune*'s CO began to collect intelligence, taking photographs through the periscope and recording noise and sonar characteristics of the Soviet vessels, the submarine was detected. The CO immediately took *Opportune* deep and moved away to the south at slow speed in an attempt to shake off the Soviet destroyers. As *Opportune* conducted evasive manoeuvres, the Skorys began to drop explosive charges – not depth charges – but the smaller explosives that were commonly used in ASW exercises to mark attacks and signal other submarines. For over thirty hours, the two destroyers hunted *Opportune* to exhaustion, until at 1130 on 23 June, when *Opportune*'s battery had almost lost all power, the CO was forced to surface the submarine. As *Opportune* broke through the waves and settled on the surface she was hemmed in by the two Skorys, at a distance of about 2000 yards on each quarter. One of the destroyers thanked *Opportune* for working with them, while the other closed in to read the submarine's pendant number and exchange identities by light. The two Soviet destroyers then escorted *Opportune* out of the area for about forty-eight hours before parting company.

The Submarine Service and Naval Intelligence conducted a detailed analysis of the incident in order to determine how *Opportune* was detected; why she was unable to break contact; whether the Soviet destroyers possessed any new detection equipment; whether they used new tactics; and at what time, if at all, they appreciated that their contact was not one of their own submarines. The analysis concluded that *Opportune* was probably detected by either radar or sonar and that tactics employed by the Soviet destroyers combined with good sonar conditions and low sea state were adequate to hold *Opportune*, which during the initial stages of the hunt used only small alterations of course and speed and remained at a steady depth. The more violent evasive actions taken by *Opportune* later in the hunt might well have been effective had they been used in the beginning during the initial period of contact and classification.[39] Because of these factors, the Deputy Director Service Intelligence (Navy) stressed 'that this incident does not present a full picture of Soviet AS capability. It must therefore on no account be quoted or used in isolation – any attempt will result in a biased or distorted assessment.'[40]

The Submarine Service and Naval Intelligence were not unduly concerned about the incident. Aside from the dropping of small explosive charges, at no time did the destroyers make any aggressive manoeuvres. One MOD official also pointed out that:

> The Russians also engage in this type of surveillance of course, and they do it on a much larger scale and less prudently than we do, using both surface vessels and submarines. We are repeatedly detecting their submarines in our habitual exercise areas; there were two cases last year when we forced them to the surface and escorted them away, for their own safety, from exercises or trials; while in June this year another submarine was escorted away submerged when we did not succeed in bringing her to the surface.[41]

However, at the time, the Navy decided not to inform the Government about the incident and when Denis Healey, the Defence Secretary, eventually found out, he took it 'very seriously' and issued firm instructions that 'the Prime Minister is to be informed at once if any similar incidents arise in the future'. When Harold Wilson eventually discovered what had happened he asked 'why he was not told about it at the time'.[42] Despite this, the Labour Government continued to authorize surveillance and intelligence-gathering operations, but only, as George Thomson explained:

> on the understanding that all possible steps will be taken to avoid a repetition of the incident when H.M. Submarine *Opportune* was detected by Soviet vessels, tracked under-water for 30 hours as she tried to withdraw, and finally obliged to surface, whereupon she was escorted from the area by two Soviet destroyers. Although there have been no political repercussions from this incident, so far as we know, I am anxious that it should not be repeated.[43]

By the mid-1960s, Soviet submarines were increasingly venturing into waters off the United Kingdom. In March 1964 and January 1965 Russian submarines were 'hunted' in the Londonderry area. The first Russian submarine was forced to the surface and escorted out of the area while the second, which was thought to be nuclear, was hunted for two days before it cleared the area. Until the beginning of 1966, Soviet submarine operations off Northern Ireland were described as 'sporadic', but a 1967 note on recent incidents with Soviet submarines prepared for Healey pointed out that 'it is clear that since January last year the Russians have kept a virtually continuous patrol NW of the UK, supported by 'W' [Whisky] class boats from the Baltic'.[44] Among the Soviets' prime intelligence targets were the transiting US Polaris submarines based in Holy Loch. In November 1965, HMS *Otus* was diverted from a surveillance exercise to follow a Soviet 'Okean' class intelligence-gathering vessel, or AGI, called *Zond*, which was patrolling in Royal Navy exercise areas off Londonderry.[45] *Otus* observed how *Zond* reacted when the USS *Daniel Webster*, a US Polaris submarine, made its way back into Holy Loch and then made four runs underneath the trawler, taking photographs of its hull fittings, and recordings of its machinery.[46] The fact that *Otus* was able to remain undetected only reinforced the view that a Soviet submarine intruder could also remain undetected in waters off the United Kingdom for a considerable time, conducting its own intelligence-gathering and surveillance operation.

The Soviets also appeared to be unconcerned about the political repercussions of their own operations. They did not suspend their operations during high-level political visits. In February 1967, the Soviet Chairman of the Council of Ministers, Alexei Kosygin, visited the United Kingdom for another high-level visit. At the time, HMS *Dreadnought*, with the C-in-C Home Fleet on board, three frigates and two conventional submarines were exercising 100 nautical miles off Northern Ireland when intelligence indicated that a Soviet 'Whisky' class submarine which had left the Baltic on 22 January was operating in the area. Initially the Flag Officer Scotland and Northern Ireland ordered the ships taking part in the exercise to close in and hunt the submarine and force it to the surface. However, when Healey was made aware of the presence of the Soviet intruder, he concluded that 'during the period of Mr Kosygin's visit, it would be best to avoid any action which might carry the risk of an awkward incident'.[47] Flag Officer Scotland and Northern Ireland was urgently telephoned and told 'that all ships and submarines were to be withdrawn from the hunt and to carry on with the exercise, and that the RAF maritime aircraft only were to continue to follow up the contact.'[48]

By the end of the 1960s, the Submarine Service was devoting so much time to surveillance and intelligence-gathering operations that in June 1968 the Admiralty Board added an additional directive to the Submarine Service's principal tasks in both peace and war: 'to carry out intelligence procurement and surveillance'. FOSM ordered that it was 'therefore desirable that Submarines exercise these tasks whenever opportunity permits and particularly against Soviet Forces when political approval for this can be given'.[49] In 1969, two standard operational orders were issued for the locating, shadowing and reporting of Soviet forces. The first, Operation 'Alfa', involved the locating and shadowing of Soviet vessels by operational submarines.[50] The second, Operation 'Bravo', was far more clandestine, covering operations involving submarines that were fitted with special equipment, specifically designed to collect intelligence.[51]

Royal Navy submarines were not only engaged in surveillance and intelligence-gathering operations against Soviet forces; in the mid-1960s they were also taking part in operations in the Far East.

INDONESIAN CONFRONTATION

Royal Navy submarines assigned to the 7th Submarine Division, operating out of Singapore, also played a crucial, yet largely unsung, role in the Far East during the 1963–6 conflict between the Republic of Indonesia and Malaysia, otherwise known as the Indonesian Confrontation. From 1963 onwards, Indonesia 'confronted' Malaysia by waging an undeclared war, conducting armed incursions across international borders with the aim of destabilizing Malaysia and dominating the region. The UK, Malaysia's strongest ally, deployed significant numbers of land, sea and air forces to deter the Indonesians and stop the conflict from escalating into an all-out war. The Indonesians operated vast amounts of Soviet equipment, including a 'Sverdlov' class cruiser, several 'Skory' class destroyers and significant numbers of MIG-15s, -17s, -19s and -21s. The Indonesian Navy also possessed one of the most powerful submarine forces in the Asia–Pacific region, consisting of twelve Soviet-built 'Whisky' class submarines, two torpedo retrievers and one submarine tender.

Up until 1965, five submarines were assigned to the 7th Submarine Division, which were supported by the immobile, converted Maintenance and Repair Craft, HMS *Medway*. Prior to the confrontation in Indonesia, the Chiefs of Staff had recommended the deployment of seven operational submarines East of Suez, including two nuclear submarines by 1970. To meet this requirement plans were subsequently made to deploy eight conventional submarines to Singapore in 1968, followed by a nuclear submarine in 1969, as well as a new submarine depot ship, designed to support nuclear submarines, HMS *Forth*. HMS *Forth* arrived in 1966, three years early, due to the outbreak of the confrontation, to support the 7th Submarine Division, which at the time comprised the Royal Navy's older 'A' class submarines: HMS *Alliance*, HMS *Ambush*, HMS *Amphion*, HMS *Anchorite*, HMS *Andrew* and HMS *Auriga*. The Captain of the Singapore Squadron, Commander John Moore, was a dynamic wartime submariner who was determined to make a submarine contribution to events. He ensured that the submarines of the 7th Submarine Division were employed on a variety of operations, including special intelligence-gathering patrols as well as operations that involved the insertion and extraction of Special Forces from 40 and 42 Commando (some of which were personally authorized by the Prime Minister).[52]

Many of the 'A' class submarines assigned to the 7th Submarine Division had their old Second World War 4-inch deck guns refitted to counter blockade-running junks, and although the class had been designed for operations in tropical waters, conditions on board were difficult and uncomfortable. They carried little water, which meant the crews rarely washed and many suffered from prickly heat because of the intense tropical temperatures. The extreme heat also influenced how COs operated their submarines. Dived sea time was governed by trying to keep battery temperatures below the ceiling of 124 degrees Fahrenheit, which meant that the batteries – which were located underneath the accommodation spaces and often pushed temperatures up to 110 degrees – were seldom used in full charge or discharge rates. When the submarines surfaced and the crew went on the bridge or the casing in inadequate clothing they were hit by a reduction in temperature of around 50 degrees. There were many cases of pneumonia. Operational conditions improved marginally with the arrival of the more modern HMS *Oberon*, which was equipped with battery cooling, and air conditioning in the accommodation spaces, resulting in more tolerable temperatures on board.[53]

In July 1965, intelligence indicated that the Indonesian Navy was planning to conduct a live firing of their Soviet-supplied Styx missile from a Soviet supplied Komar fast patrol boat somewhere in the Java Sea, and HMS *Ambush* was deployed to observe the firing. As Richard Channon, who was on board *Ambush* at the time, describes:

We cleared the Singapore Strait to the eastward and got out of sight of land, then turned right, dived, and started snorting south-eastward through the Karimata Strait between Borneo and Sumatra. It was an unusual passage, thundering along at 6 knots in a flat calm sea with never anything in sight, and the echo-sounder running continuously with the trace flickering up and down to show the bottom seldom more than 100 feet below the keel. Charted soundings were few and far between. Having turned south-west off Belitung we were on station off Djakarta a couple of days after sailing, and settled into our billet.[54]

HMS *Ambush* was fitted with special communication intercept equipment. 'We used to creep about during the day, observing the movements of naval and merchant ships, and recording "interesting" buildings on the littoral. Some "sneaky beakies", specialized secret communications ratings, were embarked with us and spent hours listening to and recording local VHF traffic,' wrote Patrick Middleton, a young engineer on board at the time.[55] 'It was interesting to put it mildly,' says Channon:

We had almost exactly 12 hours of daylight and 12 of darkness, and the sea remained glassy calm, which dictated that during the day we patrolled very slowly (one shaft, group down) up and down a line parallel to the coast keeping an eye on what was going on, and snort charged at night. By day the horizon was continually filled with the sails of trading and fishing *praus*, so that the Contact Evaluation Plot resembled a spider's web and periscope drill with the attack periscope perforce became immaculate for all watch keepers. Preventing *praus* getting too close as they drifted past was often a considerable problem, and of course they were almost always noiseless, so being forced deep was another hazard to be avoided at all costs.[56]

At night, when *Ambush* was snorting to charge batteries, the poorly lit little boats, sometimes numbering as many as 300, erratic in course and speed, were almost impossible to keep track of. Only the man on periscope watch stood between the submarine and a collision. Channon, who was looking through the periscope, recalled:

As I swung round, I thought I fleetingly caught a light, almost at the top of the field of view. Continuing the swing right round, and elevating the lens a touch, there it was, a man swinging a lantern to illuminate his sail, so close that I could see the seams in the cloth. So close that he must have heard the engine bearing down on him, and been not a little scared by the fact that he could not see what was making the noise. By the Grace of God we were already turning, otherwise we might well have collided, with neither having sighted each other. Hearing the noise receding, he promptly dowsed his lantern and was no more seen. I have often wondered since what tall stories he told in the coffee shop in Tanjong Priok, and whether his unlikely tale caught the attention of the authorities. If it did, there was never any evidence of their following it up.[57]

The submarines of the 7th Division were also involved in inserting and extracting Special Forces undertaking reconnaissance missions in both Borneo and Sumatra.[58] During the Second World War, the CO of the 7th Submarine Division, Commander John Moore, had experienced first hand the difficulty of landing men from submarines and the submariner's dislike of entering shallow water. When he was posted to the Far East during the Confrontation, he took a special interest in the operations of the Special Boat Service and was determined to do something about the problems. The 'A' class, with its flat-topped tanks and low casing, was ideal for the role. By 1965, Moore had led the development of 'Goldfish', an underwater method of leaving and re-entering submarines. The technique, which was based on the Second World War X-craft, was adapted by Moore, who invented a homing device called 'Trongles' to allow swimmers to find a submarine at night. Moore also converted a Mark 20 torpedo into an underwater delivery vehicle called Archimedes and adapted a Polaroid camera to take reconnaissance photographs through submarine periscopes, as well as perfecting a technique whereby Special Forces could be parachuted to a waiting submarine, collect their equipment and proceed to their target. Moore was so dedicated and determined to see these new techniques succeed that he dropped by night to a waiting submarine to prove it could be done, disobeying the orders of his Commander-in-Chief, Vice Admiral Sir Frank Twiss, who had forbidden him from partaking in 'aerial activities'.[59]

Moore also understood the importance of securing political support for his ideas. During a visit to the Far East in the summer of 1966, the Defence Secretary, Denis Healey, asked to see what the Special Boat Service was up to. He watched as a group of Marines 'parachuted from an aircraft into the sea and swam to meet me on the submarine' – one of them was a young officer called Paddy Ashdown.[60] When Sam Fry became Commander SM, 7th Submarine Squadron, in September 1967, he 'found these gadgets very dangerous, and when we endeavoured to show them to the Commander in Chief they failed miserably. So much so that I was able to discontinue this dangerous waste of submarine time.'[61] Moore's pioneering tactics and equipment were later developed by the SBS and the Royal Navy in the 1970s.

TRANSFORMATION

When the Confrontation in Indonesia ended, Harold Wilson's Labour Government, under substantial economic pressure, started to look objectively at the future of British commitments in the Far East. The eventual withdrawal of British forces from East of Suez in the late 1960s and early 1970s had a dramatic impact on the disposition of the Submarine Service, as the various submarine squadrons that comprised the post-war fleet were steadily disbanded. The process began when in August 1960 the 1st Submarine Division in Malta was rebranded the 5th Submarine Squadron and combined with the 108th Minesweeping Squadron to form the composite command Submarine and Minesweepers Mediterranean (SMS MED). The 1st Submarine Squadron title was then transferred to HMS *Dolphin* in Gosport. The 4th Submarine Division in Sydney and the 6th Submarine Division in Nova Scotia were discontinued in the mid-1960s after the Royal Australian Navy and the Royal Canadian Navy acquired their own British-built 'Oberon' class diesel electric submarines (see p. 274). In 1967, the remnants of the Malta-based 5th Submarine Squadron were disbanded when the Royal Navy's Mediterranean Fleet was abolished, and following the disbandment of the Far Eastern Fleet in the early 1970s the 7th Submarine Division in Singapore was also abolished. This left the Royal Navy with just four Submarine Squadrons: the 1st Submarine Squadron at HMS *Dolphin*; the 2nd Submarine Squadron in Devonport; and the 3rd Submarine Squadron at Faslane, which shared the new operating base with the recently formed 10th Submarine Squadron, comprising the four new Polaris submarines.

The 1966 Defence White Paper considerably reshaped the Royal Navy, phasing out the three large carriers, HMS *Victorious*, HMS *Eagle* and HMS *Ark Royal* by the 1970s and cancelling the new replacement carrier CVA-01.[62] The role of the submarine had played a crucial part in the debates surrounding the future of aircraft carriers, with many, such as Joseph Mallalieu, the Parliamentary Under-Secretary of State for Defence, believing that submarines, not aircraft carriers, were the future of the Royal Navy. In a note to the Minister of the Navy, Christopher Mayhew, Mallalieu wrote that 'my own hunch is that, over the next 15–20 years, the role of the submarine is likely to be paramount. Should we not, therefore, be thinking of submarines, not only as hunter-killers and replacements for strike aircraft, but as all-purpose ships of the RN?'[63] The announcement that the proposed new carrier was to be cancelled led Mayhew and the First Sea Lord, Admiral Sir David Luce, the first submariner to become First Sea Lord, to resign. It also led to a major reconfiguration of the Royal Navy, into a primarily anti-submarine warfare specialist force, whereby the Navy's primary offensive and strike capabilities were transferred to the Submarine Service.[64]

These increased responsibilities required a different kind of submariner. There was little room for the hard-drinking culture that had characterized the early post-war years. Long gone were the days when 'Some Commanding Officers suppressed their ambition and lived for the day, often imitating their perception of the wartime heroes. Inevitably this involved alcohol. There was an awful lot of it about.'[65] Changes had already started to occur in the mid-1950s, brought on by the so-called 'William Tell' incident, when two submariners (one of whom was the duty officer) in the middle of a heavy drinking session, decided to liberate a pistol from the onboard weapons locker and take shots at a half-full whisky glass perched on a submariner's head while standing in the Control Room. One of the officers missed the whisky glass and grazed the skull of his fellow officer.[66] In the new Submarine Service, where submariners were the custodians of the deterrent, as well as the primary offensive and strike capabilities of the Fleet, 'the drinking was decidedly moderate, everyone wore passable uniforms, and there was a general air of serious endeavour, some distance from the cheery piracy of the past'.[67]

The rest of the Royal Navy also struggled to come to terms with, and indeed understand, the new nuclear 'black art'. One submariner took to the pages of a 1967 edition of the Naval Review to complain that the 'last three issues, of a publication expressing a representative cross section of naval thought, make not even a passing reference to the nuclear attack submarine – a ship that is beyond doubt the most potent package of offensive power at sea today'.[68] With HMS Dreadnought already in the Fleet, and HMS Valiant and HMS Warspite due to commission in the mid-1960s, the rest of the Royal Navy struggled to integrate the Navy's new nuclear-powered submarines into the surface fleet. It chose to describe them not as SSNs, attack submarines or hunter-killers, but as 'Fleet Submarines', in order to emphasize that 'this new and revolutionary form of submarine was a Fleet asset' that 'would not threaten the surface Fleet'.[69]

For much of the 1960s the Royal Navy's first nuclear submarines were operated in direct support of surface units, much to the chagrin of submariners, who repeatedly complained that the Fleet submarines, which had proven themselves as the 'most promising and effective anti-submarine vehicles found so far', were being operated in a purely defensive role. 'I believe a better name is Nuclear Attack Submarine (or SSN) which not only differentiates it from Polaris and conventional submarines, but lays emphasis on its primary ability – attack,' argued one submariner in the *Naval Review*.[70] Another anonymous submariner, also writing in the *Naval Review*, argued for 'acceptance of these submarines for the magnificent, exciting, offensive ships that they really are' as well as 'Acceptance by the Navy as a whole that the Submarine Service has grown from a subordinate arm (dashing but rather scruffy) to a responsible maturity where it should be our major tactical offensive force.' He urged 'Acceptance by the Submarine Service itself, which must be partly to blame for the existing situation, by having been the silent service for too long.'[71]

When Harold Wilson's Labour government entered office in October 1964, it inherited an SSN programme that had been postponed by the demands of the Polaris programme. Once the new Government clarified its policy towards Polaris – continuing with the programme, but cancelling the planned fifth submarine – it resumed the SSN programme.[72] The fourth, and first post-Polaris SSN, HMS *Churchill*, the first of the so-called 'Repeat Valiant' class, was ordered from Vickers in Barrow in October 1965. Virtually identical to the original *Valiant* design, the 'Repeat Valiant' class included a number of minor improvements mainly consisting of 'certain equipments not previously available' that could be accommodated without detriment to the SSBN or other SSN development programmes.[73] HMS *Churchill* was followed by two more SSNs: HMS *Conqueror*, ordered from Cammell Laird in August 1966, and HMS *Courageous*, ordered from Vickers in March 1967. *Churchill* differed from her two sister ships in that she was chosen to test a new experimental propulsion system, the culmination of a lengthy R&D programme at the Admiralty Research Laboratory in Teddington.[74] Instead of a traditional propeller, *Churchill* was equipped with a pump jet, similar to a water turbine, consisting of a rotor and a stator, both with a large number of blades, surrounded by a carefully shaped duct.[75] *Churchill*'s pump jet enabled the flow of water to be carefully controlled in velocity and pressure and could be either more efficient or quieter than a propeller or a bit of both.[76] The trials in *Churchill* were so successful that the Navy later decided to fit pump jets to the next class of British nuclear submarine, what became known as the 'Swiftsure' class. It remains a standard feature of all British nuclear submarines.

With HMS *Churchill*, HMS *Conqueror* and HMS *Courageous* under construction, the Navy slowly began to embrace its underwater future. At the institutional level, the importance of the SSN was clearly demonstrated by the depiction of an SSN on the cover of volumes one and two of the 1967 Future Fleet Working Party report, which referred to the SSN as 'the Navy's main offensive weapon', saying further that 'no other type of warship holds so much potential for the future' and that it 'has brought a new dimension to naval warfare'.[77] However, it was still unclear how many SSNs the Navy required or indeed how many the country could afford. When the Defence Secretary, Denis Healey, began to examine the overall size of the submarine fleet in October 1967, he refused to commit himself to any rolling programme without 'fuller information about the case for a force of 12 or more of these submarines'.[78] In response, the First Sea Lord, Admiral Sir Varyl Begg, presented Healey with a detailed paper – a classic exposition of the attack submarine as the capital ship of the future – that illustrated the military and political value of an SSN force in the 1970s and beyond and outlined the role the SSN could play in the context of the Navy's overall concept of naval operations, on which it was now basing the general shape of the Royal Navy's future Fleet.[79]

This new concept of naval operations foresaw a Fleet that was capable of conducting operations both inside and outside the NATO area. The main force consisted of relatively lightly armed frigates which were intended to form a military presence and conduct anti-submarine warfare, backed up by ships of higher capability which would provide punch and power projection. Without aircraft carriers, nuclear-powered submarines were now considered to be the most important higher-capability vessels where they would 'provide an important part of the Navy's anti-submarine capability' and 'provide the main independent offensive power of the Navy against enemy surface ships'. The paper also highlighted the way in which a nuclear submarine could be employed in a number of additional roles, any one of which could assume primary importance, depending on the circumstances. These included surveillance and intelligence gathering; special operations, such as the landing of intelligence agents or Special Forces; offensive mine-laying; the training of Allied surface and air forces in anti-submarine warfare; and the protection and escorting of Polaris submarines.

In terms of future threats, all of these tasks, the paper explained, were important:

> In the Atlantic the main threat to the NATO naval alliance will come from increasing numbers of Soviet nuclear submarines, and the Soviet missile-firing surface ships. The contribution RN submarines can make to NATO capability against both of these will therefore be of major importance. Elsewhere a relatively small number of modern submarines can pose a very powerful deterrent to hostile naval operations by any power who might wish to interfere with our maritime interests. Any country with naval pretensions contemplating local warlike operations contrary to British interests might be tempted to undertake maritime operations against a thinly-spread British surface fleet. It could suppose that even a small number of ex-Soviet missile-firing ships, including KIL-DIN and KOMAR types, or submarines, would provide an effective counter to interference by British surface units and conclude that it would have an advantage in local operations. The SSN, with its speed, endur-ance and immunity from air or missile attack, redresses this balance, and this is known now and will become more widely known by the Navies of possible opponents as well as those of our Allies.[80]

The paper also delicately demonstrated the advantages of nuclear submarines over conventional submarines. While it acknowledged that the 'Porpoise' and 'Oberon' class submarines, 'with their silent oper-ation under water, are the finest conventional submarines in the world', it also highlighted two principal limitations: their need to surface at regular intervals and their slow sustained submerged speed. 'Whatever further development and modernization is built into conventionally powered submarines in the future,' the paper concluded, 'the balance of the advantage will remain in favour of nuclear powered submarines and in favour of A/S forces armed with improved detection and quick reaction AS weapons, unless these fundamental limitations are over-come.'[81] The advantages of the nuclear over the conventional submarine were overwhelming:

> The outstanding characteristics of nuclear-propelled submarines are their ability to remain under water for very long periods, where they

enjoy virtual immunity from air attack and from the effects of the weather; their self-sufficiency for long periods; and their ability to move under water at speeds which equal that of most surface ships and exceed that of many, even in calm conditions. In rough weather, they may often have a speed advantage over all surface vessels. Nuclear submarines are not limited, as conventionals have been, to operating in focal areas, waiting for targets to pass. Nuclear submarines have the ability to perform in all roles now undertaken by conventionals, and to achieve, ship for ship, far more than any conventional can hope to do against other nuclear submarines and modern ASW defences. Furthermore they can keep up with the enemy, shadow him, attack and re-attack at will; they can also keep up with, and act in direct defence of surface units of the fleet. Nuclear submarines are much less vulnerable than conventionals and can operate successfully in a hostile environment in which conventionals would be at considerable risk. As air and satellite reconnaissance of the sea becomes more extensive and precise, the advantages of nuclear over conventional submarines will increase.[82]

The paper then went on to examine the range of tasks that nuclear submarines could undertake. The first was the anti-submarine role:

Nuclear-powered submarines have introduced a new dimension in to the concepts of submarine and anti-submarine warfare. They have multiplied many times the problem of surface ship defence against submarine attack; and all the means at our disposal now – surface ships, helicopters, LRMP [Long-Range Maritime Patrol] aircraft and submarines are barely sufficient to deal with the threat which they can pose. On independent ASW patrol our own Fleet Submarines can exploit their unique characteristics effectively to deter enemy submarine operations, at very long distances from their bases.

In the close A/S role they provide a passive sonar capability superior to that of a surface ship or helicopter and thus provide early warning. In certain conditions of water and weather they can choose their depth of operations to provide a better active capability than surface ships. They can, while submerged, keep pace and co-operate with friendly surface and submarine (including POLARIS) forces and thus provide a submarine detection capability unaffected by the weather, and undisturbed by air attack. As they do not need any air defence when submerged they relieve the task of maritime air defence. In addition they can increase the tactical flexibility of surface ships which they are supporting. Lastly they are the only vehicles which can match the speed and endurance of another nuclear submarine in all weathers.

Fleet submarines will therefore make an important (and unique) contribution to our anti-submarine deterrent forces which will complement surface ships and aircraft, whether for independent deterrent operations outside Europe or as part of an Allied force in the Atlantic or elsewhere.[83]

The paper also emphasized the nuclear-powered submarine's anti-surface ship role:

The characteristics described earlier also make fleet submarines exceedingly powerful anti-surface ship weapon systems. They can be deployed overtly or covertly and quickly over very large distances. This means that they can be sent to carry out patrols of one to three months' duration, unsupported, in any part of the world, sailing from and returning to the United Kingdom. In this way, they can be used at long range to deter enemy ships from leaving harbour; to trail and shadow them; and to attack and sink them. They can be switched immediately from a deterrent to an offensive role with or without alerting the enemy and without exposing themselves to pre-emptive attack.

In considering how best to provide the Fleet in the late 1970s with an independent strike capability against missile-firing surface ships, and bearing in mind the contribution of RAF aircraft, the Admiralty Board decided to rely on the Fleet submarines. They therefore put forward only modest proposals for the arming of the surface fleet against the surface threat. No surface-ship SSGW is to be introduced – at least during the 1970s; and the only surface to surface capability which the Fleet itself will have will be armed helicopters, the 4.5" gun and such capability as surface to air weapons such as SEA DART offer against surface targets.[84]

Thus in the later 1970s, the Fleet submarines will be a key element in the future Fleet 'package', The Fleet will have, in its surface ships, some high quality defence against aircraft, missiles and submarines, but without its own nuclear submarines, it would have only a small offensive or deterrent capability against an enemy on the surface.[85]

Begg then went on to explain how the nuclear submarine could participate in surveillance and intelligence-gathering operations:

The surveillance of enemy naval forces in peace-time can give a valuable indication of the future intentions, and can therefore make a vital

contribution to the control of a crisis and consequently to the prevention
of war. Intelligence gathering gives a long-range forecast of the enemy's
future capability, and is therefore important to the means of countering
future threats.

Fleet submarines are not the only means of carrying out these surveillance and intelligence tasks. But whenever these tasks have to be
performed clandestinely – as is frequently the case – for either political or
practical reasons, Fleet submarines with their long underwater endurance will be the best, and sometimes the only means to employ.
Both these tasks were, for example, carried out by submarines off Indonesia during the period of confrontation. Patrol submarines [conventional
submarines] were then used: but Fleet submarines would have been
greatly superior, and in any more sophisticated hostile environment, even
more so. Moreover, Fleet Submarines have a great advantage over conventional submarines for operations in northern waters where periods of
darkness are short for half the year.[86]

Finally he emphasized the training role of the nuclear submarine:

Fleet submarines will also be required for training our own ASW forces,
including surface ships and aircraft, if we are to maintain an effective
capability in face of the growing nuclear submarine threat. Conventional
submarines and synthetic training aids can, and must, be used for basic
training; but there is no substitute for a nuclear submarine for the more
advanced aspects. Although they are now obtaining their own patrol submarines, Commonwealth Navies have in the past relied heavily on RN
submarines for their ASW training; and they are likely to continue to do
so for the more advanced training in tactics against nuclear submarines,
which they cannot yet afford themselves.[87]

Before Healey made a decision on the overall size of the SSN fleet he
wanted to 'form judgments on the position of the submarine in an escal-
atory situation, and the extent to which we need SSNs as part of the
conventional deterrent'. Outside of NATO, he wanted to 'see settings
in which we could assess the likelihood and scale of SSNs being used
both in an anti-shipping role and also in an offensive role against enemy
shipping and military installations. In the latter case, we would need to
judge the capabilities of the SSN's weapons and possible alternative
ways of taking such offensive action.'[88] In order to aid the development
of a quantitative assessment of the requirement for SSNs, Healey established a Working Party to 'provide material to assist in making an

assessment of the nuclear submarine programme' and to 'consider possible political and military circumstances at various levels of activity in which maritime forces, including nuclear submarines whenever appropriate, may become involved, and prepare suitable scenarios for examination'.[89]

The Working Party, chaired by Sir Alan Cottrell, Deputy Chief Scientific Adviser (Studies), presented its interim report in December 1967. It outlined a dozen scenarios, illustrating the kinds of operations and incidents, both inside and outside of NATO, which 'might conceivably arise' during the next 20–30 years. There are two examples which are very much a reflection of the concerns of the time; the first involved the protection of the US and Royal Navy Polaris/Poseidon submarines operating out of Scotland:

NORTH ATLANTIC

Background

The general build-up in Soviet maritime force in the late 60s and early 70s is accompanied by a gradual increase in single-ship and squadron passages and exercises on the high seas as well as in 'domestic' waters. The periodic detachment of ships from the Northern Fleet to the Mediterranean gives opportunities for either North Sea or Atlantic observation of British maritime activity. A more or less regular pattern of movement is established usually at a fairly low level of activity.

Polaris

Since 1968 the British Polaris squadron as well as the American Holy Loch squadron has been operating out of the Clyde. The arrangements which were made to evade any Russian attempts to track the submerged SSBNs worked adequately and to the best of our knowledge the Russians have not successfully pinpointed SSBN operating areas.

In the summer of 1972 shortly after COMSUBRON 14 gets his first POSEIDON boat 'Resolution' comes out of her first refit and following the appearance of a small Russian squadron in British waters on their way home from Algiers, intelligence indications show unusual activity off the north of Scotland. There follows a series of associated incidents – trawlers fishing, some sonar contacts, aircraft sightings – which show clearly that a co-ordinated attempt at the surveillance of SSBNs on passage is being mounted. It is not clear whether American or British submarines are the prime target, but this is a threat to the NATO deterrent and a co-ordinated response is necessary.

Development

It becomes clear that NATO SSBNs are being shadowed by Soviet SSNs with active and passive sonar.

Role of the SSNs

(a) Deter the shadower by overt surveillance.

(b) Attempt to decoy the shadower by producing sonar traces and noise interference.

(c) Trace the shadower through a sanitizing area in which the surface ships might take active measures to interfere.

(d) Hold the shadower when his contact with the SSBN has been broken, and report his movements if approaching an SSBN operating area.

Working Party's provisional assessment of SSNs on task

1 possibly 2.[90]

Another scenario, this time set in northern waters, envisaged the Soviet Union taking advantage of détente and a perceived lack of American interest in NATO and European affairs:

Finmark [a county in the extreme northeastern part of Norway]

By 1977, a prolonged period of détente and renewed prosperity have relaxed NATO attitudes. At the same time the European members of NATO have grown closer together in many ways: some people think that as a consequence the new administration in the United States will drift away a little bit more openly.

The Scandinavian members of NATO have remained active, after their traditional fashion: no foreign troops, no nuclear weapons, but loyal and otherwise fully paid-up. Norway, for example, organizes an army exercise in the Spring, involving British, Canadian and German troops besides her own in Arctic manoeuvres.

Russia unexpectedly reacts, and protests to Norway about an alleged frontier violation. She demands assurances that Norway will not allow foreign troops to use her facilities, and shows some indication of wanting to use the affair to test Western political solidarity.

The Russian Northern Fleet undertakes some ostentatious exercises off North Cape. Maritime aircraft patrols are set up, and a limited amount of troop movement towards the Norwegian frontier is observed. All of these military gestures are made unprovocatively and there is no very clear evidence that the Russians intend to push the situation to the extremes. The Norwegians are badly rattled nonetheless and the affair has the result of forcing the US Government's hand. They renew their undertaking to meet their obligations, but make it known that they will not 'force the pace'. Quite what they mean is not clear.

Development
NATO decides therefore initially to make a limited response by deploying ships and aircraft to the area to demonstrate a measure of support and to keep an eye on Russian dispositions. The forces operate from the UK and Germany, but could be moved quickly to Norway. Preparations are made for land reinforcements which may be called for if the Russians persist. But finally, a form of words is found which satisfies the Russians, leaves the Norwegians unscathed and NATO formally intact.

Roles of the SSNs
(a) Close A/S support of the NATO Task force.
(b) Surveillance of Soviet major units and of any further deployment from Northern Fleet bases.
(c) Off shore patrols of possible landing areas.

Working Party's provisional assessment of SSNs on task
1 in support of each of several groups
3 for surveillance and offshore patrols
TOTAL between 5 and 10, probably nearer 10.[91]

The Working Party also outlined a series of additional scenarios in the North Atlantic, principally the Greenland Sea, northern waters and the Barents Sea. There were also Mediterranean scenarios off Malta and Hellenic Thrace as well as general cooperation with NATO and the United States in the region. Finally came scenarios in the Far East involving Indonesia/Australia and Hong Kong. Although the report was useful, the Working Party concluded that it 'would be unproductive to try to base any further quantification of the SSN requirement on more detailed development of scenarios' and instead recommended that a detailed study should be carried out of the number of SSNs to meet the expected training and other peacetime commitments.[92]

A judgement of the general balance of capability required by the Royal Navy and of the proportion of limited financial resources that the Navy could afford to devote to SSNs began to crystallize at the end of 1967. For financial reasons, the figure was unlikely to be less than twelve and no more than twenty. The Navy eventually concluded that in order to fulfil the peacetime tasks of the Submarine Service, as well as make a worthwhile contribution to NATO's maritime forces in war, it required between fifteen and eighteen submarines by the early 1980s, when the first of the 'Oberon' and 'Porpoise' class submarines reached the end of their projected operational lives.

Initially the programme settled down to a planned order rate of one SSN every twelve months. This would have led to a fleet of eighteen SSNs by the time HMS *Dreadnought* retired from service at the end of 1982. However, as a result of the financial cuts imposed by the Wilson Government in January 1968, following devaluation in November 1967 (see p. 253), this was reduced to one SSN every fifteen months. While this elongated the SSN programme and produced the required immediate, but not long-term, savings in cost, militarily there were a number of consequences. These included a reduced NATO and national SSN order of battle throughout the 1970s, which was reflected in UK force allocations to NATO (sixteen SSNs by 1982 as opposed to the earlier eighteen) and a reduced and delayed capability for the SSN fleet to assume the role of the Navy's anti-ship strike capability in which it had been cast after the phase-out of the carriers in the early seventies. It also meant the possible life extension of some of the patrol class submarines in order to make good the shortfall in general submarine capability as well as a reduced ability to provide realistic anti-SSN ASW training for the fleet.[93] The MOD also looked at ways of substantially increasing the productivity of SSNs on certain peacetime tasks.

One of the biggest problems affecting the size of the Royal Navy's SSN fleet was British industrial capacity. Since 1966 the Treasury had been pressuring the Navy to slow down the SSN building rate to the point where just one shipyard could cope with the Navy's requirements, rather than the two (Vickers and Cammell Laird) that were still involved in both SSBN and SSN construction. After examining the economics of the proposal, considering the return on capital investment at each yard and the effects on the SSBN building programme of a decision to place no more nuclear work at one yard, the MOD concluded that it would not be more expensive to revert to one shipyard and could in fact be cheaper. Vickers was the obvious choice. As we have already seen, Cammell Laird did not have the design capacity or the experience to act as a lead yard and their performance over the Polaris programme compared unfavourably with that of Vickers.

Although Cammell Laird had been awarded the contract for the fifth SSN, HMS *Conqueror*, this was only done to avoid what is known in the shipbuilding industry as 'last ship syndrome' – whereby the workforce take longer on the last unit of a contract – having an impact on the Polaris programme. By the time HMS *Revenge* was finally accepted into the Navy in December 1969, Cammell Laird and its workforce had, in the eyes of the MOD, worked themselves out of the nuclear-submarine business. The final straw came shortly before *Conqueror* was completed when an individual put a handful of metal objects in the submarine's gearbox, which later exploded during testing, delaying completion by many months and costing the taxpayer millions.[94] Had the Navy's fourth Polaris submarine, HMS *Revenge*, suffered the same fate as *Conqueror*, a Cammell Laird employee would have been guilty of delaying the Polaris programme and sabotaging the United Kingdom's strategic deterrent.

In 1969 the government announced that in future all nuclear submarines would be built in Barrow.[95] The nearly 3000 highly specialized men taken on during the Polaris years were now surplus to the company's needs.[96] Throughout 1969, as 3000 'horrified' workers at Cammell Laird looked on, the 'Polaris birds came home to roost'.[97] The pay and productivity deals awarded during the Polaris contract were viewed as inappropriate in the harsher world of competitive pricing for merchant shipbuilding, in which technology had moved on. Managers and men struggled to relearn habits more appropriate to the competitive environment of commercial work.[98] The 'easy regime' of Polaris work had been 'neither conducive to wise overall management of resources, nor to the cultivation of good practices'.[99] Added to this was a realization that many of the fixed-price contracts that the company had entered into for merchant vessels to be delivered from 1969 to 1972 would lead to crippling losses. By spring 1970, losses on shipbuilding work were running at an annual rate of £10m and the Labour Government, worried about the forthcoming general election, stepped in to bail out the company.[100] As Kenneth Warren has noted, 'From these drastic changes of summer 1970 may be dated the beginning of the last phase of Cammell Laird's business career.'[101]

The decision not to continue constructing nuclear submarines at Cammell Laird had a significant and longstanding impact on the UK's industrial capacity to design and build them. Thereafter, Vickers in Barrow-in-Furness was the only shipyard in the United Kingdom capable of designing and constructing nuclear submarines for the Royal Navy. Vickers had accepted the nuclear-submarine programme as a challenge and, as a 1967 report on its impact noted, 'despite some serious and unexpected setbacks, its purpose has so far been achieved'.[102]

The programme also had some unexpected benefits across British industry such as a marked increase in the standards of materials and workmanship, improved performance in programme evaluation and discipline, and a greater awareness of the need to get a contract defined more specifically before it was awarded, as well as the elimination of some traditional demarcation problems between particular trades. Vickers concluded that 'the Nuclear Submarine Programme has served a serious purpose of putting certain companies in a position where they will be more competitive in the international environment'.[103] It also estimated that the fallout from having access to the American nuclear-submarine programme, on which vast sums of dollars had been spent on development, 'could well represent in money, figures comparable to the whole cost of our programme'.[104]

But the nuclear-submarine programme, which was described in a 1967 briefing to the Minister of Defence (Equipment), Roy Mason, as 'comparable in sophistication to the most intricate items in the American Space Programme', was making extraordinary demands on the Royal Navy and British industry.[105] Not without reason, in the nuclear-submarine business SSN was said to stand for 'Saturdays, Sundays and Nights' and SSBN for 'Saturdays, Sundays and Bloody Nights'.[106] Since April 1963, the Royal Navy's first SSN, HMS *Dreadnought*, had been in near constant use and the operational part of the first commission, which involved proving and modifying equipment destined for the 'Valiant' and Polaris submarines, was described as an outstanding success. *Dreadnought* participated in many major Fleet exercises where her speed, endurance and sonar introduced a new dimension into submarine warfare in the Royal Navy. In 1966, the Captain of the 3rd Submarine Squadron reported that *Dreadnought* was 'well led, highly trained and fully operational'.[107]

But there were problems. In October 1966 the first in a planned series of sea trials was abandoned when a number of cracks were discovered inside *Dreadnought*, initially in two frames that formed part of the submarine's hull. The Prime Minister, Harold Wilson, was concerned about safety but also because 'the costs involved are bound to lead to acid comment by the C. and AG [Comptroller and Auditor General] and by the Public Accounts Committee'.[108] Wilson wanted to set up an independent public inquiry, but was convinced not to by Healey due to concerns about the impact on the Anglo-American relationship. The Americans discovered from their own experiences that higher specification steel was required to construct submarine hulls, but they did not tell the British. 'We now know that they have had similar problems to which they claim to have found the solution,' Healey wrote to Wilson in October 1966. 'But they told us nothing until we approached them about our cracks; and even now they remain very reticent, possibly because of the attention the matter has aroused in the British Press, and the cloak of security they have cast around their own difficulties.'[109] When looked at in perspective the problem was aggravated by an improvement in testing techniques: as more sensitive tests were conducted, more cracks became evident.[110] A number of operational penalties were accepted, but as Healey later told the House of Commons, the 'cracks are so small that they are only discernible by the most modern techniques of ultra-sonic testing. They represent an additional maintenance task, but they do not involve the crews of the boats in any degree of risk.'[111]

Cracking problems continued to plague *Dreadnought*. When the submarine entered Rosyth for the first refit of a nuclear submarine in 1968, investigations revealed yet more significant cracking in the valves and some of the pipework associated with the American S5W reactor system.[112] Sir Solly Zuckerman was once again drafted in and asked to raise the problem with Admiral Rickover. Rickover 'admitted to having had similar experience in American nuclear submarines, expressed his belief that the cracking was due to thermal stress and suggested remedial action. He did not regard the problem as "extraordinarily important" although he has been replacing the valves and pipe sections as and when opportunities arose.' Rickover was clear that had the cracking constituted a serious safety hazard he would have informed the Royal Navy about it when US Navy submarines first experienced it. He 'reaffirmed his intention to honour his obligation to inform the Royal Navy of any major safety problem concerning the *Dreadnought* plant, but he was equally clear' that he did not consider himself obliged to inform the Royal Navy of every US technological development in the field of nuclear propulsion. He advised the British to look for similar cracking elsewhere in the plant which may not have been observed by the refitters.[113]

Experience with the first two all-British SSNs, HMS *Valiant* and HMS *Warspite*, was far better. Both had shown that the class of SSN the Royal Navy was building could travel long distances at high speed, free from detection by anything but a combination of good fortune and the most highly developed detection equipment. The designers, manufacturers and installation test teams had succeeded in ensuring that the

PWR1 plant was extremely safe and reliable. In a fourteen-month period in *Valiant*, between July 1966 and October 1967, the reactor compartment required less than one man-hour per day on defect rectification, half of which was 'unskilled work'. The high standard achieved on the reactor plant became the datum against which all other equipment and systems were compared. During one period on *Valiant* the reactor compartment only required 3 per cent of total man-hours on defect rectification while the remaining propulsion system took 40 per cent and hull systems 25 per cent. Fifteen per cent of the propulsion work was steam leaks, which took up a considerable amount of time and skilled labour.[114] But while the Submarine Service was finally producing powerful platforms to fulfil the Royal Navy's future concept of operations, there were serious problems with the weapons systems that were intended to give these new capital ships teeth.

As we have seen, at the beginning of the 1960s, the only available anti-surface ship weapon available to the Submarine Service was the Second World War era Mark 8 torpedo, which had first entered service in 1940. Although the Mark 8 had been repeatedly modified, these modifications did little to alter its basic characteristics and by the 1960s it was increasingly suffering from age-related defects. Operational limitations were imposed on its use, which reduced its reaction time and forced the attacking submarine to carry out noisy tube operations close to a target, which increased the chances of detection. The relatively short range also limited a submarine's freedom of movement during the final stages of attack. The ineffectiveness and unreliability of the Mark 8 was demonstrated in June 1967 when HMS *Dreadnought* was ordered to sink the *Essberger Chemist*, a German tanker that had caught fire and was considered a hazard to shipping. *Dreadnought* took two days to cover the 1000 miles from Gibraltar to the Azores area, where the derelict tanker lay. Of the four Mark 8 torpedoes fired at the tanker one missed at a range of a mile, while the other three hit but failed to sink the small (280 foot) long vessel. Embarrassingly for the Submarine Service, the tanker had to be sunk by gunfire from a frigate. The *Daily Telegraph* later highlighted the implications of this dismal performance for the Navy's claim that its future strike power lay with the Submarine Service and proclaimed that 'so long as these vessels are armed with the Mark 8 torpedo such a claim is ludicrous'.[115]

Although the Mark 8 could also be used to attack submarines, the primary anti-submarine torpedoes in use with the Submarine Service were the Mark 20 'Bidder' and the Mark 23 'Grog'. The Mark

20 medium-speed, unguided, passive-acoustic homing torpedo was considered obsolete by the late 1960s as it was slow and could be easily countered by decoys. It was never deployed on the Navy's new SSNs and SSBNs. However, it remained in service in the 'Oberon' and 'Porpoise' class conventional submarines as it was the only torpedo capable of being discharged from stern tubes. The Mark 23 wire-guided anti-submarine passive homing torpedo was also obsolete. It was neither designed to home in on modern Soviet SSNs, nor capable of doing so, and while it had enough speed to deal with conventional Soviet submarines, it could only home in on the 45 Khz radiated noise produced by a snorting/cavitating Soviet 'Whisky' class submarine, which was expected to be relegated to second-line duties by the mid-1970s. The torpedo, which had an overall weapon system effectiveness of just 33 per cent, also 'had a very bad reputation among submariners and would have been entirely useless in a war situation against Soviet submarines'.[116] 'It was a rotten torpedo, being just a Mark 20 with a guidance wire dispenser attached,' complained Sam Fry. 'Quite how it was ever accepted into service is beyond me.'[117]

There was only one conclusion: 'In the 1970s even conventional submarines cannot be hit by current weapons.'[118] The Submarine Service set its hopes on a new wire-guided torpedo known as the Mark 24, which promised 'a marked advance on any previous weapon', and was capable of attacking the latest Soviet submarines. The scientists and engineers in charge of the Mark 24 development were so confident of their claim that their new torpedo would be 'the end of the line for torpedo development' that they named the development programme Project Ongar, after Ongar station, which was then the last stop on the Central Line of the London Underground. However, in the early 1960s the Mark 24 development programme suffered from setback after setback, aggravated by the closure of the Torpedo Experimental Establishment at Greenock in Scotland in 1959 and the subsequent transfer of its staff to Portland in Dorset. In the early 1960s, wideranging reviews, one of which was entitled 'Whither ONGAR?', resulted in the torpedo performance specifications being reduced considerably to achieve the in-service target date of 1969. The torpedo's propulsion system was changed from an internal combustion engine to an electric motor with a silver zinc battery, reducing the planned speed of the torpedo from 55 knots to 24 knots, with a short final-attack-phase capability of 35 knots. The Soviets had also started building submarines that could operate well below the floor depth of the Mark 24.

'SSNs cannot be hit by current weapons,' concluded the Admiralty assessment. 'They can be attacked only by the MK.24, provided that the target is not too deep (over 1200 ft).'[119] The original version of the Mark 24 was also incapable of hitting surface ships. It had no anti-surface capability.

Unsurprisingly, the Admiralty's torpedo experts were not highly regarded. Even the Superintendent of the Royal Navy's Torpedo Experimental Establishment, Captain G. O. Symonds, admitted that the reaction of any officer who has dealings with torpedoes or the Experimental Establishment was 'almost inevitably one of horror'. 'Nothing new ever comes out of that place, except promises and fanciful project names' was the usual reaction from submariners in the late 1950s and early 1960s. Symonds admitted that 'there has been justification for such remarks in the past' and he blamed changes in policy, his establishment, the Admiralty, delays and former project leaders and scientists 'trying to attain a ridiculously high Staff Requirement, which he has regarded as a challenge'.[120] According to Symonds, Staff requirements for the Royal Navy's torpedoes had often been 'based on an individual's pipe dream, who, finding the future very hazy, has doubled the speed, added a couple of noughts to the run and one nought to the depth and then sat back thinking "Well, that should take care of the next fifteen years!" or, possibly, "That should take care of my turn in the Admiralty." '[121]

In February 1969, Admiral Michael Pollock, Flag Officer Submarines, sent a paper entitled 'Submarine Weapons in the 1970s' to the Controller of the Navy, Admiral Sir Horace Law, in which he expressed his frustration and anger:

> The attached paper analyses the total inadequacy of the weapons at present available to the Submarine Command in the anti-submarine role and forecasts that a similar situation will occur in the anti-warship role within the next five years.
>
> It illustrates the degree to which the viability of the submarine force depends upon the timely introduction of a Mk.24 torpedo of specified performance and proved reliability, and to the further development of this or another weapon to restore anti-surface ship capability.
>
> It records the unsatisfactory effect upon morale which so inadequate a capability would cause in war and the dangerous use to which such a credibility gap in our armoury could be put in disputing the submarine.
>
> The situation portrayed is unsatisfactory in the extreme, but I should

be failing in my responsibilities, both as an operational commander and as your advisor on submarine matters, if I failed to draw your attention most urgently to it.[122]

Pollock concluded that it was 'rapidly becoming more and more evident how ineffective it is to go on building modern capital ships and arming them with the equivalent of a bow and arrow'.[123]

This truly dismal record of torpedo development was to get worse. Acceptance Trials for the Mark 24 began in November 1968, with the aim of issuing the new torpedo to the fleet by January 1970. However, the trials ran into 'significant initial problems' and by 1969 they were 'some months behind schedule'.[124] By the end of 1969, as development of the Mark 24 continued, the Submarine Service was so desperate for an effective weapon that it explored a number of alternative options, including the purchase of US weapons such as the Mark 45 ASTOR torpedo, which was capable of carrying a nuclear warhead, the Mark 40 Mod-1 torpedo or the SUBROC (SUBmarine ROCket) anti-submarine rocket. Pollock also raised the possibility of fitting a WE 177A nuclear warhead to the Mark 8 torpedo 'as a way to achieve a large increase in the effectiveness of our torpedoes within a relatively short period'.[125] He argued that such a torpedo would be 'much superior to any present British submarine weapon' and that 'if urgent action is not taken, we shall find that the "main striking power" of the Fleet consists of an excellent vehicle with no punch'.[126] However, the proposal was never taken seriously. The short range of the Mark 8 put the firing submarine well within the destructive range of a nuclear torpedo.[127]

Despite the fact that the Royal Navy's newest and most capable SSNs lacked effective torpedoes, the Navy was anxious to send them into northern waters to gather intelligence on the Soviet Fleet and provide strategic warning in the event that the Cold War ever heated up. By the 1960s, the assumption was that prior to a Soviet attack on NATO, the Soviet Northern Fleet would leave its bases and head for open waters, and there would be a marked increase in communications activity on the Soviet side. By collecting information about levels and patterns of Soviet military activity, which could then be used to form a detailed database, major deviations from the normal pattern could be detected and serve as a warning of attack. As in the 1950s with the early conventional submarine operations, the northern waters of the Soviet Union such as the Barents Sea became an important region for surveillance and warning.

UP NORTH

For the Commanding Officers of the Royal Navy's newest SSNs, these intelligence-gathering operations were exceptionally demanding. At least one former diesel submariner turned down the opportunity to command an SSN because he had had such a bad time up against the Soviets, who hunted his diesel submarine to exhaustion.[128] Such were the demands on COs that the Navy would assign an extra command-qualified officer to a submarine conducting an intelligence-gathering operation to ease the burden. In an SSN, the complexity and continuous twenty-four hours a day operations in often hazardous conditions were now well beyond the reach of a single person. While the CO generally set the tone, technically and tactically he needed to have some competent men at his side in every department. The success of a submarine, particularly a nuclear submarine, was the result of a team effort. 'In the case of an attack,' wrote one SSN Commander, Sam Fry:

the estimation of the target's course, speed and range will be made by the Captain. He will be aided by his observations and experience but the capability of the team, the tubes' crew, trimming Officer of the Watch and many others will have all contributed to the successful [outcome]. All the same factors applied to a Cold War patrol, and one should never forget the technicians that have been responsible for propelling the submarine, and ensuring that all the complex equipment works.[129]

When Royal Navy COs were recognized with awards for their exploits, often in the form of an OBE, the crews were rightly proud, but sometimes referred to the OBE as 'Other Buggers' Efforts' or 'Other Bastards' Efforts'. 'Whilst this is a truism,' acknowledged Fry, 'someone should show off the reward for all their hard work, and I have always hoped that my crew members shared that view.'[130]

Like the 'T' class submarines that conducted the first intelligence-gathering operations in the mid-1950s, Royal Navy SSNs were extensively prepared before they sailed north. Marine Engineers such as Patrick Middleton ensured that the submarine was 'extre-e-mely quiet':

This involved checking that all the rotating machinery was in balance and sitting on its rubber mountings so that no noise was transmitted to the hull and onwards. No easy task. Despite rubber or flexible pipework, machinery sitting on springy rafts and endless examination, it was all too possible for a bucket to slip down in some cloacal corner and short out the insulation.[131]

SSNs were then put through a Static Noise Range in Loch Fyne, where the submarine would be secured fore and aft to large buoys and lowered to a depth where it was surrounded by hydrophones. A scientist sitting in a hut ashore would then listen as the submarine's crew ran all sorts of machinery in different combinations to see the result. 'A noisy pump could blank our own sonar, and provide a beacon for the enemy; distinctive combinations of frequencies, unique to [the submarine], could make a powerful and unwelcome identifier,' says Middleton.[132] In some cases the forward escape hatch and the indicator buoys were welded shut in order to reduce the risk of any unwanted noise. The type of equipment fitted to SSNs for intelligence-gathering patrols had also changed considerably. In the days prior to departure there would be a constant stream of scientists and engineers installing some 'special fit' equipment. Often the equipment would arrive accompanied by its inventor 'of the wild-haired kind, who sometimes had little idea of the practicalities of installing or operating his precious baby,' recalled Middleton.[133]

The Navy's first SSN, HMS *Dreadnought*, was too noisy to be considered for sensitive intelligence-gathering operations in northern waters. The first SSN to head north was therefore HMS *Valiant*, under the command of Commander Peter Herbert. Before *Valiant* departed Faslane at some point in 1968, Herbert was warned by Denis Healey: 'Don't you bloody well get detected.'[134] Herbert quickly found that the operation 'was surprisingly easy because we were quiet and we could wander and watch things going on'.

'I watched, from about 1000 yards behind a cruiser, watching its missile launch and those sorts of things and following a submarine and going underneath a submarine and looking at its bottom, those sorts of things. But mostly ... at night when nothing was going on, then you went up and listened to communications ... hoovering up as much as you could.'[135]

Herbert was 'bloody careful not to get detected', but he had 'to get close several times in order to get any information'. At one point during the operation *Valiant* was sitting underneath and taking photographs of a Soviet submarine when it unexpectedly started to dive. Herbert came away from the Navy's first SSN intelligence-gathering operation with a healthy respect for the Soviet surface fleet.[136] Their submarines 'were less worrying', he says. Although Soviet surface ship sonars were very poor – which explained why *Valiant* and other SSNs were able to do what they did and remain undetected – Soviet surface ships' weapons capability was 'very very impressive'. *Valiant* observed Soviet surface ships 'shoot down their targets in quite a spectacular way. The number of missiles they fired was fantastic.' 'It was an interesting time,' said Herbert, 'all of us were bloody tired at the end of it . . . you didn't get much sleep really.'[137]

In the mid-1950s, the Soviet Union had embarked on an intensive anti-ship missile programme to counter the threat of US aircraft carriers. As Khrushchev explained in his memoirs, 'The Americans had a mighty carrier fleet – no one could deny that. I'll admit I felt a nagging desire to have some in our own Navy, but we couldn't afford to build them. They were simply beyond our means. Besides, with a strong submarine force, we felt able to sink the American carriers if it came to war. In other words, submarines represented an effective defensive capability as well as reliable means of launching a missile counterattack.'[138] The P-6 (SS-N-3a) Shaddock cruise missile, armed with either a conventional or a nuclear warhead and with a range of up to 245 nautical miles, could be fed guidance information while in flight by a video data link, enabling long-range reconnaissance aircraft and later satellites to relay radar information from great distances to the launching submarines. These anti-ship cruise missiles were carried by the first-generation Soviet nuclear cruise-missile-carrying submarine, the Project 675/'Echo II' class and the conventionally powered Project 651/'Juliett' class, which housed eight launch canisters fitted in the deck casing. Twenty-nine of the class were produced between 1963 and 1968.[139]

On one occasion in the late 1960s, HMS *Warspite* under Commander John Hervey entered into northern waters and started to trail a Soviet 'Echo II' cruise-missile-carrying submarine.[140] When submerged, *Warspite* depended on its Type 2001 bow sonar to determine what was taking place. The submarine was also fitted with a very short-range passive sonar in the rear of the fin and a line/flank array on both sides of the hull known as Type 2007. Unlike in the surface world, where radar could be used to determine the range, course and speed of a target, passive sonar only gave the bearings of a contact, not its range, course and speed. This could be obtained by using active sonar, but during a covert intelligence-gathering operation the use of active transmissions was unacceptable as any use would immediately be heard and betray *Warspite*'s presence.

Warspite's attack team were 'closed up' with the bearings of all contacts, including the Echo II, plotted together with additional annotations such as characteristics and classification on a Contact Evaluation Plot (CEP). The maintenance of a comprehensive tactical picture required experience and judgement and very good liaison between a submarine's Sound Room (which gathered the sonar information) and Control Room. Although *Warspite*'s crew was good at interpreting the Contact Evaluation Plot, using it to build up an accurate picture of the location, range and course of the Echo II, it was very difficult to determine the depth at which the Echo II was operating. The best *Warspite*'s crew could do was estimate based on occasional machinery noise emanating from the Soviet submarine. Maintaining an accurate track of its range by following how its bearing movement changed over time was also challenging. Little change to the bearing could mean either that the Echo II was too far away to register any movement, or that it was on a collision course at very close range.

As *Warspite* continued to trail the Echo II, the Soviet submarine shut down one of its two propeller shafts, which reduced its speed by around 3 knots as well as its noise signature. *Warspite*'s crew detected none of this; nor did they appreciate that the bearings remaining relatively steady meant that the distance between *Warspite* and the Echo II was closing fast. At around 0025 *Warspite* slammed into the ballast tank of the Soviet submarine. 'There was an awful bang, and crushing, and scraping and we were pushed right over in our chairs, about 74 degrees, so we were almost on our backs,' said Frank Turvey, *Warspite*'s Engineer, who was walking through the Control Room at the moment of impact. 'There were alarms on all of the panels, bells ringing, red lights flashing.'[141] The initial impact caused *Warspite* to violently heel to starboard. She then swung back, passed under and thereupon collided with the Echo II again, causing *Warspite* to roll to starboard for a second time, between 65 and 70 degrees. *Warspite*'s CO, John Hervey, was standing at the far end of the Control Room at the moment of impact. As he clambered to the ship control area he found his Executive Officer, Lieutenant Commander Tim Hale, standing on the almost horizontal face of *Warspite*'s normally vertical Systems Console, where he had initiated the emergency surface procedure.[142]

Once on the surface *Warspite*'s engineering team prevented the nuclear plant from shutting down, which would have severely limited the available power and the ability to withdraw from the area. They set about calmly and methodically restoring the various items of machinery that had developed faults during the collision and checked for casualties. No one was injured. One man on the planes had fainted and had to be relieved. Within five minutes, *Warspite*'s Navigator, Lieutenant Tom Le Marchand, despite the failure of all the submarine's compasses, managed to get an accurate fix of *Warspite*'s location and was able to plot a course out of the area.[143] A large chunk of *Warspite*'s fin, at the top port forward corner, where submariners would normally stand when the submarine was running on the surface, was damaged. *Warspite* did not require immediate repairs but was now restricted in its speed and too noisy to complete the operation. The Echo II was also damaged, though, like *Warspite*, not critically. It reportedly returned to its base, where repair crews discovered a hole in the outer hull so large that a Soviet officer said 'a three-ton truck could easily' have driven through it.[144]

Hervey dived *Warspite* and set course for Faslane, later rendezvousing with a Royal Navy frigate, HMS *Dundas*, near the entrance to a Scottish sea loch. The frigate then escorted *Warspite* to a sheltered anchorage, where a team of waiting shipwrights carried out a series of cosmetic repairs to the top of *Warspite*'s fin to conceal the damage from Soviet reconnaissance aircraft and spy trawlers as well as unwanted media attention.[145] By the time *Warspite* reached Faslane the press had somehow caught on to the fact that there had been an incident involving a Royal Navy SSN. The government promptly issued a cover story in order to preserve operational security. Officially, *Warspite* had hit an iceberg. The 19 October 1968 edition of *The Times* ran a story on its front page: 'Ice on the sea has damaged the 3,500-ton Warspite, one of the Navy's five nuclear-powered submarines. There was no risk of radio-active leakage, the department said. Warspite, which was delivered to the Navy last year, returned to Faslane, on the Clyde, for repairs. The Warspite, on exercise in the North Atlantic, cleared the obstruction with slight damage to her conning tower and other parts of the superstructure.'[146]

A number of former submariners, shipyard workers and former senior ministers have unofficially revealed what actually happened.[147] 'It's become known now that during my tenure one of our submarines was quite badly damaged,' says the then Minister of the Navy, Lord Owen. 'Its conning tower was not quite destroyed, but very seriously bent. Of course we lied about how it had happened. But now it is known it had hit a Soviet submarine because it was shadowing it. Now, boys will be boys. There was an element of just going off and doing this for bravado and that we need to curb some of it. And I think some of it was curbed.'[148]

The repairs at Barrow took twenty-eight days. *Warspite*'s badly damaged fin was removed and replaced with HMS *Churchill*'s, then building at Vickers. A strike by the Barrow workforce helped maintain the cover story as management, who were used to dealing with sensitive information, completed most of the work. One shipyard worker, on seeing *Warspite* as it entered Barrow, is reported to have said: 'First iceberg I've heard of with antifouling.'[149] Although Hervey suffered the severe displeasure of the Admiralty Board, he remained in command of *Warspite* and conducted a number of other successful Cold War patrols. As he wrote:

> The moment one committed a submarine to cold war type ops, one had to do so in the certain knowledge that to perform the task effectively – especially in an SSN – the CO was being encouraged to operate – quite deliberately – close to Soviet submarines – sometimes very close – but without any depth or space separation guaranteed – nor any knowledge of the other submarines' future intended movements. So arguably, those ordering the operations were accepting that collisions, if not inevitable in the short term, were certainly very likely to occur, sooner or later, in the long term. But the value of the intelligence to be gained and the – in your face type – pressure which such ops put on the Soviet high command were probably thought to justify the risks. We certainly never doubted this. Moreover, it was excellent training for war.[150]

HMS *Warspite*'s second intelligence-gathering operation under Hervey's command was very successful. Only one member of the submarine's crew asked to be relieved. 'I had already agreed that anyone who wanted to go should be allowed to do so,' recalls Hervey. 'All the rest bravely swallowed their fears and just got on with the job, for which I think they deserve praise.'[151]

In 1969, Hervey handed over command of *Warspite* to Commander Sandy Woodward. *Warspite*'s programme involved working with surface ships, activities that required high speeds and vigorous manoeuvring. Not long after taking command Woodward discovered that some of *Warspite*'s crew were still suffering from the incident with the Echo II:

'It was my first week at sea, before Christmas in '69 I suppose, I was given a week to work myself into the submarine and we were at periscope depth taking the wireless routine and I said to the third hand who was on watch when the wireless routine is finished, ten degrees bow down, onto twenty knots, two hundred feet down to the deeper areas to do our business. I went down to the Senior Rates mess on the deck below to meet all the Chiefs and Petty Officers. We were all having a beer, cheerful, chatting away, getting to know them. Then the angle came on and there was a deathly hush, absolute quiet. Then we pulled out and as we levelled off, slowly conversation started again. I thought "Fuck me", my entire Senior Rates Mess is twitchy about a ten degree angle, which should be as normal as blueberry pie in an American boat. I went back up to the Control Room and told the First Lieutenant what had happened and he said: "Oh didn't you know?" I said: "Oh God, what haven't I been told" and he told me the whole story.'[152]

In order to demonstrate to *Warspite*'s crew that he was capable of handling any situation that might occur and to reassure those still unnerved by the more severe angles associated with surface ship exercises, Woodward spent a week taking *Warspite* through an extreme version of 'Angles and Dangles', putting the submarine through its paces, diving, rising, twisting and turning.

Warspite's first intelligence-gathering operation under Woodward's command was directed against one of the Soviet Navy's newest vessels, an 18,000 ton helicopter carrier. These were 'a new development and are of particular interest' and their role was 'uncertain'.[153] Intelligence indicated that one of the new helicopter carriers, the *Leningrad*, was on its way from the Soviet Black Sea Fleet to join the Northern Fleet and that 'something' large was projecting 50 feet below her hull. Woodward was ordered to take *Warspite* into the Atlantic, find the *Leningrad* and gather intelligence on its general behaviour. Not long after departing Faslane, *Warspite* intercepted the *Leningrad* and its escorts off the west of Ireland, heading north at a speed of around 10 knots. This was ideal for an important manoeuvre in the submariner's intelligence-gathering craft, the so-called 'underwater look', which would see *Warspite* move directly underneath the *Leningrad*.

Great care and attention were taken when preparing for this complex manoeuvre. The submarine would close up at action stations with extra men placed underneath all the hatches to listen for the target and report when it was overhead. In the Control Room, the Operations Team would be working hard to try and identify the course, speed and range of the target, gleaning as much information as possible from the sonar. Meanwhile the First Lieutenant and the Captain would agree on the depth the submarine should maintain while directly underneath the target. This was a complex judgement that took into account the estimated height of the swell which led to the pitching up and down of the target, the height of the periscope when raised, the target's draught, the depth separation from the bottom of the hull to the top of the periscope and an added safety margin. Once these had been calculated the submarine would move down to the required depth, check the visibility was good enough and then begin the approach.[154]

The approach phase of the manoeuvre took time as the submarine sneaked up astern of the target, attempting to establish its exact speed. In the Control Room there was silence as everyone listened out for the words of the Sonar Controller who, using a large-scale plan of the submarine, with the position of all the sonar sets and the fans of the bearings emanating from the positions, plotted, triangulated and established a rough set of ranges. Ship Control then increased or reduced the revolutions in stages while planesmen concentrated on maintaining the course and depth required. One lapse of concentration could lead to disaster for both the submarine and the ship above. As the submarine moved in closer to the objective, the periscope was raised and the CO looked for any evidence, such as the wake of the ship, that would indicate that the submarine was close to its target. A report from men waiting in the fore-ends of the submarine that they could hear the target was eventually received, followed by the appearance through the periscope of the murky outline of the hull. The Captain provided a running commentary of what he saw while a camera, fixed to the periscope, was used to take photographs of whatever was required of the underwater fittings, and in the Sound Room recording tapes were switched on to suck in as much sonar data as possible.

Woodward recalled that *Warspite* 'went straight in, lined up astern and slowly overtook her while looking ahead and above underwater through my periscope. First I saw her wake, then her huge rudders appeared and then her screws. Now for the fifty-foot projection, I thought, which for all I knew could well write off my periscope if I didn't see it soon enough. Sure enough, a large dark shape appeared ahead and with a quick check to see that my periscope was going to go safely underneath, we continued towards it. At about fifteen feet, it became clear that it was only a large sonar dome, probably less than fifteen feet deep, and similar to what would be required to house a long-range, high-powered, active sonar array such as I had in my bow. I went a bit deeper, made some sketches and decided to stay underneath the *Leningrad* to await developments – it seemed as safe a place to be as anywhere else – who would, or even could, look for me there?'[155]

Underwater looks required a great deal of concentration from all involved, especially the planesmen, who were responsible for ensuring the submarine did not wander off the required depth. Often, when they were relieved, they would be soaked in sweat. Intelligence-gathering operations were exhilarating for *Warspite*'s crew. Patrick Middleton described how:

> The boat thrummed with excitement and with the water whooshing past. We were dead level, only the smallest angle on the planes was needed to keep depth, and the machinery all seemed well matched at what was, after all, its design condition. The only downside was that if something did go wrong it would go wrong a lot more quickly, and more profoundly, so we were all kept on our toes. The Command team were all metaphorically peering ahead, trying to keep their eyes, well their ears actually, on the quarry.[156]

Warspite remained under the *Leningrad* for the next few hours. All seemed well until the Soviet ship suddenly increased speed. *Warspite*'s crew then detected some Morse code letters on her own underwater telephone. 'I didn't know quite what to make of that at the time but thought it might be some kind of identification or interrogation message – perhaps they were expecting one of their own nuclear submarines,' says Woodward. Rather than move away from the *Leningrad* he decided to wait and see. Half an hour later *Warspite*'s sonars indicated that one of the *Leningrad*'s escorts was off the starboard beam pinging away at *Warspite*. Woodward ordered full helm away and with 45 degrees of heel in the turn *Warspite* reversed course at 21 knots. The escort quickly lost contact and *Warspite* slipped away and continued to shadow the surface group. 'I had no difficulty shadowing from about twelve miles away, watching *Leningrad*'s masts over the crystal-clear horizon, listening to and recording her big sonar banging away, very similar to my own, if I cared to use it' recalled Woodward. 'Things seemed fairly settled for a while, the *Leningrad* was doing about ten knots and making it easy for me to keep up at periscope depth.'[157]

But then it all started to go wrong. An American Long-Range Maritime Patrol aircraft appeared and spotted *Warspite*'s search periscope. Mistaking *Warspite* for a Soviet submarine, the American aircraft started dropping smoke floats to mark the position of the submarine. An exasperated Woodward attempted to call the American pilot off on the radio but received no answer. But it was too late, the *Leningrad* decided to investigate what was happening and turned immediately towards *Warspite*. 'We were getting excellent recordings of her big new sonar at work,' said Woodward. 'I let her get contact, indicated by a predictable change of operating mode, giving more useful intelligence on its performance, before going deep and evading at speed.' Woodward moved *Warspite* astern of the *Leningrad* and continued to shadow from the other side, before breaking off and returning to Faslane with what was referred to as 'the take', all the intelligence collected on the *Leningrad*.[158]

While Soviet cruise missile submarines and surface ships remained important intelligence targets, by the mid-1960s the West was increasingly concerned about the development of a new class of Soviet strategic missile submarine that appeared to be comparable to the US Polaris system. The Soviets had started work in this area in 1958 and by 1962 the Soviet Navy had approved the design of the Project 667A ('Yankee I') missile submarine, which carried sixteen R-27 (SS-N-6) ballistic missiles, each with a range of 3000 kilometres, capable of reaching targets in the eastern United States and Canada within minutes from launching positions in the Atlantic. Construction of the first Yankee I began in 1964 with the first of class entering service in 1967, eight years after the completion of the first US Polaris submarine.[159] Thirty-four 667A submarines were eventually built.[160] By 1969, twelve 'Yankee I' class submarines were in service with the Soviet Navy. Due to the relatively short range of the missiles (3000 kilometres) the Yankees had to deploy to patrol areas relatively close to the US coast. As the bulk of the Soviet strategic submarine force was based at Zapadnaja Litsa, a naval base 45 kilometres from the Norwegian border, the Yankees had to transit south, through the passages between Greenland, Iceland, the Faroe Islands and the Shetland Islands, to reach their patrol areas in the Atlantic off North America. While on transit the Yankees were vulnerable to detection in areas such as the Norwegian Sea.

On a later occasion in the late 1960s, Woodward was ordered to take *Warspite* and intercept a newly refitted 'Yankee' class submarine that intelligence indicated was fitted with long missile tubes, capable of carrying missiles with increased range, thus allowing them to be fired further from the US mainland. Woodward was ordered to remain undetected and to measure the height of the Yankee's missile tubes as accurately as possible. The patrol started badly. Shortly after departing Faslane one of *Warspite*'s 'Submarine Sunk' indicator buoys broke loose and repeatedly banged against the side of the hull, making a noise that could be heard for miles. The buoy eventually separated from *Warspite* and fortunately did not transmit the standard 'Subsunk' distress signal which would have alerted anyone within the area that a submarine had been lost. Then an electrical mechanic, conducting maintenance on one of *Warspite*'s masts, was injured when the mast was lowered onto his head. Fortunately a doctor had been embarked on the patrol and was able to stitch the mechanic's scalp back together.[161]

As *Warspite* moved towards its target, Woodward began to consider how he was going to measure the length of the submarine's missile tubes. 'I did know enough about the construction of our own Polaris submarines to realize that the tubes would extend from virtually the bottom of the keel to the top of the after deck – the tube length effectively defined that part of the design. If I could measure his draught accurately and then quickly after that get a photo of how much of his hull was above the water, we'd have as good an answer to our top intelligence target as anyone was likely to get without his tape measure.' But he had no time to test the technique. 'It had to be right first time.'[162]

After locating the Soviet submarine, which was conveniently sitting on the surface, Woodward, peering through the thin Attack Periscope, slowly moved *Warspite* in directly astern of the target. At around 400 yards, he ordered an increase in depth to 80 feet in order to take *Warspite* underneath the Russian to conduct an underwater look. 'We had some initial difficulty getting down in time but all went well,' recalls Woodward. 'There was no wake behind me to warn him of my arrival, he was still dead in the water, but rudders, after-planes overhead told me I'd got the run-in right and we took shots of his underwater fittings as we went along his whole length.'[163]

Warspite returned to periscope depth. Woodward planned to calculate the height of the Yankee's missile tubes by taking photographs through *Warspite*'s periscope, using the height of the Yankee's fin as a yardstick. However, when *Warspite* moved in to take the first round of photographs Woodward discovered that the top window of *Warspite*'s periscope had become fouled and the initial photographs were blurred. In order to obtain better photographs Woodward was forced to leave the periscope up to dry in the sun as *Warspite* crossed the stern of the submarine. Vision and photograph quality improved steadily as *Warspite* proceeded to photograph the target's starboard quarter, stern and port quarter. While this gave Woodward what he needed, it also gave the Soviet lookouts on the conning tower ample opportunity to spot

Warspite's exposed periscope. The first in the series of photographs from this particular stage of the operation shows a single figure on top of the fin. The last shows a group of individuals with binoculars, pointing directly at *Warspite*'s periscope.

'Obviously, I had been counter-detected,' said Woodward, 'contrary to my prime patrol order. Not good. I thought I'd try to make the best of it by passing up his port side with two masts still up so that he could estimate my course . . . and, as I knew he would, report it as such. I then pulled the masts down, went deep, turned away 180 degrees, right round to the south . . . the direct opposite of what might be expected . . . I stayed deep for fifteen minutes and returned to periscope depth . . . I continued slowly south and over the next three days observed a large part of the Soviet Northern Fleet and Air Force conducting a totally ineffective ASW search for me in a huge expanding semi-circle twenty miles away . . . Much useful intelligence was gained both of the ICBM's tube length as ordered and of Soviet Northern Fleet ASW operations as a bit of profit.'[164]

Although Woodward achieved his primary intelligence objective, he had contravened his primary patrol directive by being counter-detected. 'Plainly, I was going to be in trouble when I got home – you do not deliberately disobey clear orders and hope to get away with it.' Rather than breaking off, he decided to continue the operation. 'I was almost unconsciously even less careful to avoid counter-detection in subsequent encounters,' he said. 'I would stand to gain much valuable tactical intelligence from further counter-detections, deliberate or otherwise. So, in approximate accordance with orders, I would continue to avoid counter-detection but not be too fussed if I failed.'[165]

According to Middleton:

> There was always a running battle between the communications specialists and their sonar opposite numbers, the former wanting us to be at periscope depth with their aerials poking above the sea surface, the sonar boys wanting us to be deep with our ears flapping. Tempers sometimes ran hot in Sandy's daily planning meeting. Occasionally we thought that the opposition had got a sniff of us, and moved delicately to avoid any questing Soviet, eager to flush us out. No more iceberg incidents, we determined![166]

Eventually *Warspite* came across another Soviet SSBN and spent two days following it and collecting intelligence about its behaviour. 'We established a clear pattern of his manoeuvres and I was close to breaking off at about 0300 one morning,' continued Woodward. 'However, I decided to hang on a bit longer since I had become interested in his behaviour, particularly how he cleared his stern arcs (blind to his sonar and why I usually kept astern of him) by doing substantial course changes every few hours. My normal response to his turn was to point directly at him and go as slowly and quietly as possible, waiting for him to resume his original course. This had happened about four times already without incident and I was perhaps becoming a bit complacent. In war, we could have sunk him at any time over the previous two days. But this is peacetime, so I turned over command to my First Lieutenant to snatch a quick nap – I hadn't had all that much sleep since arriving in area.'[167]

Twenty-two minutes later, *Warspite*'s First Lieutenant, John Coward, knocked on Woodward's cabin door and said, 'Captain, I think you should see this.'[168]

'I got to the periscope in about ten seconds flat and looked astern to see an array of submarine masts close behind us, leaning over slightly as he followed us in a sharp turn to starboard,' said Woodward. 'He was actually turning inside us as well, which was further bad news and presented a risk of collision if he got it wrong – which seemed all too possible. I pulled our masts down, reversed our helm and increased to my self-imposed maximum speed of sixteen knots on a not quite straight course to get away from him.'[169]

Woodward was confident that *Warspite* had a speed advantage over the Soviet submarine, and although the Russian had a firm, solid active sonar contact, Woodward reckoned he knew quite a bit about how Soviet sonars behaved because 'they had reportedly copied it from our own, courtesy of some traitors at our Research Establishment at Portland some years before. From my own experience,' he said, 'I probably knew more about what it could and could not do than he did.' In order to find out, he came up with a plan: 'I believed that while he could hold me on his active sonar out to about four miles in the water conditions prevailing if he kept his speed below eight knots, if I ducked beneath the layer at about that range he would lose all contact immediately unless he too ducked beneath the layer. If he did that, I would know soon enough to go back above the layer again and stay out of contact from him. I calculated that if I made sixteen knots and he only eight, I could open the range to four miles within half an hour, go deep below the layer, turn ninety degrees off track and lose him.'[170]

After half an hour *Warspite* turned 135 degrees and pointed straight at the Russian. The crew waited and watched on passive sonar as the Soviet submarine went by. 'As soon as [I was] satisfied that he was not using his active sonar in the search mode towards me, I returned to periscope depth to try and monitor his enemy report,' Woodward went on. 'He obliged, reporting losing contact exactly where I had predicted and gone deep at a range of four miles and giving my earlier course and speed while he trundled on up that course, pinging away to no effect. Disappointing for him, satisfactory for me and allowing me to potter off and think about it all.'[171]

Subsequent analysis revealed that the Soviet submarine had probably begun to suspect the presence of *Warspite* up to twenty-four hours earlier. He had developed the habit, unnoticed by *Warspite* at the time, of taking a single high-powered sonar ping back down his course when he cleared his stern arcs. 'As it was,' concluded Woodward, 'I think he must have got a sniff there was someone lurking behind him and that last time he turned to clear his stern arcs, instead of turning some ninety odd degrees to clear his arcs and then resuming his original course, he had pointed straight back towards us and held that new course until we nearly met – physically.'[172]

Two days later, *Warspite* came across a Soviet torpedo-firing exercise and Woodward once again moved his submarine in to observe: 'Suddenly, there were two torpedoes in the water and to my slight alarm as we watched their bearing on sonar, coming directly my way ... they were most probably "practice" torpedoes without warheads, but they might just not be. I had detected no TRV [Torpedo Recovery Vessel] and the possibility that they are not practice torpedoes due to be recovered at the end of their run had suddenly become a bit more real. Could I have got this badly wrong? Should I evade them, making a lot of unavoidable noise in the process, or should I cross my fingers, wait and see and trust meanwhile that they were only "practice"? As I had neither the speed nor the countermeasures to have much chance of getting clear of them, the only sensible thing was to point between them and trust that firstly they were indeed "practice" weapons and secondly that neither of them would physically hit us. As the torpedoes approached, it became apparent that at least they were not active homers or I would have been hearing their transmissions. And if they were passive homers, I had remained quiet enough for them not to pick me up. Fingers began to come uncrossed, but not until they went safely past me either side. Little intelligence gained but I know to be careful next time.' *Warspite* returned to Faslane with a vast amount of technical and operational intelligence on Soviet submarines, surface ships and weapons systems.[173]

Herbert, Hervey and Woodward were the first commanding officers in the Royal Navy to take what was then the most advanced of Her Majesty's warships into the frontline of the deep Cold War. Just as other COs had done with conventional submarines in the 1950s, under minimal supervision, they operated with a degree of unparalleled independence and flexibility which had much in common with that given to their eighteenth- and nineteenth-century predecessors, who often took risky and sometimes controversial decisions, independent of any senior authority, while operating on the far side of the world. The advent of nuclear-powered submarines raised the Submarine Service to a central position within the Royal Navy.

6

'No Refuge in the Depths': The Cold War in the 1970s

Why did we need covert surveillance operations in the Cold War? We had to demonstrate that we could prevent the Soviets breaking out into the Atlantic from their Northern Fleet bases. Could we and our allies have stopped them? Answer, probably not. There were just too many of them, even if our weapons had been effective against their very-deep-diving hulls and high speeds. But at least we could have given them a bloody nose, and at a higher level they knew it.

Captain Richard Sharpe, Royal Navy, 2010.[1]

We are going to have relatively few SSNs. We have got to use our SSNs in the most cost effective way – and that is to put them on the tails of the most threatening of the enemy submarines. If we can achieve this it is probably the best way of protecting our forces. If all of them can be attacked approaching our forces, we should achieve our objective.

Captain Peter Herbert, Royal Navy, 1973.[2]

THE DECADE OF THE PASSIVE

Throughout the 1970s new doctrines and tactics, combined with steady material advances in sonar, communications, navigation and weapons systems, transformed the Submarine Service and moved it to the front line of a largely unnoticed and unacknowledged undersea confrontation between the West and the Soviet Union. Thanks to a steady combination of construction, machinery and operating techniques, the Royal Navy's nuclear and conventional submarines, designed for maximum silence and stealth, exploited a crucial sonar advantage over the Soviet Navy and, alongside the United States Navy's submarine force, conducted a major surveillance effort against the Soviet Union, identifying likely deployment routes and tracking, trailing and shadowing Soviet nuclear submarines, with the ultimate aim of attacking and destroying them on the outbreak of war. Throughout the 1970s, surveillance and shadowing of Soviet forces became part of an acceptable deterrent posture.

In April and May 1970, some 200 ships, submarines and land-based aircraft of the Soviet Navy conducted Exercise 'Okean' (Ocean), the largest naval exercise to be held by any navy since the Second World War. Conducted simultaneously in the Atlantic and Pacific Oceans, and the Barents, Norwegian, Baltic, Mediterranean, and Philippine Seas, the exercise involved the rapid deployment of forces, anti-submarine warfare exercises, a simulated anti-carrier exercise followed by amphibious landings. During the anti-carrier phase, simulated air strikes flown against Soviet task groups in the Atlantic and Pacific struck the simulated US Carrier Groups in the North Atlantic and North Pacific within a few minutes of each other, a remarkable achievement of planning and execution.[3] According to the then Soviet Minister of Defence, Marshal Andrei Grechko, 'The Okean maneuvers were evidence of the increased naval might of our socialist state, an index of the fact that our Navy has become so great and so strong that it is capable of executing missions in defense of our state interests over the broad expanses of the world.'[4] But Exercise 'Okean' also revealed some troubling realities about the Soviet nuclear-submarine fleet. On 8 April 1970, a 'November' class submarine participating in 'Okean', the K-8, suffered engineering problems when a spark ignited a fire in the flammable chemicals of its air regeneration system, while submerged at a depth of 395 feet in the Atlantic, off Cape Finisterre, Spain. The submarine was able to reach the surface, where smoke and carbon dioxide forced most of the crew onto the deck. On 12 April, after drifting for three days and being battered by strong gales, the K-8 sank with the loss of fifty-two of her men, including the Commanding Officer.[5]

SOSUS

Early Soviet nuclear submarines also had one other major weakness: they were noisy. Unlike diesel electric submarines, which radiated much submerged noise only when snorkelling, the first Russian nuclear submarines generated noise all the time, because machinery, such as reactor coolant pumps and generators, flow noise (the passage of the hull through the water), propeller noise (cavitation) and transient noise (noise generated by activity within a submarine), could all be distinguished from own-submarine self-noise and the ever-present ambient sea noise.[6] In the early 1950s, the US Navy developed a system of underwater surveillance listening arrays designed to provide a regular check on the movements of Soviet submarines. Hydrophone arrays were installed on the ocean bed in deep water and signals were carried ashore by submerged cables into processing stations known as Regional Evaluation Centers. The US Navy realized that monitoring the different frequencies generated by Soviet submarines could not only be used for detection purposes, but also, through signal processing, could be broken down by frequency into their component parts, displaced and used for classification purposes. Just as communications transmitted can be fingerprinted, so can a maker of ship noise be recognized by the sort of sound it makes.[7] High-grade machinery, for example, radiated low frequencies that could be recognized as characteristic of different classes of ship or submarine. These low frequencies could not be determined by the human ear but were exhibited on paper plots, known as LOFARGrams, which traced out lines drawn out by sensitive sonar sets. Submarines and warships had distinguishing lines that could then be interpreted and used to identify individual vessels.[8]

Operationally, the US Navy intended to use a fixed underwater-listening system, the Sound Surveillance System (SOSUS) to provide early warning of hostile submarines entering the North Atlantic or Eastern Pacific, as well as generating 'cueing' information which could be used to direct ASW forces towards targets. The US built a number of SOSUS chains in the Western Atlantic and north of the Iceland–Faroes Gap, each terminating at regional SOSUS Evaluation Centers (later called Naval Oceanographic Processing Facilities, NOPFs) that correlated contact information with other intelligence information.[9] The West Atlantic array terminated in the US and the Iceland–Faroes Gap array (the Norwegian array) in Iceland. By 1968, 'improved cable technology and processing methods had enhanced the capability for cross fixing of contacts' and the US decided to extend SOSUS. It approached the British government about the possibility of participating in a joint Anglo-American project.[10] Initially, the US Government proposed a single new Eastern Atlantic array known as SDC2 to cover the greater part of the Eastern Atlantic and terminating in the UK at a specially built SOSUS evaluation facility. The Americans also proposed a third North Atlantic system (SDC3), which would terminate at the same facility. When combined, the two systems would provide coverage of the entire North Atlantic.[11]

The British had also experimented with fixed passive sonar arrays. In May 1958, the Director of Naval Intelligence had emphasized 'the vital importance it is to this country that we should know, in advance of a war, the movements of enemy submarines'.[12] The Admiralty Board 'strongly endorsed' involvement in the SOSUS project, seeing the hosting of facilities as 'an economical means of obtaining an entry to a costly USN system'.[13] Due to the mid-Atlantic ridge, sonar coverage in the Eastern Atlantic was very poor and as a result:

> Intelligence is unable to track submarines on patrol in the eastern Atlantic, particularly SSBNs in holding areas east of the Azores, and submarines in transit to the Mediterranean and West Indies. This information is of the utmost importance for the long term study of deployment patterns which would provide a warning of any significant change in Soviet posture, and for the day to day tracking of individual submarines for marking [within weapon range with a Fire Control Solution of sufficient accuracy to support weapon discharge] in times of tension and prosecution in the event of hostilities.[14]

The British Government recognized that 'Active British participation in the provision of sites in the UK should ensure that we derive serious benefit from the system' and that 'refusal to provide what the Americans may regard as a modest contribution to a joint effort may well result in our being declined valuable information'.[15] Access to SOSUS intelligence was seen as 'invaluable to the RN/RAF ASW capabilities' as well as the MOD's Defence Intelligence Staff, which required 'the best obtainable information on the type and disposition of the Soviet nuclear strike forces . . . The SLBM force is increasing and the proposed extension to SOSUS, with a terminal in the UK, should add considerably to the total knowledge on the dispositions of the Soviet Navy.'[16] The RAF also supported British involvement as its new Nimrod Maritime Patrol Aircraft were equipped with 'underwater acoustic detection and location capability which could be used if required to enable it to exploit the SOSUS information'.[17] These sophisticated new aircraft would put the UK 'on the same footing as the US and Canadian LRMP [Long-Range Maritime Patrol Aircraft] force'.[18]

UK participation in SOSUS was first considered in the spring of 1968 and a Joint US/UK SOSUS Project Team was established later that year to oversee the proposals. Tortuous negotiations continued well into the 1970s over the location of the Regional Evaluation Centre.[19] The US eventually settled on an airbase in Pembrokeshire, West Wales, RAF Brawdy. Construction of the new facility began in 1970 under a deep veil of secrecy, with the UK providing considerable assistance in the form of a cable ship and RAF Shackleton Long-Range Maritime Patrol Aircraft to conduct an acoustic propagation survey of the northeastern Atlantic, codenamed Project Neat, to determine the most suitable location for the SOSUS arrays. The US began surveying the UK continental shelf for the location of the undersea cables in May 1971.[20] As far as the public was concerned, the new facilities were part of an oceanographic research programme designed to provide the US with detailed information on oceanographic and acoustic conditions off the continental coasts.[21] Throughout much of the Cold War the true purpose of the facilities was kept on a strict 'need to know' basis and the secrecy surrounding the base sparked off a number of conspiracy theories, ranging from a secret nuclear-weapons storage facility to a UFO tracking station.

When the first US SOSUS arrays were activated in 1961, the USS *George Washington* was tracked across the North Atlantic on her first transit from the United States to the United Kingdom. In June 1962, SOSUS achieved contact with its first Soviet diesel submarine, followed a month later by the first detection of a Soviet SSN in waters north of Norway. SOSUS also played a significant role during the Cuban Missile Crisis, detecting the small number of Soviet nuclear-armed 'Foxtrot' class diesel electric submarines that escorted Soviet merchant vessels to Cuba. In 1968, SOSUS successfully detected Soviet 'Charlie' and 'Victor' class submarines. It also played a key role in locating the wreckage of the USS *Scorpion*, a 'Skipjack' class nuclear submarine which sank 400 nautical miles southwest of the Azores on 22 May 1968 with the loss of ninety-nine crewmen, as well as the discovery and retrieval years later of a Soviet 'Golf' class submarine that sank in 1968 north of Hawaii.[22] When the SOSUS facilities at RAF Brawdy opened in 1974, an emblem was adopted with an appropriate Welsh motto: 'Dim lloches yn y dyfnder': 'No Refuge in the Depths'.

SOSUS did not provide the exact position of a contact, nor its identity, or its range. Passive sonar was unable to produce the sort of instant picture generated by radar. What it did provide was bearing information, which over time could be used to determine the bearing movement, thus giving a rough idea of where a contact was.[23] These areas, known as SOSUS Probability Areas (SPA) could vary in size from between 750 sq. nm to over 3000 sq. nm. At the very least SOSUS would tell the

Navy the areas where something was not. As one submariner later wrote, 'At best it [SOSUS] can only be taken as a good guide and at worst it can be totally misleading.'[24]

SOSUS was of unique value in maintaining a plot of the general deployment status of Soviet nuclear submarines. Once SOSUS had detected a contact, the SPAs would be relayed to Maritime Patrol Aircraft such as RAF Shackletons and later Nimrods, which would then fly out to the area and attempt to narrow in on the contact using sonobuoys. This information was then used to guide submarines and surface ships into intercept positions, where they would attempt to maintain contact with and track Soviet submarines. Once Brawdy was operational, the Royal Navy developed SOSUS-aided intercept procedures which it used to aid its detection and tracking efforts. In order for these to work properly, when good SOSUS information was available it had to be passed to the RAF and the Submarine Service in a timely manner. When this was in the order of 2–3 hours the information was of great value and often resulted in detection. When it exceeded four hours its use was minimal, particularly if the intelligence didn't contain a track update to confirm where the submarine should have been. While SOSUS provided a good steer, it was also geographically limited to the Iceland–Faroes Gap. It could not, in large areas of the Norwegian Sea, give positions accurately enough for submarines to effectively plan a search. Apart from gaps in coverage, SOSUS suffered from periodic unserviceability caused by trawling damage to seabed cables; in war, the shore stations would be vulnerable to attack and the system would become liable to acoustic interference. For all of these reasons there was a need to supplement SOSUS with other detection systems.[25]

Overusing SOSUS also caused problems. While the Soviets knew of the existence and location of the SOSUS chain on the US eastern seaboard, where US and Canadian LRMP aircraft regularly prosecuted contacts, the Soviets did not, in the early 1970s, know about the new Eastern Atlantic SOSUS chains.[26] In order to keep the existence of the arrays secret the US prosecuted SOSUS contacts covertly, via passive means. It avoided overtly prosecuting contacts via active means and the same restrictions applied to the RAF.[27]

Soviet submariners also eventually developed several methods to escape detection and reduce the effectiveness of SOSUS and Western ASW systems. Soviet submarines would sometimes stay in direct proximity to merchant or naval ships with noise levels that were high enough to obscure the noise signature of the submarine; they would also move at slow, quiet speeds when in areas known to have a SOSUS system. But the Soviets had to strike a balance when transiting to their patrol areas, travelling at speeds that were covert, but quick. We now know that Project 667A 'Yankee' class submarines were regularly deployed in the Atlantic from the end of the 1960s onwards and that these missions normally lasted for sixty days, with each submarine spending four and a half weeks in their patrol areas. As the average speed of a 'Yankee' class submarine crossing into the Atlantic from Soviet Northern Fleet bases was 12–14 knots, for the 11–13 days it spent transiting to its patrol area, it was vulnerable to detection.

During Exercise 'Okean' in 1970, six 'Yankees' were on patrol in the Atlantic, representing all but one of the worked-up hulls available at the time. Outside of exercises, two 'Yankee' class submarines could normally be found on station at any one time in the early 1970s, but the UK's Defence Intelligence Staff expected the number to increase as more of the class entered service.[28] Early 'Yankee' patrols did not operate in areas that were within missile range of North American targets, and by 1970 only one Yankee had ventured into range of North America, during a brief foray on its homeward transit from a patrol.[29] Fortunately for the West, the 'Yankee' class stuck to a strict patrol schedule, which was one of the reasons why US/UK anti-submarine systems were so effective at tracking them in the 1970s.

By 1971, the JIC assessed:

> that the Soviet leaders are probably well satisfied with the measures available to them to protect the Soviet Union from strategic strikes mounted from aircraft carriers. Their reconnaissance aircraft and submarine and surface shadowing ships, supplemented by sophisticated intelligence gathering facilities, make them confident that they will know the location of all aircraft carriers within or approaching strike range of the Soviet Union. Their attack capability, again by aircraft as well as by submarine and surface units, also gives them the capability of mounting rapid strikes with nuclear or conventional weapons against the carrier. Their ship-borne SAM also give [sic] them some capability against any aircraft which manage to take off.[30]

But, largely due to the existence of the US Navy and Royal Navy's Polaris/Poseidon force, the JIC did 'not believe that the same confidence exists in respect of the Western SSBN threat to the Soviet Union'. The JIC assessment concluded that the Soviet Union had 'no direct counter to the Polaris or Poseidon submarine force within her grasp' and predicted that the Soviet Navy would attempt 'to provide one using every element of anti-submarine warfare available to her'.[31]

In the late 1960s, the Soviets started to focus considerable effort on detecting US Navy and Royal Navy SSBNs in strategically important Western chokepoints, such as the Gibraltar and Sicilian Straits, the Iceland–Faroes Gap, and off Western SSBN operating bases such as Faslane and Holy Loch. They also developed a new class of nuclear submarine designed primarily for anti-submarine operations, to hunt down US Navy and Royal Navy SSBNs. Known as the Project 671, or 'Victor I' to use its NATO nomenclature, the 'Victor' was one of the Soviet Navy's first second-generation nuclear submarines and its bow sonar – thanks to the Portland spies – appeared to be loosely based on that used in the first-generation Royal Navy SSNs. Although the Soviets made a major effort to reduce the self-generated noise of the 'Victor' class, to lower the possibility of detection, the first variant, the 'Victor I', of which a class of fifteen were built until 1974, were still relatively noisy compared to their US Navy and Royal Navy counterparts.[32]

By the 1970s, the Royal Navy and US Navy's anti-submarine warfare advantage over the Soviets afforded by intelligence measures such as satellite reconnaissance, ELINT and SOSUS also forced the Soviet Union to divert resources towards protecting their own missile-firing submarines and away from hunting Western SSBNs. The 'Victor Is' were used to escort the 'Yankee' class SSBNs while on patrol in order to protect them from attack by US and Royal Navy submarines. Although this was clearly intended to complicate Western ASW efforts, in reality the practice unintentionally contributed to the success of US/UK ASW techniques as the 'Victor I' was so noisy that it often revealed the location of the 'Yankee' it was escorting.[33]

Despite the increased ability to detect Soviet ballistic-missile submarines afforded by SOSUS, there was little complacency. The 'Yankee' class was still seen as a serious threat that needed countering. A 1971 JIC assessment concluded that the 'Soviet submarine-launched ballistic missile force [was] already a highly credible component of their strategic nuclear armoury' and that:

> 'Y' class submarines are now regularly deployed in the Atlantic and, to a lesser extent, on the Pacific seaboard to the North American continent. The building of this class appears to be top priority in a generally highly active submarine construction programme. The Russians could

numerically match the number of Western SSBNs with 'Y' class submarines by 1975, and are likely to do so, although there is nothing to suggest that they will stop there. Soviet ballistic missile submarines are at a disadvantage compared to their Western counterparts in that they mount shorter-range missiles and lack foreign operating bases. Thus their time in transit is longer and time on station relatively shorter. This may well make it more difficult for them to pose a threat to the United States than vice versa and may well affect the eventual size of their force (assuming no ceiling has to be placed on it as a result of the current United States/ Soviet strategic arms limitations talks). The introduction into service of a 3,000 nautical-mile range missile, which we believe to be under development, but for which no submarine launch vehicle has yet been identified would increase the flexibility and credibility of the Soviet seaborne nuclear strategic force.[34]

What the JIC did not know was that in 1965 the Soviets had started work on a new class of ballistic-missile submarine, the Project 667B 'Delta I', which was designed to carry the single-warhead R-29 (SS-N-8) missile, with a range of up to 7800 kilometres. From 1972 onwards, the increased missile range of the new 'Delta' class SSBNs significantly increased the survivability of the Soviet SSBN force. The first 'Delta I' submarine entered service in 1972 and by 1977 eighteen were in service. In addition four 667BD 'Delta II' class submarines, each carrying sixteen R-29D (SS-N-8 Mod 2) missiles also entered service in 1975. With the increased range afforded by these new long-range missiles, the 'Deltas' would in theory be able to patrol in seas closer to the Soviet Union and, unlike their 'Yankee' predecessors, they would not have to cross the Greenland–Iceland–UK Gap in order to fire their missiles at targets in the US.[35] As a result, the new Delta SSBNs would be far less vulnerable to Western ASW operations.

Far more worrying was the possibility that the Soviets would develop a missile with such increased range that it could station its strategic-missile submarines in waters close to the Soviet Union and, more specifically, in the Arctic Ocean, over half of whose shores were Soviet. The world's fourth largest ocean, nearly landlocked with land-free central waters, depths of roughly 1500–2000 fathoms, permanent ice cap with seasonal growth and diminution, as well as a complex pattern of water flows determined by the combined effects of the great currents of the Pacific and Atlantic Oceans, and by the rotation of the earth, the Arctic was to become an area that neither side could permit the other to dominate.[36] Soviet submarines deploying in the Arctic from their bases in Murmansk had no chokepoints to transit and were operating in areas with which they were considerably more familiar than was the West. Conversely, Western submarines seeking to track Soviet SSBNs had themselves to transit the chokepoints of the Davis and Denmark Straits and to operate in areas that were great distances from the US and UK bases, out of the range of other Western air and surface assets. If the Soviets deployed their submarines in the Arctic, they could potentially erode the ASW advantage enjoyed by the West.[37]

UNDER THE ICE

The Americans recognized this danger early on and after the launch of Sputnik in 1957, the US Navy sent the USS *Nautilus* underneath the Arctic ice cap, where, in 1958, she became the first submarine to surface at the North Pole.[38] In 1960, the USS *Sargo* managed to pass the Bering Strait and the USS *Seadragon* the Davis Strait; in 1962, the USS *Skate* traversed the Northwest Passage and rendezvoused with USS *Seadragon* at the North Pole. But from 1962 until 1969, US nuclear-powered submarines largely avoided the Arctic ice pack due to safety concerns brought on as a result of the loss of the USS *Thresher*, which sank in the Atlantic Ocean in April 1963 with the loss of all 129 men on board.[39] However, the results of these early research trips were incorporated in the new US 'Sturgeon' class nuclear attack submarines, which could operate year round in the Arctic.[40] The Soviets, too, had taken their first steps in the Arctic. In 1962, the Soviet 'November' class nuclear submarine K-21 made the first Russian Arctic voyage, firing four torpedoes to create a hole in the ice through which it surfaced. On 17 July 1962, the Soviet nuclear submarine K-3 surfaced at the North Pole, and in September 1963 the Soviet ballistic-missile submarine K-178 travelled from the Pacific Ocean across the Arctic to the Northern Fleet, completing a total of 167 miles submerged under the ice.[41]

The Royal Navy's knowledge of under-ice operations in high latitudes was limited and primarily restricted to two sources: small amounts of second-hand information from the US – the use of which was limited due to differences in submarine and equipment design – and the Royal Navy's own patrol reports from conventionally powered submarines, many of which had conducted operations up to the ice edge. In March 1963, the 'Oberon' class submarine HMS *Opossum* and the 'Porpoise' class submarine HMS *Finwhale* took part in Exercise 'Portent' to obtain experience operating in fringe ice, under ice and close to the ice edge. *Opossum*'s CO, Lieutenant Commander W. L. Owen, was 'Struck by the way in which conditions at the edge of the Arctic ice-pack can change from year to year, from month to month, and even from day to day.' As Owen took *Opossum* 600 miles from Jan Mayen to Spitsbergen he was surprised at how much the ice edge changed and 'its position was found, in places, to be more than 100 miles from the "assumed boundary" shown on the Meteorological Office Ice Chart of only two weeks previously'.[42]

Owen was forced to conclude that 'the ice edge is an unfavourable area for transiting by a battery-powered submarine in war'.[43] SSKs faced a number of operational difficulties operating under the ice: their speed was comparatively slow and they had to operate mainly on the surface as snorting was unacceptable due to the risk of losing a mast or even propellers on a block of ice. But operating on the surface was also challenging. At night the submarine ran the risk of damaging its torpedo tube shutters as in very rough weather and even in daylight, blocks of ice were almost impossible to see. The risk of visual or radar detection from the air was also high and Owen recommended 'deploying battery powered SSKs in other areas, where they would be more likely to encounter non-nuclear targets, and to restrict ice edge deployment to SSK(N)'s'.[44]

In December 1970, the then FOSM, Sir John Roxburgh, concluded that it was 'Of considerable importance that we should extend our knowledge of the problems associated with operating under ice with a view to re-designing equipment if necessary. This knowledge,' he wrote, 'will be of major benefit to Commanding Officers who may be called upon to detect an enemy in, or follow him into, this environment.'[45] Roxburgh proposed that HMS *Dreadnought* should carry out a short Arctic patrol in the spring of 1971, codenamed Exercise 'Sniff'. *Dreadnought* would be accompanied by the 'Oberon' class submarine HMS *Oracle*. Planning and preparation were compressed into a very short timescale as a trip in February/March provided the right balance between availability of daylight and the southerly movement of the ice. But those months were also when winter in the Arctic was at its most severe, with regard both to the amount of ice present and to the temperatures likely to be experienced on the surface. Extensive preparations were put in hand and *Dreadnought*'s CO, Commander Alan Kennedy, and his Navigating Officer travelled to the US to discuss under-ice operations.

Crossing the Arctic Circle at 1300 on 22 February 1971, *Dreadnought* rendezvoused with HMS *Oracle* and on 23 February both submarines moved towards the ice pack, where the sea water temperature dropped to 29.7 degrees Fahrenheit. *Dreadnought* and *Oracle* then spent two frustrating days trying to find the ice edge. Once they did so *Oracle* became separated and *Dreadnought* spent ten hours trying to relocate her. At 1630 both submarines rendezvoused again and carried out an exercise to detect each other under the ice, which Kennedy described as 'a complete failure'. *Dreadnought*'s sonar was 'swamped with ice contacts' and was unable to locate *Oracle*. The two submarines then went their separate ways with *Oracle* withdrawing to the ice edge for independent trials, while Kennedy took *Dreadnought* north. *Dreadnought* spent the next few days under the ice until 1042 on the morning of 28 February, when Kennedy identified what looked like a suitable area for surfacing through the ice. He took *Dreadnought* to diving stations and attempted to bring the submarine into a stationary hovering position just beneath the ice. However, the first attempt to surface was a failure and *Dreadnought* 'sank to the depths'. Kennedy decided to try again and positioned *Dreadnought* in a hover underneath a thin layer of ice, varying the ascent of the submarine between 10 and 15 feet per minute. He wrote how:

> At 220 feet, I raise the periscope and see a fantastic view of the edge of the Polynya [an area of unfrozen sea within the ice pack] with one small crack of water showing. The submarine is moving slightly relative to the ice. This is corrected by a kick astern and the ascent continues. At 1129 with a rate of ascent of 15 feet per minute the fin touches the ice and fails to break through. Gentle blowing with the two main vents open, moves the submarine up a fraction. More short blows with the vents shut breaks the submarine through. Further blowing lifts the submarine to 39½ feet and the upper lid is opened. After confirmation . . . that the top of the fin is clear the search periscope is raised to reveal the beautiful arctic day with the sun just above the horizon. We are in a frozen lead which conforms to the shape plotted prior to surfacing. Full surfaced arctic routine is instituted.[46]

As *Dreadnought*'s crew packed onto the submarine's casing they were confronted with an air temperature of -13 degrees Fahrenheit. After a few minutes on the surface, *Dreadnought* began to list again, this time by up to 2¾ degrees to starboard. Kennedy had been aiming for a bigger area of ice, but had inadvertently surfaced *Dreadnought* in a smaller skylight that was just over the length of the submarine and only around 100 yards wide. A small ridge of ice separated the two skylights and as the ice began to move *Dreadnought* came perilously close to the thicker ice. After three and a half hours on the surface, its Search Periscope had frozen and could not be rotated. Fearing damage to the hydroplanes, Kennedy noted that 'we may have to move shortly':[47]

In retrospect I obviously should have continued to the larger of the two areas but one can never be certain of returning to the same point. The penalty for this soon becomes obvious as the noise of ice movement can be heard until the list on the submarine increases. At 1330 I consider we must get out of this place and the sooner the better. The dive commences at 1346 by which time the list has increased to 6°. The main difficulty is breaking the bows free. By 1348 the fin is just about under but we have a bow up angle of 7½ degrees. The angle is reduced slightly by a half second blow in the after main ballast group and a thousand gallons of water is flooded into 'D' tank. I am reluctant to put any more weight in or else when we drop clear we will go like a stone. It seems likely that we will slowly come out. The main requirement is patience.[48]

Ten minutes later, Kennedy's patience was rewarded and *Dreadnought* passed 110 feet, eventually levelling off at 250 feet. The submarine continued north, surfacing again the next day for eight hours, before diving and heading towards the pole.

The 120 miles to the Pole were beneath some of the thickest polar ice that *Dreadnought*'s crew had ever seen and Kennedy was 'not too hopeful of finding somewhere to surface'. At 0800 the submarine passed under the Pole and immediately turned back south on the 175°W meridian, searching for an area to surface. At 0855 *Dreadnought*'s crew identified what looked like a small area of thin ice and Kennedy attempted to surface, but the attempt failed after heavy ice drifted over *Dreadnought*. Kennedy attempted to bring *Dreadnought* to the surface on two further occasions, but both failed. 'It looks as if we are not going to be able to get up,' wrote Kennedy at 1230, 'a great disappointment felt by all on board.' But Kennedy was determined to surface at the Pole if he possibly could. 'I decide to have one more go by going round in a square of 7 miles,' wrote Kennedy. 'Having got this organised I turn in, in a pessimistic frame of mind.' As *Dreadnought* moved forward a large stretch of thin ice appeared and Kennedy decided to try and surface again:

The submarine is positioned and an ascent commences, again made more difficult by the echo sounder reporting thick ice over the fin but not on the bows or the stern. This seems unlikely as we had no evidence of thick ice on the final approach. We continue upwards and I am unable to see any light through the ice before putting the periscope down at 100 feet. The fin touches at 68 feet and stops at 67 which is again odd, since the top of our fin is at a keel depth of 58 feet. The situation looks much more rosy on the radial gauges and I decide to blow. A short blow on 2, 3, 4 and 5 Main Ballast and the fin breaks through at 1718.[49]

Dreadnought surfaced four miles from the North Pole. The ice was so thick that only the submarine's fin broke through the ice. 'The view is quite breathtaking in the half light of an arctic dawn,' wrote Kennedy, 'with great blocks of ice lying against the side of the fin where they have had to be pushed off the top of the fin by the surfacing OOW [Officer of the Watch].' 'Sniff' was an appropriate name for the exercise. It was, Kennedy wrote, 'breathtakingly cold', −38 degrees Fahrenheit.[50] *Dreadnought* remained on the surface for nine hours, while the majority of the crew went out on to the ice to play football and take photographs. Most were awed by the spectacle. Photography was very difficult. It was pitch dark and so cold that the camera shutters froze.

After nine hours on the surface *Dreadnought* dived and after surfacing on two occasions altered course to pass beneath the Pole again before steering south down the Greenwich meridian on a course back to Faslane. On the return passage strange noises, a scraping along the hull followed by one loud clang, were heard in the Control Room and the submarine became increasingly difficult to handle, and the helmsman struggled to keep *Dreadnought* from steering to starboard. When *Dreadnought* arrived in Faslane at 0900 on 11 March, Kennedy trimmed the submarine so that the propeller was just visible outside of the water. As it slowly rotated a small bend on one of the blades was identified, most likely caused by a piece of ice which had fallen off the casing when diving. *Dreadnought* also had two small dents in the bows and a piece of the fin was missing – the latter lost in heavy seas on the return voyage.

Dreadnought's patrol, which was made public, proved that a Royal Navy nuclear submarine could operate successfully under the ice in all latitudes up to the Pole. Kennedy confirmed that there were sufficient areas of thin ice to make under-the-ice operations entirely feasible. But he warned that:

The presence of areas in which DREADNOUGHT was able to surface daily should not be taken to imply that Patrol Submarines could also operate far into the polar pack. There were long stretches of ice where no suitable surfacing area was available and it would be extremely hazardous to allow a Patrol Submarine to go anywhere under the ice other than at the edge.[51]

Only a nuclear submarine could stay under the ice for extended periods and while the patrol confirmed that the standard equipment fitted to *Dreadnought* was satisfactory, provided a few minor modifications and additions were made, Kennedy pointed out that if there was a requirement to operate under the ice where icebergs were present, more sophisticated iceberg detection equipment was required. He also highlighted the peculiar sonar conditions. There was a great deal of background noise from the rumbling of the ice overhead as well as severe temperature and salinity gradients, all of which were antipathetic to sonar detections. Kennedy was well aware that 'the ability to detect enemy submarines in such an environment is largely unanswered' but:

> From experience gained during SNIFF it is probable that under ice detections will be difficult to achieve either actively or passively due to the large number of non submarine contacts which are always present. It would also appear that Sonar Type 2001 is reverberation limited when operated actively under the ice. At the ice edge bathy conditions [the underwater equivalent to topography] were far worse than they were further into the ice. This was undoubtedly caused by the warm Gulf Stream waters sliding in under the cold arctic currents. In such conditions DREADNOUGHT found it almost impossible to detect ORACLE actively or passively, when the latter was operating at below cavitation inception speed.[52]

Kennedy urged the development of more-advanced sonars as well as further trials as the Arctic environment could 'offer very real cover for our own forces such as SSBNs if ever required. It is not my province to comment on targets for SSBNs, nor indeed do I have the knowledge to do so; but it is relevant that all places in the USSR North of Latitude 50°N from Leningrad in the west to the Sea of Okhotsk are theoretically in range from the arctic basin.'[53]

A SPECIALIST SERVICE?

Dreadnought's patrol to the North Pole was yet another first for the Submarine Service. It was now responsible for the Royal Navy's most advanced capital ships as well as the United Kingdom's independent nuclear deterrent. Yet many submariners continued to feel that the Royal Navy as a whole was failing to accept the new position of the submarine in maritime warfare. In June 1972, Flag Officer Submarines, Vice Admiral John Roxburgh, warned his successor, Rear Admiral Tony Troup, that while 'a good deal of lip service is paid' to the increased role of the submarine, 'I am left with the feeling – below Board level I must say – that many officers have not really hoisted in where we are going. A sort of mental blockage.'[54]

This did not stop the Submarine Service from highlighting their increased importance. In July 1971, a new submarine badge was issued to all qualified submariners to signify their specialist role in the service. Between 1958 and 1964, a submariner's badge was available and could be worn on the right sleeve of rating and senior rating uniforms. But the badge, which was never issued – it had to be purchased by the wearer from naval stores – was never popular and due to its ugly design quickly acquired the nickname 'sausage on a stick'. In June 1964, in advance of the arrival of the Royal Australian Navy's (RAN) British-built 'Oberon' class submarines, the RAN established a submarine project team in Canberra. One of the officers assigned to the team, Commander Alan McIntosh, designed a distinctive version of the US Navy Submarine Insignia, otherwise known as 'Dolphins' or 'Fish', a breast pin that was worn by enlisted men and officers to indicate that they were qualified in submarines. The RAN design included a set of gold dolphins, facing each other, separated by the British and Commonwealth Crown.

In 1968, the crew of HMS *Trump*, the last Royal Navy submarine in the 4th Submarine Squadron in Melbourne, was given the badge to wear for a year. Reception and uptake were overwhelmingly positive, and although Roxburgh was 'not in favour of the introduction of a submarine badge' when he took up his post as FOSM, he 'quickly became aware of the desire for a badge in the Command, particularly amongst junior officers and ratings'. In July 1971 the Royal Navy issued a variation of the Australian submarine badge, known as 'the Dolphins', which to this day is worn 'above or in the position of medal ribbons'. 'The badge has been very well received and is worn with pride,' noted Roxburgh in June 1972. 'I believe it will only continue to be held in high regard if its wearing is guarded.'[55] To this day, only qualified submariners are permitted to wear the Dolphins, which they are awarded (after passing a series of rigorous practical exams) in a ceremony that involves drinking two fingers of rum and catching the Dolphins with their teeth.

Yet in 1971 a worrying episode demonstrated that standards in the Submarine Service were deteriorating. On 1 July 1971, one of the last of the Royal Navy's 'A' class submarines, the 25-year-old HMS *Artemis*, was alongside at HMS *Dolphin* in Gosport undergoing maintenance before a planned deployment to the West Indies and sank after undocking. The sinking was caused when the submarine, which was in a stern-heavy trim, became unstable by the filling of its external fuel tanks with 178 tons of compensating water. The addition of compensating water brought the after escape hatch to the waterline. The hatch was not shut and uncontrolled flooding quickly overwhelmed the various compartments inside the submarine. The duty watch on board and twelve men managed to jump onto the uprights of the jetty while others dived into the water and swam to safety as the submarine went down. It took just sixty seconds for the submarine to sink. Three men trapped on board ran from the flooded main accommodation area to the torpedo compartment, shutting a watertight bulkhead door. *Artemis* developed a 45 degree list on the muddy harbour bed and there were fears that the escape hatches would not function properly unless the submarine was righted.

As soon as the incident occurred, the Subsunk procedure went into operation and diving teams and salvage vessels quickly arrived. A rescue operation began, with a salvage tug putting a line under the submarine's stern, while another was placed under the bow. After working through the night, communication was eventually established with the three men stranded in the forward torpedo compartment, who reported that the air inside was starting to foul. After ten hours, the most senior of the trapped men, Coxswain David Guest, was ordered via radio to flood the hatch and escape. At first light, the ten-hour ordeal ended as all three men bobbed to the surface, one after the other, in front of hundreds of officers and men on the quayside. 'There was a slight leak and the next compartment was flooded,' recalled one of the trapped men, Tom Becket. 'It was coming up through a hole and the air got very, very thick. We had trouble breathing.'[56]

The Board of Inquiry concluded that the fault lay entirely with the

submarine's personnel, who 'were unaware of the circumstances prevailing in their submarine at the time, and of the dangers towards
which they were heading'.[57] From the moment *Artemis* arrived alongside at Gosport, a number of apparently disconnected events began to
build up to disaster. They had one common thread: bad submarine
practice. Poor communication, the incorrect issue of and failure to
carry out orders all contributed to the accident. The CO, Lieutenant
Commander Roger Godfrey, who was not on board at the time of the
sinking, was charged with negligence in failing to take personal charge
of his submarine or to ensure that she was entrusted in his absence to
an officer competent to take charge. In his absence, and the absence of
the submarine's First Lieutenant, who was on leave, Godfrey arranged
for Lieutenant John Crawford to take charge of the undocking. Crawford had little experience of conventional submarines, having served in
an SSBN for two years before joining *Artemis*. Other officers' submarine experience varied between two years for the Engineer Officer, to
four months for the sixth hand. The senior ratings all had at least five
years' conventional submarine experience.

One of the most striking features of the resulting court martial was
the inability of both senior and junior officers giving evidence to answer
questions on subjects such as the differences in weight between fuel and
water in fuel tanks and tons per inch of immersion, the height of hatches
above the waterline and the angle of inclination and draught of submarines on the surface and at different stages of flooding and fuelling.
Lieutenant Commander Michael Everett, defending Crawford at the
court martial, said that the accused had insufficient training and experience as a duty officer and that his 'actions were a clear indication
of his lack of training or awareness of the situation. His brain was not
programmed to recognise the danger signals.' Everett argued that since
the war the Submarine Service had forgotten many valuable lessons.
'The sense of urgent self-preservation kept alive by the constant contact
with the enemy is falling off,' he said. 'Officers with wartime service
are now senior officers no longer at sea. Nuclear submarines, too,
meant that many experienced senior ratings have left conventional
patrol submarines.'[58] Although the CO, Godfrey, was cleared of failing
in his duty, Crawford was severely reprimanded for negligence. The
Board of Inquiry, in reviewing the standards and practices of the Submarine Service, made over sixty-five recommendations and highlighted
a number of areas that it was concerned about and which it strongly

suggested needed further study, including a lack of emphasis during training on basic submarine practices.

The Board also suggested that the impact of the nuclear-submarine programme on the conventional Submarine Service needed further investigation. In its view, the SSN and SSBN programme had contributed to the accident and had had a direct effect upon standards in conventional submarines. Although Roxburgh adhered to the principle that 'there should be no question of a "second eleven" and people should move freely between Polaris [SSBNs], Fleet [SSNs] and Patrol submarines [SSKs]', he recognized that:

> because of the more demanding task in the nuclears it has been inevitable, particularly in the technical field, for the brighter officers to be selected for nuclear training and the lesser brethren have tended to gravitate to the Patrols. Over the years the level of competence in Patrols has thus reduced and this has led to the development of bad practices and disregard for safety rules.[59]

'The only answer', wrote Roxburgh, 'is for Captains SM and their COs to insist on the maintenance of proper standards and basic submarine practices. I have been at pains to ram home to Captains SM their prime responsibilities in this respect.'[60]

There were other problems with the Submarine Service in the early 1970s; one of the biggest concerned operational command and control. While the SSBNs were controlled centrally from Commander Task Force 345 based in Northwood, the Royal Navy's SSNs and SSKs were under control of the remaining individual submarine squadrons, who would surrender control to other authorities such as Flag Officer Scotland and Northern Ireland or Flag Officer Portland during exercises. There was no centralized command and control. Communications and intelligence remained independent, and little consideration was given to clearly defining areas in which submarines could safely operate, a system known as water space management. In February 1972, confusion between operational authorities and a serious lapse of liaison over interaction between submarines and their operating areas caused an incident which was later labelled the 'great submarine chase'. HMS *Conqueror* was exercising its active sonar in the Clyde outer areas when it gained contact with an unknown submarine. After hearing *Conqueror*'s active sonar, which at high power was easily identifiable, the submarine contact took off at high speed. *Conqueror*'s crew, believing they had detected a Soviet SSN, gave chase. In fact, they had detected a friendly NATO submarine that did not want to identify itself. For several hours, as *Conqueror* chased the contact, shore staffs failed to realize what was happening and the charade continued with commendable enthusiasm.[61]

A VICTOR PENETRATES THE CLYDE

Eleven months later, in January 1973, a Soviet submarine was detected in the sensitive Clyde approach areas, where it appeared to be waiting for Royal Navy and US Navy SSBNs to sail from Faslane and Holy Loch.[62] For six days, the Soviet submarine, which was eventually classified as a 'Victor' class SSN, remained sixty miles northwest of Donegal and fifty miles off Colonsay, where it was kept under twenty-four-hour surveillance by RAF Nimrods operating out of RAF Kinloss.[63] At 0400 on 27 January 1973, Commander Chris Ward, the CO of HMS *Conqueror*, which was based in Faslane and acting as what was known as the 'Ready Duty SSN' – at short notice to deploy – received the following orders:

IMMEDIATE   270400Z
                                             JAN 73

FROM ...
TO: HM SUBMARINE CONQUEROR
INFO: FLAG OFFICER SUBMARINES
COMMODORE CLYDE
CAPTAIN SM3

FROM FIRST SEA LORD
A SOVIET NUCLEAR POWERED SUBMARINE HAS BEEN DETECTED IN THE NORTH WEST APPROACHES. YOU ARE TO SAIL FORTHWITH AND SWEEP THE RUSSIAN FROM OUR WATERS.[64]

Eight hours after receiving the signal *Conqueror* sailed from Faslane to find the Soviet intruder, which was still approaching UK territorial waters. 'It was an unusual situation,' recalled Ward.[65] 'Normally we would be pursuing the opposition as it were in his general area. But this Soviet submarine was quite plainly determined to act as bait. What were they trying to prove?' Roger Lane-Nott, a young Navigator on board *Conqueror*, was 'horrified that a Soviet submarine should consider let alone be able to enter the Clyde Inner Areas, which were absolutely our backyard'. 'How dare he come in here,' he said, remembering the operation. 'We had to get rid of him. And yet I took it as perfectly normal that we could and would operate in international waters in the Barents Sea, in the Soviets' backyard. It was just we were not used to this type of aggression in our home waters. Nor was it expected, even though we knew that the Soviets wanted to keep track of our SSBNs.'[66]

It was left to Ward to determine exactly how to 'sweep' the Victor from a busy shipping environment, where noise from merchant vessels significantly reduced the chances of detections on passive sonar. 'My task as I knew it was to get this bloody submarine away from the Firth of Clyde and rattle my brain as to how to bloody do it,' he explained.[67] 'What could I do to seduce this bastard? Talk to him on the underwater telephone with the few words of Russian that I still remembered?'[68] Ward concluded that the best means of luring the Russian away from UK waters was to disguise *Conqueror* by adopting the silhouette and lights of a Polaris SSBN and move out in the Atlantic, with the Soviet submarine trailing behind.

At first the Victor refused to participate in Ward's plan. 'It proved incredibly difficult to get him to get into contact with us,' he said.[69] *Conqueror*'s crew resorted to making the submarine as noisy as possible to increase the chances of the Victor being able to detect it. 'It was ridiculous. We were at action stations for hours trying to move the submarine into a position where he was bound to pick us up. But he didn't.'[70] After consulting with *Conqueror*'s officers, Ward decided to use *Conqueror*'s Type 2001 active sonar. 'We were all quite nervous about that,' recalled Lane-Nott.[71] Active sonar would almost certainly alert the Victor to *Conqueror*'s presence – Type 2001 active sonar transmissions were unmistakable – but they would also give away *Conqueror*'s exact position, which the Victor could then exploit to acquire a fire control solution and potentially carry out an attack. 'What I wanted to do was try and make it like we were really going out on patrol. That we had been doing trials on our sonar, hence all the active stuff,' said Ward.[72]

Ward took *Conqueror* to 235 feet and ordered the use of the Type 2001 Active Sonar, the noise of the transmissions reverberating throughout the submarine. 'After the first transmission we got him at 5000 yards, then 3000 yards, then 500 yards. I can remember it now,' said Lane-Nott.[73] With the Soviet submarine now aware of *Conqueror*'s presence Ward increased speed and set course to the southwest, out into deep water. 'I felt that if we got about a half day's steaming away with him stuck behind me then that would be absolutely fine. That would leave enough clear water for our submarines without any interference from him.'[74] It worked. The Victor followed *Conqueror* as it moved south at full speed on 100 per cent reactor power. Once clear of the Clyde areas and into deep water, Ward took his submarine deep and then stopped, 'like a log in the water', and switched off all non-essential machinery and equipment, attempting to stay as quiet as possible. Tension in *Conqueror*'s Control Room then reached fever pitch as the Sound Room provided update after update of the Victor's possible position, counting down the range of the Soviet submarine as it moved closer and closer to *Conqueror*. 'Eventually we got this incredible noise of the submarine going over the top of us,' said Ward. The Victor passed right over *Conqueror*. 'The boat rocked slightly,' remembered Lane-Nott, 'it was a strange situation because we all felt it, but we could not define what it was we had felt.'[75]

Conqueror continued to quietly track the Victor as it moved away, before handing over responsibility to an RAF Maritime Patrol Aircraft. Ward then signalled the Admiralty:

IMMEDIATE   311605Z
                                                       JAN 73
FROM: HM SUBMARINE CONQUEROR
TO: ...
INFO: FLAG OFFICER SUBMARINES
COMMODORE CLYDE
CAPTAIN SM3
FIRST SEA LORD'S 270400Z JAN 73

MISSION ACCOMPLISHED. SOVIET SUBMARINE HAS BEEN LURED INTO THE ATLANTIC AND DISENGAGED. OUR WATERS ARE SANITIZED. AM PROCEEDING IN ACCORDANCE WITH PREVIOUS ORDERS.[76]

The incident marked the first known penetration of the Clyde area by a Soviet submarine, and a number of procedures were later developed to deal with any future intruders.[77]

Both events involving HMS *Conqueror* highlighted the inadequate state of SSN/SSK submarine command and control and the potential for mutual interference with SSBN operations. In 1972, the wartime submariner Admiral Anthony Troup was appointed Flag Officer Submarines. Troup, who at the age of twenty-one had been awarded command of the training submarine H32, was a distinguished submariner who recognized that a more professional, competent operational command and control organization was required. By the early 1970s the waters in the North Atlantic were becoming increasingly crowded as more and more submarines from many nations began to put to sea. Soviet SSKs and SSNs would transit through the Norwegian Sea and the North Atlantic to the Mediterranean and elsewhere while Soviet 'Yankee' class SSBNs continued to transit from their Northern Fleet bases to patrol areas off the United States' eastern seaboard. US Navy, French and Royal Navy SSKs, SSNs and SSBNs also deployed from their European bases to patrol areas in the North Atlantic. Surveillance operations continued as well, in the Barents Sea, the Mediterranean and the Baltic, and SSKs and SSNs acting on intelligence were increasingly diverted at short notice to intercept Soviet forces. This was all 'an amalgam of activity that required very careful control, both of measures to avoid mutual interference and of aggressive moves to place our submarines closely alongside those of the potential enemy', wrote one submariner.[78]

Troup recommended the establishment of a new command and control organization located alongside that for Polaris, CTF 345 at Northwood, and the relocation of FOSM and the transfer of the submarine command from HMS *Dolphin* in Gosport. However, when Troup retired in 1974 his successor, Vice Admiral Iwan Raikes, whose father, Admiral Sir Robert Raikes, had been FOSM from 1936 to 1938, adopted an 'over my dead body' attitude towards the proposed move to Northwood and vowed to stay put at Fort Blockhouse, fearing that FOSM would ultimately be absorbed into the Fleet staff and lose his identity and independence.[79] While Raikes held out at Gosport, Troup's new twenty-four-hour organization, known as CTF 311, was nevertheless established in Northwood in 1974/75 to control Cold War submarine operations. Led by a submarine-qualified Commander, with a twenty-four-hour staff, it was responsible to C-in-C Fleet's Assistant Chief of Staff (Operations). FOSM allocated submarines to the new organization and was kept informed of developments, but because of his location at Gosport and the lack of twenty-four-hour, seven-days-a-week staffing at HMS *Dolphin*, he was largely isolated from day to day operations.

The twenty-four-hour operational arrangement at CTF 311 brought many benefits. All the Northwood communications, real-time intelligence and SOSUS information was channelled into one location, analysed, and made available to submarines for the first time. The Americans based in Northwood were also informed about when and where Royal Navy submarines would be operating, in order to avoid any conflict with US submarine operations. An innovative and complex water space management system known as Area Allocation was also established to avoid mutual interference and to manage and coordinate the large number of submarines, both friendly and unfriendly, operating in the vast oceans. US Navy and Royal Navy SSNs and SSBNs were allocated to geographical patrol areas in different sections of the ocean, and were sometimes depth-separated. This meant that an SSN or SSBN operating in a designated patrol area could be certain that any contact it detected was not a friendly SSN or SSBN.

A NEW CONCEPT OF OPERATIONS

The Submarine Service had also started to rethink how it used its SSNs, particularly in the support and escort of surface ships. In the early 1970s, the introduction of Soviet submerged-launched anti-ship missiles considerably altered anti-submarine warfare. Until the early 1970s the United States Navy and the Royal Navy followed separate paths in anti-submarine warfare. The United States Navy independently deployed its SSNs away from a surface force, using passive sonar as the primary means of detecting enemy contacts. In contrast, the Royal Navy concentrated on what was known as the Link Ship/SSN combination, which involved an SSN sitting in front of a surface force and using its high-powered active sonar to search for contacts. Once a contact had been detected the SSN would report to the link ship (surface ship), which in turn would launch a helicopter to carry out an attack. This divergence of view between the two navies arose because the Royal Navy developed the Type 2001 long-range high-powered active sonar, which was fitted in its SSNs; whereas the US Navy developed low-frequency passive sonars whose effectiveness as an ASW sensor was degraded in the close vicinity of surface ships because of the high ambient noise in the proximity of surface forces. By the 1970s the Royal Navy had amassed a considerable amount of data and experience of the use of the SSN in ASW operations and had concluded that the linked-SSN concept, while a useful addition to the ASW armoury for defence against torpedo-firing submarines, was not the final answer to the threat of Soviet submarines.[80]

With the vast increase in the size and composition of the Soviet SSN fleet, in particular the entry into service of missile-firing submarines, which were regarded as the Soviet Navy's main offensive weapon, Royal Navy surface forces could no longer expect to encounter short-range torpedoes fired from patrol submarines, but medium- and long-range missiles. In other words, an opposing submarine no longer always needed to approach a surface ship to make an attack. As a result, detection opportunities for an SSN operating in the linked role, using active sonar ahead of a surface force, decreased. Experience at sea had also shown that when used in the wrong circumstances the linked combination was highly vulnerable due to its comparatively slow speed, the beacon effect of the Type 2001 active sonar out to some fifty miles, and the intelligence disclosed by the insecure means of communication between submarine and surface ship. Experience had also shown that an attacking SSN was often able to avoid the linked unit by using sonar intercept equipment to work its way around the force, shadowing it and then closing to attack. One Royal Navy SSN that was lying in wait for a linked SSN and surface force during an exercise reported that 'they sounded just like a circus coming over the hill'.[81]

In order to counter Soviet short-range tactical-missile-firing submarines, as well as their long-range counterparts, both of which would attack from well inside the maximum effective missile range, the United States Navy developed a form of unlinked operations in which US SSNs were placed in what the Americans termed 'the deep-field', some fifty or sixty miles from a friendly surface force, operating on passive sonar. The Americans also spent vast sums of money making their submarines quieter and they developed highly capable passive sonars to go into them. They also developed and continuously improved anti-submarine torpedoes, such as the Mark 48. In 1974, after observing these American developments, the Royal Navy concluded that 'Studies have shown that the use of a nuclear submarine in this [linked] role is not an effective employment.'[82]

The Submarine Service now recognized, as Commander Richard Sharpe, a submariner on FOSM's staff, explained in a 1973 presentation to an annual anti-submarine warfare conference that:

The ASW battle has moved away from the area of water surrounding a surface force. The mobility, endurance and weapon systems of modern

Soviet submarines can primarily be countered by mounting a major surveillance effort across likely deployment routes and subsequently by vectoring the Fleet submarine to a shadowing position. In the NATO area a state of confrontation exists today and surveillance and shadowing are part of an acceptable deterrent posture.[83]

The Submarine Service still played an important role in the observation of Soviet maritime deployments, a key requirement that in the run-up to war was considered essential to political and military decision-making. But it now sought to play a more active role in the Cold War at sea. As Sharpe pointed out, 'Confrontation is the name of this game.'[84]

Yet this confrontation bore little resemblance to past undersea confrontations. As Gary Weir has noted:

> With the constant threat of mutually assured destruction placing erroneous but strict limits both on the use of weaponry and the traditional expectations of battle, Cold War submariners formulated strategy and tactics, and calculated victory and defeat in a very different way. Since using even conventional offensive weapons could easily precipitate horrible and nearly uncontrollable geopolitical consequences, undersea warriors measured victory in terms of surveillance, detection and constant monitoring. If you knew the enemy, his vehicle or ship, his location and capability and you could follow or 'shadow' him without betraying yourself, you claimed victory by Cold War standards.[85]

Thus the undersea conflict during the Cold War that the Royal Navy fought alongside the US Navy was redefined away from physically destroying targets in encounters that would almost certainly escalate with unthinkable consequences, towards surveillance, detection, submerged capability and destructive potential. As Weir has again noted, 'With this victory achieved, a national leader might anticipate any threat with sufficient time to destroy an adversary's ability to act. *Control* over the opponent became the operational objective and precise knowledge of him the means to that end.'[86] By demonstrating that the West was in control over what was happening at sea, Royal Navy submarine Captains hoped their actions would serve to deter the Russians. As Sharpe explained:

> To contain these unambiguous threatening capabilities, we had to confront (both on land and at sea) this massive military power in order to deter the Soviet hierarchy from dangerous initiatives, which could trigger a real war. In other words, in the maritime world we had to demonstrate by covert

strategic nuclear deterrent patrols and surveillance operations that we were better than they were, and together with the United States and other navies could confront any expansionist ambitions by countering them with conventional or, as a last resort nuclear weapons ... The obvious question: 'How could covert surveillance operations deter?' The answer is that they were only covert in time, place and detail but in the sure knowledge that the Soviets at a higher level would be aware of our superior capabilities.[87]

From the early 1970s onwards, Royal Navy submariners put to sea with the aim of developing in the Russians an inferiority complex, the thought that whenever they went to sea, they would know that there was going to be a Royal Navy or US Navy submarine around that could probably hear what they were doing and thus attack at any moment.[88] The Royal Navy set out to ensure that every submarine operation contained, as Sharpe put it, 'an underlying element of threat, such as to make the use of force an unattractive option'.[89] As Peter Herbert, another submariner on FOSM's staff in the early 1970s, explained to the conference on ASW in 1973:

> We are going to have relatively few SSNs. We have got to use our SSNs in the most cost effective way – and that is to put them on the tails of the most threatening of the enemy submarines. If we can achieve this it is probably the best way of protecting our forces. If all of them can be attacked approaching our forces, we should achieve our objective.[90]

This doctrine of controlled confrontation, or flexible response, depended on Royal Navy submarines being in the right place at the right time with sufficient force either to de-escalate tension or be able to initiate action or counter-action effectively with the minimum force necessary. The SSN, with its ability to shadow a target using an intermittent contact policy or, depending on sonar conditions, the ability to establish a firm trail on a contact and mark it with a weapon system, ready to fire on command or on pre-ordered rules of engagement, was the ideal platform to fulfil this doctrine. Royal Navy policy was to always remain covert, and conduct operations in a passive manner. As one submariner, Captain Tom Le Marchand, has explained, 'The one thing that you make sure above all else is that you are not detected by the enemy. If you are, you break that detection as aggressively as you can. A number of our crews have been detected. It is a pretty tense experience. If you think you are being tracked, you turn round and get out as quickly as you can.'[91]

While identifying, trailing and marking Soviet submarines sounds easy, both the US Navy and the Royal Navy had to overcome a number of different problems in order to meet this aim. First, how to find a contact? US Navy and Royal Navy SSBN patrols had already shown the virtual impossibility of finding a modern nuclear submarine in vast oceans. Second, in order for shadowing and trailing to be effective, it needed to be covert, which demanded a significant sonar advantage so that, in effect, Royal Navy submarines could hear the target, but the target could not hear them. Third, a secure two-way communications system was required in order to keep Maritime Headquarters in the picture, and to receive, quickly, the directive to fire and take out the target. Finally, a quick and responsive weapon system was required. The Royal Navy recognized that it was no good having a weapon system that took twenty minutes to reach its target; by that time the enemy, such as a Soviet SSBN, would already have fired its missiles.

The Royal Navy had some of the answers to these problems and a large part of the submarine development programme in the early 1970s was concentrated on improving performance in these areas. The best place to gain initial contact with Soviet submarines was during the deployment and transit to a patrol area. By concentrating all long-range detection sensors (airborne, seabed, SOSUS, ship, submarine or other intelligence) on the transiter the Navy hoped to provide sufficient data to vector in its SSNs towards a target using a new concept of anti-submarine warfare operations, known as 'Aided Intercept'. In order to obtain initial contact with Soviet submarines during their deployment and transit to patrol areas, SOSUS, satellite intelligence and various other forms of intelligence were combined, analysed and relayed to Maritime Patrol Aircraft, surface ships and submarines. This new concept of ASW also demanded centralized command and control, backed up by reliable all-source intelligence as well as good, fast communications. It also demanded significant advances in navigation, as intercept information was largely based on geographical data rather than a target's position relative to a unit in contact with it.

The new concept also required significant advances in submarine tactical development. In 1966, a Submarine Tactical Development Group (STDG) was formed at the Clyde submarine base, Faslane, to develop submarine tactics. In 1974 the Submarine Tactics and Weapons Group (STWG) was formed from STDG and a weapons trial group, to take over submarine tactical development and in-service tactical-weapon-firing evaluations. The group changed the tactical picture and was responsible for developing many of the submarine tactics that enabled the Submarine Service to maintain superiority over the Soviet Navy. The STWG also developed a very close relationship with the US Navy's tactical-development squadron, Submarine Development Squadron Twelve (SubDevRonTwelve), whereby a Royal Navy Exchange Officer was involved in the development of the US Navy's newest submarine concepts and tactics.[92]

The US Navy and the Royal Navy also continued to send submarines to observe Russian naval movements and to gather as much information as possible on new Soviet ships, submarines and weapon systems. Two operations could be initiated at short notice: Operation 'Larder' – shadowing and intelligence gathering by surface ship, submarine on the surface or aircraft; and Operation 'Lacquer' – covert intelligence collection by submarine.[93] The US Navy and the Royal Navy also continued to cooperate on intelligence-gathering operations in northern waters and by 1973 the Submarine Service was running two operations a year under American auspices as part of a programme known as the Special Naval Collection Programme, or SNCP.[94] The Royal Navy could not afford to install the special intelligence-gathering equipment and modifications into all of its SSNs and throughout the Cold War Royal Navy policy was to keep two operational SSNs equipped for SNCP work.[95] These submarines were more commonly known as 'special fit' submarines and were employed on intelligence-gathering operations for the majority of their operational time at sea.[96]

One of the first submarines to receive 'special fit' equipment in the early 1970s was HMS *Courageous*. The coxswain's, engineer's and ship's offices were stripped out and transformed into what was known as the 'L Shaped Room', in which the very latest radio 'listening' and recording gear was fitted. Up-to-date electronic-warfare equipment was also added to the Radar and Electronic Warfare Office and new equipment was installed in the Control Room. The old radar mast was also removed and replaced with a completely new one, known as the telegon mast, which provided the 'L Shaped Room' with loose radio chatter and telemetry. A video system with a playback facility was also incorporated into the periscopes, known as the 'periviz'. A digital time system was incorporated in all operational areas and microphones were fitted to various compartments to record what was said during sensitive operations onto a large eighteen-track tape recorder for later analysis. *Courageous*'s sonar equipment was also upgraded, including prototype active sonar intercept equipment which provided the source and range of active sonar transmissions. It was nicknamed 'Donald' after its inventor, a scientist from the Navy's Underwater Detection Establishment.[97]

'Special fit' submarines employed on these operations would carry specialists in ACINT, SIGINT and COMINT, known as 'riders'. The submarines would remain under British control, but the US would provide training, support and guidance, as well as a liaison officer who would often serve on board during operations – another example of the intimate and close relationship that existed between the Royal Navy and the United States Navy beneath the waves. Royal Navy officers also served on board US Navy submarines on similar operations. One such US Navy Exchange Officer who served on board *Courageous* during a special intelligence-gathering operation was Lieutenant Commander Bruce Schick, the Assistant Intelligence Officer on the staff of COMSUBLANT. In the early 1970s, Schick spent two months on board *Courageous*, where, because of his intelligence experience, he was appointed as the CO's personal advisor, on twenty-four-hour call. He was also responsible for drafting the patrol report and this created some security problems, as Schick explains:

> At the end of the patrol, when requisite copies of the final product were produced, the skipper took a razor blade to the one which was destined for the U.S. He literally cut out sections which the British 'Intel weenies' had declared NOFORN (no Foreign Dissemination). This little ploy has always rankled submariners on both sides of the pond. The skipper was obviously embarrassed. After all, I had written the thing. Then he produced another undefiled copy. Inside the front cover was a little envelope with razor blades inside. This is yours, he said. Put in the US diplomatic pouch and cut it up when you get home. What a gentleman.[98]

There could be few better illustrations of the trust which the underwater Anglo-American relationship carried. But, as the Flag Officer Submarines, John Roxburgh, pointed out, the close relationship with US Navy submariners needed to be 'fostered by continuing attention and effort' and would 'only persist as long as we remain effectively up to date with our nuclear propelled submarines and have ourselves something to offer'.[99]

Aside from these special operations, the Submarine Service was also increasingly intercepting and trailing Soviet submarines. One of the most important intelligence priorities in the 1970s was to detect the sea-lanes, the routes that the Soviets used to transit to and from their patrol areas. Once these had been identified, Royal Navy submarines would lie in the centre of the lanes and wait for contacts to appear. One of them, the Soviet Northern Fleet homebound lane, was identified in the Mediterranean and in March 1975 HMS *Courageous*, under the command of Captain Richard Sharpe, then operating as part of the UK contribution to US deterrent forces in the Mediterranean, was ordered to intercept a transiting Soviet 'Echo II' class submarine. *Courageous* travelled over 300 miles in seventeen and a half hours, and moved into the centre of the Northern Fleet homebound lane. Initial detection of the 'Echo II' was achieved using Aided Intercept procedures based on SOSUS track assessments. Long-Range Maritime Patrol Aircraft narrowed down the location of the enemy submarine and transmitted the information to *Courageous*. Initial contact was made at a range of eight miles and *Courageous* trailed the Echo II, covertly for 320 miles, remaining on the Soviet submarine's quarter at an average range of between ten and fifteen miles and at speeds of up to 15 knots.[100] By the 1970s, the Echo II was an old noisy submarine and Sharpe recorded just some of the sounds the submarine made in *Courageous*'s log: 'Target clearing stern arcs to port. Compressed cavitation. Changing propulsion mode. Rattles and bangs. Ringing tones. Buzz saw noise. All typical of Type I nuclear.'[101] As a result of another successful operation, *Courageous* won the coveted US Navy 'HOOK EM' award for excellence in anti-submarine warfare.

Detecting and remaining in contact with Soviet submarines was extremely difficult and was nowhere near as simple as is often portrayed in Cold War fiction or Hollywood films. Establishing viable estimates of a target's course, speed and range when only provided with passive sonar bearings was, as Sharpe later explained, the 'most important and obscure of the submariner's black arts':

It is difficult enough when the noise source is constant, as in a cavitating surface ship propeller, but achieves a whole new plane of obfuscation when the contact is irregular. A simple analogy is that it is like being in a field with a herd of cows in pitch darkness. You can hear munching, the swish of tails, footfalls and the occasional seismic contribution to global warming, but only a fool would claim that he knows the exact PIM (position and intended movement) of any individual animal. Part genius or pure 'con job?' The answer is, a bit of both, and to an extent the dynamics of each encounter are variable and uncertain. You really do need first-hand experience of submarine versus submarine operations to understand what happens and . . . what doesn't happen. This takes years for Commanding

Officers to learn. The trouble is that if you are looking in the wrong direction, or the noise is shielded by acoustic interference, or even by the target's aspect, any fire control solution is not guaranteed. Even if it is, you still have the bearings only/range rate analysis computation, which can go wrong and go wrong and go wrong. Active sonar can help in exceptional circumstances but instantly gives away your own position. Hence the collisions and near misses in this era of NATO–Soviet confrontation, for example, *Warspite* in 1968 and, in later years, *Sceptre* were involved in actual scrapes with nearly disastrous consequences.[102]

The 1970s saw the introduction of new technology which enhanced the Royal Navy's ability to track Soviet submarines. One of the most important came from HMS *Warspite*'s former CO, Sandy Woodward. Before assuming command of HMS *Warspite*, Woodward had been appointed as Teacher to the Perisher course. During his week of preparation, he spent time with the then incumbent Teacher, Commander Sam Fry. One lunchtime, as the frigates hunting the submarine opened up for the next run against the Perishers, Fry said, quietly, without even looking up: 'They've turned.' Woodward asked Fry how he knew, but Fry couldn't answer. 'I questioned him carefully but to no avail,' said Woodward. 'He had no idea how he knew. He just did.'[103] After months of pointed curiosity Woodward eventually discovered how it was Sam Fry knew the frigates had turned. The vital clue was a change in the note of the frigates' sonar transmissions as heard over the loudspeaker in the submarine's Control Room.[104] Woodward realized that the attacking frigates' active sonar transmissions jumped in frequency as the frigates turned at the end of a run. This was due to the Doppler effect – the common acoustic phenomenon, heard every day in the changing pitch of an ambulance's siren as it goes by.

Woodward realized that the change in frequency could be exploited passively for tracking purposes and could help avoid the occurrence of unintentional close-quarters encounters, such as *Warspite*'s collision in 1968. If the frequency was higher than the base frequency, then the range was closing. If the frequency was lower, then it was opening. If submariners were armed with such information then they should be able to prevent a collision as it would be evident immediately that a range was closing regardless of the bearings. Woodward approached the Submarine Tactical Development Group for technical assistance, where a mathematician, Lieutenant Guy Warner, got to work on developing tactical guidance. Mathematics was then used to determine the base frequency and in the early 1970s various calculators were introduced into submarines to assist submariners to extract tactical information from frequency data.[105]

During any normal period in the early 1970s, the Royal Navy estimated that there were two 'Charlie', one 'Victor', one 'Juliett', ten 'Foxtrot' and two 'Whisky' class Soviet submarines operating in the Mediterranean. These submarines, in particular the 'Juliett' class, were thought to lurk in the basin between Crete and Libya/Egypt, waiting for a US Carrier Group. They were also reported as operating within thirty miles of the coasts of Crete, Libya and Egypt, where the bottom was recognizable, making navigation easier, the sonar sea state was higher, and counter-detection by active sonar was very unlikely. In times of tension NATO expected the Soviets to reinforce their Mediterranean submarine force by the covert passage of submarines from the Atlantic Ocean and the Black Sea. In order to locate and quantify these reinforcements NATO deployed air, surface and sub-surface ASW forces on surveillance and barrier operations in the chokepoints off Sardinia, Sicily, Malta and eastern Crete as well as the Straits of Gibraltar and the vicinity of the Dardanelles through which the Soviet submarines had to transit.[106] To find the submarines, there was a coordinated ASW effort employing US P3 Maritime Reconnaissance Aircraft, flooding the area with radar, as well as submarines using overt active sonar.

In July and August 1978, HMS *Dreadnought* was ordered to establish and maintain an active trail of Soviet 'Juliett', 'Tango' and 'Foxtrot' conventional submarines and Soviet nuclear submarines, and to conduct surveillance of high-interest Soviet surface ships and exercises.[107] *Dreadnought* had considerable difficulty locating its primary target, the 'Juliett' submarine. A coordinated radar flood from a US P3 Orion Maritime Patrol Aircraft and an active search using *Dreadnought*'s sonar on 25, 26 and 27 August revealed nothing and *Dreadnought*'s CO, Sam Salt, took his submarine into the Sollum anchorage to conduct surveillance of a number of Soviet vessels. As conditions in the anchorage were ideal, Salt decided to conduct hull surveillance on a Soviet modified 'Kashin' class destroyer, making two underwater passes during which Salt's observations through the periscope were recorded as detailed sketches, recording in-line sonar arrays on the under-hull and a main sonar dome suite, which reinforced the impression that this modified 'Kashin' class possessed a very-low-frequency active sonar capability. *Dreadnought* remained undetected throughout the operation.[108]

Trailing Soviet submarines sometimes led to incidents. In 1977, HMS *Valiant*, under the command of John Coward, was trailing a Soviet 'Echo II' class submarine off the coast of Syria. Coward was in *Valiant*'s Control Room when the Engineer entered and said: 'I think you ought to come and see something.' Coward went back to the re-inforced tunnel over the top of the reactor compartment in the bottom of which was a glass window that could be used to see into the unmanned reactor compartment. 'It looked like a huge cathedral of machinery, brightly lit, lovely and quiet, but you can't get in it because of the radi-ation,' recalled Coward. 'But we couldn't see any of that because five feet below the window was the shimmering surface of the Mediterra-nean. We had embarked god knows how many hundreds of tons of sea water into the reactor compartment.'[109] A salt-water services pipe situ-ated in the reactor compartment had fractured, causing sea water to flood into the compartment. Coward returned to the Control Room and slowly reduced *Valiant*'s speed, which created problems with the trim of the submarine as the extra weight began to take hold. 'If we'd have slowed down without knowing that we were several hundred tons heavy, we could have sunk it,' said Coward. 'But fortunately we were going extremely fast, so we hadn't noticed the stuff coming in.'[110]

Valiant was surfaced and with the reactor shut down the diesel gen-erators were started to provide power and limited propulsion. As *Valiant*'s engineering team pumped water out of the reactor compart-ment Coward was instructed to wait for further instructions once a technical analysis had been carried out, and the arrival of a chemist from Rolls-Royce and Associates, who was later flown out and winched onto the submarine by a helicopter from a County Class Royal Navy destroyer, HMS *Fife*. However, the noise of surfacing and the diesel engines alerted the Soviet submarine to *Valiant*'s presence. With the possibility of Soviet warships detecting *Valiant* on the surface, Coward made the decision, backed up by the advice from his own engineering staff, to restart the reactor. At the subsequent Board of Inquiry into the incident Coward was heavily criticized for restarting *Valiant*'s reactor after only a limited clean of the salt-water-drenched pipework. The Flag Officer Submarines, John Fieldhouse, became increasingly irri-tated by the committee's patronizing tone and intervened in the proceedings: 'Gentlemen, those of us who serve in submarines live closer to the consequences of our actions than any of you could pos-sibly imagine.'[111] In the end the Board of Inquiry condemned Coward's decision on engineering grounds, but a top-secret annex to the Board's report acknowledged that Coward had faced a bleak choice between possibly serious engineering problems in the submarine, many months, if not years later; or causing grave political embarrassment if *Valiant* was found on the surface surrounded by Soviet forces. The Admiralty accepted the Board's conclusion and Coward went on to have a highly successful Navy career, becoming an Admiral and Flag Officer Submarines.

Such incidents were rare and the fact that *Valiant* had continued to operate while under such conditions was a testament to the designers of the PWR1 nuclear reactor that powered the submarine. 'It was so well constructed, so well insulated and so well clad that it didn't mind running in sea water,' said Coward. 'We'd gone absolutely ape to stop any drop of sea water getting in there before, the smallest drop in a thousand gallons of beautiful fresh water we'd have ditched it and started again . . . And then we flooded the whole blinking thing with sea water!'[112]

These Cold War operations were very demanding on the crews. Once clear of a patrol area the submarine's company would be a hive of frantic activity as the crew prepared for returning alongside for leave. Once back after an operation, the effects of a long patrol became all too apparent on the crew. Everyone smelt, their clothes were impregnated with the submarine odour of air conditioning, hydraulics, carbohydrates, sewage, the Galley, diesel and natural body odours. Married men on meeting their wives again were quickly told to get in the shower. Daylight also caused problems. The crew was used to fluorescent lighting and distance perception was very poor due to the maximum visual distance inside a submarine being about twenty or thirty feet.[113]

Prolonged patrols also took their toll on the fitness of the crew. The watch-keeping system messed up individual body clocks. If a crewman had been on the second watch, normally from 0100 to 0700 and 1300 to 1900, then at home he would be wide awake when everyone else was sleeping, or feeling tired in the morning when everyone else was wide awake. Knowledge of current events was also poor and if the weather had been bad on the way back with the submarine pitching about the crew would feel a rolling sensation for a couple of hours once back on land. The crew were also discouraged from driving for a day or possibly two as the levels of CO or CO_2 and other gasses may have had an effect. This was compounded by the fact that most had not had anything alcoholic to drink for eight weeks. A couple of drinks ashore could have a powerful effect and most had built up enough money while away at sea to live like kings for a few days.

To alleviate the demands on their crews, SSN COs would try and make as many port visits as possible, but because of the complex safety regime surrounding nuclear submarines, obtaining a nuclear licence for berthing in various ports was complicated. On 2 May 1976, HMS *Warspite* was berthed in Seaforth Docks, Liverpool, for celebrations for the Battle of the Atlantic weekend. *Warspite*, then under the command of Terry Woods, had been heavily worked because of a shortage of SSN availability. While the submarine was alongside, with its nuclear reactor shut down, a serious fire erupted in her Diesel Generator Room. While in harbour, nuclear submarines usually received their power from a shore generator, but due to problems with the facilities in Liverpool *Warspite*'s own diesel engines were operating to generate power. The cause of the fire was a sudden spray of hot, high-pressure lube oil from the engine of one of the diesel generators and its ignition on hot exhaust fittings on the engine. The fire started as a violent conflagration giving off intense heat and vicious choking black smoke. By the time the diesel generators had been shut down, after about one minute, the blaze, akin to a flame inside an oil-fired boiler, had created extremely high temperatures in the higher levels of the Diesel Room on the port side of the submarine. Conditions were such that the compartment quickly became an inferno. Attempts by teams of firefighters from the ship's company to locate and extinguish the fire were hampered by their inability to move freely around the Diesel Generator Room wearing the standard emergency breathing apparatus.[114]

Within seconds of the first report of the fire, *Warspite*'s Manoeuvring Room was filled with dense, choking black smoke and visibility was reduced to just a few inches. The effects of extremely acrid toxic smoke and heat, together with difficulties experienced with the emergency breathing system, made effective watch keeping of the machinery almost impossible and *Warspite*'s crew struggled to read meters and identify alarms. Electrical supplies began to fail and dense smoke quickly filled the after-end of the submarine. The civilian fire brigade was summoned as soon as the fire was reported and arrived within six minutes of receipt of the call. Woods returned on board at 1615 and found that the fire was beyond the stage of being extinguished by portable appliances. He asked the Chief Fire Officer of Liverpool's Fire Brigade to fight the fire with members of the ship's company available as guides. Four and a half hours later, the fire was out. It had caused severe damage to *Warspite*'s Diesel Generator Room; all electrical supplies aft were lost and the nuclear plant was reduced to emergency cooling. Four senior ratings, including a trainee reactor panel operator in the Manoeuvring Room, who remained at his post until he was eventually overcome by fumes, were treated in hospital for the effects of smoke.[115]

Woods was well aware that if the fire had not been controlled he would have had to face a very difficult decision. 'I would have had to shut both doors and shut all the hatches,' he said. 'I had thought out what to do if we really started a nuclear incident. We had at least 20' water beneath us and I was going to evacuate the submarine and use escape flooding valves forward and after.'[116] Preparations were made for initiating a reactor accident signal but after discussion with the Chief Staff Officer Engineering and the Flag Officer Submarines it was decided that there were no indications that a reactor accident was likely to take place; those indications available demonstrated that the reactor was being cooled satisfactorily. Radiation monitoring, both internally and external to the submarine, gave no readings above background level and tests conducted after the incident showed no abnormalities, indicating that no release of fission products to the primary circuit had taken place. Later detailed analyses showed no sign of core damage. The Marine Engineering Officer, Lieutenant Commander Tim Cannon, was awarded the Queen's Gallantry Medal for his role in maintaining the reactor during the fire.[117]

This was the first major fire in an SSN or SSBN. No British nuclear submarine had previously had to deal with such an incident in terms of difficulty and complexity. Although there had been frequent exercises, realism was necessarily limited due to the possibility of accidents in poor visibility and the toxic nature of the smoke. Many valuable lessons were learned. The incident resulted in the upgrading of much firefighting equipment and systems on all British submarines and a revision of some of the procedures for fighting fires. In the long term, there was little impact on future port visits by nuclear submarines and at the end of 1983 HMS *Courageous* visited Liverpool.[118]

THE PERILS OF SPECIAL OPERATIONS

Aside from Cold War operations, the Submarine Service was increasingly involved in a number of other roles, stimulated by the Northern Ireland Troubles. The first was somewhat puzzling. During the 1974 Ulster Workers' Council strike, the Prime Minister, Harold Wilson, in desperate need of radical solutions to break the blockade of the

Ballylumford power station, asked the Ministry of Defence to report on the possibility of using nuclear submarines to power Belfast. Officials in the Ministry of Defence looked into the idea 'as a matter of urgency' and concluded that an SSN could leave for Belfast within forty-eight hours. However, it would not have been able to make a worthwhile contribution to the electricity supply, as the electricity generated by a nuclear submarine was not compatible with the national grid. Only a few places, such as the Royal Dockyards and one or two shipbuilders, had the suitable conversion machinery. Even then, due to the design of the submarines' electrical system, only a few hundred kilowatts could be fed into the grid.[119] The idea was shelved and the strike eventually contributed to the collapse of the power-sharing Sunningdale Agreement.

The Submarine Service was also involved in the Troubles in Northern Ireland, when submarines were involved in intelligence-gathering operations and in support of Special Forces.[120] The British Government suspected that arms, ammunition and explosives were being smuggled into Northern Ireland by sea using containers or car ferries. Since 1967 the Royal Navy had been conducting Operation 'Grenada', providing minesweepers or similar craft for coastal patrolling in order to intercept provisional IRA clandestine arms smuggling. However, these vessels often suffered from the extremely poor winter weather off the east Irish coast, which made interception hazardous. The overt nature of the patrols and the minesweepers' slow speed often meant that smugglers were able to avoid or outrun them. In June 1975, the UK's military authorities in Northern Ireland requested that a submarine be assigned to conduct covert surveillance of the east coast to help intercept vessels involved in smuggling. The submarine was attractive because of its relative immunity to bad weather, the fact that it could, if equipped with image intensifiers and low-light television, identify and follow suspect ships. It could then direct the minesweepers to take over the trail of suspects into ports, or to carry out an intercept at sea. Submarines could also move close inshore and covertly observe landings, gathering intelligence. The psychological value was also attractive.[121] In July 1975, HMS *Osiris* was ordered to conduct a patrol in Red Bay, in Northern Ireland, in order to test the feasibility of using SSKs in covert surveillance. Little suspicious activity was observed, and it was concluded that future operations should only be mounted if the submarine had specific intelligence about smuggling activities.[122] Operations continued well into the 1970s and in September 1976 HMS *Onyx* and HMS *Otus* were also involved in 'Grenada' patrols.

Royal Navy submarines were also involved in deploying and extracting SBS Royal Marines in support of operations in Northern Ireland. In January 1975, HMS *Cachalot* infiltrated two SBS teams by canoe in the area between Torr Head and Garron point as part of Operation 'Aweless', an anti-gun-running operation.[123] Such operations were not confined to Northern Ireland. Although Sam Fry had discontinued a number of Special Forces practices while Commander SM, 7th Submarine Squadron in the late 1960s, by the mid-1970s submarine Special Forces operations had developed into something of an art form and Special Forces were involved in training for a number of operations using submarines. One such exercise, codenamed Exercise 'Cold Shoulder', was conducted off Norway in the early 1970s and saw Special Forces carry out a submerged attack on an enemy ship at anchor in a Norwegian fjord. According to an SBS Marine who took part in 'Cold Shoulder', one of the Royal Navy's 'Porpoise' class submarines, HMS *Walrus,* entered the Malangan fjord, just south of Tromsø on 14 March 1973, and conducted a reconnaissance of a suitable landing point using the submarine's periscope.

Exiting and re-entering a submerged submarine was a lengthy and complex procedure that involved the SBS leaving through the submarine's forward escape tower (SET). Prior to departure the SBS would pack stores – wooden skis, pulk (a toboggan), rucksacks and an inflatable Gemini called a Capella – and place them in a stores box located under the casing of the submarine. As *Walrus* moved in towards the landing point, the CO ordered his crew to remain silent. Speed was reduced to less than half a knot and the SBS began to move outside. Wearing specialist breathing apparatus, the SBS Marines, followed by a casing diver, entered the escape tower, shut the lower lid and waited while the compartment was flooded. Once the pressure had been equalized the casing diver would signal that he was ready by using a 'knockometer', a hammer to hit the bulkhead. The upper hatch was then opened and the casing diver exited and began to prepare the stores for release. Meanwhile, the rest of the team, each wearing a Rechargeable Air Breathing Apparatus (RABA) proceeded to exit the submarine using the same procedure. 'Initially in the SET, we breathed from the submarine Built-in Breathing Supply (BIBS) to conserve the air in the RABA' recalled one Marine involved in the exercise:

> The cold water slowly flooded into the tower and, when it reached about head height, the pressure inside equalized with the outside water-pressure.

The upper hatch started to lift and cold seawater tumbled in, sometimes knocking your facemask off. Once the hatch was fully open, we changed from BIBS to RABA. Grabbing a wire along the casing, we then exited the tower and swam, or rather the slipstream pushed us, aft to the area of the Forward Torpedo Loading Hatch and got under the casing. We called this space the 'lurking area'. As each team member made his exit, he went in to the lurking area to get out of the slipstream and waited with the RABA plugged into a large 150 cu ft air bottle.[124]

Once the SBS team was safely in the 'lurking area', the casing diver hit the Torpedo Loading Hatch five times to signal to the submarine that the team was ready. Inside *Walrus*, one of the crew informed the Control Room and the CO gave permission to release the stores. Outside, the casing diver waited for permission to cut a ring main to allow the stores to float to the surface. On hearing taps from inside the submarine, he replied with his hammer, and then cut the ring main with a diving knife. The stores, with pent-up buoyancy, shot towards the surface, still linked to the submarine by a long towline. With their stores floating the SBS team moved out of the lurking area and began to swim to the surface:

In pairs, we went up the towline and, whilst the Casing Diver waited in the lurking area, we inflated the Capella and got in and stowed the rucksacks and thankfully got the engine (a Johnson 40hp) running. We gave a red torch signal to the periscope and the Casing Diver was given the signal to cut the towline; he then re-entered the submarine.[125]

Once separated from *Walrus* the SBS team moved to the side of the fjord, where they lifted their Capella ashore, capsized it and stowed the engine and unwanted diving equipment underneath. They then made their way towards the target, which was located in an anchorage. The Marine continued:

We watched the vessel come in and anchor and, during the evening prepared our diving sets. The CO_2 canisters had been compressed a bit during the submarine's passage but apart from that all was well. We donned our wooly bears and dry suits ... and whilst our two 'mules' [SBS1] and [SBS2] waited on land, [SBS3] and [SBS4] entered the water shortly after dark and dived. It was not a long swim and having placed our limpets on the hull we then returned to the shore and got into our skiing clothes ready to exfil. Rather than carry our diving kit, we cached it and made our way on skis throughout the night back to a position close to the cached inflatable.[126]

The SBS team then waited until nightfall before they prepared to row back to *Walrus*. The Marine described how:

> We had the painful business of donning very cold dry-suits and wet-suit hoods and gloves that had frozen solid. Amazingly, again the outboard motor started and we headed towards the submarine RV in Bals Fjord. Using our pinger, once we were in the estimated position, the dived submarine homed in on us. We snagged the periscope and, shortly afterwards, the Casing Diver came to the surface and in pairs we dived down the fin along the casing to the lurking area. Recharging our RABA both in the fin and in the lurking area, in turn we re-entered the submarine via the SET.[127]

With the SBS having re-entered the submarine the casing diver stored the remaining equipment, slit and sunk the Capella inflatable and then re-entered the submarine.

These operations required extensive training and there were plenty of opportunities for incidents. One of the Marines taking part in 'Cold Shoulder' had a particularly close call on another exercise in waters off Gibraltar while standing on the casing of a submarine. Suddenly an Iranian frigate began to head straight for the submarine. The submarine CO ordered an emergency dive, leaving the SBS soldier on the casing. He was eventually ripped away, passed the propellers, which he narrowly missed and was dragged into their vortex. Eventually he stopped spinning and was left floating underwater. Fortunately the exercise was taking place during daylight hours and the Marine was able to swim to the surface, where he found himself facing the bows of the frigate, which passed him by yards.[128]

In January 1977, the same Marine was also involved in a tragic accident with HMS *Orpheus*, which was fitted with a new five-man chamber specially designed for exit/re-entry operations. On 15 January, *Orpheus* was conducting exercises with the SBS in Loch Long, in blustery and cold conditions. *Orpheus* was dived to 73 feet with her W/T mast raised and three SBS divers in the special chamber. Just as the SBS team prepared to exit and begin the exercise, *Orpheus* cruised into a patch of fresh water and the sudden change in water density caused the submarine to dive uncontrollably and speed up to maintain depth.[129] As *Orpheus* continued to dive the safety diver, who was attempting to hold himself in position on the submarine's casing was pulled off like a puppet by the tension on his communications line. On the surface, an SBS team sitting in a Gemini inflatable was pulled vertically downwards by *Orpheus*'s W/T mast as the submarine increased depth. The bowline then tore off completely.

The first Marine had already left the chamber when *Orpheus* hit the patch of fresh water. He attempted to return to the chamber to warn the other two members of his team but soon realized that he stood little chance of making it. He reluctantly pushed off the submarine's casing and inflated his specialist suit. *Orpheus* was now so deep that for the first 15–20 feet of his ascent, the first Marine's suit inflation had no effect whatsoever. It was only through his own vigorous swimming effort that he managed to start moving upwards. When he eventually surfaced he was quickly recovered by the SBS in the Gemini Inflatable. When *Orpheus* surfaced after blowing ballast to give the submarine positive buoyancy there was no sign of the other two Marines. After a surface search had been conducted it was clear that the two divers were missing. As the search continued most people believed that the two divers were alive, that they had swum ashore to find a telephone or make their presence known to someone. But the next day, they were found 10 yards apart in 221 feet of water. One had managed to get rid of some of his equipment, but in a frantic attempt to swim had lost one of his fins. Both Marines had died from drowning following HMS *Orpheus*'s dramatic dive.[130]

A naval Board concluded that the two Marines probably left the casing when the submarine was at approximately 105 feet, two minutes after the upper hatch indicated shut. It appeared that they were free of the chamber and on the casing breathing off their special RABA sets. They then tried to make it to the lurking area to recharge their breathing apparatus but when *Orpheus*'s ballasts blew they were swept off the casing as the submarine suddenly accelerated towards the surface. A diver could comfortably move about the outer casing, but only if the speed of the submarine did not exceed half a knot. Any faster and he would need to concentrate on holding on. If the speed exceeded 2 knots, he would no longer be able to hold on because of the force of water pushing against him. If he let go, there was a danger that he would pass through the propellers. Both divers had opened up their inflatable suits, but because they were so deep there was little air. They were also negatively buoyant. Lessons were learned. After the incident a special life jacket was designed for SBS divers that, when pulled, inflated a massive air bag capable of pulling a man in full operational equipment from a depth of 100 feet. The maximum speed for exit and re-entry was also reduced.[131]

THE 'SWIFTSURE' CLASS

On 27 November 1976, Vice Admiral Sir Iwan Raikes stood down as Flag Officer Submarines. His successor, Admiral Sir John Fieldhouse, was the first 'post-war' officer to hold the post, and the first to have held a nuclear command. Fieldhouse was very much a product of the Cold War, and the nuclear age, having served as First Lieutenant on board HMS *Totem*, which as we have seen conducted the Royal Navy's first submarine intelligence-gathering operation against the Soviets. He had also commanded HMS *Acheron*, HMS *Tiptoe*, HMS *Walrus* and HMS *Dreadnought* and was responsible for the Polaris force during its formative phase in the late 1960s. He also held various posts on the Staff of Flag Officer Submarines at HMS *Dolphin*, all of which meant he was well versed in submarine matters and well prepared to take up leadership of the service.[132]

Fieldhouse recognized that if he did not move to Northwood he could lose control of his submarines. In early 1978, he therefore re-located the Headquarters of the Submarine Service from its traditional home in Fort Blockhouse, Gosport, to Northwood and assumed direct control of Commander Task Force 311 (CTF 311). There was now complete control over the submarine operational broadcasts; CTF 345 was next door and water could be divided between SSN and SSK operations and Polaris patrols. In order to test the new combined and integrated submarine operations centre Fieldhouse ran an exercise called Operation 'Gratitude' and ordered Royal Navy submarines to target any Soviet submarines that happened to be at sea at the time.[133]

Fieldhouse also supervised the introduction of an entirely new class of nuclear submarine: the 'Swiftsure' class. Work on the new boats began in the early 1960s, while 'Dreadnought' and the 'Valiant' class were building. However, as we have already seen, design work was interrupted by around three years because of the Polaris programme and it did not resume until the mid-1960s.[134] This additional time allowed the naval architects 'to consider the new design more carefully and to feed in the lessons learnt from operating *Dreadnought*, building the "Valiants" and the first SSBNs'.[135] Known as the 'Swiftsure' class, its design team was led by a brilliant naval architect, the Assistant Director of Warship Design at Bath, Norman Hancock. Hancock was so determined to produce a submarine that was faster, stealthier and deeper diving than any contemporary vessel that disbelieving naval staff in London had to repeatedly revise their expectations. Hancock rejected the American teardrop hull form and used one of equal diameter throughout, which combined a longer pressure hull within a shorter overall length. According to Hancock, the class was 'probably the biggest single step forward we have made in nuclear submarine design'.[136] His design philosophy was to:

> make all improvements which were practicable within existing technologies, with existing materials and without extensive research or development. This necessarily precluded large changes in reactor technology, but permitted redesign of the hull and of the layout of the weapons in the front-end as well as of the machinery in the back-end. Our aims were to simplify wherever possible by cleaning up and removing all unnecessary duplication and to introduce the necessary improvements which would produce a faster, stealthier, safer, deeper diving and more easily maintainable weapons platform.[137]

The design condensed the machinery to such a degree that in order to maintain Swiftsure's longitudinal centre of gravity, Hancock had to introduce an empty space aft of the engine room, known as 'Hancock's hole'.

With a submerged displacement of 4922 tons, and a deep surface displacement of 4478 tons, the overall length of the 'Swiftsure' class was 271 feet with a maximum diameter of 32 feet 3 inches. The final design was 14 feet shorter and one foot less in diameter than the 'Valiant' class and the diving depth was increased from 750 to 1250 feet. The external shape of the submarine was cleaned up and compacted to reduce hull resistance. The bridge fin was also reduced in size and the forward hydroplanes made retractable. The first of class, HMS *Swiftsure* was fitted with a series of highly secret special sophisticated swept-back conventional propellers designed to reduce cavitation, which gave a submerged speed of approximately 28 knots, and a surface speed of approximately 12 knots. Subsequent submarines of the class were fitted with a pump jet driven by two steam turbines, which contributed to the much fuller and blunter stern of the submarine design.[138] The class was powered by the same PWR1 nuclear reactor used in the 'Valiant' and 'Resolution' classes, but it featured a new, more advanced and longer-lasting core, known as Core B, which had double the life of and a maximum power rating approximately 20 per cent greater than the Core A.[139] The class was also significantly quieter than earlier British SSNs. Propulsion turbines, turbo generator sets and main gearbox were all mounted on a moveable raft which was noise-isolated from the pressure hull. At full power radiated noise in the 'Swiftsure' class was reduced to a level comparable with the quietest main machinery mode of the 'Valiant' class.

The class was also fitted with an improved version of Type 2001 sonar and the main array was repositioned at the fore end of the submarine in a downward position to exploit what was known as the bottom-bounce technique, a method of using the ocean bottom to increase the range of sonar. Other sonar outfits included the Type 2007 long-range low-frequency passive sonar and Type 2017 frequency analysis equipment, as well as the normal range of echo sounders, U/W telephones and cavitation indicators. The class also differed from previous Royal Navy SSNs in that it was fitted with five torpedo tubes in a modified arrangement, instead of the six in the 'Dreadnought', 'Valiant' and 'Resolution' classes, with a total of twenty torpedo stowage positions – all of which were capable of accepting the new Mark 24 torpedo.

The first of class, HMS *Swiftsure*, commissioned on 17 April 1973, and was followed by five others: HMS *Sovereign* in July 1974; HMS *Superb* in November 1976; HMS *Sceptre* in February 1978; HMS *Spartan* in September 1979; and HMS *Splendid* in March 1981. Once again there were construction problems. HMS *Superb* was to have been built with the new QI(N) high-yield steel. However, as in previous programmes the British steel industry struggled to meet the demanding build schedule and the Royal Navy was once again forced to turn to the Americans for supplies of the American-made HY80 steel. Strikes, overtime bans, lockouts, the energy crisis of 1973–4, and demarcation disputes, as well as a lack of steelworkers, electricians, outfitters and pipe fabricators, contributed to other delays. However, as more of the class entered service, they came to be universally recognized as one of the most successful of all post-war nuclear submarines. Most operated out of the 2nd Submarine Squadron in Devonport.

The 'Swiftsure' class was also the first Royal Navy nuclear submarine designed with under-ice operations in mind. Since *Dreadnought*'s patrol to the North Pole in 1970, the Royal Navy had kept a close eye on the icy waters of the Arctic Ocean. In the early 1970s, the Americans placed acoustic arrays in the ice cover, which provided regular scientific and operational intelligence and demonstrated that the Russians were deploying their submarines under the ice. The polar ice cap of the Arctic Ocean was now an area of utmost importance to both East and West, an area that neither side could permit the other to dominate. The US Navy recognized that scientific data, operational experience and simple presence in the Arctic had to be accrued, developed and maintained so as to understand the geographical and strategic aspects of the environment. In 1973, the first ever war game between two US Navy SSNs, USS *Hawkbill* and USS *Seadragon*, was conducted in the Bering Sea. 'The game was on,' wrote the chief architect of the US programme, Dr Waldo Lyon, 'two of us in the shallow water under the ice, and that's when we learned . . . no way . . . we don't know how to do this.'[140] In order to address this deficiency the US Navy started to send an SSN under the Arctic ice cap every two years.

The Royal Navy was still relatively inexperienced with under-ice operations and in October 1976 Fieldhouse decided to send one of the new 'Swiftsure' class submarines to the North Pole. HMS *Sovereign*, under the Command of Commander Michael Harris, departed Devonport on 1 October and in cooperation with HMS *Narwhal* took part in Exercise 'Brisk', travelling to the North Pole in order to conduct geophysical survey work and collect data on the underwater profile of the ice cap.[141] The first stage of the operations centred on the ice edge, where the extraordinary and unstable water conditions, combined with the high background noise level in anything other than flat calm weather conditions, made it an ideal area for remaining undetected. However, for nuclear submarines, the ice edge was a more awkward area in which to operate than either the open ocean or under the Arctic pack ice and *Sovereign* spent a considerable amount of time dodging growlers and icebergs. Conditions under the Arctic ice were far more stable and when *Sovereign* surfaced at the North Pole on 23 October, the submarine was wedged into a comfortable and secure berth in the ice sheet. While on the surface *Sovereign* was unable to establish contact with the UK due to faults with the submarine's communication masts and the now well-known high-latitude communication problems first encountered by HMS *Dreadnought* in Exercise 'Sniff'. After three days, *Sovereign*'s failure to establish communication created something of a media storm back in the United Kingdom as the press got hold of the story and incorrectly concluded that the submarine was lost beneath the ice.

Despite these relatively minor problems the patrol confirmed that the 'Swiftsure' class could easily handle the operational aspects of under-ice operations. The Captain of the 2nd Submarine Squadron in Devonport, Richard Heaslip, concluded that 'any "Swiftsure" Class SSN can be deployed at very short notice for under-ice operations'.[142] This, according to Fieldhouse, was 'highly significant' as the 'S' class could now be sent to conduct an ice patrol without special training or modification.[143] *Sovereign*'s CO recommended deploying a Royal Navy SSN to the Arctic at least once every five years in order 'to keep the art alive' and that on the next occasion at least two SSNs should take part in operations in order to improve knowledge of how Royal Navy sonars performed under the ice.[144] Fieldhouse agreed and in August 1977 he opened discussions with the United States about participating in combined US/UK under-ice exercises, using HMS *Sovereign* and a US SSN in April 1979.[145]

The 'Swiftsure' class was also fitted with new technology that significantly enhanced the capability of Royal Navy submarines to track Soviet submarines. In the early 1970s, the Admiralty's Underwater Weapons Establishment (AUWE) at Portland was tasked with producing the first computer-assisted Tactical Data Handling System for the 'Swiftsure' class. The first system, known as the Tactical Data Handling System, or DCA, was capable of accepting and displaying information generated by active and passive sonars and was fitted in HMS *Swiftsure* in 1973, with tactical trials in 1975. In combination with advanced mathematics, these new computers were able to automate the manual processes for tracking contacts, removing the need to depend on a myriad of manual plots, slide rules, calculators, stopwatches and position-keeping devices that previously placed big demands on a submarine's manpower. One of the most important advances was the realization that the same complex equations used in US guidance and navigation systems developed by an American control engineer, Rudolf Kalman, could be adapted and applied to more accurately determine the movement of a submerged submarine contact.[146] The DCA system was subsequently fitted in the next two 'Swiftsure' class submarines, HMS *Sovereign* and HMS *Superb* and it was also used as the Command part of a new Target Tracking and Fire Control system known as DCB, which was first fitted to HMS *Sceptre*.

In theory, the new computerized integrated command and fire control system could handle twenty-five submarine or surface ship contacts simultaneously and produce accurate fire control solutions for each one. However, submariners soon found that the poor bearing information provided by the main sonar and a lack of computer memory meant that the solutions displayed were sometimes unreliable. There was an understandable tendency to assume that because solutions were displayed in plan form on a computer screen they were necessarily accurate, when they were in fact portraying the best information on a contact available at the time.[147] COs who were capable of recognizing when the information was incorrect would say their operators were suffering from a fictional set of symptoms known as 'Kalman Syndrome'. As with most computers, the principle of 'rubbish in, rubbish out' applied and operators would often see a contact hurtling across the screen at vast speeds, until it was deleted. It took a great deal of effort by the operators to control the picture, to the continual frustration of command, and many submarine officers hankered after the old paper plots, which they understood better.[148] Some even resorted to maintaining clandestine versions.[149]

The 1970s also saw considerable advances in submarine communications, an area that was notoriously difficult and beset by constant problems such as mast heights, immersion and ingress of water. Alongside the existing very-low-frequency (VLF), medium-frequency (MF) and high-frequency (HF) communications, three additional means of communication, some technical, some procedural, were introduced into Royal Navy submarines. The first was a process known as 'Postboxing'. It involved a submarine arriving at a prearranged area, raising its wireless mast and transmitting and receiving communications from a nearby RAF Nimrod aircraft, which would then relay information to the relevant authorities. While innovative, the system depended on both submarine and aircraft sticking to complex schedules and being in the right place at the right time. The second means of communication, known as 'Burbling', was more rudimentary and intended for use during sensitive, time-critical operations when a submarine was unable to come to periscope depth and communicate using traditional means. In order to transmit a message a submarine CO would order a series of revolution movements, which corresponded with prearranged messages such as 'in trail', 'breaking trail' etc. These revolution movements would be picked up by SOSUS arrays, analysed and the appropriate messages extracted.

The third means of communication, the Submarine Satellite Information Exchange System, or SSIXS, revolutionized submarine communications and allowed almost instantaneous and secure two-way communications between submarines and operating authorities.

Provided the appropriate satellite was free, communications traffic could be transmitted at any time and all outstanding traffic held by the shore station transmitted to the submarine in a matter of seconds. Developed by the Americans in the mid-1970s, SSIXS was fitted to all USN submarines in 1977 and a number of sets were made available to the Royal Navy by COMSUBLANT. Following successful trials a complete inventory was acquired and fitted to all Royal Navy submarines. SSIXS greatly aided the trailing of Soviet submarines as it allowed Royal Navy submarines to send reports to operational headquarters without breaking high-frequency radio silence. It also significantly enhanced UK/US inter-operability as both nations used the same system.

While these advances significantly enhanced the capabilities of the Royal Navy's submarines, their weapons systems were still, to use the words of the then Flag Officer Submarines, Michael Pollock, in a 'parlous condition'.[150] As we have seen, the Mark 24 torpedo, on which the Submarine Service depended, had failed its acceptance trials and its designers were forced to return the torpedo to the development stage. In 1970 the former head of the Dreadnought Project Team, Rowland Baker, was called back from retirement and appointed head of a new organization known as the Torpedo Project Executive, to manage a 'get-well programme'. While this intensive programme succeeded in resolving many of the torpedo's defects, by the middle of 1970 there were still two major problems with the Mark 24: it consistently failed to pass within the required 15 feet of a target when it entered the final phase of an attack, which had significant implications for the lethality of the torpedo; and it had a tendency to roll when it was discharged from the firing submarine, severing the control wire. The Marconi Space and Defence Systems Company (GEC-AEI) had to employ some 270 men in order to resolve these problems.[151] After development and engineering problems the Mark 24 Mod 0 torpedo finally entered service in 1974.

In any case, the Mod 0 variant of the Mark 24 had other shortcomings. At first, it had no anti-surface ship capability. When the Admiralty provided one, as the Mark 24 was so old, the design and engineering modifications required to update what was then late-1950s-era technology meant that the result, the Mark 24 Mod 1, was essentially a new torpedo. But it too suffered from a number of shortcomings. It was limited to speeds of up to 24 knots and thus had no speed advantage over surface ships. At higher speeds the torpedo's flat nose created so much cavitation that it either lost what contact it had with a target or was incapable of gaining contact at all.[152] The torpedo's designers were also unable to guarantee that its warhead, which was only 300 lb compared with the 800 lb warhead on the old Mark 8, would sink a surface ship. Indeed, Admiralty studies indicated that the probability of sinking a Soviet 'Kynda' or 'Sverdlov' surface ship 'with a single hit from either weapon is low'. The Mark 24 was later modified to explode underneath the keel of a surface ship, creating a whipping effect that would break the ship's back.[153] The combined anti-submarine and anti-surface Mark 24 Mod 1 finally entered service in October 1980, but it too suffered from unreliability and shortcomings.

The Mark 24 Mod 1 was unable to reach modern Soviet deep-diving submarines. The Navy developed a modification kit which, when applied to the original torpedo, enabled it to go below its depth floor of 1150 feet to a crushing depth of 1450 feet, once it had gained acoustic contact. However, scientists and engineers in the Admiralty's Underwater Weapons Establishment (AUWE) in Portland were still concerned that it was:

> glaringly obvious that we are still a long way from meeting the threat. On the submarine side, the [Mark] 24-1's 1450 ft will go nowhere near the limiting diving depth of the Soviets. Against surface targets, the [Mark] 24-1's slow speed is inadequate against a fast ship, and of course the firing submarine is still tied to a bit of wire and so is restricted in manoeuvre; if faced with an A/S weapon such as the [Soviet] SUW-N-1 the situation becomes distinctly unhealthy.[154]

AUWE looked at further ways and means of modifying the Mark 24 so as to give it an adequate capability against deep submarine targets, by making its hull of a new and stronger material and by employing new welding techniques. AUWE hoped that this would increase the crush depth of the Mark 24 to 2500 ft. Early research had also started on the successor to the Mark 24 series, which the AUWE hoped to have ready by the mid-1980s.

The Navy was also exploring new weapon systems. The first was a private venture by Vickers, a Submarine-Launched Air-Flight Missile (SLAM) system designed to take out helicopters involved in anti-submarine warfare. In 1972, one of the last of the Royal Navy's 'A' class submarines, HMS *Aeneas*, was loaned to Vickers to test a new weapon system consisting of six Short Blowpipe missiles mounted on a stabilized launcher, housed in a pressure vessel. The submarine would surface its fin, so that it was protruding out of the water, raise the launcher hydraulically, acquire a target by periscope and fire a missile, the whole procedure taking around 20 seconds. Trials indicated that the system had good capability against a helicopter out at 3 kilometres and some capability at 5 kilometres, particularly if the helicopter was hovering. But the system required manual guidance, the target had to be visual throughout the firing sequence and missile flight and there was no night capability. The requirement to expose some of the submarine for a short period was also a distinct disadvantage and many submariners regarded the SLAM as 'over-rated . . . very ingenious, but it is a clumsy short range, daylight only system. We are not about to fit it to our submarines'.[155]

The second new weapon system was a submarine-launched sea-skimming missile, known as the Under Sea Guided Weapon (USGW), intended for use as an offensive anti-surface ship weapon. By the mid-1980s, NATO expected in the event of war to have substantial difficulties countering the vast Soviet surface fleet, of which it expected over thirty heavily armed units to be deployed in the Eastern Atlantic.[156] The requirement for the USGW rested primarily on the SSNs' ability to operate against Soviet surface forces in areas where the enemy enjoyed air superiority, but it was also envisaged as a complement to the anti-ship capabilities provided by other maritime forces. Its covert deployment was regarded as of considerable tactical significance. Of the two possible contenders, the Hawker Siddeley Dynamics Sub Martel and the McDonnell Douglas Astronautics Company Sub Harpoon, the Royal Navy preferred the US Sub Harpoon on grounds of reduced development risk, timescale and cost. Sub Harpoon also had a number of operational advantages such as longer range, larger warhead, advanced state of development and a more assured in-service date.[157] Ministers endorsed the proposals in September 1973 and negotiations with the US Government were opened to procure a UK variant known as the Royal Navy Sub Harpoon.

Royal Navy Sub Harpoon was a small rocket of about 15 feet, with retractable wings. It was housed in a capsule about the size of a normal torpedo which was embarked on board a submarine in the traditional way and stored in the Torpedo Compartment. The missile itself used a dangerous liquid fuel known as JP10 and special precautions had to be introduced and enforced for those tasked with handling the weapons, as well as those living in the Torpedo Compartment, as strict regulations had to be observed. Much thought and redesign of the existing weapons system in the submarines was also required. When firing Sub Harpoon, the submarine had to be kept level within very tightly controlled limits for pitch and heel and be steady on course. The weapon capsule was discharged from the torpedo tube by a slug of water, as if the submarine were firing a torpedo. The capsule would then shoot towards the surface and, on breaching, small charges would blow the capsule nose cone off, allowing the rocket engines to ignite and the rocket wings to deploy. The weapon would then hurtle up into the sky before returning down towards the sea, where it would skim along towards its target at a speed of about 550 knots using active radar to home in. A salvo of Sub Harpoons would normally be fired at high-value targets such as a task group's replenishment ship, tanker or aircraft carrier such as the *Kiev*. Although reliable, a salvo was fired in order to saturate a target's close-range defences.[158]

The 1970s also saw the introduction of new sonar equipment. The primary means of detecting Soviet submarines was through passive broadband sonar. The performance of Type 2001 sonar in the passive role was downgraded by the noise of water flowing over the dome that housed the sonar array. This meant that the submarine was limited to a low speed when it was required to detect or hold targets. A fibreglass dome was eventually fitted over the array, which markedly increased the ability to hold contacts at higher speed. But passive broadband sonar ranges were relatively limited (out to 30 miles was exceptional) in the North Atlantic, even in favourable isothermal conditions found in the winter months. This meant that in order to detect any contacts Royal Navy submarines had to be relatively close to them. In poor sea and weather conditions this could be notoriously difficult.

The problem was illustrated in October 1977, when HMS *Superb*, under the command of Commander David Ramsay, took part in Operation 'Crusader', a covert ASW patrol in the north Norwegian Sea designed to assess the operating environment and ASW coordination in the area, and to initiate trailing operations on transiting Soviet SSBNs should the opportunity arise. During the 23-day patrol *Superb* attempted to intercept four SOSUS contacts assessed as Soviet nuclear submarines, including one 'Delta', on patrol in the north and three – one inbound and two outbound – 'Yankees', transiting to their patrol areas. Despite setting up thirteen separate barrier tracks ahead of and across the predicted tracks of the four transiting Soviet SSBNs, *Superb* with its standard, but domed, Type 2001 sonar fit failed to make any definitive classifications.[159] Like British and American submarines, Soviet submarines and surface ships were also becoming quieter. If the Navy was to maintain its lead in sonar detection further improvements would be needed.

In the late 1970s a new item of technology was introduced that dramatically improved the capability of the Royal Navy's submarines to detect, classify and trail Soviet surface ships and submarines. The towed array consisted of a long length of neutrally buoyant, flexible snake-like wire with hydrophones embedded in it at intervals of up to 30 metres or more, which was towed at the same depth as the submarine.[160] The arrays were sometimes kilometres in length in order to intercept long wavelengths of very-low-frequency noise radiated from Russian vessels and submarines.

In 1977, recognizing the need to bring towed-array sonars into service quickly, the Navy developed Sonar Suite 2024, consisting of a UK towed array feeding into an off-the-shelf signal processor of US manufacture. Although this interim sonar provided a step forward, it had limited all-round surveillance capability, poor data display facilities and rudimentary connections with the Submarine Command Team and the Action Information System. A separate project to develop an improved towed array and advanced signal processor known as Sonar Suite 2026 was initiated. As well as exploiting advances in processing technology, it was specifically designed to match UK Tactical Data Handling Systems.[161]

Early US towed arrays were flushed out from a sheath in a submarine's hull, but this limited their length to that of the submarine. Later towed arrays, such as those used by the Royal Navy, were clipped on with assistance of a tug. Once the array was clear of the submarine, and thus the submarine's self-noise, towed arrays could detect, locate and classify noise emissions from Soviet ships and submarines at long ranges. Emphasis shifted away from detecting broadband frequencies to narrowband frequencies, which the towed array could detect at great distances. Emissions from equipment inside submarines, such as reactor coolant pumps and generators known as 'tonals', were then used to aid the classification of a target. They could also be used to determine the course, speed and range of a contact by a complex technique known as target motion analysis.[162] Mastering the use of towed arrays and target motion analysis required considerable skill and expertise. Submariners with backgrounds in mathematics, physics and chemistry thrived when using the complex mathematical formulae associated with narrowband tracking, but others found the new science difficult to master. 'We got it,' recalled Mark Stanhope, who in 1975 was a relatively junior officer in HMS *Swiftsure* with a background in Physics, 'but it took a lot of others a long time.'[163]

The towed array significantly enhanced the performance of the Royal Navy's nuclear submarines and gave the service a powerful new means of detecting Soviet forces. The first towed-array systems, the Type 2024, were installed on the new 'Swiftsure' class submarines at the beginning of 1977. One of the first to gain operational experience with the new equipment was HMS *Sovereign* during Exercise 'Agile Lion', which took place between 28 January and 3 March 1977. After exercising with HMS *Walrus* and HMS *Churchill*, *Sovereign* sailed to take part in a joint US/UK trail of a Soviet nuclear submarine transiting south from the Northern Fleet, later identified as a 'Charlie' class SSGN. The towed array allowed *Sovereign* to make a long-range detection of the Soviet submarine, well before it took over from the US submarine that was tracking it. *Sovereign* settled down for what turned out to be a 62-hour, 707-mile trail in a position just abaft the Charlie's starboard beam and started experimenting with tactics to improve the trail and gain information on the target using the then largely untested towed array. *Sovereign*'s crew quickly discovered that the accuracy of narrowband bearings was a great improvement on that obtained by the standard sonar fit.[164] The post-patrol analysis warned that:

> despite the apparent simplicity of the target's transit the effort required to simply maintain contact for hour after hour must be recognised. Continuously aggressive trailing techniques were absolutely necessary to keep up and out of the target's stern nulls [immediately behind the submarine] yet clear of his bow arcs whenever he zigged or cleared stern arcs. It was considered vital that MSN UK-22 [*Sovereign*] was not counter detected and the desire to obtain acoustic intelligence at close range was initially suppressed for this reason. As familiarity with the target increased and his pattern of operations developed so did the target's speed of advance. Once a classification of a Mediterranean bound 'Charlie' class submarine was the most probable and deep water reached the opportunity to close up for ACINT [Acoustic Intelligence] was lost along with the target. Considerable quantities of data were recorded which may provide a

fund of useful information, and the ship's command team gained a lot of experience.[165]

The towed arrays were so effective that Royal Navy submarines testing the new equipment tended on occasion to pick up Soviet submarines while they were conducting exercises. In January 1978, HMS *Swiftsure* was conducting an evaluation of the Type 2024 Towed Array, codenamed Exercise 'Six Bells', in an area west of Ireland with HMS *Churchill* acting as a target. Not long after starting the exercise *Swiftsure* detected a probable Soviet conventional submarine and interrupted the exercise to conduct a covert trail for a period of eighteen hours before intentionally breaking contact.[166]

The US Navy was also enjoying considerable success with its own towed arrays. On 17 March 1978, the USS *Batfish*, equipped with a 1000-foot towed array, intercepted a Soviet Yankee SSBN in the Norwegian Sea and trailed it for fifty-one hours, before losing contact on 19 March during a severe storm. On 21 March, firm contact was re-established in the Iceland–Faroes Gap after a US P3 Orion Maritime Patrol Aircraft was dispatched from the US Keflavik airbase in Iceland, to home in on a SOSUS contact. Once the *Batfish* regained contact it trailed the Soviet Yankee for a staggering forty-four continuous days, the longest trail of a 'Yankee' class yet conducted by a US submarine. The *Batfish* observed the Yankee travel 8870 nautical miles, including a nineteen-day 'alert' phase, much of it 1600 nautical miles from the US coast. Just outside the range of its sixteen RSM-25/R-27U missiles.[167]

OPERATION 'AGILE EAGLE'

A few months later, HMS *Sovereign*, under the command of Commander Richard Farnfield, surpassed the *Batfish*'s record by conducting what remains perhaps the longest trail of a Soviet submarine during the Cold War. In September 1978, *Sovereign* was ordered to track down what was then known as the 'Eastern Atlantic Victor', with a secondary aim of locating any Type II or Type III Soviet nuclear submarines in the Eastern Atlantic. Departing on 24 September with two other SSNs, one British and one American, *Sovereign*'s command team had little experience of trailing using the new towed array and forty-six hours was allocated to practise trailing techniques. The exercise, codenamed 'Agile Budgie', was disappointing. *Sovereign* only achieved six hours of tracking before it came to an end on 28 September.[168] The next day, *Sovereign* began Operation 'Agile Eagle' and arrived at her designated patrol area, over 28,800 square miles of ocean, 200 miles southwest of Rockall. For the next five days, *Sovereign* and the two other SSNs searched the Eastern Atlantic for the Victor, but they found nothing. On 4 October, the other Royal Navy SSN, operating in a nearby area, concluded its patrol and *Sovereign* was allocated an additional 24,000 square miles to the east. The same day, SOSUS updates indicated that a homeward bound 'Yankee' class SSBN was closing from the southwest. *Sovereign* manoeuvred in to establish contact, but again failed to intercept the Soviet submarine.

Two days later, on 6 October, a further intelligence report indicated that a 'Delta' class SSBN was operating 190 miles to the west of *Sovereign*'s position. At 2330 that evening *Sovereign* located the 'Delta' and spent three days slowly closing the range. However, just as she approached within forty miles of the 'Delta' the complex computer equipment that interpreted the data from the towed array crashed and it took thirteen hours for *Sovereign*'s crew to get the system working again. During those critical hours, without information from the towed array, Farnfield was forced to switch to *Sovereign*'s far less capable hull-mounted sonar. *Sovereign* was unable to regain contact. On 10 October, a UK Maritime Patrol Aircraft indicated that the Delta was continuing along a southwesterly course at a steady speed of 7 knots, and a few hours later a US Navy P3 Orion Maritime Patrol Aircraft reported a firm contact 290 miles to the east. Farnfield ordered an intercept course and *Sovereign* sprinted 320 miles at an average speed of over 25 knots to re-establish contact with the Delta at 0845 on 11 October, at a range of 63 miles. SSIXS had dramatically improved the way intelligence was communicated to *Sovereign*. SOSUS information, as well as real-time intelligence, could now be transmitted in time for it to be useful during operations. According to Roger Lane-Nott:

> In 1976 when I was in fleet headquarters running special submarine operations, we reckoned real time was twenty minutes. That was the time in which you got a piece of intelligence, whether it came from an aircraft or some other form of surveillance, and you were able to turn that around and relay it to our submarines. Remember, we were transmitting on teletypes on very-low-frequency radio to submarines at that time. Assuming they were at periscope depth or had a wire out to receive it, twenty

minutes was a pretty amazing timescale to get a piece of information that the submarine could use when they were chasing another submarine.[169]

Farnfield was able to inform headquarters that he was trailing the Delta at 1529 on 11 October and an acknowledgement was received on VLF some forty minutes later, 'a very impressive example of the effectiveness of the SSIXS Communications system,' he wrote.[170]

Sovereign settled in to the trail of the Delta some 650 miles off Cape Finisterre. It continued southwards until 16 October, when it altered course to the southwest, and then again southwards. On the morning of 20 October, Farnfield broke off the trail temporarily and ordered *Sovereign* to periscope depth in order to transmit and receive signals. At 1043, after *Sovereign* had dived back down to continue the trail, the Delta unexpectedly began to clear its stern arcs, by conducting a routine manoeuvre practised by all nuclear-submarine commanders in order to ensure that they were not being followed by another submarine. In Royal Navy submarines, the Type 2001 sonar only provided main sonar coverage about 120 degrees either side of the submarine's bow. This meant that 120 degrees through the stern and 60 degrees on either side were not effectively covered by broadband sonar. In order to check these blind spots submarines would slow down and clear stern arcs at infrequent intervals. In the Royal Navy this manoeuvre was carried out in two ways. The first was to swing 70 degrees one way, sweep the arc for contacts before swinging 140 degrees the other way, clearing that arc before resuming course. This was often used when the submarine was on a fast passage and did not want to lose speed. The second method involved slowing down and swinging 140 degrees either to port or to starboard, clearing both stern arcs before resuming the original course.

Soviet submariners practised another method, the so-called 'Crazy Ivan', which involved reversing course, swinging around 180 degrees and hurtling back down their original path at full speed while searching for any trailing submarines using active sonar. This was exciting and testing for the trailing submarine, and Command teams had to take swift action in order to avoid a close-quarters scenario, counter-detection or, even worse, a collision.[171] As the Soviet Delta cleared stern arcs and proceeded to head back down its previous course, it passed approximately 800 yards down *Sovereign*'s starboard side. Despite the extremely close range, *Sovereign* remained undetected. There was a complete lack of subsequent evasion or reaction by the Delta and it quickly resumed course. This was not unusual. The Royal Navy enjoyed such superior sonar advantage over Soviet submarines that throughout the 1970s its own boats were often able to conduct their patrols without being counter-detected. The counter-detection capability of the 'Victor' class on a Royal Navy 'Valiant' class was estimated as a factor of 0.5 or less.[172] Given the improved silencing techniques used in the 'Swiftsure' class, *Sovereign* most likely enjoyed a far superior capability over the Delta.

By 25 October, the Delta was some 400 miles west of the Cape Verde Islands, which lay 350 miles off the coast of West Africa. Back in the United Kingdom, news of the trail was circulating throughout Northwood. Both the Commander-in-Chief Fleet and Flag Officer Submarines informed *Sovereign* via personal signals that the patrol, which had been due to end on 3 November, was being extended by forty-two days, and would probably end no later than 15 December. 'This was because of the exceptional interest being shown on both sides of the Atlantic in the Delta's unusual deployment,' wrote Farnfield. After the initial shock had worn off, *Sovereign*'s crew reacted in a relatively philosophical manner. Everyone on board was inconvenienced in one way or another; one crew member was due to be married in November and four ratings had paid for their Christmas holiday in Miami. Food rationing was introduced, which resulted in a bread and soup lunch, no choice for dinner, and progressively lighter breakfasts. The Chefs carefully designed the menu around remaining stocks, calorific values and presentation. Milk, sugar and coffee were steadily reduced and canteen stores soon ran out, with matches and toothpaste the two most needed items.

The patrol was now so important that *Sovereign* was assigned a new patrol area, the entire Atlantic south of 10°N. But just as she was settling into the trail, events took a turn for the worse. On 26 October, *Sovereign*'s unreliable communications mast became defective. With no second mast with a UHF capability, Farnfield was forced to reverse *Sovereign*, open the range up to 40 miles and surface to repair the mast. After seventy-five minutes on the surface, the defect was remedied and communications were re-established. *Sovereign* dived and sprinted to restore contact with the Delta, which had once again altered course. When contact was made again, the Delta was conducting what *Sovereign*'s crew assumed was a bottom contour survey of the seabed over an area within a radius of 10 miles. This was valuable intelligence, as it appeared the Delta was carrying out a survey so that the position could be used as an SSBN reference point for bottom contour navigation. This proved to be the southernmost point of the patrol, as the Delta soon turned 180 degrees and headed north.

By 2 November, *Sovereign* had continuously (with one short interruption) trailed the Delta for twenty-two days. She remained in contact for another ten, after which the patrol entered its most difficult period as acoustic conditions deteriorated when a Force 6–7 gale chopped up the seas. On 16 November, *Sovereign* unintentionally passed between 3000 and 4000 yards from the Delta. As conditions continued to deteriorate, contact with the Delta was lost at 1323 on 20 November. *Sovereign* then spent the next three days searching, without success, attempting to determine the Soviet submarine's likely route. It looked as if the Delta would cross the George Bligh Bank, around 100 miles northwest of Rockall, but after searching for twenty-two hours *Sovereign*'s crew concluded that the Delta had chosen a different track. By reviewing all the available intelligence, as well as the alternative routes available, Farnfield concluded that the Delta would probably keep to the west of the Iceland–Faroes Gap. *Sovereign* moved seventy miles southeast of Iceland and waited.

Sovereign had now been on patrol for over two months. To pass the time the various available distractions became very well tried, including quizzes, card games, domino competitions, films and reading, although Farnfield complained that *Sovereign*'s 'library was substantially comprised of paperbacks which were rarely man enough to endure the re-reading incurred' and he later urged that it was 'imperative to put the procurement of a ship's library onto a firmer basis'.[173] *Sovereign* had also developed a number of defects, the most serious of which concerned atmosphere control. High freon levels had started to build inside the submarine and by 26 November there was insufficient nitrogen pressure to start up *Sovereign*'s second electrolyser after it had been stopped to repair a defect (the other had been defective early in the patrol). For three days Farnfield was forced to bring *Sovereign* to the surface for forty minutes and ventilate to keep the amount of oxygen at a reasonable level.

After she had waited patiently for twenty-two hours, the gamble paid off. At 1103 on 24 November *Sovereign* regained contact with the Delta and after careful manoeuvring resumed the trail, following the Soviet submarine from 15–20 miles away right into the Arctic Circle, observing it alter course to the northeast and remaining in contact as it loitered in the Barents Sea. In order to avoid ventilating while trailing near Russian waters, *Sovereign*'s crew started to burn the hundred emergency oxygen candles carried on board. As the Delta crossed longitude 25°E in the Barents Sea, *Sovereign* broke off the trail after receiving orders from the UK and withdrew to the southwest. She continued through the Shetland–Faroes Gap to the west of the UK and surfaced in the southwest approaches on the evening of 5 December. A day later, at 0930 on 6 December, she berthed alongside at Devonport, concluding what Farnfield described as 'a most challenging, testing, wearying and successful operation'.[174]

From initial detection on 6 October until the trail was broken on 1 December – a period of eight weeks – HMS *Sovereign* travelled 10,724 miles and remained in contact with the Delta for a total of forty-nine days, forty of which were spent continuously in the trail. The Royal Navy now held the record for the longest trail of a Soviet submarine. 'It was hard work,' wrote Farnfield, 'and for all the Forward Watchkeepers who were keeping watches one in two for over seventy days it was physically and mentally tiring, particularly for the Officers and Ratings most closely involved in the trail.'[175] The Patrol was also important for Northwood. 'This proved once and for all that [CTF] 311 was totally capable and set up ready to do this and could support any type of operation,' recalled Lane-Nott, 'the fact that we had the communications there and the intelligence and weather people upstairs and 345 down the road, we had the ability to share all the information, that was very important.'[176]

It was now clear that a Royal Navy SSN, fitted with the new towed array and narrowband frequency analysis equipment, was capable of maintaining a trail on a quiet Soviet nuclear submarine for a sustained period of time. 'The passive sonar towed array ... may well be the most important single development in ASW sensors since 1945,' commented Rear Admiral Hill in a 1984 assessment of ASW.[177] *Sovereign* obtained valuable electronic and acoustic intelligence, particularly about how a Soviet commander operated a 'Delta' class SSBN while it was on patrol. Perhaps most importantly, the Royal Navy and the US Navy had now obtained a detailed track of a patrolling Soviet 'Delta' class. They knew the speed and depth at which it operated, where and when it adjusted course (by 40 degrees every 24–36 hours using a long-leg zig-zag), when the Soviet commander carried out a check of his stern arcs (at least every four hours, sometimes as often as every two hours), and when the submarine came up to periscope depth (between three and six times daily), as well as the various navigational methods employed (bottom contours for navigation). These patrol observations were not treated in isolation. They were fed into an overall picture that the West was building about Soviet naval movements and tactics and in particular about Soviet SSBN operations.

ASSESSMENTS OF THE SOVIET NAVY

By the late 1970s, that intelligence picture had allowed the West to assess the strengths and weaknesses of the Soviet submarine fleet as a whole. Despite its weight in numbers there were many differences between the Soviet and Western philosophies and approaches. In October 1978, the MOD's Sea/Air Warfare Committee conducted an audit. The Soviets' strengths were:

a. <u>Command and Control</u>. The Soviet Navy has developed a centralised command and control system, with secure and rapid communications, enabling close control to be exercised over submarines at short notice . . .

b. <u>Satellite Targeting</u>. Soviet reconnaissance satellites, using radar, EW and photography, make it progressively more difficult to conceal the presence of multi ship formations at sea. They still have a major data handling and collation problem to solve, but if they can do this they will have a command and control system offering a much increased ability to direct their submarines. It is now a realistic Soviet aim to develop real time satellite targeting for their SSBN/SSG missiles.

c. <u>Speed</u>. Soviet design has put a premium on speed. With the top speed of the VICTOR at 32 knots and the YANKEE/DELTA at 27 knots most Soviet nuclear submarines have a speed advantage over their Western opponents.

d. <u>Resistance to attack</u>. Soviet submarines are designed with double hulls. Together with their thick deep diving pressure hulls, this considerably reduces their vulnerability to impact and proximity explosions.

e. <u>Anechoic coatings</u>. Most Soviet submarines are fitted with either an 8cm or 10cm anechoic coating which may in some conditions reduce the homing range of active torpedoes by up to 40% and the detection ranges of active sonar by up to 30%.

f. <u>Anti Ship weapons</u>. Soviet submarines have effective and reliable anti ship missiles and torpedoes with large warheads.

g. <u>Minelaying</u>. All Soviet submarines have the capability to lay mines.[178]

Despite these apparent strengths, Soviet submarines had a number of weaknesses that could be exploited by Western submarines:

a. <u>Noise</u>. Soviet radiated noise levels are generally higher than for corresponding Western nuclear SMs. They are estimated to be some 10 years behind the SWIFTSURE Class in noise reduction, which makes them considerably more vulnerable to detection by passive systems.

b. <u>Sonar</u>. Sonar and noise reduction are closely linked, giving Soviet submarine passive sonar a low level of performance. Analysis of encounters between UK and Soviet submarines indicates a sonar range advantage of 2 or 3 to 1 to the UK.

c. <u>Navigation</u>. It is believed that the Soviet SINS needs frequent independent navigation checks; that the Soviet SATNAV system is inferior to the US system; and that Soviet submarines rely heavily on bottom contour navigation which necessitates the use of echo sounders liable to detection.

d. <u>Communications</u>. Although the Soviet command and control is generally a strong point, Soviet submarines do not appear to use floating wire aerials and are only just beginning to use communication buoys. This equipment deficiency constrains the submarines to frequent periods at periscope depth. Furthermore the rigidity of the Command and Control system, particularly its probable requirement for exchange of communications during tactical missile firing, may offer detection possibilities. Soviet information transfer systems have, of course, the same inherent vulnerability to jamming as similar Allied systems.

e. <u>Training</u>. 60–70% of Northern Fleet submarine ratings are conscripts serving only 3 years. It must be a problem of some magnitude to produce and sustain the high operating standards needed in nuclear submarines; and this tends to be borne out by the simple nature of many exercises carried out with the minimum of freeplay.

f. <u>Experience and Tactical Proficiency</u>. Some Soviet SMs have recently demonstrated an improvement in tactical expertise, but with only 15% of any class on patrol at one time and few major Fleet exercises, general operational competence is not high by Western standards.[179]

The overall intelligence picture also led to some troubling conclusions. Throughout the 1970s, the US and UK developed just one type of nuclear submarine each: the 'Los Angeles' class and the 'Swiftsure' class. In the same period, the Soviets deployed eight different classes, many of which departed from mainstream design. One of the most advanced was the Project 705, later designated the 'Alfa' class, whose appearance puzzled Western intelligence agencies. In the UK, the JIC was certainly aware of the existence of what a 1972 assessment of the Soviet Navy described as the 'A' class SSN and that it had been delivered to the Northern Fleet. But the 1972 assessment stated that the 'Alfa' had 'not yet been evaluated' and that it was 'expected to have a performance better in some aspects, than the V'.[180] The United States mounted a considerable intelligence effort against the submarine. Many analysts refused to believe that the Soviets had departed from a basic philosophy that underpinned Soviet submarine development, one based on steady evolutionary advances in design and steadily improving tried technologies. The West believed that when it came to submarine design, the Russian's adhered to a basic philosophy: 'to build and create things good enough to do what they were meant to do was considered wise; to make them better than necessary was a waste of energy and precious resources'.[181] Indeed, an old Russian proverb, *Better is the enemy of good enough*, was apparently inscribed on a plaque on the desk of the chief architect of the Soviet Navy, Admiral of the Fleet Sergey Gorshkov.

While the Victor was a relatively conservative design, a direct evolution of the first-generation Project 627, 'November' class, SSN, the Alfa marked a radical departure from previous Soviet submarines. Using a liquid-metal reactor plant instead of the PWR systems common to both US and RN submarines, the 'Alfa' was a small, high-speed ASW submarine designed to seek out and destroy Western missile and attack submarines. Highly automated, with a crew of just thirty, the Alfa had a streamlined, visually striking teardrop-shaped hull designed to maximize underwater speed; and fabricated not out of steel but titanium, which allowed the submarine to dive to depths of 1300 feet (400 metres) – well out of range of both UK and US torpedoes. The first of class was laid down in 1968, just as the 'Victor' class SSNs entered series production. The first Alfa, K-64, commissioned into the Soviet Navy in December 1971. However, severe production and engineering problems plagued K-64 and in 1972 the submarine suffered a major reactor incident when the liquid metal in the reactor's primary coolant hardened. The submarine was taken out of service and cut in half and the next of class did not appear until 1977, with a further five completing between 1978 and 1981.[182]

As both US and UK intelligence communities looked ahead to the mid-1980s and beyond they assessed that although the Soviet Navy was not expected to increase in numbers of ships and submarines, its potential for war and its political effect worldwide would increase significantly. Intelligence assessments indicated that the Soviets would develop a ballistic missile and a submarine to launch it with capabilities as close to those of the US Trident programme as they could achieve. Such a submarine, known as the 'Typhoon', would probably appear around 1980 and by 1990 constitute around 18 per cent of Soviet SSBNs. The 'Yankee' class, which would be twenty-five years old by 1990, was expected to be replaced by a new class of SSBN. The Soviets periodically updated the Delta design, evolving the basic Project 667B into the Project 667BD, 'Delta II' SSBN. Between 1967 and 1977, the Soviets produced 56 SSBNs: 34 'Yankee' class; 18 'Delta I' class; and 4 'Delta II' class.[183] In 1972, they also started work on the 'penultimate manifestation' of the 'Delta' design, the Project 667BDR 'Delta III': the first Soviet SSBN capable of carrying the first Soviet SLBM with Multiple Independently Targeted Re-entry Vehicles (MIRV): the SS-N-18 Stingray as it was known in NATO. This powerful new missile was capable of delivering a single warhead to a target 4320 miles away; or three or seven MIRV warheads up to a range of 3500 nautical miles.[184]

Assessments also indicated that by the end of the 1970s the Soviets intended to deploy a new class of nuclear-powered cruise-missile-firing submarine, equipped with new submerged-launched anti-ship missiles with a range of up to 100 miles. Assessments also indicated that by the mid- to late 1980s, due to the phasing out of the by then obsolete 'Echo II' and 'Juliett' class SSGNs, a further new class of SSGN could also enter service. The Navy expected these two classes to constitute around 35 per cent of the Soviet SSGN fleet by 1990.[185] As for SSNs, assessments concluded that a new class of SSN, codenamed SSNX-1 with improved ASW capability, a reduction in self-noise, sonar improvements and a longer-range weapon system, could be expected to enter service in the early 1980s. Assessments also correctly concluded that at least one and probably more units of the 'Alfa' class were still under construction. An entirely new class of highly capable attack submarine, codenamed SSNX-2, was also expected to enter service by the end of the 1980s, and along with SSNX-1, both classes were expected to constitute up to 39 per cent of all Soviet SSNs by 1990.[186]

The Navy's assessments cautiously concluded that the Soviet Navy would continue to lag behind Western sonar performance. Despite speculative assessments there was little technical evidence that suggested the

Soviets had an effective noise reduction programme. Evaluations indicated that the new Soviet SSN classes would only have a noise signature that was comparable to the 'Valiant' class as it was in 1977. Indeed, the new 'Alfas' were reportedly 'noisy as a freight train and could be detected thousands of miles away'.[187] However, intelligence assessments acknowledged that it was 'reasonable to assume that, given the inclination, the Soviets could possess the technical capability of achieving a steep noise reduction with their new submarine classes which could compare' to future Royal Navy submarines.[188] It was predicted that by 1985 the Soviets would be likely to achieve a small improvement in counter-detection ranges, which would have an important effect on Royal Navy submarine philosophy as its putative attacks depended heavily upon attaining tactical surprise for the subsequent launch of quiet but relatively slow torpedoes. Any extensive noise reduction programme was, however, expected to result in a weight penalty that would adversely affect the speed and depth of Soviet designs.[189]

The Royal Navy remained cautious but confident that it would continue to enjoy superiority over Soviet submarines, but it acknowledged that:

> The general increase in Soviet ship or weapon effectiveness across a broad spectrum of activities will pose new problems to our forces. Improved sensors, either sonar or non acoustic systems, backed up by anti-submarine weapons with greater range and improved homing techniques will extend the area within which it is dangerous to approach the enemy, enhanced noise reduction will reduce our own detection capability, and anechoics and advanced decoys or countermeasures will degrade our own sensor and weapon effectiveness.[190]

While the Royal Navy was still occupied with the Soviet threat, it became involved in the first of many operations in the South Atlantic, thousands of miles away from Cold War waters, against an entirely different enemy.

7

Hot War: The Falklands Conflict

For my generation of submariners it was everything we would expect to have done, except it was the wrong enemy. My generation had been totally brought up on the Russians and the Soviets and it was quite extraordinary that my generation had spent all its submarine career, all thirty years of it, dealing with the Russians, but the only thing we've sunk is an Argentinian. I find that very ironic.

Roger Lane-Nott,
CO, HMS *Splendid*, 1979–83.[1]

When the dust has settled on the Falkland Islands Campaign it will be seen that the single most significant Naval Event, after the arrival of the Task Group itself, was your sinking of the Cruiser BELGRANO. That action brought the Argentinian Navy up with a round turn and sent it scurrying to the Twelve Mile Limit, there to stay for the duration while we got on and fought the Air War. That cool and determined attack was typical of your whole patrol. Bon Voyage. Take a well earned break.

Rear Admiral Woodward to
HMS *Conqueror*, 16 June 1982.[2]

OPERATION 'JOURNEYMAN'

In December 1976, the Royal Navy's Antarctic ice patrol ship HMS *Endurance* discovered that an Argentinian detachment consisting of approximately twenty men, apparently in military uniforms and led by a major, had occupied and established a weather/scientific station on Southern Thule Island, a Falkland Islands Dependency. Amid protests from the Foreign Office, the Argentinian Foreign Minister told the British Government that the Argentinian detachment would leave the station at the onset of the Antarctic winter in March/April 1977. In June 1977, the British and Argentinian governments opened negotiations over the future of the Falkland Islands, with a first round of exploratory talks in Rome, followed by a second and third round of ministerial talks in November and December 1977 in New York.[3] The then Foreign Secretary, David Owen, was concerned about the hard-line attitude adopted by the Argentinians over the future sovereignty of the Falkland Islands and the possibility that the final round of talks in December would either break down or end in deadlock. Owen asked the Cabinet Office's Joint Intelligence Committee for an assessment of the possibility of direct Argentinian action against the Islands. Although the JIC assessed that a full-blown invasion of the islands was unlikely, it concluded that action against British shipping in the surrounding waters was a more likely response. The British Government eventually adopted a policy of 'non-provocative preparedness' and towards the end of November 1977 dispatched a small naval Task Force to the South Atlantic to protect British shipping and against the eventuality of an Argentinian invasion. The Task Force consisted of two frigates, HMS *Phoebe* and HMS *Alacrity*, and two Royal Fleet Auxiliary vessels, RFA *Olwen* and RFA *Resurgent*. This surface force would remain at a considerable distance from the Falkland Islands to avoid provoking the Argentinians and triggering a serious incident.[4] An accompanying SSN would move in closer to the Falkland Islands, where it would remain hidden and be called on should it be required.[5]

In November 1977, HMS *Dreadnought*, under the command of Commander Hugh Michell, was ordered to interrupt a planned eight-month deployment to Australia and divert to Gibraltar. Arriving at 1200 on 23 November, Michell received orders to embark stores for seventy-five days and to deploy a full 'war load' of torpedoes. In order to accommodate the extra stores and weapons, *Dreadnought*'s crew was forced to offload large amounts of the extra naval stores and personal gear that they had previously stowed for their Australian deployment. The second and third decks of the submarine were crammed with extra stores – tins layered on top of tins – a process known as 'false decking', and the twenty bunks in the upper level of the *Dreadnought*'s fore-ends, which were usually allocated to ratings, were stripped out in order to accommodate the thirteen extra Mark 8 torpedoes. This had severe consequences for *Dreadnought*'s company as it limited the number of available bunks to just 90 and Michell was forced to reduce his normal sea-going complement from 113 to just 99. *Dreadnought* was provisioned with so many extra stores and torpedoes that a check-trim dive of the submarine, carried out shortly after it left Gibraltar, revealed that it was impossible to achieve a state of neutral buoyancy at periscope depth, as the submarine was 4000 gallons too heavy. This left little if any flexibility for any ocean density changes that *Dreadnought* might encounter while sailing to the South Atlantic.

Seventy-two hours after arriving in Gibraltar, *Dreadnought* departed for the South Atlantic to begin what would amount to nine weeks of covert surveillance off the seaward approaches to the Falkland Islands, while HMS *Endurance* and an accompanying RFA, *Cherryleaf*, remained in Port Stanley and the two frigates and two Royal Fleet Auxiliaries remained in a holding area 1000 miles to the northeast of the Islands.[6] There is still a considerable debate about whether the deployment of the Task Force was an exercise in deterrence or merely a precautionary measure. David Owen, Foreign Secretary at the time, later claimed that *Dreadnought* was operating under specific Rules of Engagement that stated: 'if Argentine ships came within 50 miles of the Falkland Islands and were believed to have displayed hostile intent, the submarine was to open fire'.[7] Other accounts have argued that Michell was under orders to find an Argentinian vessel and deliberately expose *Dreadnought* in order to demonstrate to the Argentinians that a Royal Navy submarine was in the area and intended to defend the Islands.[8]

In fact, *Dreadnought* was under very clear orders 'to establish a presence in the area of the Falkland Islands and their Dependencies to protect British lives and property by deterring or countering Argentine aggression'. *Dreadnought* was issued with Rules of Engagement that were based on the minimal use of force and in the event of deteriorating relations between the UK and Argentina any vessels approaching the Islands were to 'be asked to identify themselves and to state their intentions'.[9] *Dreadnought*'s First Lieutenant, Martin Macpherson, is also 'absolutely categorically certain that we were not told to expose ourselves'. When *Dreadnought* arrived 4–5 miles off Port Stanley on 12 December, Michell ensured that the submarine remained at periscope depth, prosecuting and evaluating any nearby contact that approached the islands. There was no Argentinian naval activity or local fishing boat sightings and over a five-day period *Dreadnought* only detected a total of nine ships. There was 'nothing untoward at all', recalls Macpherson, 'by itself it was a deadly dull patrol'.[10] The submarine was detected, but not by the Argentinians. During a practice covert panoramic photograph sweep on the RFA *Cherryleaf*, *Dreadnought* was sighted by the well-trained lookouts who had just completed an intense anti-submarine exercise off Portland. Around forty hours later, a signal arrived from Northwood asking: 'Was this you?' 'We'd clearly caused a bit of a stir,' recalls Macpherson.[11]

Questions have also been raised about whether or not the Argentinians were directly or indirectly informed about the presence of the Task Force. Clive Whitmore, Margaret Thatcher's Principal Private Secretary, said: 'Argentina did not know of this action at the time.'[12] However, when the former Prime Minister, Jim Callaghan, appeared before the post-Falklands Franks Committee in 1982, he was unable to say whether the existence of the Task Force was conveyed to the Argentinians or not: 'I do not know and therefore I think we should presume that it was not conveyed to them, although I do not know. I don't know, that is the simple answer,' he insisted. But he did admit that 'my own belief is that they did know'.[13] Callaghan expanded on this point in his 1987 memoirs by recalling a discussion he had with the then Chief of the Secret Intelligence Service, MI6, Maurice Oldfield:

> I told Maurice Oldfield of our plans to send a naval task force to stand off the Falklands at a discreet distance, but I did not make a direct request to him to inform the Argentineans [*sic*] of our purpose. We discussed the future prospects against the background of possible hostile action by Argentina, and I am clear that he understood that I would not be unhappy if the news of our deployment reached the Argentinean Armed Forces. I did not question Maurice Oldfield subsequently about what action he had taken so that remains speculative.[14]

In his memoirs, Lord Owen did 'not believe that Maurice Oldfield . . . would have disclosed the naval deployment as a result of a discussion with the Prime Minister, at least not without talking to me first'. He did admit that 'in some delicate areas' Oldfield would have been 'entitled to respond only to the Prime Minister' but did 'not believe this was one of them'.[15]

According to the official historian of the Falklands conflict, Sir Lawrence Freedman, 'there were no indications at the time to suggest that "C" did anything as a result of this conversation or that Argentina was aware of this deployment or allowed it to affect its behaviour'.[16] However, there is some evidence to suggest that Oldfield did, in fact, take matters further. Had MI6 wanted to inform the Argentinians that a naval Task Force – in particular an SSN – was operating close to the Falklands, one of the most appropriate and credible means of doing so would have been through Navy-to-Navy channels. The British Naval Attaché in Argentina at the time was a Royal Navy officer named Daniel Leggat. Leggat allegedly discreetly informed his Argentinian counterpart about the presence of the Task Force. Interviews conducted in 1992 for a television programme told of a 1977 conversation between Admiral Juan José Lombardo, Commander at the time of the Argentinian Navy's submarine force, and Admiral Jorge Anaya, the then Fleet Commander of the Argentinian Navy, in which Anaya asked Lombardo if Argentina's new German-built diesel submarines could find and attack a British SSN, to which Lombardo replied 'No.'[17]

Whatever the truth, Owen was 'very grateful to the Royal Navy for mounting this operation so quickly'.[18] The negotiations with the Argentinians continued and the Callaghan Government successfully 'avoided any immediate risk of dangerous confrontation with the Argentines'.[19] *Dreadnought* returned to Faslane in March 1978 with a vast amount of information about the communication and oceanographic conditions in the waters surrounding the Falkland Islands, which in almost every respect were different to those in the North Atlantic. Her crew were ordered not to reveal where they had been. When the officers walked into the Wardroom at Faslane they were met by a jubilant group of wives wearing T-shirts emblazoned with 'HMS Dreadnought Magical Mystery Tours'.[20] Negotiations between Argentina and the United Kingdom continued over the next two years as the two governments attempted to find a solution that met the concerns of both the Argentinians and the islanders, the most promising of which involved a proposal to transfer sovereignty to Argentina, which would then lease the Falkland Islands back to the United Kingdom. However, support for lease-back faltered. In June 1981, as part of its Defence Review, the Thatcher Government announced that the Royal Navy's Antarctic patrol ship, HMS *Endurance*, the most important symbol of the UK's commitment to the Falkland Islands, was to be withdrawn from service. In December 1981, a new junta took power in Argentina led by the commander of the army, General Leopoldo Galtieri. On 18 March 1982, a party of Argentinian scrap metal dealers landed on another Falklands dependency, South Georgia, where they raised the Argentinian flag and damaged property belonging to the British Antarctic Survey. Six days later, on 24 March, a detachment of Argentinian

Marines landed on South Georgia and the Argentinian junta secretly advanced plans for a full-scale invasion of the Falkland Islands.

OPERATION 'CORPORATE'[21]

The Royal Navy had foreseen the possibility of using submarines in a scenario in which the Argentinians invaded the Falklands. In December 1967, during the Healey-initiated studies to determine the size of the nuclear submarine fleet, one such scenario offered a remarkable glimpse into the future:

> In the later 1970s, after a prolonged diplomatic stalemate, an Argentinian party – which is thought to have covert official backing – lands and occupies the Falkland Islands. Their success forces the hand of the Argentinian Government. The United States vacillates. HMG decides it cannot accept the Argentinian action without protest, partly because of the casualties which occurred during the incident. Argentinian sterling balances are frozen, a diplomatic initiative in the United Nations and Washington is set in train but, in addition, it is decided to reoccupy an island in the group as a gesture before allowing the whole affair to go to arbitration. A force is therefore required which can undertake this duty; and also to provide appropriate reconnaissance.[22]

Ten years after this scenario was written, it looked as if it could become a reality.

The Argentinians were certainly aware of the importance of the Royal Navy's nuclear submarines. On 26 March 1982, it was widely reported in the UK press that HMS *Superb*, under the command of James Perowne, had sailed from Gibraltar. The Argentinian press assumed that *Superb* had sailed for the South Atlantic in response to the events in South Georgia and the MOD was happy to neither confirm nor deny the accuracy of the reports.[23] 'We never disabused them of that idea,' wrote Sir Henry Leach, the Royal Navy's First Sea Lord, 'because it did not profit us to do so.'[24] However, in the minds of the Argentinian junta, the perceived deployment of a Royal Navy nuclear submarine to the South Atlantic confirmed the need to press on with the invasion as soon as possible: 'it did no more than confirm a decision already made'.[25] In fact, *Superb* had sailed north, to pass west of Ireland before proceeding to the Shetland–Faroes Gap to conduct Operation 'Sardius', in order to detect and track Soviet submarines

northwest of the United Kingdom.[26] A 'Victor II' submarine had
deployed on 15 February and was sighted on the surface by Maritime
Patrol Aircraft on 20 March. *Superb* was tasked with conducting surveillance against the Victor and to establish whether it had interacted
with another SOSUS contact, a possible Victor III which arrived in the
vicinity of Porcupine Bank, an area off the continental shelf approximately 120 miles west of Ireland, on 22 March and then moved into the
Royal Navy's Northern Fleet Exercise Areas.

Superb sailed from Gibraltar on 26 March and quickly detected the
Victor II. It manoeuvred onto its starboard quarter as it moved slowly
north. For five days, *Superb* remained in contact with the Soviet submarine before handing over responsibility to an RAF Nimrod. *Superb*
moved off to the south, to the Northern Fleet Exercise Areas to search
for the Victor III, which was now thought to have penetrated the Clyde
Exercise Areas.[27] Perowne hoped to intercept the Soviet intruder as it
withdrew through the North Channel, between Kintyre and Northern
Ireland, but it proved elusive. *Superb* then detected and trailed a Soviet
'Delta' class SSBN that was transiting to either the Mediterranean or
the South Atlantic, collecting intelligence, and broadband and narrowband acoustic recordings, before covertly returning to Faslane in the
dark on 16 April in order to maintain the fiction that she was in the
South Atlantic.

On 29 March, with Argentinian forces preparing to invade the Falkland Islands, the Secretary of State for Defence, John Nott, informed
the Prime Minister, Margaret Thatcher, that he had:

> today instructed that the nuclear submarine (SSN) HMS *Spartan* should
> be sailed covertly to the South Atlantic. She has been taken off exercises in
> the vicinity of Gibraltar and will proceed there to stock up with suitable
> weapons and provisions. She will sail from Gibraltar early on 31 March
> and should be in vicinity of the Falklands by 13 April. This is the quickest
> available means of deploying an SSN suitably equipped to the area. We are
> planning on the basis that a second SSN will be earmarked.[28]

In mid-March HMS *Spartan*, under Commander James Taylor, had
been off Lisbon in Portugal taking part in a naval exercise codenamed
'Springtrain' after returning from an intelligence-gathering patrol
in which the submarine had come into close contact with a Soviet
'Alfa' class SSN. Taylor received an underwater telephone call from
John Coward, the CO of HMS *Brilliant*, a frigate *Spartan* had been
exercising with. Coward told Taylor to go to Gibraltar as quickly

45. HMS *Repulse*, HMS *Revenge* and HMS *Resolution* sailing up the Clyde – a rare photograph, as the patrol cycle requires that when one submarine is on patrol, the others are in either deep or routine maintenance.

46. One of the principal tasks of SSNs was to provide protection of the strategic deterrent. Here a 'Valiant' class SSN escorts a 'Resolution' class SSBN out to sea.

47. HMS *Warspite*, the Royal Navy's third SSN. In October 1968 she hit what was officially described as an 'iceberg' but was in fact a Soviet 'Echo II' cruise-missile-carrying submarine.

48. HMS *Warspite*'s second CO, John Hervey, in 1967. His strong leadership ensured that *Warspite* returned home safely following the collision.

49. *Warspite* returns to Vickers Shipyard, Barrow-in-Furness, for repairs after the collision. Note the black tarpaulin covering the damage to the conning tower.

50. (*top left*) *Warspite*'s third CO, Sandy Woodward, 1969.

51. (*top right*) An underwater photograph of HMS *Antelope* taken through HMS *Conqueror*'s periscope.

52–53. (*middle left*) These photographs of a Soviet 'Yankee' class ballistic-missile-carrying submarine were taken through *Warspite*'s periscope. Note the individual on the conning tower in 53.

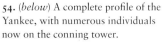

54. (*below*) A complete profile of the Yankee, with numerous individuals now on the conning tower.

55. After the completion of the Polaris programme, construction of SSNs resumed. Pictured here is the improved 'Valiant' class HMS *Courageous*.

56. In the 1960s, the Royal Navy also continued to construct diesel electric submarines. Pictured here is HMS *Oberon*.

57. (*above*) The 7th Submarine Division deployed throughout the Far East during the Indonesian Confrontation, 1963–6. Pictured here is HMS *Alliance*, complete with camouflaged paint scheme and Second World War-era surface gun.

58–60. Techniques to deploy and extract Royal Marine Special Forces from submarines were pioneered during the Indonesian Confrontation. (i) Royal Marines prepare to disembark from a submarine; (ii) Royal Marines dropping by parachute prior to being picked up by a submarine for transportation to their destination; (iii) Royal Marines prepare to deploy from an 'Oberon' class submarine in folding kayaks (folboats).

61. The faster, stealthier, deeper-diving 'Swiftsure' class SSN, introduced in the early 1970s, played a leading role in Cold War operations.

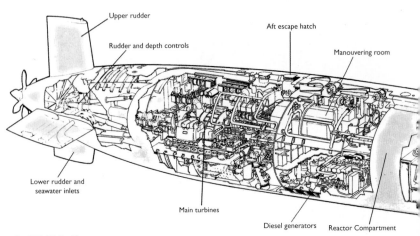

Upper rudder

Aft escape hatch

Rudder and depth controls

Manouvering room

Lower rudder and
seawater inlets

Main turbines

Diesel generators

Reactor Compartment

62. HMS *Swiftsure*.

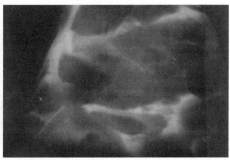

63–64. Diesel submarines did venture under the ice, but they were unable to stay dived for long. (i) HMS *Grampus* in the Arctic during Exercise 'Skua' in February and April 1965; (ii) looking through *Grampus*'s periscope while under the ice.

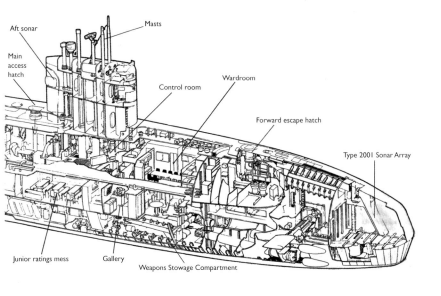

Aft sonar

Masts

Main access hatch

Control room

Wardroom

Forward escape hatch

Type 2001 Sonar Array

Junior ratings mess

Gallery

Weapons Stowage Compartment

65. (*above*) Commander Alan Kennedy took HMS *Dreadnought* under the Arctic ice and surfaced at the North Pole on 3 March 1971.

66. (*left*) In July 1971 the Submarine Service was issued with its own badge, the Dolphins.

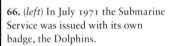

67. (*bottom left*) In the early 1970s there were worrying signs that the demands of the SSN and SSBN programmes had resulted in the neglect of the conventional submarine fleet. In July 1971 HMS *Artemis* sank while alongside at HMS *Dolphin*. She is pictured here being raised shortly afterwards.

as possible. When he asked why, Coward simply told him to 'Just go, now.'

As Taylor withdrew HMS *Spartan* from the exercise and made for Gibraltar, the Royal Navy's newest SSN, HMS *Splendid*, under the command of Roger Lane-Nott, was in the North Western Approaches, trailing a Soviet SSN. As *Splendid* came up to periscope depth to receive a broadcast and the latest intelligence update, one of its radio operators came into the Control Room and told Lane-Nott that 'There's a Blue Key message for you, sir.' Blue Key messages were heavily encrypted and could only be deciphered by the Captain. Lane-Nott retired to his cabin, where he used the special crypto keys held in a secure safe to decode and read his orders. They instructed him to 'Proceed with all dispatch' back to Faslane and store for war. 'I'd waited my entire career for one of those,' recalls Lane-Nott. 'It was wonderful.'[29]

As *Splendid* left the Victor and returned to Faslane, Taylor's HMS *Spartan* arrived in Gibraltar to take on stores for a 75-day war patrol as well as seven old Second World War vintage Mark 8 torpedoes, which were offloaded from the diesel electric submarine HMS *Oracle*. *Spartan*'s crew ransacked Royal Navy ships docked in Gibraltar for spare crypto, spare old-fashioned decoders and spare hoist wires for *Spartan*'s wireless communications mast. Taylor was given a limited briefing on events in the South Atlantic and was ordered to sail for South Georgia as soon as possible. After recalling the small number of *Spartan*'s crew that had remained in Lisbon on leave, the submarine sailed on 1 April and set course for the South Atlantic. 'We were doing 28, 29 knots solid,' remembers Taylor, 'completely deaf, completely blind, just a bullet.'[30]

In London, Mrs Thatcher was briefed by the Ministry of Defence on further military measures. She was told that:

The great advantage of using these SSNs is that their passage can remain covert until we wish to reveal it. Even if media speculation focuses on SSNs the Argentinians themselves will be left guessing. It will, of course, be necessary to draw up precise rules of engagement for the submarine Commanding Officers. While on patrol they would carry out covert surveillance and would be available to afford protection to HMS ENDURANCE. If need be, their presence could be declared in order to deter the Argentinians from any precipitate military action. In the worst case they could carry out formidable retaliation against the Argentine Navy.[31]

The section of the text that says 'The great advantage of using these SSNs is that their passage can remain covert until we wish to reveal it' was underlined by Thatcher, who scribbled in the margin that 'It was on the news this morning.'[32] Indeed, news of *Spartan*'s departure from Gibraltar made headlines after Richard Luce, the Minister of State for Foreign Affairs, who later resigned following the invasion of the Falklands, implied that the Government had dispatched a nuclear submarine after he was baited by backbenchers during a Conservative Party meeting in the House of Commons about why the Government was not doing more about South Georgia. Thatcher was 'not too displeased' about the leak. 'The submarine would take two weeks to get to the South Atlantic,' she wrote, 'but it could begin to influence events straight away. My instinct was that the time had come to show the Argentines that we meant business.'[33]

The possibility of sailing a third nuclear submarine to the South Atlantic was also discussed and, while it was identified, it was not given orders to sail because, as the Minister of Armed Forces explained to the Prime Minister, 'There would be significant operational penalties which would, among other things, adversely affect joint operations with the Americans.'[34] The Flag Officer Submarines, Vice Admiral Peter Herbert, was already deeply unhappy with how events in the South Atlantic were disrupting his regular submarine operations. He wrote in his war diary that 'with twelve scrap iron merchants creating a stir in South Georgia it is difficult to believe that it is necessary to disrupt *Spartan*'s exercises with FOF1 [Flag Officer 1st Flotilla, at that time Rear Admiral Sandy Woodward] in SPRINGTRAIN and send her to the South Atlantic as MOD requires'.[35] With only six or seven operational SSNs available at any one time, Herbert had to ensure that the crisis did not detract from the Royal Navy's prime Cold War commitments, of intelligence-gathering operations and the safeguarding of the UK strategic deterrent.[36] 'We must accept certain operational penalties as a result of the deployment of 2 SSNs to the South Atlantic,' noted an MOD brief for Nott. 'SSNs have a crucial role both in the conduct of anti-Soviet intelligence-gathering tasks, which are important for our intelligence relationship with the United States, and in safeguarding the deployment of our strategic deterrent. The deployment of a third SSN would considerably exacerbate the operational penalties.'[37] Indeed, the Navy was already complaining about the disruption caused to 'an important counter-intruder operation' – the trail of the Soviet Victor – by the allocation of HMS *Splendid*.[38]

On Friday, 2 April, eighteen hours after arriving in Faslane, Lane-Nott's HMS *Splendid* sailed for the South Atlantic. As the submarine cleared the Clyde and dived off the Isle of Arran the crew tuned into Radio 4's *The World at One* and learned that Argentina had invaded the Falkland Islands. Margaret Thatcher later said that news of the impending invasion was 'the worst ... moment of my life'.[39] Two days earlier, on Wednesday, 31 March 1982, the general consensus in Whitehall had been that if they were taken, recapture of the Islands was all but impossible. 'You'll have to take them back,' said Thatcher to the Defence Secretary, John Nott. 'We can't,' he replied, giving the MOD view that the Falklands could not be retaken once they were seized. The First Sea Lord, Sir Henry Leach, hurried to the House of Commons but the ushers in Central Lobby were reluctant to let him in. After waiting for quarter of an hour Leach was eventually rescued by a whip and taken into the Prime Minister's office. Leach believed that the Royal Navy should do everything it could to respond to the invasion and that meant assembling a Task Force and sailing it with orders to recapture the Falkland Islands. All he needed was political clearance. 'Can we do it?' Thatcher asked. 'We can, Prime Minister,' Leach said, 'and, though it is not my place to say this, we must.' 'Why do you say that?' asked Thatcher. 'Because if we don't do it, if we pussyfoot ... we'll be living in a totally different country whose word will count for little.' This was what the Prime Minister needed to hear. 'Before this, I had been outraged and determined,' wrote Thatcher in her memoirs. 'Now my outrage and determination were matched by a sense of relief and confidence. Henry Leach had shown me that if it came to a fight the courage and professionalism of Britain's armed forces would win through. It was my job as Prime Minister to see that they got the political support they needed.'[40] As Thatcher's official biographer has written, 'Her instincts told her to fight, but she could not do so in defiance of all expert advice. Leach gave her the necessary countervailing expertise' and he left the House of Commons with the authority to assemble a naval Task Force.[41]

With two Royal Navy submarines already sailing towards the South Atlantic, the third nuclear submarine that had earlier been identified, HMS *Conqueror*, under the command of Commander Chris Wreford-Brown, was now earmarked for South Atlantic operations. In the early hours of 1 April, Wreford-Brown, who had only taken command of *Conqueror* three weeks earlier, received a signal from Flag Officer Submarines ordering him to store for war. As *Conqueror*'s crew prepared the submarine for departure, a bus marked *Royal Marines Sky Diving Team* containing an SBS team arrived alongside at Faslane. 'These SBS guys came in two sizes,' recalled one of *Conqueror*'s crew, 'five foot six and scrawny, or six foot seven and size twenty-five-inch neck, or no neck. They were a tough bloody crew.'[42] Their presence was so secret that they were not allowed on *Conqueror*'s casing unless it was dark and as they were so unsure of what exactly they would be called on to do in the South Atlantic, they brought all their equipment, including limpet mines, knives, handguns, rifles, heavy machine guns, plastic explosives, hand grenades, Gemini inflatables, outboard motors and skis, all of which amounted to over nine tons of equipment to be stowed in *Conqueror*'s torpedo compartment alongside the torpedoes.

Nott was 'firmly of the view that SAS/SBS deployed from SSNs will be the answer to winning back the Falklands'.[43] When the Chiefs of Staff met on the morning of 4 April, Leach informed them that preparations were under way to embark Special Forces, both the SBS and the SAS, on board SSNs sailing south.[44] However, the Navy argued that using submarines for Special Forces operations would detract them from their primary role of preventing Argentina reinforcing the Falklands. Instead a list of diesel electric submarines was drawn up for possible deployment and the use of SSNs was restricted to transporting SBS teams to the South Atlantic. A second SBS Squadron had already flown into Ascension Island and the possibility of diverting HMS *Spartan* to pick them up was briefly considered but later disregarded, as it would have delayed *Spartan*'s arrival off the Falklands by twenty-four hours.

HMS *Conqueror* sailed from Faslane on the afternoon of 4 April, dived in the Irish Sea and proceeded south at full power. With three submarines now speeding towards the South Atlantic, the Flag Officer Submarines, Vice Admiral Peter Herbert, known officially as Commander Task Group (CTG) 324.3, devised a policy for submarine operations in the South Atlantic. 'There were two things that drove it,' recalls Herbert. 'The first was to have barriers, north and south of the Falklands, so that the Argentineans [*sic*], particularly the 25 *May* [the Argentinian aircraft carrier] and the *Belgrano* [Argentinian cruiser], and their forces would be detected on their way towards the Task Group. The second one was to react to intelligence.'[45] Herbert's plan involved deploying three SSNs to cover a possible breakout by the Argentinian Navy from its bases on the Argentinian mainland, with a fourth SSN operating behind, acting as a sweeper and providing defence in depth.

For this Herbert considered two additional SSNs: HMS *Courageous* and HMS *Valiant*. He had intended to sail HMS *Courageous* in advance of HMS *Valiant* but *Courageous* was confined to the Admiralty's large floating dock in Faslane, receiving a much needed hull preservation and essential maintenance programme, following a 26,000 mile, 302-day deployment. Despite the best efforts of the *Courageous* crew, the maintenance programme took far longer than expected and Herbert was forced to send HMS *Valiant* instead. HMS *Valiant*, under the command of Commander Tom Le Marchand, was on a deep and fast passage across the Atlantic when the 'Prepare for War' signals arrived. At first Le Marchand thought he 'could go straight down' to the South Atlantic, but his submarine had been at sea for two months. There was not enough food on board. 'As ever the human machine is the limiting factor in a nuclear submarine,' he later wrote. 'So the plan evolved: get home fast, top up with stores and torpedoes, and deploy for a long trip, which in the end was to last exactly 98 days under water.'[46]

With *Valiant* on its way back to Faslane, Herbert identified an additional SSN for South Atlantic operations. There was some discussion about sending HMS *Sceptre*, under the command of Commander Doug Littlejohns, to the South Atlantic. The submarine had been en route to northern waters to conduct an intelligence-gathering patrol, but at the time of the invasion had been forced to divert to Faslane to repair a small defect. 'I went into Faslane and there was *Conqueror* and *Splendid* loading torpedoes and I went ashore and saw the Captain of the Squadron,' recounts Littlejohns. 'I said: "I'm ready, I'll go." I was stored for war, but I was told to get back in my box. Apparently there had been some discussion about sending me but [Caspar] Weinberger [US Defense Secretary] said: "Absolutely not because I don't have an American boat to fill that gap" . . . I just said: "I want to go to war" and was told: "Get in your box."'[47] *Sceptre* went north. The only other available SSN was HMS *Warspite*, which was in the final stages of a three-year refit in Chatham. The refit work was accelerated and *Warspite* sailed on 21 April for an emergency workup and the incorporation of the new Royal Navy Sub Harpoon missile.

On 5 April, with three submarines already en route to the South Atlantic, the aircraft carriers HMS *Hermes* and HMS *Invincible*, the first ships of a Royal Navy Task Force that would ultimately consist of 115 ships, set sail from Portsmouth on Operation 'Corporate'.

Submariners were intimately involved in many aspects of the operation. The overall commander of Operation 'Corporate' was the Commander-in-Chief Fleet (CINCFLEET), Admiral Sir John Fieldhouse, a submariner by profession. His command headquarters were located in Northwood, where he worked alongside the Flag Officer Submarines, Vice Admiral Peter Herbert, who commanded Task Force 324, the SSNs deployed to the south. At the operational level, the senior officer with and in command of the Task Force was Rear Admiral Sandy Woodward, another submariner.[48] One naval historian has suggested that the 'generally successful working relationship which obtained between Fieldhouse and Woodward in 1982 owed much to their earlier acquaintance as submariners'.[49] Indeed, Commodore Michael Clapp, commander of the Royal Navy's Amphibious Task Force, concluded that Woodward's experience as a submariner made him ideal to lead the Task Force: 'Sandy was clever, mathematical, analytical and decisive,' he wrote. 'The surface Navy used to call his a "periscope mentality" – as, unlike on a ship, the submarine captain is the only one to know what is going on on the surface and in the air around . . . he was an excellent fighter, and in many ways in the right place at just the right time.'[50]

As the Task Force left Portsmouth, the Chiefs of Staff were conscious that when HMS *Spartan* arrived in the South Atlantic on 11 April, the submarine would need 'clear rules of engagement or other instructions'.[51] The SSNs had sailed with a set of Rules of Engagement that placed specific emphasis on remaining covert while on passage to the South Atlantic. Once they arrived in the vicinity of the Falkland Islands they were to conduct surveillance of Argentinian forces and collect intelligence about naval movements. They were permitted to use minimum force in self-defence, and if Argentinian forces attacked the ice patrol ship, HMS *Endurance*, they were 'to return fire to the minimum extent necessary to prevent further attack'.[52] On 6 April, the Cabinet Secretary, Sir Robert Armstrong, briefed Mrs Thatcher about what should be done during 'the period when the SSNs are on station ahead of the main force, should they be confined to a reconnaissance role; should they be authorized to attack Argentinian naval forces (and, if so, within an announced area or wherever they may be?)':

> There will be an interval of nearly a fortnight between the arrival (we hope secret) on station of the first SSN and the arrival, inevitably public, of the task force. If early in that period the SSN were to sink an Argentine warship, that could have military advantages, and would be seen as

a success by domestic public opinion. On the other hand it could lead to reprisals against the Falkland Islanders; and it could turn international opinion against us, leading to resolutions in the Security Council (which we should have to veto) calling upon us not to use force, and possibly to the withdrawal of logistic support by some of those countries now providing it. These consequences could have implications for the use of the task force even before it arrived on station.[53]

Leach was worried that the Argentinian Navy was 'clearly engaged in a rapid resupply operation, and his fear was that, unless prompt action was taken, the Argentine navy might complete their resupply operation and then return to home ports'.[54] Nott tabled a memorandum in which he proposed that an SSN should launch an early attack on an Argentinian warship, essentially launching a surprise opening shot in the conflict. This action would be followed by the declaration of an Exclusion Zone in which SSNs could sink Argentinian warships as well as support vessels involved in resupplying the Falkland Islands.

When the War Cabinet, officially called the Oversea and Defence, South Atlantic committee (OD(SA)), consisting of Thatcher, Nott, the Home Secretary, William Whitelaw, the Foreign Secretary, Francis Pym, and the Chancellor of the Duchy of Lancaster and Paymaster General, Cecil Parkinson, met for the first time the next day, 7 April, Nott's proposal was considered too radical and was dropped.[55] Had it been approved the task of finding an Argentinian vessel and sinking it would almost certainly have fallen to Taylor and the crew of HMS *Spartan*. 'It was very frustrating,' recalls Taylor: 'There were targets there, there were warships and other ships flying Argentinian flags and I think the straightforward removal of one or some of them, which I would have been content to do, that's the job, I think would have saved a great deal of bloodshed later on, I really do. Had John Nott's view prevailed then it would have been the same end result, but I think there would have been rather less loss of British life.'[56]

Instead, a revised proposal was tabled in the form of an announcement of a 200-mile Maritime Exclusion Zone (MEZ) around the Falkland Islands to come into effect from midnight on 11/12 April. The next day Nott stood up in the House of Commons and announced that:

> From the time indicated, any Argentine warships and Argentine naval auxiliaries found within this zone will be treated as hostile and are liable to be attacked by British forces. This measure is without prejudice to the right of the United Kingdom to take whatever additional measures may

be needed in exercise of its right of self-defence, under Article 51 of the United Nations Charter.[57]

The War Cabinet approved 'an à la carte menu' of Rules of Engagement that had been agreed by the Chief of Defence Staff, Admiral Terence Lewin, in consultation with officials of the Foreign and Commonwealth Office, including the Legal Adviser, Ministry of Defence and Cabinet Office, from which certain rules could 'be selected and activated on Ministerial authority as and when the situation requires'.[58] The rules, which applied to *Spartan* and all other Royal Navy submarines as and when they arrived on station, were as follows:

a. Any vessels positively identified inside the Exclusion Zone as being Argentine warships, submarines and naval auxiliaries may be attacked.

b. After the first successful attack, withdraw from scene of action and report. Having reported or, if unable to clear report after 12 hours, continue patrol.

c. Situation Reports are to be made at your discretion as soon as possible after any subsequent successful attacks and on all Argentine units detected.

d. Additionally, if attacked you are authorized to retaliate as necessary for your self-defence both inside and outside the Exclusion Zone.[59]

The Prime Minister also added that 'The commanding officer of a submarine would naturally have the right to exercise his discretion in any matter relating to the safety of his vessel.'[60]

Nott later admitted that 'Nearly every decision we had to take on the rules of engagement was difficult.'[61] In the opening stages of the conflict, two aspects in particular caused problems. The first concerned the requirement to positively identify Argentinian submarines before carrying out an attack, a requirement that carried very high risks of counter-detection and attack from the small number of submarines in the Argentinian Navy. The Argentinian Navy possessed four conventional diesel submarines, two old former US Navy 'Guppy' diesel electric submarines constructed during the Second World War – *Santiago del Estero* and *Santa Fe* – and two relatively modern German-built Type 209 diesel electric submarines: *Salta* and *San Luis*. While the US 'Guppy' submarines were probably too old to be of any use in offensive operations, very little was known about the more modern 209 submarines, which were very quiet and had the potential to pose a serious threat to the Task Force as well as the SSNs operating in the South Atlantic. Lack of intelligence about other nations' submarines operating in the area, especially Soviet boats, also caused problems.[62] What should a Royal Navy submarine do if it encountered a Soviet submarine?

In this context, the incorrect press reports about HMS *Superb* were relevant. Following the invasion of the Falklands the Argentinian press continued to report that a Royal Navy submarine had been located 250 miles off the coast of the southern Argentinian city of Mar del Plata. At the time, the British Air Attaché in Washington, Air Vice Marshal Ron Dick, was chairing a meeting of the United Nations Military Staff Committee, 'the most moribund committee ever devised by man'. Before he left for New York, he:

> was told that the Buenos Aires newspapers had been headlining a report that a British nuclear submarine had been detected operating off the coast of Argentina. I knew that to be wrong, but it was good news because, if they even *thought* a nuclear submarine was in the offing it was almost as good as having one there. Back in the UN, we dragged ourselves through the motions of our dreary meeting and I then stopped to speak to my French colleague near the conference room door. The Soviet representative that day happened to be an admiral and he brushed my shoulder on his way out. He did not stop or even look at me. He just kept going through the doorway, but a question floated back over his shoulder: 'Are our submarines being of any help?'[63]

It is not clear whether this was a reference to the Soviet 'Victor' class SSN that HMS *Superb* was ordered to intercept prior to the invasion or to additional Soviet submarines operating off the Argentinian coast in the South Atlantic. The latter was very much on the minds of the Chiefs of Staff on 6 April, when a Foreign Office official, Peter Wright, asked Leach 'to what extent our SSNs would be able to distinguish Argentine from other, including Soviet, submarines'. Leach's response was not encouraging:

> He said that he could give me no watertight assurance unless submarines were to fortuitously give away, by underwater telephone conversation, which nationality they were. There might also be indications from the nature of sonar use, and he commented that Soviet submarines were likely to remain passive while the Argentines might have an operational need to transmit. In brief, it is possible, but by no means certain, that

reliable identification could be made. I said that this was a point which was likely to be of considerable interest to Ministers.[64]

The Soviets, as well as other South American states, were eventually warned to keep their submarines away from the Falkland Islands.[65]

The second aspect that caused difficulties was highlighted on 10 April, when concerns were raised that HMS *Spartan*, operating under the new Rules of Engagement and thus with authorization to attack Argentinian vessels, would arrive in the MEZ just as the US Secretary of State, Al Haig, was beginning various diplomatic initiatives. Any attack by a Royal Navy submarine could have damaging political repercussions and jeopardize any diplomatic breakthrough. Haig urged the British Government to suspend enforcement of the MEZ, and as a precaution ministers instructed *Spartan* and the other three SSNs to 'Remain covert. Carry out surveillance of area allocated. If detected evade. If unable to evade, prosecuting units may be attacked in self defence.'[66] For Lane-Nott and the crew of HMS *Splendid*, these alterations were extremely frustrating. 'We had national rules of engagement and they decided to change those,' he later said. 'I had some difficulty with that. We had a set of national Rules of Engagement and at the moment we were supposed to use them somebody decided to ditch them and produce some new ones, which I found extraordinary, because all our training had been on the other ones.'[67]

Aside from the normal recognition publications, such as *Jane's Fighting Ships and Aircraft* as well as NATO publications, the submarine crews knew little about the capabilities of Argentinian forces. 'There was virtually nothing coming out of Northwood or the Ministry of Defence as to what these people had,' remembered Taylor.[68] 'The lack of detailed information on Argentinian forces at the start of Operation CORPORATE was disappointing,' wrote Lane-Nott.[69] 'It was a very strange feeling to go out not knowing what you were going to do. I mean, we were used to that, we were flexible, that's part of submarining, getting out there and changing role three or four times, we're capable of doing that. But the information on Argentina was so small, there was precious little on the intelligence side apart from *Jane's*, I mean it got better, but at the beginning there was nothing. Which I found astonishing really.'[70] Other than HMS *Dreadnought*'s Operation 'Journeyman' December 1977 patrol report, very little was known about the waters surrounding the Falklands. Extracts from the report were studied in detail, alongside other oceanographic and navigational publications that were communicated to the SSNs as they sped south. But with so little information on the nature of the threat from Argentinian forces, the crews of the SSNs were anxious. 'It was the first time that I began to feel the burden of responsibility,' recalled Lane-Nott on *Splendid*. 'I'd never felt it before. I'd done all sorts of dangerous things, I was the Captain of a nuclear submarine and suddenly I was going to war. If I'd been going against the Russians I wouldn't have minded at all, because I knew what the enemy's capability was. I had no idea what the capability of these people was, at all.'[71]

On 12 April, HMS *Spartan* arrived in the Exclusion Zone and set course for the Falkland Islands. 'There was a sense of being on the far side of the world,' says Taylor, 'the assurance of once we got close to the Falklands that if you did get a contact it was probably the enemy. It was rather like being under the ice, if there's a noise there, it's probably theirs.' *Spartan*'s crew were good-humoured, but the possibility of a shooting war weighed heavily on those with families back home. 'It didn't really dawn on people until we heard on the media broadcast of John Field-house saying: "This will be a sad and bloody business,"' recalls Taylor. 'I said to the ship's company that this is what we're trained for, it's not a big deal, except hopefully there will be a bang at the end of it. We're just going to do what we've always done. This is the job, this is who we are, this is what we do.' Attendance at the on board Sunday church service improved, and although *Spartan* had never been a drinking submarine, alcohol was restricted. Taylor's biggest challenge was to ensure that his crew, many of whom had been at sea since January, did not get 'stale'.[72]

Taylor positioned *Spartan* off Port Stanley and then moved three nautical miles off Cape Pembroke lighthouse to conduct visual reconnaissance of Argentinian movements, taking photographs through the periscope of Argentinian aircraft taking off and landing on the nearby airstrip. *Spartan*'s crew quickly obtained a vital piece of intelligence that fed into the Task Force's plan to reoccupy the Islands. According to Northwood, the Argentinians did not possess any minelaying capabilities. However, on 15 April *Spartan* observed two Argentinian vessels, one the landing-ship *Cabo San Antonio*, laying two minefields near the entrance to Port Stanley Harbour. Taylor wanted to attack both vessels, but he was unable to sink them due to the restrictive Rules of Engagement. All he could do was watch. *Spartan* remained in the area until 21 April, providing regular updates on Argentinian activity in and around Port Stanley.[73]

HMS *Splendid* arrived in the Exclusion Zone on 15 April, and took up a position in the northwest, between the main Argentinian ports and the Falkland Islands. The Royal Navy's nuclear-powered submarines had allowed the British Government to establish a 200-mile Maritime Exclusion Zone around the Falkland Islands just ten days after they had been invaded. Argentinian naval operations in and around the immediate Falklands area virtually ceased. The Argentinians knew, as one submariner later put it, that 'the only way to know for sure that there is a submarine is when one starts losing ships – and that's a very expensive way to find out'.[74] Lane-Nott and James Taylor remained concerned about the Argentinian Navy's submarines. Although intelligence assessments suggested that at least one of Argentina's 'Guppy' class submarines had been stuck in harbour for some time and was probably non-operational, other assessments indicated that by 18 April Argentina's three remaining submarines had already been at sea for about a week, and although there were indications that both Type 209 submarines had problems with their periscopes and torpedo tubes, at least one was suspected of operating inside the MEZ, while there was some evidence that the *Santa Fe* had been dispatched to South Georgia.[75]

Lane-Nott started to prepare his crew for a possible engagement with either the *San Luis* or the *Santa* Fe as soon as *Splendid* arrived in the MEZ. He ordered two Tigerfish Mark 24 torpedoes loaded into *Splendid*'s nos. 3 and 4 tubes, which were then equalized and the bow caps opened. He also trained his crew hard. 'So we trained and we trained and we trained and we trained attack teams, we trained different types of scenarios, it was constant,' recounted Lane-Nott. 'We had intelligence, we had recognition tests, we got the blokes to make models of the masts so that people knew what they could expect so see when on the periscope.'[76] That training was put to the test almost immediately when *Splendid*'s Active Intercept sonar alarmed on precisely the frequency that a US-built 'Guppy' class submarine would use when transmitting on active sonar.

'The immediate reaction was that we've got an Argentinean [*sic*] "Guppy" diesel submarine transmitting on active sonar on us, to get a range before he fires,' said Lane-Nott. 'We got a sort of bearing, but it wasn't very good and we had pre-transmissions and it was exactly what you'd get in that situation. We went to action stations and the quiet state that was required, but I didn't really have a bearing so I didn't know where the hell the problem was coming from. Really I was waiting for the torpedo . . . I was just waiting for the torpedo. We were looking at all the things, we were shutting off for counter-attack, we were looking at what the best depth we should be at, where were we now, have the water conditions changed, all those things we were thinking about all the time, what's your evasion depth, where do you go to? It was a very, very tense time, because we just didn't know. So I changed direction and changed depth a little bit, but I basically stayed where I was and tried to keep as quiet and alert as possible and damn me if it didn't happen again. But we still didn't have a bearing. This was a very crude system as it was then, as unless you get a very strong transmission it won't give you a bearing, but the frequency was absolutely spot on. It was quite clear to me that we were in an attack situation and the ship's company got the message as well.'[77]

Splendid's crew spent the next fifteen minutes attempting to locate the source of the contact until someone realized that one of the submarine's ballast pump hull valves was making exactly the same frequency as the Argentinian submarine and triggering the alarm. 'Suddenly it became deadly serious,' Lane-Nott went on. 'They knew it before but it hadn't been crystallized in their minds, "Bloody hell, we're at war here" . . . everybody took it seriously.'[78] *Splendid*'s crew had started to split into Hawks and Doves. 'There were more Hawks than there were Doves,' said Lane-Nott, 'but there were certainly Doves and the Doves were very much the older Senior Rates, who didn't like what all this was about, they hadn't spent all their time in the Navy to be killed in the last five minutes.'[79]

As HMS *Conqueror*'s crew sailed towards South Georgia, they too reflected on the implications of what was to come. 'I thought about it on the way down there,' said *Conqueror*'s Chief Engine Room Artificer, Edward Hogben, 'I mean, when you do patrols up north the Russians would deter you from staying there by dropping various bits of armament at you. That was to get you to go away. I thought, this time, these buggers want to kill us. But, if I go, I shall be going in good company and we'll all go down together. Once you've got over that bit, I found it relatively simple to just get on with life. There's no point screaming and making a fool of yourself. That would be embarrassing. So, you know, life was normal, as normal as it can be.'[80]

Conqueror's CO, Chris Wreford-Brown, was also reflective. 'In my own heart I hoped that our Task Force would be sufficient deterrence to persuade the Argentinians to withdraw from the Falklands without a fight,' he wrote, 'but it soon became clear that we would be involved in some form of conflict.'[81]

Conqueror was part of the first significant engagement of the Falklands conflict, Operation 'Paraquat', the reoccupation of South Georgia. *Conqueror* entered the Argentinian 200-nautical-mile Exclusion Zone around South Georgia on 18 April 1982 and arrived at the island at 0630, where she conducted a sonar sweep along the coastline to check for contacts as well as to give the SBS the opportunity to see the coastline. There was little activity at the two inhabited areas on the island and Wreford-Brown decided to 'open from the coast. Set up an Anti-Shipping Patrol, knowing there is "nothing over my shoulder".'[82] As the reoccupation of South Georgia continued with *Conqueror* patrolling offshore, two helicopters from HMS *Antrim* were lost while trying to land SAS troops on a glacier.[83]

Conqueror's crew were immediately reminded of 'the truth and drama of the situation' when on 21 April, during the forenoon watch, they 'detected the classic signature of a submarine running on diesel engines'. As *Conqueror*'s Navigator, Jonty Powis, recalled:

> The bearing rate was high enough for a snorting submarine to be close. The Captain was summoned and we rushed to periscope depth at action stations with tubes ready. Nothing was in sight so we assumed that the submarine was dived and snorting at slow speeds just outside visual range. We returned to the depths to approach the firing position. As we left the layer we lost contact and never regained it. We tried all sorts; going shallow again then deeper, active sonar, springing beyond supposed maximum range and looking back at the target actively and passively; all fruitless.[84]

After searching, tracking and even going active failed to produce a solution or indeed confirmation, *Conqueror* came shallow again. On 24 April, *Conqueror* was ordered to find the Argentinian 'Guppy' class submarine *Santa Fe*, which was reportedly operating in the area, but before she could do so, on 25 April the *Santa Fe* was caught on the surface by helicopters from HMS *Endurance* and HMS *Antrim* which disabled her as she attempted to enter Cumberland Sound. 'The event brought us to the realization that we were actually at war and could have fired real torpedoes at a real target full of real people,' wrote Powis. 'Furthermore they would probably have a go at us too if we were careless. We became sharp.'[85]

Thereafter *Conqueror*'s contribution to Operation 'Paraquat' was limited and in its final days the decision was taken to offload 6 SBS and their equipment to HMS *Antrim* via helicopter. Bad weather complicated the transfer and at one point two men and a full load of equipment were washed off *Conqueror*'s casing by rough weather before being recovered by helicopter. *Conqueror*'s crew were sorry to see the Royal Marines go. 6 SBS had integrated into life on board the submarine. 'I will have an everlasting memory of their officer's amazing ability to eat food any time it was put in front of him,' recalled Wreford-Brown, 'he was always appearing in the wardroom in the wrong uniform, which normally resulted in the mess president fining him a bottle of port.'[86]

With the Royal Marines disembarked, *Conqueror* dived and resumed course for the Falklands to join HMS *Splendid* and HMS *Spartan*. During the passage south in very rough weather, *Conqueror*'s communications wireless mast had been damaged: bent forward slightly by the harsh waves, while the submarine was at periscope depth attempting to receive a broadcast. Thereafter the communication mast repeatedly caught the top of *Conqueror*'s fin when it was lowered. This made communication with Northwood next to impossible and was extremely frustrating. 'I have had enough,' wrote Wreford-Brown on 19 April. 'Isn't communicating FUN!?' SSIXS reception was so poor that a number of copies of each signal had to be patched together in order to get a fair copy, a lengthy process that on occasions took up to one and a half hours.[87] Recognizing that the position was untenable if *Conqueror* was to continue to play an active part in Operation 'Corporate', Wreford-Brown kept *Conqueror* at periscope depth and at a slow speed of 7 knots with the mast slightly raised until nightfall, eventually surfacing the submarine to allow two of *Conqueror*'s crew to clamber out of the fin and into the freezing conditions to remove the antenna from the mast.[88] The antenna was then taken inside the submarine and repaired while two crew members attempted to fix the mast so that it would once again retract into the fin. Although communications were eventually restored, they were very intermittent.

While Argentinian submarines preoccupied the SSNs operating in the South Atlantic, the War Cabinet was far more concerned with the Argentinian Navy's sole aircraft carrier, the ARA *Veinticinco de Mayo* (*25 May*). Although old – the carrier had been built for the Royal Navy and commissioned in January 1945 as HMS *Venerable* – it posed a considerable military threat to the British Task Force. It could carry between seven and nine Skyhawk and up to five Super Etendard fighter aircraft, both of which were capable of mounting air-to-surface and air-to-air attacks from a distance of around 400 miles. The carrier also

possessed helicopters, which provided a limited anti-submarine warfare capability, as well as six Tracker aircraft that could undertake
radar surveillance operations at distances of up to 500 miles, and able
to direct carrier aircraft as well as other air and naval units into attack
positions.

On 21 April, after a meeting of the War Cabinet, Lewin told Mrs
Thatcher that it had been possible to identify the location of an Argentinian naval force, including the carrier, in a patrol area between the
Argentinian coast and the Maritime Exclusion Zone. Intelligence
reportedly indicated that the carrier was operating a few miles off the
Argentinian coast, some way south of her Puerto Belgrano base. Fieldhouse had already ordered HMS *Splendid* to leave the MEZ and
proceed in the direction of the area in which the Argentinian naval
force was patrolling in order to reduce the time which it would take the
submarine to carry out an attack if ministers decided that this is what
it should do. He told Thatcher that it would take *Splendid* about two
days to reach the carrier and as it was sailing outside the MEZ it would
be under 'high seas' Rules of Engagement – meaning *Splendid* could
not attack it except in self-defence. Fieldhouse proposed that any submarine detected by *Splendid* and not classified nuclear should be
presumed to be Argentinian and attacked.[89] Mrs Thatcher agreed.

HMS *Splendid* was operating off Port Stanley when orders from
Fieldhouse arrived. Lane-Nott immediately set a northwesterly course
at maximum speed in order to intercept the carrier. But back in London, as *Splendid* sped north, the Foreign Secretary, Francis Pym, who
had not been consulted, was concerned that *Splendid*'s orders increased
'the risk that an Argentine submarine may be engaged and sunk when
it is neither in the MEZ itself nor in the direct path between the British
task force and the Falkland Islands'. Pym, who was about to visit
Washington, was nervous about the political ramifications of any possible engagement and protested that 'The option of an attack on the
Argentine naval force outside the MEZ and in its current patrol area
would be a major policy decision which has not yet been considered by
Ministers.'[90] As a result of Pym's political negotiations in Washington,
surveillance of the Argentinian Task Group was no longer considered
acceptable and the next day Thatcher was persuaded to suspend the
orders. *Splendid* was instructed not to leave the MEZ and if she was
already outside it she was to change course immediately and return
within it.[91] When Lane-Nott received the recall *Splendid* was 'within a
few hours/miles of our quarry'. 'It was an extremely frustrating

moment,' commented Lane-Nott. 'I really thought I had her.'[92] Rather than allowing *Splendid* to remove the primary threat to the Task Force, the Government simply conveyed a warning to the Argentinians through the Swiss Government that any approach on the part of Argentinian warships which could amount to a threat to interfere with the mission of British forces in the South Atlantic would be regarded as hostile and was liable to be dealt with accordingly.[93]

On board HMS *Hermes*, Admiral Woodward observed all of this with dismay. He was deeply unhappy with the command and control arrangements for the submarines operating in the South Atlantic. Although he was the overall Commander of the Task Force, he had no authority to command and control the submarines, which comprised the separate Task Force, TF 324.3, commanded by Herbert back in Northwood who reported directly to Fieldhouse. 'It was my opinion,' wrote Woodward, 'that I should take control of them myself, rather than have them run directly from Northwood by the Flag Officer Submarines':

I felt there were several good reasons for this:

a) I had Captain Buchanan on my staff, and one of the main reasons he was with me at all was to act as the local Submarine Force Coordinator.

b) It made more sense, to me at least, that the submarines should be under my command locally in case it became necessary to deal with a quickly changing set of circumstances which required very early action.

c) It might be alleged that I knew something about the subject of submarine warfare in my own right since I had been appointed, admittedly only for a week or two, to command the Submarine Flotilla in 1981.

d) *Hermes* was fully equipped with all the necessary submarine communications channels to do the job.

Above all, I wanted to change the operating methods – make them better suited to the conditions prevailing in the south.[94]

Woodward certainly knew a great deal about submarine warfare, but his confidence in *Hermes*' ability to handle submarine communications was misplaced. Although *Hermes* was equipped with all the necessary submarine communications equipment and channels, as the Task Force proceeded south communication demands increased steadily as the conflict intensified. Woodward later noted that his Operations

Officer was reading over 500 signals a day and, with communications under such considerable strain, taking responsibility for the submarines would have only added to the problems. There was also the distinct possibility that *Hermes*, a high-value target, would at certain periods of time have to go radio silent in order to avoid attack by Argentinian forces.

Woodward had also never experienced the comparatively new, but highly successful, arrangements of centralized operational control introduced in the 1970s with the creation of CTF 311. Northwood had transplanted the system it used to run submarines when in the North Atlantic to the South Atlantic, carving up the areas around the Falklands into three separate grids; Northeast, Northwest and the South. An SSN was allocated to each grid and under the Rules of Engagement was forbidden from crossing into another grid even if it was pursuing an important target. This method of operating submarines was in part a reflection of tried-and-tested Cold War missions in the busy North Atlantic, where, with a large number of submarines patrolling, there was always the risk of so-called 'blue-on-blue' incidents. Woodward wanted to give the submarines in the South Atlantic free rein to search for Argentinian surface units and then, if necessary, sink them once authorization had been granted. As far as he was concerned, given the limited number of Argentinian submarines in service with the Argentinian Navy it was 'no longer necessary to confine our SSNs to separate areas, provided they were forbidden to engage submerged contacts. By releasing the SSNs from the constraints of separate areas, I could attach any of them (or they could attach themselves as the chance offered) to any group of Argentinian surface ships, ready to attack the moment they got final clearance from London.'[95]

Lieutenant Commander Jeff Tall, Admiral Woodward's submarine Staff Officer, also shared this view:

> We were subdividing the South Atlantic into just three areas. It took away all flexibility. I asked the question 'What if something happens over here and we need more than one submarine?' I didn't get an answer. So we began to understand at a very early stage that Northwood had no feeling for either the fight in hand or what was needed to fight a fast-moving scenario. I suggested that we could have a hot pursuit scenario, where if one of our submarines was in hot pursuit of a high-value surface unit they could cross into the other sub's water. I got a reply saying that 'The Mark 24 [the new sonar-guided torpedo on the submarines] can't distinguish

between friend or foe.' Well that was nonsense. What you do is the submarine in contact with the surface target sets a floor depth to the torpedo and the other submarine stays below it.[96]

But Fieldhouse and Herbert were determined to retain control of the submarines. Prior to taking up his position on board *Hermes*, Tall witnessed some of the early discussions at Northwood between Fieldhouse and his staff about command and control arrangements. 'It was obvious to me there was disagreement,' recalled Tall. 'It was clear that he was not going to give away command and control of his submarines.'[97] This was how it had always been throughout the Cold War and, as Herbert has pointed out, 'The whole philosophy in the Iceland–Faroes Gap and so on was based on this and so everybody understood it. Why change that philosophy suddenly and chuck all that away. It was so simple. I was determined there was going to be absolutely no possibility of blue on blue. They were going to always be separated.'[98] With so many SSNs operating within the vicinity of the Falklands, the possibility of a blue-on-blue, friendly fire incident was very real and one of which Herbert had to take account. On 26 May, HMS *Valiant*, which was equipped with a towed array, held a contact that later turned out to be HMS *Splendid*. 'There's no doubt in my mind,' wrote Tall, 'that the overriding consideration was do not lose a reactor in the South Atlantic. Because the political ramifications of that, I think, would have been extremely serious.[99] The Royal Navy was and continues to be distinctly aware that it is always one accident away from the end of the nuclear-submarine programme.

Woodward recognized that he was in no position to argue, especially against two such seasoned senior submariners as Fieldhouse and Herbert. 'I retired from the debate with as much grace as I could summon, which, as I recall, was not all that much.'[100] Operationally, there were mixed views about the command and control system. The CO of HMS *Valiant*, Tom Le Marchand, concluded that the standard of command and control was 'very impressive'. 'I felt that I was in touch with a Controller who was in full possession of all the facts and who was managing up to four SSNs and one SSK as a team, without straying towards being oppressive,' he later wrote.[101]

Taylor in *Spartan* was also content with the system. 'I had the assurance that Northwood was still going to be there in the morning and not some burning bit of wreckage,' he says. 'When we actually got into the South Atlantic I was sending daily sit reps for Operation "Corporate". I just started sending my daily reports in the format we used in the North Atlantic and in the Norwegian Sea. To me this was just another submarine operation. Why would you invent something new? This is what we do, this is where I am, where I'm going to be, where the opposition is, any new problems, defects. Basically it was a format. My operational authority knows what that is. I didn't even know if Admiral Woodward and his staff had a copy of this document. What did they want me to tell them every day apart from "I'm here". What could they offer me in terms of material support? I wanted the whole package . . . How did I know that they were going to be on the end of that circuit twenty-four hours a day, for all three submarines, all four submarines, eventually five and six submarines? You can't do it. You just can't do it. At the end of the day I had no absolute assurance that my command and control headquarters was still going to be floating.'[102]

Lane-Nott was also satisfied that the 'division of the operating areas was found to be manageable', but this was partly because both Lane-Nott and James Taylor had come to an arrangement. 'I promise not to attack a nuclear submarine, if you do too.' *Spartan* and *Splendid* both strayed into each other's respective patrol areas, exercising the degree of flexibility and ambiguity that was expected of submariners.[103]

As the Task Force neared the MEZ, the SSNs attempted to find the main Argentinian surface task groups. On 26 April, *Spartan* was ordered southwest to cover any move by the *Belgrano* group towards South Georgia and was tasked with intercepting, and reporting and gaining intelligence on, the MV *Río Carcarañá* which was believed to be engaged in resupply operations. The *Río Carcarañá* was far further west than anticipated, but before *Spartan* could reach its target the search was abandoned in favour of pursuing one of the Type 209 submarines which intelligence indicated may have been approaching the Falklands from the north. By 28 April, *Spartan* had moved north of the Falkland Islands, clear of the 100 fathom line and established an ASW patrol, 'searching along likely enemy track with finger on the trigger'.[104] But due to poor weather conditions significantly reducing detection ranges *Spartan* found nothing. 'Whether he was there or not I haven't the faintest idea,' says Taylor.[105] On 29 April, *Spartan* was ordered to patrol to the west of the Falklands to search for a far more worrying target: the still elusive Argentinian carrier group.

On 24 April, intelligence had indicated that the carrier had left its home port of Puerto Belgrano with two escorts, and sailed south. HMS *Splendid* was also ordered to locate it and, if possible, sink it. But the carrier continued to elude Lane-Nott. 25 April proved to be a 'quiet day with very few contacts . . . no Argentinian military activity at all' and the next day, 26 April, *Splendid* was facing the possibility of 'being dragged South again' to search for the Argentinian submarine *San Luis*.[106] Lane-Nott was deeply unhappy and 27 April was yet another 'frustrating day with no contact whatsoever'. On 29 April, his luck changed. After spending the night and early morning searching, *Splendid* finally detected something promising.

'I found myself,' recalls Lane-Nott, 'in a situation where I had in one sweep of the periscope, within 14,000 yards, that's seven miles, I had two Type 42 frigates, identified as Argentinian, and three A69 frigates. So effectively I had the sum total of the escort forces of the Argentinian Navy, all outside of the twelve-mile limit, all within one sweep of the periscope, all visual. I worked on the principle that if the carrier was going to come out, then this was their escort, there was no other escort, nobody else could do it. They weren't doing anything in particular They were just wandering around, all within sight of each other, there was nothing structured about it. They weren't using active sonar, they were doing very little transmitting on radio, there very few people on deck. It was an extraordinary situation.'[107]

Lane-Nott was 'euphoric', but he was once again hampered by the Rules of Engagement, which did not permit him to sink enemy vessels. 'The Rules of Engagement had started to become specific, you can sink this but you can't sink that, you can attack this, but you can't attack that, which I didn't really understand. There didn't seem to be much sense to it to me. If it was an Argentinian ship we should be sinking it. The whole idea was to put the fear of god into them with nuclear submarines, if you get out there, you get sunk and here we were, they were all within sight of the periscope and we didn't have the [authority under the] Rules of Engagement.' Lane-Nott was 'absolutely convinced that if I stayed and shadowed this lot that sooner or later they would lead me to the carrier'.[108] *Splendid* stuck with the three A69s and chased them southwest, and in the process discovered that the destroyers *Comodoro Py* and *Hércules* were patrolling nearby.

As *Splendid* continued to shadow the gathering Argentinian ships, in London the War Cabinet met to discuss a paper on 'the military threat posed to our forces in the South Atlantic by the Argentinian aircraft carrier "25th of May"' and to formulate 'options for eliminating that threat'. Lewin told the War Cabinet, OD(SA), that 'Militarily the Argentine aircraft carrier could amount to such a [military] threat from virtually any position on the high seas; it would not always be known where she was, she was capable of covering 500 miles in a day, she could carry aircraft with an operating radius of a further 500 miles and the supply line for British forces was strung out between Ascension Island and the Falklands.' The paper outlined that action against the carrier could be taken by surface ships, Harriers or submarines, but concluded that:

> On balance, a SSN would give the greatest chance of success with the least prejudice to other operations. Following a torpedo attack from an SSN, the '25th of May' should be disabled. It is possible that she might sink quickly, but this is unlikely given good damage control practices . . . If the carrier did sink, it is possible that a large proportion of her crew numbering 1,000 would have time to abandon ship with life rafts. Her escorting vessels should be able to pick up survivors.[109]

After considering the legal, military and political issues involved, OD(SA) agreed that 'British forces should be authorized to attack the Argentinian aircraft carrier as soon as possible wherever she was on the high seas.' No further warning need be given, but only if the carrier was north of latitude 35°S and west of longitude 48°W.[110] 'On the high seas' was defined as outside the internationally recognized territorial waters limit of 12 nautical miles. The 200-nautical-mile Maritime Exclusion Zone was now redefined as a Total Exclusion Zone and Argentina was informed that:

> Any ship and any aircraft, whether military or civil, which is found within this zone without due authority from the Ministry of Defence in London will be regarded as operating in support of the illegal occupation and will therefore be regarded as hostile.[111]

At that time intelligence indicated that the carrier was not heading towards *Splendid*'s position, but was in fact operating to the south. Northwood ordered Lane-Nott and *Splendid* to leave the escorts and go and search for it. Lane-Nott was furious. 'There's an old submarine adage, which I've certainly lived by: "attack what you see, not what you think might be there." If you've got something in front of you, you attack what you see.' He 'turned a Nelsonian blind eye' and ignored three or four of the broadcast routines in which the orders were transmitted, hoping that the carrier would appear and meet up with the escorts: 'I just could not believe that after I'd sent that report and we were still in contact with them, that they were telling me to go somewhere else. I just couldn't believe it. I refused to believe it effectively. I decided not to get into a slagging match. I decided first of all to ignore it and hope that by staying with them the carrier would suddenly appear over the horizon. It was twenty-odd hours by this stage. The weather was relatively balmy, there was a swell, but it wasn't much of a sea and we were able to stay at periscope depth perfectly satisfactorily and keep track of these guys, and keep out of their way, and just move around and keep track of everything that was going on. I mean it seemed absolute nonsense to me to be leaving them for some intelligence report.'[112]

Eventually, Herbert sent Lane-Nott a personal signal that said, 'I understand. Go now.' But Lane-Nott was still not convinced. He talked it over with his First Lieutenant, and said: 'I don't like this at all. This goes against every principle of submarining, everything we've ever been taught about how we deal with the enemy, and deal with an attacking situation. This is totally against my instincts to leave what you see.' But he had no choice. 'Eventually I did break off, with great reluctance and a lot of bad grace from me,' recalls Lane-Nott. But when *Splendid* arrived at the expected rendezvous point on the morning of 1 May it found nothing. 'I was not a happy boy,' he says. 'And we started chasing shadows again.'[113]

The Argentinian aircraft carrier was, in fact, far further northeast than expected. At 2307Z on 1 May, the Argentinian Operational Commander, Rear Admiral Jorge Allara, issued orders to prepare for an all-out attack. He ordered the carrier group to deploy to a safe position, locate the British Task Force and launch an air attack at first light. He also ordered his second task group to deploy south of the Exclusion Zone, into a position to attack any British warships that attempted to flee the carrier attack. The third task group, consisting of the *Belgrano* and two escorting destroyers, was ordered south to Burdwood Bank, in order to deal with any British vessels operating to the south of the Falklands, using the destroyers' deadly Exocet missiles.[114] At 0113Z on 2 May, the Argentinian Chief of Naval Operations, Vice Admiral Juan José Lombardo, sent a further signal to Argentinian units ordering them to close on the Task Force and attack. This information was eventually communicated to *Spartan* and *Splendid*, but by the time they received it they were over 100 miles away from the supposed location of the carrier.[115] Taylor wrote of an 'Indication of a lunge to the SE by enemy CV [Carrier] group, which then withdrew to own coastal waters. Sprinted to intercept, but no interaction.'[116] *Splendid* also attempted to intercept, but like *Spartan* failed to locate the carrier.

With the carrier still eluding the Task Force, Woodward hoped that *Conqueror*, which was operating in the south, would have more luck locating the *Belgrano*:

> My hope was to keep *Conqueror* in close touch with the *Belgrano* group to the south, to shadow the carrier and her escorts to the north with one of the S Boats up there. Upon word from London, I would expect to make our presence felt, preferably by removing the carrier, and almost as important the aircraft she carried, from the Argentinian Order of Battle.[117]

On 29 April, Chris Wreford-Brown in HMS *Conqueror* received orders to locate the *Belgrano* group and was told of its possible location. She began to head towards it. The next day, Woodward was given permission to proceed inside the Total Exclusion Zone and to start the process of recapturing the Falkland Islands. The same day, *Conqueror* was given permission to attack the *Belgrano* group, but only if it was within the Total Exclusion Zone. On 1 May, *Conqueror* located the *Belgrano* and its escorts. 'A good day – in contact with the Enemy at last!' wrote Wreford-Brown.[118] But the *Belgrano* was outside the Exclusion Zone. *Conqueror* could not attack it.

Woodward now had a problem, as Freedman has explained. 'To the north he had permission to attack the carrier but no contact, while to the south he had a contact with no permission.'[119] Woodward was convinced that his carriers were about to become victims of a classic pincer movement. With both 'S' boats struggling to find the carrier, Woodward launched his Sea Harriers to investigate reports that the radar of an Argentinian S-2E Tracker (ASW carrier-launched aircraft, used by the Argentinian Navy) coming from the northwest had been detected by the aircraft carrier HMS *Invincible* and the destroyer HMS *Coventry*. At 0330 on 2 May, one of the Harrier pilots reported that he had found several radar contacts, indicating a group of four or five ships, including what could have been the carrier, 200 miles to the northwest. Woodward concluded that 'We could expect a swift thirty-bomber attack on *Hermes* and *Invincible* at first light ... he might also have Exocet-armed Super Etendards to add to our problems.' Once the *Belgrano* was factored into the equation, Woodward quickly reached 'the worst possible case ... *Belgrano* and her escorts could now set off towards us and, steaming through the dark, launch an Exocet attack on us from one direction just as we were preparing to receive a missile and bomb strike from the other.'[120] Woodward knew that the *Belgrano* was approaching the Burdwood Bank, an area of shallow water that would have made it difficult for *Conqueror* to continue to shadow her:

> Deep down, I believe she would continue to creep along the back of the Bank, and then when she is informed that the carrier is ready to launch her air strike, she will angle in, on a north-easterly course, and make straight for us, the Exocets on her destroyers trained on us as soon as they are within striking range. I badly need *Conqueror* to sink her before she turns away from her present course, because if we wait for her to enter the Zone, we may well lose her, very quickly.[121]

Withdrawal was out of the question; such action was, Woodward said, 'scarcely in the traditions of the Royal Navy'. There was only one option remaining. 'I cannot let that cruiser even stay where she is, regardless of her present course or speed. Whether she is inside or outside the TEZ is irrelevant. She will have to go.'[122]

SINK THE *BELGRANO*

Without a change in the Rules of Engagement, Woodward could not order *Conqueror* to sink the *Belgrano*. In order to provoke the Government into a quick decision Woodward exceeded his designated authority and directly ordered *Conqueror* to sink the cruiser. On board HMS *Hermes*, Jeff Tall was responsible for communicating the order to *Conqueror*: 'I said to him, "Well, you haven't got the ROE, Admiral." And he [Woodward] said, "Are you disobeying an order?" It was getting serious. I said, "No."' Tall sent *Conqueror*, 'From CTG 317.8 to *Conqueror*, text priority *flash* – attack *Belgrano* group' but he delayed sending the signal by four minutes so that *Conqueror* would not receive it on the scheduled 0600 broadcast to submarines.[123] Woodward was confident that Herbert, his former Commander in HMS *Valiant*, would intercept the signal and 'would know, beyond any shadow of a doubt, that I must be deadly serious.' Woodward hoped that the signal would get as far as the Commander-in-Chief, Sir John Fieldhouse, and that 'he would personally recommend that it should be left to run, given the urgency of my message, while he negotiated with the MOD and the Cabinet'. I had quite clearly exceeded my authority by altering the ROE of a British submarine to allow it to attack an Argentinian ship well *outside* the TEZ,' wrote Woodward.

Such a breach of Naval discipline can imply only two things – either Woodward has gone off his head, or Woodward knows exactly what he is doing and is in a very great hurry. I rather hoped they would trust my sanity, particularly because there is always another aspect to such a set of circumstances – that is, should the politicians consider it impossible for the international community to approve the sinking of a big cruiser, with possible subsequent great loss of life, I had given them the opportunity to let it run and then blame me, should that prove convenient.[124]

As soon as Herbert saw Woodward's signal appear on the SSIXS satellite he immediately removed it in order to stop *Conqueror* downloading it. 'I was in the Ops Room in Northwood,' recalls Herbert, 'and was very, very cross. I took it off the broadcast, obviously it had to go right up to the Prime Minister.'[125] Herbert passed Woodward's intercepted order to Lewin, who at 1045 assembled a meeting of the Chiefs of Staff, who quickly agreed to ask the War Cabinet to extend the altered Rules of Engagement permitting attack of the carrier to all Argentinian warships (not auxiliaries) operating outside the TEZ.[126] An ad hoc meeting of the War Cabinet then convened at 1245 at the Prime Minister's country retreat, Chequers. There, Mrs Thatcher, Whitelaw, Nott, Parkinson, Havers, Admiral Lewin and the Force Commander and the Permanent Under-Secretary, Foreign and Commonwealth Office, were all briefed on the latest developments. Mrs Thatcher's memoirs offer the most vivid account of the meeting:

Admiral Fieldhouse told us that one of our submarines, HMS *Conqueror*, had been shadowing the Argentine cruiser, *General Belgrano*. The *Belgrano* was escorted by two destroyers. The cruiser itself had substantial fire power provided by 6 inch guns with a range of 13 miles and anti-aircraft missiles. We were advised that she might have been fitted with Exocet anti-ship missiles, and her two destroyer escorts were known to be carrying them. The whole group was sailing on the edge of the Exclusion Zone. We had received intelligence about the aggressive intentions of the Argentine fleet. There had been extensive air attacks on our ships the previous day and Admiral Woodward, in command of the Task Force, had every reason to believe that a full scale attack was developing. The Argentinian aircraft carrier, the *25 de Mayo*, had been sighted some time earlier and we had agreed to change the rules of engagement to deal with the threat she posed. However, our submarine had lost contact with the carrier, which had slipped past it to the North. There was a strong possibility that *Conqueror* might also lose contact with the *Belgrano* group. Admiral Woodward had to come to a judgment about what to do with the *Belgrano* in the light of these circumstances. From all the information available, he concluded that the carrier and the *Belgrano* group were engaged in a classic pincer movement against the Task Force. It was clear to me what must be done to protect our forces, in the light of Admiral Woodward's concern and Admiral Fieldhouse's advice. We therefore decided that British forces should be able to attack any Argentine naval vessel on the same basis as agreed previously for the carrier.[127]

For the War Cabinet, it was a straightforward decision. As Nott later wrote, 'The military decisions on the rules of engagement were easy and the Belgrano was the easiest of the lot.'[128]

As soon as the decision had been taken, one of the officials in the meeting, Robert Wade-Gery, immediately rang up Northwood and said: 'Sink it.' When he asked: 'Do you want that confirmed in writing?' The answer was 'No. There won't be time. They'll have sunk it by the time it arrives.'[129] The MOD signalled CINCFLEET with the change in ROE at 1207Z and within half an hour Herbert was transmitting the new orders to his SSNs.

Back in the South Atlantic, *Conqueror* was still trailing the *Belgrano*. Wreford-Brown had at least some idea of what was going on back in London:

> During quiet moments, as I lay on my bunk in my cabin, I considered the next moves. We were the first submarine to maintain contact with enemy units and although the rules of engagement in force did not allow me to attack, I was in no doubt that those back in the UK would realise the tactical implications of what we were doing and form the conclusion that it would be militarily sensible to engage these enemy units before they threatened our task force.[130]

Conqueror was carrying two different types of torpedoes. The first was the Mark 8, which dated from the Second World War (see pp. 74–5). It was accurate at close ranges, was reliable and had a sizeable warhead more than powerful enough to penetrate the hull of the Second World War vintage *Belgrano*. *Conqueror* also carried the more modern wire-guided Mark 24 Tigerfish torpedo, but, as we have seen, the torpedo had many weaknesses and its unreliability, particularly its repeated control wire breaks, were a cause for concern. Indeed, Lane-Nott on *Splendid* had received a number of signals from Northwood about 'several problems' with the Tigerfish torpedoes, one of which was later diagnosed as a propulsion battery defect.[131] He had little 'Faith in the Weapons'.[132] After consulting his officers, Wreford-Brown decided that if he received orders to sink the *Belgrano*, he would do it with the old Mark 8 torpedoes. 'It was not a long debate,' recalled Powis.[133]

The order to sink the *Belgrano* arrived late in the afternoon on 2 May. In order to maintain radio contact with the UK, *Conqueror*'s radio mast had to be raised while travelling at slow speed at periscope depth. This was incompatible with maintaining a relative position and trail of the *Belgrano*. *Conqueror*'s communications also continued to cause problems and although it quickly became apparent that the ROE had changed, how exactly they had changed remained unclear. *Conqueror*'s second-in-command, Tim McClement, was forced to piece together message fragments from seven different copies of the signal. 'I got it all together and said: "I can prove that you have got it" because there's overlap here and there's overlap there and I've typed it out for you here,' recalls McClement. 'It took me about an hour, but there it was.'[134] *Conqueror* had permission to sink the *Belgrano*. McClement was relieved. 'It felt like a logical thing to do,' he said.[135] Wreford-Brown checked himself and then got the Navigator to recheck. By 1710Z he was satisfied that he had authority to attack. 'Traffic now received. COR [Coded Order] 177 gives me permission to Attack,' he wrote in *Conqueror*'s log. 'Delay due to confusion in that COR 174 cancelled an instruction (not received) from CTG 317.8 to attack the Cruiser. DTG of COR 177 not initially received and so awaited reruns to be absolutely certain.'[136] 'I have ROE to attack,' he continued. 'Aim now is to close TG 79.3 and then work into a firing position. Preferred weapon MK 8 Mod 4. If a good attacking position cannot be achieved because of the Escorts, then I shall use MK 24.'[137]

By the time the signal from Northwood had been pieced together *Conqueror* had fallen seven miles astern of the *Belgrano* and her escorts. Wreford-Brown ordered three Mark 8 torpedoes loaded into *Conqueror*'s torpedo tubes, along with three Mark 24s in the remaining tubes, and started the lengthy and time-consuming approach to a firing position. 'It was still daylight. The visibility was variable,' remembered Wreford-Brown. 'It came down to 2,000 yards at one time. I kept coming up for a look – but when at periscope depth we were losing ground on them – and then going deep and catching up. I did this five or six times. They were not using sonar – just gently zigzagging at about 13 knots.'[138] As *Conqueror*'s crew closed up at action stations to prepare for the attack 'The atmosphere throughout the boat was extraordinary,' wrote Powis. 'Everybody had a role even if it was to sit tight and await damage to repair, we were all concentrating intently on the task at hand.'[139]

An hour later, *Conqueror*'s crew had succeeded in achieving a firing position on the port beam of the cruiser with the escorting destroyers on the other side. At 1813Z, Wreford-Brown wrote in *Conqueror*'s logbook:

> In position 55 23.1S 61 21.0W. I am on the port quarter of TG 79.3 and my target, the Cruiser G BELGRANO, is on the left wing. My intention is to close to a firing position such that I shall fire a salvo of 3 × Mk 8 Mod 4's from 1800 yds on a Torpedo Track Angle of 100° and a zero gyro angle. I shall then evade to the SE leaving the datum between me and the destroyers for as long as possible to assist my escape.[140]

For the next thirty minutes *Conqueror* went 'deep and fast', diving in order to close the range and coming up to allow Wreford-Brown to look through the periscope every few minutes. At 1851Z, *Conqueror* came up to periscope depth and Wreford-Brown sighted the *Belgrano* again, moving at 11 knots. He ordered torpedo tubes 1, 2 and 6 into standby mode and opened *Conqueror*'s torpedo tube bow caps. Two minutes later, at 1853Z, *Conqueror*'s crew started the final approach, with Wreford-Brown conducting a final all-round look through the periscope at 1854Z. 'Nothing else in sight,' he wrote in *Conqueror*'s log book, 'M-04 BELGRANO is not aware of my presence. Intend to wait until the target's relative bearing is G13 [Green 13 degrees, very fine on the starboard bow] when it will be a zero gyro angle shot.'

The Mark 8 torpedoes relied on a very old mechanical gyro that turned the torpedo when it was in the water. But submariners preferred not to rely on the gyro, preferring instead a straight shot. The ideal position for discharging the Mark 8 was on the beam of a target with angle on the bow at discharge of 90 degrees plus an angle equal to the speed in knots at a range of 1200–1500 yards, with the attacking submarine's own course equal to the mean torpedo course. This meant that the weapons had a nearly zero gyro angle and, provided the target's speed and course remained correct, the firing solution was independent of the target's range.

'The attack drill was conducted as if we were taking part in a demonstration for a training film,' recalled Powis. 'It was not a difficult attack, we were all well practised and the mood in the control room was tense but professional.'[141] With only seconds to go before Wreford-Brown ordered the release of the torpedoes, McClement noticed that the Torpedo Course Calculator indicated that the three Mark 8 torpedoes were going to leave *Conqueror* at 17 degrees to the right, not the zero angle that Wreford-Brown was hoping for. At the moment of firing McClement shouted from the corner of the Control Room: 'Do not fire! – gyro angle improving.' Wreford-Brown lowered the periscope and waited for the *Belgrano* to reach the optimum position.[142] Eventually, McClement signalled that it was okay to fire and Wreford-Brown raised *Conqueror*'s periscope again. 'Do you mind if I fire now?' he asked McClement. At 1857Z, HMS *Conqueror* fired three Mark 8 torpedoes at an interval of three seconds, from a range of 1400 yards.

Conqueror's crew continued to plot the movement of the torpedoes as they streaked towards their target. When the Fire Control Officer's stopwatch indicated 15 seconds to first impact, Wreford-Brown raised *Conqueror*'s periscope and focused on the *Belgrano*. The first torpedo missed and allegedly lodged in the bows on one of the *Belgrano*'s escorts. It did not explode. The second torpedo hit the *Belgrano* near the middle, while the third hit near the stern. 'Orange fire-ball seen just aft of the centre of target, in line with the after mast, shortly after the first explosion was heard,' wrote Wreford-Brown in *Conqueror*'s log. 'Second explosion heard about 5 seconds after, I think I saw a spurt of water aft, but it may have been smoke from the first. Third explosion heard but not seen – I was not looking!'[143] When the first torpedo impacted a cheer erupted in *Conqueror*'s Control Room. 'There was a lot of jubilation,' recalled Wreford-Brown, 'when I looked around there were a lot more people than there should have been there, on the outskirts, hanging around the corners and things. I think everyone was very happy that we'd achieved what we'd aimed to achieve, what we'd been instructed to achieve.'[144] But within seconds, silence descended once again as *Conqueror*'s men realized what they had done. On board *Belgrano* the first two explosions killed around 200 men while another 850 took to life rafts as the cruiser sank. In all, 321 men would lose their lives.

In the Control Room of every Royal Navy submarine is a piece of equipment that carefully records significant events during a patrol for later analysis. The 'big brother' recording of HMS *Conqueror*'s attack on the *General Belgrano* is harrowing to listen to. Three loud clanks, followed by an intense, humming machinery noise can be heard as each of the Mark 8 torpedoes leaves *Conqueror*'s torpedo tubes. 'Weapon discharge,' says a crackling voice. There then follows around 40 seconds of a noise not unlike an underwater recording until the sound of the torpedoes steadily intensifies. Fifty-seven seconds after the first Mark 8 was discharged there is a loud bang, followed by a cool calm voice: 'Explosion.' Three seconds later, a second loud, crashing thud follows. 'Second explosion,' says the calm, collected voice, before an almost identical third noise erupts. 'Third explosion,' says the voice, which is quickly followed by a fourth bang, which later analysis revealed was the direct path and bottom bounce of the first torpedo impact, which by chance arrived at the same time cadence as the firing interval.[145] There is an eerie silence, calmness, the sound of water rushing down the side of the submarine's hull, followed by a faint, tinkling, cracking sound, similar to someone running their hand through a chandelier. Only later did the analysis reveal that these were the sounds of the *General Belgrano* breaking up as the ship sank beneath the waves. Wreford-Brown finds it 'remarkable how like an ordinary drill it was. It sounded like a good attack in the Attack Teacher at Faslane, everything tidy, no excitement. I'm not an emotional chap and I had been concentrating the whole time on getting into a good position. It was tedious rather than operationally difficult.'[146]

After watching the final torpedo strike the *Belgrano*, Wreford-Brown lowered the periscope, ordered *Conqueror* to a depth of 500 feet and set a course south, away from the scene of the attack at a speed of 22 knots. As the submarine moved away 'banging' noises could be heard in the distance. What they were was not immediately apparent. 'If this is the Destroyers chasing me they are either lucky in their choice of direction, or there happened to be a NEPTUNE [Argentinian anti-submarine warfare aircraft] in the local area ... that has directed them,' wrote Wreford-Brown. 'The "Bangs" reported by the Sound Room sound like gun fire – is it the BELGRANO's ammunition going off?' he speculated, 'perhaps the "Big Bang" was her magazine.' Later analysis revealed that the explosions might have been depth charges from the *Belgrano*'s escorts but by the end of 2 May *Conqueror* was well clear of any counter-attack. 'I was both relieved and exhilarated that it had been successful,' recalled Wreford-Brown. 'I had no doubt about our capabilities and we had spent countless times practising; nevertheless, there was a certain sense of relief that the team had got it right.'[147] 'The Royal Navy spent thirteen years preparing me for such an occasion,' he later said. 'It would have been regarded as extremely dreary if I had fouled it up.'[148]

'It had been by any standards, a textbook operation by Christopher Wreford-Brown and his team,' said Woodward, 'which is probably why it all sounds so simple, almost as if anyone could have done it. The best military actions always do.'[149] 'Mixed feelings of elation and envy,' wrote Taylor, in HMS *Spartan*'s Report of Proceedings.[150] 'I was envious because he'd done it and we might still yet. There were all these mixed emotions. Did I think for one second about the poor bastards from the *Belgrano* floating around the Southern Ocean? No, I didn't. I think that's a luxury you can't afford. You look back on it later and say that a bunch of kids got killed, but that's what it's all about, I'm afraid. I think the older wise guys thought about it quite deeply. The young guys were the hawks and they were cheering happily. They regarded this as just part and parcel.'[151]

Lane-Nott on *Splendid* was 'thrilled to bits that we'd sunk it, but would admit fully that I was professionally extremely jealous that it was Chris that had sunk it and not me. That was my reaction at the time.'[152] Now, on reflection, he is 'extremely grateful . . . that it wasn't me that sank the *Belgrano*, that it was him and not me. And I say that because I saw the pain that he had to go through and the examination and scrutiny he went through when he got back.' When *Conqueror* returned to the United Kingdom, the press hounded her crew and many of their families. 'The intrusion was disgraceful,' says Lane-Nott. 'People saying that he needed psychoanalysis and how much counselling has he had and all this nonsense. He's a submarine Commanding Officer, for god's sake, he doesn't need any counselling. He did his job.'[153]

Accusations were later made that the Prime Minister had deliberately ordered the sinking of the *Belgrano* in order to scupper an American-backed Peruvian peace plan. Much of this controversy arose due to the complexity of the day's events and a subsequent TV interview in which Thatcher was grilled by a 56-year-old teacher from Cirencester, Diana Gould, who accused the Prime Minister of 'sabotaging any possibility of a peace plan succeeding'. In denying this accusation, Thatcher responded that 'one day, all of the facts, in about thirty years' time, will be published'.[154] Those facts now show that on 2 May GCHQ intercepted Argentinian signals that indicated that Argentinian military chiefs had in fact ordered the *Belgrano* to reverse course. But although the order had been intercepted by GCHQ, it took time to process and the new intercepts containing the information were not distributed throughout Whitehall until 3 May, long after the decision to sink the *General Belgrano* had been taken. They did not influence events.[155] Even if the latest intelligence intercepts had been available, those responsible for assessing them would have had to ask themselves a series of questions: was the intercept accurate? Was it a deception? What followed it? Was it followed by another signal rescinding the orders that had not been intercepted?[156]

There is so much information now in the public domain that it is possible to piece together what happened. The *Belgrano* did alter course at 0811Z on 2 May and *Conqueror* reported the alteration to Northwood at 1400Z, just as the signal with the ROE alterations came in. The information was not passed on to the MOD or to Woodward.[157] It changed little. As Freedman has noted, 'Concluding that an Argentinian attack was not imminent was not the same as presuming that it had been postponed indefinitely ... If it was not attacked that day then it would be returning to the fray the next time the Argentinian Navy positioned itself for battle, by which time *Conqueror* could well have lost contact.'[158] Quite aside from the possibility that the *Belgrano* would disappear over the shallow Burdwood Bank, the periods when *Conqueror* was actually trailing the *Belgrano* were considerably shorter, as were the periods when she was marking the cruiser with a fire control solution. These all absorbed time. Herbert 'believed that the Task Force had to take its chances when it could, because the next day the chances might fall the other way', and, as Freedman has noted:

> Even if he could have gained political authority at this point to rescind the new ROE and transmit them back to *Conqueror*, which was highly unlikely, there is no reason why he would have thought that sensible. The change in ROE was seen as a necessary step that would have to be taken at some point, to enable the Task Force to engage an enemy that was clearly geared up for battle.[159]

Indeed, in 2003, the Captain of the *Belgrano*, Captain Héctor Bonzo, ended thirty years of silence and admitted that his change of direction was a temporary manoeuvre. 'Our mission in the south wasn't just to cruise around on patrol but to attack,' he said. 'When they gave us the authorization to use our weapons, if necessary, we knew we had to be prepared to attack, as well as be attacked. Our people were completely trained. I would even say we were anxious to pull the trigger.'[160]

There were few regrets on the British side. As Lewin put it:

> A catastrophe like the sinking of a major unit was bound to happen sooner or later once the Argentines had invaded and the Task Force sailed. They

must have realized the risks, we certainly did, and it was my job to ensure that if possible it didn't happen to us first.[161]

Among *Conqueror*'s crew there was 'a range of reactions to the events of the day', according to Powis. 'Some became rather introspective, some seemed unaffected, most were pragmatic. We were at war in all but name and assumed that the Argies would have had a go at us if they had detected our presence.'[162] Wreford-Brown had mixed emotions. 'Afterwards I had a certain amount of regret about the loss of life. I did not know the numbers involved, but one presumed it was considerable. But I feel we did just what we were invited to do and I would have no hesitation in doing it again. It is a fact of life that if you want to go to war, you must expect losses.'[163] McClement holds a similar view:

> They started it, so all the lives lost are the Argentinian government's fault. In war, tough decisions have to be made and people die. The Argentinians had invaded our country aggressively and that's war. We were protecting the carriers and if we had lost them we would never have won the war. It was the right thing to do and I believe it saved more lives being lost because their ships never came out again. It had a very aggressive psychological impact on the Argentinians. Admiral Lord Fisher said in 1903 "the art of war is violence, moderation in war is imbecility." Once someone has started the war and once the other side has decided to go to war with them, then the best thing you can do is to do it as hard as you can. The sooner they realise there's only one option for them – and that's surrender – the better.[164]

None of this was lost on the Prime Minister. Mrs Thatcher later wrote in her memoirs that:

> As a result of the devastating loss of the *Belgrano*, the Argentinian Navy – above all the carrier – went back to port and stayed there. Thereafter it posed no serious threat to the success of the taskforce, though of course we were not to know that this would be so at the time. The sinking of the *Belgrano* turned out to be one of the most decisive military actions of the war.[165]

Following the sinking the Argentinian Navy was reluctant to run what was now a gauntlet of SSNs which had formed a barrier on the high seas. Herbert referred to the Argentinian Navy as the 'Stay at home Navy'.[166] That reluctance eloquently demonstrated the deterrent effect of the nuclear submarine, and as James Taylor commented at the time, 'it was that, as much as any other factor, which permitted CTG 317.8 to win the crucial war of attrition in the air unmolested by the enemy surface fleet'.[167] The sinking of the *Belgrano* was, as Woodward recorded, 'one of the more riveting days in the history of the submarine service', and he later remarked that 'if *Spartan* had still been in touch with 25 *de Mayo* I would have recommended in the strongest possible terms to C-in-C that we take them both out that night.'[168]

FRUSTRATION

As *Conqueror* carried out the attack on the *Belgrano*, HMS *Spartan* and HMS *Splendid* continued to search for the Argentinian carrier, which by 3 May, intelligence indicated, was some sixty miles south of Puerto Deseado (the Argentinian mainland).[169] *Spartan* sprinted to intercept but found nothing.[170] Lane-Nott also moved *Splendid* to the southwest until at 0330Z *Splendid*'s sonars detected between three and five ships around seven miles inside Argentinian territorial waters (twelve miles from the coast), close to Puerto Deseado. 'We were right on the 12 mile limit, just outside it,' recalled Lane-Nott:

> with what we had convinced ourselves, not from intelligence reports, although some of them were vaguely in that direction, but from acoustics that what we were hearing on sonar, that we'd got the carrier and that it was running down the coast inside the 12 mile limit. And we spent an extraordinary night when we were at . . . action stations, waiting for the morning and the daylight, going along the 12 mile limit, following this vessel, which we were pretty convinced was the carrier, gradually going south at about 10 or 12 knots. It was an extraordinary night to the extent that Ian Richards [XO of HMS *Splendid*] and I didn't go to sleep at all, we just sat in the Control Room the whole time and stayed with it. We normally shared things, I did a bit, he did a bit, he'd call me if something came on which allowed me to get some sleep. But we were so absorbed in it neither of us was prepared to go to bed and the adrenalin was such that we didn't want to go to bed.[171]

Lane-Nott was convinced that he had the carrier. 'There are numerous contacts off Deseado,' he wrote. 'Not surprising as it is a port with some facilities. However, I am sufficiently suspicious of some of these contacts to believe that they may be warships. Ranging manoeuvres indicate that they may be outside the TML [Twelve-Mile Limit] but I am not certain. We will have to wait until daylight.'[172]

Back in London, the reported position of the carrier raised 'a potentially tricky problem'. The sonar contacts appeared to be moving north to 'a vulnerable position off the coast of Argentina ... outside what Britain would recognize as territorial waters but in a bay which the Argentinians claimed was inside the territorial waters baseline.'[173] The War Cabinet was forced to seek the advice of the Attorney General, Sir Michael Havers. Although Havers advised that the sonar contacts were approaching an area that was outside Argentinian territorial waters, he stipulated that the carrier could only be attacked if it was more than 12 miles from the nearest point of land.[174]

As London debated what to do, on board *Splendid* excitement was building as the crew waited for dawn. In the submarine's Manoeuvring Room, the engineering crew had linked up with the Sound Room so that reports could be broadcast to the Engine Room. 'At least four contacts are heading North West towards Cabo Blanco,' wrote Lane-Nott at 0947Z in the morning. 'Intend following them North outside the 30 fathom line until sunrise when we will close to identify and attack a selected target. It is possible that they are forming up into two groups.'[175] As dawn approached Lane-Nott took *Splendid* to action stations and up to periscope depth in order to see if he could visually identify the carrier.

'The ship was at action stations, we went to periscope depth,' recounted Lane-Nott, 'and you won't believe it, the visibility was about ten yards, we were in solid fog. You couldn't see a damn thing. The visual aspects had gone, ok the electronic aspects were there, but the visual aspects had gone. You couldn't possibly do a visual attack on a carrier with ten yards' visibility, you could do it deep if you were convinced. But one of the things that the Rules of Engagement talked about was visual identification, so we had to have the visual identification. If we hadn't had that, I might well have fired on it.'[176]

By the end of the morning intelligence information, possibly gleaned from US satellite reconnaissance, indicated that the 'Argentine aircraft carrier group was in-shore ... and enjoying close air cover from the mainland' and that the 'Argentine commander had apparently questioned his orders to move forward and launch air attacks and had pulled back for reasons of weather and military prudence'.[177] The carrier then reportedly moved away to the north.[178]

'A day of frustrations' was how Lane-Nott summed up 4 May. He moved *Splendid* east, just outside the Argentinian twelve-mile limit, hoping to intercept the contacts as they continued north to Cabo Blanco, a chokepoint for coastal traffic from Deseado and the Golfo San Jorge. 'They have certainly not come outside the TML or run the gauntlet across the Eastern edge of Golfo San Jorge,' he wrote. 'By standing off to the East we should be better placed to detect the one who does do something out of the ordinary and is therefore worth prosecuting.'[179] But by 5 May, as Herbert noted in his war diary, 'At present, the Argentinian units do not appear to have any intention of moving . . . outside the TML.'[180] But this did not stop the Argentinians from mounting other air assaults on the Royal Navy Task Force and on the afternoon of 4 May HMS *Sheffield* was attacked.

With the conflict intensifying, further intelligence on the location of the Argentinian carrier prompted the War Cabinet to revisit the matter. On 5 May, Francis Pym explained that:

> we have good reason to believe that she is in, or very close to, Argentinian territorial waters, heading North and so away from the Falkland Islands and our Task Force, and that her attack capability is reduced to six aircraft with a radius of only 200 miles. Thus the carrier in her present posture can hardly be regarded as a direct or imminent threat to our Task Force.[181]

In a two-page memo to the Prime Minister, Pym argued that 'there is not in my judgment an immediate military need to attack the carrier in its present posture'. He urged the Prime Minister to 'have possible political consequences of an attack on the carrier at the front of our minds' due to the ongoing Haig negotiations, which the Cabinet had endorsed. 'I am in no doubt whatever,' he wrote:

> about the political consequences of an attack upon the carrier in that posture while we were waiting for the Argentine response to the proposals which Cabinet endorsed this morning. International opinion would be outraged. We should, I believe, forfeit much of our Parliamentary and public support at home. We should make it impossible, at least in the short term, for Argentina to accept an agreement of the kind envisaged. And most important, we should be thought by the Americans, and by Mr Haig in particular, to have deliberately destroyed the prospects of an initiative to which we had just given our conditional agreement. I believe that the consequences of that for American public opinion and the American Government's support could be incalculably grave.[182]

As Freedman has shown:

> This was the most dovish point in British thinking. The risk of political
> isolation and the pressure from the Americans was palpable. The War
> Cabinet and then the full Cabinet had been in discussions all morning
> about an unsatisfactory American plan that might have to be accepted.
> A cease-fire might be imminent. In these circumstances it seemed folly
> to push Britain's luck by attacking the carrier close to the Argentine
> mainland.[183]

Pym recommended 'that the instructions to the submarine concerned
should now be modified, at least until we know whether the Argentines
are going to accept the Haig proposals to the effect that the submarine
should not attack the aircraft carrier so long as she continues on a
northerly course in or close to Argentine territorial waters'.[184] Pym recommended that 'the Submarine Commander might be authorized to
attack the carrier only if she has moved out of the vicinity of Argentine
territorial waters and has changed course in a direction which clearly
implies hostile intent'.[185] The Attorney General backed Pym's recommendation and insisted that the 'legal defensibility of an attack on the
Argentine carrier would be reduced if she were not within 12 hours'
steaming of a point from which her aircraft could threaten British
forces.'[186] Havers also pointed out that 'even outside the 12 mile limit,
an attack on the Argentine aircraft carrier might be difficult to justify
legally if as a result of moving northwards she was much too far away
to pose an immediate threat to British forces'.[187]

Nott later recorded that:

> One of our nuclear submarines found the Argentine aircraft carrier lurk-
> ing within Argentinian territorial waters ... and the rules of engagement
> did not permit an attack within Argentinian territorial waters. The Navy
> sought a change in the rules, although the shallow water would have
> posed a hazard to our submarine. Margaret Thatcher was keen to agree
> the change, on the basis that the aircraft carrier would present a continu-
> ing threat to our ships and to take the Falklands even after we had
> recaptured them. I opposed the change, arguing against her and Terry
> Lewin on the grounds that action in South American territorial waters
> could bring in other countries on the Argentinean [sic] side just as we
> were about to achieve a victory. We did not agree the change.[188]

Although the War Cabinet decided that 'no immediate changes were
needed in the Rules of Engagement', they were amended on 6 May and

British forces were restricted from carrying out attacks on the carrier when it was more than twelve hours' steaming from the Task Force.[189] Haig was also informed that Britain was exercising restraint in its search for the carrier, while Argentinian commanders were not operating under similar restrictions.[190]

The next day, 7 May, the MOD objected to the restrictions on the grounds that 'the range of the carrier was not fixed but dependent upon the relative weights of the weapons and fuel being carried as well as their ability to refuel in-flight'.[191] The War Cabinet once again issued a warning to Argentina through the Swiss explaining that any Argentinian warship or aircraft found more than 12 nautical miles from the Argentinian coast would be regarded as hostile and dealt with accordingly. This effectively extended the Total Exclusion Zone right up to the Latin American coastline. A public announcement and formal notification to the Argentinian Government were made later that day and, after allowing twenty-four hours for the Argentinians to act on the warning, the ROEs were once again altered to remove the previous restrictions on SSNs attacking the carrier.[192]

Back in the South Atlantic, *Splendid* was no longer in a position to search for the carrier. The submarine was suffering from a number of mechanical defects, the most serious of which concerned one of its two electrical generators. To repair it, *Splendid* would need to snort, or to run its diesel engines while on the surface. On 6 May, Lane-Nott decided to withdraw from his operating area to well outside any areas of Argentinian activity to carry out repairs. From 6 May until 14 May *Splendid*'s engineering team battled to repair the faulty generator. While they were able to repair the machinery, in order to bring full power back on line *Splendid*'s reactor had to be scrammed and restarted, a procedure that required a great deal of power, more than was available in the submarine's batteries. To provide that power, *Splendid* would have to run on diesel engines.

'We came up to periscope depth to find ourselves in the worst weather that we had in the whole of the time we were down there,' explained Lane-Nott. 'It was storm force twelve, mountainous seas, I mean staying at periscope depth is difficult enough in those weathers, yet alone snorting in it . . . We managed to get a snort on, we then full scrammed and we were into the process of doing this and we lost it and we couldn't stay at periscope depth, we couldn't keep the thing open. We had no alternative but to go deep and try and hang on . . . "reactor is critical", "normal electrical supplies restored" were two phrases that will remain ingrained in my heart because I honestly thought we were going to run out of battery. I won't tell you what was left at the end, but it was absolutely touch and go. I mean, another thirty seconds I reckon and we would have been out of gas.'[193]

VALIANT ARRIVES

While *Splendid* was carrying out repairs a fourth SSN, HMS *Valiant*, entered the South Atlantic. The day before *Conqueror* sank the *Belgrano*, HMS *Valiant*, the Royal Navy's oldest SSN, under Commander Tom Le Marchand, sailed into Faslane after conducting an operational workup with HMS *Torquay* and HMS *Oracle*. Once in Faslane she embarked a full war load of weapons and topped up on stores and victuals. 'Earnest preparations at base ensued,' recalled Le Marchand. 'Our weapon system was upgraded to the most recent Tigerfish development, stores and provisions for 95 days were stuck down below, and one member of the ship's company married his fiancée. From the outset, we were utterly convinced that this was going to be a shooting war, and that our task was going to be to take the Argentinian navy out of the equation.'[194] *Valiant* sailed for the South Atlantic on 2 May and arrived in the hostilities zone at 1800Z on 15 May after a transit of 7000 miles at over 500 miles per day.

Although *Valiant* had only just arrived in the South Atlantic, Le Marchand described how:

Keeping the highly sophisticated but 21-year-old nuclear submarine at peak performance for night after day was a fantastic achievement. Heroes daily dealt with steam leaks, hydraulic bursts and even the odd fire; one person, the smallest man on board, had to slide 12ft down between the pressure hull and the port main condenser (a space of 9 by 18 inches cross-section) to replace a flange from which steam was leaking. Unrepaired we would not have been able to use full power – a crucial get-away requirement. Discipline and morale were outstanding.[195]

On 17 May, two days after arriving on station, Northwood received further intelligence which indicated that the Argentinian submarine *San Luis* was transiting to Puerto Belgrano and *Valiant* was ordered to intercept and sink it. Although weather conditions were flat and calm, and sonar conditions were excellent, a fishing trawler 'remained in the general area during the rest of the day, often confusing the towed-array picture with bursts of cavitation, flutter, diesel signature and tonals'. Also, the shallow water and no fewer than 180 biological contacts, such as whales, prevented *Valiant* from achieving a firm classification on the 'plethora of contacts' that she detected over a six-hour period, from 1200Z to 1800Z.[196] In order to remain vigilant, Le Marchand kept the bow caps on *Valiant*'s torpedo tubes open all afternoon. This was a typical entry in his log:

> At 1452 another promising [submarine] contact was gained (S92) to the East moving left. A rapid ranging manoeuvre gave 1936 range of 1600 yds, and VALIANT was on the point of cutting the firing bearing when the 'contact' split, passing down either side at 200 yds. Further attempts at relocation produced intermittent detections, but none with sufficient confidence or collateral to justify firing, and subsequent careful re-examination of the LOP confirms that no meaningful solution was obtainable. Nevertheless much disappointment was felt, and it will never be known for sure whether S92 was a missed opportunity, although on board analysis points yet again to a BIO contact. Most pundits would agree that a great deal of luck would be necessary for an SSN to hold an S209 class on main motors long enough to attack it, even without the unpredictable distractions caused by fish, fishermen and amorous sea mammals.[197]

With the arrival of *Valiant* the SSNs were all, by 17 May, involved with providing full defensive cover for the Amphibious Group. Three of the four SSNs were tasked with intercepting and attacking Argentinian surface ships moving east from their bases on the coast to strike the UK amphibious force, with the fourth SSN, HMS *Valiant*, positioned behind, acting as a sweeper to provide defence in depth. At first *Valiant* was allocated an area to the northeast of the Falkland Islands, in effect a centre billet, with *Spartan* to the west covering the northern ports, *Conqueror* to the east and *Splendid*, which had returned 'into the CVA [carrier] chase again' after successfully repairing its faulty turbo generator on 16 May, covering the Tierra del Fuego area to the south-west.[198] However, *Conqueror* was unable to take up its position due to yet more communication problems. Instead, *Valiant* was ordered to a new patrol area to the northwest of the Falklands, covering the port of Comodoro Rivadavia and the direct route between Argentina and the Islands.

Conqueror's communications equipment had in fact deteriorated to such an extent that it proved impossible to transmit or receive any communications traffic. The communication problems were now so severe that Wreford-Brown was forced to withdraw her sixty miles to the east and surface at night in order to attempt to replace the wire. 'This has most probably been the most frustrating day of the Patrol,' he wrote in the *Conqueror*'s patrol log:

> We were all set to move West into water where the ARG Warships are thought to be, when reception on the HF tail on the floating wire started to show signs of deteriorating. I therefore withdrew 60 NM to the East to surface in order to replace it with one of the other two I carried. On surfacing these were both found to be damaged by water ingress – Why? Brought all three below to work on them, but successful repair seems unlikely. I then had to watch the water to the West of me be reallocated to VALIANT, so that I now sit with an SSN buffer between me and any chance of some action. Present indications (confirmed at 190050) show that all 3 wires are just as useless. The result – I have wasted at least 36 hours operational time, lost the water where I might do something useful and now have to sit in a passive posture, not the best employment of an SSN; all because of the inability to repair a wire.[199]

Conqueror dived and after taking advice from Northwood the crew attempted to adapt all three wires. When *Conqueror* surfaced late on 21 May, in rough weather, and streamed one of the newly built wires it became wrapped around the submarine's propeller. This only became apparent when *Conqueror* dived and cavitated heavily at speeds above 7 knots, with obvious detection opportunities for the enemy. 'I sound as though I am trailing a metal dustbin,' wrote Wreford-Brown.[200] Attempts to dislodge the wire by putting *Conqueror* through a series of vigorous manoeuvres failed. The only option was to surface again and put a diver in the water to cut the wire, but the weather had deteriorated to such an extent that it had become too rough to surface again. 'It is too rough to put Divers down at present,' wrote Wreford-Brown, 'I shall have to wait for the weather to improve ... it is very restrictive and gives excellent opportunities for counter detection.'[201]

The weather finally abated on the afternoon of 23 May and Wreford-Brown called for volunteers to make up the dive team. Petty Officer Graham Libby responded, knowing that if *Conqueror* was detected by aircraft it would have to dive. The weather conditions were so bad that the likelihood of his lifeline parting on the edge of a propeller were very real. In both cases, his chances of recovery would have been negligible.[202] 'I was a single man,' recalls Libby. 'I was quite happy to go out there because I was pumped up. We had just sunk a blooming great warship – this could be the icing on the cake, you know. It's just something exciting that I might never ever get another chance to do.'[203] For twenty minutes Libby was battered by heavy waves in dark, freezing and terrifying conditions. At one point, while he was in the water, cutting away the wire, the Diving Supervisor on *Conqueror*'s casing was washed overboard and recovered by use of his lifeline at 2207. Libby succeeded in clearing most of the obstruction, cutting two long 'tails' of wire away from the propeller, leaving one six foot tail and an unknown amount still around 3–5 of the blades. Although cavitation was still present at some speeds, it was no longer operationally limiting and *Conqueror* could rejoin operations. 'I was absolutely done in, totally exhausted,' he said, 'I slept for eight hours straight after that.'[204] Libby received the Distinguished Service Medal for his 'cold, calculated courage and willingness to risk his life for the benefit of his ship far beyond any call of duty'.[205]

By the time *Conqueror* returned to operations, the Falklands conflict had entered its final phase. On 21 May, British forces began the invasion of the Falkland Islands, with 3000 troops and 1000 tons of supplies landing at San Carlos on East Falkland. This first phase of the invasion came at great cost. The Royal Navy frigate HMS *Ardent* was sunk by Argentinian aircraft, leaving twenty-two dead, while HMS *Argonaut* and HMS *Antrim* were hit by Argentinian bombs that failed to explode. Three days later, HMS *Antelope* was lost after a bomb disposal expert was killed attempting to defuse an unexploded bomb. The next day, the Royal Navy destroyer HMS *Coventry* was also sunk with the loss of nineteen men after it was attacked by Argentinian aircraft. The Merchant Navy container ship *Atlantic Conveyor* was also hit by Exocet missiles and later abandoned with the loss of twelve crew and vital supplies.[206]

On board the SSNs, the crews were all frustrated by the absence of any contacts, feelings that were deeply exacerbated by the news of British losses in the amphibious operating area in and around the Islands. HMS *Splendid*'s crew took the losses very hard. '*Ardent* and *Alacrity* and *Sheffield* really hit us hard,' recalls Lane-Nott, 'very hard indeed because we were frustrated that we were not being used properly, we weren't getting the information and they were getting through to these guys and why weren't we in there. I would have liked to have operated inside the Falkland Sound. I felt that we could have been a great support to the landings.'[207] Taylor on *Spartan* felt like 'some sort of detached visitor' compared to those in the ships sitting in bomb alley. 'We were out there tucking into our soup and sarnies in a shirt-sleeved environment and getting really quite frustrated.'[208] 'Throughout the deployment all were acutely interested in how the "real battle" was going,' recalled Le Marchand, 'and determined to do all possible to contribute.'[209] At the time of the landings HMS *Valiant* was 'ideally placed' in the centre of the Estrecho – straits – de le Maire to attack two Argentinian destroyers it had sighted. However, the two destroyers were firmly within the Argentinian twelve-mile limit and *Valiant* was unable to attack without a change in the Rules of Engagement. After identifying the destroyers, Le Marchand signalled Northwood, 'seeking an ROE change in order to effect retribution for British losses within the British territorial waters of the Falkland Islands'.[210]

Back in London, Le Marchand's request triggered an intense debate about how to respond to British losses and the Argentinian warships lurking inside the Argentinian twelve-mile limit. On 27 May, the Chiefs of Staff argued that there had been 'a considerable change in both political and military circumstances' since the previous discussions about the Rules of Engagement in Argentinian territorial waters on 7 May. 'Negotiations are no longer taking place,' they said. 'The military conflict has escalated dramatically with four of our warships having been sunk. The Argentines have not imposed any parallel restrictions upon their own freedom of action. On the contrary they continue to occupy much of the Falkland Islands and are attacking from safe bases British forces within the Falkland Islands (i.e. British) territorial waters.' As two such losses had occurred in the British territorial waters, the Chiefs of Staff believed that the time had now come to 'remove the sanctuary' which the British Government had 'allowed up to now for Argentine forces within 12 NM of their own coast'.[211]

At a meeting of OD(SA) on 28 May, the Chiefs put their case forward in a paper that sought 'discretion for the Argentine Navy to be attacked inside the 12-mile limit'.[212] During what Mrs Thatcher described as 'difficult' discussions, ministers pointed out that 'Britain had lost 4 ships in the past week, two of them in British territorial waters' and that 'The Exocet missiles with which a number of Argentine navy ships were armed constituted a serious threat to current operations, and the possibility of a sudden sally by such ships could not be ruled out. In these circumstances,' they argued, 'it was unreasonable

and dangerous that the Argentine navy should be allowed sanctuary if
within 12 miles of the coast of Argentina.'[213] Despite ministerial concerns that 'Public opinion would not understand if opportunities . . .
were missed and later a major British ship such as the SS *Canberra*
were in consequence sunk', ministers were concerned about the lack of
'legal justification for operations within what Britain herself regarded
as Argentine territorial waters unless war had first been declared'. Ministers also argued that the 'threat posed by the Argentine navy, while
within these waters, was not direct enough to justify action under the
right of self-defence' and that 'Force could not be used if it was disproportionate.' Ministers were also concerned about wider arguments
against what would be seen as 'the equivalent of an attack on the
Argentine mainland' and the possibility that such attacks would
increase the danger of other Latin American countries joining the conflict on Argentina's side.[214] The War Cabinet postponed a decision as to
whether or not to allow the attack of Argentinian Navy ships within
the twelve-mile limit from 28 May until the following week.[215] They
had two further meetings, but failed to reach a decision.[216]

Mrs Thatcher was well aware of how frustrated the submarine commanders were:

> We remained grievously concerned at the naval losses and aware that the
> surface Argentine navy had retreated into 12 mile territorial waters. As
> two of our ships had been sunk in <u>our</u> territorial waters we tackled
> the A. G. [Attorney General] – could we <u>not</u> sink theirs in their territorial
> waters. There were parts of the coast where the water was deep enough
> for our submarines to operate. But time and time again the A. G. said No.
> Once they moved out & into the direction of the Falklands – yes – but not
> unless. Our submarine commanders were left prowling up & down the
> line, very frustrated.[217]

There is little doubt that, had the Cabinet allowed attacks within
Argentinian territorial waters, *Valiant* would have been able to sink the
destroyers. 'Weather conditions in the area were ferocious,' recorded
Le Marchand, 'although visibility was generally very good, and predicted sonar ranges gave a high probability of detecting the enemy.'[218]
That afternoon, *Valiant* also detected a 'Ton' class minesweeper, but as
with the destroyers was unable to attack it. On 31 May, *Valiant* was
ordered to locate the hospital ship *Bahia Paraiso*, which was suspected
of abusing her hospital status in carrying supplies or reinforcements to
the Argentinian land forces. Le Marchand had 'severe misgivings'

about taking *Valiant* into the declared destination of the hospital ship, the Bay of Harbours in the Falkland Islands, and after carefully considering all the navigational data for the harbour, parts of which were shallow and badly surveyed, he decided not to take *Valiant* inside.[219] On 6 June, Le Marchand then sighted *Hipólito Bouchard* well inside the twelve-mile limit south of Tierra del Fuego. As Le Marchand wrote:

> This was a particularly frustrating event because enemy action in the Falkland Islands was at its height and already had brought the devastating loss of four of Her Majesty's Ships and many lives, but Rules of Engagement specifically precluded an attack on Argentine forces within their own territorial waters. An understanding of the immense political implications of such an attack did nothing to allay my own and my Ship's Company's disappointment as the destroyer had to be permitted to pass a '90-track' at a range of two thousand yards, and proceed unmolested to Ushuaia.[220]

HMS *Conqueror* was also operating off the Argentinian coast, just outside the twelve-mile limit, reacting to intelligence which indicated that two Argentinian Task Groups, the first consisting of the Type 42s and American-built destroyers, and the second of the Type 69 frigates, were both moving down the Argentinian coast to an area between Puerto Belgrano and Puerto Deseado. But on 4 June intelligence indicated that one of the Argentinian Type 42 destroyers, *Hércules*, had damaged a shaft and had retreated to Puerto Belgrano with *Santísima Trinidad* in company. After briefly patrolling off Delgado Point, *Conqueror* then moved north to search for *Hércules*, which, on 5 June, intelligence indicated was creeping along the coast as far inshore as possible. The Gulf of San Matías presented an opportunity for *Conqueror*. The entrance to the bay was greater than twenty-four miles, so *Hércules* would have to leave safe water unless she continued to hug the coast and went around the inside of the bay to remain in territorial waters. *Conqueror* waited, but detected nothing. It looked as if the Argentinian destroyer had taken the long way round inside the bay. Wreford-Brown decided to take *Conqueror* inside to have a look. 'This was another interesting navigational challenge,' noted Powis. 'The area of chart on which we were to navigate was a little smaller than a post card: approximately 5 miles to the inch.'[221] *Conqueror* moved into the bay, which in places was so shallow that the depth beneath *Conqueror*'s keel was between 130 and 34 feet, and spent until 8 June searching for the *Hércules*. But by 8 June Wreford-Brown was forced to reluctantly conclude that *Conqueror* had missed the target. He withdrew *Conqueror* from the bay and later that day intelligence indicated that *Hércules* had arrived at Belgrano, seventeen hours ahead of schedule. 'Therefore we missed her before even entering Gulf San MATIAS,' wrote Wreford-Brown. 'However, it was an interesting experience!'[222]

HMS *Spartan* was also searching for targets. Intelligence indicated that *Veinticinco de Mayo* was operating off Cabo Blanco.[223] At one stage *Spartan* came slowly but inexorably – and inexplicably – to a complete halt. Eventually it became clear that the 4900-ton submarine had been arrested by a giant bank of kelp, a reminder that the sea still had some surprises to spring.[224] *Spartan* was eventually propelled free after varying her depth by pumping and flooding. 'Very bizarre experience,' noted Taylor, who made extra efforts to avoid the 'water gardens'. Eventually *Spartan* 'Intercepted contact heading North across mouth of Gulf.' *Spartan*'s operators detected two shafts, five blades. 'This looked very hopeful,' wrote Taylor, 'all swing into action well at the prospect of a target at last.' But as with the other SSNs there was nothing to shoot at. 'A terrible anti-climax to find a modern 9000 tonne M/V on arrival at P/D,' he recorded. *Spartan* continued to hunt, but with little success. 'No other contacts, no intercepts,' continued Taylor on 4–5 June. 'A depressingly quiet billet, and the enemy shows little sign of wishing to run the gauntlet.'[225] 'Not enough targets' was how Able Seaman Graham Wrigley later summed up his view of HMS *Spartan*'s Falklands tour.[226]

Back in Northwood, Herbert was distinctly aware of how tedious the SSNs were finding the restrictions they were operating under. On 6 June, he signalled *Conqueror*:

> I am sure you are frustrated by lack of Targets and last minute programme changes. However, during this critical period when we are unclear of ARG intentions, we must remain ready to support TG 317.8 with all available assets. The overall programme is being kept under continuous review and I will keep you informed once plans firm up.[227]

By early June, Northwood started to look again at its long-term plans to maintain a force of SSNs in the South Atlantic. Wreford-Brown was asked how long *Conqueror*'s crew could survive with the supplies on board. He advised that there were enough supplies for a further twenty-eight days on normal diet and forty-two days on a restricted diet. On 3 June, in anticipation of a prolonged stay in the South Atlantic, Wreford-Brown introduced food rationing just in case the submarine was required to remain at sea into July. 'Enforced Dieting' was implemented with a typical day's meals on board *Conqueror* looking like this:

Breakfast	One egg
	One piece of bacon OR One sausage
	Baked Beans OR Spaghetti
	Arries
	'Special' every other day
	Bread/Toast – as normal
	Jams/Marmalades – as normal
	Butter/Margarine – will be rationed
	Tea/Coffee as normal
Lunch	Thick home-made soup with bread
	Cold meats/pickles three times a week
	Sundays – Normal full Roast and duff
Supper	Normal 2 choice main course with 2 veg + spuds
	Duff
	Sundays – Rice/Curry/Chinky/Stroganoff/Pot Mess[228]

Taylor also introduced a restricted diet for HMS *Spartan*'s crew, after forty-five days at sea, which consisted of a single-choice cooked breakfast, lunch of soup, bread and a pudding, and single-course evening meal. This diet was further restricted after sixty-four days by the introduction of a 'continental breakfast' consisting of coffee or tea and a bun on alternate days. The diet proved less unpopular than might have been supposed, and in many cases led to a welcome weight loss.[229]

REINFORCEMENTS

As the battle for the Falklands continued the beginning of June saw the withdrawal of the first SSN, HMS *Splendid*, which began its transit back to Devonport on 28 May. *Splendid* was replaced by HMS *Courageous*, which had arrived back in the United Kingdom in early May after a 26,000-mile, 302-day deployment. The submarine was put into the floating dock in Faslane, where 'manic' much-needed hull preservation and essential maintenance were rushed through in order to get the submarine in a fit and ready state to deploy south. *Courageous*'s

CO, Commander Rupert Best, was concerned that 'the leave and maintenance, harbour training and Index/Calibration period might be barely sufficient to rectify the many accumulated defects as well as complete the programmed maintenance', but thanks to the support of the Faslane Base and assistance from other submarines the submarine sailed for the Falklands on 12 May, ten days earlier than originally planned, with a full war load of thirty-one weapons, including Tigerfish, Mark 8 torpedoes and the new Royal Navy Sub Harpoon missile system – the first time it had been deployed operationally – which the crew were eager to use against the Argentinian carrier. 'To 25 *de Mayo*, from *Courageous*' was inscribed on one of the capsules that housed the deadly missiles.[230]

Courageous arrived in the waters surrounding the Falklands on 28 May, the day *Splendid* left for the UK, and was positioned to the west of the Islands, transiting down the 100-fathom line to the twelve-mile limit off Puerto Deseado. 'Depending upon where we were, our task was to patrol and look out for aircraft and surface ships,' recalled Best. 'We were shifted about a bit, first north and then north-west of the Falklands before returning to the Argentinian coast and the airfields at Comodoro Rivadavia, Puerto Deseado (where the story behind the *Rime of the Ancient Mariner* originated) and San Julian.' Best was 'keen that the Task Force should recognize the unique capability of RNSH [Royal Navy Sub Harpoon] as an anti-ship weapon and therefore position us so that we would be in the right place to take advantage of this long range missile should any opportunity to use it occur'. But aside from contact with the Argentinian hospital ship *Bahia Paraiso*, which Best also allowed to proceed on her way, *Courageous* saw little of the enemy. 'On occasions we were going quite close in, to about 20 miles off the coast,' recalled Best. 'The Argentinian fleet was in harbour, the *Hércules* . . . and one or two others came out occasionally, but well inshore and basically scuttled from one place to the next. We were there listening out and waiting for any intelligence that we could gain on their carrier 25 *de Mayo* or anyone else.' Disappointingly for *Courageous*'s crew, the carrier failed to appear. 'On the whole we were doing our own thing and rather independent of all that was happening around the Falkland Islands,' said Best.[231]

The SSK HMS *Onyx* also arrived in the waters surrounding the Falklands, bringing the total number of Royal Navy submarines in the South Atlantic to five. *Onyx*, under the command of Lieutenant Commander Andy Johnson, was equipped with a new five-man chamber for deploying and recovering Special Forces. The submarine had been recalled from Devonport on 18 April and ordered to proceed to Gosport to store for war. An intense week at Portsmouth followed, while the crew prepared for departure, carrying out essential maintenance, storing supplies and loading weapons, including 10 Mark 24 torpedoes, 11 Mark 8s and 2 Mark 20s. 'There was also a need for briefing, planning and the assimilation of large quantities of data concerning both our own and Argentine units, not least of which was a geography lesson,' remembered Johnson, 'Few of us had any idea of the size or exact location of the islands we were off to recover!'[232] A week later, on 26 April, *Onyx* departed from HMS *Dolphin* in Gosport and set course for the South Atlantic.

On 16 May, the submarine arrived at Ascension Island to restock stores and carry out what was probably the first submarine refuelling at sea from a tanker in over forty years. During the refuelling one of *Onyx*'s external fuel oil tanks was punctured, but the leak was only discovered after the submarine had left Ascension and it was forced to return and find sheltered waters so that divers could carry out a quick repair. *Onyx* finally departed Ascension on 19 May and set course for the South Atlantic. Diesel electric submarines were considerably slower than their nuclear-powered counterparts so travelling to the Falklands was a feat in itself, a journey that was expected to take just over a month. 'We found the passage a trial,' admitted Johnson, 'particularly after we left the tropics and entered the South Atlantic winter. Surfaced at night to travel faster was wet, cold and uncomfortable, dived in the day was quieter, but it seemed we would be travelling for ever.'[233]

Onyx arrived in San Carlos Water on 31 May. Although intended to insert and extract Special Forces, the submarine was only used in this role once operationally. 'It was a great pity that she had not been dispatched in time to reach us before D Day for the four nuclear boats in the South Atlantic were not ideal for SBS insertions in shallow water,' wrote Michael Clapp and Ewen Southby-Tailyour.[234] *Onyx* was to have played a pivotal role in an aborted Special Forces assault, codenamed Operation 'Kettledrum', on the Argentinian mainland, designed to destroy Argentina's Exocet missiles before they could be launched by the Super Etendard fighter-bombers that carried them. *Onyx* was to have delivered a team of Special Boat Service Marines to attack a naval airbase at Puerto Deseado, which Northwood suspected was being used by Argentinian pilots following missions against the Task Force and the Falkland Islands.[235] However, the day before HMS *Onyx* arrived in San Carlos, the Argentinians had expended the last of their Exocet missiles and due to the considerable efforts of the British Secret Intelligence Service there was little prospect of the country obtaining any replacements. Thus the need for the operation was questionable. The Officer Commanding, SBS, Major Jonathan Thomson, was so concerned about the operation 'inside the borders of a country with which the UK was not, in any legal sense, at war' that he insisted on leading the operation himself.[236]

On 1 June, *Onyx* berthed alongside HMS *Fearless* and collected a team of SBS Marines and their associated equipment. The submarine sailed from San Carlos Water at 0700 to prepare for Operation 'Kettledrum', with a lot of what Johnson called 'unanswered questions'. It was still unclear whether the operation was to be 'in-and-out' or one way. Regardless of the answer, merely conducting a landing at Puerto Deseado would have been fraught with difficulties, due to the seabed profile and the fact that the Gemini inflatables would have 'to be launched from as far as 20 miles offshore, where the water begins to shoal sharply from 230 feet to 100 feet'.[237] Accommodating the additional SBS personnel within the cramped confines of the small submarine was not easy. As Johnson recorded:

> When we stored prior to our departure, we had laid a complete 'false deck' of canned food and stores throughout the submarine, reducing headroom from six to four feet in some places. Even the showers were full of stores. I can remember the three occasions in the entire 116 day patrol on which I was able to have a proper wash, rather than just a dip in a bucket! Living with eighty-four people in a space designed for sixty-eight, as well as all the extra equipment, was the epitome of 'cheek by jowl'. Remarkably, however, there was no friction, and our 'visitors' soon became fully integrated into the crew, taking a full part in the day-to-day operation of the boat.[238]

Onyx continued to prepare for Operation 'Kettledrum' throughout 2 and 3 June, conducting a series of 'wet drills' with the SBS: surfacing the submarine, inflating two Geminis on the casing and diving beneath the craft, before surfacing again to quickly recover them. But at 1800 on 3 June, Northwood cancelled the operation. 'Puerto Deseado was such an insignificant air base that it did not warrant a major military and diplomatic failure – and a failure it would have been.'[239]

Onyx was re-tasked to conduct a far more pressing operation. The

British suspected that the Argentinians had landed on Weddell Island, the third-largest of the Falkland Islands, and that an Argentinian reconnaissance team was involved in directing Argentinian aircraft towards the archipelago. If this was the case, the reconnaissance team had to be taken out. On 5 June, *Onyx* attempted to insert an SBS team at Chatham Harbour on Weddell Island, but at 1314, while transiting to the area, the submarine struck an uncharted rock pinnacle south of Cape Meredith, damaging the bow shutters on two torpedo tubes and trapping a Mark 24 torpedo in one of the tubes. 'The SBS Operations Officer was sitting in the seat nearest the wardroom door when this happened – we were having lunch,' recalled Johnson. 'He was trampled by the rest of the wardroom who literally ran over him to get into the control room. He arrived some minutes later wearing a lifejacket and enquiring politely if everything was all right!' Concerned, Johnson asked the SBS officer if he would don his diving gear and exit the submarine and carry out an inspection. The SBS officer politely declined.[240] *Onyx* resumed the operation but was unable to land the SBS team at Chatham Harbour due to poor weather. A second attempt, at a different southern location, Pillar Cove, was carried out the next day. *Onyx* went to action stations at 2000 and approached to within 3 cables, 0.3 nautical miles, of the coast, where the SBS team was successfully inserted. They found no evidence of Argentinian activity.

PICKET DUTY

With British land forces involved in fierce fighting as they advanced across the Falklands towards the capital, Port Stanley, and the reluctance of the Argentinian Navy to leave Argentinian territorial waters, the SSNs sought to contribute to the conflict in a way that demonstrated the flexibility of submarines if the crews were well trained and capable of unrestricted operations in shallow water.[241] While HMS *Valiant* was waiting west of the Falklands for two Argentinian destroyers, the sister ships *Hipólito Bouchard* and *Piedra Buena* to run the gauntlet from the mainland, *Valiant*'s crew realized that their submarine was between forty and sixty miles off Rio Grande Naval Air Base, the home of the Super Etendards, Naval Skyhawks and other Argentinian aircraft and sitting directly under the flight path the Argentinian air force was using to reach the Falklands. *Valiant*'s crew started to send flash enemy locating reports to the Task Group of all the Argentinian aircraft approaching the Falklands, by simply looking through the periscope and counting the number of aircraft that flew over the submarine. This seemed a relatively safe and easy way to contribute to the lack of early warning available to the Task Force.

At 1815 on 23 May, *Valiant*'s crew suddenly became aware of a number of explosions, each getting progressively louder, until the fifth and last was loud enough to shake the submarine. *Valiant* had taken up position on the flight path of Argentinian aircraft adopting the shortest route back to the mainland dropping their bombs while being chased by the Task Force's Sea Harriers. 'The moral was get off track from their return route,' noted Le Marchand, 'We were lucky, but a few feet closer and it might have been something of a bad luck story.'[242] Later analysis revealed that some of the bombs used by the Argentinians had a short delay in their fusing, which coupled with the channelling effect of the shallow water explained the heaviness of some of the detonations. 'FOSM seems to be impressed with our ability to act like this,' wrote Le Marchand.[243] Thanks to SSIXs communications equipment, *Valiant* was able to detect, classify, report and receive acknowledgement from the Task Force within about five minutes and in some cases as little as two minutes. 'This is the first time anything like this has been done as it would have been impossible with H. F. Communications,' wrote *Valiant*'s Navigator, David White.[244] As Le Marchand noted, while sitting in *Valiant*, counting Argentinian aircraft, 'a new dimension to the Concept of SSN Operations was born'.[245]

Just how valuable this information was to the Task Force was demonstrated on 8 June 1982, when a Chilean long-range radar that had been providing vital intelligence to the Task Force about Argentinian aircraft taking off and landing from airbases in Argentina failed.[246] With the radar out of action, Argentinian aircraft took off late on the morning of 8 June. Flying with their radar permanently switched on, the aircraft were easily detected by *Valiant*'s electronic-warfare mast. *Valiant* immediately signalled Northwood via satellite to circulate an air-raid warning to the Task Force. As this could be completed in around two minutes and the distance to East Falkland was 440 miles there should have been plenty of time to notify the ships in San Carlos Water and the ground troops to prepare for the attack and take all necessary precautions. However, due to communication problems *Valiant*'s messages did not get through to the Task Force in time.[247]

At San Carlos Water, the first indication of any attack was when the Harriers on Combat Air Patrol were drawn to four Mirages from the Argentinian 8th Fighter Group, which had conducted a low-level flight to the north of the Falklands. However, these were decoys while the main flight went for the landing ships. The Harriers then diverted to the Type 12 frigate HMS *Plymouth* in Falkland Sound, which was attacked by Daggers and hit by four bombs which fortunately, due to insufficient altitude, failed to detonate. The Harriers were then forced to return to the carriers to refuel, leaving the landings at Bluff Cove unprotected. Five Skyhawks flew on to their targets in Bluff Cove. The first run over Fitzroy was unsuccessful, but during the second run the Skyhawks sighted *Sir Galahad* and *Sir Tristram*. *Sir Galahad* was attacked from the east, taking three direct hits with conventional 250-kilogram bombs. A cluster of bombs then hit *Sir Galahad* again before all five Argentinian aircraft turned south and returned back to base. Fifty-six British servicemen were killed and over 150 were wounded. In his book on naval intelligence, Nigel West concludes that 'The central tragedy of Bluff Cove is that the warning transmitted by *Valiant* on the morning of the raid never reached the Welsh Guards at Bluff Cove, who were completely unaware of the urgent need to disembark quickly.'[248]

Despite the Bluff Cove tragedy, early-warning duty was now one of the most important tasks for the Royal Navy submarines stationed around the Falklands. HMS *Onyx* and the SSNs were ordered to join the air picket patrol. *Onyx* was stationed off the west of the Falklands, where she spent fifteen days monitoring Argentinian aircraft movements, while *Conqueror* moved to early-warning picket duties off Puerto Deseado from 11 to 14 June, while *Spartan* was in the north, observing mainly transport aircraft. The early-warning information the SSNs were providing was so valuable to the Task Force that *Valiant* was ordered to turn back while sprinting to intercept an Argentinian destroyer, the *Piedra Buena*, as the 'job of ELINT picket off RIO GRANDE considered to be more important'.[249] This was possible despite the fact that *Valiant*'s interception equipment had a number of shortcomings: an inability to identify detections rapidly, an inability to determine radio frequency, and poor bearing discrimination. The 'Valiant' class was fitted with what Powis on *Conqueror* described as 'a truly ante-diluvian' ESM outfit:

> Barely capable of reliable operations for a periodic return to periscope depth it was a thing of cathode ray tubes and valves. It warmed up like a

1950s television set but was not half so easy to use. More than 4 or 5 radars in a band and it was swamped. In coastal waters it was at its limit: off an airbase with 2 dozen angry aircraft testing their radar before launch it was all but useless. We knew something was coming our way but until we saw its classification, it was guesswork.[250]

Equipment shortcomings meant, as Le Marchand noted, that 'any meaningful attempt at Target Motion Analysis in the time scale was a forlorn hope'. However:

as operator experience increased, considerable skill was developed and much confidence held in the early identification of emitters, even with the very few sweeps that could be intercepted before an aircraft had passed. At the same time a highly efficient periscope watch was maintained, and on several occasions raids were observed as contrails heading East, whilst low-flying aircraft could be identified shortly after take off or landing.[251]

Valiant's record speaks for itself. From 17 May to 10 July, *Valiant* detected, classified and reported a total of 263 contacts to the Task Group, a feat greatly appreciated by Woodward and the ships that made up the Task Force. Before *Valiant* departed the South Atlantic on 11 July, Woodward sent the following signal to Le Marchand:

1. OVER 300 ENEMY REPORTS TESTIFY TO THE EFFI-CIENT CONDUCT OF YOUR LONELY VIGIL OFF RIO GRANDE. YOU CAN BE PROUD OF YOUR PART IN CORPORATE AND I AM GRATEFUL FOR ALL YOU HAVE DONE
2. BRAVO ZULU AND BON VOYAGE.

Chris Craig, the CO of HMS *Alacrity*, later wrote of the surveillance provided by the submarines:

Such notice would enable us to shorten our readiness at appropriate times and significantly reduce the potential for surprise attack, even though we had just lost one of our three air-defence destroyers. The early warning that the submarines provided by watching the enemy's doorstep saved the lives of many men. That the achievement of *Valiant*, *Splendid* and others were not recognized more emphatically after the war, I found a travesty.[252]

DEPARTURE

On 14 June 1982, British forces advanced to the outskirts of the Falk-lands' capital, Port Stanley, where General Mario Menéndez, in command of 9800 troops, surrendered to Major General Jeremy Moore, Commander of British Land Forces. The conflict was over. As we have seen, HMS *Splendid* was the first to depart. She arrived back in Devonport to awaiting families on the jetty and a low-key reception. Her crew had been expecting to return to the South Atlantic, but the Argentinian surrender put a stop to that.[253] For Lane-Nott, Operation 'Corporate' was a 'demanding, stimulating though often frustrating experience' but he was impressed with how the Royal Navy's latest nuclear submarine had performed. 'It will be four months to the day since the submarine left Devonport,' he wrote. 'During this time SPLENDID has steamed over 30000 miles without either an AMP [Assisted Maintenance Period] or SMP [Short Maintenance Period]. The Swiftsure Class continues to prove itself ... For me personally, Operation Corporate has provided the culmination of a unique experience. I have taken SPLENDID from being hoops on the slip to war – something few are fortunate or privileged to achieve.'[254]

James Taylor's HMS *Spartan* started its homeward passage on 9 June and entered Devonport on 24 June, after steaming a total of 24,320 miles. The submarine was greeted with an emotional reception as she came alongside. The third submarine to depart the South Atlantic was HMS *Conqueror*. Before departing, *Conqueror* signalled *Courageous* and *Valiant*: 'Goodbye and Good Luck. We will spin a few dits in the Back-Bar for you. For COURAGEOUS – I look forward to being as well rigged as you on my return!' – a reference to RN Sub Harpoon. HMS *Courageous* responded with:

> VMT for your signal
> When they hear you've gone away
> We trust the Argies come out and play
> For we can't accept there are no
> Targets like the Gen BELGRANO
> As to Rig, we plan to say
> Performance matches bold display!
> Meanwhile, I know VAL will agree
> Fair shares for all 'mongst SM3

Go Conquering Heroes bathed in glory
Regale the Bombers with your Story

Well done. Good Luck. Happy Homecoming.

HMS *Conqueror* arrived in Faslane on 4 July 1982, flying the first Jolly Roger with an atomic symbol as well as a symbol denoting a sunken warship and a dagger for the special operations transfer off South Georgia. It was the first Jolly Roger flown from a nuclear submarine.[255] Herbert met each of his submarines as they returned from the war and the 'low-key sort of feeling about things' on board *Conqueror* surprised him. 'I didn't understand it at the time,' he says, 'when I went down, I thought everyone would be you know: "Super, we've done it." But it wasn't like that and it took me by surprise.'[256]

Operation 'Corporate' did not end with the surrender of Argentinian forces on 14 June 1982. HMS *Valiant* and HMS *Courageous* remained on patrol in the South Atlantic and continued to keep a close watch on the Argentinian Navy. On 29 June, *Valiant* sighted *H. Bouchard* and *P. Buena* off the coast near Rio Grande. But once again the Rules of Engagement prevented Le Marchand carrying out an attack. 'I have no doubt,' he wrote, 'that had the Rules permitted it I would have been able successfully to have attacked the two destroyers.'[257] By early July, supplies on board both submarines were running low. On *Courageous*, a lack of cigarettes forced many of the crew to start rolling and smoking pipe tobacco. *Valiant* departed for home on 11 July and arrived back in Faslane in early August after spending eighty-seven days at sea and travelling 26,400 nautical miles. 'There was one single frozen chicken in the ship's fridges,' recalled Le Marchand.[258] 'There is no doubt that the ultimate test is War,' he wrote, in the official summary of *Valiant*'s patrol, 'and whilst there is disappointment in not having a scalp to hang at our belts, I consider a real contribution to the collective CORPORATE effort has been made, and my Ship's Company deserve the highest praise for the manner in which they have done so.'[259] HMS *Courageous* left the Falklands on 29 July after a total of sixty-two days in the area. In total *Courageous*'s Operation 'Corporate' patrol lasted ninety-three days.

The last submarine to arrive home was HMS *Onyx*. Prior to departure *Onyx* had been awarded the unpleasant task of sinking the landing ship, RFA *Sir Galahad*, which had been damaged by bombing at Bluff Cove on 8 June. *Onyx* lined up in ideal conditions and Johnson took his boat to action stations at 1215, firing the first Mark 24 at 1235. It failed to explode. At 1254, *Onyx* fired another Mark 24, but it also failed. The crew were unable to determine what caused the failure, but they suspected there was a problem with the torpedo batteries. At this stage Johnson gave up on the Mark 24s and resorted to the Mark 8s. At 1330, a Mark 8, the same torpedo Wreford-Brown had used against the *General Belgrano*, was fired from a range of 1500 yards. It struck RFA *Sir Galahad*, which later sank. Had Wreford-Brown used the Mark 24, could he have expected similar results?

Onyx's transit homeward started on 17 July and continued on the surface from 22 July without incident until 11 August, when the starboard main generator developed a fault which could not be repaired. Passage continued at reduced speed until the submarine finally arrived in Gosport on 18 August 1982. The patrol covered 20,000 miles in four months with minimum support. The submarine was immediately put into a floating dry dock in Portsmouth, where damage to the hull and the Mark 24 torpedo trapped in the tube could be examined. Johnson explained how:

> Both bow tubes were damaged while the torpedo in one tube was cracked like an egg, with the safety range clock 'wound off' as the battery had partially energised ... the experts in the armament depot had no idea how to dismantle the torpedo while it was still in the tube – and we couldn't move it forward or back. In the end, the dockyard staff cut away the area of the torpedo tube around the warhead, then an engineer from the dockyard, with myself and one of the 'fore-ends men' hacked the sonar head off the torpedo with drills and crowbars in the middle of the night. The area around the floating dock had to be evacuated and the cross-channel ferry terminal closed while we did so. So I guess from that we can conclude that the torpedo was in quite a dangerous state. One expert assured us that it could certainly explode at any time – not that we knew that until we entered Portsmouth. Ignorance is bliss![260]

That the torpedo had failed to explode during *Onyx*'s 8000-mile journey back to the United Kingdom was a minor miracle.

AFTERMATH

There was a near constant SSN presence around the Falkland Islands for years after the conflict. This was almost entirely the duty of the older SSNs, HMS *Valiant*, HMS *Warspite* and HMS *Courageous*.

These submarines were known as the South Atlantic Runners. Before departing on patrol the boats would enter the floating dock and be fitted with a special Falklands propeller. HMS *Warspite* was the first submarine to conduct a post-conflict patrol. Fresh out of refit from Chatham after completing an emergency workup in April and May, *Warspite*, under the command of Commander Jonathan Cooke, sailed on 25 June for a 75-day patrol, 43 days of which were spent on station in the Falklands. *Warspite* was armed with a full weapon load of 20 Harpoons, 5 Mark 8 torpedoes and 6 Mark 24s. In the aftermath of the fighting the period on station was extremely uneventful, which was something of an anti-climax for her crew, who had spent a considerable amount of time rushing the submarine through the refit. *Warspite* was assigned to an area to the north of the Falklands, on the edge of the Protection Zone. She reported twenty Polish trawlers and one merchant vessel before returning to Faslane on 21 September.

Warspite's second 'Corporate' patrol in the South Atlantic was far more eventful. The submarine sailed from Faslane on 25 November. She was due to return in February but because of a defect in the submarine designated to take over from her the patrol was extended. *Warspite* completed what was at that stage the longest submarine patrol in Royal Navy history, 111 days (77 days on station) between 25 November 1982 and 15 March 1983 and 57,085 kilometres (30,804 nautical miles), which remains in the *Guinness Book of Records* as the longest submarine patrol in history. *Warspite* sailed from Chatham for the first time without the Mark 8 torpedoes on board as a result of conversion to the Tigerfish Mark 24 Mod 1, along with sixteen RNSHs. The Commanding Officer, Jonathan Cooke, was due to marry his fiancée the weekend before *Warspite* was now expected to return from sea. The crew even prepared a wedding cake for their CO. But *Warspite* spent so long in the South Atlantic that by the time she returned to Faslane there was little food remaining on board. 'We were living on tinned food,' recalled Cooke, 'including steak and kidney and tomatoes. If you looked in the deep freezer just now, I think you'd find about three herrings and two lemons. Another week and my cake would have gone. We did not anticipate we would be out so long and I believe it is the longest trip by quite some time. The crew kept going very well. I think what kept them cheerful was thinking about the problems I had with my wedding.'[261] 'We have asked them to make an extraordinary effort and they have responded magnificently,' said the Secretary of State for Defence, Michael Heseltine, after he was winched on board off Ardrossan in the Clyde approaches on 15 March.[262]

HMS *Valiant* deployed for yet another patrol on 1 August 1983 until 29 October, while HMS *Warspite* conducted another patrol between 5 January and 3 April 1984. South Atlantic patrols involved operations off the Argentinian coast, where teams of specialists would monitor Argentinian radio frequencies. Commanding Officers were also told to occasionally let the Argentinians know that there was still a UK submarine presence in the area by deliberately exposing the submarine's periscope. This did not always produce immediate results, as one crew member on board HMS *Courageous* described:

> So we were at periscope depth amongst a fishing fleet and steamed up and down waiting for them to sight us, but no radio traffic. So we put up another mast and steamed up and down, still no radio chatter to indicate that we had been noticed, so Rupert [Best, CO of HMS *Courageous*] kept sticking up more and more masts, but still nothing! In the end getting slightly annoyed with the Argentine fishing fleet he told Ship Control to broach and expose the fin. This duly happened and the ether went mad with radio chatter from the fishing fleet.[263]

In 1985 Dan Conley took over command of HMS *Courageous* and conducted the first of three South Atlantic patrols:

> The three patrols were aimed at collecting general intelligence of Argentinian air and military activity. Most of the time *Courageous* conducted her patrols close to the Argentinian coast, mainly cruising at periscope depth just outside the 12 mile territorial limit during daylight hours, which was the only time any military activity was detected. We had embarked two Spanish speaking 'spooks' who with their specialist equipment were able to tune into Argentinian radio military circuits, particularly aircraft control frequencies. Our torpedo tubes were loaded with three Tigerfish Mod 1 torpedoes and three anti-ship Sub-Harpoon missiles. The Rules of Engagement directed that we attack and destroy any submarine detected within a 150-mile radius of the Falklands (the Exclusion Zone), but two years after the end of the war I doubt whether our politicians or the British people would have welcomed news of such an engagement. In the event Argentinian military activity was at a very low ebb; the only submarine detected was a German-built 209 Class testing its radar alongside in its base in the city of Mar del Plata. The only Argentinian surface vessel sighted was a coastguard cutter.[264]

Diesel submarines were also deployed to the South Atlantic. HMS *Opportune*, for example, conducted a South Atlantic deployment between 14 January 1985 and 28 July 1985, often acting as an 'enemy' submarine against British surface forces during exercises.[265] HMS *Sealion* also conducted a South Atlantic patrol in 1987 and returned to the United Kingdom flying a Jolly Roger, with two daggers, suggesting that the submarine was involved in a Special Forces-related operation. But, as HMS *Courageous*'s CO recalled, 'as time went on, it became equally clear that Northwood's priority and interest was now firmly focused back on the USSR even though those in the Falklands were still conscious of a residual threat. The reality was that there was not much left to be done by submarines in the South Atlantic.'[266]

THE DETERRENT

There is one other aspect of the Falklands crisis that requires clarification. After the conflict there was a widespread belief that in the early stages of the crisis the Government deployed one of the Royal Navy's Polaris submarines close to Ascension Island. A story later appeared in the *New Statesman* which alleged that the submarine was sent south in order to provide the option of launching a demonstrative nuclear attack against Córdoba in northern Argentina in the event of the loss of a major Royal Navy capital ship, such as one of the Task Force's aircraft carriers.[267] The details of the deployment were apparently 'given in a series of highly classified telegrams sent to the British Embassy in Washington'. According to Freedman, who 'found no such telegrams', the source for the *New Statesman* article was Alan Clark.[268]

The suggestion that a Polaris submarine was involved in the Falklands conflict first appeared in the early days of the crisis. HMS *Resolution*, under Commander Toby Elliott, was a few days into a standard 72-day patrol in the North Atlantic. At the same time as Elliott learned that the Argentinians had invaded the Falklands, he was somewhat surprised to hear BBC World News report that HMS *Resolution* was stationed off Buenos Aires. 'I sent for the Navigating Officer and told him this good news,' says Elliott.[269] He thought no more about it, particularly as *Resolution*'s crew never heard the report again. However, back in Faslane, the families of the sailors were so taken aback by the report that they sought clarification from Faslane, who quickly reassured them that it was inaccurate.

Elliott and *Resolution*'s crew were far too occupied with maintaining the UK's independent nuclear deterrent. Three to four weeks into the Falklands conflict the Soviet Northern Fleet decided to take advantage of the fact that a significant portion of the Royal Navy's submarine fleet was occupied in the South Atlantic. Seven SSNs from the Soviet Northern Fleet deployed into the Norwegian Sea, through the Iceland–Faroes Gap and into the North Atlantic. Some went down into the Bay of Biscay while the others continued to hunt for HMS *Resolution*. US and Canadian Maritime Patrol Aircraft, as well as the RAF's Nimrods were deployed to keep track of the Soviet SSN breakout and at one point an RAF Nimrod thought it had detected *Resolution* in an area that was less than two miles away from a Soviet SSN. Northwood was immediately alerted.[270]

Despite the best efforts of the Soviet Northern Fleet, HMS *Resolution* remained undetected throughout its 72-day patrol, although Elliott was forced to carry out a number of evasive manoeuvres in order to avoid a Soviet 'November' class submarine during the initial stages of the SSN breakout. 'The one thing which I knew we were so keen to preserve was the very straightforward claim that no SSBN had been detected,' he said.[271] The Soviets continued to search for *Resolution* for the remainder of its 72-day deterrent patrol.

Both Elliott and Herbert are clear that there was 'absolutely no thought of redirecting an SSBN down to the South'. The reports, says Herbert, are 'Rubbish. Absolute rubbish ... I think I would have known ... A crazy idea ... Not on my watch.'[272] 'I have to say,' said Elliott, 'that at no time did I ever get the feeling that we were going to be needed. We remained well out of maximum strike range. We did not even feel the urge to break out the charts which would have been required for the long voyage south.'[273] However, speculation that a Polaris submarine was sent to the South Atlantic was given further credence with the publication in November 2005 of the diary of François Mitterrand's psychiatrist, Ali Magoudi. According to Magoudi, on 7 May 1982, a few days after an Exocet missile fired from a Super Etendard aircraft had struck HMS *Sheffield*, Mitterrand, who was late for a meeting, said:

> Excuse me. I had a difference to settle with the Iron Lady. That Thatcher, what an impossible woman! With her four nuclear submarines in the South Atlantic, she's threatening to unleash an atomic weapon against Argentina if I don't provide her with the secret codes that will make the missiles we sold the Argentinians deaf and blind.

The Prime Minister was apparently so livid that she blamed Mitterrand 'personally for this new Trafalgar' and he was 'obliged to give in. She's got them now, the codes':

> One cannot win against the insular syndrome of an unbridled Englishwoman. Provoke a nuclear war for a few islands inhabited by three sheep as hairy as they are freezing! But it's a good job I gave way. Otherwise, I assure you, the Lady's metallic finger would have hit the button.[274]

Exactly what 'codes' Magoudi is referring to is unclear. What is clear, as the official historian of the Falklands campaign, Sir Lawrence Freedman, has noted, is that 'the nuclear option was never seriously discussed in the War Cabinet and nor were preparations made for its implementation by the Royal Navy'.[275] However, we do now know that Margaret Thatcher does appear to have considered the use of nuclear weapons against Argentina under certain extreme circumstances. 'She certainly took it very seriously, to the point certainly where she would have been willing to face up to the real eventuality of use,' recalled the former Permanent Secretary at the Ministry of Defence, the late Sir Michael Quinlan:

> I do recall an occasion after the Falklands War when she said something to me which suggested that she would have been prepared actually to consider nuclear weapons had the Falklands gone sour on her. I found that a terrifying suggestion, but she undoubtedly made it . . . The very fact that she could say it raised my eyebrows and my hair slightly.[276]

Submarines played an important role in the Falklands conflict, maintaining a successful blockade of the Falkland Islands and forcing the Argentinian Navy to remain within its own territorial limits from the time the Maritime Exclusion Zone was established until the end of the conflict and sinking one of the Argentinian capital ships. But, as Rear Admiral Frank Grenier, Flag Officer Submarines between 1987 and 1989, later noted:

> Submarines did not win the Falklands War. Their contribution was a combination of achievement and threat. Performance, as ever, came from the inheritance of fine breeding, sound experience and a wonderful response from our people. The machines were a tribute to the technicians, both on the spot and in support. We were fortunate that the political resolve and backing were in place.[277]

But could the Submarine Service have prevented the conflict? During a conference on the Falklands conflict at the Centre for Contemporary British History in 2005, John Nott and Admiral Woodward clashed when attempting to answer that very question:

John Nott: 'The big "what if" of history is what would have happened if we had one or two nuclear submarines there. Is anybody seriously suggesting that that would have deterred the Argentinians?'

Admiral Woodward: 'I am damn sure of it.'

John Nott: 'Right, okay, well that's a big "what if" of history.'

Admiral Woodward: 'I'm a submariner.'

John Nott: 'You are a submariner. I have no doubt that if Margaret Thatcher had immediately agreed rules of engagement to allow us to sink Argentine merchant ships approaching the Falkland Islands, then a nuclear submarine could have sunk them. But what I am saying is – if we had got together and tried to decide whether we would give rules of engagement for the sinking of approaching Argentine merchant ships, it seems to me extremely doubtful whether we would have agreed them before a landing had even happened. Admiral, you have to live in the real world!'

Admiral Woodward: 'They are not privy to our rules of engagement. They are deterred by our being there, they didn't know that our politicians were saying "keep your hands tied behind your backs". As far as they are concerned we have got two SSNs who will sink their ships, by which means nearly all their Falklands forces had to travel to go. It is quite clear from the interview with Admiral Anaya that was exactly so. That is what made him do it earlier.'

John Nott: 'I see. So we would have had SSNs going backwards and forwards to the Falkland Islands?'

Admiral Woodward: 'If we had done that it wouldn't have happened.'[278]

Ultimately, the Falklands was a war nobody expected to have to fight. It was an unusual, old-fashioned contest in which the Royal Navy's submarines displayed remarkable professionalism, flexibility and endurance, deploying a mixture of old tried-and-tested skills acquired in the First and Second World Wars alongside those newly developed in the Cold War. 'I remain very proud of the fact that we were able to store for war and get cracking very, very quickly with really very little fuss,' says Herbert. 'I told them what to do and they did it. They were a super bunch of COs at that time.'[279]

The Falklands conflict was only a temporary respite from the main confrontation between East and West, but it helped the West win the

Cold War. Years after the conflict, a Soviet General told Mrs Thatcher that 'the Soviets had been firmly convinced that we would not fight for the Falklands, and that if we did we would lose. We proved them wrong on both counts and they did not forget the fact.'[280] Indeed, John Lehman, President Reagan's Navy Secretary, has argued that:

> Britain's actions in the South Atlantic, with support from the United States, made a major contribution to breaking the will of the Soviet Union. Prior to the Falklands episode, Moscow had considered Europe a paper tiger. In the event of a Cold War conflict, the Soviets had assumed the Europeans would have neither the will nor the stomach to fight. Much of their contingency planning proceeded from that assumption. Moscow watched developments in the South Atlantic with great interest. And I can say, based on highly classified documents that came across my desk, that Margaret Thatcher's decision to fight for the Falklands came as a real shock to the Soviet leadership. It forced them to rethink their assumptions for Western Europe and begin to take NATO forces far more seriously.[281]

'Normality returned swiftly' after the conflict, recalled HMS *Conqueror*'s Navigator, Jonty Powis. *Conqueror*'s crew had been promised a long maintenance period and some much-needed rest. 'However,' recalled Powis, '5 weeks later we were at 75 North chasing shadows on SOSUS.'[282] It was back to business as usual against the Soviet adversary in the cold northern waters.

8

Maintaining the Deterrent:
From Polaris to Trident

He [President Carter] was lying on his bed in his swimming trunks when I walked in – we all had grass huts or something equivalent – and I just walked across ten yards of grass, with the blue sea shimmering almost beneath our feet and woke him up and said, 'Jimmy, before we resume tonight, on our next session, I want to have a word with you about the possible replacement of Polaris.'

Jim Callaghan recalling the Guadeloupe summit of January 1979 in conversation with Peter Hennessy in 1988.[1]

When it came to choosing the modernisation of Polaris, we went into everything very carefully, including Cruise as a possible alternative. Not on! You need far more submarines, far more weapons. They do not get through . . . they have not the range . . . We went into it very thoroughly.

Margaret Thatcher, 17 November 1986.[2]

The spendthrift and bellicose Nott
Is involved in a nuclear plot;
He's impaled on the fork
Of his Tridential talk
And Howe has to finance the lot.

A limerick scribbled by Lord Carrington,
Foreign Secretary.[3]

IMPROVING POLARIS

In August 1982, two months after victory in the Falklands, the Prime Minister, Margaret Thatcher, boarded HMS *Resolution* as it was returning to Faslane. The submarine, under the command of Paul Branscombe, dived, while the Prime Minister lunched on board, met most of the crew and was photographed, sensibly trouser-suited, working the periscope.[4] The Prime Minister also sat in the Missile Control Centre, directly in front of the trigger that was used to launch the Polaris missiles and, with headphones on, witnessed *Resolution*'s crew go through a practice missile launch. Branscombe also debriefed Thatcher about *Resolution*'s patrol, which had started in the final weeks of the Falklands conflict. 'She was concerned with the bigger picture, that we were maintaining the bigger picture because that is what it was all about,' recalls Branscombe. 'I think she realized that if indeed difficult decisions were going to have to be taken, they would be taken, subject to what she had written, by people who were not only ordinary, but stable.'[5] This is evident in a letter Thatcher wrote to John Fieldhouse a few days after the visit:

> It was a marvellous experience – made wonderful by the superlative and yet modest qualities of the commander and crew. The feeling of comradeship and yet discipline and respect were marvellous to see. We are fortunate indeed in the high personal qualities of our ordinary folk – if ordinary is the word to use: they all seem so able to demonstrate extraordinary qualities when called upon to do so.[6]

The Falklands conflict was a distraction from the Submarine Service's primary mission, that of fighting the Cold War and the challenges of keeping pace with an evolving Soviet threat. The 1980s represented the most intense period of activity in the deep Cold War. As we shall see in the next chapter, throughout the 1980s the Royal Navy's submarines were increasingly involved in sensitive and highly dangerous operations against more modern, advanced Soviet submarines. But they were also responsible for maintaining the United Kingdom's strategic nuclear deterrent, deployed on continuous Polaris patrols, twenty-four hours a day, 365 days a year.

Deciding to become or remain a nuclear-weapons power has always been a very prime ministerial business, and the detailed casework underpinning such policy-making is very much like the submariners' craft – something to be kept concealed and deep below the surface of events until it's judged to be the moment to come up into the light of public attention. The deterrent-carrying submarines have always been intensely political boats. We have already seen how much this was so when the UK deterrent first went underwater (see Chapter 4). It was so, too, albeit to a lesser degree, when the Chevaline Improvement to the Polaris front end became known, as we shall see. It was palpably true of the leap from Polaris to Trident, as it is today, more than fifty years after the Polaris Sales Agreement was signed.

When the Prime Minister visited HMS *Resolution* in August 1982, the Polaris missile system on board the Royal Navy's four 'Resolution' class submarines was about to be overhauled by an extensive and highly secret modernization programme that had its origins in the late 1960s. When Polaris first became operational in the late 1960s, one submarine was capable of destroying 7–10 Russian cities, including Moscow and Leningrad.[7] When two submarines were at sea there was a 96 per cent per cent chance of destroying twenty cities, assuming a 70 per cent reliability of the Polaris missiles (see above, p. 259).[8] Before the 'Resolution' class went to sea in 1968, the United States and the Soviet Union had started to conceive of ways to destroy offensive warheads either in space or after they had re-entered the atmosphere. In the late 1960s, the possibility of the Soviet Union deploying a defensive shield, known as an anti-ballistic-missile system, around its major cities became a very real possibility. If the Russians deployed an ABM system to defend Moscow, British Polaris warheads would be unable to penetrate it if they were not hardened and redesigned. This left the British with a difficult decision. Unless it modernized Polaris, modified the warheads, or altered and reduced its deterrence criteria, the British independent nuclear deterrent would, it was argued, lose its credibility.

In September 1967 the United States responded to the perceived threat by announcing that it also intended to deploy its own ABM system. It also developed two means of countering the new Soviet ABM system. The first was codenamed 'Antelope'. This upgrade to the front end of the Polaris missile system involved sacrificing one of the three warheads in favour of penetration aids designed to get through the Soviet shield. The second was an entirely new missile system known as Poseidon, a missile with far greater range and a Multiple Independently Targeted Re-entry Vehicle (MIRV) system, capable of delivering up to ten warheads, smaller than those in Polaris in terms of yield, with an aerodynamic design such that they re-entered the atmosphere at great speed. Technically, Poseidon was available under the terms of the Polaris Sales Agreement but the British Government decided against asking the Americans for it.[9] According to Denis Healey, this was 'partly because of its immense cost, and partly because we would be responsible, as with Polaris, for producing the nuclear warheads, and we could not expect to master the MIRV technology except at a disproportionate cost in our scarce scientific manpower'.[10] Wilson's Labour Government was unable and unwilling to move forward with anything other than studies and a very limited amount of improvement work such as the purchase of hardened missiles that protected the missile electronics from electromagnetic pulse effects.[11]

Studies into possible Polaris replacement and improvement programmes continued following the election of Edward Heath's Conservative Government in June 1970. In May 1972, the Anti-Ballistic-Missile Treaty restricted the United States and the Soviet Union to two ABM sites: one around their capital and another to defend a single ICBM site, with a total of 200 ABM missiles spread between the two sites. The Treaty had important implications for the British as it stopped the Soviet Union from deploying vast ABM systems around the Soviet Union and possibly negating the ability of the Royal Navy's Polaris force to meet the Government's deterrent criteria. The treaty threshold of 100 ABM missiles around Moscow was crucial as British intelligence estimates indicated that if the Soviets deployed 128 ABM interceptors around Moscow, Polaris would be incapable of penetrating and upholding the Moscow Criterion.[12] Despite the ABM treaty, and a later 1974 protocol that further restricted Soviet ABMs to one site around Moscow, the need to improve or replace Polaris to ensure that it could penetrate Moscow remained.

In the preceding years a number of options had materialized. The first involved adapting elements of the American Antelope programme into a UK project known as 'Super Antelope'. The second involved purchasing a fully MIRVed Poseidon system from the United States. The third, essentially a hybrid system, envisaged incorporating elements of Super Antelope on top of de-MIRVed Poseidon missiles which later evolved into Option M, a de-MIRVed Poseidon front end which was still at the design stage in the United States.[13] A UK MIRVed Poseidon purchase again was ruled out in September 1973 for a variety of political, economic and strategic factors. This left two practicable alternatives, Option M or the Super Antelope programme. Option M was eventually dismissed due to its expense compared to Super

Antelope and the likely Congressional difficulties expected to arise should the Nixon administration attempt to request a sale. US attitudes and behaviour towards the United Kingdom deteriorated throughout 1973, including the termination of some elements of US intelligence cooperation as a result of the Yom Kippur War. It also led many British officials to conclude that it was imperative to ensure that the nuclear deterrent retained a greater measure of independence.[14] The Heath Government was also anxious to avoid taking any action that could be interpreted as reaffirming the UK special relationship with the United States so soon after joining the EEC.

CHEVALINE

The decision to proceed with Super Antelope was taken in October 1973. The programme, which was later renamed Chevaline, involved a complete redesign of the payload of the Polaris missile to enable it to penetrate Russian anti-ballistic-missile defences. It contained a number of significant improvements over the original Polaris system. It carried re-entry vehicles which were hardened to resist the effects of ABM warheads and two warheads, one mounted on the second stage of the missile and the other mounted on a Penetration Aid Carrier (PAC) which contained a large number of penetration aids designed to confuse the ABM radars stationed around Moscow. Chevaline did not provide an MIRV capability. The PAC was a highly sophisticated spacecraft which, after separation from the second stage of the Polaris missile, manoeuvred itself in space so that its payload, with a second warhead, could be correctly deployed.

The Chevaline project covered a vast range of technical, scientific and engineering disciplines and demanded an in depth understanding of almost every major field of defence technology – scientific intelligence, radar and electronic systems, nuclear technology and effects, battle modelling, system modelling, defence philosophy, materials science, explosive, propellant and pyrotechnical technology, aerodynamics, aero-ballistics, space ballistics, control systems, mathematics, computers, underwater technology, analytical chemistry and so on. The integration of such diverse technologies and the engineering design of a total system to provide both the performance required to meet the threat and the levels of reliability deemed necessary to achieve a credible deterrent system posed major problems and required dedicated teams of very high calibre as well as large supporting groups of technologists, trials engineers and production engineers.[15]

The early years of Chevaline were troubled. The politics surrounding the programme meant that crucial decisions were postponed while alternatives were explored. The technological challenges involved, combined with questions about the way in which the programme was managed resulted in delays and increased costs. This had consequences. By November 1975, the Chiefs of Staff had considerable doubts about the ability of Polaris to penetrate Soviet ABM defences.[16] By the mid-1970s, intelligence assessments indicated that the Soviet Union had deployed sixty-four ABMs around Moscow.[17] By the end of 1975, the original Polaris system, which had only been at sea for around six and a half years, was judged no longer capable of satisfying the criterion of inflicting unacceptable damage to the Russians through the destruction of Moscow.[18] The Chiefs of Staff were forced to alter their criteria for deterrence, and in March 1976 the Chief of Defence Staff, Field Marshal Sir Michael Carver, recommended to the Defence Secretary, Roy Mason, that the UK alter its National Retaliatory War Plan and target either ten Russian cities, excluding Moscow, or operate the 'Resolution' class submarines in the Mediterranean to launch at Moscow in an attempt to outflank the Soviet ABM radars.[19]

Against the prospect of management deficiencies, technical risks, significant delays, and spiraling costs, a major independent review of the programme reported in 1976. While it concluded that Chevaline should continue, it recommended that overall control should be placed in the Navy's hands. The Deputy Controller (Polaris), Rear Admiral David Scott, a widely respected Submariner in the Royal Navy, Submarine Service and the United States, was nominated to head a recreated Polaris Executive, by the Chief of the Naval Staff, Admiral Sir Edward Ashmore. Scott, who had been Deputy Controller (Polaris) since 1973, had doubts about Chevaline. According to his unpublished memoir:

As I carefully read through the details of the extremely complicated design of the new re-entry system, then known as Super Antelope, I began to have doubts as to the rationale of proceeding down the course on which we had embarked. My worries initially stemmed from the following considerations:

1. The introduction of highly toxic, highly reactive, liquid fuels into a system which had been primarily designed to use only solid fuel as a submarine safety measure.

2. With its new front end, the missile system would become a hybrid one, being a mixture of U.S. and U.K. technology.

3. The system relied on the use of decoys to achieve penetration of the defences. These decoys would burn up at about 250,000 feet. Furthermore, their radar signatures could be compromised during test firings. Besides having specially equipped ships to observe firings in the vicinity of the trials area, there were Russian radar stations on Cuba.

4. The shape of the warheads, which had appeared in publications in the U.S., was such that during descent they became sub-sonic in speed at a height of about 90,000 feet, so becoming an easy prey to the Russian terminal defence missiles and even to the Russian SAM 10, a surface-to-air missile.

5. The reduction in warheads from three to two per missile.

6. The reduction in range, which resulted in a massive reduction in the size of the patrol area available to the submarines. I am not, even at this time, at liberty to give figures for this, but I can say that the size of these areas was much less than that available for the unmodified system, and minute compared to the size of the areas available for the Poseidon and Trident missile systems.

7. The accelerating cost escalation of the project. First estimated in November 1970 at £85 Million, it had risen by November 1973 to £235 Million, by March 1976 to £594 Million and by completion in 1980 to over £1 Billion.

8. The inexperience and lack of knowledge of our scientists in the field of Gas Dynamics. This became apparent in 1974 and 1975 when RAE Farnborough, who were responsible for the design of the penetration aids and the penetration aid carrier, had, on a number of occasions, to seek the advice of the Lockheed Missiles and Space Corporation on both designs and calculations.

9. The large number of pyrotechnic devices incorporated in the new front end, amounting to well over a hundred per missile.[20]

Many of the problems with Chevaline concerned the highly sophisticated Penetration Aid Carrier (PAC) at the very heart of the system. This complex piece of engineering equipment posed mechanical, electrical, chemical, explosive, propellant and pyrotechnic problems alongside major problems in ballistics and dynamics.[21]

Scott's misgivings were also a reflection of wider concerns about Chevaline that existed within the Royal Navy, which had always favoured Poseidon to maintain commonality with the US Navy. The introduction of a liquid-propellant system for the main motor which enabled the PAC to meet its various velocity increments before deployment of its payload caused many worries among submariners, particularly from a safety point of view. The probability of a missile accident leading to the loss of a submarine was judged to be no greater than that which applied to the original Polaris system, but submariners, who recalled the fate of HMS *Sidon*, remained distinctly uneasy about the prospect of working with liquid propellants, despite the introduction of numerous safety systems.[22]

Submariners were also concerned about the size of the Chevaline operating areas. These depended on a variety of factors such as Russian activity, the seasonal variation depending on ice limits, the disposition of friendly forces at the time, and communications and navigational support requirements. Chevaline missiles had a reduced range of 1960 miles compared with Polaris maximum range of 2460 miles. This entailed an almost 60 per cent reduction in sea room/operating areas compared with un-Chevalined Polaris, as Scott had warned.[23] A new lightweight warhead which promised to extend the range of the missiles, and provide more sea room and thus greater insurance against improvements in Soviet anti-submarine warfare capability, was tested in 1978.[24] However, according to one estimate these lightweight warheads only led to a modest improvement in range, from 1960 miles to 2030 miles, still a 50 per cent reduction in sea room compared with that afforded by Polaris.[25]

In April 1976, Scott assumed control as Chief Polaris Executive. Accepting that the Royal Navy was stuck with Chevaline, he led a new management team which undertook a complete review of the project, established tighter controls, significantly increased its cost estimates to £495m (at autumn 1972 prices) and extended the programme to allow additional flight trials. Participation from industry was also strengthened and British Aerospace was appointed to coordinate the various main contractors. When the Public Accounts Committee reviewed the Chevaline programme towards the end of 1981 it confirmed that significant management and control weaknesses had existed for some time before the autumn of 1976 – before Scott assumed control and the Royal Navy took over – during the period when heavy costs were incurred, and that the changes in management could and should have been made earlier.[26]

Work continued throughout the late 1970s and into the early 1980s. By the time the existence of the programme was disclosed to Parliament by the Defence Secretary, Francis Pym, on 24 January 1980, its cost had risen to £1000m (£530m in 1972 prices).[27] Did the Callaghan Government ever consider cancelling Chevaline? In his retirement, Lord Callaghan said:

> When I came to office as Prime Minister I could then have said, 'Well, all right, we'd better cancel it.' But it's awfully difficult, unless you have the virtue of hindsight, when something is going on, has been going on for three or four years, and you're told, 'Oh, it's going to be pretty soon now, can we have another hundred million or fifty million?' to say 'No, put it all on one side,' to be so certain you're right that it's not going to succeed. In fact, it did succeed; but it did cost a lot more than everybody expected. And every time they called for a new tranche, I used to write 'agree' on the minute . . . because one always thought it was just around the corner.[28]

In spite of technical problems with the first submarine firings in November 1980 and delays to the trials programme in 1981 because of a Civil Service dispute, the Chevaline Approval Firings from HMS *Renown* in early 1982 were an outstanding success. HMS *Renown*, equipped with a full outfit of Chevaline missiles, became operational in late 1982, followed by HMS *Revenge*. HMS *Resolution* and HMS *Repulse* came next, in 1985 and 1987, after refits.

The Royal Navy's inventory of Polaris A3 missiles also required upgrading. No quantity of US missiles, however cheaply acquired, could carry the Royal Navy into the 1990s – almost all US Polaris missiles were appreciably older than the Royal Navy's and by the late 1970s many had started to show a disconcerting failure rate in tests.[29] The central factor was motor life. In the early 1980s a major problem was discovered in the first stage rocket motors that had the potential to jeopardize 'the continuing serviceability of the UK Polaris First Stage Rocket Motor inventory'. As Rear Admiral John Grove, CSSE, explained to the House of Commons Defence Committee in February 1985:

> Over the years we have had three types of fault come up on these. You get detectable and repairable faults which, of course, you can repair, and maintain your stockpile. We have had detectable faults that are not repairable so that diminishes the stockpile. The really worrying one, of course, that caused us to go into this programme were faults that were not detectable by non-destructive testing means. They were due to flight failures – the Americans had some failures and we had some failures – and, of course, it is a difficult situation. That was the circumstance that drove us to a remotoring programme.[30]

A Joint Motor Life Study Co-ordination Group concluded that 'unless a repair technique for the defect is developed and qualified the required number of motors cannot be maintained until 1994 without the purchase of substantial numbers of motors to replace those rejected. A conservative estimate indicates that at least 88 motors would be needed even if fall out from the additional motors was at a somewhat lower rate.'[31] The Royal Navy was once again forced to turn to the United States and purchase a total of eighty-two new rocket motor sets, the product of a complex Polaris re-motoring programme.

The Polaris re-motoring programme and the Chevaline front end were both designed to ensure that the Polaris force remained effective and credible – able to penetrate Moscow – until about 1994, a date determined primarily by the hull and machinery life of the 'Resolution' class submarines.[32] After that date it was expected to become increasingly difficult and costly to maintain the submarines, with longer and more frequent repair and refits required and the risk of a break in the cycle and the continuity of deployment rising markedly. Operationally the submarines would fall steadily further and further below likely future standards of quietness, and would thus be increasingly vulnerable. If the United Kingdom was to continue to operate a credible independent nuclear deterrent into the twenty-first century a new class of submarine and a new missile system were required.

TOWARDS TRIDENT

Ever since the creation of the Campaign for Nuclear Disarmament in early 1958, and its brief capture of the Labour Party Conference in 1960–61, a particular politico-emotional equation has tingled in the air at every ministerial meeting on the Bomb during a Labour Government:

Nuclear Weapons = Political Neuralgia

The politico-emotional charge was especially pronounced when entirely new systems, as distinct from improvements to existing ones, were under consideration. For Labour governments, the sensitivities are particularly acute and for them the current and future nature of the Submarine Service has been intimately bound up with this politico-emotional calculus. The Royal Navy has continually been aware of this since the Nassau Agreement in December 1962, when the Labour leadership, as we have seen, made the Polaris programme a matter of party competition during the run-up to the 1964 general election. Naturally, when Labour governments have been in power, it has been difficult for submarine Britain to track the political mathematics as it was calibrated inside the tightest nuclear circles. The sensitivity was probably at its highest in the late 1970s, when a small group of ministers, under Callaghan's chairmanship, moved through a series of meetings to consider the possibility of flouting Labour's unambiguous October 1974 manifesto pledge not to commission a new generation of weapons, the technical choices if they did so and the public spending price tag that would be attached to each option. For historians of government, too, this patch of nuclear history has particular fascination, for Jim Callaghan broke with post-1945 precedent and kept the making of nuclear-weapons policy away even from the Cabinet Committee structure where it usually lay.[33]

It was the Cabinet Secretary, Sir John Hunt, who urged Callaghan to keep the top-secret 'Nuclear Studies' on options for Polaris replacement away from the full Cabinet – at least for the time being – in December 1977:

I hope . . . that you will allow me to say why I have considerable doubts about the wisdom of doing this at the present stage . . . For example, I do not think you could possibly tell the Cabinet that we are considering whether the Moscow Criterion [a capacity to penetrate the anti-ballistic-missile screen around the Soviet capital] is necessary. This is not however to say that a <u>decision</u> should be taken behind the backs of the Cabinet. We are however at least one or two years away from any decision and the need to know principle ought therefore to operate strictly in the meantime.

My second doubt is a political one. You have rightly directed that these studies are to be conducted by a very few people on a strictly in-house basis and without any political commitment whatever. If however the Cabinet are told the risk of a leak must be greater. This would not only alert the Russians to what is going on but could create a political problem for you within the Labour Party with both those in favour and those against us staying in the nuclear game exploiting the situation for their own purposes. And what if some members of the Cabinet refused to agree that the studies should go ahead?[34]

Callaghan concurred, partly to avoid embarrassing his number two, the Deputy Leader of the Labour Party, Lord President of the Council and convinced unilateral nuclear disarmer, Michael Foot, who was a member of the Cabinet's Oversea and Defence Policy Committee but was kept away from Callaghan's minimalist inner group on nuclear-weapons policy (minimalist in the sense of being confined to only those ministers who had to be there – the Prime Minister, plus the Chancellor of the Exchequer, Denis Healey, the Foreign Secretary, David Owen, and the Defence Secretary, Fred Mulley).

Callaghan's team gathered initially as simply 'a Ministerial meeting' in 1976–7 but, at the end of 1977, it mutated into what was described as the 'Restricted Group' on 'Nuclear Defence Policy'. At Callaghan's 'meeting of ministers' in No. 10 on the morning of Friday, 20 October 1977, the discussion of a successor system to Polaris – or not – flickered between Chevaline, the progress of talks on a Comprehensive Test Ban Treaty and Strategic Arms Limitation Treaty II (SALT II), and the prospects for Enhanced Radiation Warheads (popularly known as the 'Neutron Bomb').[35] The minutes, taken by John Hunt, indicate that:

> In discussion, there was general agreement on the desirability of maintaining an independent UK nuclear deterrent but the view was strongly expressed that the criterion on which the effectiveness of our existing deterrent was judged – namely its capacity to penetrate the ABM defences round Moscow and destroy 40 per cent of the Moscow region – should be re-examined. It was argued that only a small deterrent was needed to deter an enemy, although a larger one might be necessary to re-assure oneself and one's friends.[36]

Minutes rarely attribute views, but one can hear the sceptical voice of David Owen in that section on the Moscow Criterion.

Polaris replacement came up in conclusion IV:

> The Ministry of Defence should prepare for consideration by Ministers, proposals for the conduct of preliminary studies – to be carried out without any political commitment – of the options for a possible successor system to the Polaris force.[37]

Immediately, the political neuralgia is infecting the cold print of John Hunt's minutes:

> In discussion of these points, some opposition was expressed to conclusion (IV). It was argued that in allowing the Ministry of Defence to

prepare proposals for the conduct of preliminary studies of a possible
successor system to Polaris, an ineluctable process could be set in train
which it would be difficult to control or conceal . . . If it were to become
known – as it would as a result of any discussions involving the Americans – that consideration of a successor system was under way, there
would be a major row within the Labour Party in view of the commitment in the Manifesto against a new generation of nuclear weapons . . .
Even a general study of the future of the UK nuclear deterrent, not
including a commitment to retain one, would cause political trouble if it
were to leak. Was it really essential to initiate any sort of study at this
stage?[38]

The road to Trident and the boat on patrol today begins five weeks
later at the first meeting of what is now described in the paper trail
as 'CABINET NUCLEAR DEFENCE POLICY', on Thursday,
1 December 1977. Callaghan summed up the discussion by saying:

> That there could be no question of the Government taking a decision in
> favour of a successor generation during the lifetime of the present Parlia-
> ment to which their Manifesto undertaking applied. They noted however
> that the Polaris system would cease to be effective in 17 years and that it
> would take about 15 years, at the maximum, to develop a successor
> system . . .
>   The majority felt . . . that a study should be undertaken now of a
> character which would enable the next government to reach decisions
> about whether a successor system should be developed, and if so, what
> system should be adopted. Such a study should be set against the back-
> ground of the strategic problems we would face and should cover the
> various options, covering the political, financial and technical
> implications.[39]

This commission led to the production of what is remembered among
the submariners and the guardians of the deterrent as the Duff–Mason
Report, at the time one of the most sensitive pieces of paper in Whitehall and, to this day, a crucial artefact in both British submarine and
nuclear history.

   Circulated on a strict need-to-know basis in December 1978, its formal title was 'Factors Relating to Further Consideration of the Future
of the United Kingdom Nuclear Deterrent' and it was divided into three
parts:

The Politico-Military work, Parts I and II, was led by Sir Antony Duff, the senior Foreign Office expert on nuclear and intelligence matters. The technical work, Part III, was led by Professor Sir Ronald Mason, Chief Scientific Adviser at the Ministry of Defence – hence Duff–Mason Report.[40]

Helpfully for the ministers on Callaghan's 'Restricted Group', the essentials of the Duff–Mason Report were presented in stripped-down summaries as well as in full, and, as instructed, give both the pros and cons of going to a new generation of nuclear weapons. Duff–Mason is one of the most important pieces of the history of the UK as a nuclear-weapons state, firstly because on Callaghan's personal instructions in his last hours in office, it was given to Mrs Thatcher, thus breaking the normal convention that new governments do not see the papers of the previous administration; and secondly because it prepared the way for over thirty years of Trident patrols aboard 'Vanguard' class submarines.

Here are Duff–Mason's own summaries:

Factors Relating to Further Consideration of the Future of the United Kingdom Nuclear Deterrent

Part I. The Politico-Military Requirement

Summary of Report

1. For deterrence to be achieved a potential aggressor must believe that his opponent has the capability to inflict unacceptable damage on him and there is a real possibility that this capability might be used. NATO's deterrent strategy depends on the link between conventional, theatre nuclear and strategic nuclear forces being maintained and the Soviet Union being convinced that, in response to aggression, the Alliance would if necessary be prepared to escalate the conflict to a level at which the consequences to the Soviet Union would outweigh any possible gains (paragraphs 1–7).

2. As the gains to the Soviet Union from eliminating the United Kingdom would clearly be less than those from eliminating the United States, the United Kingdom can expect to deter aggression by posing a smaller deterrent threat than that posed by the United States. There can be no absolute certainty that, following a massive nuclear

attack on the United Kingdom a Government would take a deliberate decision to order a retaliatory strike by the British deterrent. But the essential thing is that the Soviet Government should believe that there is a real possibility of their doing so. Provided our deterrent was perceived to have the capability, the Russians could not rule out this possibility. This is sufficient for deterrence (paragraphs 8–15).

3. Over the next 30–40 years, our planning need not be geared to any nuclear threat beyond the Soviet Union. We can assume that European links with the United States in the North Atlantic Alliance will continue though the credibility of American nuclear retaliation in defence of European interests could be weakened (paragraph 16).

4. The case for and against British strategic nuclear force can best be discussed in terms of the purpose which such a force would serve:

   (i) <u>A minimal contribution to NATO's assigned nuclear forces.</u> The British deterrent represents a significant proportion of NATO's assigned nuclear forces. The importance of this should not be exaggerated since our deterrent represents only a very small proportion of the total nuclear forces of the Alliance, including American strategic forces (paragraphs 17–20)

   (ii) <u>A second centre of decision making.</u> This is the distinctive nature of our contribution. It complicates Soviet calculations and means that not all nuclear decisions in the Alliance are left entirely to the United States President. Two situations are envisaged: First, a decline in the credibility of the American nuclear guarantee to Europe. A British nuclear force could provide, with the French, the nucleus of a European deterrent and thus reduce the risk that Germany might seek to develop a nuclear capability. Second, hesitation by the United States to use her nuclear weapons in support of NATO. Neither super power could exclude the possibility that, in this situation, a British Government might act to make good the weakness of American resolve. On the other hand, it might be argued that the existence of a second centre could imply lack of confidence in the American guarantee and thus undermine its credibility. Moreover the Russians might not believe that the United Kingdom would ever act independently of the United States, especially over an issue not directly affecting United Kingdom territory (paragraphs 21–7)

(iii) <u>A capability for independent defence of national interests.</u> The British deterrent provides an ultimate option for national defence should collective security fail, which would assist us to counter politico-military pressure or to deter aggression itself. The question is whether it is necessary or credible for us to seek to provide against such a contingency (paragraphs 28–9)

(iv) <u>Political status and influence.</u> To give up our status as a Nuclear Weapons State would be a momentous step in British history. It gives us access to and the possibility of influencing American thinking on defence and arms control policy and has enabled us to play a leading role in international arms control and non-proliferation negotiations. But a decision to embark on a new generation of the British deterrent might be seen by many Non Nuclear Weapon States as inconsistent with our declared arms control and non-proliferation aims and thus reduce our capacity to exercise influence in these fields (paragraphs 30–35).

5. The cost of a successor system would be high and funds spent on the deterrent would not be available for our conventional forces. But we would be buying a unique capability which could not be provided by our European allies. On the other hand, it could be argued that, from the Alliance point of view, conventional forces had a higher priority than the maintenance of the British deterrent as a measure of ensuring continuing American commitment to the defence of Europe.

Factors Relating to Further Consideration of the Future of the United Kingdom Nuclear Deterrent

Part II: Criteria for Deterrence

Summary of Report

1. Of the purposes discussed in Part I, the key ones are the second centre of decision making and a capability for independent defence of our national interests. We should need to deploy a capability which the Soviet Union would regard as being able to inflict un-acceptable damage and to be used independently (paragraphs 1–5).

2. 'Unacceptable damage' is essentially a matter of judgement. It is sug-gested that it could be achieved either by the disruption of the main government organs of the Soviet State or by causing grave damage to a number of major cities involving destruction of buildings, heavy loss of life, general disruption and serious consequences for indus-trial and other assets. An attempt is made (in Annex A) to quantify this judgement (paragraphs 6–10).

3. Three options for creating unacceptable damage are identified:

   (i) Destruction of the main Government centres (both the above and below ground) within the Moscow outer ring road and, outside it, a selected number of alternative bunker locations which are associated with the centralised system of command and control of the Soviet Union at national level.

   (ii) Breakdown level damage to Moscow as a city <u>and</u> Leningrad <u>and</u> two other large cities.

   (iii) Damage to a number of cities, but excluding Moscow. Two variants are suggested –

      (a) Breakdown level damage to Leningrad and about 9 other major cities;

      (b) grave damage, not necessarily to breakdown level, to 30 major targets, including Leningrad and other large cities and possibly selected military targets.

   Option (i) could inflict a greater penalty than the others and would therefore provide a greater certainty of deterrence. But it would be more difficult and expensive to achieve. Option (ii) has the advantage (over Option (iii)) of involving the destruction of Moscow as a city: on the other hand Option (iii) (either variant) might avoid the need to penetrate Moscow ABM defences. It is considered that any one of these options would constitute unacceptable damage, and that, if the United Kingdom capability fell short of meeting one of them or its equivalent, there would be room for significant doubt about its adequacy (paragraphs 11–14).

4. Other criteria include the retention of sole national control over the order to fire our nuclear weapons, ability of our deterrent to survive a pre-emptive attack, continuous deployment at early readiness to fire and a substantial probability that the damage threatened would be achieved. Moreover, if our strategic deterrent is to be credible, it should be seen as complementing other levels of defensive capabilities, i.e. there should not be any major gaps in our spectrum of response (paragraphs 15–17).

The Chiefs of Staff had considerable 'reservations' about criteria option 3b, which required thirty hits on thirty cities. This was inserted to appease the Foreign Secretary, David Owen, who argued that the nuclear capability able to inflict a million fatalities identified in the three options might still meet UK deterrent requirements.[41] Owen favoured a scaled-down deterrent, using nuclear-armed cruise missiles based on board the Royal Navy's SSNs.[42]

Reading the detail of Duff–Mason in its section on what would, for deterrent purposes, constitute unacceptable damage to the Soviet Union brings to mind, in terms of its near unimaginable human impact, Whitehall's own assessment, in the Strath Report of 1955, which was only declassified in 2003, of what ten Russian 10-megaton hydrogen bombs would do to the United Kingdom.[43]

Unacceptable damage

1. As we point out in Part I of the study, in attempting to define an effective deterrent, we must make assessments of probable Soviet attitudes which cannot be founded on precise data. The judgements made by the super-powers about the scale of damage which they need to threaten against each other are no guide to our own requirements, since the scale of deterrence is related to the gains foreseen by the potential aggressor, and the gains from eliminating the United Kingdom would clearly be less than those from eliminating the United States. There is in our view no unique answer as to what would probably constitute unacceptable damage. Some of the options may be preferred as being more likely than others to make the Soviet Government reappraise its intentions. But the choice must weigh cost and other aspects.

2. It has been UK policy not to say exactly how we would use our nuclear capability: the Soviet Union itself is left to draw its own conclusions from what it can see of the capability. We assume that this policy will be maintained. There is little practical risk of the Soviet Union's so misreading the scope and character of the capability that deterrent value would be lost; and our target options need not therefore be constrained by the problems which would arise if we had to make our intentions public.

3. We believe that a deterrent threat of unacceptable damage might be posed in one or both of two potentially overlapping but distinct ways:

    a. if the general level of destruction likely to be suffered by the Soviet Union was such as to outweigh the benefits from removing the UK from the international scene and/or appropriating her resources

    b. if the damage were likely to undermine, at least for a considerable period, the Soviet Union's ability to compete across the whole range of her capabilities as a super power with both the United States and China.

4. Broadly, our deterrent might be designed to threaten capabilities of key importance to the Soviet state; or cities as a whole; or a combination of the two. The extent to which threatened damage against particular capabilities and/or against cities might be perceived as unacceptable by the Soviet leaders is discussed in detail in Annex A; our judgement is that they would find unacceptable:

a. the disruption of the main Government organs of the Soviet state; or

b. grave damage to a number of major cities, involving destruction of buildings, heavy loss of life, general disruption and serious consequences for industrial and other assets.

5. In order to illustrate this judgement, we have identified three broad options for creating damage which, compared with the gains from aggression against the United Kingdom, would be considered by the Soviet Union to be unacceptable. Possible future Soviet defences against strategic attack are considered in Part III of the study. But to illustrate the possible implications of these options for the characteristics of a future ballistic missile force (while not ruling out a non-ballistic solution), we have placed them in the context of current assessments of Soviet Anti-Ballistic Missile (ABM) defences. As to the total Soviet ABM capability we have assumed continuing SALT I restrictions. Current assessments suggest that the Soviet Union may in future have the capability for the endo-atmospheric close defence of the Moscow area as well as for the exo-atmospheric defence of a much wider area around Moscow (as in the current GALOSH system).[44] For illustrative purposes, we have assumed a coverage for the exo-atmospheric system based on the observed performance of GALOSH (shown in map at Annex B). But the development of a longer range missile providing ABM cover over more cities cannot be ruled out. (Part III of the study discusses the technical feasibility of such a development, and the implications of alternative judgments on this point).

6. We have identified three broad options for creating unacceptable damage:

(1) Option 1 would be to threaten disruption of the main Government organs of the Soviet state. It would require the capability to penetrate ABM defences (both endo- and exo-atmospheric) and to destroy hardened targets. On the assumption that a

British strategic force would be used only after a period of tension, and probably only as a retaliatory strike, we must assume that the Soviet Government would have implemented protection measures against nuclear attack. In order credibly to threaten disruption of the Government, we would therefore need to be believed to be capable of destroying the main government centre (both above and below ground) within the Moscow outer ring road and, outside it, a selected number of alternative bunker locations which are associated with the centralised system of command and control of the Soviet Union at national level (see paragraph 1 of Annex A). Attack on the Governmental capability would additionally carry with it widespread destruction of Moscow as a city and in the area around Moscow thus enhancing the deterrent value of this option. (As it would be necessary to ground burst the nuclear weapons targeted against the bunkered centres, the threat posed would include a significant hazard from residual fall-out.)

(2) <u>Option 2</u> would be to threaten breakdown level damage to a number of cities including Moscow. It would require the capability to penetrate both endo- and exo-atmospheric ABM defences, but not that needed to destroy hardened targets. We do not believe that to threaten breakdown-level damage solely to Moscow as a city (leaving the Government capability probably seriously impaired but still functioning) would constitute unacceptable damage. We believe it would be necessary to threaten breakdown-level damage to Moscow as a city (defined as the area within the outer ring road) <u>and</u> Leningrad <u>and</u> two other large cities. A capability on this scale could threaten damage beyond repair to nearly half the buildings in four major cities of the Soviet Union and the possibility that more than 5 million people might be killed and a further 4 million injured. It would involve the destruction of the Soviet capital and the centre of the Soviet bloc, of cities which are major centres of military research, development, and production (RD and P), and of areas within these cities which are of major importance for Russian history and culture.

(3) <u>Option 3</u> would be to threaten damage to a number of cities, but excluding Moscow. It would require neither the capability to attack hardened targets nor, on current assessments, to penetrate ABM defences (see paragraph 10 above). We believe that

the option would constitute unacceptable damage if it threatened either of the following:

(a) to inflict breakdown-level damage on a significantly larger number of cities than under Option 2, which effectively lie outside the present Moscow ABM defences. The cities which might be selected are illustrated in the map at Annex B which shows all cities west of the Urals with a population of more than 500,000 people (numbered in terms of their population) and the limit of ABM coverage. We believe that the selection of cities for targeting under this option should be based primarily on population size, but should also take account of military RD and P facilities, industrial importance, and of historical and cultural importance. We therefore consider the cities chosen should include Leningrad which is of major importance in all these respects. For the remainder of the 'package', (while taking account of any limitations imposed by the range of the delivery vehicle) a wide geographical spread might be chosen to heighten the threat of psychological shock and nationwide dislocation. As to the number of cities to be threatened, we believe that this should be calculated to put at risk a broadly similar total number of people as under Option 2. This would require attacks on Leningrad and about 9 other major cities;

(b) to inflict grave damage not necessarily to breakdown level on 30 major targets, including Leningrad and other large cities and possibly selected military RD and P targets (such as submarine building facilities), which effectively lie outside the present Moscow ABM defences. The major impact of a threat of this sort would be in terms of general shock and dislocation plus the risk to sensitive military facilities not associated with major cities.

7. We believe that any one of these options would constitute an unacceptable level of damage. Option 1 would, in our view, inflict a greater penalty than the others, involving as it would both large scale civilian destruction and very severe damage to the government capability; and it would therefore provide greater certainty of deterrence. It may, however, be more difficult and expensive credibly to threaten damage of this sort since it would involve the ability both

to penetrate endo- and exo-atmospheric ABM defences around Moscow and to destroy structures hardened against nuclear attack.

8. As between Options 2 and 3, the former has the advantage of involving the destruction of Moscow as a city, with inevitably some resultant effect on the working of central Government. It is also arguable that if the Soviet leadership believed that we were content permanently to abandon any attempt to penetrate the Moscow defences, there might be a distinct reduction in the psychological effect of our deterrent since, for the first time since we attained nuclear status, there would be an important part of Western Russia which was free from the risk of our attack. These considerations would need to be weighed against the advantages of a package under Option 3 which includes a larger number of cities and which might avoid a requirement to penetrate the ABM defences around Moscow.

9. These three major options (and the alternatives indicated within Option 3) clearly do not exhaust the possibilities; they are intended only to illustrate levels of threat which would be sufficient for our purposes. In our judgement, if UK capability fell short of meeting one of these options or its equivalent there would be room for significant doubt about its adequacy.

Independence

1. We believe that to satisfy the key purposes identified in paragraph 4, we must retain sole national control over the order to fire our nuclear weapons. (This would be qualified only to the extent of our present arrangements for consultation with other members of the Alliance if time permits.) This view carries implications for possible co-operation with another state or states in the procurement and maintenance of a strategic capability. We must be able to sustain our capability nationally for a period of time, to guard against the risk that a partner might seek to neutralize our capability for independent action by cutting off his support during a crisis. A judgement cannot be reached in isolation from the cost and other considerations, but we believe we should aim to be able to maintain an independent capability for a period of at least a year.

Other Criteria

1. We have also considered the other major criteria for the characteristics of a strategic nuclear capability. These are discussed in Annex C. For the reasons stated there, we believe a UK capability should:

   (a) offer a high assurance that it will survive a pre-emptive attack;
   (b) preferably be continuously deployed at early readiness to fire;

(c) offer a substantial probability that the full damage threatened
     would be achieved. We believe that an aggressor would be
     deterred if there was a one in two chance that the threatened
     damage would be achieved in full (and a higher probability of
     some lesser damage).[45]

The CDS's Duff–Mason file also contains two maps, the first of the
Moscow region; the second of the anti-ballistic-missile screen that
Chevaline was designed to penetrate; the trajectory of a Chevaline
missile fired from a Royal Navy Polaris submarine in the Faroes area
of the North Atlantic; and the information that the yet-to-be fitted
Chevaline warheads and decoys are 'designed to give 50% probability
of attaining the number of penetrations needed in order to achieve the
existing damage criterion when facing an exo-atmospheric ABM
defence . . .'[46]

Jim Callaghan convened a meeting of his 'Restricted Group' nine
days after the Duff–Mason Report was circulated. The Prime Minister
upped the pace of its work not just because Duff–Mason was now
available but because he was due to meet the US President, Jimmy
Carter, at the Guadeloupe quadripartite summit (US, UK, France and
West Germany) in the Caribbean in early January 1979. This would
present an opportunity to sound Carter out about the possibility of
extending Nassau and the Polaris Sales Agreement to cover a possible
British purchase of the C4 Trident missiles the United States was planning to fit to its new 'Ohio' class submarines.

The Cabinet Secretary, John Hunt, briefed Callaghan to this effect
on 20 December 1978, the day before the 'Restricted Group' was to
meet:

the only decision required now is whether the stage has come when there
should be talks with the Americans. It is very difficult to take the study
any further without knowing what co-operation from them we can get,
their view on the cost of the various options, etc. If Ministers agree that
there ought to be such talks, they could best be initiated by your talking
to President Carter at Guadeloupe.[47]

Hunt used a minute from David Owen to Callaghan circulated the day
before, pressing the case for further study of the cruise-missiles-on-
hunter-killer-submarines option which, wrote Owen, 'is intrinsically
more attractive than the officials' paper [Duff–Mason] allows', to give
Callaghan his (Hunt's) views on the next-generation UK deterrent.[48]

If the group thinks there is merit in Dr Owen's arguments further work can of course be done on the options he advocates. I have however myself considerable doubt about a bargain basement deterrent. The case for having a British independent deterrent is not an overriding one but rests on a balance of arguments. If however we are to have one, it has to be credible both to ourselves and our potential enemy. If it is not so credible, it would be better to do without it. This does not mean that it has to have the ability to destroy Moscow. Indeed, I would not recommend going for Option I in the criteria study. It does however mean the assurance of creating other unacceptable damage.[49]

The MOD was also sceptical. In a letter sent to the Private Secretary of the Defence Secretary, Fred Mulley, on 18 December 1978, Michael Quinlan, the Deputy Under-Secretary of State, Policy, at the Ministry of Defence, described Owen's arguments as 'unexceptional' and 'superficial' and argued that an effective deterrent required 'options an order of magnitude higher than this' of up to 10 million dead. The Soviet Union's 'threshold of horror', he wrote, was different to the UK's because it had lost more than 20 million people in the Second World War. 'In this field nothing is provable, but it is far from clear that they would regard less than half of 1% of their population as an unthinkable price for contemplating a conquest of western Europe,' wrote Quinlan. 'Still more to the point, it is far from clear that they would rate highly the probability that a country would choose incineration rather than subjugation in order to inflict a strike of such relatively modest proportions.' The UK had to possess the ability 'to run the whole course right up to unacceptable strategic damage – to finish what we start', argued Quinlan.[50]

Mulley put in his own paper too, on 20 December 1978, in response to Owen's of the previous day. The 'concept of small packages of cruise missiles on existing SSNs has further implications which we should note', he wrote:

Once we get away from a 'dedicated' force, we are in effect changing other criteria as well as the damage criteria – we break away from year-round readiness, we increase vulnerability (because boats have other things to do besides hide), we may increase reaction time (because boats may be in the wrong place for firing and out of communication). In addition, we detract from the other and very important roles of the SSN – the pattern of deployment and operation for the strategic deterrent roles is incompatible with that for the existing roles. We have none too many SSNs now for these roles, and I note the Foreign and Commonwealth Secretary's view that we should have more. His proposal would have the opposite effect unless we built more boats – which in logic should then, as the paper by officials notes, be charged to the deterrent role.[51]

Mulley also stressed that the cruise missile option would seriously affect the undetectability – and hence the invulnerability – of the UK deterrent as 'for technical reasons about forty minutes must elapse between salvoes. The chance of CMs penetrating to the target – already a matter of some concern unless one is in the business on the United States' scale – would not be improved by coming at the defences in penny packets. In addition, the submarines themselves would be more vulnerable, since the first salvo would disclose their positions.'[52]

The meeting in No. 10 on the morning of Thursday, 21 December 1978, opened with Jim Callaghan explaining that the Duff–Mason teams had 'been asked to advance the completion date of the study so as to give Ministers a chance to consider it before the Guadeloupe meeting'.[53] There was more thought on the general uncertainty in the world in the thirty years to 2009–10, including the notion of a Britain standing alone and in serious peril. John Hunt's minute records the 'standing alone' contribution to the 21 December discussion like this:

we could not rule out the possibility that within the timescale we had to consider we might find ourselves having to face alone Soviet political pressure or military threats. In this situation a British deterrent would provide us with the basis for resistance.[54]

Short of that desperate last contingency, Britain's remaining a nuclear-weapons power was seen as a stabilizing factor 'particularly in relation to Germany, and also as a balance to France, who would otherwise remain the only nuclear power in Europe'.

Tony Duff's politico-military assessment for the Callaghan Group was that for the next 30–40 years (i.e. up to 2009–19) 'UK deterrent planning need not be geared to any nuclear threat beyond that posed by the Soviet Union ... [and that] [w]e should base our policies on the assumption that much the same adversary relationship will continue with the Soviet Union as we have today.'[55] But among the ministers in No. 10 that morning there was at least one who sensed just how precarious were their and everybody else's predictive powers:

The changes which had taken place over the last 30 years in international affairs, such as the Sino/Soviet quarrel, Yugoslavia's withdrawal from the

Cominform [the Soviet bloc co-ordinating body], the development of fusion weapons [hydrogen bombs] and the rapid progress in decolonization, had not been predicted in advance. Developments in the next 30 years were likely to be equally uncertain. The Atlantic Alliance [NATO] had lasted longer than might have been predicted 30 years ago, but no one could say what would be its future or the future of the United States/Soviet relations in the period to which we had to look forward.[56]

The Group ranged over the destructive force needed to deter the Russians, the cruise missile option and the Trident C4, as well as the possibility of collaborating with the French not being foreclosed at this stage. They agreed to meet again on 2 January 1979 to decide whether or not Callaghan should approach Carter at Guadeloupe. At that meeting 'there was general agreement that the Guadeloupe meeting presented an ideal opportunity to broach the matter with President Carter privately . . .'[57]

The 2 January session was significant for another reason. In his summing up, Jim Callaghan came out to his colleagues as a Trident man. John Hunt's minute records that the Prime Minister:

> said that the Group were not yet ready to take a decision even in principle, though he himself favoured the Trident C4 option. They agreed however that he should raise the issue privately and without commitment with President Carter in Guadeloupe. To some extent he would have to feel his way, but if President Carter showed readiness to be helpful he would work out with him the best way of exploring the matter more thoroughly.[58]

There was no more sinuous a politician than Callaghan at feeling his way and he already possessed a good working relationship with Jimmy Carter. According to the note Callaghan circulated to the 'Restricted Group' on his return, it was at 3.30 on the afternoon of Friday, 5 January 1979, that he roused the President from his afternoon nap.

It's all too easily forgotten that the one person in the world who, alone, could bring to an end Britain's status as a nuclear-weapons nation is the US President. If the beach hut conversation had ended with a presidential 'no', the Polaris boats and the RAF's WE 177 gravity bombs could have run on for a decade and a half (and no doubt something could have been done within the UK to replace the WE 177). But the country could well have ceased to be a top-of-the-range nuclear power in terms of equipment. We now know that in just over a year Mrs Thatcher would be in No. 10 and Ronald Reagan in the White House and both were nuclear-minded, to put it no higher. But the calculus inside the Nuclear Defence Policy Group would have been entirely different if Callaghan had returned from Guadeloupe with a piece of paper recording an 'I'm sorry; no more' reply from Carter.

Instead Callaghan's note for his nuclear group read:

> I woke the President up and said I wanted to talk to him about something important. I then explained to him the ground we had been over in considering the next generation of nuclear weapons. I explained to him that we might have to replace some of the Motors of the Polaris missiles in the middle of the 1980s and that we had made a detailed in-house study ourselves about what would be involved if we decided on a new generation and we now wished to carry our studies further on a confidential basis. I wanted to know what his reaction would be. Our approach was that any successor system should be cost effective and that it should add to total security. For us as a nation the balance of advantage was only marginal and it could well be that we could use the resources to better effect in more conventional directions. This was why I had put the question to [Helmut] Schmidt [West German Federal Chancellor] this morning and had received the same answer that he had given me on an earlier occasion, namely that he wishes us to remain in the nuclear field.[59]

Carter, as Callaghan knew, had considerable nuclear and submarine experience in his pre-political life. In the 1950s he served under Rickover and personally witnessed the Royal Navy's early attempts to foster collaboration with the US Navy's nuclear-propulsion programme. Carter's reaction was favourable:

> The President said that he too was glad that we were in the nuclear field and that he hoped strongly that both we and the French would remain. He did not wish the United States to be the only country that confronted the Soviet Union. What kind of system were we thinking of? I said we ruled out the GLCM [Ground-Launched Cruise Missile] for the time being. At this stage we were basically attracted to a submarine launched missile and for my part if the cost could be properly apportioned what I thought would be best would be the Trident C4. Did he see any objection? He said that there was no objection at all.[60]

The pair of them danced a strange little quadrille of names and systems:

I pointed out to him that it [the C4] was MIRVed [carrying Multiple Independently Targeted Re-entry Vehicles] and that we did not have a MIRVed missile at the moment. He said, 'Well, so is the [Soviet] SS20 MIRVed.' Incidentally what I called the Trident C4 he called the Trident I. I asked if they were the same and he said they were. He said it was the C5 (which I was calling the D5) which was still on the drawing board. At any rate, it was quite clear that both of us were speaking of a MIRVed warhead. He said that the United States had always got the greatest benefits out of co-operation with Britain . . .[61]

Callaghan's report to his nuclear group on Guadeloupe noted 'President Carter's offer in relation to the Trident missile was welcomed.' The two men agreed that Callaghan could send a couple of officials to Washington to talk further about systems and cost. At John Hunt's suggestion, Sir Ron Mason and Sir Clive Rose, head of the Cabinet Office's Oversea and Defence Secretariat and other minute taker at the 'Restricted Group', were chosen for this mission.[62]

The group agreed to extend the proposed bilateral talks to cover the other options being considered, i.e. cruise missiles and a modernized Polaris A3 as well as the C4. It was noted that President Carter had told the Prime Minister that the SLCM [Submarine-Launched Cruise Missile] could not be regarded as a serious option because of major technical difficulties which it had not yet been possible to overcome. It was, however, pointed out that President Carter's own experience was 'likely to give him a natural bias in favour of ballistic-missile options and that he had a need to get the Trident programme through Congress'.[63]

It should be remembered that all this nuclear diplomacy and the work of the Nuclear Defence Policy Group was being conducted against a very substantial political, economic and industrial crisis – the 'winter of discontent' – as the majority-less Callaghan Government struggled with a spiralling series of strikes that hit the essentials of life, absorbing copious quantities of ministerial time and energy and siphoning away support for the Government in the country. Indeed, it was on his return from the Caribbean at Heathrow Airport on 10 January that the well-tanned Callaghan made the still-remembered political gaffe which is misremembered as him saying 'Crisis? What crisis?' (which was a newspaper headline in the *Sun* rather than something he actually said).[64]

The Labour Party's nuclear neuralgia was still running through the nervous system of Callaghan's nuclear group and a cover story was concocted lest it be needed: 'Provided participation in the bilateral talks was strictly limited on the American side, as we assumed it would be, even if this meant some loss in terms of detailed information, the risk of leaks should be reduced to a minimum. But it would be advisable to have ready a defensive line and it was suggested that we could say in the event of a leak that the talks concerned our Polaris missile motors and our position in relation to the SALT negotiations in preparation for SALT III.' Callaghan undertook to write to Carter about the Mason–Rose mission and the ground he wished the talks to cover.[65]

The 'Restricted Group' did not meet again, but they corresponded on a form of words for the next Labour general election manifesto that would give them a degree of flexibility.[66] Callaghan's letter to Carter was still being tweaked when his Government entered what proved to be its terminal phase as a result of a no-confidence motion in the House of Commons at the end of March 1979. On the day before the crucial vote Callaghan finally wrote to Carter a hand-delivered letter 'For the President's Eyes Only':

> As you know, the Government here faces a crucial Vote of Confidence on Wednesday, and no one can predict which way things will go. But either way an election cannot be far off, and I want therefore to put on record what I told you privately in Guadeloupe about the future of our nuclear deterrent.

He indicated that he would wish to dispatch the Mason–Rose team to Washington but that decision might be with his successor.[67]

Hunt minuted Callaghan the same day asking his authority to show Mrs Thatcher the letter should she become PM. As a result, on 4 May Callaghan gave the written instructions that 'The incoming Prime Minister should be briefed on the need for replacing Polaris (or otherwise – as she thinks!) and should decide whether to make her own approaches to President Carter.'[68] In the meantime, Carter sent a personal manuscript letter to Callaghan (which is not in the file but is quoted in part in a Private Office exchange with John Hunt on 6 April 1979) extending good wishes for the coming election (the Government had lost the confidence vote by one) and stating that the US administration 'will be glad to talk to your people as suggested, recognizing that there are no presumptions about the result of the talks'.[69]

One of the many differences between Jim Callaghan and Margaret

Thatcher as Prime Ministers was that she was not afflicted by a Conservative equivalent of Labour's nuclear neuralgia. This (unlike so much of the Polaris to Trident story) is quite visible from a simple reading of the respective parties' manifestos for the May 1979 general election. Compared to that of the Conservatives, the Labour manifesto's entry about the Bomb is prolix and laden with caveats:

> *Labour*
> In 1974, we renounced any intention of moving towards the production of a new generation of nuclear weapons or a successor to the Polaris nuclear force, we reiterate our belief that this is the best course for Britain. But many great issues affecting our allies and the world are involved, and a new round of strategic arms limitations negotiations will soon begin. We think it is essential that there must be a full and informed debate about these issues in the country before the necessary decision is taken.[70]

> *Conservative*
> The SALT discussions increase the importance of ensuring the continuing effectiveness of Britain's nuclear deterrent.[71]

Mrs Thatcher's ad hoc ministerial group on 'Nuclear Defence Policy' (MISC 7 in the Cabinet Secretariat's lexicon) was one of the first Cabinet Committees she created. Its first meeting took place in No. 10 in the late afternoon of Thursday, 24 May 1979, exactly three weeks after the general election polling day which brought the Conservatives back into power with an overall majority in the House of Commons of 43.[72]

PURCHASING TRIDENT

Following her victory in the general election, the new Prime Minister pressed on as swiftly as she could on the matter of replacing Polaris. Her Government revealed the existence of Chevaline. On 24 January 1980, Francis Pym said to the House of Commons:

> Without breaching the provisions of the 1972 treaty on anti-ballistic missile defence, the Soviet Union has continued to upgrade its ABM capabilities, and we have needed to respond to that upgrading so that we can

maintain the deterrence assurance of our force. The previous Conservative government, therefore, pressed ahead with a programme of improvements to our Polaris missiles, which our immediate predecessors continued and sustained. The House will, I am sure, understand that I cannot go into detail, even to correct the widely mistaken assertions that have sometimes appeared in public, but I think the programme has now reached a stage where I can properly make public more information about it.

The programme, which has the codename Chevaline, is a very major and complex development of the missile front end, involving also changes to the fire control systems. The result will not be a MIRVed system, but it includes advanced penetration aids and the ability to manoeuvre the payload in space. The programme has been funded and managed entirely by the United Kingdom with the full co-operation of the United States . . . It has been a vital improvement. I do not think the House will be surprised that it has been costly. The programme's overall estimated cost totals about £1000 million.[73]

This was done in part to minimize opposition to proceeding with Trident from the Labour Party. The former Foreign Secretary, David Owen, criticized Pym for making a 'cheap party point' and the announcement on the grounds that it had been made 'for purely party political reasons to justify the decision to buy Trident missiles and to embarrass the Labour Party for the fact that it had put the interests of the country first and had been ready at all times to modernize the deterrent – and it was right to do so'.[74]

Margaret Thatcher took nuclear-weapons matters very seriously. She carried out her own Nuclear Release Procedures exercise in the Cabinet Office's Nuclear Release Room (then next to COBRA) in October 1979. In what circumstances would she have been prepared to use either Polaris or Trident? This was a question the veteran BBC broadcaster Sir Robin Day asked the Prime Minister in the run-up to the 1987 general election. Mrs Thatcher replied:

'The nuclear weapon is a deterrent. NATO has said that we are only a defensive organisation; that we only use any of our weapons in response to an attack. If there is no attack, there will be no war. If there is a nuclear deterrent, I believe that there will be no attack. I would not have that confidence if there were no nuclear deterrent. After all, Europe was full of weapons when Hitler went to war, and if you look on the side of the allies, there were probably more than Hitler had. Russia was full of weapons when Adolf Hitler attacked her. It did not stop a war. The nuclear deterrent has been so powerful that it has stopped it. And that is the argument.'[75]

In an earlier exchange in the House of Commons the Prime Minister left little doubt. 'Yes, of course if you have got a nuclear deterrent you have to be prepared to press the button because that deters anyone from using nuclear and also from crossing the NATO line on conventional.'[76] Enoch Powell, never a believer in the nuclear deterrent, seized on this statement and expressed his belief that the Prime Minister would not 'take a decision that would consign a whole generation to destruction in any conceivable circumstances whatsoever'.[77] Indeed, early on in her Premiership she told one of her Foreign Office experts on the Soviet Union, Sir Rodric Braithwaite, 'that she was not at all sure that, in the event, she could press the button: "I want grandchildren too . . ."' she explained.[78]

Those responsible for operating the deterrent during Mrs Thatcher's long Premiership were equally committed to carrying out the Prime Minister's instructions should they be called on to do so. Paul Branscombe and Toby Elliott both commanded HMS *Resolution* during the first Thatcher Premiership. Did Elliott think he would ever have to launch? 'No,' he says. 'In those days I was so convinced that the deterrent was working so well, and because we were operating these things with precision, and the Russians knew that we were operating them with precision, they knew that if we were told to fire, we would fire. I didn't really need to think about it too deeply. It most certainly didn't cause me any sleepless nights and it most certainly didn't stir my conscience, at all.'[79]

Did Elliott ever wonder what Mrs Thatcher had written in her letter of last resort? 'I think I knew,' he says. 'She would have wanted me to execute the target plan that I had been allocated.'[80]

How would he and *Resolution*'s crew have handled the task?

'I would imagine that it would have been quite a challenge to psych everyone up and say: "Come on guys, we've got to do this, this is what we've been instructed to do." There would have been some who would have been overcome with the implications not only for themselves, but their families, particularly those that lived around the base. They would have been very emotionally disturbed by it. Then there would have been those who you would have had to control their anger and throttle them back. It would have been quite difficult, particularly after they stopped issuing rum. I suppose you could have had a medicinal tot but it's not quite the same as a steadying tot of rum. I can't imagine that there wouldn't have been a period of intense, or building international tension, certainly during my period. I can't believe it would have been the bolt from the blue. We would have followed the increase in tension intensely and I would have told the ship's company about this and I would have started to gear them up to think about the need for us to step up our game if we needed to. We would have done more drills, not necessarily rehearsing the weapon system readiness because that was exercised sufficiently anyway . . . You would start exercising the attack team. You would spend time with your officers doing table top stuff, discussions about what was going on, we'd be listening to the news and so on. We would have got used to the very much heightened tension that would exist on board as well as what was going on outside. So our own little world would become very tense from that point of view. In that situation, assuming the punch-up would have started, one would like to think conventional war would have started, then tactical nuclear weapons might have been used in the ground war, so you would have gone through that stage. One would be thinking it's not too long before it's going to happen, before you had to press the button. So if you woke up the following morning, if you could get any sleep, and instead of being told to press the button overnight, you woke up and Britain wasn't there, then it wouldn't be too much more of a shock, but the longer you had to wait before you were actually instructed to retaliate, I think it would be more difficult to keep your guys pulling together.'[81]

Branscombe's crew used to debate in a jocular fashion where they would go after unleashing their deadly arsenal. Some favoured Florida, others the southern hemisphere, somewhere like Perth. They also used to debate how many Polaris missiles they would keep back, as a bargaining chip. 'It's all pretty grim stuff,' he says. 'Living in a submarine is pretty grim in all circumstances. If you didn't have an advanced or what one might say warped sense of humour you probably would not survive. So being able to think of silly things and make sure everybody was engaged in that, not only does it relieve tension, but it meant that we survived doing what we did.'[82]

For all Mrs Thatcher's sense of purpose and legendary work-rate, the trail from Polaris to Trident took over three years to complete. There were a number of reasons for this. One of the most important was a change of President from the Democrat Jimmy Carter to the Republican Ronald Reagan in January 1981. Carter was, as he had indicated to Callaghan in their beach-hut meeting, willing to provide Trident under the terms of the 1963 Polaris Sales Agreement but wanted Mrs Thatcher to delay completing the deal until he, Carter, had sorted out his difficulties in getting ratification of the Strategic Arms Limitation Treaty II through Congress.[83] With the change of presidency, the Reagan administration decided to move to the more sophisticated and powerful D5, causing the question to be reopened in Whitehall. At every stage Mrs Thatcher, to her credit, practised the 1950s Churchill model of nuclear decision making by doing the detailed work in a small Cabinet Committee before taking the final decision to the full Cabinet.[84]

At the first meeting of MISC 7 in late May 1979 she set the tone and pitch for what was to follow and conducted herself with characteristic briskness. First of all, she summoned a streamlined set of ministers: her deputy, the Home Secretary, Willie Whitelaw; the Foreign Secretary, Lord Carrington; and Francis Pym, Defence Secretary. There was no place for her Chancellor of the Exchequer, Geoffrey Howe. Interestingly – and unusually for Cabinet Committee records – the MISC 7 minutes reveal that these three were her Nuclear Deputies for the purposes of authorizing nuclear retaliation if the UK was under nuclear attack and she was dead or beyond reach. It is plain that Willie Whitelaw was second in the chain as the Cabinet Secretary had already sent him a note on the drills on 18 May. John Hunt, Mrs Thatcher told them, 'would arrange a briefing for them to explain what was involved. It was important that Ministers themselves took part in exercises to practise nuclear release procedures so that they became familiar with them.'[85]

Her nuclear ministers had received the Duff–Mason Report, all three sections, as part of their briefing material for the meeting. The minutes contain no hint that she told them about the Callaghan bequest – though it would have been plain that such a substantial document could only have been prepared under the previous Government. Characteristically, Mrs Thatcher opened with no verbal foreplay:

THE PRIME MINISTER said that the starting point for their discussion must be that the Government was fully committed to maintaining an effective strategic deterrent. The question for consideration was what system should be the successor to Polaris: and to enable them to decide this they needed more information about the costs and the other implications of the alternative options.

Once more the Callaghan bequest was apparent in Mrs Thatcher's preamble:

> Much of this information could only be obtained from the Americans and it was proposed that a small team of senior officials should visit Washington. There were good reasons for thinking that President Carter would agree to such a visit.[86]

The customary choreography of the nuclear question, before 1979 and subsequently, was followed 'in discussion' (as Cabinet and Cabinet Committee minutes always put it) when:

> it was noted that Trident C4 came out clearly in the officials' study [Duff–Mason] as the preferred solution. But this would be a very expensive option and we would need to look very carefully at the possibility of going for something cheaper. It was essential therefore that the options should be examined and presented by officials without any implied ministerial backing for the C4 so that all the factors, including cost, could be taken into account when the decision was reached.[87]

When MISC 7 convened for the second time, on 10 July 1979, the Chancellor of the Exchequer, Geoffrey Howe, was included. Perhaps he had been excluded from the first because Mrs Thatcher was not going to appoint him as one of her Nuclear Deputies for retaliation purposes. It would have been odd if he had remained off MISC 7 given that cost – as always – is a potent factor in any nuclear-weapons procurement decision. The MISC 7 meeting on 19 September concluded that the Submarine-Launched Cruise Missile option, championed by David Owen in the previous Government, would not suit British needs. By the time MISC 7 was convened on 5 December 1979 ahead of Mrs Thatcher's visit to President Carter in Washington, the factors in play surrounding a successor system to Polaris were caught vividly in a briefing prepared by the Cabinet Secretary, Sir Robert Armstrong, for the Prime Minister. The underlinings on the text are the Prime Minister's, not the Cabinet Secretary's.

Armstrong's opening sentence carried a great deal of freight:

BACKGROUND

This is a <u>key decision</u>, which will affect our most important means of defence over the next 40 years and thereby the basis of our international military posture, and will have major implications for the defence budget, and indeed for public expenditure, for at least the next decade.[88]

In terms of spending, Armstrong reminds Mrs Thatcher that Howe and the Treasury have agreed with Pym and the Ministry of Defence that up to 1983/84 (the planning horizon) the cost of Polaris replacement will come out of the Treasury's Contingency Reserve rather than the Defence Budget (though the Cabinet did not yet know this, nor did the other members of MISC 7).

The Conservative Government which came into office in 1979 under Margaret Thatcher faced a number of serious economic problems. From the Government's point of view, in defence terms, the country was trying to do too much, with the certainty of not doing it well enough. Sustaining the entire conventional- and nuclear-weapons programme was projected to need at least £300m a year above NATO aims for a 3 per cent increase in real terms up to 1987/88. The Conservative Government set out to reshape the defence programme to a more sustainable and relevant structure and to assign to the new structure resource levels with sensible headroom to absorb inevitable cost growth and other such pressures.[89]

Would the cost of Trident after the planning horizon had concluded in 1983/84 fall on MOD (as is the case at the time of writing for the 'Successor' programme)? Armstrong's next paragraph implies that if the Defence Secretary, Francis Pym, succeeded in bringing defence spending under control, the UK's nuclear-weapons capability would avoid any serious cuts:

> MISC 7 <u>cannot resolve that point now</u> [of where the cost will fall] and <u>need not do so</u>, provided that its members are prepared to agree that Polaris replacement is our <u>top defence priority</u>, and in consequence that, if we do not have the resources to sustain all four 'pillars' of the Secretary of State for Defence's strategy, as discussed at OD [the Cabinet's Oversea and Defence Committee] yesterday, this pillar will be the last to go.[90]

Armstrong outlined the six decisions ministers were facing, which have a timeless quality for anyone involved with the British bomb from the 1940s onwards:

> Do we retain our strategic deterrent?
> What should it be capable of doing?
> Which weapon should we choose?
> Number of boats
> Foreign policy factors
> Timing of announcement

He reminded Mrs Thatcher that in public she had committed her administration to replacing Polaris and that 'It was MISC 7's starting point at its first meeting in May that the Government was fully committed to doing so. But the Chancellor [Geoffrey Howe] was not present at that meeting and the seriousness of the issue is such that you may wish your colleagues to reaffirm that we do wish to stay in what is, for us, a big league . . .' The next sentence has been redacted as has the entire following paragraph, which suggests they deal with the detail of what damage a 'big league' deterrent should be capable of inflicting. Armstrong backs the opinion of the Defence Secretary, Pym, that the C4 missile topped by MIRVs 'clearly emerges as <u>the best option</u>, on both <u>military</u> and <u>financial</u> grounds. Details in "the revised Mason Report" [which has not been declassified] show that only this weapon will adequately meet the damage criteria. <u>Cost</u> is relatively <u>low</u> and <u>reliable</u>, because we should be acquiring a weapon which the Americans will be continuing to procure themselves, not one specifically designed for us.'[91]

As we have seen, the Duff–Mason Report (see pp. 476) identified four targeting options that would constitute an unacceptable level of damage to the Soviet Union. Assuming the Soviets remained within the limits of the 1972 ABM Treaty, a fleet of five submarines that could maintain two at sea at all times armed with a total of thirty-two Trident C4 missiles was required to meet the damage criteria identified in Option 1 (destruction of the main governmental organs of the Soviet state), while one submarine at sea, from a fleet of four, could meet the damage criteria identified in Option 2 (breakdown-level damage to a number of Soviet cities, including Moscow) and Option 3a (breakdown-level damage to a larger number of cities than in Option 2, but without Moscow or any other city protected by an ABM system) and 3b (grave, but not necessarily breakdown-level, damage to thirty targets without AMB protection).[92] The Government was clearly aiming for a Trident force that could meet the criterion set out in Option 1, the targeting of specific sites and facilities within Moscow, such as underground command centres. The importance which the Soviet leadership attached to maintaining their administrative centre unimpaired was shown by its positioning of their ABM system around Moscow and the construction within the city of shelters hardened against nuclear attack for the hierarchy of the Party, the Government and the armed forces and their key staffs; and of alternate bunkered offices for redeployment if sufficient warning time was received, in an area some 600 kilometres from

Moscow. Some ninety alternate bunkered offices had been identified, twenty-seven of which were for the major national and military leadership and those responsible for the operational control of the armed forces.[93]

Robert Armstrong's section on the size of the Trident force and whether it should consist of four or five submarines is revealing. Five boats would push up the cost from £7bn to £8bn over twenty years and Geoffrey Howe was arguing 'for the cheaper solution'. But the Cabinet Secretary nonetheless argued for five as:

> A five boat force would give us some <u>hedge against accidents</u>; and barring these, it would enable us to have two boats on patrol at all times (your Nuclear Release exercise in October pointed up the disadvantages of only having one boat on patrol). The French are building their sixth ballistic missile submarine.[94]

In the flurry of memoranda exchanged between the MISC 7 ministers ahead of the December meeting there had been concern from Lord Carrington in the Foreign Office about the developing anti-submarine warfare threat from the Soviet Union. Pym, as the Cabinet Secretary reminds the Prime Minister, had pointed out that, in Armstrong's words:

> a five boat force (with two always on patrol) offers a far better bet against this threat because it is almost inconceivable that the Russians in this timescale will develop a capability to find and sink two submarines <u>simultaneously</u>. The fact is that the fifth boat would double the operationally available strategic deterrent, and diminish its vulnerability by a much greater factor, than two, for relatively modest extra cost.[95]

Robert Armstrong then addressed the question of the degree of UK dependence on the USA, which was a matter of some concern to the Prime Minister (and which will be covered in greater detail later in the chapter):[96]

> A decision to go for C4 MIRV will keep us <u>totally dependent</u> on United States cooperation over a very long period. As with Polaris, once we have our boats and weapons, we shall have full operational independence in a crisis. But as with Polaris we shall be relying on the Americans not only for initial supply but also for <u>continuing logistic support</u>. If the latter were cut off at any time – and it would be dependent not only on successive Administrations but also successive Congresses – we could not keep going on our own for more than 6–12 months.[97]

The power of a US President or a US Congress to finish off the UK as a 'big league' nuclear-weapons state, almost certainly for ever in both practical and political terms, was clearly laid out. Sir Hermann Bondi, Chief Scientific Adviser to the Ministry of Defence, 1971–7, made the same point even more tersely when interviewed in 1988:

> If the Americans were to tell us at one stage, 'we will go on for another twelve years but not a day longer', we can adapt. If the Americans say tomorrow, 'All we do now for you will stop', then it won't be many months before we don't have a weapon.[98]

In the intense privacy of his brief for Margaret Thatcher in December 1979, Robert Armstrong did not exhibit steel-clad confidence in the United States as a nuclear provider. 'We cannot foresee,' he told her:

> how Anglo-United States relations will develop over the next 40 years. It is impossible to be as confident of continuing support for a quarter of a century ahead in 1979 as it was in 1949. But they have not so far either let us down or used our dependence as a means of pressure. In any case, we have no real alternative. Going it alone would be prohibitively expensive. That only leaves co-operation with the Americans or co-operation with the French.[99]

The French always put in an appearance when the nuclear-weapons question is under discussion in Whitehall, only (so far) to be dismissed as a deterrent partner for a variety of reasons. In late 1979 they were these:

> <u>The French.</u> To avoid later recrimination it is important for your colleagues to be clear that they are choosing the <u>American</u> rather than the <u>French alternative</u> and why. In the light of your decision with <u>President Giscard</u> on 19 November our preference will come as no surprise to him. Our basic reason for not choosing the French alternative is that it would almost certainly give us <u>a less effective weapon at greater cost</u>. If we were convinced that we should base our long term decisions on the hypothesis that the American connection was likely to decline, and the French connection to become our <u>predominant international link</u>, then we should arguably go into partnership with the French. Politically and economically it would be a more evenly balanced partnership, but it would seriously worry the <u>Germans</u>, it would pose great problems with the Americans, on whom we remain dependent for keeping Polaris going through the 80s. And is France's <u>long term reliability</u> inherently greater than the Americans'?[100]

The Armstrong memorandum underscores the degree to which a nuclear-weapons procurement decision is like no other ministers make, not only because of their awesome power, but also because, given the lifespan of nuclear-weapons systems, nothing (apart from civil nuclear decisions) matches its long-term nature.

Armstrong concluded by pressing C4 MIRV on a five-boat force as the preferred option and the sequence (MISC 7; then full Cabinet) of consulting ministers (he urged, interestingly, that the full Cabinet be told Mrs Thatcher was approaching President Carter 'but in the interests of security not of the choice of system').[101] It appears that ministers agreed to defer a decision on four or five submarines. We don't have the detailed minute of the MISC 7 discussion on 5 December as it was 'recorded separately, and . . . retained by the Secretary of the Cabinet' and is retained still.

Just as in Polaris days, heated discussions about the size of the Trident force continued throughout 1980. The Treasury was deeply opposed to a fifth submarine and in June 1980 Geoffrey Howe wrote to Thatcher and questioned the ASW case for the fifth submarine.[102] The Chiefs of Staff were also divided. The Chief of the Defence Staff, Sir Terence Lewin, a naval officer by background, recalled the long and tortuous arguments over the fifth Polaris submarine in 1964 and argued strongly for a five-boat Trident force. He was firmly backed by the Chief of the Naval Staff, Admiral Sir Henry Leach, while the Chief of the General Staff, Sir Edwin Bramall, and the Chief of the Air Staff, Sir Michael Beetham, supported five in principle but remained deeply concerned about costs.[103] The Chiefs feared 'that the government would eventually settle, given all the financial problems, for a three boat force, which would stretch operating cycles to the absolute limit in order to keep one on patrol, and allow no margin at all for the slightest unserviceability if the continuity of deterrence – and therefore its credibility – was to be maintained'.[104]

In November 1980, Ronald Reagan was elected President of the United States. Among the new administration's first decisions was to accelerate the development of US strategic nuclear forces, one of which was the production of the new Trident D5, a three-stage solid-fuel ballistic missile. It is 13 metres long and over 2 metres in diameter, weighs 60 tonnes and has a range of up to 7000 nautical miles. In flight its length is increased by the activation of a two-metre aerospike which reduces drag during its journey through the Earth's atmosphere. Each missile is capable of delivering up to twelve Multiple Independently

Targeted Re-entry Vehicles. The D5 was due to phase into the US Navy so that by the middle of the 1990s the Trident C4 would have been removed from the US armoury, almost exactly the moment when the Royal Navy's Trident fleet was due to enter service. This left the British with a considerable dilemma. The UK now 'faced a choice between two unattractive alternatives; C4 would have all the penalties of uniqueness, while D5 would be better and costlier than we needed, would involve the financial risks of an untried system, and would increase the dollar content of the overall programme'.[105]

Trident D5 offered superior performance over both Chevaline and the Trident C4. The Chevaline system operating in the Atlantic could only attack targets in the Soviet Union west of the Ural Mountains. The greater range of the D5 would allow the UK to attack targets across the Soviet Union, even when operating from the Clyde Estuary and the Norwegian Sea, and targets west of the Ural Mountains from as far away as the US eastern seaboard.[106] The greater range also gave the Royal Navy a vast increase in operating areas in which it could hide its submarines. Very roughly, the difference between the areas in which Trident D5 and Polaris could be fired was 'something like a ten to one ratio'.[107] 'The additional sea room which the long range of Trident gives us is a very useful property for the system,' said Rear-Admiral Grove. 'We look forward to the increased sea room.'[108]

But the cost of the Trident programme continued to overshadow any discussions. By 1981 Thatcher had concluded that the Defence Secretary, Francis Pym, had failed to grip defence expenditure and seemed unable to take the difficult decisions that were required. In January 1981 Pym was removed as Secretary of State for Defence and replaced by the former Trade Secretary, John Nott. Nott inherited a defence overspend from the previous financial year of £200m and a projected overspend for 1980/81 of £400m out of a total defence budget of £11.2bn. The question of cost was captured particularly well during the 1972–82 decision-taking cycle by Nott's memorandum prepared for MISC 7 in late 1981 ahead of the D5 decision, a document that supplements the Armstrong memorandum of two years earlier:

> In the midst of a recession with no economic growth, low confidence and an understandable preoccupation with our short term economic difficulties it would be all too easy to wash our hands of this commitment. Our predecessors in both major parties also had economic problems which seemed equally pressing to them but they have kept this country with an

independent deterrent ever since the 1950s. So far the present Government has a good story to tell in relation to our election commitments on defence. On an objective analysis of the threat it would be a startling moment to change this policy as the 'window of vulnerability' [in terms of Soviet forces] opens wider in the mid-1980s; as Mitterrand modernizes his independent deterrent; and in the light of all our statements over the past 2½ years to the effect that our independent strategic deterrent is essential for the protection of this country and the maintenance of peace.[109]

This appeal to the doctrine of unripe time, always a powerful factor, was followed by Nott's version of what Michael Quinlan (at that time his chief adviser on nuclear matters inside the Ministry of Defence) later called 'the gut instinct'. Nott asked his MISC 7 colleagues:

Would the Conservative Party forgive us – and more important would the nation do so – if the dangers of the world increase? By foregoing Trident we would be abandoning our stake in the only available technology likely to meet both the requirement of invulnerability and penetration up to 2020 . . . to decide not to proceed with Trident would be in effect to opt out of the nuclear business. It is also probably inevitable that within our lifetime other small countries will acquire a nuclear capability. In the eyes of our allies, and of our enemies, we would seem quite a different nation (and the Conservative Party quite a different party).[110]

The overall cost of the Trident programme beyond 1983/84 (the planning horizon) remained a significant issue, not only for the Royal Navy, but the country in general (see p. 495). On 10 February 1981, Nott informed Thatcher that 'two-thirds of the Party and two-thirds of the Cabinet were opposed to the procurement of Trident. Even the Chiefs of Staff were not unanimous.'[111] The Foreign Secretary, Lord Carrington, who was also present in the meeting, supported the purchase but thought 'that the Ministry of Defence is guilty of gold-plating' and hoped 'that a cheaper way might be found'.[112]

The Royal Navy argued that the overall cost of the Trident programme should be spread equally over the three Services. The Army and the Royal Air Force were unwilling to give up any of their planned and costed programmes to finance the project. The Navy argued that because Trident had not, for political reasons, been included in the Long-Term Costings, and because it was a national politico-military requirement outside the normal business of defence provision, it should be paid for by a special subvention separate from the general defence budget.[113] Nott refused and in the spring of 1981 ruled that the Navy would have to bear the cost of Trident alone. He explained the reasons for doing so in his memoirs:

> First, the Royal Navy was responsible for the operations and running cost of Polaris, the existing deterrent. Secondly, as a pure question of financial control and responsibility, it had to be part of a single budget. We could not have the Royal Navy running Trident – and then the Army and the Air Force carping about the management and the cost. Thirdly, the nuclear deterrent had always been a single-service responsibility – the Air Force being in control of the early air-launched deterrent. And finally, as Trident was the United Kingdom's most valuable defence resource, I wanted the high quality and management dedication of the Royal Navy to continue; I would not have had the same confidence in the Army or the Air Force.[114]

After two further meetings, MISC 7 agreed in January 1982 to go for Trident D5 and three submarines, with the question of a fourth to be considered later.[115] The superior accuracy and destructive power of the Trident D5 missile compared to the Trident C4 had weakened the arguments put forward for a force of five submarines. Unlike the Trident C4, which required two submarines carrying a combined thirty-two missiles to meet the damage criterion identified in Option 1 of the Duff–Mason Report, the more advanced Trident D5 was capable of penetrating hardened targets such as command bunkers and destroying them with a single detonation which was the equivalent of four Trident C4 detonations.[116] The fifth boat was now seen as 'extravagant', to use Howe's words, and it was 'knocked out'.[117]

Once MISC 7 had reached its conclusions, the full Cabinet was briefed on the D5 decision on 21 January 1982. Nott, who was responsible for organizing the brief, noted in his autobiography that:

> We explained to the full Cabinet how much we knew about Soviet nuclear, biological and chemical capabilities, where their command and control bunkers were situated, and how the development of anti-ballistic missile defences bore down on the requirements for a credible deterrent. This was an era when the extent of satellite photography and electronic and signals intelligence was not much known to those outside a small circle . . . My colleagues were fascinated; but the Chancellor had come from a good lunch and slept during the briefing.[118]

Negotiations of the terms on which the D5 could be made available took place in Washington on 8, 9, 24 and 25 February 1982 with an American team drawn from the White House (the Deputy to the National Security Advisor), the Pentagon and the State Department, the British side from the Cabinet Office, Ministry of Defence, and Foreign and Commonwealth Office. In a meeting of MISC 7 to review the negotiations on 4 March 1982, Nott argued that 'the terms negotiated by officials represent an extremely favourable deal which we should have no hesitation in accepting'.[119]

During that MISC 7 meeting, discussion turned to the size of the Trident fleet. The question now remained, should the fleet consist of four or three submarines? As was pointed out in MISC 7 discussion:

> On the question of the size of the Trident force, the point was made that a force of three D5 boats had equivalent hitting power to a force of four C4 boats. The fourth boat was required solely as an insurance against failure of one of the other three. No such failure had occurred with the Polaris force. A fourth Trident boat, at a cost of £1 billion, would be an expensive form of insurance against a contingency which experience showed was unlikely to arise in practice. To keep open the option of three or four boats would also provide a margin against cost increase and help to counter criticism of the increased cost of the programme since 1980. The American Trident programme would not increase. Against this, the point was made that the greater effectiveness of the D5 missile system was not the main factor bearing on the choice between three and four boats; the issue was determined by the need to be sure of maintaining at least one boat on continuous patrol. This could not be achieved with a force of only three boats unless accidents to the boats themselves could be totally avoided. Many of the Government's supporters would argue cogently for a five boat force and would certainly not be content with less than four. Above all, a force of three boats would not be a credible deterrent so far as the perception of the Soviet Union was concerned. If the Government were seen to be contemplating the possibility of having only three boats, and therefore of giving up the ability to be sure of a continuous deterrent patrol, the case for the maintenance of an independent deterrent would be questioned.[120]

In summing up the discussion, the Prime Minister said that 'four submarines should be built, each with 16 missile tubes, improved tactical weapons and propulsion system, with a view to the first boat entering service in 1994'.[121]

On 4 March 1982, the full Cabinet also agreed the D5 for up to a four-boat force – the system in operation today. Nott opened with a distillation of MISC 7's accumulated deliberations. He said in a 'most confidential record', kept separate from the normal Cabinet minutes, of which only three copies were made:

> that the strategic nuclear deterrent was central to the defence of the United Kingdom. No one could foresee what might over the next 30–40 years happen to the North Atlantic Treaty Organisation or to the United States attitude to the defence of Europe. A strategic deterrent under British national control was therefore essential.[122]

In retrospect, we can see that Nott's opening remarks about the unpredictability of the world were prescient. Within a month the Falklands had been invaded by Argentina and the UK found itself in a war that none round the Cabinet table in March 1982 were expecting. Nott fleshed out his argument for Britain purchasing the most sophisticated missile in the world. Though he did not mention the Chevaline experience, the perils of not operating the same system as the US were almost certainly in his mind:

> The Polaris force would be 30 years old by the 1990s and its credibility would be declining. Only a four-boat Trident force could provide a Successor which would be credible in Soviet eyes and remain operational well into the 21st Century. The D5 Trident 2 missile would be more cost-effective than the C4 Trident 1 version, because it would preserve commonality between Britain and America.[123]

At this point Nott presented the cost and the French factors together:

> It would also be cheaper during the years immediately ahead. Its total cost over fifteen years would average £500 million a year or just over three per cent of an annual Defence Budget of over £14,000 million. By contrast France's nuclear deterrent was costing twenty per cent of her defence expenditure.[124]

Concern about overall costs of the programme also led the Government to search for other savings. One of the most significant concerned the initial assembly and periodic refurbishment of the Trident missiles. The Royal Navy's Polaris missiles had been serviced and maintained at the Armaments Depot at Coulport on Loch Long. Now, the United States Navy offered to support and maintain the Royal Navy's Trident D5 missiles at the US Navy Strategic Weapons Facility

Atlantic (SWFLANT) at Kings Bay, Georgia. Due to the greatly increased reliability, safety and lifespan of components used in the Trident system it was no longer necessary to build specialist missile-processing facilities in the United Kingdom. Whereas the Polaris A3 missiles had to be regularly removed from SSBNs between patrols for routine maintenance, Trident D5 missiles could be retained on board an SSBN for up to ten years. All the vital control, guidance and electronic packages placed in the missiles could be readily exchanged, at sea or in harbour, without offloading them. Use of Kings Bay also reduced the task at Loch Long to that of storing and processing warheads, and mating them with the missiles aboard the submarine. Under United States law British warheads could not be stored or processed in the United States, but had to be removed from the missiles before the submarines crossed the Atlantic to offload their missiles for processing. Servicing at Kings Bay would have the added advantages of reducing the opportunities for anti-nuclear campaigners to seek to disrupt the operation of the deterrent through demonstrations or industrial action. (The local planning authority had also refused to cooperate with the Ministry of Defence over the possible expansion of the Coulport base.)[125]

This considerable reduction in capital works meant that once Polaris was phased out, only a few hundred civilian staff would need to be employed at Coulport, compared to the approximately 2000 working there in the 1980s. The Government estimated that the arrangement would lead to savings in capital expenditure worth some £500m over eight years with significant savings in running costs throughout the lifetime of the system. The most serious problem concerned the impact on the independence – real or perceived – of the UK deterrent by any decision to rely on the US for missile processing. The MOD argued that the UK would still be independent. As Nott explained in a paper to MISC 7 in July 1982:

In reality, given the advance in technology which Trident represents, we would be more independent with US processing than we are now with Polaris with British processing should US assistance be cut off. We shall still depend on US to supply and repair individual components of the Trident strategic weapons system and missile as we do now with Polaris. But the reliability of Trident will be greater and we shall be able to replace components in the submarine from our own stocks of spares in UK so maintaining our capability for an extended period. And, most crucially, our independence will be assured since, at all times, a high proportion of our missiles will be held in our submarines: two boatloads of Trident missiles will always be available and under HMG's absolute control; and there will be three boatloads for much of the time.[126]

The summary concluded:

Given that a high proportion of our missiles will be held in our submarines, operationally available, and under HMG's sole control at all times, the UK ability to use its deterrent if necessary is maintained. It is, of course, true that with Trident, as with Polaris now, the UK will be dependent on the US to undertake the refurbishment, repair and logistic support of individual components of the strategic weapon system and missiles. But this is unaffected by whether we use US processing and facilities or set up our own in Scotland. Moreover, in the inconceivable case of a future US Government deciding to cut off supplies of components and no longer to repair them, there would be no sudden effect on the effectiveness of our deterrent. We hold considerable stocks of spares, and given the greater reliability of Trident, and that components in the missile can be exchanged without removing it from the submarine, we would be better able to continue to maintain the deterrent than would be the case now with Polaris, although over a period of two or three years its effectiveness would inevitably begin to fall off. In reality, therefore, our independence is not reduced if we use US processing facilities for our Trident missiles, and indeed is enhanced by comparison with Polaris in view of Trident's improved technology.[127]

Mrs Thatcher, in summing up the discussion on missile processing, concluded that:

the political and financial advantages of carrying out missile processing in the United States outweighed the marginal reduction in the independence of the Trident system and the eventual loss of job opportunities in Scotland. The initial decision to adopt Polaris, and now Trident, as the United Kingdom's strategic nuclear deterrent already involved a considerable degree of dependence on the United States. The use of American processing facilities was merely an extension of the arrangements for the supply of the missiles.[128]

The Cabinet was also told on 4 March 1982 that in the event that 'such support were ever cut off, the success of the Chevaline programme suggested that Britain would not be technologically unable to replace it on a national basis'.[129] The Polaris Sales Agreement was later extended to allow a 'mingled asset' ownership and management system for the D5 missiles. Missiles embarked on UK submarines are randomly selected from the inventory of missiles at SWFLANT at Kings Bay, Georgia. The submarines then go to the Royal Naval Armaments Depot at Coulport, where the missiles are fitted with UK-designed and UK-manufactured warheads. Initial plans expected the UK to be at all times in possession of two boatloads of Trident missiles, under British control, and three boatloads for around 80 per cent of the time.[130]

As well as the practicalities, the Thatcher Government was clearly concerned about the ethics underpinning the Trident decision. Open Government Document 80/23, 'The Future United Kingdom Strategic Nuclear Deterrent Force', published in May 1980, represented a break in traditional UK policy of not discussing targeting policy and plans. Nott believed that it was no longer 'feasible to stand publicly on so bland a refusal of discussion, and that something a little less unforthcoming, particularly on the potentially contentious issue of "city bashing" will be required'.[131] This was due to 'the public concern which the difficult ethical issue of nuclear deterrence naturally attracts'.[132] The final document hinted at the targeting options identified in the Duff–Mason Report.

> Successive United Kingdom Governments have always declined to make public their nuclear targeting policy and plans, or to define precisely what minimum level of destructive capability they judged necessary for deterrence. The Government however thinks it right now to make clear that their concept of deterrence is concerned essentially with posing a potential threat to key aspects of Soviet state power. There might with changing conditions be more than one way of doing this, and some flexibility in contingency planning is appropriate. It would not be helpful to deterrence to define particular options further.[133]

According to Quinlan, 'The phrase was intended to imply targeting concepts which, while still countervalue [targeting of cities and civilian populations] and not promising to exempt cities or in particular Moscow, would not be exclusively or primarily directed at the destruction of cities. The impulse behind this was ethical, and reflected in some degree vigorous public debate in Britain on the moral tolerability of striking at populations.'[134]

Moral and ethical debates about Britain's status as a nuclear-weapons state intensified in the mid- to late 1980s. In 1983, CND successfully captured Labour's policy-making apparatus and influenced the party's

1983 manifesto, memorably described by Gerald Kaufman as 'the longest suicide note in history'.[135] In 1983, Labour's then Defence Spokesman, John Silkin, said that a Labour Government would begin ridding Britain of its nuclear weapons within days of coming to power, with the ultimate aim of removing every nuclear base, British or American, from the United Kingdom within five years. The new Labour leader, Neil Kinnock, was equally clear about his attitude towards the UK nuclear deterrent. 'There are no circumstances,' he wrote in 1983, 'in which I would order or permit the firing of a nuclear weapon.'[136] Two years later he reportedly told a group of American Congressmen that if he ever became Prime Minister he would 'never authorize the use of nuclear weapons even if Britain itself was under nuclear attack'.[137]

The prospect of an anti-nuclear Labour Government obtaining power echoed that of 1964, when Harold Wilson's Labour Opposition campaigned on an apparent pledge to cancel Polaris. Once again, the Conservative Government considered making it as difficult as possible for a future Labour Government to cancel the programme should they win office. The Cabinet was told that 'it would not be possible to devise penalty arrangements which would preclude a future Government from abandoning' Trident, 'But politically it might not prove disadvantageous that comparatively little would have been spent on the programme by the time of the next General Election; the Government could not be accused of pre-empting the issue and in practice many of their opponents in Parliament would if they came to office be forced to recognise that the decision now being taken was the only possible one.'[138] The Conservatives won the seat of Barrow in both the 1983 and 1987 general elections.

How would the Chiefs of Staff have reacted to a Labour victory? The veteran BBC broadcaster Sir Robin Day asked Mrs Thatcher that very question in the run-up to the 1987 general election. 'What do you think is the duty of the Chiefs of Staff?' he asked. 'To resign if they disagree? Or obey orders of the democratically-elected first Minister?' 'The Chiefs of Staff have to make up their own minds,' replied Thatcher:

> Each person is responsible for what he decides. It would be for the Chiefs of Staff to decide whether, in their view – it would certainly be mine – that the damage done to NATO; the damage done to liberty, because Britain has always stood for liberty; the damage done to Britain's defences would be so deep, so fundamental that they could no longer be responsible for

carrying the burden of defence, or for being in charge of our Armed Forces without a nuclear weapon of any kind, when those Armed Forces faced an adversarial attack. I know what I would do. I just could not be responsible for the men under me under those circumstances. It would not be fair to put them in the field if the other people have nuclear weapons. I know what I would do, but they are free to make their decisions. That is the fundamental part of the way of life in which I believe.[139]

The Prime Minister was very clear that a Labour Government following through with its policy of nuclear disarmament 'would do untold damage':

Britain is not just another country; it has never been just another country. We would not have grown into an Empire if we were just another European country with the size and strength that we were. It was Britain that stood when everyone else surrendered and if Britain pulls out of that commitment, it is as if one of the pillars of the temple has collapsed – because we are one of the pillars of freedom and, hitherto, everyone, including past Labour Prime Ministers, have known that Britain would stand and Britain had a nuclear weapon.[140]

The electorate, it seems, agreed. Throughout much of the early phase of the Trident programme, the Labour Opposition remained committed to unilateral nuclear disarmament. By the time of the 1992 election, the Labour leadership had reversed its position and adopted a clear commitment to multilateral rather than unilateral disarmament. The party leadership pledged to keep Britain's independent nuclear capability and continued to advocate multilateralism, despite Annual Conference votes in favour of scrapping Trident in 1993 and 1994.

THE TRIDENT PROGRAMME

The particular responsibility for managing the Trident programme fell to Rear Admiral John Grove, CSSE. A former submariner, Grove was one of the first Electrical Officers to serve in submarines in the 1950s, on board HMS *Tally Ho*, HMS *Turpin*, HMS *Porpoise* and HMS *Dreadnought*. He was appointed Chief Polaris Executive with responsibility for running the 'Resolution' class submarines and the original Polaris system, but also the continuing development of Chevaline, which would maintain the credibility of the UK's nuclear capability until the new 'Vanguard' class submarines and the Trident system entered service. When the Trident programme was announced in July 1980 Grove sought permission to change his title to Chief Strategic Systems Executive, or CSSE, a generic term that was more reflective of his responsibilities for Polaris, Chevaline and Trident. He also looked at how Polaris had been managed and structured his management system accordingly.

A top-level body, known as the Trident Group, comprising the Chief of the Naval Staff, Chief of Defence Procurement, Chief Scientific Adviser, Controller of the Navy, the Chief of Fleet Support and Deputy Under Secretary Programmes, was heavily involved in the early days when major policy decisions were taken. It met fairly regularly every two months, but as policy decisions were taken it became clear to the Controller that as the project management task grew in importance it was sensible also to have a lesser group, known as the Trident Watch Committee, consisting of the Controller of the Navy, the Chief of Fleet Support and the Controller R&D Establishments, Research and Nuclear. Close liaison was also intensified with the US Special Projects Office in Washington. Grove and the then Director of the Strategic Systems Project Office (SSPO), Admiral Glenwood Clark, together supervised a jointly agreed programme. 'I have to say that the cooperation and the support we get from the Americans' Strategic Systems Project Office is very good indeed. They regard our programme as their programme almost as much as their own and I get tremendous support from them,' said Admiral Grove.[141] A small Royal Navy team had a permanent office within the Strategic Systems Project Office, headed by a Royal Navy Captain reporting to both Admiral Clark and Admiral Grove. An American cell was also established in London.

The foundations of Britain's Trident submarine programme were laid by Ministry of Defence staff in Bath in preparation for the award of the contract for developing the 'Vanguard' class as a whole to Vickers in July 1982. Unlike the 'Resolution' class, the 'Vanguard' class was designed from the outset as a purpose-built ballistic-missile submarine, incorporating a number of successful design features from other British nuclear submarines. At 149.9 metres long, with a 13-metre beam and a 15,980 tons submerged displacement, the 'Vanguard' class are almost twice the size of the 'Resolution' class, most of which is attributable to the sixteen-tube Trident D5 missile compartment. The increased size allowed the 'Vanguard' class to be fitted with a fourth deck, which (together with a reduced complement of 132) allowed more spacious living quarters and better working conditions for the crew. The hull contains approximately 21,000 lengths of electrical cable totalling some 300 miles and around 1.5 miles of ventilation trunking.

Aside from the D5 missile compartment, the most significant other improvement was a new pressurized water reactor, known as PWR2, built by Rolls-Royce and Associates. By the mid-1970s, it was apparent that the PWR1 plant had reached the limits of its development and future improvements would require a radically new design. In April 1976, Rolls-Royce began to define a new reactor plant, known as PWR2, with improved military characteristics – increased power, lower noise, improved shock resistance and increased core life. The Admiralty Board also placed heavy emphasis on enhancing safety margins and improving the plant design for in-service inspection. Although the design and safety criteria used in the mid- to late 1950s and early 1960s represented the state of the art at the time, the original PWR, based on the US S5W plant, did not allow for easy inspection and by the 1980s plant experience, major advances in mechanical engineering design processes, and a greater understanding of the mechanism of crack initiation, as well as knowledge gained from land-based PWRs and successive refits of submarine plants, demonstrated that inspection of components vital to safety was possible as non-destructive testing and remote control techniques improved.[142]

To meet the design objectives, Rolls-Royce proposed a plant that was significantly bigger than the PWR1 with a larger reactor pressure vessel and pressurizer, and more powerful main coolant pumps. Manufacturing the plant presented a number of challenges as all the major components, including the reactor pressure vessel, the steam generators, the pressurizer, and the main coolant pumps, as well as the reactor core and the control and instrumentation systems, were new designs. Operational proving was therefore necessary before the PWR2 entered service in a submarine. A second shore test facility was built on the same site as the existing Dounreay Submarine Prototype, in time to provide sufficient testing before the first core for HMS Vanguard was ordered.[143] Given the large size of the 'Vanguard' class a new secondary propulsion plant compatible with PWR2 was also required. The design evolved from the machinery used in the 'Swiftsure' and 'Trafalgar' classes, but took maximum advantage of the latest technology to meet requirements of increased power density, weight reduction, longer operational life between overhauls, improved equipment reliability, improved access for maintenance and operation, and reduced noise.[144]

Alongside significant improvements in safety, stealth, reliability, and ease of operation and maintenance, the 'Vanguard' class was also the first Royal Navy submarine equipped with an advanced computerized Submarine Command System (SMCS). This multi-screen command system employed eleven highly advanced computers, tied together by a high-capacity fibre-optic link, and would eventually replace the DCB system on all Royal Navy submarines. SMCS had more than twenty times the processing power of previous submarine combat systems, at lower cost and with greater reliability.[145] The class was also fitted with Sonar 2054, a system unique to the 'Vanguard' class, employing an array of hydrophones and transducers twice the size of any others in operation with the Royal Navy at that time.

The Vanguard construction programme was the biggest, most complex and expensive construction programme in Western Europe. All four submarines were pieced together in a new purpose-built indoor shipbuilding facility – the largest in Europe – situated at the northern end of the Devonshire Dock in Barrow-in-Furness. Construction of the Devonshire Dock Hall, as it is known today, started in 1982. The land on which the Hall stands was reclaimed by filling in the existing Devonshire Dock with 2.5 million tonnes of sand from Morecambe Bay, creating a barrier on which concrete foundations were laid. The hall, which is approximately 269 metres long, 67 metres wide and 51 metres high – twice the size of Manchester United's Old Trafford ground – dominates the Barrow-in-Furness skyline and, with its associated workshops and amenities, covers an area of 25,000 square metres. Breaking with the tradition of ship construction on exposed slipways, the hall provides a warm, dry environment which contributes towards improved standards in work quality and efficiency. The facility also contains a 24,300-tonne capacity ship lift – the largest of its kind in the world – capable of lowering completed submarines into the water as well as returning them either to the hall or sideways to an adjacent hard standing.[146]

To aid the submarine designers a one-fifth-scale model of the submarine was built in a dedicated storage facility and model human figures were used for basic ergonomic evaluations. The Chairman of Vickers, Lord Chalfont, compared Vanguard as a piece of engineering as 'something like putting a man on the moon in terms of advanced concepts and engineering'.[147] Construction of HMS *Vanguard* began in September 1986 at a special keel-laying ceremony attended by the Prime Minister. 'HMS *Vanguard* is important,' said Mrs Thatcher.

'It's important to the defence of our country and when you judge the strength of a country, when you judge its spirit, one of the tests you use is to say, well are those people willing to defend their country if they believe in it? Yes, we are willing to defend it. We have to be right up front in nuclear technology because if a potential aggressor has that technology we could not possibly deter him unless ours was as good as his and preferably better. My generation is constantly grateful for the fact that at the end of the last war it was the free world that got the secret of the nuclear weapon first because had it been Hitler who had got it first we perhaps would not be here today enjoying the freedom of life we do. So yes, we have to have nuclear submarines, we have to have the nuclear weapon because that is the very latest technology and we must have it for that reason. So that we may continue to enjoy in future years the peace we have enjoyed for the last forty years. And so I come to HMS *Vanguard*. I have had the privilege of going on the Polaris submarine for a short time. The people who command them, the people who go to sea in them are quite outstanding. They take their duties to us seriously, we take our duties to them equally seriously. And so I am glad and proud to be associated with laying the keel of this first HMS *Vanguard* ship which will carry Trident because I believe that it will keep the peace which is the greatest desire of all of us.'[148]

Construction of two further submarines, HMS *Victorious* and HMS *Vigilant*, started in December 1987 and February 1991, with a decision on the fourth postponed until after the 1992 general election.

In order to accommodate Trident, a major works programme was required to expand, enhance and modernize facilities at the Clyde Submarine Base. The Trident Works Programme, like the Polaris Works Programme in the 1960s, became one of the largest and most complex building tasks ever undertaken by the Ministry of Defence and the Government's Property Services Agency, and one of the biggest construction projects in Europe. Redevelopment of Faslane and Coulport began in 1985 at a final cost of £1.9bn. The overall scheme consisted of some 110 individual projects, many of them linked, and created more than 4000 jobs, with Scottish firms providing much of the planning and technological expertise. At Faslane a covered ship lift, the height of an eleven-storey building and as long as Wembley Stadium, was constructed to raise a 'Vanguard' class submarine clear of the water for repairs and routine maintenance. A new power generator facility, with the capacity to serve a town of 25,000 people, was built. At the Royal Naval Armament Depot in Coulport a new floating explosives-handling jetty, as tall as Nelson's Column, as long as two football pitches, and weighing four times more than the aircraft carrier HMS *Invincible* was built to load and unload Trident missiles and warheads. A new refit facility in Rosyth dockyard, outside Edinburgh, to refit and refuel the 'Vanguard' submarines was also planned but later abandoned on cost grounds.

The infrastructure programme was beset with problems. Plans for the facilities at Faslane changed radically following the Chernobyl incident in Ukraine in April 1986 as MOD officials worried about the impact of an earthquake on the west coast of Scotland. Many of the new facilities were at that point redesigned to even more rigorous standards in order to withstand earthquakes and other extreme environmental conditions. The sudden changes put the Ministry of Defence in a weak negotiating position with contractors and building work was allowed to begin before designs had been finalized. The ship lift alone required 7200 alterations and at the peak of its construction programme 1000 consultants from sixty-seven firms were engaged on it, many on open-ended contracts. A hardhitting 1995 report from the House of Commons Public Accounts Committee heavily criticized the Ministry of Defence for its management of the refurbishing of Faslane, where costs increased from an initial budget of £1.1bn to £1.9bn.[149]

New facilities were also required at the Atomic Weapons Establishment at Aldermaston, where warhead production facilities built in the late 1950s and early 1960s were approaching the end of their life. The UK purchased the Mark 4 re-entry vehicle from the US for the Trident D5 missiles. As with Polaris, AWRE was responsible for designing a warhead that met the specifications of the re-entry vehicle in terms of weight, size, shape, centre of gravity and centre of inertia. The Trident warheads, known as Holbrooke Mark 4s, were based on the US-designed W76 warhead and were tested three times at the US Nevada test site. The warheads were to have been manufactured in a series of new facilities at AWRE, but costs spiralled from initial estimates of £250–£300m to over £1bn because of deficiencies in planning, control and liaison with the main Trident programme. Manufacture of the first warheads began in 1987, initially in AWRE's existing facilities to supplement capacity until the new facilities became available.[150]

Debates about the fourth submarine continued into the late 1980s and early 1990s. In 1992, Grove's successor as Chief Strategic Systems Executive, Rear Admiral Ian Pirnie, told the House of Commons Defence Select Committee that it was possible to show 'on paper' that three submarines could be sufficient to keep one on patrol at all times, but three submarines would allow for 'no contingency at all or insurance against any form of material defect, refit delays, or anything else'. Opponents of the fourth submarine seized on Pirnie's evidence and the Admiral was forced to return to the Committee and explain that his remarks had been misinterpreted. Pirnie's advice to ministers had always been that four submarines were needed and that cancellation of a fourth boat would only save around £400m, around 4 per cent of the total cost of the Trident programme. 'For that 4 per cent we increase our submarine availability by 50 per cent,' said Pirnie.[151]

By the early 1990s, the fourth submarine had become tied up with the Labour Party's retreat from unilateral nuclear disarmament. It featured heavily in the 1992 general election campaign, where the Prime Minister, John Major, declared during a speech in April that a future Conservative Government 'will order, build, deploy and arm that fourth Trident submarine that our armed forces tell me they need': 'We will not take any risk with that crucial shield. But I tell you who would. Labour would. They don't say, can't say, won't say what their attitude on Trident would be. Because they don't know. First Labour say they will build it. Then they say they won't build it. They even say they might build it and send it floating round the world devoid of arms. What would they call it? HMS *Spineless*? HMS *Witless*? HMS *Clueless*?'[152]

The decision to proceed with the fourth 'Vanguard' class submarine was taken in July 1992, two months after the Conservative election victory. The keel of HMS *Vengeance* was laid down in February 1993. HMS *Vanguard* commissioned into the Royal Navy in August 1993, followed by HMS *Victorious* in January 1995, HMS *Vigilant* in November 1996 and HMS *Vengeance* in November 1999, long after the Cold War conflict in which they had been conceived had ended. As the submarines slipped away from the wall at Faslane and Devonport, their crews never lost the sense that they sailed into their own peculiar limbo somewhere between peace and war. But when Mrs Thatcher and her colleagues on MISC 7 slogged through the options between 1979 and 1982, there was no sign that, as the Duff–Mason Report had put it, a superpower Soviet Union would cease to pose a nuclear threat to Britain for decades to come.

9

The Silent Victory:
The Cold War in the 1980s

Our [RN] nuclear powered hunter killer submarines, and those of the United States, are the platforms best able to operate well forward and threaten the whole range of Soviet submarines and high value surface units.

Admiral Sir William Staveley, First Sea Lord, May 1986.[1]

'I'm under no illusions. I was at war. I felt that when I was at home. We were all incredibly secretive in the Submarine Service. We didn't talk to people outside our own group.'

Rear Admiral Roger Lane-Nott, former CO HMS *Splendid* and former Flag Officer Submarines, March 2014.[2]

'The Forward Maritime Strategy depended entirely on getting nuclear-powered submarines up there early. Outcome unknown. [In the event of war] I do believe that we would have been quite good at it.'

Commander James Taylor,
former CO HMS *Spartan*, April 2014.[3]

THE COLD WAR HEATS UP

By the 1980s, the Soviet submarine fleet had made significant advances in worldwide deployment, submarine performance, and weapons. Submarines were now active in all oceans and backed by efficient worldwide command and control. The Russians had also made impressive technological advances, constructing submarines that were generally faster and deeper diving than their Western counterparts. They were also equipped with an impressive array of weaponry, and Soviet submarines led the West in their acquisition of tactical anti-ship missiles. However, it was weight of numbers rather than individual quality that was the strength of the Soviet submarine threat: by the 1980s the Soviets were completing one nuclear submarine every six weeks, well in excess of the combined NATO building rate. The UK's Defence Intelligence Staff (DIS) estimated that by 1985 the Soviets would have some 320 submarines in service, of which 210 would be nuclear-propelled. By comparison, in 1980 the Royal Navy submarine fleet consisted of 4 SSBNs, 12 SSNs and 16 conventional submarines. The DIS also estimated that at the start of a conflict there could be 70 Soviet nuclear cruise missile and attack submarines posing a threat in the Eastern Atlantic areas and, in addition, the Soviets could maintain around 45 SSBNs on strategic deterrent patrols from the Northern Fleet. In general, the Soviet submarine fleet constituted a formidable anti-ship force, being fast, well armed, and well directed. But it had a much lower anti-submarine capability, which gave US Navy and Royal Navy submarines, with their noise and sonar advantages, individual superiority.[4]

In the late 1970s, the Royal Navy developed an anti-submarine warfare concept which looked into the 1990s.[5] Because of the UK's position on the maritime flank of Europe the Royal Navy was well aware that in the event of war it would have to provide the main contribution of maritime forces immediately available to the Alliance in the Eastern Atlantic as the US had ceased allocating surface ships to the area, and France had decided to place more emphasis on the Mediterranean at the expense of the Atlantic. UK forces were therefore configured to provide long-term surveillance of Soviet forces, to withstand an initial Soviet attack, and to sustain operations through all levels of conflict. The concept was designed to cover Royal Navy tasks in war, periods of tension and periods of peace:

Missions in War. Soviet Submarines will have to be countered in a wide variety of settings in war. Furthermore, if hostilities are extended, attrition of Soviet submarines will become an increasingly important task. However, it can be seen that the main NATO maritime missions whose successful completion will depend on mastering the Soviet submarine threat are:

a. Protection of NATO SSBN deployments. UK and some US strategic missile submarines deploy from the restricted waters of the Clyde. It can be expected that Soviet Submarines may attempt to detect and attack departing SSBNs in the Clyde approaches, and our maritime forces must be disposed and employed accordingly.

b. <u>Support of the NATO Strike Fleet Atlantic.</u> The importance of the NATO Strike Fleet is unlikely to diminish in the timescale considered and is such that the Soviet Navy can be expected to make determined efforts to neutralize it by co-ordinated air, surface and sub-surface attacks. UK maritime forces would play a part in ensuring the Strike Fleet's safe transit to and security within its operating areas, which could be anywhere in EASTLANT.

c. <u>Protection of Amphibious Operations.</u> A prime NATO requirement is for the reinforcement of Northern Norway and the Baltic area. It will be the aim to commence such reinforcement before the outbreak of hostilities, but in any case the amphibious forces will require ASW protection in transit and whilst operating in their landing areas.

d. <u>Protection of Reinforcements and Resupply of ACE [Allied Command Europe] from North America and the United Kingdom.</u> In a prolonged build up or in extended hostilities the survival of Europe would be largely dependent upon the safe arrival of war reserves and reinforcements, and a major Soviet onslaught on the Atlantic and [English] Channel lifelines can be expected, particularly from submarines.

<u>Tasks in Tension.</u> In a period of tension all the main maritime missions referred to, and which require ASW protection in support, may have to be initiated. Otherwise Allied strategy remains essentially one of deterrence. Amongst other things this requires an increased emphasis on intelligence gathering and surveillance to acquire immediate and detailed information on Soviet submarine deployment. It will also be a requirement to trail or mark Soviet submarines, to allow NATO to take the offensive should hostilities occur.

<u>Tasks in Peace.</u> The Soviet perception of the UK's maritime capability is formed in peace. By conducting a range of activities in EASTLANT/ ACCHAN [Allied Command Channel], UK maritime forces aim to demonstrate a broad span of defensive and offensive capabilities. In addition a number of national maritime tasks are undertaken, which in anti submarine operations include intelligence and surveillance to provide the basis for future action and planning, and the training and exercising of our own ASW forces to maintain their efficiency.[6]

As always, the priority for SSNs was the protection of the UK Strategic Deterrent, the 'Resolution' class and later 'Vanguard' class submarines deploying from the Clyde areas, where they were potentially vulnerable to interception by Soviet submarines. SSNs were also used for 'delousing' SSBNs to ensure that they were not being tracked by Soviet submarines.

Aside from this protection role, the Royal Navy also developed additional strategies designed to take advantage of long Soviet submarine transit routes, Soviet weaknesses in radiated noise levels and sonar, and the strength of NATO and UK passive detection technology that had been developed. The first such strategy, termed Forward Operations, involved anti-submarine warfare activity in northern parts of the Eastern Atlantic, with Soviet submarine bases in the Barents Sea and their approaches as key targets. In war, the Royal Navy intended to attack Soviet submarines immediately on leaving their bases, by either mines or torpedoes, or both. The strategy concluded that 'nuclear submarines must be the means of delivery' and that 'only the SSN could conduct covert operations with a good chance of survival in such a hostile environment'.[7] Other Forward Operations aimed at intercepting Soviet submarines while they were transiting to their patrol areas. These would take place early in any conflict, along identified Soviet submarine transit routes in the general area of the Greenland–Iceland–UK Gap, or at other positions indicated by intelligence or the circumstances of a particular operation. At such points UK forces, including Maritime Patrol Aircraft and towed-array frigates, would mount area or barrier ASW operations with Royal Navy SSNs, with their noise and sonar advantage, locating, trailing and attacking Soviet SSNs.

In any of these operations the SSNs' roles would increase in proportion to the expected Soviet air threat, and to the degree of covertness required. The SSN was also 'the only practicable vehicle' capable of conducting a further type of Forward Operation, providing a counter to Soviet 'Delta' class SSBNs in the Barents Sea, an area of ocean that would be dominated by Soviet forces in the event of any conflict. As the only European country possessing SSNs (France did not commission its first SSN until February 1983), countering Soviet SSBNs was seen as an important UK role. But the Royal Navy also recognized that in a period of tension or conflict some of the large number of Soviet SSBNs could be deployed further south in the Eastern Atlantic and targeted against European nations. In either scenario, the Royal Navy expected counter-SSBN operations to absorb many SSNs and that would inevitably affect other ASW operations adversely. Operations in the approaches to Soviet bases also had their place in periods of peace and tension, for intelligence gathering and surveillance.

The Royal Navy also intended its submarines to play a key role in supporting so-called Open Ocean Warfare operations in the Eastern Atlantic. Where SOSUS coverage existed, the Maritime Patrol Aircraft would be the primary means of following up on individual contacts, but SSNs were also required to covertly trail Soviet SSNs and SSBNs in open ocean waters. While Forward Operations and Open Ocean Warfare were expected to take their toll on Soviet submarines, the Royal Navy recognized that inevitably some Soviet submarines would penetrate to their targets. Alongside other ASW units, such as towed-array frigates, helicopters and Maritime Patrol Aircraft, SSNs were also involved in direct support of main NATO maritime missions in the Eastern Atlantic such as the NATO Strike Fleets, Transatlantic Reinforcement and resupply shipping to Europe, as well as amphibious operations on the northern flank. In the event of war SSNs would deploy at ranges of 50–100 miles from high-value targets and provide ASW protection against the large number of Soviet cruise missile submarines seeking to launch torpedo or missile attacks against NATO forces.

In periods of tension, the Royal Navy's ASW activity was 'designed to deter the Soviets from increasing their pressure or resorting to open hostilities', as the Operational Concept explained:

> To this end, increased emphasis will be placed on intelligence gathering and surveillance. An ability to trail or mark Soviet submarines will allow NATO to take the offensive should hostilities occur; this ability has been demonstrated by SSNs and MPA acting in accordance with SOSUS, and can be provided to the extent that numbers will allow. A primary consideration will be the protection of deploying SSBNs, in which MPA and large helicopters have useful capabilities. SSNs can be used to 'delouse' SSBNs, and towed array fitted SSKs might play a part. Fixed Arrays are expected to provide valuable protection for SSBNs in the Clyde Approaches.[8]

In the early 1980s the Royal Navy's submarines, particularly the new 'Swiftsure' class, continued to conduct covert 'live' operations against the Soviets exercising on operations in war deployment areas and to gain intelligence.

In July 1980, HMS *Spartan*, under the command of Commander Nigel Goodwin, was diverted from Mark 24 Tigerfish deep-water test firings in an area between the Outer Hebrides and Rosemary Bank (a seamount just over seventy miles west of Scotland) to take part in

Operation 'Ephebe', to obtain 'current intelligence concerning the movements, missions, tactics, doctrine and operational patterns of Soviet and Warsaw Pact submarines'.[9] *Spartan*'s primary target was a Soviet task force comprising the helicopter carrier *Leningrad*, two Soviet 'Krivak' class destroyers and a 'Uda' class tanker, *Lena*, all of which had entered the Norwegian Sea.[10] Northwood wanted to determine whether or not Soviet SSNs were operating in direct support of the Soviet task group.

Shortly after embarking on the operation *Spartan* quickly detected the task force and moved in behind it. *Spartan* also found two submarines operating approximately seven miles astern of the *Leningrad*, one of which had previously been picked up on SOSUS and was assessed as a Type II Soviet nuclear submarine.[11] *Spartan* spent the next forty-eight hours loosely trailing the task force as it proceeded northeast at a speed of between 12 and 16 knots collecting intelligence from sonar transmissions, as well as the noise signatures of the various vessels. While in the trail *Spartan* made eight separate detections of the two Soviet submarines classified as a 'Charlie' class and 'Victor' class. To observe how the submarines were operating and interacting with the task force, *Spartan* then closed the range to five miles. But there was little evidence that the submarines were operating in a direct support role and *Spartan*'s crew concluded that they were simply exercising against the Soviet task force.[12]

After collecting intelligence on the various vessels comprising the task force, *Spartan* moved away and returned to a central search position in the Norwegian Sea to intercept submarine contacts detected by SOSUS. The submarine spent the next forty-eight hours investigating two possible contacts, the first of which was a 'Yankee' class SSBN that appeared to be transiting back to its Northern Fleet base. *Spartan* closed the Yankee to an estimated fifty miles before breaking off and opening to the south in a planned withdrawal from the operation. While moving south *Spartan* detected a second submarine that was classified as a possible Type II/III Soviet nuclear submarine. *Spartan* closed the range until it was approximately five miles on the Soviet submarine's port bow, before moving in and establishing a trail on its port quarter at a range of 6000 yards. *Spartan* spent the next hour following the submarine, which was continuing on a steady course at a speed of between 14 and 20 knots. After an hour in the trail the Soviet submarine suddenly started to clear stern arcs, turning quickly back towards *Spartan*, which was forced to evade to reduce the risk of counter-detection. After completing the evolution, the Soviet submarine returned to its original course and speed, while *Spartan* slowly moved away until it was beyond narrowband trailing range.[13]

As *Spartan* continued south, Northwood extended the operation and ordered Goodwin to intercept a possible 'Alfa' class submarine which was detected operating to the northwest of North Cape, possibly waiting to delouse the inward-bound Yankee that *Spartan* had detected at the beginning of the patrol. *Spartan* sprinted to outflank the Alfa and head off the Yankee's attempts to return to Russian waters. While moving east at high speed *Spartan*'s crew had considerable difficulty holding the Alfa on sonar. Goodwin slowed *Spartan* at periodic intervals in order to re-establish contact, but determining the range of the Alfa remained difficult. As the range closed Goodwin summoned his XO, Lieutenant Commander Dan Conley, to *Spartan*'s Sound Room. 'The towed array displays were indicating a very confused picture,' recalled Conley, 'almost saturated by noise from the "Alfa", but curiously there was no contact on the hull sensors.'[14] *Spartan*'s sonar operators concluded that the Alfa was between thirty and forty miles away, but as more information became available it was obvious that this was wrong. Then a contact with a fast-moving bearing was detected. The Alfa was not 30–40 miles away, but around 10,000 yards off *Spartan*'s port beam. The range continued to close, until at one stage the Alfa passed approximately 3000 yards from *Spartan*.

Fearing counter-detection, Goodwin attempted to manoeuvre *Spartan* behind the Alfa into its blind sonar arcs. However, it soon became clear from the continued bearing movement of the Soviet submarine that the Alfa was suspicious. It also began to circle and both submarines turned through approximately 240 degrees. 'The Russian commander had clearly effected counter-detection and a situation had developed akin to an underwater dogfight, the two opposing submarines manoeuvring less than two miles apart and risking collision,' wrote Conley.[15] Goodwin waited for the range on the Alfa to open and then ordered *Spartan* away to the south at high speed, making frequent course alterations, hoping to increase the distance between the two submarines. But as *Spartan* sped away the Alfa started to transmit on active sonar, clearly attempting to confirm if there was indeed another submarine in its area.[16] 'Sonar transmissions now came from astern at regular intervals,' said Conley, 'a clear and unambiguous indication that "the opposition" was well and truly in active contact and that, quite simply, *Spartan* was now the Russian's quarry.'[17]

Spartan continued south at high speed, but with the Alfa capable of speeds of over 40 knots, compared to *Spartan*'s 30 knots, withdrawing was difficult and the Alfa remained on *Spartan*'s quarter, 'very close and clearly having no trouble in keeping up', remembered Conley.[18] 'There was little that could be done, other than hope the Russian would find it increasingly difficult to maintain sonar contact in the high-speed chase.'[19] After an hour *Spartan*'s evasive manoeuvres succeeded in confusing the Alfa. Goodwin reported the possible counter-detection to Northwood, but instead of the submarine being ordered back to port, the Alfa was such a high intelligence priority that *Spartan* was ordered to relocate it. *Spartan* returned to the vicinity of the Alfa's last known position, but after searching for thirty-six hours failed to detect anything of interest. *Spartan* returned to the UK with a significant amount of acoustic, electronic, and tactical and operational intelligence about the Alfa, including its broadband and narrowband noise signatures, as well as considerable evidence that it was fitted with some sort of propulsor and enhanced passive sonar capability.[20]

Aside from intelligence, such patrols provided opportunities for submarine crews to experience prolonged periods at sea confronting the Soviet Navy in war-like scenarios. The training value of operations in the north Norwegian Sea, where there were copious opportunities to confront Soviet submarines in a highly operational environment, were self-evident and it eventually became accepted policy that each CO should have at least six weeks conducting operations against the Soviet opposition during his time in command. These operations also demonstrated that the Royal Navy's latest submarines were reliable and able to meet the demands placed on them during intensive periods at sea.

By the early 1980s SOSUS was regularly tracking over 200 submarine contacts a year, from which information was disseminated to Royal Navy and US Navy submarines, which would then attempt to track them. In May 1981, Commander Michael Boyce took HMS *Superb* on patrol on Operation 'Monsea' and over a period of twenty-six days detected and encountered twelve Soviet submarines.[21] *Superb* departed Devonport and initially went north to intercept a Type II nuclear submarine that had been detected in the north Norwegian Sea, designated L-045. As *Superb* transited towards its target, a separate SOSUS contact, previously classified as L-041, a probable Type I nuclear submarine, was reportedly loitering in an area 150 miles north of the Iceland–Faroes Gap. Northwood ordered *Superb* to break away from L-045 and instead attempt to find L-041 in order to establish the nature of its operations and whether or not it was interacting with any other Soviet forces. *Superb* quickly obtained contact with L-041, closed, and fell in behind it to start a trail. There was no indication that the Soviet submarine was operating with any other vessels, but it was very noisy, and Boyce suspected that it might have been either on a simulated anti-shipping patrol, or acting as a confusing contact to decoy SOSUS/MPA detection efforts while quieter Soviet submarines passed in and out of the Atlantic.

As Northwood had instructed *Superb* to remain in contact with L-041 only if it was conducting anything other than a standard patrol, Boyce moved away in order to prosecute another SOSUS contact, a 'Yankee' class reportedly operating 300 miles to the northeast of *Superb*'s position, designated L-048. Before *Superb* could obtain contact with L-048, Northwood cancelled the prosecution order and ordered *Superb* to intercept another contact, a Type III nuclear submarine that appeared to be bound for the Western Atlantic via the Iceland–Faroes Gap, and designated L-046. *Superb* spent two days searching for L-046 in difficult conditions, but due to sparse SOSUS tracking information, which was in the region of 75 nautical miles, and a large amount of merchant traffic which helped to mask the submarine's noise signature, *Superb* first detected the Soviet submarine at a range of only 2500 yards, and although it later passed about 1000 yards down *Superb*'s side the Soviet submarine did not show any signs of having detected *Superb*, which then manoeuvred into a trailing position and maintained both narrowband and broadband contact on it for the next forty-eight hours.

After collecting a considerable amount of intelligence *Superb* was tasked with intercepting a possible 'Delta' class submarine that had been picked up on SOSUS and designated L-052. However, after searching for the Delta, *Superb* was once again ordered to break away and intercept a Soviet task group led by the nuclear-powered guided-missile cruiser *Kirov*. The *Kirov* had deployed from the Barents Sea in company with a modified 'Kashin' class destroyer, a Soviet support ship and a possible 'Charlie' class SSGN. HMS *Glasgow*, a Royal Navy Type 42 destroyer, and a support ship, RFA *Olwen*, were already tracking the small task force; *Superb* was ordered to assist in the surveillance job and in particular to conduct underwater looks on all the Soviet vessels. *Superb* remained in contact with the *Kirov* group for three days with four principal objectives: to track the submarine consort – a 'Charlie' SSGN; to achieve close monitoring of *Kirov*, including an underwater look; to gather data on *Kirov*'s sonar; and to conduct an underwater look on the Soviet oceanographic research ship, now identified as *Vladimir Kavrayskiy*.

Superb failed to detect the 'Charlie' class submarine, as the level of background noise from the various ships in the area was far too high. The *Kirov* appeared to be carrying out sonar trials as it was tracking around a 40 nautical miles box with the 'Kashin' class destroyer and HMS *Glasgow* in company while *Vladimir Kavrayskiy* remained in the centre monitoring sonar transmissions. Due to these constant transmissions *Superb* was unable to approach *Kirov* and conduct an underwater look, but it did manage to conduct one on *Vladimir Kavrayskiy*, which was lying static in the water well clear of *Kirov*. Although *Superb* passed directly underneath, very little was observed due to poor water visibility. However, *Superb* was able to obtain a vast amount of information about the task group, particularly the *Kirov*, which did not stop transmitting while *Superb* was in the area. After a number of days in contact with the task group, Boyce was ordered to prosecute the 'Delta' class SSBN designated L-052.

While *Superb* was conducting surveillance against the *Kirov* group SOSUS had lost the 'Delta' and had only recently regained contact. *Superb* had little difficulty detecting the Soviet SSBN but quickly determined that it was extremely quiet and very well handled by the Soviet CO, who employed a number of tactics that led *Superb* to frequently lose contact. Despite this difficulty, *Superb* continued to track the Delta but with only weak and tenuous contact it proved very difficult to establish a firm trail. Eventually *Superb* reduced the range to around 3000 yards, and remained in the Delta's stern arcs. *Superb* was sitting directly behind the Soviet submarine when it suddenly began to turn through 270 degrees and passed 600 yards astern of *Superb*. Boyce and *Superb*'s crew held their nerve and concluded that the manoeuvre was just a coincidental turn as part of the Delta's erratic course alterations. The Delta showed no sign of having detected *Superb* and it continued to open up to the south, causing *Superb* to lose contact as the Soviet submarine was in its stern null. Twenty hours after initial contact *Superb* was ordered to withdraw to assume a new role in the Northern Fleet Exercise Areas.[22]

Not all trailing operations were successful. Towards the end of 1980 HMS *Sceptre* was forced to abandon an operation when it suffered extensive damage after being involved in one of the most fraught moments of the deep Cold War. Officially *Sceptre* hit an iceberg.

Unofficially, the submarine collided with what was probably the 'Delta III' class SSBN *K-211*, which at the time believed it had collided with an unknown US Navy 'Sturgeon' class submarine.[23] *Sceptre* had been trailing the Delta III, one of the Soviet Navy's most advanced and quietest SSBN classes and an important intelligence target, but had lost contact. After attempting to regain contact for thirty minutes, *Sceptre* suddenly started to shake. 'There was a huge noise,' recalled one of *Sceptre*'s officers. 'It started very far forward, sort of at the tip of the submarine, and it trailed back. It sounded like a scrawling. We were hitting something. That noise lasted for what seemed like a lifetime. It was probably only a couple of seconds or so. Everybody went white.'[24] 'I knew we'd impacted something straight away,' said Chief Petty Officer Michael Cundell. 'I ran out of the mess and alerted someone to shut one of the bulkhead doors, number 29, even though I knew there were people inside in the sleeping accommodation – I had to do it, because at that point we didn't know if the hull had a leak and not doing so would have compromised the rest of the submarine.'[25]

According to accounts from those on board, *Sceptre* was then pursued for two days by a Soviet attack submarine. 'We just made a sharp exit and escaped under the ice without trace,' said Cundell. After shaking off the Soviet submarine *Sceptre* surfaced so that her crew could inspect the damage. There was a long tear along the outer hull and the conning tower and mangled pieces of material from the Soviet boat's propellers were embedded into the hull. 'It took a big chunk out of our casing,' he said. The Soviet submarine had driven across *Sceptre*'s bow and its propellers had ripped into the submarine. 'The tear started about three inches from the forward escape hatch', said Cundell. 'If that hatch had been hit or damaged – it's about 2 foot 6 in diameter – if that had been ruptured, then the fore-ends would have shipped water which would have made the boat very heavy. We would probably have sunk.' *Sceptre* returned to Devonport and was put into dry dock for repairs. 'I remember digging a lot of the propeller out of our hull, and although most of it went to the Ministry of Defence, I still have a piece,' said Cundell.[26]

Doug Littlejohns assumed command of HMS *Sceptre* on 21 September 1981 and was given responsibility for bringing the boat back into the fleet. 'The submarine was broken and so was the crew,' recalled Littlejohns. 'My main objectives were to get her out of dockyard hands as quickly as possible while also ensuring that the crew was fully trained and ready to venture back to sea. There were major challenges on both fronts.'[27] When *Sceptre* finally put to sea Littlejohns found that, like all submariners, his crew were highly resilient:

> The crew had been badly shaken but within 48 hours of leaving Plymouth you would not have known it. We took her out into the Western Approaches, dived with a remarkably good trim, given the structural and material changes which had taken place, and then proceeded gently to cross the 100 fathom line. We conducted a slow and careful dive to maximum diving depth – there were quite a few nervous and white faces as we got even deeper – and remained there for some time while the engineers did thorough checks all over the boat before pronouncing that all was well. Following that I changed tempo and ordered 20 knots, 20 degrees bow up and a depth of 200 feet. That caused the first set of badly secured crockery to smash and there was more to come when I ordered the planes reversed and at high speed plunged back down deep. I carried on like this for an hour or so at ever higher speeds, tighter turns and bigger angles and dangles until the crew realized that this was how it was going to be from then on! By now all the white faces and nervous looks had disappeared and were replaced by grins and laughs.[28]

Although *Sceptre* returned to the fleet and was declared fully operational, a legacy of the incident stuck with the submarine for the rest of its operational life.[29] During the collision a piece of the Soviet submarine went through *Sceptre*'s propulsor and at certain speeds *Sceptre* would emit a noise signature which could be easily detected.

Intelligence collection operations never ceased (except in the weeks before general elections). At the end of 1981, HMS *Superb* detected the Soviet Union's first third-generation nuclear submarine, the Project 949/'Oscar' class SSGN. At 144 metres in length, with an 18.2-metre beam, and displacing 22,500 tons, the 'Oscar' was one of the world's largest submarines. Designed to counter US aircraft carriers, the submarine was very fast and capable of speeds in excess of 30 knots. It was armed with twenty-four new P-700 Granit anti-ship missiles, known in NATO as the SS-N-19 – three times as many cruise missiles as the previous Soviet SSGNs, the 'Echo II' or 'Charlie' class. Twelve missiles were placed in angled launch canisters between the pressure hull and the outer hull on each side of the submarine. The speed and impressive firepower of the 'Oscar' posed a major threat to surface ships; it was also extremely quiet in comparison with previous Soviet submarines. This was the vessel on whose beam HMS *Superb* fell in at about 5000 yards and watched as the submarine carried out a test firing.

The Royal Navy's conventional submarines were also involved in collecting intelligence against the Soviet Navy. In July 1982, HMS *Opportune* was involved in an incident with the Soviet factory ship *Rybachiy*. *Opportune*, under the command of Lieutenant Commander Richard Burston, was tasked with conducting surveillance of the *Rybachiy* in an area 300 miles west of Lands End. *Rybachiy* appeared to be engaged in a bottom contour survey, using an echo sounder and an unidentified towed-array body. Burston's orders were to try to ascertain what the *Rybachiy* had been doing since moving down from the Porcupine Bank, an area approximately 120 miles west of Ireland. *Opportune* spent five hours conducting two underwater looks on the *Rybachiy* in modest visibility, which produced no clear photographs of the hull domes or towing arrangements. The only unusual features were an unexplained fitting protruding two feet from the hull just aft of the port bilge and a probable towed-array body. Prior to departure Burston was briefed on the possible appearance of 'lights near the waterline'. After completing the underwater looks, *Opportune* withdrew to snort and recharge batteries overnight. However, at 2232 on 11 July four white lights were indeed sighted on *Rybachiy*'s waterline. Burston decided to approach and investigate with the intention of making a close 400-yard pass up the port side attempting night photography for the exact locations.

As *Opportune* moved astern of the *Rybachiy*, Burston's last visual look indicated that his submarine was on a safe course. However, just as *Opportune* was closing on *Rybachiy*, sonar information indicated that the Soviet vessel had altered course towards *Opportune*. Burston quickly raised the periscope to take another visual look and immediately ordered an emergency dive to take *Opportune* deep. But it was too late. At 0030 on 12 July, at approximately 62 feet, with 10 degrees bow down angle, the top of *Opportune*'s fin made contact with the underside of the *Rybachiy* and ran into its port screw and rudder. Alarms sounded all around *Opportune*'s Control Room as the submarine continued down to 140 feet, then 180 feet. Water was coming in from the bottom of the radar mast and neither periscope could be raised. Initial reports indicated that there were no serious breaches of watertight integrity or damage to propulsion or major systems and Burston decided to open up the range to the northwest at a speed of 8 knots for one hour, surfacing before twilight to conduct a visual inspection. After running for eight hours, *Opportune* surfaced. The damage was extensive. The bridge area had been pushed aft approximately 5 feet and compressed 2–3 feet on top of the attack and search periscopes. The remaining top of the fin area from the Officer of Watch position to the W/T mast was pushed back 2–5 feet and compressed downwards. The top 3 feet of the forward periscope was also bent back at 80 degrees and the top window had snapped off the aft periscope and lodged in the fin. *Opportune* returned to Portsmouth. Attempts to conceal the incident largely failed due to the media focus on the Navy in the aftermath of the Falklands conflict.[30]

The incidents involving HMS *Sceptre* and HMS *Opportune* were two exceptions to mostly highly successful operations against Soviet naval forces that yielded vast amounts of operational, acoustic and electronic intelligence on the movement and tactics of Soviet submarines.

THE 1981 DEFENCE REVIEW

With such increased emphasis on the SSN in the Royal Navy's ASW operations, it was only a matter of time before someone started to question the need for the rest of the Royal Navy. As we have seen, the Conservative Government which came into office in 1979 faced a number of serious economic problems. When John Nott succeeded Francis Pym as Secretary of State for Defence in January 1981, he inherited a defence overspend from the previous financial year of £200m and a projected overspend for 1980/81 of £400m out of a total defence budget of £11.2bn. In trying to bring the defence budget under control, Nott also sought to understand naval strategy. The Navy saw its principal function in the event of war as the convoying of US reinforcements across the Atlantic and to support the American Carrier Battle Groups which were to move into Greenland, Icelandic and Norwegian waters to contain the Russian naval threat in the Atlantic. Nott was deeply unsatisfied with the quality of the briefings that he was receiving from the Naval Staff. 'Again and again I saw that I was being briefed in a way that fortified the traditional naval interest without getting down to the real nitty-gritty,' wrote Nott. He felt that he received 'one inadequate briefing after another' and was 'asked to accept ideas which even a layman . . . could see were nonsense'.[31]

Of great concern to NATO was the possibility that the Soviets would mount a ground assault on northern Norway to seize aircraft bases to provide air cover to Soviet naval forces attempting to dominate the Norwegian Sea. NATO developed elaborate contingency plans for the rapid reinforcement of Europe in the event of war which were demonstrated in periodic exercises in which each plan would be tested and various NATO forces given opportunities to familiarize themselves with the affected waters and territories. The Atlantic campaign, designed to protect the transportation of Allied reinforcement and resupply across the Atlantic, was periodically practised in a series of exercises known as 'Ocean Safari'. A shallow-seas campaign designed to prevent the exit of the Soviet Baltic Fleet into the North Sea and to protect Allied convoys in the North Sea and the English Channel was practised in exercises codenamed 'Northern Wedding', and a Norwegian Sea campaign designed to prevent the exit of the Soviet Northern Fleet into the Norwegian Sea and the North Atlantic, and to provide sea-based support to Allied air and ground operations in Norway, was practised as Exercise 'Teamwork'.

Although these exercises demonstrated the solidarity of the NATO Alliance and its determination to defend against any attack, the Royal Navy realized the vulnerability of the convoy system. In the NATO 'Ocean Safari' exercise of 1981, the direct defence of shipping carrying troops and supplies across the Atlantic was abandoned in favour of an experiment with a new 'defended lanes' concept, an attempt to use the potential of modern long-range passive sonars to sanitize an area through which independently routed ships or lightly escorted merchantmen could traverse.[32] By the 1980s, as one naval historian has put it, 'doctrinal thought was moving even further away from the ideas confirmed by past experience'.[33]

Until the late 1970s, analysts and strategic planners in the United States assumed that Soviet Admirals would act and react in the same way as their American counterparts, that in the event of war the Soviets would aim to bring about fleet-to-fleet actions on the high seas, with the aim of obtaining control of sea lines of communication, the so-called 'SLOCS', on which the resupply of Europe depended.[34] As one naval historian noted:

From the early 1960s, when the growth of Soviet naval power became evident, the pre-dominant view in America was that the Soviets were building a naval force with many capabilities similar to the United States Navy. Most importantly, the existence of a blue-water Soviet Navy seemed to emphasize, in American minds, the capability for peacetime power projection, the facility for wartime attack on U.S. and Western naval forces and sea lines of communication, as well as the ability to launch strategic nuclear strikes from the sea. Increasingly, Americans worried about the Soviet Navy as a sea-denial force that could deprive the West of the free use of the sea, thereby creating political, economic, and military disaster. In short, Americans tended to view the new Soviet naval capabilities in terms of mirror-imaging and refighting World War II.[35]

However, beginning in the late 1970s, the United States obtained specific high-level intelligence about Soviet naval plans and capabilities. The intelligence indicated that the Soviets regarded the interdiction of sea lines of communication as a far less urgent task than the United States had previously assumed. The nature of this intelligence was predominantly SIGINT, but also 'some very significant HUMINT penetration of senior echelons of the soviet leadership', which remains highly classified.[36] It is likely that at least one of the sources was a combined US Navy/National Security Agency top-secret programme to tap into Soviet seafloor communication cables in the Soviet Far East and Arctic areas, codenamed Operation 'Ivy Bells'.[37]

Regardless of the source, the highly classified intelligence had a profound impact on US intelligence assessments of Soviet intentions and capabilities:

> The insights gained from these sources allowed the U.S. Navy, led by Naval Intelligence, to totally reassess how the Soviets would fight a war, where their strengths and vulnerabilities were, and how their perceptions and prejudices caused them to view us [United States]. This enabled Naval Intelligence to stimulate and participate not only in a complete rewrite of U.S. naval strategy and the war plans which governed how the U.S. would fight a war with the Soviet Union, but also to plan and conduct meaningful perception management.[38]

In November 1981, a US interagency intelligence memorandum on 'Soviet Intentions and Capabilities for Interdicting Sea Lines of Communication in a War with NATO' expressed the general agreement of intelligence analysts that Soviet military planners regarded the wartime interdiction of NATO sea lines of communication as a secondary mission and that only a few submarines would be employed in attacking commerce in the North Atlantic in the opening stages of a NATO–Warsaw Pact war. The US concluded that the majority of Soviet forces would instead be deployed close to the USSR to defend the Soviet SSBN force and to protect the Soviet Union from NATO's nuclear-armed strike force.[39] This view was confirmed a year later in a 1982 US National Intelligence Assessment, which concluded that:

> The Soviets view SLOC interdiction as a less urgent task than providing combat stability for their SSBNs and defeating the West's nuclear-capable naval strike forces. They believe that Warsaw Pact forces would defeat the main grouping of NATO forces in Central Europe or the war would escalate to theater nuclear conflict before NATO's seaborne reinforcement and resupply of Europe or US forces in the Far East became a critical factor. Only a few forces – primarily diesel submarines – would therefore be allocated to open-ocean SLOC interdiction from the outset of hostilities.[40]

Put simply, the Soviets appeared to have concluded that 'increased interdiction effort would be at the expense of SSBN protection and the defense of the Soviet homeland'.[41] Exactly why they had shifted their naval strategy remained a mystery.

When Nott visited Norfolk, Virginia, in March 1981 for a briefing by Admiral Harry Train, the Supreme Allied Commander Atlantic, he 'began to grasp the true scenario'. SACLANT's briefings 'were of a much higher intellectual and practical quality than anything that I obtained from the Naval Staff in Whitehall'.[42] Significantly, he found that the American convoy plans for the resupply of Europe 'did not support the Naval Staff's Second World War type convoying proposals, even if the war turned out to last six months rather than a week'. The US Navy's new convoying plans took account of the reduced threat to the reinforcement and resupply of Europe by sea and 'involved a series of hectic, dispersed single sailings to avoid the Soviet submarine threat, rather than Second World War escorted convoys'.[43]

Nott was also heavily influenced by his belief that any conflict with the Soviet Union would end up being a short rather than a long war. Whenever a battle on the central front was 'gamed', NATO was in such a desperate position after seven days that it was forced to initiate the use of tactical nuclear weapons.[44] This meant that once reinforcement by shipping across the Atlantic arrived it would be too late to be of any use. 'I wanted to have an open debate with the Royal Navy about these issues,' recalled Nott, 'but they [the Naval Staff] stuck to the long-war thesis on the basis that this justified the traditional naval role.'[45] Nott wanted 'fresh and original thinking to meet the Soviet submarine challenge':[46]

Not only were these Soviet submarines (some hundreds in number) moving faster and diving deeper, but they were also being equipped with anti-ship missiles of increasing sophistication and range, targeted by satellite, which could be released by the Soviet submarines from under the sea. Clearly these missiles posed a very great threat, when coupled with the Backfire bombers, to our surface ships in the North Atlantic. This raised the whole question of whether we should not put more emphasis on our submarine fleet.[47]

As a result of the deadlock, Nott turned to his non-naval advisers, in particular to the MOD's Chief Scientific Adviser, Sir Ronald Mason, who initiated his own studies to determine the size and shape of maritime forces in the Eastern Atlantic. Studies from the Defence Operational Analysis Establishment indicated that land-based maritime air power and submarines were the vital assets in any conventional deterrent force for the Eastern Atlantic. The reports rejected the logic behind the navy's re-equipment programme and argued that it failed to deal with the increasing anti-ship missile threat to warships conducting anti-submarine warfare. Nott believed that the way out was to strengthen NATO's conventional forces on the ground in Europe and to ensure that naval forces in the Atlantic were strong enough to safeguard reinforcement.[48] He wanted to shift the Royal Navy's emphasis away from the traditional escorted convoy towards nuclear submarines and Nimrod Maritime Patrol Aircraft operating in the Atlantic, and away from expensive ships to cheaper ones such as the Type 23 frigate. This would allow deep cuts to be made to the Royal Navy's destroyer and frigate force, as well as a reduction of the carrier force to just two ships and the eventual phasing out of the Navy's amphibious capability, while the resources devoted to escorting surface forces in the Eastern Atlantic could be diverted to the Greenland–Iceland–United Kingdom Gap chokepoints. The Royal Navy's top priority would remain the provision of the UK strategic nuclear deterrent, initially with Polaris/Chevaline and later the new Trident force. But for general maritime warfare the Royal Navy would concentrate first upon the provision of a powerful submarine force to exploit the country's position on the flank of the Soviet Navy's main exit to the Atlantic, defending the Greenland–Iceland–United Kingdom (GIUK) Gap through which Soviet submarines would have to travel if indeed they planned to interdict NATO sea lines of communications.[49]

The First Sea Lord, Sir Henry Leach, was so incensed by the proposed reductions in the Royal Navy's surface forces that just nine months before the Falklands conflict he wrote personally to the Prime Minister:

> The proposal has been devised ad hoc in two months. It has been neither validated nor studied in depth. No alternative options have been considered. It has all been done in a rush. Such unbalanced devastation of our overall Defence capability is unprecedented . . . We are on the brink of a historic decision. War seldom takes the expected form and a strong maritime capability provides flexibility for the unseen. If you erode it to the extent envisaged I believe you will undesirably foreclose your future options and prejudice our National Security.[50]

Leach's letter had little effect. 'In the end, we agreed to an expansion of the submarine programme at the expense of the surface fleet,' wrote Nott.[51] In the introduction to the long-awaited and partially leaked White Paper 'The Way Forward', published on 25 June 1981, Nott declared that:

> Our basic judgment . . . is that for the future the most cost-effective mix – the best balanced operational contribution in our situation – will be one which continues to enhance our maritime-air and submarine effort, but accepts a reduction below current plans in the size of our surface fleet and the scale and sophistication of new ship-building, and breaks away from the practice of costly mid-life modernization.[52]

The White Paper emphasized that:

> Our most powerful vessels for maritime war are our nuclear powered attack submarines (SSNs), soon to be equipped with the anti-surface ship guided missile Sub-Harpoon. There are 12 at present in service, and the fleet will build up further to 17. An order worth £117 million is now being placed for the next Trafalgar-class boat to be built by Vickers (Barrow). We intend also to proceed with the new class of diesel-powered submarines (SSKs) – which may have considerable export potential – and will if possible introduce these at the rate of one per year. Both SSNs and SSKs will be equipped later in the 1980s with a new heavyweight torpedo of high performance; we are considering the choice of design.[53]

In short, the Royal Navy's hunter-killer submarines were to be the modern Dreadnoughts – *the* capital ships of the deep.

Although the Falklands Crisis enabled the Navy to hold on to certain surface assets, the conflict had little impact on the Submarine

Service. A foreword to the 1982 Defence White Paper, issued in June 1982 and written by Nott, stated that:

> The events of recent weeks must not, however, obscure the fact that the main threat to the security of the United Kingdom is from the nuclear and conventional forces of the Soviet Union and her Warsaw Pact allies. It was to meet this threat that the defence programme described in Command 8288 ['The Way Forward'] was designed. The framework of that programme remains appropriate.[54]

Indeed the 1982 Statement on the Defence Estimates went on to stress that 'In the field of anti-submarine warfare, we attach particular importance to increasing the size of the nuclear-powered submarine force as rapidly as resources will permit.'[55]

In Nott's mind, the future belonged to the SSNs, but there was a constraining factor of which the Royal Navy was well aware. In June 1977 a paper had been submitted to the Fleet Requirements Committee on the size and shape of the Submarine Flotilla. The paper argued that because of industrial and support constraints, the maximum number of SSNs that the Royal Navy could build and support was less than that required to meet operational tasks. SSNs were expensive in terms of capital and running costs, and the analysis by the MOD revealed that the Navy could afford to build and support no more than twenty-four nuclear hulls (including the four Polaris submarines). The possibility of replacing the SSBNs also limited the number of SSNs that the Royal Navy could build – only two or three extra (compared with pre-1981 plans) could be in service by the year 2000 and even that was very much dependent on a smooth ordering programme. One way of increasing numbers and meeting the various operational tasks assigned to the Royal Navy was to use conventional submarines to fill the resultant shortfall, operating a mixed fleet of SSNs and SSKs, with the SSKs carrying out a number of tasks ordinarily assigned to SSNs.[56]

THE 'UPHOLDER' CLASS

This represented a clear turn in naval policy. In 1962 the Macmillan Government had decided not to order a new generation of conventional submarines to replace the 'Porpoise' and 'Oberon' classes, the first of which was due to be decommissioned in 1984. However, by the late

1960s, the views of a number of senior Royal Navy submariners began to shift. When Vice Admiral John Roxburgh took up the post of FOSM in 1969 he 'felt there was a case for another generation of conventionals on [grounds of] cost'. However, by December 1971 his views had changed, as he explained in a letter to a former FOSM, Vice Admiral Sir Ian McGeoch:

> Having now lived with the problem for two years and discussed it at all levels I have become convinced it would be a grave mistake to dissipate our limited resources by going for more conventionals. I have very much in mind the time scale – the earliest we could in practical terms produce a new class would be by 1979–80. Whereas I am happy with their effectiveness now I cannot agree that they will be an effective warship on into the late 1990s, albeit correctly armed. The disadvantages of immobility and lack of endurance compared to Fleet submarines [SSN] in both anti-surface and anti-submarine warfare need no emphasis from me to you. I am also happy that the comparative noise advantage of the conventional is disappearing.[57]

Any new class of conventional submarine would need to be large to accommodate modern sonars and computer systems, all of which required large amounts of electrical power, both for running the equipment and maintaining a suitable environment in which it could operate. An SSK equipped and armed with the sensors and weapons of the 1980s and 1990s and powered by diesel electric propulsion was also judged to have little chance of survival against future ASW tactics, given the need to regularly expose a snort mast to take in air. 'The conventional has little hope of survival, in the 1980's and beyond, off a Russian Base,' noted Roxburgh.[58] Other forms of propulsion designed to prolong underwater endurance, thus giving some of the advantages of nuclear power, would have been very expensive in research and production, eliminating many of the cost advantages. As far as Roxburgh was concerned, 'the case for a submarine which can fully exploit the advantages of endurance and mobility afforded by nuclear power far outweighs anything that can be provided by conventional power'.[59]

However, there was plenty of evidence to reinforce the case for a new class of conventional submarine. Nuclear submarines, due to the complex machinery required to maintain nuclear reactors, tended to radiate far more noise than conventional submarines. A modern SSK operating at slow speeds on battery power was far quieter and a very difficult target for passive sonars. In the 1970s, the lack of detections on the Soviet 'West of UK Whisky', a conventional Soviet submarine that was almost always stationed off the United Kingdom, and the difficulty which very considerable NATO ASW forces had in detecting Soviet diesel-powered 'Juliett' and 'Foxtrot' class submarines in the Mediterranean, bore witness to the SSK's ability to avoid discovery. Royal Navy and Dutch Navy SSKs frequently demonstrated the ability to conduct surveillance patrols in the vicinity of Soviet ASW vessels without being counter-detected. Major NATO exercises also highlighted just how difficult it was to detect diesel submarines. During the 1976 'Teamwork' exercise the tactics used by four diesel submarines awaiting a NATO Strike Fleet in the Shetlands–Faroes Gap resulted in all of them being undetected. Technological advances, such as the fitting of higher-performance diesel engines and high-capacity batteries, had also considerably reduced the period which SSKs had to spend snorting to as little as 5–10 per cent of a patrol.

In November 1977, following the appearance of its June 1977 paper on the size and shape of the submarine flotilla, which concluded that because of industrial and support constraints, the maximum number of SSNs that the Royal Navy could build and support was less than that required to meet operational tasks, the Fleet Requirements Committee endorsed an Outline Staff Target for a new SSK to exploit its main advantage – quietness – and complement the Royal Navy's SSNs.[60] In times of tension and war the Navy envisioned deploying the new SSKs on area ASW surveillance patrols in the GIUK Gap, the Norwegian Sea, IBERLANT (Allied Forces Iberian Atlantic Area) and the northwest and southwest approaches to the UK. In times of minimal tension they would undertake the larger portion of the training and trials tasks, releasing the SSNs to concentrate on shadowing, trailing, and intelligence-gathering and surveillance operations. Preliminary feasibility studies began in December 1977, when the operational roles of the new SSKs were defined in more detail.

The Navy required a submarine that had good endurance, was very quiet, and had excellent sensors, good communications and adequate weapons, with the potential for continuing weapon development to match improvements in SSNs. These characteristics had to be balanced against cost-effectiveness and great care was taken to avoid designing a submarine which would simply duplicate all the roles of an SSN, but, where it could, would do it more cheaply, both in capital and in support costs. The new SSKs would be appreciably better than the 'Oberon' class as a result of improved technology in all fields, particularly the greatly enhanced endurance of modern batteries. The aim was to produce a submarine capable of performing operational roles in the Eastern Atlantic and Channel areas, but also able to conduct low-intensity operations in other areas. ASW operations, both surveillance and offensive, were judged to be the most important and this required an ocean-going submarine with good sonar, good Action Information Organization, good communications and sufficient endurance to spend six weeks on station in the Iceland–Faroes Gap. Other operational roles included intelligence gathering, protection of the strategic deterrent, and coastal work, including mining and special operations as well as ASW training and exercises. The size of the new SSK had to be kept to a minimum in order not to degrade its shallow-water coastal performance and, most importantly, to keep the cost within the region of £52m each, about one third the cost of an SSN.

A number of different designs were produced. The first was Design A1, a submarine 62 metres in length and 7.3 metres in diameter, with a submerged displacement of 1960 tons. Design B1 was a slightly larger submarine, 67 metres in length and 7.6 metres in diameter, with a submerged displacement of 2250 tons. Design D1 was even bigger, 75 metres, with a hull diameter of 7.6 metres and an overall displacement of 2650 tons. It quickly became clear that available design resources, both for the submarine and for its weapons systems only permitted further detailed study of one of the options. Design A1 was quickly discarded, and discussion centred on B1 and D1. While B1 was the preferred design of the Naval Staff, Vickers preferred D1 and its greater range and weapons load capacity because it had ambitions to market the new SSK design overseas and believed potential customers would be interested in larger ocean-going SSKs. However, D1 was definitely too big for the Navy's operational requirements, and it would have proved too expensive to maintain the cost-effectiveness argument against the SSN.[61]

As a compromise solution between the Navy's operational requirements and the demands of the export market, the Director General Ships proposed a B1-based design, with a slightly rearranged internal layout and an increased submerged displacement of 2400 tonnes. This would allow the submarine to carry sufficient fuel for a patrol range of 24,000 miles and twenty-eight days on station. The design could also be stretched to include more fuel if some customers considered this necessary, and an even greater weapon load.[62] This was considered acceptable by Defence Sales, British Shipbuilders and the Royal Navy, which accepted a small number of minor penalties, such as reduced

68. The Submarine Service expanded rapidly throughout the 1970s. Pictured here in 1976 are 'Porpoise' and 'Oberon' class diesel electric submarines at HMS *Dolphin*.

69. Vice Admiral Sir Tony Troup, FOSM, 1972–4. A hardened Second World War submarine ace, Troup developed a far more aggressive posture towards the Soviets.

70. Admiral Sir Iwan Raikes, FOSM, 1974–7, feared that his post would ultimately be absorbed into the Fleet staff and lose its identity and independence.

71. Admiral Sir John Fieldhouse, FOSM, 1977–81. Fieldhouse completed the centralization of Submarine Command and Control and ensured the service was ready should the Cold War heat up.

72. 2 May 1982: the *General Belgrano* sinks after being struck by two Mark 8 torpedoes fired from HMS *Conqueror*.

73. (*above*) HMS *Conqueror* returns home from the South Atlantic flying the Jolly Roger, 4 July 1982.

74. Mrs Thatcher visited HMS *Resolution* shortly after the Falklands conflict. She is pictured here in *Resolution*'s Missile Control Centre witnessing a simulated firing of a Polaris missile. The Commander-in-Chief Fleet, Admiral Sir John Fieldhouse, is standing behind her.

75. In the early 1980s the Thatcher Government decided to replace Polaris with Trident. Pictured here is a Trident D5 missile launching during HMS *Victorious*'s DASO, witnessed by the authors in October 2012.

76. A new class of SSBN was required to carry Trident. HMS *Victorious*, the second of the 'Vanguard' class SSBNs, at the Vickers shipyard, Barrow-in-Furness, September 1993.

77. The northern end of Faslane in July 1994. The entire base underwent a vast expansion to accommodate Trident. New berthing facilities and the new ship lift are pictured in the foreground. The old Polaris berths and the Admiralty floating dock is in the background.

78. The Armament Depot at Coulport also underwent expansion. Pictured is the new Explosives-Handling Jetty. In the background is the old Polaris storage area and the old Explosives-Handling Jetty.

79. The new Explosives-Handling Jetty at Coulport, with a 'Vanguard' class submarine receiving an outload of Trident D5 missiles.

80. FOSMs gather every ten years for a reunion. Pictured here in 1985 in front of HMS *Andrew*'s 4-inch gun at HMS *Dolphin* are, left to right: (standing) Admirals John 'Sandy' Woodward, Iwan Raikes, John Roxburgh, Ben Bryant, Tony Troup, 'Tubby' Squires, Dick Heaslip; (seated) Admirals 'Rufus' Mackenzie, Peter Herbert, Guy Grantham, Sir John Fieldhouse, Horace Law, Sir John 'Baldy' Hezlett, Ian McGeoch. Fieldhouse was at that point Chief of the Defence Staff, the most senior submariner in British naval history.

81. In the 1980s submariners found themselves increasingly isolated from their surface colleagues, giving rise to the 'Black Mafia', who among other escapades painted this submarine silhouette on the stern of HMS *Scylla*.

82. The Royal Navy Sub Harpoon, introduced in 1982, enabled submarines to strike at surface ships from a range of sixty miles.

83. HMS *Trafalgar* during sea trials in 1982. The 'Trafalgar' class was the culmination of British Cold War submarine design and a triumph for the naval architects in the Royal Corps of Naval Constructors.

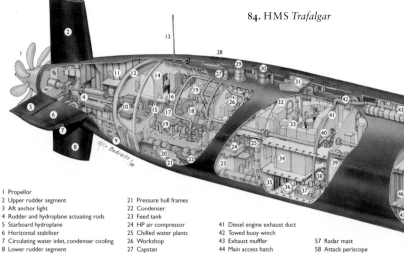

84. HMS *Trafalgar*

1 Propellor
2 Upper rudder segment
3 Aft anchor light
4 Rudder and hydroplane actuating rods
5 Starboard hydroplane
6 Horizontal stabiliser
7 Circulating water inlet, condenser cooling
8 Lower rudder segment
9 Retractable secondary propulsion motor
10 Rudder and hydroplane hydraulic actuators
11 Aft ballast tanks
12 Hull aft pressure dome
13 Portable ensign staff
14 Technical office
15 Propeller drive shaft thrust block and
   flexible coupling
16 Emergency propulsion motor
17 Combining gearbox
18 Main turbines
19 Turbo generators
20 Lubricating oil tank

21 Pressure hull frames
22 Condenser
23 Feed tank
24 HP air compressor
25 Chilled water plants
26 Workshop
27 Capstan
28 Hinged cleats
29 Engine room hatch
30 Aft escape hatch
31 Towed buoy
32 Watertight bulkhead, typical
33 Manoeuvring room
34 Switchboard room
35 Hydrogen storage bottles
36 Air treatment unit
37 AC/DC motor generator
38 Diesel generators
39 Reactor compartment
40 Main steam valve

41 Diesel engine exhaust duct
42 Towed buoy winch
43 Exhaust muffler
44 Main access hatch
45 Commanding Officer's cabin
46 Cooling tank
47 Diesel oil tank
48 Air purification plant
49 Junior ratings' mess
50 Air system main supply fan
51 Mast well
52 Mast hydraulic actuators
53 Aft sonar
54 Diesel exhaust
55 Snort induction mast, fresh air
56 Communications antennae
   masts

57 Radar mast
58 Attack periscope
59 Search periscope
60 Surface navigation position
61 Forward Sonar
62 Diesel fuel-oil expansion
   tank
63 Conning tower access
64 Control room
65 Galley
66 Garbage ejector
   compartment
67 Data distribution racks
68 Battery compartment
69 Torpedo handling gear

85. (*top left*) By the 1980s, the Submarine Service was starting to resolve the problems with its torpedoes. Here a salvo of three practice Mark 8 torpedoes approach a Type 21 frigate. The torpedoes are set deep to run under the frigate.

86. (*top right*) HMS *Berwick* sinking after being struck by a Mark 24 Tigerfish torpedo during a live-fire exercise. The torpedo was designed to explode underneath a target, in this case the resultant shockwave breaking the back of the ship.

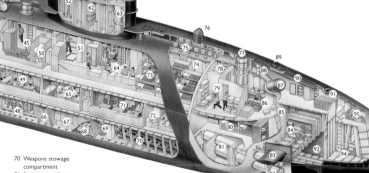

70 Weapons stowage compartment
71 Senior ratings' mess
72 Junior ratings' mess
73 Senior ratings' bunks
74 Sonar equipment room
75 Hull sonar glands
76 Sonar housing
77 Forward escape hatch
78 Coxswain's stores
79 Dry provisions store
80 Junior ratings' bunks
81 Torpedo tubes
82 Torpedo tube bow caps
83 Retractable forward hydroplane
84 Hydroplane hydraulic actuator
85 Hull forward pressure dome
86 CO2 absorption unit
87 Weapon embarkation hatch
88 Forward capstan
89 Hinged cleats
90 Removable casing plates
91 Retractable fairlead
92 Number 2 Main ballast tank
93 Forward sonar housing
94 Anchor stowage and cable locker
95 Anchor windlass
96 Number 1 main ballast tank

87. The end of the Cold War. The Chief of General Staff, Soviet Armed Forces, General Vladimir Lobov, visits HMS *Revenge* at Faslane, 4 December 1991. Astern of HMS *Revenge* are another SSBN and an 'S' class SSN.

88. (*above*) The 1990s also saw the introduction of the 'Upholder' class diesel electric submarines to replace the ageing 'Porpoise' and 'Oberon' classes: HMS *Upholder*, HMS *Unseen,* HMS *Ursula* and HMS *Unicorn.*

89. (*left*) The end of an era. HMS *Valiant,* HMS *Olympus* and HMS *Upholder*: the 'Valiant', 'Oberon' and 'Upholder' classes were all retired in 1993.

maximum dived speed and reduced submerged endurance. The final design, known as the Vickers Type 2400, represented a significant increase in capability over the designs of the 1950s 'Porpoise' and 'Oberon' classes. The hull design, machinery mounting and hull coatings all benefited from three generations of SSN design. The result was an SSK that promised a more up-to-date weapon fit, greater endurance and quietness, and a complement of 47 compared with the 69 in the 'Oberon' class. It was also anticipated that maintenance and refit times would be reduced by 20 per cent.[63]

With a submerged displacement of 2465 tonnes and a surface displacement in diving trim of 2205 tonnes, the pressure hull of the submarine was constructed using high-tensile steel, with high-tensile steel frames, and extensive use was made of glass-reinforced plastic in the casing and bridge fin. The class had a deep-diving depth of 250 metres. Its single propeller, of an advanced noise-reduced design, allowed it to travel at speeds of 20 knots dived and 12 knots surfaced, with a range of some 8000 miles. Course and depth were controlled using an advanced new autopilot. It was also equipped with the French bow passive, intercept and active sonar 'Triton' which was branded Sonar 2040 in the Royal Navy. The first of class, HMS *Upholder*, was fitted with a Sonar 2026 towed-array system, while later submarines featured the more advanced and cheaper 2046.[64] Because it would not have an anti-ship role, the new SSK would only carry wire-guided torpedoes, such as the Mark 24 Tigerfish, usually only fired singly or in pairs, allowing a reduced weapon load.[65]

To ensure that the required minimum of submarine hulls was maintained as the 'Oberon' class reached the end of their service lives, it was decided that the first of the 'Upholder' class should be in service by 1986. A Project Order contract was placed with Vickers in 1980 and a significant percentage of the Vickers design and drawing office staff were engaged on the project, for which a one-fifth scale model and part mock-up were produced. HMS *Upholder* was laid down in 1983 and launched in 1986, while three more of the class, HMS *Unseen*, HMS *Ursula* and HMS *Unicorn*, were laid down at Cammell Laird (which had by then been purchased by Vickers Shipbuilding and Engineering Ltd) in 1986. HMS *Upholder* was meant to commission into the Royal Navy in 1988, but there were delays and increased costs, and the submarine was only delivered in June 1990.[66] HMS *Unseen* followed in June 1991, HMS *Ursula* in May 1992 and HMS *Unicorn* in June 1993.

THE 'TRAFALGAR' CLASS

The Royal Navy's twelfth SSN, HMS *Splendid*, the last of the 'Swift-sure' class, was commissioned into the Royal Navy in March 1981. Development work on the next class of Royal Navy SSN had started as far back as 1968 under the codename SSNoY when Rolls-Royce and Associates began development work on a new SSN primary plant, with increased core life and reductions in the pumping power and noise of the main coolant pumps. Work on the design of a new, longer lasting core, known as Core Z, was completed in 1971 and a prototype was fitted at Dounreay in 1973/74. This new design significantly reduced the low-frequency radiated noise and self-noise of the plant and offered an extended operating life over that of Core B used in previous SSN/SSBN classes. The new core was installed in most of the 'Trafalgar' class sub-marines, and retrofitted to the 'Swiftsure' class.[67]

As in the 'Swiftsure' class, the 'Trafalgar' class was fitted with five forward torpedo tubes, all capable of accepting the Mark 24 torpedo, and RN Sub Harpoon. Twenty weapons stowage positions were fitted and salvo firing capability was improved. These adjustments resulted in a small increase in both the weight and space demands of the design and it was necessary to lengthen the 'Trafalgar' design by several frame spaces. The 'Trafalgar' class thus had a submerged dis-placement of 5210 tonnes and a deep surface displacement of 4740 tonnes, and a speed reduction of around 1 knot compared with the 'Swiftsure' class. The overall length was 85.4 metres and the pres-sure hull diameter of 9.83 metres was the same as the 'Swiftsure' class. Deep-diving depth also remained the same, at 381 metres, and in spite of the increase in length the maximum submerged speed of the class was nominally 29 knots with a maximum surface speed of 12 knots – again similar to the 'Swiftsure' class. The first of class, HMS *Trafalgar*, was fitted with a seven-bladed Skew propeller, but subsequent submar-ines were fitted with a new Pre-Swirl Pump Jet (PSPJ). The PSPJ was designed to decrease the range of detection from blade noise and to permit the replacement of individual blades.

The 'Trafalgar' class was also designed to meet lower noise-level targets in order to reduce detection risk and interference with on board sonar systems. The class was fitted with the latest technology including principal hull-mounted active/passive sonar known as sonar Type 2020, to replace the 1950s-era Type 2001. Based on more modern technology, most notably computer processing, Type 2020 provided improvements in range, accuracy and target handling ability and gave submarines an improved capability to identify and track several ships and submarines simultaneously. It could also obtain fire control solutions earlier, allowing the class to use its weapons to greater advantage.[68] From 1983, the 'Trafalgar' class was also fitted with the latest towed-array system, an all UK equipment to replace the less capable US/UK sonar 2024, which had been installed on thirteen SSNs and four SSBNs. As well as exploiting advances in towed-array/processing technology, the new sonar suite Type 2026 was specifically designed to match UK submarine command and control arrangements.[69] It also had a great deal of 'stretch potential' – its capacity to adapt to future threat changes – and the Navy anticipated that alterations to its software would maintain its in-service life until the end of the twentieth century.[70] The 'Trafalgar' class was also fitted with Type 2007, a long-range passive flank array; Type 2019, an intercept, forecasting array, as well as other types.

The Project Order contract for the 'Trafalgar' class submarines was placed with Vickers in April 1972, with the first of class, HMS *Trafalgar*, due on 7 September 1977. A shortage of Grade 'A' Welders and a thirteen-week industrial dispute at Vickers meant that it was not launched until July 1981.[71] Commander Martin Macpherson took HMS *Trafalgar* out of the Vickers building yard and was immediately impressed by its capabilities. 'I suppose the thing I remember most, being most confident about, was how quiet she was,' he said. 'Unbelievably quiet, which gave you huge confidence in terms of operating against the Soviets.'[72] During trials to test for underwater noise transmissions, US sound engineers on board a US monitoring vessel reportedly asked Macpherson when he would be starting his run past the monitoring ship, to which the CO replied: 'We have just completed it.' As well being exceptionally quiet, the 'Trafalgar' class were also fast.[73] 'You could run at maximum power about 27 or 28 knots in *Trafalgar* with a pencil standing on the desk and it wouldn't fall over,' recalls Macpherson. 'You go maximum speed in *Valiant* and you really know it. It's like driving an MG TC [a classic car produced in the 1940s] over a bumpy road. In *Trafalgar* it's absolutely smooth.'[74] HMS *Trafalgar* was followed by six other submarines of the same class: HMS *Turbulent* in April 1984; HMS *Tireless* in October 1985; HMS *Torbay* in February 1987; HMS *Trenchant* in January 1989; HMS *Talent* in May 1990; and HMS *Triumph* in October 1991.

SSNOZ AND THE FOLLOW-ON SSN

The Nott Review (see p. 529) also influenced the Navy's plans for the successor to the 'Trafalgar' class. The main offensive and deterrent tasks that were anticipated for the SSNs by the end of the twentieth century had important implications for the construction of the Navy's SSN fleet. Whereas in the past it had always been held that quality in SSNs was paramount, the projected burden devolving on future SSNs shifted the balance towards quantity. The Royal Navy had intended to follow the US Navy, which had started to develop a brand-new class of highly capable submarine known as the 'Seawolf' class. A revolutionary new submarine would break the evolutionary philosophy that designers had followed since *Dreadnought* first went to sea in the 1960s. Studies were first put in hand in 1969 to establish the kind of attack submarine the Royal Navy would require beyond the 1980s. An Outline Staff Target (OST 7052) for an entirely new design, known as SSNoZ, was endorsed in May 1978. But to meet the Naval Staff Target, the Fleet Requirements Committee required further studies into another five options with reduced costs. These studies reported at the same time as the Thatcher Government embarked on the Trident programme. All work on SSNoZ ceased in the autumn of 1980 when work on the SSBN successor absorbed the available resources.

SSNoZ aimed to achieve improvements in every operational aspect of SSN performance, an increase in speed, diving depth and firepower, propelled by the new PWR2 nuclear reactor designed for the 'Vanguard' class SSBNs, and equipped with an integrated sonar system, and a more capable combat system.[75] This resulted in numerous designs ranging from 6500 to 7300 tonnes with an average cost increase of 38 per cent compared to that of the *Trafalgar* design. As Rear Admiral Marsh, Assistant Chief of the Naval Staff (Operational Requirements), pointed out, 'in the face of insistence on maintaining the greatest capability in all areas, the design had become large and expensive'. The Navy attempted 'to trim the design to bring down costs but it had proved generally insensitive to cost-cutting attempts.' The design became 'over-ambitious' and the Navy, faced with a realistic view of the country's economic position and its effect on resources available for SSN procurement, concluded that 'it would have been difficult to argue such a costly weapons platform through Committees'.[76] The design was shelved and further work effectively postponed.

Not everyone was happy with the demise of SSNoZ. During a meeting of the Fleet Requirements Committee, it was pointed out that:

> Work on the SSNoZ design had been properly carried out on the basis of threat assessments and should not therefore be lightly dismissed. It would be more appropriate for the design to be suspended pending the outcome of the further studies rather than abandoned completely; this would allow future judgments to be made on the complete range of options available.[77]

Indeed, Rear Admiral Anthony Whetstone, Assistant Chief of the Naval Staff (Operations), who was on leave when the paper went to the Fleet Requirements Committee, observed that:

> the aim of the SSNoZ design study was not to produce a submarine costing 38 per cent more than SSNoY [the 'Trafalgar' class] but to ensure that our SSNs (which must inevitably number less than the Soviets) maintained superiority of quality. Unless the studies were dishonest or incompetent the 38 per cent cost increase was considered to meet this aim.
>
> If we design down to SSNoY + 10% cost what guarantee have we that the operational capability will meet the requirement? A submarine that will encounter its Soviet opponent at an inherent disadvantage must be of questionable value. Having two or three more won't affect the issue as (1) submarine encounters are almost invariably one against one and (2) losing ten encounters is no better than losing eight.
>
> Therefore if we are to keep an SSN force at all, its members must be designed to out-detect and out-fight the Soviets. Whether they need also out-run, out-dive and out-communicate is perhaps open to argument.
>
> The conclusion I draw from the above is that in the SSN case we should design up to operational requirement not down to cost. If savings are needed they should be made by identifying the areas of performance which are not critical to operational capability and accepting lower standards (and use of existing equipment) in these areas. If this means a slightly smaller force then I submit this is the lesser evil.[78]

The implication of the postponement was a trough in the SSN building programme. While there would be eighteen SSNs by 1991, they would then fall to just fifteen by 1999. If force levels were not to fall dramatically a follow-on SSN would need to be ordered by 1992, requiring an Official Staff Target to be endorsed by 1985 at the latest. It was therefore imperative that studies be redirected as soon as possible, particularly if an SSN substitute were needed in the event of the cancellation of the SSBN successor. Was quantity the only yard-stick? The cheapest, quickest solution would have been to reopen the 'Trafalgar' production lines. However, the Navy concluded that while the full SSN0Z design was too expensive for the follow-on SSN requirement, extending into the twenty-first-century the 'Trafalgar' submarine system, whose propulsion and tactical weapon systems as well as numerous other components stemmed from the 1950s, would be a false economy.

The Navy eventually settled on a follow-on SSN programme that exploited the advances made up to the start of the SSBN construction programme, with considerable savings on the estimated cost of SSN0Z. A Steering Group chaired by the Director of Naval Operational Requirements was tasked with ensuring that the optimum balance between operational requirements and costs was maintained. Consideration was given to building small nuclear submarines in the range of 2800–3000 tonnes. These submarines, described as Nuclear Patrol Submarines, were based on the Type 2400 design but stretched as necessary to include a new low-powered nuclear-propulsion plant. Although these submarines would have had a restricted performance compared with the 'Trafalgar' design, such disadvantages were offset by affordability.[79] However, the Navy eventually concluded that to develop a new and indigenous Nuclear Steam Raising Plant was not achievable in the timescale of the project. Ways of adapting existing plant components, with significant penalties in terms of noise signature, available space for weapons, sensors and reactor safety constraints, were also explored as was the possibility of international collaboration. But it quickly became apparent that the most cost-effective solution was to use the new PWR2 plant, at that stage in development for the 'Vanguard' class. The PWR2 promised considerable advantages over the PWR1, including reduced noise signature, improved reactor core life, improved shock protection and more generous reactor safety margins. It could also be scaled to provide power to a secondary machinery set producing approximately 50 per cent of that in the Trident submarine.

However, if the Navy selected the PWR2 it would have to abandon the idea of designing a much smaller submarine, as the plant implied a hull size of a minimum of approximately 5000 tonnes.[80] While this would undoubtedly increase the cost of the new class of submarine, it would also give considerable scope for the incorporation of a capable weapons/sensor system. This would be significant: many of the war games run by the Royal Navy revealed that in war its SSNs would be unable to make the most of what would be a 'target-rich environment' because they quickly ran out of torpedoes. Increasing the weapons complement on board any future SSN was therefore seen as very important.[81] In late 1982, work was tentatively begun on a submarine that was more like 'Trafalgar' but with a cost increase of 10 per cent, rather than 38 per cent as with SSN0Z, and a proposed maximum displacement of 5000 tonnes submerged. The aim was to complete initial studies by 1985, when the naval architects involved in the Trident programme would become available, place the order for the first of class in 1992 and aim for an in-service date of 1998.

THE WALKER SPY RING

The Royal Navy's decision to partly abandon its quality-over-quantity submarine philosophy could not have occurred at a worse time. The Navy was well aware that its considerable success against its Soviet adversary depended 'largely upon the performance of passive acoustic sensors keeping pace with Soviet submarine noise reduction'.[82] Submarine noise reduction was a complex task, involving a high quality of detailed engineering, and requiring an iterative process of trials analysis, design and refitting. The Navy estimated 'that if the Soviet Navy pays unremitting attention to noise reduction their newer submarines might be approaching the performance of our Swiftsure class by 1990'.[83] If that occurred, to preserve the Royal Navy's passive superiority over the Soviets, the Navy would have to rely on the continuing presence of older Soviet submarines with relatively high noise levels, and the growth of passive sonar technology.[84]

For now, the Royal Navy still enjoyed considerable advantage over the Soviets, particularly when it came to trailing Soviet submarines. Since the introduction of the first towed-array systems in the late 1970s, a number of Royal Navy COs had become experts at exploiting narrowband information to detect and track Soviet submarines. (The CO of HMS *Courageous*, Dan Conley, earned the nickname 'Dan, Dan the Narrowband Man' due to his extensive knowledge of narrowband and towed-array operations.)[85] Whereas in the 1970s, simply trailing Soviet submarines at long ranges was considered satisfactory, by the 1980s the Royal Navy no longer believed it was enough merely to maintain contact with Soviet submarines for extended periods. With the introduction of the Mark 24 Tigerfish torpedo, many in the Submarine Service argued that they now needed to demonstrate the ability to acquire and destroy Soviet submarines in warlike scenarios. Royal Navy submarines started to close Soviet submarines at the end of a trail and transmit on active sonar to see how they reacted. They also started to 'mark' Soviet submarines regularly, acquiring fire control solutions and carrying out simulated firings. 'We all knew that at the end of the day we had to get a fire control solution,' recalls Lane-Nott. 'It was all very well trailing something, but you weren't effective unless you could get a fire control solution and that you knew that if the torpedo worked properly you could fire and sink that submarine.'[86]

In November 1985, HMS *Conqueror* spent eighteen days on a covert intelligence-gathering and ASW-training war patrol, codenamed Operation 'Sheikh', in a 160,000-square-mile area in the Norwegian, Greenland and Barents Seas. *Conqueror* quickly detected a 'Delta' class SSBN deploying to the Greenland Sea and moved in to conduct two simulated Tigerfish firings, which were later judged to have been successful.[87] *Conqueror* then broke away and began to track yet another 'Delta' class SSBN, which appeared to be fixing position using bottom-mapping topography. *Conqueror* closed to investigate, but by the time it arrived the Delta had moved off to the west, outside *Conqueror*'s operational area. Instead of pursuing the Delta, *Conqueror* moved southwest of Bear Island and after four days in the area detected a homebound 'Yankee' class SSBN. Once again, *Conqueror* moved into firing range off the Yankee's starboard quarter, but was forced to break away when a small merchant vessel bound for Spitsbergen crossed *Conqueror*'s bows and disrupted the narrowband sonar picture. By the time *Conqueror* regained contact with the Yankee, it had increased speed and it took *Conqueror* over seven hours to regain a fire control solution. 'This is a nerve wracking game,' wrote *Conqueror*'s CO, 'trailing a 12kt target is known to be quite difficult but trying to get to a firing position from astern is murder. There is no doubt that there is a chance of counter detections (in wartime being sunk) particularly when one has no idea when he will slow down to CSA [clear stern arcs].' As *Conqueror* closed the range, the Yankee suddenly slowed and began to clear stern arcs. *Conqueror* quickly simulated firing two torpedoes, but due to the close range and the possibility of the Yankee deploying decoys and other countermeasures in a wartime scenario, the attack was only judged to have had a fair chance of success.[88]

Like the Delta, the Yankee remained unaware that it had just been under simulated attack. It continued to the southeast, while *Conqueror* altered course to the west and moved into a new search area southwest of Bear Island. There *Conqueror* briefly made a long-range detection of another nuclear submarine patrolling to the northwest, but lost contact after a short time. *Conqueror*'s CO was forced to surface his submarine and shut down the reactor in order to carry out repairs to a steam leak. As engineers repaired the problem *Conqueror*, which was snorting, detected the Type III nuclear submarine again.[89] After pressing on with the repair *Conqueror* dived again and vigorously pursued the Soviet submarine, closing in and acquiring a fire control solution and firing two simulated Tigerfish torpedoes, which were later judged to have been successful. *Conqueror* then withdrew to the southeast and detected another homeward bound 'Yankee' SSBN travelling at unusually high speed, possibly due to a mechanical defect or a casualty on board. As with the previous submarine contacts, *Conqueror* closed the range to a firing position and simulated firing two Tigerfish torpedoes at the fast-moving Yankee. In just eighteen days at sea, *Conqueror* had conducted four simulated firings at four Soviet submarines.[90]

Despite these successes, starting in the late 1970s Western Intelligence had been puzzled by the appearance of a number of different Soviet submarine designs. Instead of quantity the Soviets started placing an increased emphasis on quality in ship construction and weapon systems. The most worrying was the culmination of the 'Victor' design, Project 671RTM, 'Victor III', of which fourteen were built between 1977 and 1993. The 'Victor III' was 4.8 metres longer than the 'Victor II', quieter and fitted with tandem propeller and improved sensors and weapons. High-frequency, narrowband noises, such as those generated from machinery inside the submarine, were significantly reduced. One of the most distinctive deviations from the previous 'Victor' variants was the inclusion of a pod, about the size of a small minibus, mounted atop the upper tail rudder. When the first 'Victor III' appeared in late 1977, Western Intelligence suspected that the pod housed a towed communication cable, a torpedo decoy system or an auxiliary propulsion system for extremely quiet underwater speeds.[91] The US eventually confirmed that the pod contained a towed array, and in November 1983 a Victor III, K-324, operating off the east coast of the United States, snagged the towed-array cable of a US frigate, the USS *McCloy*, 282 miles west of Bermuda and was forced to surface, allowing US photography of the inside of the pod.

The 'Victor III' signalled the beginning of the end of the acoustic advantage enjoyed by both the Royal Navy and the US Navy. 'The Victor Is, Victor IIs, the early Deltas, Hotel, Echo, Novembers, all those were easy,' recalls James Perowne, 'and then suddenly, they turned out the Victor III ... Life got much harder.'[92] Soviet submariners also appeared to be getting better, finding new ways to counter SOSUS. The 1980s saw older, noisier first-generation Soviet submarines sent to waters surrounding the Iceland–Faroes Gap to act as noisy, confusing contacts, to decoy Royal Navy and US Navy SOSUS and Maritime Patrol Aircraft, while newer, quieter Soviet submarines passed in and out of the Atlantic. It was almost as if the Soviets started to realize just how susceptible their SSNs and particularly their SSBNs were to detection and trailing by Royal Navy and US Navy submarines. All of this was deeply puzzling until May 1985 when the FBI uncovered the biggest espionage leak in US naval history, with the arrest of the former senior US Navy warrant officer John A. Walker.

John Walker's espionage started in October 1967. He was deeply dissatisfied with aspects of his naval career, the American political system and, as he later wrote, 'The farce of the Cold War and the absurd war machine it spawned.' He was also in financial difficulties after entering into a number of failed business ventures.[93] In exchange for money, he decided to spy for the Soviet Union. When he first walked into the Soviet Embassy with a batch of photocopied documents and asked to see security personnel the KGB immediately recognized the importance of the information Walker was offering. His documents contained highly classified details about US Navy submarines, as well as cryptographic information, including the month's settings for the American KL-47 encryption machine. For the next eighteen years Walker provided cryptographic information, operational orders, war plans, technical manuals and intelligence digests to the Soviets, by selectively photographing small numbers of the thousands of documents that he had access to and depositing them in carefully prearranged spots known as 'dead drops', where he would collect cash and further instructions.[94]

To ensure his espionage activities continued when he retired from the Navy in 1975, Walker recruited an old Navy friend, Senior Chief Petty Officer Jerry Whitworth, who re-enlisted in the Navy in 1974 and continued to deliver information to the Soviets until he also retired in 1983. Whitworth left his last posting, on the US aircraft carrier USS *Enterprise*, with a foot-high stack of documents, including cable traffic and photographs of various US cryptographic systems. With Whitworth retired, in 1983 Walker also recruited his son, Michael Whitworth, a yeoman in the Navy, who copied over 1500 documents for the KGB while serving on board the carrier USS *Nimitz*. Walker's older brother, Arthur L. Walker, a retired Navy Lieutenant Commander working as a defence contractor, was also enlisted, after he owed money to John.[95]

In 1984 the seeds of the ring's destruction were sown, first by Jerry Whitworth, afflicted with guilt, who opened an anonymous correspondence with the FBI. However, he failed to follow through and the FBI was unable to identify him. It was Walker's wife, Barbara, who eventually revealed to the FBI that her husband had been spying for the Soviet Union. The ring was finally broken on 20 May 1985, when the FBI arrested John Walker. In exchange for a reduced sentence, Walker made a deal to reveal all and plead guilty. He was sentenced to life in prison. Arthur Walker received three life sentences and a $250,000 fine. Michael Walker received a 25-year sentence, while Jerry Whitworth was fined $410,000 and given a 365-year prison sentence. John Lehman, the US Navy Secretary, was deeply angered by what in his view were lenient sentences and by the Navy's not being able to court-martial Walker for espionage, which would have entailed the death penalty. Michael Walker was released in February 2000 for good behaviour.[96] John and Arthur Walker were eligible for parole in 2015, but they both died in prison in 2014.

At first, the Americans attempted to play down the impact of the spy ring. The US Chief of Naval Operations, Admiral James D. Watkins, declared that the US Navy had the problem 'bounded and can leave it in the dust behind us'.[97] However, over time the US was forced to admit that Walker–Whitworth had caused a great deal of damage. In July 1985, a high-ranking Soviet KGB defector, Vitaly Yurchenko, told the FBI that 'the information delivered by Walker enabled the KGB to decipher over a million messages'. The man responsible for first handling Walker, the former KGB station chief in Washington, Boris Solomatin, also admitted that 'Walker's information not only provided us with ongoing intelligence, but helped us over time to understand and study how your military actually thinks.'[98] In April 1987, the US Defense Secretary Casper Weinberger admitted that 'We now have clear signals of dramatic Soviet gains in all areas of naval warfare, which must now be interpreted in the light of the Walker conspiracy.'[99] The Director of US Naval Intelligence, William O. Studeman, later said that 'the Walker–Whitworth espionage activity was of the highest value to the intelligence services of the Soviet Union, with the potential – had conflict erupted between the two superpowers – to have powerful war-winning implications for the Soviet side'.[100]

Walker provided a vast amount of secret US Navy documentation to the Soviets, including operational orders, war plans, technical manuals, intelligence digests, sonar systems, operating procedures, operating areas, signals intelligence, electronic intelligence, and communication intelligence. Suddenly, the new Soviet behaviour, the change in tactics and overall naval strategy, made sense. Through information furnished by the Walker Ring, the Soviet Navy began to understand its doctrinal inferiority, how vulnerable its submarines were to acoustic detection and just how effective the US Navy and the Royal Navy were at trailing Soviet submarines. 'The real thing they told them was that they weren't as good as they thought they were,' recalls Martin Macpherson.[101] Armed with this information they had embarked on a series of programmes designed to close the technological gap between themselves and the West.[102]

Acoustic-quieting technology had already been introduced into the 'Victor II' design, significantly reducing machinery-produced noise and US intelligence indicated that the 'Victor II' already had the same noise levels as a US 'Sturgeon' class completed five years earlier, in 1967.[103] But only seven Victor IIs were constructed between 1972 and 1978 and the programme was interrupted in the mid-1970s when the Soviets embarked on the 'Victor III'. In the late 1970s, the Soviets had sought out Western technology to make their submarines quieter. The Toshiba Machine Company, a subsidiary of Japan's giant electronic company Toshiba, and the Kongsberg Våpenfabrikk, a company wholly owned by the Norwegian Government, sold highly advanced propeller-milling equipment to the Soviet Union in direct violation of Export Control rules that 'led to more efficient production of Soviet-designed submarine propellers' and the fabrication of smoother thus quieter submarine propellers.[104] A skewed propeller, very similar to that used in HMS *Swiftsure* in the mid-1970s, was fitted to the first Soviet third-generation nuclear submarine, the Project 945 'Sierra'. Like the 'Alfa', the 'Sierra' was constructed out of titanium. Double-hulled, its single reactor employed natural convection at slow speeds, alleviating the use of pumps, which contributed to its quietness at slow speeds. The 'Sierra' was equipped with a towed array and was capable of carrying up to forty torpedoes. Advances in automation also allowed for significant reduction in crew size; 31 officers and warrant officers and 28 enlisted men compared to the 75 men on the 'Victor' class.

The first 'Sierra' commissioned into the Soviet Navy in September 1984, but progress on subsequent submarines, including a modified design incorporating additional quieting features, Project 945A 'Sierra II', was slow. The Soviets intended to build a fleet of forty Sierras, but production difficulties associated with welding titanium led to significant production delays and by 1993 only four of the class had been built. Due to the delays the Soviets pursued a steel hull version of the design, which eventually evolved into an entirely new design known as the Project 971, 'Akula'. Slightly larger than the 'Sierra', the 'Akula' used the same propulsion plant, but contained a number of major innovations intended to reduce both weight and more importantly noise, including modular isolated decks and a revolutionary active noise cancellation system. The first 'Akula' was commissioned into the Soviet Navy in December 1984 and series production continued well into the 1990s. The 'Akula' was so quiet that when the first of class began its sea trials in 1986, the US Navy was 'aghast to learn that its "quiet level" approached that of the 688 [class] submarines we were building just a few years before'.[105] Through much of the later years in the Cold War the US Navy described the 'Akula' as the 'Walker' class.

The information furnished by the Walker Spy Ring also finally explained the fundamental shift in the strategic and operational policy of the Soviet Navy. The Soviets recognized that they would have to bring about a substantial weakening of NATO's potential to protect its sea lines of communication before they could expect much success in a large-scale, conventional campaign against the reinforcement and resupply of Europe. They concluded that the only way to do this was by gaining control of the Norwegian Sea, by neutralizing NATO's anti-air and ASW capabilities based in northern Norway, significantly reducing NATO's air and ASW defences in the GIUK Gap, while at the same time degrading NATO's wide-area ASW surveillance capability in the North Atlantic either by attacking or sabotaging SOSUS terminals. The Soviets calculated that if they could achieve all of this then they could dictate the time and place of any campaign and disrupt the reinforcement and resupply of Europe. They knew that if they failed, they would have to fight a campaign on NATO's terms, with minimal capabilities.[106] Aware of these vulnerabilities, the Soviets realized that what they lacked materially and personally they could make up for tactically through the adoption of a series of techniques designed to frustrate the West's superiority over the Soviet Navy.

The Soviets improved their knowledge of the underwater environment and began to exploit the different thermal/temperature fronts in the seas in which they operated. Different sea conditions could wreak havoc with trails. The contrast between long-range detections in the Atlantic and short-range detections across a front was stark. Ocean fronts – the boundary between two distinct masses of water, where properties can change markedly over a short distance – such as the southeast Iceland, Norwegian and Jan Mayen fronts could disrupt trailing operations and severely reduce detection ranges against submarines. The Soviets also started to deploy older noisier submarines near SOSUS arrays, as decoys, while modern, quieter Soviet submarines passed in and out of the Atlantic undetected. Soviet SSBNs deploying from the Northern Fleet also started to behave as if they knew they were being trailed. They began to visit a strong frontal area as part of their deployment process in order to shake off and possibly counter-detect any trailing SSNs.

The Soviet Navy also developed countermeasures to disrupt US Navy and Royal Navy trailing activities. One such approach was the Shkval rocket-powered underwater-unguided torpedo with a nuclear warhead, designed on the assumption that the first warning a Soviet submarine would have of an impending attack at the hands of a Royal Navy or US Navy submarine would be the sound of an incoming torpedo. The Shkval was fired back down the bearing of the attacking torpedo's transient noise with the aim of destroying both the attacking submarine and the incoming torpedo with a nuclear explosion.[107] Shkval was later described as 'the physical embodiment of the Soviet high command's realization' that the Royal Navy and the US Navy could 'sink Soviet ballistic missile submarines at will' in the 'northern Atlantic and Pacific seas'.[108]

This awareness of increased vulnerability also finally explained the earlier intelligence that indicated that the Soviets had abandoned the wartime strategy to intercept and disrupt the NATO sea lines of communication in the Atlantic. Instead, the Soviet Navy now intended to withdraw its vulnerable SSBNs, armed with long-range missiles with sufficient range to reach the United States, to waters such as the White Sea, the Sea of Okhotsk and shallow territorial waters around the Kamchatka Peninsula and Novaya Zemlya. Once stationed in these areas Soviet SSBNs no longer needed to transit through the choke-points and acoustic barriers of the GIUK Gap to attack targets in the United States and Europe. This new approach, sometimes termed the 'bastion strategy', allowed Soviet SSBNs to conduct patrols in friendly home waters, close to the Soviet Union, where it was difficult for Royal Navy and US Navy submarines to remain undetected. It also allowed the Soviet submarine fleet to exploit the protection of the Arctic ice cap and by the early 1980s there was increasing evidence that more and more Soviet submarines were operating under it.

'BEARDING THE BEAR IN ITS LAIR' – US MARITIME STRATEGY[109]

This shift in Soviet naval strategy posed a considerable challenge to both the Royal Navy and the US Navy. The US Navy responded with a naval strategy designed to seek out and destroy Soviet SSBNs and the forces protecting them at the outset of a major conflict. Confident that the Soviet Navy now intended to confine the majority of its forces to northern waters (see p. 529), the new US strategy envisaged US submarines moving towards the naval and airbases of the Soviet Northern Fleet on the Kola Peninsula, followed by US Navy Carrier Battle Groups. The Americans hoped this would allow the US Navy to gain the initiative and keep the Soviet Navy occupied far away from NATO territory, shipping and military operations, forcing the Soviet Union to 'disperse its forces into areas and activities' where they could do NATO least harm.[110]

The process of reevaluating and rewriting war plans began with the US Navy submarine force in 1982, and was eventually distributed in classified form throughout the US Fleet in 1984.[111] In 1982 the strategy was endorsed by NATO's three military commands and NATO's highest political authorities as NATO's Concept of Maritime Operations (CONMAROPS), the first aim of which was described by the former US 2nd Fleet Commander, Admiral Henry C. Mustin, as 'to contain and destroy the Soviet Northern Fleet'.[112] However, in 1984, the British CINCFLEET, Admiral Sir John Fieldhouse, and NATO's SACLANT, Admiral Wesley MacDonald, launched an initiative to alter NATO's naval rules of engagement so that they were far more aggressive. They aimed at more 'flexibility' for naval commanders, including the ability in certain circumstances to fire first in self-defence.[113] Other NATO countries refused to endorse the alterations on the grounds that they were not consistent with the fundamental principle that NATO was a defensive alliance.

The US was determined to proceed and in 1985 took the unusual decision to publicize and promote the strategy in order to secure public support. In a widely distributed 48-page supplement to the US *Naval Institute Proceedings*, the US Navy's Chief of Naval Operations, Admiral James D. Watkins, argued that 'the Soviet Navy's role in overall strategy suggests that initially the bulk of Soviet naval forces will deploy in areas near the Soviet Union, with only a small fraction deployed forwards'. He insisted that 'Soviet exercises confirm such an interpretation' and that 'the option some advocate, of holding our maritime power near home waters, would inevitably lead to abandoning our allies. This is unacceptable, morally, legally, and strategically. Allied strategy *must* be prepared to fight in forward areas. That is where our allies are and where our adversary will be.'[114]

The Maritime Strategy called for offensive operations at the very outset of war. 'We have to move up north of the GIUK ... gap,' said Watkins:

> We have to control the Norwegian Sea and force them back into the defensive further north, under the ice, to use their attack subs to protect their nuclear missile submarines, to use their attack subs to protect the Kola and the Murmansk coasts, and similarly their Pacific coast as well. If we try to draw a 'cordon sanitaire' and declare that we are not going to go above the GIUK gap or we are not going to go west of such and such a parallel, then obviously they have the capability to use their attack subs offensively against our SLOCS.[115]

As President Reagan publicly put it in January 1987, the purpose of the Maritime Strategy was to 'permit the United States to tie down Soviet naval forces in a defensive posture protecting Soviet ballistic missile submarines and the seaward approaches to the Soviet homeland, and thereby to minimize the wartime threat to the reinforcement and resupply of Europe by sea'.[116]

The Royal Navy played an important part in the strategy's aggressive, forward aspects. 'Our nuclear powered hunter-killer submarines, and those of the United States, are the platforms best able to operate well forward and threaten the whole range of Soviet submarines and high value surface units,' declared the First Sea Lord, Admiral Sir William Staveley, in 1984.[117] During a public speech to a naval conference in 1986, Staveley argued that:

> Better sensors, weapons and communications would today allow an unhindered enemy to wreak havoc in a very short timescale and, in that

same timescale, to cause an effect out of all proportion to the effort required. It is therefore vital in peacetime that our submarines develop, maintain and demonstrate their warfighting ability in order to deter an aggressor and prevent war. And, equally important, that if and when required, we are capable in war of inflicting heavy losses on enemy submarines very early in battle. We must do so to maintain our supply lines in order to support the conventional battle on land and at sea, such that war stopping leverage can be applied without recourse to the use of tactical or strategic nuclear weapons. NATO's submarines would be in the vanguard of any war effort. Their task, in conjunction with other NATO fleet elements, is critical. By their covertness, and their independence upon material support once deployed, they suit the concept of Forward Defence. They multiply the options available to politicians and commanders in peace, tension or war.[118]

At the same 1986 naval conference, the Commander-in-Chief, Naval Home Command, Vice Admiral Sir Peter Stanford, outlined the Royal Navy's role in any future war:

It will be essential, to conduct forward operations with attack submarines, as well as to establish barriers at key world chokepoints using maritime patrol aircraft, mines, attack submarines, or sonobuoys, to prevent leakage of enemy forces to the open ocean where the Western Alliance's resupply lines can be threatened. As the battle groups move forward, we will wage an aggressive campaign against all Soviet submarines, including ballistic missile submarines.[119]

This would contain Soviet submarine forces in a defensive campaign and prevent them from surging out into the Atlantic to threaten NATO lines of communications.

The Royal Navy recognized that one of its key battlefields in any future conflict would be the Norwegian Sea. Losing control of it would, Staveley argued, put 'at risk the sparsely populated region of North Norway, then Iceland and the Faeroes and thus placing the North Sea and the United Kingdom so much closer to the front line of Soviet forces, needlessly exposing ourselves to a greater threat which would make war fighting a much more daunting prospect for NATO'.[120] According to Staveley, the defence of Norway and control of the Norwegian Sea were both:

indivisible parts of what must be seen as a sub-strategy for the entire Northern region, which must itself form a coherent part of NATO's overall strategy. Soviet domination of the NATO territory-bounded area would enable them to fulfil their maritime objectives more easily. If their increasingly capable submarine-based Northern Fleet can be contained by NATO forces and attacked in depth through the judicious forward deployment of our submarines, ships and aircraft, then I believe we should be able to deny the Soviets their objectives while achieving our own. Our goal must be to achieve the timely arrival of reinforcement and resupply shipping, not just across the Atlantic but to discharge their sustaining cargoes safely in whichever European port it is required.[121]

Alongside other European navies, the Royal Navy would engage in a series of complex operations designed to 'contain' the Soviet Northern Fleet, deploying aircraft and submarines forward to engage targets. Behind them, submarines and surface ships would conduct barrier operations at chokepoints in the North Atlantic and Norwegian Sea such as between the Bear Island–North Cape line and the GIUK Gap. Behind these chokepoints Soviet forces would encounter defences around British and other sea assets such as the reinforcement shipping coming across the Atlantic. The last line was the defence of British sea and air space. 'The whole point of NATO's naval strategy (and of the Royal Navy's place in it), in short, is to offer defence in depth,' wrote the naval historian Geoffrey Till. 'There can be little doubt that serious failure in this policy of containment would gravely disrupt almost anything NATO would seek to do in a serious and sustained conflict.'[122]

To carry out this containment role the Royal Navy needed, as Staveley put it, 'a regular presence and a bank of experience in those areas in peacetime. Right now.'[123] In April 1987, Sir Nicholas Hunt, CINC-FLEET, reportedly stated that 'forward defence' effectively takes Royal Navy submarines 'behind enemy lines', and that this must be carried out in peacetime.[124] The 1986 Statement on the Defence Estimates declared that 'Enemy attack submarines are successfully to be held at arm's length from the critical Atlantic routes. Defence against these submarines would begin when they sailed.'[125] The 1987 Statement outlined four tasks for the Royal Navy: 'the interception and containment of Soviet forces in the Norwegian sea'; 'direct defence of reinforcement, re-supply and economic shipping, in conjunction with US and European maritime forces, and supported by the RAF'; 'anti-submarine defence of the NATO Striking Fleet Atlantic'; and 'protection and deployment of the combined United Kingdom/Netherlands Amphibious Force to reinforce the Northern Flank of NATO'.[126] The first three highlighted the Royal Navy's integration into the new US Forward Maritime Strategy. 'We were right there in every respect, not just lip service to it,' recalls Paul Branscombe.[127] Instead of using Royal Navy destroyers and frigates as convoy escorts, the vessels, together with the Royal Navy anti-submarine carriers HMS *Invincible*, HMS *Ark Royal* and HMS *Illustrious*, would attempt to fight the next war along with SSNs in the Norwegian and Barents Seas with the aim of destroying the Soviet Northern Fleet, the Soviet means of second strike, and of launching attacks on Soviet territory.

In many ways, these were the glory days of the post-1945 Submarine Service, the years when they really stared the increasingly sophisticated Soviet adversary in the face and the East–West front line was on occasion reduced to a few cold and watery yards. They also represented the Anglo-American alliance at its tautest and fraughtest. From the mid-1980s onwards, Royal Navy and US Navy warships increasingly operated in northern waters, shadowing Soviet ships and submarines in the Barents Sea, east of the North Cape of Norway and near the highly sensitive Kola Peninsula, to determine Soviet patrol areas, operating techniques, submarine noise signatures and other characteristics.[128]

As the professional head of the Submarine Service, Flag Officer Submarines was responsible for ensuring that the service was integrated into US plans. 'I went across several times to COMSUBLANT to talk about the future and organization and our part . . . about what we were prepared to do and what we could do to fit in. You were always fitting in to the Americans rather than the other way round,' recalled Toby Frere. 'The Americans and ourselves seemed to be absolutely in step as operators in what we were going to do, and what we could do. There was no divergence, no "you are going to go and do this and we are going to go and do that", we were absolutely in step.'[129] 'They were totally honest,' recalls Martin Macpherson, FOSM's Submarine Operations Officer. 'From Northwood, we used to play war games with them, which were totally bilateral submarine war games. How we would surge deploy all the SSNs, because they had fifty or so SSNs in the Atlantic and we had something like fifteen. How in times of tension we were going to get them into the Barents without hitting each other. We were going to bottle them up.'[130]

Throughout the 1980s successive FOSMs found themselves working more closely with the US Navy's Submarine Service than to their own Commanders. When Richard Heaslip was appointed FOSM in 1984, operational day-to-day control of US Navy and Royal Navy submarines was routinely shared between both navies. HMS *Swift-sure,* for example, was attached to the US Navy under US operational control and stationed in the Pacific.[131] Royal Navy submarines were also involved in exercises designed to simulate pacifying and locking down Norwegian fjords to allow US carriers to anchor and launch attacks.[132] Relations between the two submarine services were so close that many submariners felt detached from the rest of the Royal Navy. 'We were so close to SUBLANT in Norfolk, Virginia, that the rest of the Navy virtually took no notice of us whatsoever,' admitted Heas-lip.[133] 'I do believe that most of the Royal Navy didn't have the faintest idea of what we were there to do,' admitted James Taylor, Chief of Staff to FOSM in the late 1980s. 'We thought the rest of the Navy had no idea what was going on,' said Frere. 'When we went and did a NATO exercise there was always that feeling that this is a bit of a chore, we are a clockwork mouse, and it was time out from doing our real job.'[134]

War plans were also drawn up, as were contingency plans. Heaslip was surprised to find that there were no contingency plans against the possibility the United Kingdom was wiped out in a pre-emptive nuclear strike. As a result, with his American counterpart, orders were issued in the form of a sealed envelope which was placed in each SSN's safe instructing commanders that if they had cause to believe that nuclear war had broken out and they had been unable to hear the BBC for three days then they were to go to a certain position, at a certain depth and course, and wait until a US Navy SSN came along and with orders. 'I felt that we've got a dozen SSNs and the Americans are still going to go on in combat, even if Britain is out of action,' recalls Heaslip, 'and it doesn't make any sense not to have any plans for these SSNs.'[135]

One of the most important, and indeed controversial aspects of the US Maritime Strategy concerned the tracking of Soviet SSBNs. US Navy and Royal Navy SSNs would search Soviet operating areas and in war attack any SSBNs that they found. This, it was hoped, would alter the strategic nuclear balance in NATO's favour as the conventional war progressed. It was also hoped that this would reduce Soviet willingness to continue to fight and force the Soviet Navy to divert many of its forces to the SSBN bastion areas to counter the NATO onslaught.[136] The US believed that by prioritizing Soviet SSBNs, the Soviet Navy would be forced to divert its most capable SSNs away from other offensive missions, particularly any attempt to intercept NATO sea lines of communications. As Watkins later said, 'Aggressive forward movement of anti-submarine warfare forces, both submarines and maritime patrol aircraft, will force Soviet submarines to retreat into defensive bastions to protect their ballistic missile submarines. This both denies the Soviets the option of a massive, early attempt to interdict our sea lines of communication and counters such operations against them that the Soviets undertake.'[137]

Since a proportion of Soviet nuclear weapons would almost certainly have been targeted against the British Isles, there was a *prima facie* case that the Royal Navy should seek to destroy them, in essence continuing its historic 'walls of England' role, defending the realm against sea-based attack.[138] Advocates of the strategy argued that being able to threaten something the Soviet Union valued would increase the deterrent and war prevention effectiveness of the Royal Navy. However, critics argued that such action against the Soviets' strategic reserve would be provocative and destabilizing and that Soviet force levels were so high that attrition was unlikely to reduce the Soviet nuclear threat to the UK in any significant way, so the whole effort would be pointless. Also, with only seventeen SSNs and around ten SSKs, the Submarine Service would have little impact, unless it was operating alongside the US Navy. The House of Commons Defence Select Committee, for example, argued that:

> At best, the forward deployment of RN ships in a *period of tension* could only monitor transiting Soviet submarines. It could not prevent them leaving the Norwegian Sea for the Atlantic, deal with those already operating there, or intercept and monitor those redeploying from elsewhere in the world. In a period of tension, forward deployment could appear provocative.[139]

If tension escalated into all-out war the Submarine Service hoped to inflict a damage ratio of three to one. Privately, Royal Navy submariners admitted that they would have around a 50 per cent chance of returning. They realized that once they went north, they would stay until they either ran out of torpedoes or were destroyed. 'We never got to the stage where we felt that we could flood their areas so much with submarines that we would know where every single one was,' recalled Toby Frere. 'On the other hand they didn't feel at all secure, because they did know that we were there and we did know where they were operating and we could work out how they were doing it and we had the weapons to take them out.'[140]

ARCTIC OPERATIONS

Although the Maritime Strategy had many critics, changes in Russian activity from 1986 onwards suggested that the Soviet Navy was attempting to counter it. The deployment of Soviet submarines and surface forces 'out of area' dropped markedly. Major Fleet exercises in 1986 departed from previous trends emphasizing far-reaching operations and were instead staged much closer to the Soviet mainland, under the umbrella of Soviet land-based aviation. As Lehman wrote:

> Overall, the Soviet Navy has continued to operate and to train, but activities have switched dramatically to their home waters. The net strategic result appears to us to be a Soviet fleet positioning and training to counter our new maritime strategy. That precisely was what we intended, to force them to shift from an offensive naval posture targeted against our vulnerabilities to a defensive posture to protect their own vulnerabilities.[141]

As the Maritime Strategy took US Navy and Royal Navy submarines still further into northern waters, the Soviets increasingly looked to the Arctic to ensure the safety and potency of their submerged deterrent. In 1983, a 'Delta II' SSBN, the K-092, conducted the world's first under-ice ballistic missile launch, when it fired two SLBMs while operating under the ice pack. They also began to use more and more of their SSNs to protect the SSBNs. In 1985, the Joint Intelligence Committee noted that:

> The ballistic missile submarine (SSBN) force is increasingly deployed to Arctic waters where the ice-cap can give extra protection from Western anti submarine warfare (ASW) forces. SSBNs can fire their missiles on the surface either where there is open water or by breaking through the ice-cap ... Nuclear attack submarines (SSNs) (of the ALFA, AKULA, VICTOR III and SIERRA classes) would be expected to operate in close support of these SSBNs, while Maritime Patrol Aircraft and surface ships would operate in distant support. The protection of these strategic forces thus occupies a large part of the Northern and Pacific Fleets.[142]

This posed a problem for both the US Navy and the Royal Navy. By withdrawing their SSBNs the Soviets effectively turned the tables on the West.[143] Whereas in the 1960s and 1970s the short range of Soviet submarine-launched ballistic missiles demanded that Soviet SSBNs take up patrol positions in the Western Atlantic, thus transiting through the chokepoints and home waters of NATO's northeastern flank where they were subject to detection and possible shadowing, Soviet submarines deploying in the Arctic had no chokepoints to transit. Their operating areas were now considerably more familiar to Soviet submariners than they were to NATO's. US Navy and Royal Navy SSNs seeking to hunt down Soviet SSBNs now had increasingly to transit through the chokepoints of the Davis and Denmark Straits, and operate at greater distances from US and UK bases, where operating experience was relatively poor. The considerable advantage the Royal Navy and US Navy enjoyed in air and surface anti-submarine warfare also all but disappeared, as neither was effective in the new strategic environment under the ice. This had important implications for the region.

The Maritime Strategy, and the Soviet response to it, conceptually transformed the Arctic from a natural scientific laboratory and region of occasional and exceptional activity into a possible battle-space on a par with the Northern Pacific or the GIUK Gap.[144] Had World War III come it would have seen combat of great ferocity. Beginning in the early 1980s, Royal Navy and US Navy SSNs were routinely deployed into Arctic waters to develop their battle plans and wartime capabilities. Numerous submarines participated in a series of joint United Kingdom/United States anti-submarine warfare operations in the Greenland Sea under varying conditions of ice cover, to test the capability of US Navy and Royal Navy SSNs to detect, intercept and shadow Soviet submarines operating under the ice. The first of these SUBICEX operations, as they were known, took place in 1979 when HMS *Sovereign* and USS *Archerfish* took part in a series of exercises. The second took place in 1981, when HMS *Valiant* went up against the USS *Silversides* in order to find out whether the Royal Navy's oldest class of SSN was capable of operating under the ice with minimal additional and material preparations.

Going under the ice was relatively straightforward, but detecting and classifying submarines operating underneath it was challenging. Sonar conditions under the ice were often excellent, meaning there was a high probability of long-range detections. However, the reflection of target noise from the huge number of ice keels often caused difficulties, and in contrast to the familiar chokepoint operations in the GIUK Gap where Soviet submarines would be transiting to their patrol areas, Soviet submarines operating under the ice were already patrolling and were therefore much quieter, more alert and unpredictable. Detecting and attacking Soviet submarines patrolling under the ice was full of what Tom Le Marchand, the CO of HMS *Valiant*, described as 'Problems, problems, problems.'

> We have established that the target is within range but he is probably too quiet for passive homing weapons; if we go 'active' there will be excessive reverberations, not to mention the distinct probability of the weapon homing on to an ice keel, or even striking one accidentally en route to the target; maybe it's better to track him until he leaves the ice cover, if ever he does; or maybe he'll fire his missiles before I can attack him. Maybe ...[145]

Le Marchand concluded that it was 'clear that many problems need to be solved before under-ice warfare can be waged with anything like the confidence that open ocean temperate ASW generates'. But he recognized that the 'strategic importance of the Arctic inner space is evident, and it is equally evident that neither East nor West can afford to resign control of that area to the other'. He concluded that 'nothing will replace or make up for the value of ... all-important operational experience. And it is here where one faces the nagging suspicion that we have some catching up to do.'[146]

Soviet interest in the Arctic region was beginning to be reflected in the latest submarine designs. In the early 1980s, the Soviets produced yet another version of the Delta design, the 'Delta IV', which was significantly larger and quieter than its predecessors and capable of carrying a larger missile, the three stage RSM-54/R-29RM (NATO designation SS-N-23 Skiff).[147] The first of class was laid down in 1981, commissioned in December 1984 and was followed by seven others. The Soviets also commissioned the first submarine designed specifically to operate under the ice. With a surface displacement of 18,797 tonnes, and a submerged displacement of 26,925 tonnes, the Project 941 'Typhoon' SSBN (as featured in book and film *The Hunt for Red October*) was the largest submarine in the world and its unique design differed considerably from that of any submarine built by the US Navy and the Royal Navy. Constructed from two full size parallel pressure hulls, each 149 metres long and 7.2 metres in diameter, with a twenty-tube missile compartment placed between the hulls, in two rows, forward of the fin, the Typhoon's twin propellers were housed in shrouds to protect them from ice damage and the submarine's large displacement allowed it to punch through thick ice to launch its powerful new missiles, the RSM-52/R-39 (NATO designation SS-N-20 Sturgeon), equipped with ten MIRV warheads.[148] The first of the

'Typhoon' submarines, TK-208, was laid down in June 1976 and commissioned in 1981. Six more were built between 1983 and 1989.[149]

The 'Typhoon' was a crucial Western intelligence target and in the early 1980s, HMS *Superb* became the second Western submarine to successfully detect and collect intelligence on it. In December 1986, another of the Royal Navy's SSNs, HMS *Splendid*, reportedly conducted the first long-term trail of 'Typhoon' when its towed array became entangled with the Soviet submarine. *Splendid* ended the trail and returned to Devonport, where her commander faced a naval inquiry. The incident, which somehow leaked to the press, was the first public demonstration of the Submarine Service's active participation in Forward Operations in peacetime against Soviet SSBNs.[150] It led to a series of exchanges in the House of Commons, when the Labour MP Tam Dalyell suggested to the Minister of State for the Armed Forces, John Stanley, 'that it would be helpful to the House if the Hon. Gentleman's speech could include some reference to what allegedly happened in the Barents Sea with the towed array sonar, so that we can discuss the matter on the basis of information rather than newspaper reports?'[151] Martin O'Neill, the Labour Opposition's Shadow Defence Spokesman, then linked the *Splendid* incident to the Royal Navy's participation in the Maritime Strategy:

> If the United Kingdom – and, I presume, the United States – have a presence in the Barents Sea, that will surely threaten the security of the Soviet Union's SSBNs, increase the threat of a Soviet first strike and thus threaten crisis stability. In terms of NATO-Soviet relations at such a sensitive time, when we are supposed to be close to reaching an accommodation with the Soviet Union – there has already been the prefacing of the Prime Minister's visit to Moscow – it is foolhardy to be blundering around in the Barents Sea.[152]

'There seems to have been a major change,' continued O'Neill. 'Until now we have not been aware of British boats being involved around the Kola Peninsula.'[153] He urged the Government to 'be more forthcoming about these issues' and 'to admit that there has been a change in British involvement in the NATO strategy'.[154]

Stanley followed the practice of previous governments and insisted that 'there is no way that we could be drawn into commenting on submarine operations'. It was left to the Social Democratic Party's Defence Spokesman, John Cartwright, to suggest that the 'idea that a Soviet Typhoon submarine would have deliberately severed the towed array sonar equipment is extremely implausible. Even if the equipment could have been recovered – which is a very big "if" – it would have been of limited benefit to the Soviet Union.'[155] Cartwright suggested that a more likely explanation was that 'HMS *Splendid* may have been operating under . . . a merchant vessel whose anchor severed the cable of the towed array sonar.'[156] He argued that:

> Those in our attack submarines, who have the vital task of tracking Soviet ballistic missile submarines, also carry out an important task on our behalf. It is vital to be sure that we are not asking our submariners to undertake unreasonable risks. It has been suggested to me that the HMS *Splendid* incident is not an isolated occurrence and that there have been similar incidents involving HMS *Spartan* and HMS *Sceptre*.[157]

Cartwright called on the Minister to 'give us an undertaking that if those men are put at risk as a result of what they are being required to do, their procedures and methods of operation will be reviewed'. 'We want that job carried out,' he said, 'but we do not want our submariners taking unreasonable risks on our behalf.'[158] No such undertaking was given. Bizarrely in 1987 British scientists studying weather patterns in the Arctic allegedly found remains of the towed array entangled in the rib cage of a 45-foot humpback whale. Experts said it must have drowned as the weight of the 250-foot-long equipment prevented it surfacing for air. It was found on Jan Mayen Island, hundreds of miles from the encounter between the submarines.[159]

THE PRINCE OF DARKNESS

The Soviets did not confine themselves entirely to northern waters, but continued periodically to enter the Atlantic and Pacific to probe the West's anti-submarine capability, attempting to detect and compromise the Royal Navy and US Navy's SSBNs. By the mid-1980s, many of these Soviet submarines were able to remain undetected because of improvements to their noise levels. In October 1986, a US Navy SSN, the USS *Augusta*, was trailing a Soviet SSN in the North Atlantic when it inadvertently collided with an undetected Soviet 'Delta I' SSBN. The *Augusta* returned to port and suffered $2.7m in damage, while the Delta was reportedly able to continue on patrol.[160] Far more worrying was an incident in late March 1987, when five front-line 'Victor III' SSNs deployed from the Soviet Northern Fleet bases and moved into the

Atlantic. As soon as the Victor IIIs departed from their ports they were immediately picked up on SOSUS and classified as L-015; L-018; L-019; L-020 and L-021. As they continued into the Western Atlantic, the US Navy and the Royal Navy mounted a major combined ASW operation in order to follow them and maintain the operational security of both Royal Navy and US Navy SSBNs. 'We had just about everything we could move out there,' recalled Toby Frere. Royal Navy towed-array frigates deployed, alongside SSKs with towed arrays and a number of Royal Navy and US Navy SSNs, including the USS *Sea Devil*, USS *Dallas* and HMS *Superb*.

Northwood was responsible for providing regular information about the location of the Victor IIIs to the on-patrol Polaris SSBN, a task that was complicated by the fact that very few people in Northwood knew its exact location. 'We knew by exception where he wouldn't be,' remembered Frere, 'but we didn't have a clue where he was. All we were doing was keeping tabs on the mob and feeding him the information in real time, where they were. Then it was up to him.'[161] The search was so intensive and so demanding that the RAF's Nimrods used their entire yearly supply of sonobuoys in the space of a few weeks. But the Royal Navy and US Navy were able to track the Soviet deployment. 'We held those submarines almost continuously from when they came round North Cape', recalled Martin Macpherson, the Operations Officer at Northwood, 'all the way down the Norwegian Sea, lost them briefly as they went through the gaps because of the water conditions, regained just south of the gaps, watched them all the way through our really high point of interest, West of Ireland and out into the Atlantic. We watched them go down to about 48 North, watched them turn west and go out to the States. We held almost continuous contact.' Throughout the deployment the US and the UK were able to maintain firm contact with four out of the five Victors. 'The fifth one,' recounts Macpherson, 'although of the same class was obviously very much quieter than the others. Now it may be that he was a particularly well-maintained, well-managed submarine. He was always known in the trade as the Prince of Darkness, because he was so difficult to detect.'[162]

The Victor IIIs continued to the east coast of the United States and attempted to locate US SSBNs. They then turned back to the North Atlantic. To guard against a possible second attempt at detecting the Royal Navy's patrolling SSBN, Northwood intensified its efforts to track one of the quieter and more elusive Victor IIIs, which may or may not have been the Prince of Darkness. HMS *Trafalgar*, then under the command of Toby Elliott, was in Bermuda at the time of the Victor deployment and was ordered by Northwood to make best speed back to Faslane. When *Trafalgar* arrived in late April, the Victors were operating to the west of Rockall in what appeared to be an anti-SSBN operation west of the United Kingdom. There was no evidence that they intended to resume their transit north. *Trafalgar* sailed and started a search of the Hatton-Rockall Basin before receiving orders to sail south to intercept one of the Victor IIIs, designated L-019, which the USS *Dallas* had been trailing for some time. After thirty-six hours' searching in cooperation with RAF Nimrods, *Trafalgar* detected the Victor III.

Trafalgar's crew immediately recognized that this particular Victor III was 'a worthy adversary'. One former Royal Navy officer who at the time analysed Soviet submarine operations could recognize the distinctive features of the Prince of Darkness, 'the competence and the quality, that you weren't just up against A. N. Other,' he said. 'It's not so much that he had a distinct predictable or forecastable style, it was just that it was of a different level to his peers.'[163] As Elliott wrote:

> The trail was never easy. L 019, when patrolling, was slow, with an extremely quiet narrowband signature with large bow and stern nulls, had virtually no detectable broadband signature and adopted a policy of random and complex stern arc clearances or local area searches interspersed with short, but slow transits between search areas. Even when transiting, and at a slightly higher speed, there was still no detectable broadband signature and the narrowband signature remained impressively sparse. Determined tactics were necessary to maintain, and regenerate, contact on this very quiet modern Soviet submarine.[164]

Trafalgar started to trail the Victor to the west of Rockall, following it for seven days before breaking contact and heading south. The Victor appeared to be in patrol mode and was moving slowly north towards the Iceland–Faroes Gap. Although the boat was quiet, *Trafalgar*'s crew quickly determined that the Victor III conducted complex stern arc clearances at four to six hour intervals. After one such evolution, Elliott, concluding that the Victor III's CO and crew might have been lulled into a false sense of security, decided to carry out a simulated attack on the Victor, converting *Trafalgar*'s trail into a firing position. 'I intend to "pick him off" when and if the opportunity presents itself,' noted Elliott, who slowly moved *Trafalgar* into a favourable position on the beam of the Victor III and brought his weapon system into Readiness State

One – ready in all respects to simulate an attack.[165] Once *Trafalgar* had obtained a good fire control solution Elliott fired two simulated Tigerfish torpedoes that later on board analysis concluded would have hit their target.

As *Trafalgar* continued to trail the Victor III, the waters surrounding the Iceland–Faroes Gap started to become increasingly busy, with a total of eight submarines: the Victors, USS *Sea Devil* and HMS *Superb*, all making for the gap. In the aftermath of *Trafalgar*'s simulated attack on the Victor III, it had become 'increasingly difficult' to maintain a constant trail of the Soviet submarine and at one stage it became 'so marginal that there was a real risk of losing contact altogether'. Detection problems were increased by the effect of the complex oceanographic conditions that prevailed in the vicinity of the Icelandic Front and technical problems with *Trafalgar*'s 2026 sonar, which had been experiencing 'lockouts', freezing for periods of up to fifteen minutes. At one stage, as *Trafalgar* searched for the Victor III, the two submarines came to within 1000 yards of each other.[166]

As *Trafalgar* moved through the colder but more stable waters of the Norwegian Sea it was able to maintain intermittent contact with the Victor III, which appeared to be conducting a slow search to the northeast. Although the technical problems with *Trafalgar*'s sonar suite had been repaired, deteriorating oceanographic conditions resulted in yet another close pass – 600 yards – an opportunity Elliott used to carry out another firing exercise. *Trafalgar* continued to trail the Victor III, but it also detected the tonals from a second submarine, which was later classified as another of the Victor IIIs. The two submarines appeared to be cooperating and conducting a submarine exercise, taking it in turns to transit and patrol. Again, due to the challenging oceanographic conditions maintaining contact with both Soviet submarines proved difficult and Elliott was forced to conduct a number of manoeuvres to regain contact. One of these resulted in yet another close pass, but in contrast to the previous two occasions one of the Victor IIIs altered course and speed, indicating that it may have counter-detected *Trafalgar*. Elliott immediately ordered *Trafalgar* to break contact and move out of the area.

The tactics employed against the exceptionally quiet, well-handled Soviet submarine unsurprisingly created close-quarters scenarios, but Elliott ensured that *Trafalgar*'s crew was prepared and able to anticipate them. He came away from the patrol with a number of conclusions. First, that 'The "Trafalgar" class has again demonstrated its effectiveness against modern Soviet submarines, both in terms of the sonar advantage and, if the evidence from the close passes is to be believed, noise quietening.' Second, 'The "Victor III" Class SSN, if L 019 is representative, is impressively quiet.' Third, 'Detection ranges on modern noise quietened Soviet submarines are reducing, making them difficult to prosecute.' Fourth, 'Determined tactics are required to maintain, and regain, contact when trailing such submarines.' Fifth, 'Such determined tactics will lead on occasions, to close quarters situations with CPAs [Closest Points of Approach] of less than 1 Kyd [kiloyards]. With preparation, and anticipation, these situations can not only be controlled but also turned to tactical advantage.' Elliott's final words were 'for the performance and conduct of all my people during a challenging patrol in which from time to time the close proximity and unpredictable behaviour of L019 has made life somewhat daunting – they have been magnificent'.[167]

It was now increasingly clear that the information from the Walker Spy Ring had enabled the Soviets to narrow the technological gap that once existed between the Soviet Union and both the US Navy and Royal Navy. The broadband radiated noise levels of modern Soviet nuclear submarines were beginning to bear comparison with the oldest US Navy and Royal Navy submarines and the quietness of the Soviet narrowband signature was decreasing long-range detection ranges and leading to only fleeting contact. Soviets tactics also eroded advantages previously enjoyed by the West. The Soviets continued to employ a number of anti-SOSUS tactics, shifting their transit routes away from SOSUS arrays in deep water. They also adopted disruptive steering tactics, zig-zagging every 30–60 minutes, attempting to reduce the uninterrupted periods that US Navy and Royal Navy SSNs could remain in contact.

All of this was evident in late 1986 when HMS *Conqueror* conducted Operation 'Beverash' in the northeast Atlantic. *Conqueror* spent six days trailing a Victor I/II over 1400 nautical miles as it conducted a high-speed transit to the Mediterranean. Although the Victor was 'a well handled submarine, employing frequent alterations of course . . . in what were, at times, poor sonar conditions', *Conqueror*'s crew were able to remain undetected and conduct two simulated Tiger-fish attacks. 'I seized the opportunity with confidence that he was in within weapon range,' wrote *Conqueror*'s CO, James Burnell-Nugent. 'He was a bit close to my bow for choice but there is no doubt that in wartime I would have been happy to fire. I assess that with accuracy of the NB bearings . . . that the attack had a good chance of success.'[168] *Conqueror* closed the range considerably on the second attack but 'After what appeared to be an excellent opportunity for a short range attack', the Victor 'completely vanished' and Burnell-Nugent was forced to admit that the attack was unsuccessful. The Victor appeared to be deliberately utilizing the then poor sonar conditions of the north-east Atlantic to protect itself. 'The tactical lesson learnt is that if better sonar conditions are coming up – wait,' wrote Burnell-Nugent. He also noted that 'The complete lack of cavitation, the probable lack of trans-missions, the clean broadband signature and the disruptive steering plan are all marked improvements in the handling of a Victor I/II over earlier years.'[169]

Conqueror was then ordered to locate a Type II/III nuclear submar-ine that had been picked up on SOSUS rounding the North Cape. After two days of searching, *Conqueror* located the Soviet submarine, which it assessed as another 'Victor' class, and started to trail it as it transited towards the Shetland–Faroes Gap. Once firmly in the trail, Burnell-Nugent carried out yet another simulated attack that was judged to have been successful. As *Conqueror* continued to follow the Victor, Burnell-Nugent concluded that it was 'generally well handled and by "Victor I" standards quiet both narrowband and broadband'. But he noted that there was a 'difference in modus operandi' between this particular Victor and the one *Conqueror* had encountered earlier on in the patrol. 'Both were VICTOR class SSNs bound for the Medi-terranean,' noted Burnell-Nugent, but whereas the first Victor did not transmit on active sonar, the second Victor did. Both submarines also made use of 'disruptive steering plans and both had clean broadband signatures'. The second Victor had an 'atypical narrowband signature', which led *Conqueror*'s crew to conclude that it was 'a non-standard "Victor" derivative with possibly significantly different auxiliary machinery'.[170] *Conqueror* eventually handed over the trail to an RAF Nimrod and returned to Faslane with a considerable amount of intel-ligence information about both submarines.

The 'Trafalgar' class continued to demonstrate a discernible acoustic and sensor advantage over the latest Soviet submarines. In February 1988 HMS *Torbay*, under the command of Robert Stevens, conducted Operation 'Links', a covert intelligence-gathering and training ASW patrol in the north Norwegian, south Greenland and west Barents Seas, as well as the Marginal Ice Zone north and south of Jan Mayen Island, in order to gain operation experience in the challenging conditions associated with the area. Over the period of one month, *Torbay* detected, trailed and reached attack criteria against four different Soviet submarines: a Yankee I, a Victor II and two Delta Is, while also detecting two other possible submarines, one of which may have been an acoustically quiet Victor III. *Torbay* left Faslane in late January 1988 and spent three days exercising in the Royal Navy's Northern Fleet Exercise Areas against HMS *Churchill* and two US submarines, the USS *Baton Rouge* and USS *Pargo*. *Torbay* then departed for patrol and after three days of searching unsuccessfully for a Victor III at the entrance to the Barents Sea, moved north to search for a possible Delta I that had been tracked on SOSUS for over seventy days and was beginning its southeasterly homeward transit.

Torbay easily detected the Delta and closed the range and after obtaining a good fire control solution simulated firing a salvo of two Mark 24 Tigerfish torpedoes, which were assessed as having hit their target. As *Torbay* established a trail on the Delta's port quarter, its sonars detected a separate submarine contact to the south. Stevens immediately suspected that this was a 'Victor III' SSN about to delouse the Delta as it returned home. *Torbay* slowed moved away and allowed the range on both submarines to open so that it could monitor the delouse operation from a distance. Once contact with the second submarine was lost, *Torbay* reestablished its trail of the Delta until breaking off to search for a 'Yankee' class SSBN that SOSUS indicated was loitering in the southwestern approaches to the North Cape. *Torbay*'s search for the Yankee was complicated by the heavy traffic noise from fishing and merchant vessels along the Norwegian coast, but it was able to use the noise to disguise its own broadband signature when closing at high speed. *Torbay* established a trail of the Yankee until it moved out of *Torbay*'s allocated operating area. Stevens was 'confident that if required to fire, the Target's solution would have supported a Tigerfish attack', but just as *Torbay* was chasing the 'Yankee', Northwood suspended the requirement to simulate attacks on Soviet submarines because of an incident on 12 February when a Soviet frigate rammed the USS *Yorktown* and USS *Caron*.[171]

Torbay was then assigned to search for another SOSUS contact, a 'Victor II' class SSN operating in the Norwegian Sea on the Vøring Plateau and heading southeast. Although the Victor II was acoustically quiet and the presence of fishing vessels made trailing difficult, *Torbay* was able to conduct yet another simulated firing of two Tigerfish torpedoes before moving to the northwest to intercept the Delta it had encountered early on in the patrol, which SOSUS now indicated was patrolling in the Jan Mayen Polar Front on the edge of the Marginal Ice Zone. *Torbay* struggled to locate the Delta, and with only fleeting contact concluded that it was 'slow, very quiet and probably close'. *Torbay's* crew eventually concluded that the Delta and *Torbay* were in fact facing each other bow-to-bow and if Stevens had not taken immediate action and ordered a course alteration a very close pass or a collision would have occurred. As *Torbay* went deep, attempting to act like a biological contact – a whale – and disappear, Stevens paused, levelled *Torbay* off briefly and simulated firing two Tigerfish torpedoes at the Delta. This close-in, 'white of the eyes' attack was considered successful and there was no indication that the Delta, which remained on a steady slow course, had any indication that *Torbay* was nearby. 'I cannot believe my luck,' wrote Stevens. 'This submarine is very quiet narrowband and broadband in comparison to my previous contacts but he appears to be very naïve tactically.'[172] Confident of *Torbay's* superiority over the Delta, Stevens moved his submarine across its stern to examine the various aspects of its acoustic signature before breaking contact and returning to the UK.

After thirty days at sea *Torbay* returned from a 7985-mile patrol with a considerable amount of acoustic and operational intelligence, including a comprehensive recording of the Yankee's acoustic signature and confirmation that the Soviets were transiting their SSBNs through noisy areas. Stevens's patrol report contained some initial conclusions. He argued that 'The reduced Soviet narrowband signature, the aspect dependency (generally only seen abaft the beam) and their disruptive steering has reduced the uninterrupted periods in contact significantly. This meant that all too regularly contact was lost permanently before bearing ambiguity had been resolved and relocations became something of a gamble.'[173] He concluded that 'The halcyon days of long range Narrowband detections are disappearing fast. The effects of the Soviet noise quietening techniques observed this patrol, even on the older generation submarines, are impressive.'[174]

THE FINAL ACT

In the late 1980s, the Submarine Service continued to collect intelligence on the latest Soviet ships, submarines, weapon systems and tactics. HMS *Conqueror* was allegedly involved in one such operation shortly after the Falklands conflict, codenamed Operation 'Barmaid'. *Conqueror*, under the command of Chris Wreford-Brown, was fitted with special hydraulic pincers which were used to cut through a 3-inch-thick steel cable attached to a Polish AGI trawler that was deploying an experimental towed-array sonar. *Conqueror* then separated the towed array from the AGI and returned it to Faslane, from where it was flown to the United States for analysis.[175]

One submariner who was involved in intelligence-gathering operations in the mid-to late 1980s, Lieutenant Charles Robinson, commented that 'The problem is that you are operating in very confined conditions. The main worry is collision. You have a number of submarines running around in a small stretch of ocean. Submarines have hit each other.' Sonar ranges on the latest Soviet submarines had decreased to such an extent that contacts could be as close as 1000 feet before they were picked up. 'You can't actually see the thing or what it's doing, you can only hear a noise. You have 30 seconds to do an initial assessment and in that time you've probably halved the distance. You rely on how the noise is shifting to get an idea of what the vessel is and where it is heading, whether it's on the surface or underneath. If it's on the surface, you're likely to pick it up much quicker.' According to Robinson, the Soviets started to 'play games like moving in a little closer to see if the other sub picks them up. They check each other's detection range. Often round there, they're doing exercises and trials, and there is a need to find out what's happening, particularly with torpedo testing. The worry then is to make sure you don't get shot at because you're hanging around in that area. Or you might even be trying to run under a surface ship or a submarine. You can't make mistakes.'[176]

Such operations were highly sensitive and each was approved at the highest levels of the British Government.[177] Mrs Thatcher visited the Submarine Operations Centre at Northwood and received personal reports from submarine commanders after they returned from sensitive intelligence-gathering operations. In 1989, the future first Sea Lord, Mark Stanhope, then the CO of HMS *Splendid*, alongside Robert Stevens (the CO of HMS *Torbay*) and members of the senior Admiralty Board personally briefed the Prime Minister. Mrs Thatcher was shown the results of various operations, including photographs of ships, submarines and missile firings obtained by Royal Navy submarines. 'That's not what I'm normally told,' said Thatcher after the briefing. Everyone in the room thought they had just killed the Royal Navy's intelligence collection programme. But the Prime Minister immediately saw the worth of the Royal Navy's SSNs. 'She absolutely recognized the value that we were achieving and more than that she recognized the politics, that these submarines are not cheap and she absolutely got in her mind how she could defend these submarines politically.'[178] Mrs Thatcher was also extremely interested in what she was agreeing to when she authorized such operations, what the risk was, both the military risk and the political risk. She totally understood the risk versus gain balance. She said she would continue to authorize operations but she wanted to be absolutely certain that she was going to be told the truth if something went wrong, telling senior Admiralty Board members at the briefing: 'You make sure I know.'[179]

To stay ahead of the Soviets, the Submarine Service was continuously seeking to enhance its capabilities. In the late 1980s sonar systems, such as the narrowband Sonar Type 2047, were upgraded and new specialist equipment that aided in the detection and tracking of Soviet submarines was installed on all Royal Navy SSNs. But by the late 1980s there was a growing imbalance between the operational capability and the acoustic vulnerability of the Royal Navy's oldest SSNs. Without further efforts to upgrade them, to reduce their radiated noise signature, they risked becoming increasingly vulnerable to new Soviet shipborne and airborne narrowband sonars.[180] In 1980, special anechoic tiles were applied to the outer hull of HMS *Churchill* in order to absorb the sound waves from active sonar, reducing and distorting its return signal and thus its effective range. The tiles also absorbed self-noise, reducing the range at which Royal Navy submarines could be detected by passive sonar. Following the fit to HMS *Churchill* they were applied to all the Royal Navy's SSNs and SSBNs.[181]

Accurate and timely intelligence was also instrumental to retaining an operational edge over the Soviet Navy. During HMS *Torbay*'s thirty-day patrol in 1988, the submarine received 798 intelligence signals. In the mid-1980s, the United States also injected new life into SOSUS by enhancing its underwater capabilities. SOSUS evolved into the Integrated Undersea Surveillance System, IUSS. IUSS included the addition of a mobile SOSUS array that provided mobile detection, tracking and reporting of submarine contacts at long ranges known as the Surveillance Towed-Array Sensor System, SURTASS. SURTASS consisted of a long, wide array towed by civilian-crewed ships, where data was uplinked back to shore via satellite communications and analysed by powerful land-based modern computers situated in Dam Neck, Virginia.[182] The first of these systems were installed on US auxiliary Ocean Surveillance Ships commissioned between April 1984 and January 1990.

There were also significant developments in the torpedo field. By 1988, the Tigerfish consolidation and development programme was, after almost thirty years, complete. The reliability of the torpedo had been improved significantly and its capability to destroy submarines hiding under the Arctic ice pack had also been enhanced. In March 1988, the Submarine Service once again returned to the Arctic to conduct a series of Mark 24 Tigerfish under-ice firings. HMS *Turbulent* and HMS *Superb* travelled north of Alaska to the Beaufort Sea, 120 miles northeast of Prudhoe Bay, to a tracking station set up by the Applied Physics Laboratory Ice Station from University of Washington State. Alongside two US Navy SSNs, USS *Lapon* and USS *Silversides*, both *Superb* and *Turbulent* fired sixteen Mark 24 Tigerfish Mod 2 torpedoes in a series of evaluation trials. Most were fired against other submarines, but some were fired at simulated acoustic targets, configured to represent a 'Typhoon' or 'Delta IV' SSBN hiding underneath the ice. According to Dan Conley, 'the weapons performed extremely well in the quiet Arctic conditions, achieving long-range passive homing detections. Even in the active mode, where the torpedoes' homing systems had the problem of resolving the real target from the contacts generated by returns from the ice features, the weapons homed remarkably reliably.' When FOSM Rear Admiral Frank Grenier briefed the Prime Minister about the successful trials she was 'absolutely spellbound' that the Submarine Service had both a reliable torpedo and the capability to attack Soviet submarines hiding underneath the ice pack.[183]

Despite these achievements, there were still doubts about the ability of the Mark 24 Tigerfish to home in on modern, increasingly silent Soviet submarines. In the mid-1980s, the Submarine Service placed a contract with GEC-Marconi to develop a more advanced and even more reliable torpedo known as Spearfish. Capable of attacking a fast-moving target at a distance of fourteen miles and a slow-moving target at thirty miles, Spearfish was guided by either a copper wire attached to the submarine or by an advanced inbuilt sonar. Spearfish carried a 660 lb explosive charge, which could either strike the hull of an enemy submarine, or explode underneath a surface target, breaking the back of an enemy vessel. Like its predecessors the Spearfish development programme suffered from technical problems, which delayed its in-service date until the early 1990s. It remains in service with the Royal Navy today.

These upgrades allowed even the Royal Navy's oldest SSNs to maintain an edge over the Soviet Union right up until the end of the Cold War. In August 1989, the recently refitted HMS *Courageous*, under the command of P. J. Ellis, conducted Operation 'Vaughan', a covert ASW-training war patrol in the Norwegian and south Greenland Seas. *Courageous* was tasked with trailing a homebound Victor I that appeared to be taking an unusual and unpredictable route back to the Barents Sea from the Mediterranean.[184] *Courageous* detected the Victor at the western edge of the Rockall Bank and established a trail through the Iceland–Faroes Gap. The Victor was manoeuvring vigorously and erratically at variable speeds, making any approach to carry out a simulated firing very difficult. Despite the best and erratic efforts of the Victor, *Courageous* eventually obtained a good fire control solution and carried out a simulated Tigerfish firing.[185]

Courageous continued to follow the Victor until late one August evening, when it suddenly conducted a 180-degree turn and shot straight towards *Courageous*, passing down the port side, at a range of 500 yards. The Victor then used its active sonar, known as 'Shark Teeth'. Fearing counter-detection, *Courageous* evaded and opened to the north, and moved away at high speed, making frequent course alterations. The Victor had probably detected *Courageous* on passive sonar as it was closing in to conduct a simulated attack. Even so, Ellis was confident that he had not been classified. 'The trail of MO1 [Victor] was at times exhilarating, and at other times frustrating,' wrote Ellis. The Victor 'proved to be an atypical transiter and was handled in an interesting, vigorous and aggressive manner with an obvious knowledge of the environment. Periods on transit were not long and the use of different speeds always made it difficult ... His random manoeuvres further complicated the issue and the lack of pattern made for a demanding trail.'[186]

Courageous continued to track the Victor, and the next day closed yet again to conduct another simulated Mark 24 Tigerfish attack. Northwood then instructed the submarine to break off the trail and conduct surveillance of the Soviet research ship *Boris Davidov*, which was suspected of conducting special operations 180 miles to the north of *Courageous*'s position. When *Courageous* arrived the next day it closed the research ship and carried out visual surveillance. There was no unusual activity. The ship appeared to be conducting oceanographic research.[187] *Courageous* moved away and attempted to locate a Yankee which had been detected by IUSS in the north Norwegian/Greenland Seas, but after searching for seven days in the vicinity of the Jan Mayen

Polar and Eastern Front, abandoned the search in favour of a 'Sierra' class SSN which was also operating in the area. Although *Courageous* detected the Sierra, the Soviet submarine quickly moved away and into the Barents Sea. *Courageous* then attempted to intercept a possible Victor III transiting out of the Barents, but a noisy mechanical fault forced the British submarine to withdraw and return to Faslane to conduct repairs.

This most secret Cold War of the deep, almost a private war involving three navies, peaked in terms of technology, continuous operations and intensity in the 1980s, the last decade of the great forty-year East–West confrontation. By the end of the 1980s, it was becoming increasingly difficult for the US Navy and the Royal Navy to detect and track Soviet submarines. In October 1989, a Yankee I, two Delta Is, a Delta II, a Delta III and a Delta IV were all unlocated, and a Victor III that had been held on SOSUS operating north of Bear Island simply disappeared. All three submarine services raised each other's games. Of course, the Walker–Whitworth betrayal gave the Russians huge technological and operational assistance. But this should not detract from the skills of the Soviet commanders in making use of the technical bonanza espionage handed them.

For Royal Navy submariners the 1980s will almost certainly remain the underwater equivalent of a permanent world cup competition, a continuous preparation for a dreaded penalty shootout should real war have come. The Russians were worthy and increasingly stretching opponents for the British and American submariners who went up against them year after year, in a confrontation that required ingenuity and nerve in copious quantities, of which the public knew almost nothing. Royal Navy submarines, in their intelligence and surveillance roles, are still the country's first line of defence. But for those who serve in the silent deep no front line has yet quite matched the NATO–Soviet one and the demanding, inhospitable conditions through which it ran.

IO

After the Cold War: 1990–Today

The Russian Navy has not gone away. As I speak a Russian intelligence gatherer is not more than twelve miles away off the coast of Northern Ireland ... So the nuclear attack submarine's traditional role of anti-submarine warfare remains, but with it has come a newer role, into littoral water, brown water, areas closer into the coastline.

> Captain John Harris, CO,
> 1st Submarine Squadron, 1996.[1]

The SSN is not a legacy of the Cold War. Its attributes and abilities reflect exactly modern maritime doctrine. The SSN has broad utility and offers a wide range of options to politicians and campaign planners, at low risk. It can deploy early and quickly, exercise full freedom of the sea, changing role and area of operation at will. This posture can be maintained almost indefinitely. It can be an instrument of diplomacy, coercion, or war fighting employed directly or obliquely. The multi-faceted capabilities of the SSN, in contributing to the overall effort in both political and military 'battle spaces', are dependent on remaining abreast of current technological advances. Failure to invest in the future has the potential to sideline a capability that offers the widest ranges of strategic, operational and tactical choices.

> Commander Nick Harrap, Staff Warfare Officer,
> Flag Officer Submarines, 2001.[2]

UNCERTAINTY AND DECLINE

The Cold War ended at just the right time for the Submarine Service. In December 1989, a month after the Berlin Wall came down, a series of material defects were discovered in HMS *Warspite*'s PWR1 reactor systems during routine tests while she was in refit in Devonport. Significant cracks were detected in the transition welds that joined the primary circuit pipework to the steam generator heads, known as 'trouser legs'. The one-inch-thick welds joined two large pipes about 14 inches in diameter and a crack extended across about half the weld. If the weld had failed, an uncontrolled loss of primary coolant would have occurred with the likelihood of a major reactor incident.[3] As the PWR1 was operational in all the Royal Navy's SSNs and SSBNs (excluding the new 'Vanguard' class SSBNs, which were powered by the PWR2 and were not yet in service), the Navy immediately recalled all its nuclear submarines to port and limited operations to those tasks it considered absolutely necessary such as protecting SSBNs from Soviet submarines. As *Warspite* was one of the oldest SSNs in the fleet, the 'trouser leg' problem was assumed to be age-related and a limited number of SSN operations, involving the newer 'Trafalgar' class submarines, continued, while the remaining 'Valiant' and 'Swiftsure' classes were inspected for similar problems. Although the SSBNs were of similar age to *Warspite*, the Government maintained Polaris patrols, sailing the 'Resolution' class SSBNs under so-called 'operational imperatives' to sustain the continuity of the UK's independent nuclear deterrent. There appears to have been some concern over this decision, which was taken at the highest levels of government. Alan Clark, at the time a junior defence minister, reportedly told Tom King, the Secretary of State for Defence, that 'If – if there is an accident, it's not just you who resigns; the Government falls.'[4]

The Navy eventually determined that the defect was caused by steel corrosion. A demanding testing regime for the submarines of the 'Resolution', 'Swiftsure' and 'Trafalgar' classes was immediately put in place but inspecting and rectifying the defect proved difficult and time consuming. The defective weld was located inside the lower part of the steam generator, an area in an operational submarine that is radioactively hot, at about an arm's length from a 4-inch access hole. Rolls-Royce developed a robot that was small enough to pass through the hole to X-ray the weld, a complex operation that was compared by one of those involved to keyhole surgery or wallpapering a hallway through the letterbox of a front door. Specialist teams worked four-hour shifts in reactor compartments to limit their exposure to radiation. One of the techniques used to treat the affected areas was to bombard the defective transition welds with small 'lead shot-like' particles to relieve stresses in the areas surrounding the welds and so minimize the likelihood of further cracking.[5]

Some loosening of operational restrictions on the Navy's oldest SSNs took place in the summer of 1990 and HMS *Courageous*, which had been alongside for six months, was able to sail. But after only a week at sea, the submarine was recalled due to fears that the 'Valiant' class SSNs were still at risk. They were once again confined to port until they had been inspected and certified as safe.[6] Having so many SSNs confined to port placed considerable demands on those still able to operate, particularly the latest 'Trafalgar' class SSNs, which were less prone to the problem because of their relatively young age. HMS *Tireless* spent eighteen months doing back-to-back patrols. 'It was very hard work, very very hard work,' recalled Matt Kemp, *Tireless*'s Weapons Engineering Mechanic (Ordnance). 'I think in those two years I managed about 5 weeks at home.'[7]

Despite the instability in the Soviet Union in the late 1980s and early 1990s, the Soviet/Russian Navy continued to send its submarines to sea. Although the Submarine Service continued to enjoy an edge over even the most advanced Soviet SSNs and SSBNs, the risk of counter-detection when operating against them was much higher than in the past, particularly when the Russians conducted 'Crazy Ivan' stern arc clearance manoeuvres. In mid-1990, HMS *Trafalgar* undertook Operation 'Mayhem' in the Norwegian and south Greenland Seas with orders to intercept an 'Akula' class submarine that had been picked up on SOSUS and designated L-032. Although up against one of the most advanced submarines in the Soviet Navy, *Trafalgar* detected the Akula, closed it and passed 4000 yards down the Soviet SSN's port side before turning onto its quarter and conducting a simulated Tigerfish attack. *Trafalgar* continued to follow the Akula until it conducted a complex stern arc clearance, which was at first interpreted on board *Trafalgar* as a reversal of course, but it soon became clear that the Akula had, in fact, altered course to the northwest and was rapidly closing on *Trafalgar*. Once *Trafalgar*'s crew realized what was about to happen they immediately altered course to starboard and the submarines came to within 1000 yards of each other. The Akula then came round in a tight turn to port and moved in behind *Trafalgar*. Aware of the possibility of counter-detection, *Trafalgar*'s CO immediately reversed course and slowly moved away to the southeast. *Trafalgar* continued to track the Akula for the next hour, during which the Soviet submarine transmitted twice on its 'Shark Gill' active sonar, clearly suspicious of the presence of another submarine. It was unclear whether or not the Akula achieved active contact on *Trafalgar*. It certainly took no aggressive action. But shortly after transmitting for a second time on active sonar, it accelerated to 28 knots and sprinted away to the south with an impressive display of compressed cavitation.[8]

Aside from operations, the Submarine Service also continued to use its 'Trafalgar' class SSNs on training exercises. On 22 November 1990, HMS *Trenchant* was operating in the Clyde Exercise Areas, acting as the training vessel for the final phase of the Perisher course. At 0217, as one of the Perisher students handed over control of *Trenchant* to another, the submarine's passive sonar detected a close contact to starboard. As *Trenchant* turned to port to avoid the contact, a series of banging noises were heard throughout the submarine. Assuming that *Trenchant* had snagged a fishing trawl the submarine immediately returned to periscope depth where two fishing vessels were easily identified. *Trenchant* then surfaced and a trawl wire was found on the submarine's casing. After attempts to contact the two nearby fishing vessels by radio failed, *Trenchant* reported the incident to Faslane, noting that although it had snagged a trawl, the two fishing boats believed to be involved were safe. *Trenchant* then dived again and continued with exercises. Back in Faslane, attempts to identify the trawler involved in the incident continued well into the morning, until it was realized that a fishing vessel, known as the *Antares*, was missing. A full-scale search and rescue operation later located the wreck on the morning of 23 November in the area that *Trenchant* had been operating in. All four members of the crew died.

Both an internal MOD report and a report by the Marine Accident Investigation Board (MAIB) found that *Trenchant* had collided with *Antares*' trawl gear at 0219 on the morning of 22 November, causing the trawler to capsize and sink. The MAIB report was highly critical of *Trenchant*'s command team, which it concluded 'had no clear appreciation of the surface contacts held on sonar during the period between the completion of the exercise and the collision'. They were criticized for focusing too much on the position of HMS *Charybdis*, a nearby frigate also taking part in the exercise and the focus of the then duty

Captain, whose concentration was 'impaired due to his conversation with the next duty captain in the minutes before the collision'. The report described the command team as having 'a false sense of security', making 'incorrect assumptions' and 'failing to properly assess what might have happened on the surface subsequent to the collision'. Attempts to establish contact with the two fishing boats following the incident were described as 'not adequate' and the decision to resume the exercise was criticized on the grounds that it was taken with 'a lack of appreciation of the reality of the situation'.[9] The Perisher student in charge of *Trenchant* at the time was later reprimanded in a formal naval court martial. As a result major changes were made to submarine operating policy and procedures in order to improve the safety of fishing vessels. The Royal Navy, in consultation with local fishing organizations, also implemented an information service to advise fisherman of dived submarine operations in the Permanent Exercise Areas around the Scottish coast.

Against the background of the 'trouser leg' crisis, the collapse of the Soviet Union and a changing strategic environment, the Government embarked on a mini-defence review, 'Options for Change', that resulted in a significant reduction in the size and shape of the submarine fleet. The changes in the international political climate led the MOD to review the central point of UK planning throughout much of the Cold War: that the Royal Navy's role in any crisis would be to provide first line defence until the full force of the Allies, and particularly the US, could be brought to bear. The MOD concluded that the increasing instability in the Soviet Union had resulted in a dramatic shift, from a matter of days to eighteen months or more, of the likely warning period in the run up to the 'sort of conflict which might require our submarines to be engaged in major operations against the Soviet Union'. The increased warning time naturally affected how long the Royal Navy would have to wait before the United States Navy significantly reinforced it.[10] With increased warning time fewer operationally ready submarines were required to meet the threat. With warning of up to eighteen months submarines undergoing maintenance or refit could be brought back into a state of operational readiness over a short period, while other submarines could be held at various states of readiness rather than being put into refit. The MOD estimated that given the eighteen months or so that it judged it would take the Soviet Union to recover itself so that it could mount a major attack on NATO, at least 10 Royal Navy SSNs out of a maximum fleet of 12 could be ready, compared to a peacetime figure of very short notice availability of 7 or 8 SSNs. In other words, with the increased warning time, an additional 2 or 3 SSNs could be readied over a period of some months if there was perceived to be the possibility of a serious conflict, leaving only those SSNs that were in deep refit unavailable for operations.[11]

The Navy's fleet of conventional submarines also underwent radical changes. By 1995 all the remaining 'Oberon' class submarines were due to be retired and replaced by the new 'Upholder' class SSK.[12] The MOD had intended to procure between ten and twelve replacement Upholders. However the 'Options for Change' review announced a scaled-back requirement of just four submarines.[13] The 'Upholder' class had been beset by delays. One of the most troublesome concerned the largely automatic system for controlling the sequencing of the valves in the weapon discharge system. Although the automated system was successful in achieving near-silent discharge and a marked reduction in manpower, in certain circumstances the hydraulic interlocks could cause the torpedo tube slide valve to open while the tube door was open, allowing unrestricted flooding into the torpedo compartment. The programme was also delayed due to problems with the submarine's powerful main engines (which were originally designed for use in railway locomotives) as the stress of stopping the rotating parts from full power at the end of snorting led to many failures. Consequently, the first of class, HMS *Upholder*, was accepted into service two years late in 1990.[14]

Despite these problems, the few Royal Navy officers who commanded the Upholders insisted that the class was highly capable. 'Having operated UNSEEN in all weathers and in demanding circumstances, I can vouch for the effectiveness and sturdiness of these fine submarines,' wrote one of her former COs. 'The boats were handy both surfaced and submerged, they were a joy to command.'[15] After periods of docking to put right the problems with the weapon handling and discharge system, which was redesigned and re-engineered, all four submarines entered service. At the time the decision to reduce the requirement was taken, a number of improvements were also intended for later submarines of the class, including increased bunk space, rotating conversion machinery with static inverters and increasing fuel capacity by using spaces external to the pressure hull. There were also long-term plans to reduce snorting time by introducing experimental fuel cells.

However, on 25 July 1990, the Secretary of State for Defence, Tom King, told the House of Commons that 'We need to take account of the decline in the size of the Soviet navy, but also of its continuing modernization, especially with the new class of submarine ... In addition to Trident, we envisage a future submarine force of about 16 boats of which three quarters would be nuclear powered.'[16] An accompanying fact sheet referred to the then submarine force of 27 nuclear and diesel submarines, reduced to 'perhaps around 16 nuclear and diesel submarines'.[17] The proposed reductions (twelve SSNs, four SSKs and four SSBNs) represented around a 40 per cent reduction (16–12 SSNs) on Cold War levels and a 30 per cent cut in planned twenty-first-century force levels.

The Submarine Service was deeply concerned. The twenty-seven or twenty-eight SSNs and SSKs in the submarine flotilla in the 1980s had been hard pressed to meet commitments. However, the Government insisted that a reduced force could carry out the same range of wartime and peacetime tasks and that no individual role had been abandoned. Ministers apparently reached the decision as a result of their appreciation of 'the changing strategic environment' rather than as a result of budgetary pressures. As one MOD official put it to the House of Commons Defence Committee:

> The size of the submarine fleet was dictated primarily by the need to have sufficient boats to contribute to Allied forward maritime operations. Our investment for the future will be concentrated on nuclear submarines with their long range endurance and good manoeuvrability. Conventional submarines are less suited to such operations, but we decided to retain four in service, which will act as force multipliers by releasing SSNs for operations in war in the most distant waters and by undertaking important training tasks in peacetime.[18]

Such explanations failed to convince the House of Commons Defence Committee, which was so concerned with the Government's proposals that it initiated an inquiry into the future size and shape of the Submarine Service.[19] When the Defence Committee's report, 'Royal Navy Submarines', was published in June 1991, it concluded that the 'fleet of 12 or 13 SSNs proposed is the barest minimum SSN force, subject to anticipated improvements in availability rates, support and reductions in time spent in refit over a submarine's life'.[20] HMS *Conqueror* was decommissioned in 1990, despite having had over £200m spent on a refit. HMS *Warspite* and HMS *Churchill* were decommissioned in 1991, and although HMS *Courageous* continued to operate under a series of 'trouser leg' restrictions, which reduced its available power to

50 per cent, and its speed to 19.5 knots, it too was decommissioned in April 1991. HMS *Valiant*, the first of the 'Valiant' class SSNs, continued in service until August 1994 until it was also decommissioned after suffering from repeated engineering problems.[21]

The Defence Committee's report also urged the Government to 'urgently reconsider its proposals to retain only four' SSKs, recommending instead that a minimum of six should be built and enter service.[22] However, just three years later the MOD decided to decommission the entire fleet. The Upholders had been designed very specifically for the Iceland–Faroes Gap and to fire at and sink anything that passed. The MOD concluded that the Soviet threat had declined so significantly that they were no longer needed. Their limited endurance also meant they were unsuited to out of area operations. Whereas the 'Oberon' class could operate at great distances without refuelling, when an Upholder was deployed to the Mediterranean it had to be refuelled once it arrived in the region. The Submarine Service struggled to accept the decision. 'Widespread dismay broke out among young command-qualified officers,' wrote Jonty Powis, *Conqueror*'s Navigator, 'because commanding an "Upholder" as a lieutenant commander was a powerful stimulus to retention and officers lucky enough to enjoy the experience became better nuclear captains.'[23] 'It was an outrageous decision,' recalled Roger Lane-Nott, the Flag Officer Submarines at the time the decision was taken. 'It was totally unrealistic; it didn't take into account any of the information that was available. It was a disgraceful decision. Brand new submarines that we'd just got working really well, they were fantastic submarines at sea and we just frittered away the money.'[24]

All four 'Upholder' class submarines were mothballed in April and October 1994 and laid up alongside Buccleuch Dock at Barrow-in-Furness, with VSEL awarded a care and maintenance contract, placing items such as the diesel engines into deep preservation while the MOD set about finding a foreign buyer. After trying to interest Chile, Greece, Pakistan, Portugal, Saudi Arabia, South Africa and Turkey, the class was sold to the Canadian Navy in 1998 as replacements for its old 'Oberon' SSKs.[25] With the disappearance of diesel submarines, the 1st Submarine Squadron was abolished and HMS *Dolphin*, the traditional home of the Submarine Service at Gosport, closed in December 1999. This left just three submarine squadrons: the 2nd Submarine Squadron in Devonport and the 3rd and 10th Submarine Squadrons in Faslane.

FROM POLARIS TO TRIDENT

The early 1990s also saw the decommissioning of the Polaris force and its replacement with the Trident carrying 'Vanguard' class SSBNs, the first of which, HMS *Vanguard,* was due to enter service in 1994. By the early 1990s, the Submarine Service was finding it increasingly difficult to maintain and operate the 'Resolution' class, which began to suffer from a series of mechanical and technical defects. 'It was not as if there was one thing,' recalled Toby Frere, FOSM during the early 1990s, 'it was just that these boats had been running almost continuously all of their lives. Of course they'd been refitted, but their use had been very high. Their secondary systems just weren't up to it.'[26] Aside from the 'trouser leg' defect, much of the 'Resolution' design was 'too complicated, skimping on suitable materials, and too reliant on outmoded technology,' recalled Patrick Middleton. 'The nineteenth-century secondary plant, the "steam swallowing" bit, was the worst culprit, it would need an awful lot of work and expense over the years to keep it going.'[27] On 9 December 1991 the Government announced that HMS *Revenge* would be scrapped.[28] HMS *Renown* had been refitted between 1987 and 1992, at the cost of around £200m, but only carried out three operational patrols before returning to Faslane three years later with a 'minor defect'.[29] HMS *Repulse* also spent considerable periods alongside from July 1990 onwards. During this period HMS *Resolution,* which was meant to decommission in 1991 after a 25-year service, became the workhorse of the Polaris fleet as the Royal Navy struggled to maintain continuous deterrent patrols while repairs to HMS *Renown* and HMS *Repulse* were carried out.

Maintaining the deterrent between 1990 and 1996 required complex planning. Before 1989 Polaris patrols had been of a regular length, lasting around eight weeks, but during 1990 and 1991 patrols varied in length between sixteen days and 109 days and HMS *Resolution* conducted one sixteen-week patrol, only four days shorter than the longest patrol ever completed by a Royal Navy nuclear-powered submarine.[30] In anticipation of future difficulties a number of contingency plans were drawn up to maintain the continuity and credibility of the deterrent if it proved impossible to keep a submarine at sea at all times. One of these involved resupplying a Polaris submarine with food while it was still at sea.[31] Another, for use in a worst-case scenario, involved moving a Polaris submarine into Loch Long, where it would dive and remain in a static location on Quick Reaction Alert.[32] None of these contingency plans were implemented. The overworked HMS *Resolution* was decommissioned in October 1994. The 'minor defect' in HMS *Renown* was far more serious than first realized and the submarine remained tied up alongside Faslane until it was eventually decommissioned on 24 February 1996, leaving HMS *Repulse* as the last remaining operational Polaris submarine.[33] By 1996, HMS *Vanguard* and HMS *Victorious* had entered service.

With two 'Vanguard' class submarines in operation, the Polaris force was formally retired on 28 August 1996, when the Prime Minister, John Major, addressed a ceremony at Faslane Naval Base. It is worth quoting what he said in full:

> We are here today to pay tribute to the work of the Polaris force. The debt we owe is very large. For the last 28 years this Force has mounted continuous patrols that have been vital to ensure this country's peace and security. Because of these patrols any possible aggressor has known that to attack the UK would provoke a terrible response.
>
> In particular, we are here today to pay tribute to the last of the four Polaris submarines, HMS *Repulse*, which returned from her sixtieth and final deployment in May.
>
> But not only *Repulse*, of course. I pay tribute, too, to the other three boats and their crews in her Class: the *Resolution* herself, *Renown* and *Revenge*. Each has made its own unique and invaluable contribution to the remarkable record of maintaining a Polaris submarine at sea, on deterrent patrol, undetected by friend or foe, every day, of every year, from 1969 until May this year.
>
> To those of you who have served aboard any of these submarines, past and present, I offer you the thanks not just of those others of us here but of people throughout the country.
>
> The years of the Polaris Force have seen some dramatic changes. In 1968, when Resolution began her first patrol, east/west tension was running high. The Soviet Union had invaded Czechoslovakia, and the Vietnam War raged. And yet, in 1994, I signed an accord with President Yeltsin agreeing no longer to target our nuclear forces at each other's territory. Today, the West enjoys a co-operative relationship with Russia unthinkable even 10 years ago.
>
> But throughout the turbulent years, the Polaris force has always been there, always ready, always prepared, always the ultimate guarantee of this country's security.

As I said, the debt is very great.

No tribute to those of you in the SSBN force, however, would be complete without a special mention of the contribution of your families at home. They, as well as you, have borne the continual strain of enforced separation. They have had to maintain the family while you were gone, relying for communication only on the forty words of the weekly 'familygram'. None of the achievements of the Polaris fleet would have been possible without their forbearance and their understanding. To them, too, I offer a very special thank you. And I am glad that so many are here today.

I would like to thank, too, all those who maintain the submarine and its deterrent away from the boat itself, whether at the base here, in Coulport, supporting the weapons system, on the tugs moving these massive submarines in and out of port, at the headquarters at Northwood, or in the design and support organizations further afield. Each of you has played your part.

Throughout the Polaris programme we have enjoyed very close co-operation with the United States. This will continue with Trident. Our two Navies have a very special trust and understanding. I am delighted that so many representatives of the United States Strategic Systems Programme are with us today, together with the officers and crew of the USS *West Virginia*.

There is naturally a tinge of sadness today. But it is the ending of a chapter only. As Trident takes over from Polaris and Chevaline, so the 'Vanguard' Class takes on the torch from *Resolution* and her sisters.

Let me say a word about our deterrent.

I have no doubt that we are right to maintain a minimum credible strategic nuclear deterrent for the United Kingdom. We will continue to do so for as long as our security needs require. It would be folly for us not to do so. *Vanguard* and *Victorious* are already fully operational and meeting all our expectations. I look forward to seeing them joined, in 1998, by *Vigilant* and, around the turn of the century, by *Vengeance*. Together, these four submarines will carry the UK's strategic and sub-strategic deterrent well into the 21st century.

In a few moments I shall unveil this plaque marking the proud achievements of the Polaris Force. And, as I unveil it here, so, at the entrance to this facility a little way away, a small stone monument is also being unveiled. This monument is to serve as a quiet and dignified reminder of the unique contribution made to peace and security by these submarines and the men who served in them.[34]

In the early days of his Premiership, John Major had written a set of instructions, the last-resort letters, to the CO of each of the four Polaris submarines as well as the Trident replacements:

> It is a shock. The first I realized that I was going to have to write post Armageddon instructions to our four Trident submarines was when the Cabinet Secretary told me. And it is quite an extraordinary introduction to the Premiership. I remember I went away over the weekend and I thought about it, a lot, and it was one of the most difficult things I ever had to do, to write those instructions, the essence of them being that if the UK is wiped out but its Trident submarines are at sea with their weaponry what should they then do with their weaponry. Eventually I reached a conclusion and I set it out.[35]

The Polaris monument now sits proudly at HMNB Clyde, Faslane, as a constant reminder to those who continue to operate Trident.

The retirement of the Polaris force served as a symbolic end to the Cold War. By the mid-1990s, with the Soviet Union confined to history, those Royal Navy submariners who had spent their working lives confronting Soviet submarines at sea in a deadly game of underwater cat and mouse began to reflect on their achievements. 'I think it mattered immensely,' said Doug Littlejohns. 'Apart from giving NATO, the Americans primarily, but NATO, assurance that the Russian's weren't deploying on a war footing . . . We gained a hell of a lot of intelligence about what they were up to and learned a lot about their people because you could identify individuals. I think it allowed us to stay ahead of the game. We didn't have to spend as much money on R&D as necessarily we might have had to because we had a lot of evidence we could feed into our thought processes. I think it was a very valuable thing.'[36]

Roger Lane-Nott agreed:

> It was critical that we had the best possible information about them so that due preparation could be made in the event of war. Specifically, with regard to their growing submarine fleet, we needed a wide range of information. We required intelligence as to their operating capability, such as the speed, diving depth, endurance and sonar capability of their SSNs and SSBNs. And we needed to know the range of their missiles. We had to know the habitual operating patterns of Russian submarines at sea. It was essential to accurately determine their ability to detect our submarines and, even more importantly, how we could detect them . . . I think that the intelligence that we brought back through aggressive, close-in

submarine operations was absolutely vital in a whole variety of ways. More often than not, it provided the final piece in the jigsaw about what you might have heard, or known, or got from other forms of intelligence about a missile system. But, this cannot compare to having actually seen or heard a missile launching – that is the sort of information that cannot be beat.[37]

Another Cold War SSN CO, James Perowne, holds similar views: 'What it was about was making certain that it remained a Cold War and that both sides knew enough about each other that we could stay one ahead of them before they did anything to us. There were two ways of doing it. One was deterrence. On our side of the house we were look-ing at their equipment, their state of play, their state of training, their noises, can we detect them, what happens if they deploy en masse, would we be able to trail them, can we give them enough uncertainty that they would never feel safe to come out? That is what we were doing and we were doing it very very well.'[38]

Much of this was largely unknown, not only to the general public, but also to the UK's other armed services. 'It wasn't the Army sitting in Germany and it wasn't the Air Force flying out of Germany or flying out of the UK. It wasn't even the Navy patrolling the South Atlantic,' said Stanhope. 'I don't want to underplay their contribution but it wasn't like being up close and personal with another submarine who you heard the bow caps open on and you weren't quite sure what the response was going to be. It was terribly terribly exciting to a twenty-year-old who was involved in all of this. And the brown adren-aline did flow. There's no two ways about it.'[39]

But there was a cost. 'The whole relentless twenty-four hours a day, 365 day a year Cold War submarining, I enjoyed it, but it was hard,' says James Taylor. 'It was hard emotionally, it was hard professionally, it was hard psychologically. That said, I look back on it with great enthusiasm. I'm very fortunate. I've been to two wars, the Cold War and the Falklands War, and we won both of them. Not many people can say that.'[40]

That for the bulk of the Cold War the Submarine Service could monitor the Soviet Navy passively by audio, visual and electronic means and generally remain undetected was perhaps the most import-ant piece of intelligence obtained about the Soviet Union and its submarines. In spite of the overwhelming size of the Soviet Navy, indi-vidual submarines did not spend enough time at sea to be an effective fighting force. A few ships and submarines were better than the average and, if war had started, some of their weapons and ship construction strengths would have proved to be at least as good as the Royal Navy's. Sheer numbers would probably eventually have been decisive. But 'weapon systems effectiveness' depended on the quality of people and training at sea, and in this area the Royal Navy had and continues to have a huge advantage.[41] 'I don't think we overrated them,' said Stanhope, 'but there was a danger at times of underrating them and that would have been foolish. We wouldn't have had a successful Cold War in the Submarine Service if we'd underestimated our enemy. But equally we wouldn't have had the success we had if we'd overrated them because we wouldn't have gone close enough.'[42]

The end of the Cold War afforded some limited opportunities to open a conversation with the Russians. In December 1991, the Soviet Chief of Staff, General Vladimir Lobov, was shown around HMS *Revenge* in Faslane, something that would have been unthinkable during the Cold War.[43] On 3 August 1993, HMS *Opossum* became the first Western submarine to sail into the home of the Russian Northern Fleet, Severomorsk, since the end of the Second World War. 'You must not ignore the significance of all this,' said Lane-Nott at the time. 'It is a very, very historic day.'[44] The two submarine nations met face to face, drank and exchanged stories and the Russians thanked the British for supplying Russia through the Arctic Convoys in the Second World War. 'We weren't sufficiently naive to think that all of this was just put on as a front for us,' recalls Paul Branscombe, the then CO of 1st Submarine Squadron, who accompanied HMS *Opossum* on the visit.[45] For Branscombe, meeting face to face with the submariners from the Royal Navy's former Cold War adversary 'was probably the most fascinating and the most salutary experience':

'All of those years, of course, we met them at sea, we never spoke, never knew their faces. But we'd been forbidden, of course, to travel to the Soviet Union or indeed any of the Eastern European countries for security reasons so I don't think I'd actually met any real Russians, let alone Russian submariners. It was really interesting to meet with them. It was emotional in a way, almost a life changer. If you had spent all your working life, as I had, practising and thinking about what we were going to do and so on, suddenly it was a year or so after the Cold War had ended, it was the first time I suppose that I really realized that it was over . . . What surprised me was that they were just like us. They weren't ten foot tall, or even six feet tall. They were just like us as individuals. They had the same kind of sense of humour. We supposed – and intelligence assessments told us – that everything was drilled and regimented with Marxist stuff everywhere. But actually these people were just like us, ordinary bog standard people who just happened to drive submarines around.'[46]

Lane-Nott also found himself questioning his view of the Soviet Union and Soviet submariners: 'I think I had a view that the Soviet submariner had almost a blind faith in their state system and the Motherland and their ability to come on top. I had no idea if this was true, but in conversations I had much later with Admiral Yerofeyev, the Northern Fleet Commander, and his Deputy, Vice Admiral Suchkoff, and a Deputy Fleet Commander, Rear Admiral Titarenko, this indicated to me that they were equally committed to their Communist system, as I was to mine. We had to counter that and all three of them told me they were fearful of the professionalism and grit of the Royal Navy.'[47]

When discussion turned to the business of fighting, the Russians had some interesting views of the Royal Navy and the Submarine Service and the part it would play in any conflict. Despite the many public pronouncements throughout the 1980s about the Royal Navy's role in the Forward Strategy, the Soviets said that while they knew they would fight the Americans, they never thought they would have to fight the British. 'I don't think they were just being nice to us,' recalled Branscombe, 'maybe they were. Somehow they perceived that . . . we weren't identical to the Americans. They were paranoid about the Americans but they didn't seem to be so concerned about us.'[48]

Lane-Nott agreed. 'They really saw the Americans as their real issue,' he recalls. 'They thought we were incredibly professional but they saw us almost like an irritant, we were a pest that was always nibbling around the edges. We were not seen as the main issue. But they knew we were there and they knew that they could never find us and they knew that we were good. They were always worried about what we had got and what we hadn't got and they knew that we were very close to America so they worked on the principle that everything that went to America came to Britain and vice versa.'[49]

These honest conversations had their limits. When Admiral Oleg Yerofeyev asked Lieutenant Commander John Drummond how close the Royal Navy had come into Russian waters, the swift reply was 'I'm afraid I cannot discuss operations that I've done in the past.'[50] Branscombe was also asked 'What's it like to be inside the three-mile limit legally?' He said nothing.[51] Lane-Nott was able to ask Vice Admiral Suchkoff about the incident in the Clyde in 1973, when HMS *Conqueror* had sailed at short notice to ward off a 'Victor' class SSN. Suchkoff knew of the operation and when Lane-Nott asked him why the Soviet submarine had acted in the way it did, by turning and speeding straight towards *Conqueror*, Suchkoff was very honest. 'The reality of life is that we didn't know what you would do,' he said. 'Our only tactic at the time was to be aggressive because we just did not have the ability to detect you before you detected us.'[52]

The Russians were surprisingly open, taking the British on tours around a number of vessels, including some diesel and nuclear submarines. 'Their equipment was visibly less sophisticated than ours,' recalled Branscombe. 'If you asked me for a professional opinion it was a generation or so behind and this was quite a new submarine that we were looking at. But it was robust. Their maintenance was clearly good, as opposed to the surface ships, which were a complete disgrace. They had clearly been neglected for some time. But that submarine would have been a submarine that I would have been happy to take to sea.'[53]

A year later, the Royal Navy returned the hospitality by hosting the first Russian submarine to visit a British port since the Second World War. A Russian diesel electric 'Kilo' class submarine, 431, spent five days alongside HMS *Dolphin*, Gosport, where the Russian Deputy Fleet Commander, Admiral Titarenko, was given a limited tour of HMS *Trenchant*. 'As a former submariner he clearly enjoyed his few hours on board,' said *Trenchant*'s CO, Commander Philip Mathias. 'But for security reasons, it was not always possible to answer his rather searching questions.'[54] The developing RUKUS (Russia–UK–US) confidence building discussions in the early 1990s also saw mutual visits first to HMS *Triumph* at Devonport and then the Victor III *Tambov* at Severomorsk.

SUBMARINES OF THE
FORMER SOVIET UNION

Did the underwater Cold War ever really end? The dissolution of the Soviet Union in December 1991 was meant to signal the end of hostility between NATO and the Warsaw Pact. But at sea both the Royal Navy and the US Navy continued to keep a close eye on the activities of the now Russian Fleet. On 11 February 1992, the USS *Baton Rouge* was tracking a Russian 'Sierra' class SSN as it entered international waters off the Kola Peninsula. The Russian submarine surfaced directly underneath the *Baton Rouge*, causing minor damage to both vessels.[55] Just over a year later, on 19 March 1993, the SSN USS *Grayling* collided with a 'Delta III' class SSBN while operating in international waters in the Barents Sea.[56] Unlike similar incidents in the Cold War, both collisions were heavily publicized, much to the frustration of the US Navy. The newly elected US President, Bill Clinton, was furious.[57] The Russian President Boris Yeltsin cited both incidents as one of the few 'irritants' in US–Russian relations during a summit in April 1993. Clinton personally apologized, described the collision as a 'regrettable thing, and I don't want it to ever happen again' and ordered a 'thorough review of the incident as well as the policy of which the incident happened to be an unintended part'. As a result US rules about when its SSNs conducted operations against the Russians were reportedly reviewed and revised.[58]

Despite Yeltsin's complaints, the Russian Navy continued to conduct its own operations. In the early 1990s, HMS *Tireless* was deployed on Operation 'Dogrose' against one of the Russian Navy's most modern operational 'Akula' class SSNs, which had deployed south of the GIUK Gap. The Akula was a source of great concern as it appeared to be operating to the west of the UK in an anti-SSBN posture, attempting to detect the Royal Navy SSBN that was on deterrent patrol. After deploying from Devonport for the Denmark Straits and briefly withdrawing to replace a faulty towed array, *Tireless*'s crew concluded that the best means of detecting the very quiet Akula was to establish a trail on a nearby 'Delta III' SSBN which was slowly transiting back into the Barents Sea. After sprinting northeast *Tireless* established a firm trail on the Delta III and followed it for forty-six hours, closing at regular intervals to carry out four simulated Tigerfish firings. While trailing the Delta *Tireless* observed four possible delouses by a second, far quieter submarine, which *Tireless*'s crew suspected was the Akula. The Delta III was equipped with third-generation noise reduction techniques but was relatively easy to trail. The Akula was reportedly fitted with advanced fourth-generation noise reduction techniques, was very quiet, very well handled and, in comparison to the Delta, very difficult to detect. Although some intelligence on the Akula was obtained as it deloused the Delta III, a fault with the replacement towed array, which increased the risk of counter-detection, forced *Tireless* to abandon any hope of prosecuting the Akula. It moved away and returned to Devonport.[59]

The Russians continued to attempt to track the Royal Navy's SSBNs throughout the early 1990s. In late 1992, two 'Akula' class submarines were deployed west of the UK during HMS *Vanguard*'s sea trials. Two years later, in December 1994, three 'Trafalgar' class SSNs, a US Navy SSN, Royal Navy towed-array frigates and US SURTASS vessels were all deployed to protect HMS *Vanguard*, which was on its first deterrent patrol, as well as HMS *Victorious*, which was preparing to sail for sea trials. A 'Victor III' SSN, designated Hull 36, had previously been tracked for a three-month period by Maritime Patrol Aircraft and towed-array frigates as it conducted operations in the North Atlantic. Although the Russian submarine had returned to its Northern Fleet base in late 1994, by early December intelligence indicated that it was due to sail and conduct a delouse operation with a 'Delta' class SSBN south of Jan Mayen. The Victor duly left its Northern Fleet base in early December, was picked up on SOSUS and designated L-031, intercepted the Delta and then moved south into the Royal Navy's Northern Fleet Exercise Areas where it appeared to loiter.

On 24 December 1994, HMS *Torbay*, the Royal Navy's Immediate Readiness SSN required to sail at short notice, set out from Devonport on Operation 'Porringer', a reactive covert ASW operation, with orders to locate and trail the Victor and determine the nature of its operations. Although *Torbay*'s crew knew that they could sail at short notice at any time, the notification to do so on Christmas Eve came as a shock and lowered morale. But *Torbay*'s company accepted their own loss of Christmas and New Year without complaint and focused on the job at hand and spent until early January sweeping the waters around the Northern Fleet Exercise Areas for the Victor III. Locating what by all accounts was a well-handled, slow and quiet modern Russian submarine was very difficult and after finding nothing *Torbay* moved out of the area to take part in a pre-planned NATO operation. 'The end of a very frustrating two week operation,' wrote *Torbay*'s CO. 'L-031 has a lot to answer for.'[60] As HMS *Torbay* withdrew, HMS *Talent*, which had taken over as the designated Immediate Readiness SSN on 29 December, sailed from Devonport on 4 January and spent eighteen days searching for the Victor III. Despite an exhaustive search *Talent* also failed to locate the Russian submarine and with so little information available on other Russian submarines operating in the North Atlantic, *Talent* sanitized waters for HMS *Sceptre* and HMS

Trenchant, which were due to leave the UK for other operations and returned to Devonport on 22 January.[61]

By mid-January, the Victor III had been at sea for sixty-one days and had managed to avoid detection for the last eleven. A US SSN, the USS *Albuquerque*, had joined the search as had a third 'Trafalgar' class SSN, HMS *Tireless*, which sailed from Devonport in late January to relieve HMS *Talent*. Northwood's worst-case fears, that the Victor III was lurking in the Northern Fleet Exercise Areas awaiting the sailing of HMS *Victorious* for sea trials, appeared to be well founded. As *Tireless* entered the waters around the Exercise Areas, SOSUS detected the Victor III operating about ninety nautical miles south of *Tireless*'s position. *Tireless* started on a cautious approach to close and intercept the Russian submarine, assisted by cueing information from SOSUS and numerous over flights from Maritime Patrol Aircraft also engaged in the search.[62] As *Tireless* moved in to intercept the Victor III it struggled to maintain contact with the extremely quiet Russian submarine. Shortly after arriving in the Victor III's reported position, *Tireless* was forced to suddenly alter course 60 degrees to port in order to avoid a potential collision. As *Tireless* swung around to port the Victor III broke to starboard. Once the risk of collision had passed *Tireless* steadied on a course of 190 degrees while the Victor III passed at around 300 yards on *Tireless*'s starboard quarter, before heading north-north-east. *Tireless* remained in contact with the Victor III for the next seven hours, collecting intelligence information, before breaking off to transmit an update to Northwood. After reporting, *Tireless* failed in its attempts to relocate the Soviet submarine and it was later picked up on SOSUS moving north having apparently been forced deep to avoid the large number of Maritime Patrol Aircraft operating in the area. The Victor III then altered north and transited back to its Northern Fleet base. Reflecting on the operation, *Tireless*'s CO concluded that 'The encounter with the "Victor III" has served as a reminder that the "Trafalgar" class SSN remains a most capable and reliable ASW platform, capable of conducting extended covert operations in adverse conditions.'[63]

Just over a year after Operation 'Porringer', another 'Victor III' class SSN was detected lurking off the west coast of Scotland. It appeared to be blocking the northern entrance into the Clyde, waiting for one of the Royal Navy's SSBNs to depart on patrol. The then Flag Officer Submarines, James Perowne, was forced to bring an inbound SSBN returning from patrol into Faslane through the northern approaches to the Clyde, while an outbound SSBN, departing on patrol, went out using the

southern route. What no one knew at the time was that a Russian submariner on board the Victor III was suffering from a severe case of
appendicitis. As the SSBNs moved in and out of Faslane, the Victor III
suddenly altered course and moved out of the Royal Navy's Northern
Fleet Exercise Areas at high speed. On 29 February, the Russian authorities made an unprecedented call to the British Embassy in Moscow and
asked for assistance. An RAF Nimrod from RAF Kinloss and two
Royal Navy Sea King helicopters eventually located the Victor III
approximately fifty miles northwest of the Isle of Lewis. A Lynx helicopter from HMS *Glasgow*, which was operating nearby in a major
naval exercise, airlifted a Medic to the submarine to assess and prepare
the casualty for evacuation. The Russian submariner was eventually
winched on board a Royal Navy Sea King helicopter and transferred to
hospital in Stornoway where he made a full recovery.

As Russia descended into financial collapse in 1992–3, it struggled to
put its own submarines, especially its SSBNs, to sea. In June 1994, the
British Government announced that it had de-targeted Moscow and St
Petersburg and that 'the guidance computers on UK strategic missiles
no longer routinely hold targeting information'.[64] The Major Government also reduced the number of operationally available warheads for
the Trident force. A year later, in 1995, the Royal Navy's heavy involvement in SOSUS was scaled back when the main UK facility at RAF
Brawdy closed and its functions transferred to a Joint Maritime Facility
located at RAF St Mawgan which was fully integrated in the US Navy's
Integrated Undersea Surveillance System. The facility continued to track
submarines and watch over the North Atlantic until 2009 when its
functions were transferred to Dam Neck, Virginia, in the United States.
By the end of the twentieth century the operational tempo of the Russian SSBN fleet significantly declined, and in 2002 the Russian Navy
failed to put a single SSBN to sea.[65] With the disappearance of much of
the Russian threat at sea the Submarine Service entered a period of vulnerability. When the New Labour Government published its Strategic
Defence Review (SDR) in July 1998 it announced further reductions in
the size of the submarine fleet so that by 2006 the force would consist of
14 submarines: 10 SSNs and 4 SSBNs.

The SDR also made a number of further changes to the SSBN fleet.
It considered radical measures such as taking the 'Vanguard' class off
patrol altogether and separating the warheads from the missiles. However, the Government concluded that both measures could undermine
the deterrent effect of the weapons system and also lead to escalation in

the event of a crisis.[66] But improvements in the strategic landscape allowed the Government to reduce the number of warheads carried on each submarine from 96 to a maximum of 48.[67] The SDR also cut the British warhead stockpile from the ceiling of 300 operationally available warheads to fewer than 200.[68] When set against the earlier reductions under the Conservative Government, overall the destructive power of the UK deterrent had fallen by more than 70 per cent since the end of the Cold War. 'This is the minimum necessary to provide for our security for the foreseeable future and smaller than those of the major nuclear powers,' declared the Blair Government.[69] The alert status of the deterrent was also reduced and the 'Vanguard' class submarines embarked on patrol with their missile systems at several days' notice.[70] The patrol cycle was also relaxed so that just one SSBN was on patrol at any one time. SSBNs also started to conduct secondary tasks while on patrol, such as exercises, 'without compromising their security'.[71] In November 1998, HMS Vanguard became the first Royal Navy SSBN to visit Gibraltar.

In 2001 the Royal Navy Submarine Service celebrated its centenary.[72] But as past and present submariners gathered, the Service was hit by an entirely new crisis, one that would once again challenge the professional skills of all those involved. In May 2000, a serious fault was discovered in the primary cooling circuit of HMS Tireless while the boat was operating in the Mediterranean. The submarine was forced to abandon its patrol and return under diesel power to Gibraltar, where the reactor was made safe and the leak temporarily plugged, while teams of engineers and specialists from Rolls-Royce assessed the defect and developed a means of repairing it. The MOD explored the possibility of returning Tireless to the UK using a heavy transporter ship, but concluded that preparations to do so would have taken much longer than the projected repair programme.[73] The safest and most practicable solution was to carry out the necessary repairs while the submarine was alongside in Gibraltar.

After initial inspection it became apparent that a flaw in the pipework was generic and had arisen from a design/manufacture fault from the original construction programme.[74] The Government carried out a full inspection of all Royal Navy SSNs and determined that the defect was evident in five, four of which were already alongside undergoing repairs or maintenance.[75] By early 2001 only HMS Triumph was 'ready for operations' while the remaining 'Swiftsure' and 'Trafalgar' class submarines were kept at ninety days' or more notice for operations while they underwent defect repair or long-term maintenance. The MOD hoped that eight submarines – HMS *Superb*, HMS *Sceptre*, HMS *Splendid*, HMS, *Trafalgar*, HMS *Tireless*, HMS *Torbay*, HMS *Turbulent* and HMS *Talent* – would be available for operations by the end of 2001, while the remaining three – HMS *Sovereign*, HMS *Spartan* and HMS *Trenchant* – remained in longer-term maintenance or refit.[76] HMS *Tireless* left Gibraltar on 7 May 2001, almost a year after first arriving alongside.

POWER PROJECTION

By the end of the twentieth century changed political circumstances and the priorities of Western nations led the Submarine Service to make a decisive break from its Cold War emphasis on anti-submarine warfare in the North Atlantic, towards what was termed 'littoral operations, power projection from the sea and the ability to influence the land battle ashore'.[77] This process began in the mid-1990s, when the Submarine Service increased its capacity to carry out two roles for which it had previously had only limited capability: to attack land targets, which, in the past it had primarily confined to assaults on coastal facilities; and the ability to deploy and extract Special Forces behind enemy lines, a capability that the service had nearly lost with the retirement of its conventional submarines.

·In 1995 the US and UK governments signed a Foreign Military Sales Agreement to allow the Royal Navy to purchase the Tomahawk Land Attack Missile, known as TLAM, which had been used successfully by the US military in the 1991 Gulf War, Operation 'Desert Storm'. With approximately 70 per cent of the world's population and the majority of the world's largest urban areas within 100 miles of the coast, Tomahawk gave the Royal Navy and the British Government the ability if necessary to reach out from the sea over greater distances than ever before, by attacking not only the maritime defences of a country, but high-value targets and political centres far removed from the traditional front line. Tomahawk gave the UK a weapon with which it could affect the whole course of a conflict with, potentially, a single shot.[78] The RAF did everything it could to prevent the Royal Navy from acquiring a land attack missile, as such a move of course threatened the future of its manned aircraft, which traditionally provided that capability. There were extensive arguments about the location of the new

Joint Operating Headquarters, the RAF claiming that it should be located at High Wycombe, and the Royal Navy, particularly the Submarine Service, arguing that it should be located at Northwood alongside CTF 345. The matter was finally resolved when the role of FOSM was incorporated into a wider function, Chief of Staff Operations, or COSOPS, based in Northwood and responsible for both CTF 345 and CTF 311 as well as surface operations and the rest of the Royal Navy.

Tomahawk, a winged, jet-powered, high-subsonic missile, includes a conventional warhead, solid-propellant rocket booster, and various underwater protection mechanisms contained within steel capsules, which also serve as smooth-bore launching devices when loaded into a submarine's torpedo tube from which it is horizontally launched underwater. At launch the missile is ejected from the steel capsule held within the torpedo tube. Once clear of the submarine, a lanyard initiates the firing circuit and ignites a rocket motor, and the missile's guidance system pitches the missile nose up. After it breaches the surface, the missile jettisons its engine inlet cover, wing plugs and the shrouds between the missile and the rocket booster, allowing the fins to deploy. Once airborne the missiles are highly accurate and use a number of different guidance systems and can fly up to 1000 miles to strike a target the size of a garage. The missiles are also very effective at evading detection due to their low-altitude flight path which is under the coverage of most radar systems.[79]

The first Royal Navy submarine fitted with Tomahawk was HMS *Splendid*. Not only did this include a sheaf of Tomahawk missiles but also the introduction of the Tomahawk Weapon Control System (TWCS), responsible for route planning, initialization of the missile, downloading of data and ultimately the launch. In 1998 HMS *Splendid* conducted the first test firing of a live Tomahawk against a simulated target off San Diego. The Royal Navy accelerated the trials so that the missile could be brought into operational service in time for *Splendid* to take part in Operation 'Allied Force' against Serbia. During the operation it provided the Permanent Joint Headquarters with an independent, stealthy, autonomous land attack platform, capable of keeping the pressure on the Milošević regime by launching TLAM strikes by night and during weather that prevented tactical aircraft from making similar attacks.[80] *Splendid* fired 20 of the 238 Tomahawk missiles launched throughout the conflict, including against a key military radar facility located near Priština airfield in Kosovo.[81] The performance of the new system was described as 'outstanding', with 17 of the 20 missiles hitting their targets.[82]

Tomahawk quickly became the 'precision weapon of choice' as both the Navy and the British Government recognized that it allowed SSNs to 'achieve the application of force or influence at a time and place of political choice at minimum political risk and with a low probability of unwanted damage'.[83] The equipping of *Splendid* was followed by that of HMS *Trafalgar* and HMS *Triumph*, which in 2000 were fully TLAM integrated with US Fleets in the Gulf and the Mediterranean. The capability came at a price, each TLAM costing over a million pounds. The Royal Navy initially purchased sixty-five Block III Tomahawk missiles and planned to equip only seven SSNs with the capability to fire them. In the 1998 SDR it was announced that all the Royal Navy's remaining attack submarines would be TLAM equipped.[84] As more and more SSNs were fitted with the TLAM system, the Submarine Service was able to provide the Government with continual SSN-based land attack capability throughout an extended crisis, an important step in its move away from the ASW orientated arena of the Cold War towards the new 'littoral warfare concept'.[85]

With the disappearance of the 'Oberon' and the 'Upholder' conventional submarines the requirement to embark and offload Special Forces transferred to the remaining SSNs, which because of their size were unsuited to covert operations in shallow waters. Despite this drawback the Navy developed a means of providing its SSNs with covert projection and insertion capability for Special Forces and their equipment. An SSN would approach land then raise its stern end out of the water to allow the aft engine room hatch to be opened and the Special Forces to leave the submarine, open and inflate their rubber dinghies on the after casing and float off as the SSN slowly submerged beneath them. While this may sound simple, it was more complex than operations conducted by the smaller and lighter diesel electric submarines. The Royal Navy also sought to enhance the capability of its SSNs to insert covertly Royal Marines without having to surface. This was achieved by using a portable structure known as a Dry Hangar, fitted to the upper casing of an SSN, with access through the submarine's main access hatch. A Dry Hanger, adapted from an American design, codenamed Alameda, was first installed on HMS *Spartan* in 2003, which provided the Royal Navy with the capability until the submarine was retired in 2006. The Dry Hanger could house a number of Special Forces personnel and divers, as well as a mini-submarine, known as a Swimmer Delivery

Vehicle. In 1999 the Royal Navy acquired three such mini-submarines from the United States Navy. These 22-foot-long, battery-powered craft can carry a pilot, a navigator and up to four Special Forces soldiers.[86]

SEVEN DEADLY VIRTUES

Although the addition of these two capabilities injected new life into the Submarine Service, there was little room for complacency. In 2001, the Flag Officer Submarines, Rear Admiral Rob Stevens, warned that:

> The Royal Navy's Submarine Service's past successes, unflinching courage and the bond between all of us could, if not properly harnessed, be the cause of our downfall as well. Our tactical and technological victory in the Cold War could lead us to think that the operational concept of covert independent operations was the only way to go. On this note, independent anti-submarine warfare (ASW) operations are but one part of our capability. The integrated operations with the carrier, the intelligence-gathering Special Forces and the Tomahawk missile capability are all key ingredients in the maritime contribution to joint operations. If we do not improve on all our capabilities we could fall into the trap of complacency. This was a point made by Mr Richard Danzig, US Secretary of the Navy, when he drew the analogy that the submariners, were like the Narcissus; we are so busy admiring our own reflection in the water that we are in danger of failing to see the world is developing around us and eventually we too could wither and die.[87]

The Service was very aware that it had to continue to build on its proven skills of anti-submarine and anti-surface ship warfare, but that if it was to survive and remain relevant into the twenty-first century, it had to integrate its existing skills effectively into joint and task group operations to ensure that the submarine became indispensable. The service began to emphasize the seven inherent qualities of the SSN, the 'seven deadly virtues' of flexibility, mobility, endurance, reach, autonomy, stealth, and punch. In 2001 Commander Nick Harrap, an officer on FOSMs staff, described these virtues in the following terms:

> *Flexibility.* The capability to change role almost instantaneously, without equipment reconfiguration and without changes in, or redeployment of, personnel. No arrangements have to be made for host nation support.

The submarine is constrained only by broadcast limitations – or communications windows when undertaking task group support operations – in that the transmission and receipt of an instruction takes a finite length of time. It is, however, true that some roles are mutually exclusive. Prime examples are the provision of indications and warnings and TLAM strike – the one inherently covert, the other necessarily overt in its execution.

Mobility. The SSN is capable of high speeds, for sustained periods, independent of the surface and with no requirement for an accompanying logistic train. This allows for a Speed of Advance (SOA) potentially in excess of 500nm per day for as long as is required. It further allows efficient and effective employment in support of task group operations, bringing real meaning to the term 'fleet submarine'. This inherent agility allows close range force protection or sanitisation operations ahead of the task group. It almost certainly means that the SSN can be the vanguard of follow-on forces to shape the battle-space at the direction of the joint force commander.

Stealth. The SSN represents the only true ability to operate independent of the surface, up threat, regardless of who exercises sea control or air superiority. In this manner it may be possible to conduct operations either alone or in conjunction with Special Forces that are exactly in keeping with the precepts of manoeuvre warfare. The impending advent of the dry deck hanger and the ability to deploy the Swimmer Delivery Vehicle (SDV) covertly represent a significant capability enhancement and force multiplier. The contribution to, and the effect on, the campaign estimates might be critical. Added to this is the psychological dimension. The mere threat of submarine presence has a coercive effect of its own and should not be underestimated.

Endurance. There is no requirement for dependence on outside authorities either for support, or for withdrawal, from an area if the situation changes. All life support service can be provided onboard on a continuous basis. In combination with mobility and stealth, this capacity serves to give the widest range of military and political choice in campaign planning and execution. The only limitations are expenditure on food and weapons.

Reach. Of itself, seapower provides ready and unique access to huge areas. The SSN is capable of taking this concept further, exploiting the environment to the full, including areas not accessible to other forces such as the marginal ice zone and under the ice canopy. This facilitates a variety of operational employments at the time and place of choice. It can range from mere physical presence to the delivery of selectively targeted ordnance, 'behind enemy lines' and regardless of who dominates the battle-space.

Autonomy. This virtue embraces the ability to operate alone and without support at the direction of whoever exercises overall command, as a self-contained unit of force – with broad utility – rather than part of a force package, if that is what is required. The SSN is capable of self-protection and offensive action without assistance from other units.

Punch. The punch ranges from the determination of the enemy's centre of gravity through intelligence gathering and the provision of indications and warnings, through force protection, to precision strikes against land targets at a range of 1,000nm and to an accuracy of 25 feet.[88]

The service also emphasized two unique contributions that it could make in joint operations. The first was in an area commonly described as indicators and warnings, obtaining intelligence that is not available to other sources, or is simply shut down and unavailable if a satellite or surface ship is known to be operating in the area and able to exploit the information. The capacity of SSNs to obtain electronic intelligence, intercept communications, exploit microwave links and platform-to-platform communication systems, conduct environmental assessments and close-range photography, obtain acoustic intelligence and pass tracking information was an enormous asset.[89] HMS *Tireless* was involved in intelligence-gathering operations of this kind during the Kosovo crisis, monitoring radio transmissions, intercepting communications and passing information to Allied forces.[90] The second unique contribution involved revisiting the concept of SSNs operating in support of a surface task group. With the advent of high-integrity, nearly real-time transfer of information, expanded bandwidth and high rate of data transmissions, the concept of SSN support operations of surface forces has become a more realistic and practised concept. SSNs could operate in cooperation with surface forces, providing protection from enemy submarines and surface ships. But both contributions depended on further advances in technology, particularly communications. 'This is an area in which the Royal Navy must really improve if we are to break away from the unsatisfactory twelve-hour delay that the submarine [VLF] broadcast brings into any Joint Commander's calculations,' wrote Rear Admiral Stevens in 2001. 'Until we make this transition, we in the Submarine Service will not be able to take advantage of the information revolution and make the submarine truly "joint".'[91]

In the early 2000s, the Royal Navy's SSNs also underwent a number of capability upgrades designed to expand the role of the SSN and turn it into a truly multi-role platform, capable of carrying out a variety of operations. On the technological front, the service devised a number of innovative techniques to ensure that the remaining 'Swiftsure' and 'Trafalgar' class submarines remained as available as possible.[92] A two-stage incremental programme to counter sonar obsolescence was introduced, designed to 'shift the balance of operational capability further towards the UK by enhancing the submarines' sonar performance and reducing the chances of counter-detection'.[93] The Initial Phase was completed in 1996 and resolved sonar obsolescence, introduced enhanced sonar capability (Sonar 2074 and 2082) and the new integrated Submarine Command System (SCMS) and delivered an incremental improvement in weapon system effectiveness to the remaining SSNs. The Final Phase enhanced the operational effectiveness of the four most recent 'Trafalgar' class submarines, principally by the introduction of a new integrated sonar suite known as Sonar 2076 – a software intensive system that represented 'a step change in both technology and military capability', as well as an upgraded tactical weapons system and a number of signature reduction measures.[94] In September 2006, an additional programme of work was also pursued with the aim of developing and de-risking the enabling technology that would allow a more affordable, timely and cost-effective means of sustaining and upgrading Sonar 2076 in the longer term, by using common modules from other UK submarine combat system projects as well as the sharing of advanced signal processing algorithms with collaborative partners.[95] These systems were installed in HMS *Trenchant*, HMS *Torbay*, HMS *Talent* and HMS *Triumph*.

EAST OF SUEZ

The decisive shift away from Cold War-orientated ASW operations can best be seen by the increasing use of Royal Navy submarines in areas other than the North Atlantic. During the Gulf War in 1990–91, the Royal Navy augmented its Armilla Patrol, established in 1979 to protect merchant shipping during the Iran–Iraq War, with 11 destroyers, 2 submarines, 10 minesweepers, 3 patrol craft and 11 RFA vessels. During the war itself, the 'Oberon' class conventional submarines HMS *Opossum* and HMS *Otus* were involved in a variety of operations, including the deployment of Special Forces. Both submarines were camouflaged in black and light blue to cope with the clear Gulf waters when operating at periscope depth. 'The biggest worry was mines,' said HMS *Opossum*'s CO, Stephen Upright. 'You simply couldn't know if they were there until too late and that certainly made having to surface twice a day concentrate the mind.'[96] Both submarines returned to HMS *Dolphin* in May 1991, flying the Jolly Roger with daggers.

Royal Navy submarines continued to visit the region in the aftermath of the Gulf War to reassure allies and gather intelligence. In the early 1990s focus began to shift to the other submarine nations in the world. In 1993, HMS *Triumph* visited the Gulf to conduct what her CO, Commander David Vaughan, described as a 'pathfinding' operation, to develop tactics for operations in warm waters very different from those patrolled in the Cold War. The Royal Navy stressed at the time that the visit was not specifically connected with the arrival of Iran's first conventionally powered Russian-built 'Kilo' class submarine. However, the Gulf States regarded *Triumph*'s presence as 'reassuring', not only in the aftermath of the Gulf War but also in the face of a resurgent Iran. 'The need to keep nuclear submarines close to home has changed,' said Captain Martin Macpherson, representing FOSM. 'The purchase of the Kilo-class from Russia is clearly a development. Whilst locally here the Iranian Kilo is obviously of strategic significance, similar problems exist all over the world.'[97] In 1997 HMS *Trenchant* accompanied a Royal Navy Task Group consisting of twenty vessels led by the aircraft carrier HMS *Illustrious*, which sailed to the Asia–Pacific region on deployment 'Ocean Wave 97'. The aim of the exercise was to demonstrate the UK's continuing commitment to the Far East in the context of the transfer of Hong Kong to China and the Royal Navy's ability to deploy operationally effective and self-sustaining maritime forces for a prolonged period.[98]

1997 also saw HMS *Trafalgar* complete the first ever round-the-world deployment by an SSN using both the Suez and Panama Canals. 'We are practising deploying more widely and in the last couple of years have deployed literally worldwide,' said *Trafalgar*'s CO, Commander Matt Parr. 'And as well as just going there we are actually doing things we didn't have the equipment and the expertise to do during the Cold War. We are operating much more closely with surface ships and are operating under tactical command of surface ship commanders on a much more frequent basis.' This took some getting used to. 'When I had envisaged leaving harbour for the first time as CO I always expected to be in

Faslane sailing into the North Atlantic. I never expected it to be Perth, Western Australia, sailing in the Indian Ocean.'[99]

British commitments East of Suez, particularly in the Gulf region, steadily increased throughout the 1990s. Royal Navy submarines were involved in supporting the invasion of Afghanistan by coalition forces in 2001. HMS *Trafalgar* and HMS *Triumph* fired Tomahawk missiles during the first wave of attacks, while HMS *Superb* was engaged in intelligence-gathering operations. HMS *Turbulent* was also deployed to the Gulf where, alongside HMS *Splendid* and twelve US Navy SSNs, the submarine took part in the opening stages of the invasion of Iraq, Operation 'Telic' to the British, destroying Iraqi military and regime targets with Tomahawk missiles. 'We were on a notional count-down to when we thought we would be conducting our first strikes,' recalled *Turbulent*'s CO, Commander Andrew McKendrick:

> We did our first strikes on the night of March 21 and 22, 2003. It was pitch black; it was night. We were keeping UK time. It was dark quite early. There was incredible concentration. It was faces in glowing screens. It's whispered orders, concentration and then that moment when the discharge system actually ejects the missile. I'll always remember my officer on the periscope, the communications officer. We had seen footage of Tomahawk firing before but of course we had never done it. This missile leaves the water in an absolute blaze of rocket motor. There was an expletive from the officer about how bright it was as it soared away into the night. I do remember somewhere deep in the submarine there was a cheer as the first one left. The training harnesses and tempers the adrenaline but it was there. I have absolutely no doubt. You feel this great thump and whoosh as the air blasts back into the submarine; it's something you can't mistake on board. It was remarkable to an extent; to find myself after that period in the Submarine Service to actually be using the submarine's weapon system was A, remarkable and B, it was all about getting it right.[100]

McKendrick was well aware that the world was watching.

> We had trained intensively for this mission and you're very aware people are watching and the importance that your strike is conducted properly. The co-ordination is so fine, both in terms of deconfliction of these missiles as they are flying but also when and where they are to arrive – that absolute focus. The thing that impressed me most was these guys had been away nearly for nine months and the concentration and profession-alism was absolute. It was all about getting it right and you perhaps think

about the more profound issues afterwards. It was the most demanding work schedule I've ever taken part in. Morning, noon and night we were exercising different scenarios. In between that we were sitting down to work out how we would meet that demand. For me it was the satisfaction on board after so long away in proving your worth. These people [the crew] had brought the submarine from the depths of Devonport Dockyard back to the Gulf and delivered – that for me was the culmination of a very long period of ops deployments. Bringing all that together is one of my abiding memories.[101]

Turbulent returned to Devonport flying the Jolly Roger after 300 days at sea.

Given the distances involved, submarines operating East of Suez have carried out record-breaking patrols. In 2002, HMS *Turbulent* spent 236 days at sea out of a total of 300 away and travelled more than 50,000 miles, equivalent to twice around the world. Sailing from Devonport in June, *Turbulent* entered the Mediterranean, passed through the Suez Canal and became the first UK submarine to visit the new naval base at Changi in Singapore, before taking part in a series of operations in the Far East, including patrols in the Pacific and a port visit to Guam, where she became the first Royal Navy submarine to visit a newly formed US Navy submarine squadron. She then continued west, passing over the Challenger Deep, the deepest part of any ocean in the world, through the Celebes Sea, Lombok Strait and on to the British Indian Ocean Territory of Diego Garcia, returning to Singapore for Christmas and New Year, before heading into the Arabian Gulf and Bahrain for operations and exercises with coalition warships.

Operating East of Suez is also expensive, financially and materially. Due to the lack of overseas bases, submarines have to be supported while in the area. They also have to transit through the Suez Canal, a process that costs a great deal of money and has to be planned and booked in advance. Materially, the distances involved take their toll on the SSNs, particularly their reactors and machinery. The 'Swiftsure' and 'Trafalgar' classes were never designed to operate in such warm waters, yet by the twenty-first century they were deployed to the region on almost continuous patrols. Typically an SSN will now stay East of Suez on operations for ten months. The core burn is considerable and due to delays in the 'Astute' programme, the service life of the 'Trafalgar' class is being extended well beyond what was originally designed. Once East of Suez, Royal Navy submarines are responsible for maintaining continuous

TLAM coverage in the region at a few days' notice to fire. They also simultaneously conduct what are termed Intelligence Surveillance Tracking and Reconnaissance (ISTAR) Patrols, which are demanding in that the submarines are required to deal with both potential enemies and friendly forces, avoiding all of them and remaining undetected.

SSNs have also been involved in recent conflicts. In 2011, HMS *Triumph* was deployed in the Libyan campaign, supporting operations to enforce United Nations Security Council Resolution 1973 through the Joint Task Force's Operation 'Odyssey Dawn', known in the UK as Operation 'Ellamy'. *Triumph* protected the refugee and humanitarian aid ships docking in Misrata while the port was under heavy fire by tracking Gaddafi-regime minelaying small boats, avoiding the mines and conducting intelligence-gathering and targeting missions off the coast. 'We are the special forces of the maritime world. We do things that would make your hair curl,' said HMS *Triumph*'s CO, Commander Rob Dunn. 'Not since the Second World War has a submarine been used this effectively and flexibly. We were sitting incredibly close to the coast, watching and listening to the battle raging around Misrata and providing the most up-to-date intelligence and early warnings of impending attack against the port and the Nato ships providing its security.'[102] For the duration of the scouting missions, artillery, rocket and missile explosions were easily heard throughout the submarine and shells repeatedly landed in the water nearby.[103] During the opening two nights of Operation 'Ellamy', *Triumph* launched a number – perhaps as many as twelve – Tomahawk missiles into Libya alongside US forces, which fired a combined total of 100.[104] They were directed against targets inside Libya, which included air and missile defence system radars, anti-aircraft sites as well as key communications nodes in the areas around Tripoli and along the country's Mediterranean coast.[105] These operations were designed both to de-risk subsequent missions flown by NATO aircraft, degrade the Libyan regime's capability to resist a no-fly zone and prevent further attacks on Libya's citizens and opposition groups. Justin Hughes, Captain Submarines, Devonport Flotilla, described *Triumph*'s missile firings as 'getting up close and personal' with Libya and 'landscape gardening from the deep'.[106]

Triumph remained at sea until 4 April.[107] It was relieved by HMS *Turbulent*, which was also involved in operations off Libya in June 2011 before moving through the Suez Canal and to the Gulf, where *Turbulent* suffered from a mechanical defect, which in the warm waters of the Gulf caused severe problems. *Turbulent* had been alongside in

Bahrain conducting repairs and preparing to sail for an ISTAR patrol. The submarine sailed in 34-degrees Celsius water temperature and 45-degree air temperature, with its coolant pumps operating at 50 per cent capacity. As *Turbulent* attempted to reach the cooler waters of the ocean, the boat's freon plants – part of the air-conditioning machinery – failed and the temperature inside the submarine started to rise even higher. *Turbulent*'s crew shut down all non-essential equipment attempting to reduce the heat, but there was no improvement and before long humidity in the aft machinery compartments reached 100 per cent with temperatures of over 60 degrees Celsius. Personnel quickly overheated and some of the crew began suffering from severe heat exposure. The submarine started to become uninhabitable.

'I genuinely thought there was going to be a loss of life on board,' recalled *Turbulent*'s CO, Commander Ryan Ramsey, who was on the bridge at the time. 'I came down below and was met with this incredible blast of heat.'[108] *Turbulent*'s crew faced a three-fold problem: people were collapsing; equipment was failing; and if conditions continued to deteriorate, there was a small risk that the weapons would heat up beyond their design state and the reactor would shut down. To avoid long-term damage and, in the worst case, calamitous failure, many of *Turbulent*'s systems were switched off. The marine engineers cycled personnel in and out at fifteen-minute intervals to maintain the reactor while the rest of the crew opened up the submarine's outer hatches to get comparatively cooler outside air into the submarine. Nevertheless, within hours, 'people were just collapsing everywhere, many at their workstations,' recalled Ramsey. 'We had casualties in the control room, the engine room, the bridge, the wardroom, cabins and the toilets and showers.'[109]

The only option was to dive *Turbulent* and get to below 60 metres where the temperature would drop by up to 10 degrees Celsius and even more the deeper the submarine went. Once deep, *Turbulent*'s crew could cool the boat and machinery and recover the crew. But with the foreplanes and afterplanes [used for steering, diving and surfacing the submarine] out of action, diving was next to impossible. 'It was touch and go before we dived as to what might happen to us and the submarine,' recalled Ramsey.[110] Eventually *Turbulent*'s crew managed to extend the foreplanes and the submarine dived. There was relief all round as the crew slowly changed all the water in various tanks and started to bring the air-conditioning plants online. The first started, then the second, followed by the third, but *Turbulent*'s crew was still unable to turn on any equipment which would run the risk of overloading them. Eventually a small amount of equipment was turned on, but when *Turbulent* returned to periscope depth for the first time the air-conditioning plants shut down again because of the rapid rise in water temperature. The boat dived again for the relative cool of the depths and after forty-eight hours *Turbulent*'s crew succeeded in bringing the submarine back to full operational state and continued with the operation. 'There's not a day that goes by that I do not think about what happened,' said Ramsey. 'The pain of seeing my crew like that. But when I think back to that time I quickly remember how fantastic they all were in dealing with the situation. We recovered from it. They did exactly what they had to do, and looked after the team.'[111]

OVERSTRETCH?

As a result of Labour's Strategic Defence Review in 1998, HMS *Splendid* was prematurely decommissioned in 2003, followed in 2006 by HMS *Spartan*. This reduced the total number of SSNs from twelve to ten.[112] In July 2004, a supplement to the 2003 Defence White Paper – entitled *Delivering Security in a Changing World* – announced that 'in the light of the reduced threat . . . an attack submarine fleet of 8 SSNs will be sufficient to meet the full range of tasks'.[113] HMS *Sovereign* was decommissioned in September 2006; followed by HMS *Superb* in September 2008, after it hit an underwater pinnacle in the Red Sea, eighty miles south of the Suez Canal. HMS *Sceptre*, the last of the 'Swiftsure' class was decommissioned in December 2010, reducing the number of in-service SSNs to just seven. Numbers were meant to remain at this level, but due to further delays with the 'Astute' programme, the size of the SSN fleet has continued to fall. HMS *Trafalgar* was decommissioned in December 2009. In 2011, the Submarine Service was required to maintain four SSNs at high readiness, known as R4+: able to deploy to sea within twenty days. This became increasingly difficult following the July 2012 decision to decommission HMS *Turbulent*, leaving only five SSNs in service with the Royal Navy. As of early 2014 two were usually in various stages of refit and maintenance, leaving only two or three available for operations, one of which is always East of Suez, while the other is preparing to relieve it. The remaining submarine, if not undergoing routine maintenance, or taking part in exercises and training such as the Perisher course, is expected to be available for other tasks. It is nearly always designated the Immediate Readiness SSN, ready to depart anywhere in the world at short notice should it be required.

With so few submarines, the Submarine Service is finding it increasingly difficult to meet a range of other commitments, acting as the host submarine for the Perisher course, providing the rest of the surface fleet with ASW training, participating in the large annual multinational Joint Warrior naval exercise, supporting NATO and taking part in joint operations, such as Operation 'Active Endeavour' in the Mediterranean (reporting on shipping movements) and Exercise 'Cougar', the annual exercise of the Royal Navy's Response Force Task Group.[114] The smaller the Submarine Service gets, the less resilient it becomes. If the unexpected occurs, the service will have little choice but to scale back or temporarily abandon some of these commitments.

The shortage of SSNs has been reflected in ever-longer patrols in the Mediterranean and East of Suez. In July 2012, when HMS Triumph returned from operations off Libya, the submarine had been on deployment for the best part of a year: 323 days on operations, with 257 days at sea, 93 of them on 'silent patrol'.[115] A year later, on 22 May 2013, Irvine Lindsay's HMS Trenchant returned home to Devonport – eleven months to the day after the start of what became the longest patrol ever completed by one of the Royal Navy's nuclear submarines. In total Trenchant spent 335 days away from the United Kingdom – 267 of them East of Suez as the UK's high-readiness front-line asset on a deployment that covered 38,800 nautical miles, the equivalent of one and three quarter times around the world. Trenchant spent over 4700 hours underwater – the equivalent of six and a half months. Of Trenchant's 170 crew (of whom 130 were always at sea), seven were on 'Black Watch' – on board for the entire deployment.

Nuclear submarines are some of the most complex machines man has ever built, and each has its own foibles. As the 'Trafalgar' class age, it is becoming more of a challenge to ensure they are safe to operate. In 2011 the Royal Navy was getting about 55 per cent availability out of each submarine and though the boats were originally designed for twenty-five years they are now trying to operate them for over thirty years after their construction while trying to get more out of them.[116] In 2013, the MOD's internal nuclear-safety watchdog, the Defence Nuclear Safety Regulator (DNSR), issued its annual report and highlighted the problems of extending the operational lives of the 'Trafalgar' class. It underlined how the five remaining 'Trafalgar' class submarines, launched between 1984 and 1991, are now expected to operate for up to thirty-three years, with at least one not scheduled for decommissioning until 2022. 'As a result,' the report concluded, 'the Trafalgar class are operating at the right hand end of their "bathtub" reliability curves.' The number of problems being encountered by the boats is increasing steadily and the 'effect has been seen in a number of emergent technical issues of the last few years' which 'can be directly attributed to the effects of plant ageing'.[117] In March 2013, HMS *Tireless* was forced to return to the UK with a minor leak to her nuclear reactor. This left just one SSN in service, a position that the retired Admiral Woodward called 'very worrying'.[118]

The Submarine Service's commitments continue despite their reduced numbers. The Royal Navy still aims to send a nuclear submarine into the South Atlantic twice every year, to remind Argentina that the UK remains committed to defending the Falkland Islands. Details of deployments are almost never made known. But in May 2012 HMS *Talent*, armed with Tomahawk missiles, arrived in Simonstown Dock in South Africa, a strategic staging post for missions to the South Atlantic, and the remarkably well-informed *Sun* newspaper quoted an unnamed source as saying 'HMS *Talent* will be dropping by the Falklands and keeping watch. That's what she is built to do – protect Britain's interests . . . and that is what she will be doing this summer.'[119] *Talent* arrived in the area in good time for 14 June 2012 – the anniversary of the day the British Task Force ended the 74-day Argentine occupation of the islands. South Atlantic deployments often involve a dash of careful attention-seeking behaviour.

The Submarine Service also continues to go north, to ensure that it retains the skills required to operate under the Arctic ice. At the time of writing the last Royal Navy SSN to operate in the Arctic was HMS *Tireless* in March 2007, under the command of Ian Breckenridge. The exercise was conducted as part of the Joint US Navy/Royal Navy Ice Exercise 2007 (ICEX-2007). *Tireless* was under the ice pack, 170 miles north of Deadhorse, in Prudhoe Bay Alaska when an explosion ripped through the Forward Escape Compartment killing two submariners, Anthony Huntrod, an Operator Mechanic, and Paul McCann, a Leading Mechanic Operator. A third submariner, Richard Holleworth, a navy stores accountant, was also injured. 'I was about a metre away when I heard a really loud bang,' Holleworth told the subsequent inquest. 'The room was instantly filled with bright white smoke. I could not see my arm in front of my face, just a glow.'[120] In *Tireless*'s control room Breckenridge was confronted with a rapidly deteriorating scenario as thick smoke spread through all forward compartments. A manual flood alarm, indicating that flooding was present, was also triggered. In line with the Navy's standard Emergency Operating Procedure (EOP), Breckenridge operated what is known as the 'Battleshort', overriding *Tireless*'s reactor safety systems in order to ensure that the submarine was not further 'jeopardized by a spurious automatic reactor shutdown'.[121]

As Breckenridge searched for a suitable ice polynya through which *Tireless* could surface, the crew quickly determined that the flood alert was a false alarm which had probably been triggered by the explosion. In the Forward Escape Compartment Holleworth located Huntrod and attempted to kick open the doors to the escape hatch. 'I thought I would just grab Tony and carry him with me,' said Holleworth. 'But it was futile. I had a moment of clarity that I was trapped in there. I knew I was going to die.'[122] He staggered to an oxygen relay point and pulled on a mask. 'All I remember is slumping to the floor. I accept that I must have just passed out.'[123] He was found forty minutes later when the ship's company breached the Escape Compartment. If it had not been for the stamina and presence of mind of *Tireless*'s crew, particularly Holleworth who used 'all available means to extinguish' the fires, the consequences of the incident could have been much worse. *Tireless* later punched a hole though the ice pack and surfaced and Holleworth was airlifted to Anchorage in Alaska by an American helicopter and a transport aircraft. The Board of Inquiry into the incident determined that the explosion was caused when a self-contained oxygen generator, known as an SCOG, used to supply extra oxygen to the submarine exploded. Although the inquiry was unable to determine beyond doubt why one of the canisters exploded, it concluded that the most likely cause was that oil had seeped into and contaminated the oxygen generator. The Ministry of Defence accepted responsibility for the incident and a series of recommendations relating to damage control, firefighting and equipment were quickly implemented and a new design of oxygen generator issued to the fleet.

Since *Tireless*'s operation in the Arctic the Navy has been unable to send another submarine up to the region. There is, however, still a need to maintain an intimate knowledge of the area and the skills required to operate in it. The Arctic is likely to become even more important than it was during the Cold War. Climate change may melt the Arctic's ice cap to such an extent that in this century it will become a navigable ocean for commercial shipping, as well as mineral and oil exploration. Territorial disputes are already taking place. Russia has already made extensive claims on the region, based on its continental shelf, way beyond the usual twelve-mile limit from the coastline. The US Navy's most recent exercise took place in March 2014 when the USS *New Mexico* fired a simulated torpedo against a simulated submarine in order to maintain its Arctic submarine skills. 'In our lifetime, what was [in effect] land and prohibitive to navigate or explore, is becoming an ocean, and we'd better understand it,' noted Admiral Jonathan Greenert, the US Navy's Chief of Naval Operations, who was in the Arctic to observe the exercise. 'We need to be sure that our sensors, weapons and people are proficient in this part of the world' so that the US Navy can 'own the undersea domain and get anywhere there'.[124] A Royal Navy officer was assigned to the US ICEX and the Navy still aspires to send a 'Trafalgar' and 'Astute' class SSN back under the ice at some point in the future.[125]

The Submarine Service also continues to keep a close eye on Russia. (In the office of a senior British intelligence figure with connections to the submarine world there stands a bronze statue of Lenin some 3 feet high.) Whether or not it continues to collect intelligence about Russian submarines, is an open question. Every living CO who has operated against the Russians would emphasize that collecting intelligence about Russian submarines was – and remains – Champions League play for the Queen's submariners. It is what gives the final sharpening to their dark-blue competitive edge. That edge depends on a multiplicity of moving parts – human, technical and industrial – all working as they should do all the time. It is expensive, precarious and, especially in stretching circumstances, fragile. Can – should – the UK devote the substantial and top-of-the-range human, financial and technical resources to Submarine Britain that its sustenance demands? What might the future hold for the patrollers of the silent deep?

And the Russians Came Too:
Today and the Future

The UK has always been a reluctant nuclear power
Sir Kevin Tebbit, Permanent Secretary to the Ministry of
Defence, 1998–2005, speaking in 2009.[1]

THE 'ASTUTE' CLASS

The future of the Royal Navy Submarine Service depends on a new
class of SSN and a new class of SSBN. By the early 1990s there had
been little progress in the programme to replace the 'Trafalgar' class.
After a lengthy period of abeyance, as a result of work on the 'Vanguard' class, a Staff Target was issued in March 1986 and Full Feasibility
Studies were carried out by VSEL between 1986 and 1989.[2] The Staff
Target was not translated into a Staff Requirement until August 1989,
and the Programmed Acceptance Date of the Follow On SSN slipped
by eighteen months to the year 2000. By 1990 the extent to which the
new class would be an improved or 'stretched' Trafalgar rather than a
genuinely new design had yet to be decided and several possibilities
were under discussion within the Ministry of Defence.[3] The Naval
Staff eventually revisited the requirement for the Follow On SSN and
renamed the entire programme 'Batch II Trafalgar'. The aim was to
design a submarine with similar capabilities to the 'Trafalgar' class, but
rather than developing new technology, as was intended with SSN0Z,
'Batch II Trafalgar' would incorporate the best of existing technology
already in use in the 'Trafalgar', 'Vanguard' and 'Upholder' submarines that had been built since the 1980s. The submarine would be
powered by the PWR2 reactor used in the 'Vanguard' class, as well as
carrying the new combat system for the 'Swiftsure' and 'Trafalgar'
classes and a number of systems from the 'Upholder' class.[4]

From the outset there were problems. The MOD felt that the owners of the Barrow shipyard, VSEL (Vickers), had made excessive profits from the 'Vanguard' class and a consensus developed in the MOD that competition would lead to lower acquisition costs. At the same time high overheads also contributed to the belief that the roles traditionally performed by the MOD could be carried out by industry more efficiently and effectively. As a result, design authority, traditionally the responsibility of the MOD's in-house naval constructors, the Royal Corps of Naval Constructors, was transferred to the submarine design contractor.[5] There was no planning for the transfer of responsibilities from the MOD to the prime contractor, management of which in any case changed when British Aerospace bought GEC-Marconi in November 1999 and created BAE Systems. There was a lack of cooperation and coordination between the prime contract office and the MOD and high turnover of personnel, especially at management level at VSEL, GEC-Marconi and then BAE Systems, led to numerous changes in leadership there and at the shipyard.[6] The MOD also lost its 'ability to be an informed and intelligent customer'.[7] During the construction of the 'Vanguard' class, fifty MOD staff had been based at the shipyard, but during the early stages of the 'Astute' programme the MOD's on-site presence was just four people. The shipyard also struggled to meet the requirements of the programme. By the time work on the 'Astute' class began in the late 1990s, it had been almost twenty years since the 'Vanguard' class was designed by the MOD and years since the last 'Vanguard' class, HMS *Vengeance*, was delivered. Employment at the shipyard at Barrow had dropped from over 13,000 during the 'Vanguard' and 'Trafalgar' builds to just 3000, with consequent loss of experience and expertise.

While most of the workers who departed were craftsmen, many highly trained engineers and constructors skilled in submarine design and construction also left. The shipyard tried to maintain a set of core shipbuilding skills by designing and building surface ships such as HMS *Ocean*, HMS *Albion* and HMS *Bulwark*, and a civilian oil tanker, but the skills involved in constructing relatively simple surface ships were very different from those required by nuclear submarines. When the 'Astute' contract was signed in 1997, Barrow was forced to split its resources to meet its existing commitments, meaning only a third of the available designers were employed on the submarine programme.[8] Both the MOD and GEC-Marconi underestimated the impact of the long gap between the design of the 'Vanguard' class and the start of 'Astute'.

Although the 'Vanguard' class was first conceived as a modest upgrade to the 'Trafalgar' design, the PWR2 it used demanded a wider and longer hull and the requirement for lower radiated noise signatures led to a more complex design. At 97 metres in length, with a beam of 11.3 metres and a draught of 10 metres, the 'Astute' design came in at 7000 tonnes surfaced, 7400 tonnes dived. The Navy fought long and hard to ensure the new design was large enough to carry sufficient weapons and reloads. The increase in capability is substantial. The 'Astute' class is fitted with six 21-inch torpedo tubes, one more than the 'Swiftsure' and 'Trafalgar' classes. More importantly, the new boats are capable of carrying up to thirty-eight weapons, Tomahawk Land Attack Missiles or Spearfish torpedoes, a 50 per cent increase compared to previous SSNs.

Although the keel of the first of class, HMS *Astute*, was laid in January 2001, the first years of the construction programme were beset with problems. One of the biggest was the introduction of 3D Computer-Aided Design (CAD) software. Although the designers at Barrow had used 3D CAD software for the design of some surface ships, these designs did not compare to the complexity of a densely packed nuclear submarine. Previous submarine designs had been produced in 2D on paper by hundreds of draughtsmen; wooden mock-ups were then used to understand the layout of the submarine and the access routes for pipes and cables. The CAD software required extensive modification before it could be used to design a submarine, which was difficult because of the shortage of UK designers with 3D CAD experience.[9] As a result of the delay, physical construction of HMS *Astute* began with very few complete drawings and those that were produced by the CAD process were to a level of detail with which the shipyard workers were completely unfamiliar. In previous construction programmes they had worked from drawings that contained little detail; they had then used their years of submarine construction practice and experience to decide how to build what was represented on the drawings. There were therefore further delays in the programme as the 'Astute' class workforce learned to work with the new drawings.[10] By August 2002 the 'Astute' programme was more than three years late and several hundred million pounds over budget.

These delays and cost overruns resulted in a revised contract in December 2003, with the MOD abandoning the 'eyes on, hands off' approach it had adopted in managing the requirements of the programme and asking the US Navy's primary submarine builder, General

Dynamics Electric Boat, for assistance. As a result over 200 experienced designers and managers from Electric Boat began to help BAE Systems with the design effort. A secure link was established between Barrow and Groton, Connecticut, to pass the detailed drawings necessary for construction. Electric Boat also helped Barrow to emulate the modular construction techniques that it had used for the US Navy's 'Ohio' and 'Virginia' class submarines, outfitting submarine rings using a vertical method rather than the traditional horizontal process used in all previous Royal Navy submarine construction, and assisted BAE in tracking programme progress; a senior Electric Boat employee was eventually based in Barrow and assigned to BAE Systems as the 'Astute' Project Director, with responsibility for all aspects of delivery.[11] The approved in-service date for HMS Astute was 2005, but this was moved back from 2005 to 2008.

As Barrow was incorporating these US techniques and relearning how to build a nuclear submarine, the construction of HMS Astute proceeded in a somewhat uncoordinated fashion. Testing and commissioning caused more delays as it had been ten years since anyone at Barrow had tested and commissioned a nuclear submarine.[12] HMS Astute was finally launched in June 2007 and left the Barrow shipyard to start operational trials in November 2009. During the sea trials, three kinds of problems were identified: flaws in design that only became apparent once testing began; equipment with poor reliability; and problems relating to construction. 'In the programme of testing over three years we have identified issues in all of those categories. And got on and fixed them,' wrote the then Director Submarines, Rear Admiral Simon Lister. 'Is this normal? Where is this on the spectrum of scandalous waste of taxpayers' money? Is this what we could expect? Is this the normal endeavour of dragging any ship out of the dockyard? You will have to make your own mind up.'[13] However, the delays to the 'Astute' programme must be set against the context of previous submarine-building programmes outlined earlier in this book. 'Point me to any submarine building yard that produces a first of class and I will show you a process that is extraordinarily challenging,' said Lister. 'The level of challenge in Astute I don't think has been any more than in the level of challenge in the first of class in other submarines.'[14]

Shortly after commissioning into the Royal Navy, on 22 October 2010 Astute was marooned on top of a sand bank off the Isle of Skye while conducting sea trials, very close to the Skye Bridge, allowing dramatic filming and photography by the media. Early attempts to free the submarine failed and *Astute* remained aground until shortly before the next high water. During the recovery there was a minor collision with the towing vessel *Anglian Prince*, which damaged *Astute*'s starboard foreplane. According to the official report of the Service Inquiry the causes of the grounding were 'non adherence to correct procedures for the planning and execution of the navigation combined with a significant lack of appreciation by the OOW [officer of the watch] of the proximity of danger'. A number of other additional factors also contributed to the incident, including some deficiencies with equipment.[15] The Panel found the failures were specific to HMS *Astute* and were 'not indicative of wider failings within the Submarine Service as a whole'.[16] As a result, *Astute*'s then CO, Commander Andy Coles, was relieved of command.

The following year, in April 2011, *Astute*'s new CO, Commander Ian Breckenridge, had docked his submarine in Southampton on a goodwill visit when one of her crew, 23-year-old Able Seaman Ryan Donovan, angry over losing the chance of a deployment on a surface ship after disobeying an order to help clean the submarine, opened fire with an SA80 rifle with which he had been issued to stand guard on the submarine's casing, and shot dead *Astute*'s Weapons Engineering Officer, Lieutenant Commander Ian Molyneux, and badly injured a Second Officer, Lieutenant Commander Christopher Hodge. Donovan's 'murderous onslaught' was only stopped when two civilian dignitaries, the leader of Southampton city council, Royston Smith, and the chief executive, Alistair Neill, leapt on him in the submarine's Control Room. Donovan admitted murdering Molyneux and attempting to murder Hodge, as well as two other men who escaped unhurt, and was given a life sentence.[17] Breckenridge is very stoical about the press portrayal of *Astute* as a jinxed boat. He knows it will take a long while to get its reputation back and that the press will seize upon all the inevitable teething troubles. He said in an interview with a US magazine:

> The media will always jump at an easy target. So we're an easy target, they're having a field day ... ask any of my ship's company, any of our supporting personnel, anyone in the Navy, any of our families, any of our friends in the U.S. submarine force; in fact, anybody that basically understands what a first-of-class submarine's got to go through, and they'll recognise that the term 'jinx' is just a very easy one that the media jumps on. Does it bother me? Used to, initially. Not bothered by it anymore. There are much more important things to deal with. It's a fantastic

submarine to deliver to the front line as soon as possible. And I've got a really good team helping me do that.[18]

As *Astute* continued on her sea trials, reports of further problems reached the press. On 16 November 2012, an article appeared in the *Guardian*: 'Slow, leaky, rusty: Britain's £10bn submarine beset by design flaws.' The article went on to list a number of alleged problems that had affected the boat, including: 'flooding during a routine dive that led to Astute performing an emergency surfacing'; 'corrosion even though the boat is essentially new'; 'The replacement or moving of computer circuit boards because they did not meet safety standards'; 'Concern over the instruments monitoring the nuclear reactor because the wrong type of lead was used'; 'Questions being raised about the quality and installation of other pieces of equipment': and 'Concern reported among some crew members about the Astute's pioneering periscope, that does not allow officers to look at the surface "live".'[19] The article also alleged that the 'Astute' class was 'doomed from the start' and that 'flawed thinking and design' had led to a submarine that lacked in both speed and capability. One 'insider' criticized *Astute*'s performance: '"The PWR2 was shoehorned into the Astute, and it meant the submarine's initial designs had to be changed," said a source. "That is why the Astute has a slightly bulbous look about it, not the clean lines that you might expect. The reactor was never meant for an attack submarine and it is supplying power to machinery whose designs have not greatly changed for 50 years. In very simple terms, it is like hooking up a V8 engine to a Morris Minor gearbox."'[20]

The Navy did its best to counter the claims. The day after the later article in the *Guardian*, Rear Admiral Lister wrote to the newspaper:

As the Royal Navy officer responsible for the delivery of the 'Astute' submarine programme, I must respond to your claims about the performance and potential safety of HMS *Astute* (Report, 16 November). All those involved in the delivery of our submarines have a duty to the submariners that serve on them to ensure that we provide a safe environment in which to live and work. As a submariner myself, I am acutely aware of the need to meet the exacting safety standards we demand and we are committed to meeting them both for HMS *Astute* and for the remaining submarines in the class.

I would never allow an unsafe platform to proceed to sea and the purpose of the extensive sea trials HMS *Astute* is undertaking is to test the submarine in a progressive manner, proving that the design is safe, that it has been manufactured correctly and that she is able to operate safely and effectively. This process reflects the nature of HMS *Astute* as both a prototype and an operational vessel. We have always known that it would be necessary to identify and rectify problems during sea trials and this is what we have done. All the issues noted in the story have either already been addressed or are being addressed. In particular, while we do not comment on nuclear propulsion issues, or the speed of our submarines, I can assure you that, once HMS *Astute* deploys operationally, we do not expect there to be any constraints on her ability to carry out her full combat role for the Royal Navy.

I invite the Guardian to spend time on HMS *Astute* with me to see at first hand the professionalism of the crew, the confidence they have in their boat and the rigour with which sea trials are carried out and problems addressed.

Rear Admiral S R Lister
Director submarines, Ministry of Defence.[21]

The *Guardian* did not take up Rear Admiral Lister's offer. Ian Breckenridge also defended his boat:

Astute is the most capable submarine I have served in during my 25 years in service. She is a step forward from her predecessors but her new design and first of class status mean that during sea trials she must prove not only the build but also the design – hence sea trials take time. I spent almost 200 days at sea in *Astute* and no submariner would ever think of going to sea in anything other than a safe submarine.[22]

There is little in the claims in the *Guardian*. As we have already seen, the decision to use the PWR2 was taken on affordability grounds, and whereas speed was a paramount requirement in Soviet submarines, it has never been so in Royal Navy submarines. '*Astute* isn't going to be a greyhound of the ocean,' admits Admiral Lister. 'But she's about half as quiet again as a "T" boat.'[23] Regardless of *Astute*'s maximum speed, what is lost in speed is gained in quietness. The reactor has indeed contributed to *Astute*'s unusual design, but the bulbous hull form aft to the fin can be explained by an additional requirement that was included in the 'Astute' class late on in the design phase. Each submarine is capable of carrying an underwater dock from which Special Boat Service Royal Marine commandos can launch midget submarines to carry out covert missions. Fitted behind the submarine's fin, the Special Forces Payload

Bay (or, as it is referred to by submariners, 'the Caravan of Death')
allows swimmers to enter and leave the submarine while it is submerged. It was developed under a secret programme known as Project
Chalfont.[24] 'That was a real design driver for the boat, and that's why
we've got a big sail,' explains Breckenridge. 'The shapes and curves [of
the hull] help the dry deck shelter sit in the right place.'[25]

At 0900 on Monday, 15 August 2011, on a clear, fresh Faslane
morning, we are piped aboard HMS *Astute* and formally welcomed on
board by Ian Breckenridge who takes us into his cabin and talks about
the perils of being the first boat (among other things, lots of training
and plenty of visitors): 'People who don't know the boat think we're
just a super 'T' boat. We're not. We're way beyond that. It's a quantum
leap in automation. We can press buttons where you needed three or
four sailors before. The optronics are exceptional. We have six torpedo
tubes instead of five. I can hold fifty per cent more weapons than a
'T' boat ... The communications are way beyond a 'T' boat. I have
three different satellite receivers.'[26] We are taken on a tour that begins
above *Astute*'s PWR2 nuclear reactor. We climb into an airlock, the
heavy doors close, the compartment is pressurized and the second
door into the compartment above the reactor, known as 'the tunnel',
slowly opens. We climb in and are instantly hit with the strange
smell ... and feeling. Below us is a nuclear reactor capable of powering
a small city for many years. *Astute* is powered by the latest PWR2 design,
known as Core H, the culmination of continuous improvement applied
to the previous five generations of PWR1 and PWR2 plant design.
Core H represents a ten-fold improvement in core life over the first US
and UK core designs. It also has a long life expectancy: the Royal
Navy does not expect to have to refuel the submarine throughout its
service life. One of *Astute*'s crew lifts the hatch and we see the metal
containers above the reactor which hold the fuel rods. The crew are
shielded from the reactor. Two diesel engines and one big battery provide back up.

Standing in the tunnel, above *Astute*'s reacting heart, we talk about
reputations. 'It's too early to tell', says Owen Rimmer, one of *Astute*'s
watch leaders. 'We are still finding our feet. It's a matter of using our
new capabilities and trusting them. The planes no longer work manually. People were sceptical at first but now they are getting used to it.'[27]
According to Simon Lister, 'the character of a boat is defined by its
history and the crew, whether it is well looked after and clean. What
it's like when it's operational is an accident of history. The "R" boats

from Cammell Laird were less well put together – that's the reputation. It shows over time.'[28]

We leave the tunnel through yet another airlock – the forward and aft ends of the submarine are separated, only accessible through the tunnel – and pass the Health Physics Laboratory where they check the radiation levels from the reactor and monitor for contamination. The Manoeuvring Room, essentially the engineering centre, is separated from Health Physics by a cramped central walkway. The room is slightly larger than the Engineering room on a 'Trafalgar' class, but it looks similar. There are a huge number of buttons flashing, digital readouts, touch screen computers. It looks much like a small version of the starship *Enterprise*. There is a large chair in the middle. The controls are always manned, in three- to four-hour shifts. It is from here that *Astute*'s engineers monitor and maintain the submarine's reactor.

One of *Astute*'s Engineers, Lieutenant Commander Rob Tantam, says that to get the 'Astute' class qualification you have to know the whole boat. 'That's what sets us apart from the General Service. Everyone has to be able to react, to tell what's wrong and know where the valves are.' Tantam has developed a sixth sense. 'I know what feels right. As a submariner and an engineer that's what you get. You subconsciously build a picture of what is normal.'[29] 'If you're an engineer being on a submarine is the bee's knees,' says Admiral Lister. 'It's the combination of high pressure, nuclear power, explosives, atmosphere. The whole existence of the submarine depends upon your competence. It's the best job if you are a UK engineer.'[30]

We pass one of *Astute*'s escape towers and walk through the bowels of the submarine, past noisy machinery everywhere, greasy and smelly, just like any other ship. We pass *Astute*'s Water Making Plant and enter the Control Room, the eyes, ears and brain of the boat. There is the cool hum of air conditioning and computer screens everywhere but there are also subtle differences from other control rooms. The most obvious is the lack of a periscope: *Astute* is the first Royal Navy submarine to dispense with the periscope and adopt phototronic masts that display optronics. 'The optronics are exceptional,' explains Admiral Lister.[31] 'Much as I loved my old periscopes, this is much better,' says Breckenridge.[32] We move over to the helm. Gone are the traditional helm controls. *Astute* is usually driven by a digital autopilot, but on the left hand side of the helm is a small joystick that controls the submarine via fly-by-wire. We are told that it is very sensitive: 'you don't feel like you are moving a 7000 ton submarine. On the older submarines you really had to exert yourself by pulling back on the helm. We are still getting a feel for it.' We ask about the computers, specifically, what would happen if they crashed? 'The computer is designed not to crash,' replies one of *Astute*'s officers. They laugh when we ask if such a thing is even possible. 'There are safety systems, backups, alternate supplies etc.,' insists the officer. 'Everything important is hard wired so if something does happen we are still in control.'[33] We are struck by how young *Astute*'s crew is. The officer in charge of the helm is in his mid-twenties. *Astute* is his first submarine so he doesn't know any different. He has no experience of how things are done on the Navy's older, Cold War-era submarines. 'It's a real privilege to be part of such a good project,' he says. 'You get a real feeling of pride. We've been through the pain, but it helps that we've all been through it together.'[34]

The best way to get an idea of what life on board HMS *Astute* is like is to sit down with some of her crew. Commander Breckenridge takes us into one of *Astute*'s messes where twelve ratings have gathered to answer some of our questions. Of the twelve, only one has served on a diesel boat. The other eleven are what the Submarine Service calls FLUNOBS – 'F****g Lazy Useless Nuclear Orientated Bastards'. We ask what the differences are coming into a lead boat such as *Astute*. 'It's hard work, very hard work,' answers one rating, 'proving all this new technology makes the whole process much more difficult.' 'It's high profile, on the same level as the new destroyers are,' says another. We ask about the living conditions and are surprised by some of the answers. 'Compared to an "S" boat the accommodation here is much smaller,' replies one rating. 'It feels like we have been forgotten about again,' says another. 'It's as if the Navy think all this new technology is much more valuable than the crew. I don't think the Navy even has enough weapons to fill all the space that is down in the Bomb Shop.' 'The Galley is better,' admits another, 'and everyone gets their own rack, but the bunks are smaller than on other submarines. You can't even read a paperback in your bunk properly. Anyone over six feet is going to stick out. The same is true if you're wide.'[35] A price has been paid in habitability because of the method of laying out the hull in rectangular spaces.

We ask about the public's view of *Astute* since the grounding and the shooting. 'We've been keeping busy,' replies one of the ratings. 'It's been non-stop, we haven't really had a chance to think about it, we've just had our heads down, we just want to crack on and get over it.' 'We tend to ignore the press,' answers another. 'It's the ship's company that show the boat in the best possible light.' 'The bad press really started with the fire in Barrow,' says another rating referring to the minor fire that broke out in *Astute*'s conning tower at Barrow in 2009. 'Since then every little snippet has been commented upon. People outside, who have no idea what went on or what is going on, will tell other people. It's not us saying all the negative things. The lads will defend the unit if they are out and . . . there have been some occasions when they have.' 'When things were going well we had forty-six days at sea, we were really pleased,' says one rating, 'then another incident completely beyond our control happened and we went right back to how it was in the beginning.' Another interrupts. 'We are having exactly the same problems as everyone else, that is what sea trials are for, they iron out all the problems, but we have such intense focus on us that our problems get noticed more.' 'We are a happy ship's company,' they insist. But they are clearly frustrated about the public perception of the submarine. 'The general public is ignorant about what is going on, but sometimes our biggest enemies are in the Navy. They don't know about the boat and some spread rumours. Then you have the so-called experts who pop up on TV when some incident happens, constantly spreading rubbish when they have no clue what they are talking about.'[36] Admiral Lister is standing in the background, listening. 'You just need to get on and put her into service,' he says. 'My sense is that once that happens she will be well known for the right reasons.'[37]

When we ask about the Submarine Service and its place in the rest of the Navy a sense of distinction emerges. 'We are a much more professional service than the Skimmers' (the submariners' name for the surface crews), says one rating. Another interrupts. 'It's more of a multitasking role,' he says. 'Our skills are spread across the whole submarine whereas Skimmers only have one task to do, one specialty, one piece of equipment to manage. Our equipment is the boat.' Another rating explains another key difference. 'On a ship, at the end of the working day you down tools and go to the mess. We don't do that. We do six hours on and six hours off. It's better that way.' 'They think serving on submarines is dangerous,' explains another. 'They think we smell, don't clean and they underestimate how much we have to do. They never seem to ask themselves why do we get retention pay?'[38] Submariners receive extra pay in recognition of their skills and the challenges of the job. 'There isn't much of a relationship with the Skimmers,' admits another rating. 'They can't afford to drink in the same bars as us.' It's not just the Skimmers. The Royal Marines also have a degree of friction with the Submarine Service. 'They are crazy,' insists one Rating and the others nod in agreement. 'We stay away from them when we go on shore. The best Marine is a submarine.' What about the other armed services? 'The Army think we are great,' they insist. 'They love the technological aspect. It's the same with the RAF, they understand the technical bits.' We are very impressed with the crew and ask if they are given any special media training. 'Yeah,' replies one rating on behalf of the others. 'We're told to shut up!'[39]

A few weeks after our tour, Ian Breckenridge took *Astute* to the United States for an extensive series of trials. *Astute* loaded Tomahawk cruise missiles at the US Navy's submarine base in Kings Bay, Georgia, and successfully launched two from the Gulf of Mexico onto a missile range at Eglin Air Force Base in Florida. 'Both flew beautifully,' said Breckenridge.[40] The submarine then made its way to a joint US–UK facility known as AUTEC: a bigger and better-instrumented range than the British Underwater Test and Evaluation Centre (BUTEC) range located off Applecross, which the Royal Navy uses for smaller post-refit tests. The relationship with the US is still extraordinarily close. There is 'very little we don't share with them', says Admiral Lister, and with the increasing US focus on the Pacific the relationship has become closer. Admiral Donald [Director, Naval Reactors, US Navy], the modern day Admiral Rickover, reportedly said: 'We don't want the Brits to drop out of the game . . . we want the Brits to do more.'[41]

While in the United States, *Astute* was put up against the most advanced submarine in the US Navy, the 'Virginia' class attack boat USS *New Mexico*. So important was the exercise, codenamed Exercise 'Fellowship 2012', that both the US Chief of Naval Operations, Admiral Jonathan Greenert, and First Sea Lord, Admiral Sir Mark Stanhope, both submariners who had reached the highest rank in their navies, were on board. 'With the Royal Navy getting together with the US Navy you have the best out there operating together with cutting edge technology and cutting edge tactics, and that's important for each of us to see what we have and see what we're made of,' said Admiral Greenert.[42] Both Admirals observed the crews of the two most advanced submarines in the US Navy and the Royal Navy spend two days in the Bahamas, trying to out-flank, out-manoeuvre and outwit each other.[43] *Astute* performed exceptionally well. 'We met and surpassed every expectation,' says Breckenridge. 'She is just better than any other submarine I have ever been in.' Even though *Astute* was not fully operational at the time of the trial, the Americans were very impressed with its capabilities. 'Our sonar is fantastic and I have never before experienced holding a submarine at the range we were holding USS *New Mexico*,' says Breckenridge. 'The Americans were utterly taken aback, blown away with what they were seeing.'[44]

HMS *Astute* was commissioned into the Royal Navy on 27 August 2010. HMS *Ambush* was commissioned on 1 March 2013. HMS *Artful* was launched on 17 May 2014. At the time of writing the fourth, fifth and sixth submarines, HMS *Audacious,* HMS *Anson* and HMS *Agamemnon* were under construction at Barrow. When *Audacious* is launched assembly will begin of the final submarine HMS *Ajax*. These seven submarines will keep the Royal Navy in the SSN business well into the 2050s. As of 2015, the seventh 'Astute' submarine was planned to enter service in 2024, two years after the last 'Trafalgar' class submarine is decommissioned. This will leave the Royal Navy with a shortfall of at least one submarine when measured against projected tasks. A 2011 National Audit Office report concluded, 'The Astute class submarines will not meet the Royal Navy's requirement for sufficient numbers of submarines to be available for operations over part of the next decade', and as a result of the delays the Submarine Service 'will have to use older boats beyond their out-of-service dates, work the smaller fleet of Astute submarines harder, or reduce scheduled activity for submarines'.[45]

CURRENT AND FUTURE THREATS

Why does the UK still need a fleet of SSNs? Nearly thirty years after the Joint Intelligence Committee's paper on 'Soviet Naval Policy' circulated in July 1985, several of the submarines identified and analysed in what turned out to be the final phase of the Cold War (the covert intelligence contest has never ceased), would still feature in any assessment of underwater Russia. As we write, a 'Delta IV' ballistic-missile boat of the Northern Fleet is almost certainly in one of the bastions where the Arctic ice and the floes girding its rim creak and groan making sonar detection especially difficult. There may be even one last Typhoon, the ageing matinee idol of *The Hunt for Red October*, staggering about somewhere under the Barents Sea. The Akulas and the Sierras, as we have seen, still come down from 'round the corner' and far into the Atlantic to test the undersea ears of NATO and to add an extra frisson to US and UK test launches of Trident D5 missiles off Florida as part of the two respective navies' DASO programmes. Though there was a break in continuous at-sea patrolling by the Russians for much of the twenty years after the Cold War, all the indications are that Putin's Russia has an ambitious programme under way to modernize and improve all three elements of its nuclear 'triad' on land, under the sea and in the air.

Since 2009, Russia has gradually stepped up the patrolling (punctuated by occasional moments of attention-seeking behaviour such as an Akula II surfacing off the US eastern seaboard in that same year) to remind the West that, despite the recent collapse in oil and gas revenues, the newly assertive Russian state can afford to indulge old habits by teasing and stretching its old adversaries. In August 2010, Royal Navy submarines were 'experiencing the highest number of "contacts" with Russian submarines since 1987'.[46] One of Russia's most advanced 'Akula' submarines was detected off the west of Scotland, waiting for a 'Vanguard' class SSBN to depart on its three-month patrol. In an operation reminiscent of the Cold War and the mid-1990s, a 'Trafalgar' class submarine was deployed to chase the Russian submarine out. 'The Russians have been playing games with us, the Americans and French in the North Atlantic,' a senior Navy commander said. 'We have put a lot of resources into protecting Trident because we cannot afford by any stretch to let the Russians learn the acoustic profile of one of our "bombers" as that would compromise the deterrent.'[47] 'We're still in the Cold War,' says one senior submariner. 'The reason we have that Immediate Readiness Submarine is because when the Russians come out we go and find them.'[48] In December 2012, a Royal Navy submariner, Petty Officer Edward Devenney, was jailed for eight years for breaching the Official Secrets Act after being caught in an elaborate MI5 sting operation attempting to sell to Russian intelligence highly sensitive details of submarine movements and operations, as well as highly classified code information that could have caused 'substantial damage to the security of the UK'.[49]

The Coalition Government's 2010 decision to cancel the Nimrod MRA4 Maritime Patrol Aircraft programme on cost grounds has left a serious capability gap in respect of the Nimrods' role in ship surveillance, search and rescue and anti-submarine warfare. To mitigate the risks the Navy has been forced to make greater use of frigates and helicopters to protect sea-lanes and prosecute possible submarine contacts.[50] The First Sea Lord, Admiral Sir Mark Stanhope, claimed that 'the risk can be mitigated under the current threat levels we are expecting to envisage' both 'in terms of the delivery of the strategic deterrent as well as in terms of force protection of deployed task groups'. Air Chief Marshal Sir Jock Stirrup, Chief of the Defence Staff, claimed in 2010 that he had 'made it clear in the defence review that if we went ahead with the decision to get rid of maritime patrol aircraft, in the circumstances of a resurgent submarine threat we would not be able to send a naval taskforce to sea unless someone else provided that capability.' 'It was not a case of taking a bit more risk; we simply would not be able to do it, should that particular threat level rematerialize. Nobody is saying that it will or that it won't, but we would have to look for somebody else to provide that capability.'[51]

Since the decision to abandon Nimrod was taken in 2010, relations between Russia and the West have deteriorated. Following the 2014 Ukraine crisis Russian submarine activity has increased significantly. In April 2015, the head of the Russian Navy, Admiral Viktor Chirkov, admitted that 'From January 2014 to March 2015 the intensity of patrols by submarines has risen by almost 50 percent as compared to 2013.'[52] In November 2014, the Government was forced to call in NATO Maritime Patrol Aircraft from France, the United States and Canada after a fishing trawler spotted an 'unknown submarine periscope' off the west of Scotland, not far from Faslane. Two US Orions, a Royal Canadian Air Force Aurora and a French Dassault Atlantique Maritime Patrol Aircraft operating out of RAF Lossiemouth spent early December 2014 searching for the mysterious submarine.[53] A month later, two US aircraft alongside the Type 23 frigate HMS *Somerset* were again involved in a search for a mysterious submarine contact lurking near Faslane. In March 2015, a Scottish fishing trawler was nearly dragged underwater after an unidentified submerged object was caught in its trailing nets. The trawler's Skipper, Angus Macleod, was forced to rev the trawler's engine to keep pace in front of the net for around fifteen minutes, until the boat's propeller finally severed the rope. What this submerged object was remains a mystery. The Royal Navy confirmed that there were no Royal Navy or NATO submarines operating in the area at the time. 'It was not a whale,' said Macleod, 'we have had whales in the nets before and the net is all twisted afterwards. Whatever it was, it was human powered – of that we are convinced.'[54]

If these probings are Russian submarines, the frequency is still far short of the Cold War patterns and are designed not just for the purposes of testing the US, UK and NATO but to play to President Putin's domestic audience, many of whom no doubt found it as difficult as the British previously did to cope with the pangs of geopolitical decline – in the Soviet case, the swiftest collapse ever experienced by a superpower – in 1989–91. The nuclear-triad renewal programme, there-fore, has national pride as its propellant as well as military reach and prowess.

This is why a 2015 assessment of what the Royal Navy Submarine Service might be facing in the coming decades would still begin with Russian capabilities, though it would now go much wider to embrace a number of other navies of varying levels of sophistication. What kind of underwater world might the 'Vanguard' class and their succes-sors and the 'Astute' class be facing with the best that submarine Britain can equip them? It's a world paradoxically of enhanced technical threats since Cold War days. Cyber attacks occur on almost a daily basis; though the UK Government shrinks from admitting it publicly; crucial parts of Britain's defence capabilities and what is known as our 'critical national infrastructure' are under attack by Russia and China or those who act as their instruments. There are some intriguing con-tinuities. The Russians still put great faith in human intelligence: there are almost exactly the same number of Russian intelligence officers in the Kensington embassy and the trade mission in Highgate as there were in the 1980s. And they continue to make great efforts to penetrate Operation 'Relentless' – the maintenance of continuous at-sea deter-rence since 1969 – whose integrity and protection are a permanent priority for UK Defence Intelligence and its sister secret services. This is what a 2015 version of that 1985 assessment might look like.[55]

Russia

Post-Soviet Russia kept a viable submarine capability throughout the lean and stretched twin decades after the fall of the Berlin Wall. Its SSBNs, guarded by a full panoply of air, surface and underwater pro-tectors in their cold bastions, remained the ultimate, last-resort defence of the Russian homeland, a constant and absolute top priority for the occupant of the Kremlin. And if, in the words of the 1985 JIC assess-ment, they are 'deployed to the Arctic waters . . . the ice-cap can give extra protection from Western anti submarine warfare forces' and 'SSBNs can fire their missiles on the surface either where there is open water or by breaking through the ice-cap.'[56]

In 2011, two years after stepping up patrolling, Vladimir Putin declared publicly that all branches of the Russian military would be funded for a 30 per cent modernization by 2015 (equivalent to £89bn) rising to 70 per cent by 2020. Western Intelligence was sceptical that on the naval side this would convert into the number of new hulls promised by 2015 and 2020 as Putin may have been aiming high deliberately.[57]

By 2015, the end of the first surge of new defence spending, the average age of the submarines of the Russian Northern Fleet was twenty-two years. As in the Cold War, Putin's naval construction programme is weighted heavily towards submarine construction. It is now at its highest level since the breakup of the Soviet Union, with three major submarine programmes receiving considerable investment:

- The 'Borey' class SSBNs (Borey is the Russian word for the Greek god of the north wind).
- The 'Yasen' SSNs (Yasen is the Russian for ash tree).
- The 'Lada' conventional submarines (Lada is the name of a Slav pagan goddess who had the lot – youth, beauty, love, harmony and merriment).

The first of the huge 24,000 ton 'Borey' class, the *Yuri Dolgoruki*, is already at sea after a programme that has been beset with delays. Originally conceived in 1982, the development of the 'Borey' class SSBN was frustrated by both funding difficulties and considerable missile development problems. When the first of class was laid down in November 1996, the plan was to use a new strategic missile, the SS-NX-28. That missile was cancelled in 1998 and replaced with a navalized version of the SS-27 Topol-M, known as Bulava 30. The Bulava's development programme has been marred by embarrassing and increasingly public test failures (missiles exploding mid-flight and violently cartwheeling through the night sky in flames). Despite these problems, Bulava entered service in January 2013 (and in September 2013, a Bulava missile failed in flight following a submerged test launch).[58] Overall, there are plans for eight 'Borey' class submarines in all, each carrying sixteen Bulava missiles with up to six warheads apiece. The six Delta IVs of the Northern Fleet have been refitted to keep them going until the 2020s but the three Delta IIIs in Vladivostok are in poor condition, as is Russia's SSBN capability in the Pacific. As a result, the Russians are thought to be giving the construction of the Boreys priority over all other programmes. Three are under construction at Russia's nuclear-submarine naval yard at Sevmash: the *Knyaz Oleg*, the *Vladimir Monomakh* and the *Knyaz Vladimir*.[59]

The first of the 'Yasen' class, the *Severodvinsk*, impressed Western intelligence analysts before, during and after its trials in the White Sea in 2012. It is a dual-purpose missile carrier with its cruise missiles a mix of the nuclear-tipped and conventional anti-ship warheads, and carrying too a substantial complement of torpedoes. The 'Yasen' may represent the pinnacle of Russian SSN design, benefiting not only from all the information from the Walker Spy Ring, but the considerable technological advances that have occurred in the years since the end of the Cold War. The quietness of the Yasen's signature makes it very hard to detect. Its sensors are very advanced but the US Navy's newest 'Virginia' class SSNs and the Royal Navy's new Astutes will still have a technical edge when and if they come up against one. Four are currently under construction at Sevmash: the *Kransnoyarsk*, the *Khabarovsk*, the *Kazan* and the *Novosibirsk*.[60] Three more are planned by 2023, although four in service by 2023 is probably a more likely figure.

The 'Lada' conventional submarines too are very advanced with greatly extended underwater endurance and will pose a stiff test to the US Navy and the Royal Navy. Like the Yasens, the first of the line, *St Petersburg*, and its sisters will represent a step change in Russian conventional submarine capability and seven of them are planned to be in service by 2020. Western intelligence reckons the Russians may go into partnership with the Chinese Navy in building the Ladas.

There are other more puzzling developments. One of the six remaining Delta IVs, K-64, has been converted into an auxiliary submarine to replace a specially converted Delta III to serve as a mother ship to an unusual titanium-hulled mini-submarine known as the Project 1851 'Paltus'. In each case the central section, where the missile compartment once sat, has been removed and replaced with a 43-metre plug into which the mini-submarine is able to dock and undock. The purpose of Paltus is unknown, but it is capable of diving to extreme depths of up to 1000 metres. It is also equipped with two forward manipulators, long arms capable of reaching out from the submarine and carrying out various tasks, which suggests it is used for a combination of oceanographic research and search and rescue operations, but also possibly for underwater intelligence gathering.[61] Here, the US 'Ivy Bells' programme to tap Soviet underwater communication cables immediately comes to mind (see p. 531).[62] As of 2014, 99 per cent of international communications are routed via 263 submarine fibre-optic cables, which are cheaper and quicker than satellite communications.[63] Many more are due to be laid down in the future. Are the Russians using this mini-submarine and its mother ship to tap underwater cables? Do they have a modern day equivalent to the 'Ivy Bells' programme?

Russian interest in small submarines does not end there. In 1990 construction of a nuclear-powered mini-submarine known as 'Losharik' started at the Soviet Admiralty shipyard in Leningrad. Construction was suspended following the collapse of the Soviet Union, but the submarine was completed in 2007. Built of titanium to withstand pressures at extreme depths, the highly classified submarine took part in Russia's Arctic expedition in autumn 2012, where it dived 2.5–3 kilometres in the Mendeleyev Ridge and remained submerged for twenty days. During the dive, the crew conducted geological surveys and collected 500 kg of rocks, which Russia probably intends to use to reinforce its claims to the Arctic region.[64] According to recent reports, construction of a second mini-submarine, smaller than the 'Losharik', has also started.[65]

The funds pouring into submarine Russia will finance a range of other significant upgrades tasked for anti-submarine warfare including three new classes of frigate – the 'Steregushchiy', the 'Admiral Gorshkov' and the 'Admiral Grigorovitch'. There is a big push too on underwater sensors, both magnetic and passive/active acoustic. The old Ilyushin 38 May Maritime Patrol Aircraft will be replaced by a new aircraft at what looks like the rate of one a year and the old Tupolev Bears will be upgraded again. Add to this modernization programmes for heavy and light torpedoes and anti-submarine missiles and mortars and mines and there is, or soon will be, a truly testing Russian-shaped underwater world in which today's Perisher graduates and their crews will have to operate on their peacetime patrols, let alone in the scarcely thinkable conditions of any other kind. The silent deep today and in the foreseeable future has the characteristics of a front line just as it did in Cold War times.

For the Royal Navy Submarine Service, the Russians will always be their Champions League opponents, their maritime benchmark. In the 2020s Russia will certainly have the capacity to mount another deep Cold War should the occupant of the Kremlin wish it. But it is important not to neglect another rising set of hulls just one or two of which might find themselves up against a Royal Navy SSN in the warmer waters East of Suez.

China

Since the mid-1990s, China has put a significant effort into building up its submarine strengths. By 2013 the inventory included 55 conventional submarines, 7 SSNs and at least 4 SSBNs. They were by Western standards relatively noisy and technologically backward. But the Chinese leadership has been making considerable investment in technical improvements. China increased military spending by 9.7 per cent to an estimated $216bn in 2014. Its anti-submarine warfare capacity would execute a generation leap if it procured the Russian-designed Ladas. China's considerable gifts in the world of espionage are also expected to help narrow the gap with Western submarine forces.

Top-level political support in Beijing for underwater China will almost certainly be sustained as the country proceeds bit by bit with its longer-term plans to acquire a serious naval presence in the Indian Ocean and possibly even the Middle East. The Chinese are well aware of the West's lead in underwater technology – hence the big investment in improving China's anti-submarine warfare capabilities. As China's maritime reach widens, there is an increasing chance that Chinese surface vessels and submarines will be operating in the same waters as the Royal Navy's 'Astute' class. Chinese developments are naturally closely monitored by Western intelligence and spark a special attention from those responsible for calibrating future possibilities of another huge rising power.

India

India has made great strides in its anti-submarine warfare. It has leased an Akula I from the Russians, which became active in early August 2013. It has developed its own 'Arihant' class SSBN. The Indians created their own sonar improvements for the old Kilos and are building an entirely new French-designed 'Scorpene' class of SSKs. In addition India has considerable strength in the air with a mix of Maritime Patrol Aircraft acquired from both East and West – the US P-8 Neptunes and the Russian IL-38s upgraded with the new Russian Sea Dragon ASW suites – an aerial abundance the UK could only envy when the 2010 Strategic Defence and Security Review took away the new Nimrods. India is also procuring a new fleet of frigates in the 'Talwar' class. If its forces can integrate all the elements they possess successfully, Western analysts rate India's chances of becoming a formidable ASW power. For all their progress with nuclear enrichment, however, it is unlikely that a democratic Commonwealth partner like India would ever use its underwater assets to put the Royal Navy under threat, unlike contemporary Iran.

Iran

In and around the tightly constrained waters of the Gulf, the Iranians have frequently exercised a range of underwater, surface and aerial capabilities in an area where there is always a Royal Navy presence often including a 'Trafalgar' class submarine. The Iranians operate coastal surveillance systems, unmanned aerial vehicles, some rather old Maritime Patrol Aircraft left over from the days of the Shah and a very substantial range of small combat vessels, rigid raiders and the like. Iran also operates five British-designed 'Alvand' class frigates.

Western intelligence has detected quite a large home-grown Iranian R & D and investment programme in underwater and ASW capabilities designed to get around sanctions. Of particular concern are the Iranian Yonos, their midget submarines. Though the threat such boats pose to Royal Navy submarines is low, that is not so for surface vessels as the Yonos have very similar capabilities to the North Korean midget submarine which sank South Korea's *Cheonan* in March 2010.

When assessing future threats to Royal Navy submarines great attention is paid to the capabilities of those countries which just might trouble her in the future. Intense concentration is also devoted to those technologies that might trigger a leap in vulnerability. One in particular has preoccupied her naval and intelligence analysts since the first big investments in nuclear boats in the late 1950s to today – will there be a technical breakthrough that makes the deep dramatically less opaque, tilting the balance away from the hunted to the hunters?

So far, this has not happened. ASW improvements have been substantial but evolutionary. And, over the next 30–40 years, the lifetime of the Royal Navy's 'Successor' boats, some of these are foreseeable and foreseen: sonar will improve; Maritime Patrol Aircraft will be able to cover bigger sweeps of ocean; satellites will achieve still better resolutions; the most advanced SSKs will pose an increasing threat to SSNs. Yet, as one insider put it, 'the revolutionary ones are the ones you worry about. But be under no illusions, the oceans are still not transparent. It's still a question of "small submarine; big ocean".'[66]

And for those who are paid to worry about such things there is one shared comfort with which they console themselves – that all the nations in the business are upgrading and acting on a common conviction that beneath the ocean is still by far the most secure place to conceal your weapons of last resort and their sister attack boats. Western intelligence knows that all top of the range (or near top of the range) submarine powers are working on this. But as one analyst put it, 'we don't see a game changer'.[67]

There are other technologies, however, which already exist that can imperil the Royal Navy Submarine Service without a dramatic breakthrough that turns the oceans transparent. Cyber is one of them. There are three ways in which the threat could manifest itself:

- A sophisticated cyber power among the nation states (i.e. Russia and China) could user cyber to siphon out submarine Britain's technical secrets, increasing the chances of countering them and/or adding to the cyber attacking nation's own technological base. The same could apply to the Royal Navy's operating patterns.
- On the operational side, though it would be very difficult, a sophisticated cyber attacker might try and either penetrate or spoof communications to the boats.
- Much more likely, and more easily done, is a wider attack on the UK's critical national infrastructure which would hit power supplies and communications more generally. Such an attack might have a serious and not necessarily intended impact on Royal Navy submarine operations.

Allied to this trio of malign possibilities is a far lesser one which nonetheless worries some on the inside of Submarine Britain (and needs to be recognized in a book that seeks to open up that world) – the price of greater openness and accessibility since the passing of the Cold War. Television cameras have been in the boats over the past fifteen years to a degree inconceivable in the sixties, seventies and eighties. Social media carries an abundance of insider 'dits'. Trade fairs exhibit highly sophisticated sonars for sale on the open market. All these factors make the guardians of national security uneasy even as they recognize the need for greater public awareness to help explain the purpose of the country's substantial investment in Submarine Britain.

Overstretch and long deployments East of Suez have also created other problems. In 2012, the MOD's internal safety watchdog said that 'There is a risk that the RN will not have sufficient suitably qualified and experienced personnel to be able to support the manning requirement of the submarine fleet' and that 'Inability to recruit, retain and develop sufficient nuclear and submarine design qualified personnel will result in an inability to support the Defence Nuclear Programme.'[68] The report warned that a 'dearth of experienced mid-career people' was also threatening the Service and that this would continue 'into the next decade'. By 2015, it estimated that the Service would be suffering from a 15 per cent shortfall in engineers and that one in seven posts for weapons officers at the rank of Lieutenant would be vacant. This is a particular problem in the SSBN world, where the distinction between crews of various SSBNs has started to blur as experienced crew move around SSBNs whenever there is a skills shortage. The Submarine Service takes its commitment to maintaining one SSBN on patrol at all times so seriously that it is prepared to move experienced crew members from SSNs to SSBNs, even if it means the SSN is unable to sail due to crew shortages. The Navy has carried out a senior officer manpower review looking at ways of improving the 'quality of life' for submariners, who, as we have seen, are being forced to deploy more frequently and do more jobs. But with patrols now lasting over four months, the Navy is struggling to attract recruits.[69]

The introduction of women into submarines might also improve the retention problem. Although women have been serving at sea in the Royal Navy since 1990 and make up nearly one in every ten of those in the service, until very recently they have been excluded from serving in submarines. This was due to concerns about the risks associated with the higher levels of carbon dioxide in a submarine's atmosphere and the potential to damage fertility. However, following extensive research by the Institute of Naval Medicine in Gosport, the Navy concluded that the risks are unfounded and that there are no medical reasons for excluding women from service in submarines. Pregnant submariners will not be able to serve at sea for health and safety grounds because of the risks to the unborn child. If a woman should discover that she is pregnant while on patrol she will be isolated in a special atmosphere-controlled environment until she can be safely removed from the submarine.

There is another threat to the now all-nuclear-propelled Royal Navy Submarine Service – a serious nuclear incident. Some insiders believe the UK 'is one accident away from the end of the Submarine Service',[70] which is a powerful extra incentive to make sure such an accident never happens. It is here that the human and the technical factors meet, as they do in all aspects of the submarine world, but in a singularly acute form. And once more the Royal Navy stresses the levels of training, technical competence and maintenance needed to keep men and women and the boats safe and Submarine Britain in being.

AND THE RUSSIANS CAME TOO

Central to remaining a nuclear-weapons nation is the need at regular intervals to show the rest of the world that it is exactly what the UK still is. And this particular quadrille in Britain's nuclear choreography is the Demonstration and Shakedown Operation, known as the DASO and pronounced 'day-so'. It takes place towards the end of the mid-life Long Overhaul Period and Refuel (LOP(R)) of the Royal Navy's 'Vanguard' class submarines. Three of the four boats – *Vanguard* herself, *Victorious* and *Vigilant* – have successfully completed it. One of us, Hennessy, was present at the run up to *Victorious*'s DASO in May 2009.[71] Both of us witnessed HMS *Vigilant*'s successful launch of a Trident missile on 23 October 2012 from very close in, as did the Russians, of which more in a moment.

DASOs are extraordinary events. There is a touch of unreality about them from the moment you arrive at Orlando Airport, which primarily exists to shovel huge numbers of holidaying families off jumbo jets and through to Disneyland. At DASO time, a pair of young naval officers, shimmering in tropical whites, wait at the gate to pluck you from the tide of holidaymakers and take you to another monument to architectural overstatement called Cocoa Beach just down the Florida coast from Cape and Port Canaveral. As you gather in the DASO hotel, and the US contingent arrive, it looks as if Submarine Britain and the special nuclear relationship are going on a kind of triennial holiday together – a mix of a jolly and a reunion with the launch of an intercontinental ballistic missile thrown in.

Appearances deceive. A DASO is an intense working session over several days and not just for the Commanding Officer and crew of the submarine. For the shadow over Cocoa Beach and the discussions at the nearby Naval Ordnance Test Unit, NOTU, housed inside Cape Canaveral Air Base, is the huge silhouette of the 'Successor' submarines to the 'Vanguard' class. For the first time ever in the 52-year history of shared delivery systems, the UK will be ahead of the USA in the procurement and building cycle, the first Common Missile Compartment of the Royal Navy 'Successors' will be built before the US Navy fits the same compartment to its replacement submarines for the 'Ohio' class. As a result, the mutuality of the discussions during the DASO days are even more intense than usual. We also notice that the Americans are particularly concerned that *Vigilant*'s shot goes without a hitch. It is quietly explained to us later that this is because this is probably the last DASO before the new UK Government, after the 2015 general election, moves towards the so-called 'Main Gate' decision on whether or not to proceed with the 'Successor' programme, and the Americans involved in the DASO are especially keen that Britain should do so.

There is another contributor to the electricity in the air at Cocoa Beach and Port Canaveral's berthing area – the Russian Navy, in the shape of a huge 'Vishnaya' class electronics and signals spy ship, the *Viktor Leonov*, which is lurking in international waters about twenty miles off the coast waiting for *Vigilant* to slip out into the launch area of the US Navy's Atlantic Missile Range. The *Viktor Leonov* has followed *Vigilant* all the way from Kings Bay, the US Naval Base in Georgia, where the submarine collected the pair of Trident D5 missiles from the shared pool that it needs for the test (one for use; one spare in case of technical trouble). It is the first time a Russian spy ship has turned up off Florida for a US or UK Trident launch since 1997. We gather that over the previous weekend Russia exercised its triad of land-, sea- and air-launched nuclear missiles for the first time since the Cold War ended.

The Russians told the US and the UK about these tests under the same treaty whereby the US and the UK also warn the other acknowledged nuclear-weapons states of their forthcoming trials (or, at least, of the period of days during which a missile is likely to soar out of the sea on the Atlantic Test Range off Florida or the equivalent Pacific Test Range off California). In March 2012 Russia had exercised the readiness of its forces up to strategic nuclear level for the first time since the Soviet Union broke up. A senior US naval officer says the Russians will already have agents placed in Cocoa Beach including in the strip joints 'listening to the talk of drunken sailors' and close to the hotels where

Vigilant's crew are billeted. On launch day, he adds, they'll be watching from the beaches too.

Naturally, Western intelligence is keeping a close eye and ear on the Russian spy ship. We hear that the vessel has been referring to HMS *Vigilant* as 'the big sausage'. Among the reasons the Russian AGI has been and will be watching *Vigilant* so closely is that the UK is not a signatory to the Strategic Arms Reduction Treaty, which is solely a US–Russia arrangement. Under its terms, each side must give the other the telemetry – the measurements taken at each missile test – as part of the confidence-building measures between the two countries. The Royal Navy is not obliged to and does not, so the Russians have turned up to do it for themselves. Their presence is felt throughout the build-up to the DASO and most acutely on the day of the launch itself when a Russian 'Sierra 2' SSN was detected some 300 miles off the US eastern seaboard, at about the time of *Vigilant*'s shot – a story which reached the press in early November through US Naval Sources.[72] 'I'm not sure the Cold War is over,' says a senior Royal Navy officer over breakfast on T minus 1 (i.e. the day before the firing). Apart from the extra pep the presence of the spy ship is giving the whole exercise, it is thought to be good for training purposes.

On the evening of Sunday, 21 October, a Trafalgar Night Dinner is held at the Tides Country Club, part of the nearby Cape Canaveral Air Force Station. The CO of *Vigilant*, Commander Mark Lister, presides, accompanied by his senior officers. They have brought the Ward Room silver from the boat. Rear Admiral Mark Beverstock, Chief of the Strategic Systems Executive and the Royal Navy officer who incarnates the UK end of the special nuclear relationship, opens the speeches, neatly linking the continuous deployment at sea of at least one 'Vanguard' class submarine with Trafalgar: 'Deterrence is all about being both credible and capable ... Maintaining a permanent presence at sea is fundamental to demonstrating this capability. Back in 1805 the only deterrent Britain had was Admiral Lord Nelson.' Admiral Beverstock then introduced Hennessy, whose job it was to speak on one of the themes of this book, 'Watching the Submariner', before leading the toast to 'The Immortal Memory' of Nelson.

As part of describing the human aspects of the submariner's craft as they strike an outside observer, Hennessy mentioned something he had observed on HMS *Victorious* during a visit to Faslane the previous May, there was an ironing board held by brackets inserted into missile tube number 16 – a visible manifestation of human beings' urge to domesticate pretty well anything if they have to live in close proximity to it. This story – harmless, enough, Hennessy thought – produced a flicker at the heart of this most special of relationships. Rear Admiral Terry Benedict, Mark Beverstock's US equivalent and the personification of the relationship in Washington, turned to his neighbour, Commander Lister, and asked: 'What are you doing to my tubes?' Lister instantly deployed weapons of mass reassurance to the effect that there was no damage done to said tubes. The 'Immortal Memory' was duly toasted and all was well. As Rear Admiral Henry Parker, Controller of the Navy, explained on the bus back, submariners, COs in particular, speak of their boats like lovers 'because you live in them. You don't live in an aircraft.'

There was a little of the Cold War in the night air at the Trafalgar Dinner and not just because of that huge Russian naval vessel waiting a few miles out. We were all conscious that exactly fifty years ago, in waters close to here, the Cuban Missile Crisis – the most perilous near-miss of the forty-year confrontation – was being played out with at least one possible outcome still too dreadful to contemplate. Among the US diners that evening was Lt General Jim Kowalski, Commander, Global Strike Command, who cut his aviator's teeth on the B52 and B1 bombers and described the Cold War as 'a war we dared not fight and couldn't afford to lose'. Submariners played a crucial part in resolving the crisis. When Kennedy privately offered to remove the Jupiter missiles based in Turkey in exchange for the Russians removing their missiles in Cuba he knew that the US Navy's recently deployed Polaris submarines could adequately cover the same targets as the obsolete Jupiters.

The next round of special-relationship dining occurred the following evening, Monday, 22 October, at the DASO Hotel on Cocoa Beach after a day of meetings at the Naval Ordnance Test Unit which included an inspection of a huge D5 missile broken down into its constituent parts and of the old underground test silos from the Polaris, Poseidon and Trident days. These are being rebuilt to test the new Common Missile Compartment.

Captain J. P. Hetherington and members of his team briefed us. Much of the discussion centred on the recent US decision to delay its Ohio replacement programme. There was a risk that this would also delay the UK 'Successor' programme. However, the US has decided to keep the critical component for the UK programme, the Common Missile Compartment, on its original schedule. The Successor programme will be different to all other purchases under the Polaris Sales Agreement. With Polaris in the 1960s and Trident in the 1980s, the British bought a proven weapons system that was already in service with the US Navy. For 'Successor' the UK will be the first to operate the system in the Common Missile Compartment. This says Rear Admiral Mark Beverstock, the man in charge of the British end, is 'an uncomfortable place for the UK to be'.

As a result, the two navies have developed what is called 'a mitigation strategy'. The US is building a complete Strategic Weapon System (SWS) Ashore facility at Cape Canaveral, which will allow the US to test the hardware, thereby reducing the risk to its 'Ohio' replacement and the UK's 'Successor'. The SWS Ashore facility will also end the practice of distributive testing across facilities dispersed over the United States by bringing everything together in one location. The Americans are impressively aiming to prove 100 per cent of the programme for the British before the first installation of the Strategic Weapon System takes place in the first UK 'Successor' submarine. They did not need to do it before the US programme 'shifted to the right', to use the jargon of military spending programmes. It is at this point that Rear Admiral Parker sharply interrupts the briefing with 'Can I just say thank you.'

The new SWS Ashore facility will be built on a site at Cape Canaveral that almost charts nuclear history in its archaeology. Complex 25 was a four-pad launch site built in 1957 to test the US Navy's SLBMs. Between 1958 and 1970 a total of ninety-four launches, of Polaris, Poseidon and Trident missiles, were conducted there. Thirty-three years later the SLBM programme is returning. Pad 25B was initially built with an underground launch mechanism known as a ship's motion simulator to simulate the roll and pitch of a submarine. Once completed the facility will consist of two test bays. The first will house two pairs of existing Trident missile tubes that are currently installed in the 'Vanguard' class and 'Ohio' class submarines. On top of them will be a simulated superstructure on which a Missile Servicing Unit will be able to mount in order to load and unload missiles. The second test bay will contain a new Ohio replacement and 'Successor' class missile tube, a section of the new Common Missile Compartment. A building that houses a Control Room and a mock-up of the new Missile Control Centre on the planned new submarines will separate the two test bays. Viewed from the surface the building will look deceptively unimpressive, as most of the structure is housed underground.

Although this briefing is for the guests who have flown in to watch the DASO, it quickly turns into a business meeting between CSSE and Special Projects (SP). Discussion centres on training and how much the UK needs to spend in the next few years on a new Trident Training Facility to prepare the crews of the Successor submarines. The US suggests CSSE looks at the possibility of using American facilities. The 1963 Polaris Sales Agreement sets the parameters, but it does not prescribe the generosity shown by the US Navy at moments like this.

The briefing session/meeting over, we drive the short distance to Complex 30, the Missile Assembly Building, where the Trident missiles are modified for the DASO. The Trident II D5 is manufactured in the United States by the Lockheed Martin Space Systems Company and is powered by a three-stage solid-propellant rocket motor and an onboard inertial guidance system which gives the missile a range of over 4000 nautical miles and an accuracy to a few metres. As we are taken around the Missile Assembly Building we see various sections of missiles at different stages of assembly, each with a tiny inspection hatch open so that it is possible to see their entrails. We are taken up to the top of the building to look down into the so-called 'equipment section' of the Trident II D5 missile, its front end. There are twelve empty slots for the Multiple Independently Targeted Re-entry Vehicles (MIRVs) that house the warheads.

Next we are taken to see a complete missile. At nearly 13 metres in length and over 2 metres in diameter, each one weighs 60 tons. In flight the missile's length is increased by a 2-metre aerospike, which reduces the missile's frontal drag by about 50 per cent while in the earth's atmosphere, greatly increasing its range.

During the tour of Complex 30, Admiral Benedict, recalling the Trafalgar Dinner the night before, says: 'One of the things about the Royal Navy is your sense of history and tradition.' He speaks as well about the mutuality of the special nuclear relationship: 'I have a great appreciation for the partnership with the British. What we get out of it is a strong ally. What the British get out of the Polaris Sales Agreement is a very cost-effective relationship . . . But the thing I take away more than anything is the personal friendships.' This is clearly true. The friendships are palpable, as is the long experience of working together. On

DASOs the two navies are, to all intents and purposes, fused. Over the food at the Trafalgar Night Dinner, Captain Al Holt, the US Navy officer who will be on board *Vigilant* when the missile launches on Tuesday to give Mark Lister permission to fire and to validate the whole test operation, says: 'It's almost like a big brotherhood or a big family.' He also reflects on the individuals who lead their crews: 'Command is like a marathon. You mustn't go out too fast or too slow . . . we try not to talk about submarine operations because our strength is in not being detected . . . The whole flavour of a crew reflects the personality of the Captain. All COs,' says Holt, a veteran of Cold War operations, 'are slightly micro-managers. If you don't concentrate on certain things you'll get buried. The hard part of submarining is figuring out what the information is telling you. Submariners concentrate on deficiencies. The submarine has tons of sensors, but it just gives you the data, not the answers. Sometimes it's two dimensional in a three-dimensional world.'

These eve of test dinners are consumed with a frisson of anticipation. Will the weather hold? Will it all work? What will that Russian spy ship try and do? Rear Admiral Steve Lloyd, Mark Beverstock's predecessor, has arrived. He was in charge of the 2009 DASO. We talk about the AGI before we sit down to eat:

Lloyd: 'The Russians are showing us they're back. They're interested. They want to see if we're still in business, viable.'

Hennessy: 'So it's a kind of attention-seeking behaviour?'

Lloyd: 'Exactly. They fired two Bulavas [new submarine-launched intercontinental ballistic missiles] over the weekend.'

Speaking at the dinner is Jon Thompson, the recently appointed Permanent Secretary at the Ministry of Defence (he had been its Director General Finance since 2008). The nuclear element, he says, 'is the pinnacle of our relationship'.

By great good fortune Hennessy was sitting next to Franklin Miller, who received an honorary knighthood for his close and sustained contribution to the UK deterrent and from his seat in the Pentagon knows every person and every fibre involved in the transatlantic nuclear membrane. We talk of the debate gathering pace in the UK as we approach its general election in May 2015, which is certain to contain an element of nuclear discussion. One myth always features in such public debates – the belief that the US, as the provider of the D5 missile, could always prevent a release of it from a Royal Navy submarine if a US President believed a UK Prime Minister should not authorize it. Frank says there is no such capability: 'If there was a switch the US could flick, I would have known about it,' he says.

This is the night of the third and final presidential debate between President Barack Obama and Governor Mitt Romney. The election is two weeks away. We have an insanely early start in the morning if we are to catch the United States Research Vessel *Waters* at Port Canaveral, which will carry us to the launch area, so the temptation to stay up and watch the showdown is resisted.

Tuesday, 23 October 2012, 4.45 a.m., Double Tree Hotel, Cocoa Beach

Rendezvous in the foyer. Everyone very jolly despite the hour.

Hennessy: 'Will the SVR [Russian Overseas Intelligence Service] be on Cocoa Beach?'

Senior US Officer: 'They will be there.'

5.30 a.m., the *Waters*

The *Waters*, technically known as the LASS [Launch Area Support Ship], is the huge surface vessel that watches over and measures all D5 launches, whether on this range or on the Pacific one off San Diego. The *Waters* is an old Cold Warrior. One of her jobs was to lay down the arrays that combined to operate the SOSUS system. The ship has been painted white since Hennessy was last on board in 2009. Numbers of guests are also greatly down this time. It's an austerity DASO – the extras have been cut by between half and two thirds.

We're welcomed aboard by Lieutenant Commander Mitch Puxley and we greet each other as old friends.[73] We're taken to the *Waters*' lecture theatre for breakfast, briefings and a seminar as we sail slowly towards the launch area.

0600, the *Waters* casts off

The VIP and Media Communication manager of the DASO, Captain Paul Dailey, tells us there is a second Russian vessel following us now, an auxiliary rescue ship, or ARS [a posh name for a tug] which can also gather intelligence.

0630

Introductory briefing from Jack Spiller of the US Navy Department. Jack fired the first ever D5 in December 1986 – the one that went wrong, performing cartwheels above the USS *Tennessee*'s telemetry mast before being safely destroyed. Jack says the presence of the second Russian vessel explains 'why there were so many Russian speakers in Walmart yesterday'. Jack has long experience in the business. 'Deterrence works every day,' he says. 'If it ever has to be fired in anger it's failed.'

0715

A second briefing from Commander Paul Nitz, US Navy. 'The AGI is out there waiting for us. We've already visualized her. She played with us last week. The worst case is that she'll get 7000 to 8000 yards in on *Vigilant*.'

Nitz is followed by *Vigilant*'s young, bright and sharp Intelligence Officer, who gives a rundown about our Russian watcher:

'The *Viktor Leonov*, AGI – in common parlance, a spy ship, carries two thirty-millimetre weapons systems and air defence. She went through the English Channel. Fleet [Northwood] thought she might be going to Gibraltar. I always tend to be paranoid wondering if she would come with us. Lo and behold she did. We were expecting her to go on to Havana or Venezuela. On 27 September she was in Havana. Lovely satellite picture. When we came out of Kings Bay she was waiting for us. She tried to contact us calling herself the "Jacksonville Sheriff" [Jacksonville is a Florida port to the north of Cape Canaveral]. She definitely declared her intention to monitor the DASO. We've done enough to protect *Vigilant*'s signature. And the *Waters* has kept her off our backs. At the dress rehearsal [on Saturday, 20 October] she came out, made her presence known, did not interfere.'

1030

We move into our seminar. Hennessy opens with a historical look at 'the long road to Successor' and touches on the nature of the current UK debate about our possession of nuclear weapons and the prospect of May 2015 seeing a nuclear-infused general election. Discussion is followed by a fascinating presentation by Dr Brad Roberts, Deputy Assistant Secretary for Defense in the Pentagon and the leading figure in the Obama administration's review of US nuclear posture, on notions of extended deterrence – in the Far East especially – in the twenty-first century.

1242

HMS *Vigilant* is some 2000 yards away from us and about to dive. Beyond *Vigilant* on the horizon is the Russian AGI with a US Coast Guard cutter close by it. We take it in turns to observe it through borrowed binoculars. The weather is fine for a launch (sea state 3). The temperature is in the seventies Fahrenheit. The water is a deep purply blue. For most of the afternoon the two US Coast Guard ships and their complement of Black Hawk helicopters attempt to prevent the *Leonov* from straying into the launch area. From the *Waters* we watch how this operation is played out. At one point, the two Coast Guard ships appear to be almost straddling the Russian spy ship as the Black Hawks fly overhead.

Later, from film taken on board HMS *Vigilant*, we can reconstruct what is happening a short distance away from us.

At around 1246 an alarm sounds three times in *Vigilant*'s control room:

'Diving stations, diving stations, diving stations. All compartments close up, all compartments check communications with DCHQ on the DC Net.'

Commander Mark Lister, the CO: 'XO, Captain, dive the submarine.'

Lt Commander David Fox, the XO: 'Dive the submarines. Diving now, diving now, diving now. Open four, five and six main vents.'

The submarine dives slowly in order to keep the propulsor under water at about a degree and a half. Once the foreplanes bite, the submarine can go deeper. On passing 20 metres the crew check the six main hatches for leaks. There is still air trapped in the ballast tanks and the casing so the submarine goes through a process known as 'rocking the bubble' which allows the remaining air to escape. The main vents are then shut and the submarine levels out at the minimum safe depth, its telemetry mast still clearly visible on the surface.

1246

Vigilant starts venting at the rear. A minute later the front tanks vent. Mitch Puxley says: 'Good. It's on time. There's always a bit of emotion.' T hour is 2 p.m.

1249

The front of *Vigilant*'s fin slams into the Atlantic. Bright good conditions. 'Pretty well perfect,' says Mitch. 'We're now into the launch area, which is a box of ten square miles.'

1250

We learn that there's going to be a twenty-minute delay for 'deconflicting' reasons. There are satellites up there which may collide with one of the D5s if it goes off at 1400. We learn later that it's a Russian Soyuz 3 carrying people in it. If the D5 struck it there would be an international incident in technicolour.

1253

All we can see of *Vigilant* now is the telemetry mast with its red tip. A Black Hawk helicopter flies over it.

1300 on the Bridge of the *Waters*

The Russian AGI is starting to come in closer and is being shadowed at about 7 knots by the US Coastguard.

Paul Nitz says: 'All I care about is the 5000 yards around the SSBN.'

The AGI is now just under seven miles away from *Vigilant*. Nitz says: 'I've tried to talk to him but he won't respond. I've been asking him to open up eight miles.'

Two minutes before the launch, *Vigilant* will send a signal to the *Waters*.

1313

Chat to the very affable Rear Admiral (Retired) John Butler of the US Navy. Like his friend Frank Miller (who is aboard *Vigilant*) John is one of the incarnations of the US–UK nuclear relationship. He now works for Lockheed Martin, which manufactures the D5. He's seen many DASOs but, quite plainly, they still excite him. He has brought a beach chair on board and has set himself up in pole position to witness the shot. 'It's showtime,' he says.

Those who have done DASOs before say this launch area, the

Atlantic Range, is particularly tricky because the Gulf Stream in these waters is fast – moving at about 4 knots. So to get *Vigilant* into the near motionless hover before the launch requires some deft seamanship on the part of Commander Lister and his crew. Just before it happens, a white puff of smoke will appear to give us her exact location.

1355

Black Hawk helicopters circle over the launch area.

In *Vigilant*'s Control Room, Commander Lister is sitting in front of the launch panel. He has a headset on which allows him to talk to the rest of the submarine. He scribbles in a notebook every few seconds. At T minus 20 minutes he takes *Vigilant* to action stations:

Commander Lister: 'WEO, instrumentation, this is command, tube one is released.'

Range: 'Weapon system is in condition one SQ for tube one for DASO launch, WEO, command, roger.'

The range has given the OK, it is clear to launch. SP205 is the senior American onboard *Vigilant*:

SP205: 'Command 205, roger, Final T zero prediction is valid.'

Commander Lister: 'Final T zero, command is valid, 205 command, roger. 205 command, all launch prerequisites have been met.'

SP205: 'Very well, SP205 concurs.'

Commander Lister: 'SP205 concurs, 205, roger.'

In the Missile Control Centre, Lieutenant Commander David O'Connor, sits holding the trigger in his hand. He says: 'Supervisor, WEO, initiate fire one, denote one.'

Denote means that information is being passed to the missile. This takes around thirty seconds.

1420

It's happening. Jinks sees the warning puff of white smoke.

Back in *Vigilant*'s Control Room Commander Lister inserts his key into the launch panel and turns it: 'WEO, Command, you have permission to fire.'

He picks up the intercom and speaks to the submarine: 'The WEO has my permission to fire. One in denote T minus one minute and counting. Ten seconds.'

In the Missile Control Centre there is complete silence. David

O'Connor is still holding the trigger in his hand. He waits and then pulls the red button on the black Colt 45 and says: 'One away!' There is a loud thud and the submarine vibrates.

Back in the Control Room Lister says: 'Missile away.'

On the surface, we see a huge eruption of white smoke as the missile emerges from the sea in its envelope of steam and gas. It pauses, drops a little, its motor ignites, it moves sideways lest a calamity occurs and it falls back on *Vigilant* and then a brilliant white light dominates the skyline, the D5's motor ignites and with an intensely bright tail of yellow-orange flame it soars off into the sky.

There is a pause of a few seconds and the roar of the launch carries across two and a half miles of Atlantic Ocean and envelops the watchers on the *Waters*. Spontaneous applause erupts from all the decks.

Through a gap in the cloud we see the first stage of the missile fall away against the light-blue sky. The D5 is well on its way. Its flight will take about thirty-four minutes and stretch about 4000 miles away down into the South Atlantic.

1420–1421 HMS *Vigilant*

Commander Lister is sitting in his chair holding a black telephone, speaking to the submarine: 'LASS is receiving telemetry. Telemetry looks good.'

An American voice comes over the radio: 'This is SST, Permission to Fire is removed. First Stage Separation. Missile is on trajectory. Nose fairing ejection. Missile is on trajectory.'

Lister: 'Stand by Second Stage Separation. Second Stage Separation. Missile is on trajectory.'

Within about two minutes, after the third-stage motor has ignited, the missile is travelling in excess of 6000 metres per second (13,500 miles per hour – nearly ten times faster than Concorde).

Back on board *Vigilant*, SP205 says: 'Command 205, there will not be a second T zero.'

Lister: 'There will not be a second T zero. 205 Command, Roger.'

It's all over very quickly. *Vigilant*'s crew starts carrying out a number of tests and drills before surfacing. The missile is still in flight. Once the boost phase is complete the laws of ballistics take over. While in the upper atmosphere the equipment section is deployed. It orients itself to view the stars and corrects its trajectory path. For each re-entry body, the process is repeated as the section thrusts, and reorients to achieve proper deployment conditions. Each re-entry body then reenters the atmosphere and splashes down at its designated target, in this case at the end of the test range off the west coast of Africa.

1425, the *Waters*

Two Black Hawks in tight formation circle low and roar overhead.

1436

Vigilant's fin appears, with a Black Hawk over her.

1440

The Black Hawks fly above us once more as if in a victory parade.

1535

We're proceeding in parallel with *Vigilant* back to Port Canaveral. Even on the surface *Vigilant* is swifter than the *Waters* so a gap is opening up.

Paul Nitz comes down from the bridge to tell us that after the shot the AGI called *Vigilant* on Channel 16 (which vessels of all kinds use to talk to each other) with a message of congratulation in English from her skipper in a heavy Russian accent out of central casting.

'Bravo November this is Russian Warship. We sincerely congratulate crew of your ship with the successful completion of the Ballistic Missile Trident II launch exercise. We also would like to thank United States Navy Research Vessel *Waters* and Coast Guard Ship and Helicopter for very well-skilled manoeuvring. This is Russian Warship . . . Out!'

One of the US VIPs aboard *Vigilant*, a Russian-speaking lady from the State Department, offered to reply in Russian. But Mark Lister decided not to reply and the Silent Service lived up to its name.

There's much talk of the significance of the Russian presence. We learn later that the *Viktor Leonov* had intruded about 500 yards into the launch area and the two US Coastguard cutters had to break the rules of the sea by interposing themselves in a way that risked being rammed. It was assumed afterwards that the AGI would return to Havana. In fact, Hurricane Sandy was preparing for its assault on the US eastern seaboard. The spy ship sought and was granted permission to put into Jacksonville. One gentlemanly act plainly deserved another.

2100, the Green House, pub, a few yards from the jetty

As you walk through the door, there are all the crests of the Royal Navy submarines that have conducted DASOs since the 1960s. Beer flows. Mark Lister arrives to much congratulation. Captain James Hayes RN, in effect the inspector of the exercise with his US namesake on board, presents Mark with an already framed photo of his launch complete with commemorative medals. Hayes also gives him a framed picture of the *Viktor Leonov* which has scrawled upon it 'Congratulations on a Successful Shot! Yrs, Ivan.' In his reply to Hayes's speech Lister says: 'This is the greatest day of my life apart from the days when my children were born.' His wife Mary is there as she had watched the launch from the *Waters*.

The First Sea Lord, Mark Stanhope, is glowing with pleasure. He had been aboard *Victorious* in May 2009 on the day the test had been postponed because of an electrical storm. So he was witnessing it this afternoon for the first time.

A successful DASO is a big rite of passage for the submariners and an experience that doesn't fall to many. The WEOs who have done it join a small outfit called 'The Trigger Club', which actually meets for dinners. Does such a thing exist for COs? We'll ask Mark Lister when we breakfast with him tomorrow.

Wednesday, 24 October 2012, 8.00 a.m., breakfast, Double Tree Hotel, with Commander Mark Lister and fellow officers from HMS *Vigilant*

In an understated, British way there is mild euphoria that things went so well yesterday. Lieutenant Commander Dave O'Connor, the latest recruit to the now ten-strong 'Trigger Club', tells us the launch was exactly at 1420. A manned Soyuz spacecraft had caused the twenty-minute delay. The D5 took thirty-four minutes to complete its flight. At the other end of the range, six re-entry vehicles were seen and the seventh one heard. Later we hear from American friends that *Vigilant*'s shot was among the most accurate ever of the D5 tests from both US and UK submarines.

Also breakfasting is *Vigilant*'s newest officer recruit, Lieutenant Chris Poley, whom we had met on HMS *Astute* in Faslane during the summer of 2011. What did he make of the three week voyage across the Atlantic to Kings Bay and life aboard a 'Bomber'? Above all, he said, it's how 'your life shrinks down into very small size and eventually a handful of rooms'.

Also breakfasting with us is Julian Miller, number two in the Cabinet Office's National Security Secretariat. Julian, who in the past ran the Assessments Staff in the Cabinet Office's Joint Intelligence Organisation, was wondering if the AGI added to a pattern of increased activity by the Russians.

Mark Lister knows about the Russians. He's had a long naval career serving in SSNs, SSBNs and also the old conventional submarines. He talks about the differences:

'In the diesel submarines the Captain was the only man who would order the submarine around . . . It was a reflection of his personality. I can't do everything on my own on a "V" boat. The CO of an Oberon would instinctively know almost everything about the boat bar the engines.'

Lister is deeply pleased to have this command. He's fifty but plainly still at the top of his powers. He's risen right up through the ranks starting off as an Able Seaman in 1978 and, with HMS *Spartan*, he was involved in the Falklands thirty years ago. Like pretty well all the COs, he confirmed that the boats he commands are (family apart) the loves of his life.

How tense was he before the launch?

'Not very because of two days rest before. I wasn't fatigued yesterday. I was fresh and apparently from T minus 30, I stopped being my normal self and became quite serious. In the control room it was dealing with the fourteen VIPs while talking to Range Command that was the real challenge. But after a while you go so vacant that you are able to tune out the guests.'

There were fifty-five extra riders on board the submarine, forty-five of whom were American.

'[The relationship is] Very solid. They are very keen to say that the community ethos is so strong. A US Commander is in tactical command when we are using their range to test the missile. He is the one who controls the launch. I think I can do everything . . . but he's the one who says that I can fire. The most gratifying thing is to hear an American who has been on eight or nine DASOs who says that this DASO was one of the best. I'm fortunate in that I've got a submarine in good condition and good people. The ship's company that you join is like a family. Reputation matters and this contributes to the reputation.'

What was it like in the Control Room once the missile was launched? Did they cheer?

'We were bracing ourselves for all kinds of things. I have done hundreds of WSRT [Weapons System Readiness Tests] in the submarine. You go through the test and nothing happens. This was with a real key. After the launch, 16,000 tons of submarine shook and oscillated. You hear the compensation water coming into the tube and feel the boat steadying out on depth after about thirty seconds. Then we have a short period of checks before surfacing.'

What about the Russians? How much of a bother was the AGI?

'A lot. They were trying to get our signature. If I was him I'd be doing the same thing. I'm an old Cold War submariner and I don't talk to the Russians. To call them the enemy is the wrong word ... they were adversaries and they still are. They are still the greatest credible threat.'

Mark Lister drives us down to *Vigilant* alongside a special jetty dully black in the bright Florida sun. Inside she looks bright and lived in and utterly transformed from the cold, dark, empty shell we saw in Devonport in February 2010 when the reactor was taken out to be fitted with its new core.

We tell Lister we'd heard that when the VIPs were shown where the launch keys were kept in the outer safe on their tour of the boat on the way out to the launch site, Frank Miller had asked him where the Prime Minister's letter was. Lister said he avoided getting into a discussion about that and had told Miller he couldn't talk about it. In fact, David Cameron's last-resort letter was placed in *Vigilant*'s inner safe only when she joined the patrol cycle in September 2013. We are delighted to find that that ironing board – the latest cause of tension in the long history of the US–UK special nuclear relationship, and quite unforeseen by the 1963 Polaris Sales Agreement – is still affixed to the missile tube.

A few weeks on, a letter arrives from Frank Miller returning on paper the answer to the question Hennessy put to him at the DASO Dinner in the Double Tree:

My dear Peter,

You asked me for my views on the independent nature of the United Kingdom's nuclear deterrent, and, specifically, whether I believed an American President had a technical capability to impede (or otherwise negate) the launch of a British Trident missile.

Let me say at the outset that I exercised direct Policy oversight of the US strategic nuclear deterrent from 1981 to 2001. From early in 1982 onwards that brief included responsibility for the US/UK nuclear relationship. I was intimately familiar during those years with US nuclear-weapons plans, policies, procedures and weapons systems (and with most British ones as well). At no time was there ever any evidence (or even a suggestion) that a US 'veto' existed over a British Prime Minister's ability to execute independently the UK's Polaris or Trident force. Indeed, had such a capability existed, I would have become aware of it as part of the intrusive nature of my oversight responsibilities.

I am therefore quite confident in assuring you that no such mechanism exists and that the UK's independent nuclear deterrent is exactly that: independent.

On returning from the USA, Hennessy spoke to David Young, a veteran of nuclear-weapons policy and procedures in the Cold War Ministry of Defence in the 1970s, about the letter and the question which prompted it. 'I would be very surprised' if the US President had the capacity to prevent the launch of a Royal Navy missile, he said. 'But they could do slow starvation. Ingenuity couldn't get you round the bend. It would be finito.'[74]

Sir Kevin Tebbit, Permanent Secretary to the Ministry of Defence, 1998–2005, whose experience of nuclear-weapons policy goes back to the early 1970s, confirmed David Young's view. If the United States decided it wished the UK out of the nuclear-weapons business, 'It would take six months at least; could be a bit longer than that. Whatever the period, it's irrelevant to the management of a crisis.'[75]

But there was no sign of that in Florida or the ocean to the east of it in October 2012. Quite the reverse. Beyond the immediate DASO, all eyes – British and American – were on the 'Successor' programme.

SUCCESSOR

For a period of eighteen months or so, the question of the 'Successor' system to Trident seemed briefly to fracture the iron law that Britain's nuclear-weapons decision-taking phases coincide with economic crisis and public spending stress. For when the Blair Cabinet met on 4 December 2006 and agreed, without a single dissenting voice, to authorize the construction of a new generation of missile-bearing submarines to sustain a UK nuclear deterrent over the period 2020–50, the magnitude of the financial crash of 2008 and the duration of the recovery period which followed was foreseen by few, and certainly not by the ministers gathered in No. 10. That morning they approved a White Paper, 'The Future of the United Kingdom's Nuclear Deterrent', which was presented to Parliament that afternoon.[76]

The White Paper declared that:

> We have . . . decided to maintain our nuclear deterrent by building a new class of submarines . . . A final decision on whether we require three or four submarines will be taken when we know more about their detailed design.
>
> We have also decided to participate in the US life extension programme for the Trident D5 missile which will enable us to retain that missile in service until the early 2040s. Our existing nuclear warhead design will last into the 2020s. We do not yet have sufficient information to know whether it can, with some refurbishment, be extended beyond that point or whether we will need to develop a replacement warhead: a decision is likely to be necessary in the next parliament.[77]

Tony Blair, in a personal Foreword to the White Paper, primarily deployed the uncertain world, unpredictable future argument to explain why he was not going to take Britain out of the ranks of the nuclear-weapons states:

> We cannot predict the way the world will look in 30 or 50 years time. For now some of the old realities remain. Major countries, which pose no threat to the UK today, retain large arsenals some of which are recognised as being modernised or increased. None of the present nuclear weapons states intends to renounce nuclear weapons, in the absence of an agreement to disarm multilaterally, and we cannot be sure that a major nuclear threat to our vital interests will not emerge over the long term.
>
> We also have to face threats, particularly of regional powers developing nuclear weapons for the first time which present a threat to us . . . And we need to factor in the requirement to deter countries which might in the future seek to sponsor nuclear terrorism from their soil. We believe that an independent British nuclear deterrent is an essential part of our insurance against the uncertainties and risks of the future . . . I believe it

is crucial that, for the foreseeable future, British Prime Ministers have the necessary assurance that no aggressor can escalate a crisis beyond UK control.[78]

Within days a carefully choreographed exchange of letters between Tony Blair and President George Bush had agreed to essentials of a deal under the terms of the 1958 Mutual Defence Agreement and the 1963 Polaris Sales Agreement.[79] On 14 March 2007, the House of Commons voted 409 to 161 to renew Trident (88 Labour MPs voted against but Conservative support gave the Blair Government a comfortable majority).[80] Earlier the Government had defeated an amendment which would have delayed the decision by 413 to 167 (95 Labour MPs voted for this).[81]

For all the force of Tony Blair's Foreword to the White Paper and the size of those majorities in the House of Commons, there was a lingering dash of uncertainty for those in the Royal Navy, the Ministry of Defence, at Aldermaston and Burghfield, BAE Systems, Rolls-Royce, Babcock International, plus a myriad of suppliers to Submarine Britain – which the Government itself made plain. Future Parliaments would have to take decisions on the number of successor submarines, their detailed design and build, and new warheads for the Lockheed Martin missiles.[82]

What did not emerge at the time was the degree to which Tony Blair had contemplated ending the enterprise begun by Attlee and his nuclear Cabinet Committees in 1945–7.[83] When the Blair memoirs, *A Journey*, appeared in September 2010 they contained an intriguing admission, written in the conversational style the former PM adopted for his recollections:

We agreed the renewal of the independent nuclear deterrent. You might think I would have been certain of that decision, but I hesitated over it. I could see clearly the force of the common sense and practical arguments against Trident, yet in the final analysis I thought giving it up too big a downgrading of our status as a nation, and in an uncertain world, too big a risk for our defence. I did not think this was a 'tough on defence' versus 'weak or pacifist' issue at all.

   On simple pragmatic grounds, there was a case either way. The expense is huge, and the utility in a post-Cold War world is less in terms of deterrence and non-existent in terms of military use. Spend the money on more helicopters, aircraft and anti-terror equipment? Not a daft notion.

In the situations in which British forces would be likely to be called upon to fight, it was pretty clear what mattered most. It is true that it is frankly inconceivable we would use our nuclear deterrent alone, without the US – and let us hope a situation in which the US is even threatening use never arises – but it's a big step to put that beyond your capability as a country.

So, after some genuine consideration and reconsideration, I opted to renew it. But the contrary decision would not have been stupid. I had a perfectly good and sensible discussion with Gordon [Brown] who was similarly torn. In the end, we both agreed, as I said to him: Imagine standing up in the House of Commons and saying I've decided to scrap it. We're not going to say that, are we? In this instance, caution, costly as it was, won the day.[84]

The nuclear-weapons people in Submarine Britain are well aware – none more so – that just one instance of what Tony Blair called 'caution' *not* winning the day, and that would be it. Britain would cease to be a nuclear-weapons state either within weeks (if the decision was to decommission the existing boats) or years (if it was decided to carry on for as long as the system was viable but no longer). For it would be, in practical terms, impossible to regenerate a substantial nuclear-weapons capacity once it had been abandoned.

The 'Successor' teams are very aware that the boats which exist only in the designs upon their computer screens, like the old 'Resolution' class and the current 'Vanguard' class, are the most politically controversial pieces of military equipment that have been, or will be, operated by the British Armed Forces. When you visit the more sensitive parts of Submarine Britain, that idea hangs in the air, especially at a period, in the words of one insider, 'when the politics of the programme meet the realities' of the time needed to build the new systems if there is not to be a gap between the ageing Vanguards and sufficient of the fledgling Successors to sustain continuous at-sea deterrence.[85]

Another insider, also reflecting on the wider ecology of the 'Successor' project in the summer of 2012, said: 'The political risk is significant. If it's a half-hearted programme, if there's a lukewarm appetite for the programme, if it's just another procurement, that's a risk. If [by contrast] it's like the Olympics [whose start was imminent then], you're going to do it.'[86] Between the 2006 White Paper and those candid conversations in the summer of 2012 there had been a number of anxiety-inducing elements in the lives of what one might call the 'Successorites', whether in uniform, in civvies or in overalls.[87] Tony Blair's successor in No. 10, Gordon Brown, took nearly everybody by surprise in September 2009 during a non-proliferation session of the United Nations Security Council when he declared that 'the United Kingdom will retain only the absolute minimum credible and continuing nuclear deterrent capability'. The Prime Minister explained that 'subject to technical analysis and progress in multilateral negotiations, my aim is that when the next class of submarines enters service in the mid-2020s, our fleet should be reduced from four boats to three'.[88] Neither the full Cabinet, the US allies, or even the members of the nuclear sub-committee of Gordon Brown's National Security, International Relations and Development Cabinet Committee, let alone the First Sea Lord, Sir Mark Stanhope, had been consulted in advance of his departure for New York. Only at the last minute did word reach the Royal Navy of what he intended to say and Mr Brown was persuaded to add the words 'credible and continuing' to his speech.[89]

There is a consensus within the Submarine Service that if the size of the Successor fleet was reduced to just three submarines, the Royal Navy would not be able to maintain one submarine continuously at sea. A fourth submarine reduces the workload and stress on the other three submarines, it also allows the Navy to plan against the possibility of accidents and other unforeseen circumstances, which history has repeatedly shown can and do happen.

The most recent occurred in February 2009, a few months before Gordon Brown's announcement, when HMS *Vanguard*, the on-patrol submarine, collided while submerged with the French SSBN, *Le Triomphant*. The French authorities initially believed *Le Triomphant* had hit a shipping container, but it later became clear that the two submarines struck each other while travelling at slow speeds. *Le Triomphant* was able to return to its base on L'Île Longue near Brest, under its own power. HMS *Vanguard* also remained on patrol until another submarine could sail from Faslane and take over responsibility for maintaining the deterrent. When *Vanguard* returned to Faslane, the submarine was put into the ship lift and raised out of the water. Royal Navy and Babcock engineers then carried out repairs before returning the submarine to the operational cycle. 'Doing it [Continuous At Sea Deterrence] with four is bloody difficult, especially with one in Devonport having its core done,' says one senior submariner, with responsibility for ensuring the Vanguards are able to put to sea. 'When the third boat goes into the ship lift, I'm left with two boats, one of which always has to be on patrol. It's bloody hard.'[90]

At the time of Mr Brown's announcement, eight months before the general election of May 2010, the original White Paper timetable was being adhered to. The so-called 'Initial Gate' of the 'Successor' procurement – the configuration of the boats' missile compartments and their nuclear-propulsion system – was to be announced in September 2010 early in the life of what would be a new Parliament (the previous general election having taken place in May 2005) to allow the first of the Successors to enter service, according to the original timetable, in 2024 at a cost of £11bn–£14bn.

The outcome of the general election in 2010 added a novel element of political risk. The question of the Bomb was affected for the first time by the politics of coalition (Churchill had kept his Coalition colleagues – and his War Cabinet as a whole – well away from atomic matters during the 1940–45 Second World War Coalition).[91] By the time of the 2010 general election, the Ministry of Defence and BAE Systems were some way into the 'Concept' phase for the Successor boats and the first sections of the intentionally elaborate organizational machinery of delivery were being pieced together. Shortly after the House of Commons vote, the Ministry of Defence created a Future Submarine Organisation to see the programme through its 'Concept' phase. In 2008 a Common Missile Compartment programme was established with the United States to cover both the 'Vanguard' and the 'Ohio' replacement programmes.

Stripped to its essentials, the 'Successor Programme' has five elements:

- The design and build of the submarines to replace the 'Vanguard' class.
- The industrial infrastructure needed to support the design and build.
- Life extension of the current 'Vanguard' boats.
- The operational infrastructure needed to support the 'Successor' system when it has entered service.
- Support for the current warheads and the capacity to design a new warhead if needed.

It is easy when observing these huge interlocking procurements to lose sight of the need to keep existing facilities in a condition where they can rise to the demands of the new. For example, John Reid, as Defence Secretary, told the Commons on 19 July 2005 that the Atomic Weapons Establishment at Aldermaston would be getting an additional

£350m per annum for three years for this purpose.[92] It was plain, too, before the Blair Cabinet took the decision to proceed with a 'Successor' system that the original 25-year operational life of the Vanguards would have to be extended beyond thirty years.

But the coming of the Coalition changed these calculations. The Bomb became a significant bargaining chip between the Conservatives and the Liberal Democrats in the Coalition's programme for government in May 2010. The form of words agreed stated that 'We will maintain Britain's nuclear deterrent, and have agreed that the renewal of Trident will be scrutinized to ensure value for money. Liberal Democrats will continue to make the case for alternatives.'[93]

How did the nuclear-weapons element of the two-party deal play out as Britain got used to its first taste of peacetime coalition government since 1939? Firstly, Continuous At-Sea deterrence was maintained throughout the life of the Coalition. In the early days of his Premiership, David Cameron, after being briefed by the Cabinet Secretary, Sir Gus O'Donnell, and the National Security Adviser, Sir Peter Ricketts, wrote his four last-resort letters and appointed his Nuclear Deputies.[94] The value for money study was completed in the run-up to the Government's wider 2010 Strategic Defence and Security Review (SDSR). The question of the Trident upgrade swirled through the nuclear policy sub-committee NSC(N) of David Cameron's new National Security Council and then, as is the way with the most sensitive intra-Coalition deals, in meetings outside that committee between the Prime Minister and the Deputy Prime Minister, Nick Clegg.[95]

The Cameron–Clegg deal was placed in a paper which went to the full National Security Council on 12 October 2010 before being announced in the SDSR document, 'Securing Britain in an Age of Uncertainty'.[96] The ingredients were:

- Defer decisions on a replacement for a current warhead.
- Reduce the cost of the replacement submarine missile compartment.
- Extend the life of the current 'Vanguard' class submarines to thirty-seven years.
- Re-profile the 'Successor' building programme. This means getting the first boat out of Barrow in February 2025 if it is to meet the target of leaving Faslane for its first operational patrol in July 2028.

- Consequently, take the second investment decision, the so-called 'Main Gate', in 2016 which will involve approving the design and settling the question of three or four 'Successor' boats.
- Reduce the number of warheads on board each submarine from 48 to 40.
- Reduce the requirement for operationally available warheads from fewer than 160 to no more than 120.
- Reduce the number of operational missiles on each submarine.[97]

The SDSR disclosed that the value for money review had yielded savings of £3.2bn and that the new boats will have eight operational missiles in a compartment of twelve in the shared US–UK design. It emerged later in the House of Commons that running the Vanguards on into the late 2020s and early 2030s (between six and eight years longer than the 'Resolution' class were at sea) will add £1.2bn to the eventual bill, largely because of the requirement for an extra refit for each Vanguard.[98]

Wherever you touch the 'Successor' programme, whether at the designers in Barrow, or with the policy-makers in the Ministry of Defence in Whitehall, the procurers at MOD Abbey Wood near Bristol, among the operators at Faslane or the refitters at Devonport, you have a sense of three clocks ticking simultaneously and relentlessly. The first counts the hours, days and years of effort needed to sustain the Vanguards into the late 2020s and early 2030s. This, it's plain, is a technological leap in the dark which all concerned wish could have been avoided. The second clock is related to the first. The Chief of the Defence Staff's directive (the wording and details of which are secret) lays upon the Submarine Service the duty of maintaining Continuous At-Sea Deterrence, which the Service has done through thick, thin and some very dicey moments since June 1969. Fulfilling that directive is bound to have some perilous moments as the 'Vanguard' class age and enter lifespans beyond those of the 'Resolution' class (some of which were pretty raddled by the time the 'Vanguard' class replaced them). The third clock is related to both the others – getting the first and subsequent Successors out of the Barrow Yard and through their sea trials and shakedowns in time to keep 'a bloody Union Jack' (Ernest Bevin's phrase during discussions on the UK bomb in 1946) on the British Bomb at all times from 2028 until deep into the 2050s and maybe even into the 2060s.

Just how difficult it will be to extend the life of the 'Vanguard' class and to maintain CASD was evident to us when, with the former Chief of the Defence Staff, Lord Guthrie, we visited HMS *Vigilant* in the final stages of her refit at Devonport dockyard on 1 August 2011. We were greeted by Gavin Leckie, Programme Director responsible for getting the 'Vanguard' class through refit and for maintenance on the Clyde, and Rear Admiral Steve Lloyd, CSSE, in a board room alongside the huge warehouse where all the spare parts are kept for the submarines. 25,000 pieces of equipment came off *Vigilant*, which was in 9 Dock, where we saw her in February 2010, but the dock was no longer dry. Leckie explained how the dock could withstand an earthquake of a magnitude likely to happen once in 10,000 years, tsunamis and extreme tides. On the pace of refits, he said 'the whole workforce is educated to CASD'. By the time *Vigilant*'s refit was completed 2.7 million man-hours had been spent on refurbishing the submarine – half industrial, half technical.

Boiler-suited and hard-hatted we climbed up the submarine's gangplank with a Babcock photographer taking pictures from the dockside. There were 108 crew and the galley had already produced the first batch from its chip pan. 'The day the chip pan works there's life back on the boat,' said Leckie. For refits, the chip pan moment seems to be the equivalent of a keel-laying or a launch ceremony. Gavin says there have been twenty-eight alterations and additions to increase capacity.

We started our tour in the Torpedo Room and moved through the Control Room and Sonar Room. The galley was sparkling. Full health and safety rules applied ('if you send a crew out with bugs you'll run out of loo paper'). We went through 67 bulkhead into the missile compartment, 'from metric land to imperial land', said Leckie as US specification is in evidence where the weapons are. 'If you blindfold an American submariner and put his hand on any bit of the equipment in here, he would know where he was.' We looked at the three-tiered bunks for the junior rates, eighteen of them, which ran alongside part of the missile compartment.

Much of the morning and afternoon briefings covered the LIFEX Programme – the extension of the lives of the 'Vanguard' submarines. As we gathered once again in the Board Room, Leckie explained that twenty-four years is the normal life of a nuclear submarine. HMS *Sceptre* – the last of the 'Swiftsure' class now laid up – did 32–3 years. He finished by saying: 'The reactor core burnt up and the hull life ended.' Of the 'Resolution' class extension, Gavin said that 'we stumbled into it; we cannot afford that now'. Getting the 'Vanguard' class to 2036, he said, is 'a challenge . . . we've got to understand what that requires – early warning'. It's a matter of maintaining CASD and managing the transition from the 'Vanguard' class to the 'Successor' class and anticipating 'the cliff edges'. Do we need to prove Successor technology on the 'Vanguard' class? There is a Deterrent Transition Programme to ensure CASD as the classes overlap.

As we walked back to the Board Room Leckie told us that the origin of the 'Vanguard' class Life Extension Programme was HMS *Victorious*'s DASO in May 2009 when he and Mark Beverstock went for a run down Cocoa Beach. Of the work so far, Gavin said: 'We've not found the show stopper – the cliff's edge [such as a serious defect or serious hull fatigue]. The "Vanguard" class were well designed, well built and have been well supported. Non-destructive examination is now so intense using ultrasound so we're away from using radiography. Structural surveys are not finding anything untoward at all but there's more work to do. You can fingerprint these platforms to make sure it is not changing.

'We're doing a complete evaluation to do something today to keep it going another six years. But one of the biggest risks we are going to stumble into is, are our suppliers going to be there in twenty years? Mind you, you'll need those suppliers for "Astute" and "Successor". That could drive change. And the way we control the nuclear-implicated work is very, very safe. There's independent scrutiny and standards are getting better.

'The real issue is that we've moved from twenty SSNs to seven, so regenerating suitably qualified and experienced personnel with less than half the platforms. The risk is making savings in this.'

Leckie was certain that we cannot afford to dip below the critical mass required to remain a submarine nation. 'Barrow lost the ability between "Vanguard" and "Astute",' he says. 'You cannot afford to lose that critical mass.'

The timing anxiety increased a month after our visit to Devonport when the September announcement of the 'Initial Gate' funding, the choice of nuclear reactor for the Successors and the configuration of their missile compartments, did not materialize. The chief cause of the delay was arguments between the Ministry of Defence and the Treasury about reactors. The Treasury, with Danny Alexander, the Liberal Democrat Chief Secretary as its lead Minister, were finally persuaded that rather than carrying on with an improved model of the latest

PWR2 installed in the 'Astute' class the expense of a new Rolls-Royce Pressurized Water Reactor 3 system (PWR3) was worth while, especially given the stringent health and safety requirements likely to apply in the 2030s, 2040s and 2050s. The Fukushima disaster in Japan sharpened the minds of all concerned to the dangers of running on reactors designed more than a generation earlier.[99]

Work on PWR3 began in 2000. In November 2006, Steve Ludlum, the then managing director of Rolls-Royce, explained the rationale behind the new reactor to the House of Commons Defence Committee:

> The biggest issue with a nuclear reactor is when you are not using the power to move around or for electricity it is still generating heat and you need to take that heat away. Largely speaking, you would do that using a pumped flow system and electricity is required for that. If you lost the electricity the pump flow is not there and it is much harder to take the heat away. So a new design of reactor would aim to avoid pumped flow systems and a more natural process of taking the heat away and, hence, it would be much safer. In doing all that, the affordability changes, too, so when we have looked at a new reactor design compared to the old reactor design, we are looking at something like, perhaps, 10 or 20% improvements in affordability through a new reactor design too, because of the way that we would remove some of the components on the plant that we could basically design out and, again, make the plant better and safer to operate. We would be aiming to make it safer and cheaper, and within that new design sustaining the industry as well.[100]

The decision to go for PWR3 was finally taken on 29 March 2011, though it was not disclosed in Parliament until 18 May 2011 as Government announcements fell into purdah until the local government polls and the referendum on the alternative vote system for elections to the House of Commons had taken place in early May. In addition to a statement in the Commons from Liam Fox, the Defence Secretary, the Coalition published an *Initial Gate Parliamentary Report*.[101] The document estimated that PWR3 would add £50m to the cost of each 'Successor' submarine. The reason given for going for PWR3 was that it would provide 'superior performance over PWR2' as it would significantly reduce periods in upkeep and maintenance as well as enhancing safety. 'As we move to a new class of submarine the requirement to continually improve our performance . . . is only met through PWR3.'[102]

The PWR3 is also the result of renewed technical exchange between the Royal Navy and the United States Navy. In the late 2000s the restrictions on technical exchange put in place by Admiral Rickover were finally lifted. Around 2004 the US decided that it was in its strategic interest to start cooperating again on nuclear propulsion. There were a number of triggers that led to the change of policy. On the British side, the Chief Scientific Adviser to the Ministry of Defence, Sir Keith O'Nions, realized that the technical programme had started to atrophy. On the American side, the former head of US Naval Reactors, Admiral Frank Bowman, had experienced similar problems when the US Navy moved from the 'Los Angeles' class submarines to the new 'Sea Wolf' class. In 2004, both men sponsored a naval exchange programme that led to increased cooperation. The UK had made enormous inroads in sustaining the life of nuclear cores. This is a real UK success story and something the Americans are interested in.[103]

Under the US/UK 1958 MDA, Naval Reactors is providing the Ministry of Defence with US naval nuclear-propulsion technology to facilitate the development of the nuclear-propulsion plant for *Successor*. As of 2013 a UK-based office comprised about forty US personnel providing full-time engineering support for the exchange, with additional support from key US suppliers and other US-based programme personnel as needed.[104] The PWR3 propulsion unit that will power *Successor* is 'based on a US design' but uses 'next-generation UK reactor technology, PWR3 and modern secondary propulsion systems'.[105]

This does not mean that the PWR1 and PWR2 reactors used in existing classes are unsafe. As Rear Admiral Lister explains:

> Designs have evolved over time, so while PWR-2 is perfectly safe, the opportunity to further improve safety performance existed. The other driving point is that the new plant is significantly simpler in design and operation than PWR-2. That brings a reduction in through-life costs, and also brings the levels of operational performance and availability that we seek. An additional supporting argument is that by exercising the design expertise in nuclear propulsion plant you sustain its competence and expertise. Designing a modern pressurised water reactor will sustain these skills for the lifetime of Successor.[106]

However, the Defence Nuclear Safety Regulator's 2012/13 report predicted that the 'Vanguard' class are 'likely to exhibit plant ageing effects' because their lives have been extended to the late 2020s and early 2030s.[107]

In 2012, a problem was found with the PWR2 Prototype at

Dounreay, after low levels of radioactivity were discovered in the cooling waters. Later tests revealed that a microscopic breach in the zirconium cladding around the reactor's core was thought to be to blame, but it was not entirely clear how this came about. The PWR test reactor at Dounreay, which was shut down after the leak was detected, was intentionally run at higher levels of intensity than those in the Royal Navy's submarines in order to pre-empt any similar problems with the reactors on board the vessels. It was restarted in November 2012 and continues to operate safely. But as a result of the discovery, HMS *Vanguard*, the first and oldest of the *Vanguard* fleet, will have to be refuelled with a new core during its next scheduled 'deep maintenance period' in 2015 at a cost of £120m. '"This is the responsible option," said the Defence Secretary, Philip Hammond, in a statement to the House of Commons on 6 March 2014. "Replacing the core on a precautionary basis at the next opportunity, rather than waiting to see if the core needs to be replaced at a later date, which would mean returning Vanguard for a period of unscheduled deep maintenance, potentially putting at risk the resilience of our ballistic missile submarine operations."'[108] A decision to refuel the next oldest submarine, HMS *Victorious*, will not be taken until 2018.

At sea now atop the D5 missiles in the 'Vanguard' class are Holbrooke Mark 4 warheads derived from a late 1970s design. Under way already at Aldermaston and the nearby Burghfield factory in Berkshire is a Holbrooke 'refreshment' programme whose purpose is to sustain the warheads into the 2040s, well into the patrolling life of the 'Successor' submarines. A new facility is under construction at Burghfield under the project name 'Mensa' for the purpose of disassembling and reassembling the Holbrookes after new fuses have been fitted and 'a few tweaks are made to the physics package'.

How long will it take to design and manufacture the new warheads to replace the Holbrooke Mark 4s? Seventeen years is the current estimate. Those seventeen years fall into three phases. First, seven years to conduct experiments on the new ingredients on the 'physics package'. These involve the ORION laser facility at Aldermaston and a shared facility in France, TEUTATES, which handles the hydrodynamic testing.[109] Second, another seven years to test the resultant design. The new ORCHARD computer system at Aldermaston enables the weapons to simulate the effects they would expect from a real explosion and relate them to the data accumulated from years of underground tests in the Nevada desert up to the last BRISTOL shot in 1991. This historical material remains vital. Third, a final 2–3 years for commissioning and approval.

On these timings, Cabinet and Parliament will have to approve the programme in or around 2016–17, not long after the 'Main Gate' decision on the 'Successor' submarines, if the 2040s into service timetable is to be met. If the world turned menacing and a Prime Minister and Cabinet instructed Aldermaston and Burghfield to produce a new warhead more swiftly, they could do so in 7–10 years provided they settled for an improved version of the Holbrooke or the BRISTOL (described by an insider as a 'Ford to Holbrooke's highly tuned Ferrari') and five flight trials instead of ten. As with the Submarine Service as a whole, Aldermaston and Burghfield must be sustained by well-motivated people and the highest level of equipment if the UK is to remain a nuclear-weapons state into the 2060s.

At the same time as it reported to Parliament on the 'Initial Gate', the Coalition announced yet another 'Trident Alternatives Review' to be overseen by Nick Harvey, Liberal Democrat Minister for the Armed Forces (on leaving the Government in the September 2012 reshuffle, he was replaced in this role by Danny Alexander in the Treasury as the Liberal Democrats now were without a Minister in the MOD). The review was led by Ian Forbes, an MOD official on secondment to the National Security Secretariat in the Cabinet Office, and was tasked to examine credible alternatives to a submarine-based system and credible postures other than CASD. On 16 July 2013, the Cabinet Office published an unclassified version of the Trident Alternatives Review. It concluded that:

> There are alternatives to Trident that would enable the UK to be capable of inflicting significant damage such that most potential adversaries around the world would be deterred. It also shows that there are alternative non-continuous postures (akin to how we operate conventional military assets) that could be adopted, including by SSBNs, which would aim to be at reduced readiness only when the UK assesses the threat of a no-notice pre-emptive attack to be low. None of these alternative systems and postures offers the same degree of resilience as the current posture of Continuous at Sea Deterrence, nor could they guarantee a prompt response in all circumstances. Whether the cruise missile-based systems amount to a credible alternative to Trident would depend on a political judgment on whether the UK was prepared to accept:

- a reduction in whom it could deter unilaterally in all circumstances (major nuclear powers might only be deterred if UK acted with its nuclear allies);
- a significant increase in the vulnerability of any alternative system compared with an SSBN (as a result of not being able to deploy covertly and/or not being able to sustain an assured second strike capability through-life);
- and significantly increased operational complexity, especially if Forward Operating Bases were required.

Choosing to operate the SSBNs in a non-continuous posture depends upon the level of political confidence that:

- a potential aggressor would not launch a no-notice pre-emptive attack when the UK was at a lower posture with no boat deployed;
- that, with sufficient warning, the UK could re-constitute back-to-back patrolling before a potential period of heightened tension arises (covering the availability of equipment and suitably trained and motivated civilian, military and industrial personnel); and
- that such back-to-back patrols could then be sustained long enough to cover any emergent crisis.

The costs of delivering an alternative system could theoretically have been cheaper than procuring a like-for-like renewal of Trident were it not for timing and the fact that the UK deterrent infrastructure is finely tuned to support a submarine-based Trident system. In particular, the time it would take to develop a new warhead (itself a costly and high-risk exercise) is judged to be longer than the current Vanguard-class submarines can safely be operated. Bridging the resulting gap in deterrence capability would involve procuring two Successor SSBNs so that a Trident-based deterrent remains available until at least 2040. Doing that at the same time as investing in the development of a new warhead, new missile, new platform and new infrastructure means that transitioning to any of the realistic alternative systems is now more expensive than a 3 or 4-boat Successor SSBN fleet.

Meanwhile the 'Successor' team at Barrow was gradually cranking up its strength of engineers and designers for what was going to be, by its completion, a 5½-year sequence of work led by Tony Johns, a former submariner who served on HMS *Splendid* with Mark Stanhope and describes himself as 'the conductor of a thousand-strong orchestra'.[110]

On our first visit to Barrow for this book in June 2011 he was still building his numbers; at that point he had 600 people in place, 270 of them professional engineers (recruiting the engineers needed was one of his greatest worries; they are working hard on getting retirees to come back as 'associates'). At that stage the United States 'Ohio' replacement programme was still just ahead of 'Successor' by six months to the relief of all those in Barrow. The US is to build twelve Ohio replacement boats at a cost of $5bn each. Each 'Successor' will come in at £2.2bn per boat.

By the time of our second visit to Barrow in June 2012, the timing had changed. In January 2012, the US Department of Defense, as part of a substantial cuts programme, announced '. . . we will delay the new Ohio Submarine replacement by two years without undermining our partnership with the UK'.[111] Though Washington made it plain that this would not involve any delay to the Common Missile Compartment design schedule and that the US would build the first one, this development did add to the anxiety levels in the UK as never before had Britain been procuring ahead of America in this way.

There is another first involved in the design of 'Successor'. This is the first time an SSBN has *not* come off the drawing boards of the old MOD Ships Department in Bath.[112] The facilities at Barrow are stunningly better. They are housed in a vast area, filled with computers, known informally as 'The Blue Lagoon' and formally as 'Sovereign' (Barrow likes to name its rooms after warships). What do we know about the 'Successor' design?

It is slightly bigger than the 'Vanguard' class, at 17,500 tons, which is surprising given it will be carrying fewer missiles in a twelve-tube, as opposed to a sixteen-tube missile compartment. The increase in size is due to the need to smooth out the design by avoiding the drop on the 'Vanguard' class between the missile compartment and the engine room. There will also be more distance between the missile tubes. The size of the PWR3 reactor, which is bigger than the PWR2, also dictates aspects of the design. The PWR3 will not, unlike the reactors on the 'Vanguard' class, need replacing in the mid-life refit. It is much safer too, being a passive reactor, which is 'pretty benign if it goes wrong'.

The 'Successor' class will also feature a number of stealth and signature improvements over the 'Vanguard' class. In 2006 the Joint Intelligence Committee's fifty-year-out assessment prepared for Tony Blair's ministerial group on Trident replacement stressed the need to increase the chance of undetectability by anti-submarine techniques likely to be available decades ahead. Achieving minimal 'signature' is, with sonar, one of the 'crown jewels' of the submariner's trade. A Barrow insider describes it as 'the last redoubt of national sovereignty' (signatures are something neither the UK nor the US share). Most of these elements are highly classified but a smooth external design and a carefully sculpted fin and the immensely secret details of the propulsion system, which will run to a reduced shaft speed, are all key elements. Taken together, this 'slipperiest of shapes' plus the 'propulsor standards' will be one of the 'big differences with Vanguard'. The 'Successor' class will also have X-planes at the back rather than the standard flat ones, altering the appearance noticeably, which will detune what are called 'wake harmonics', thereby reducing 'signature' still further.

The indispensability of US know-how is everywhere apparent. There are thirty Brits working at Electric Boat in Groton, Connecticut, and thirty Americans working at Barrow. The memory is still fresh of how indispensable Electric Boat people sent from Groton to Barrow were in providing what a senior BAE figure called 'a flying buttress' for the yard when it was struggling to fill the knowledge and skills gap which grew between HMS *Vengeance* putting to sea and work starting on HMS *Astute*.

Tony Johns reckons the task needs, in addition to US expertise and collaboration, 'Germanic adherence to process, Gallic flair and English bloody-mindedness'. And Tony, understandably like most involved with UK nuclear-weapons programmes, has faith in what he's doing: 'I fundamentally believe that we need a national deterrent. That's what gets me up in the morning and you only get one crack at this in a generation.'[113]

Barrow, Britain's premier submarine town, is a remarkable repository of craftsmanship. The boats they build are among the most technically demanding and advanced artefacts in the world and each time one enters the great Devonshire Dock Hall you can feel and hear it. The skilled workforce is, however, hugely understated about the specialness of what they do. A senior figure tells us that this understatement drives one of the top admirals, who visits regularly, 'nuts – "sell yourselves," he says'![114]

The overlapping anxieties, though constant and admitted to in conversations, never somehow dampen the ardour for the task or the absorption and pleasure that comes from being part of a manufacturing enterprise of the most stretching kind.

What, in addition to the political uncertainty, are the risks to getting the first 'Successor' out on patrol in July 2028? The main companies involved, BAE Systems, Babcock, etc. must perform and enough 'suitably qualified and experienced personnel' must be available to work on the programme. The UK supply chain also requires careful attention. Deindustrialization of the British economy has made this a growing problem ever since the submarines became the carrier of the British Bomb. For example, with the decline in steelmaking at Motherwell Bridge only Barrow itself has the capability to build inner hulls after buying Motherwell Bridge's machines. And the highest-grade steel for the Successors will be imported from France. The 'Astute' experience too is no end of a lesson for the designers and builders of 'Successor'. As a senior Barrow figure put it: 'I'm fabulously proud of *Astute*. A fabulous piece of engineering but a bugger to build.'[115] Only about 20 per cent of the 'Astute' design was completed when the yard began to cut steel for it. The comparable figure for the first 'Successor' will be 70 per cent, when, if 'Main Gate' is passed, it starts cutting steel in 2016. There have been considerable advances in 'modularization' since *Astute* was begun.

'The shadow of *Astute* is everywhere,' admits a senior Barrow figure. If anything, it has made them pessimistic in their calculations. It has inoculated them against the so-called 'conspiracy of optimism' between the MOD purchaser and the industrial supplier which has been the curse of so many weapons procurements. Another of *Astute*'s problems was that its highly secret Special Forces capability was added relatively late as part of the renegotiation of the contract in 2003. 'Rule one of procurement,' says a BAE insider. 'Once you've decided what you want, don't fiddle around with it.'[116] Rear Admiral Lister and his successor Rear Admiral Mike Wareham have also ensured that both BAE and the Navy feed the lessons of 'Astute' into 'Successor'. As Simon Lister explained it, 'My policy is to take every lesson I can from every quarter I can find and feed it into the design of Successor and its manufacturing plan. I am not saying "Astute has been a failure we are not doing that again". I am saying that we must learn from our experience on a daily basis in how we put Successor together.'[117]

Listening to the 'Successor' conversations makes it clear that the top figures on all sides seem genuinely to like and admire each other. But they all share the scars of the 'Astute' procurement and are extremely sensitive about the possibility of time and cost overruns for 'Successor'. They are, however, bound together by hoops of the highest grade steel, because Submarine Britain is dominated by a monopoly customer – the MOD, and a sheaf of monopoly suppliers. This adds an extra dimension to the notions of 'intelligent customer' and 'smart supplier'. And they both know that any flaws are going to be exposed either by the National

Audit Office and the House of Commons Public Accounts Committee or by the unforgiving deep ocean. Yet John Hudson, the Managing Director at Barrow, who has twenty-eight years' experience working in the shadow of the hulls, is unflustered. So, it seems are all those involved in the pinnacle of all this activity, Operation 'Relentless'.

OPERATION 'RELENTLESS'

Operation 'Relentless' is the oldest standing order across the UK's Armed Forces. It began with the first Polaris patrol in 1968 – to be precise, when the Royal Navy took over the deterrent role from the Royal Air Force in June 1969 – and has carried a variety of codewords of which *Relentless* is the latest. It embraces the UK's nuclear-weapons firing chain from the Prime Minister and his special authentication codes to the 'Vanguard' class submarine on patrol in the silent deeps of the North Atlantic. Operation 'Relentless' is signed off by the Chief of the Defence Staff, who lays upon the Tridentine parts of the Royal Navy, its boats, berths, support staff and suppliers, the duty of round-the-clock Continuous at-Sea Deterrence. It is also the most frequently exercised of the standby requirements laid upon the UK's Armed Forces.

New Prime Ministers are swiftly indoctrinated into its instruments and procedures – nuclear release being, with the secret agency briefings, the most sensitive part of the induction of every new Head of Government. This is not a world of collective Cabinet government or House of Commons votes on the use of force, but is as exclusively prime ministerial as it gets (though each Prime Minister chooses two or three Nuclear Deputies, appointed personally rather than according to the Cabinet hierarchy, lest he or she is wiped out by a bolt from the blue). It is an operation that if, heaven forbid, the UK came to launching a Trident missile, would take just under an hour to propel 60 tons of D5, as we witnessed at the DASO, out of the Atlantic.

The Prime Minister does not have to be in a bunker to do this. He can set the nuclear release procedures in train from anywhere provided he has the right people with him for briefing purposes, together with two nuclear-authenticated officers and the right cryptography in which to encode the message to the Nuclear Operations Targeting Centre (NOTC) beneath the Ministry of Defence's Main Building in Whitehall.

To reach NOTC you descend several floors beneath the MOD to well below the water level of the nearby Thames. First you pass through the heavily guarded entrance to the Defence Crisis Management Centre (codenamed PINDAR after the ancient Greek poet, for reasons nobody seems to know) and then through two huge bulkheads designed to protect PINDAR's equipment and people from the electromagnetic pulse created by any nuclear assault on the heart of government in Whitehall above. PINDAR's corridors glow with artificial light; the air conditioning hisses; the atmosphere has some similarities to that of a submarine. Then, through a navy blue door, you enter the extended domain of the Submarine Service. Once inside NOTC's three-room suite you meet the head of this extraordinary branch of the wider tribe, Lieutenant Commander Barry Wells. His chief, Rear Admiral John Gower, is present to divulge what can be shared from this world of multiple codenames, straplines and secret drills. 'We are rather evangelic about need-to-know,' he says with Delphic charm.

Gower chairs the Operation 'Relentless' Assurance Committee (ORAC) on which sit (in the spring of 2014) his co-Rear Admirals, Rear Admiral Matt Parr, COMOPS, and Rear Admiral Mark Beverstock, CSSE. They meet several times a year and produce an annual report for the Prime Minister on the maintenance of Continuous at-Sea Deterrence (though if it were ruptured, the PM would be told straightaway). 'It's all about the management of risk. Nothing in government is funded to the level of desire. It's funded to the level of necessity.' On the morning we visit, NOTC possesses an air of calm, as one would hope. The Prime Minister sets its level of readiness and it is currently the lowest. The Crimean crisis is raging away on the rim of the Black Sea but its international ripples are not affecting the pitch of 'Relentless'.

The first of NOTC's rooms is about 16 feet by 8 with light blue walls. Here sit the computer terminals and keyboards which link NOTC to the Prime Minister, to Task Force 345 in the Chiltern chalk at Northwood and with NATO's Supreme Headquarters Allied Powers Europe at Mons in Belgium. It is to here that the PM's National Firing Chain Message, with its coded authentication at the bottom, would come to be fashioned into the PM's National Firing Directive for onward transmission to Northwood.

From the moment the Prime Minister's instruction reaches NOTC's screen, it takes forty minutes to reach the submarine on patrol. Once the submarine has completed its authentications and drills, it takes (as we have seen) a further fifteen minutes for the first Trident to burst out of its tube. So a UK Prime Minister can change the world – and our country – for ever inside an hour. Add on about a further thirty minutes and the missile can reach targets over 4000 nautical miles away from the point where it roars out of the sea. The four clocks on the wall of NOTC's Communications Room give no hint of possible targets. They are local time, Zulu time, SHAPE time and Omaha time (where the US Strategic Command, STRATCOM, is based).

NOTC's second room, its registry, is even smaller. Here are kept the super-secret codewords of the firing chain. The third part of the NOTC suite carries the prosaic name of Room 128. This 30 by 30 foot space is the domain of the UK's nuclear targeteers and contains all the cryptography needed to produce the targeting plans, as directed by the Prime Minister's message, and to conduct nuclear-weapons release. This material is the most sensitive and most heavily protected in Whitehall and that the UK Armed Forces possess. Despite the calmness of Barry Wells's team, the lucid matter-of-factness of John Gower's explanations, the artificial light and the hiss of the air conditioning that makes the submariners feel at home, this understandably creates an atmosphere unlike any other.

Like Tony Blair, Britain's current Prime Minister, David Cameron, is determined to see the United Kingdom continue as a nuclear-weapons state. As Leader of the Opposition, Cameron visited the Royal Navy's Devonport dockyard in June 2008, and as part of his tour, spent over two hours on the Trident submarine, HMS *Victorious*, shortly before it sailed for the first time after its mid-life refit. The firing system was explained to him, including the inner safe and the purposes of its contents in the boat's control room.[118] Very shortly after becoming Prime Minister, David Cameron had to write his letters of last resort. On Thursday, 3 October 2013, we went to interview the Prime Minister in his office in No. 10 Downing Street and asked him how he approached the task:

'I asked John Major in and asked for his advice and I talked to him about it. I also talked to the Chiefs of Staff, I talked to CDS. But then, in the end, it is you in the office on your own. I sat at that chair and there's a great big shredder that was placed right here and you write . . . and then you seal it up. Hopefully nobody will ever see these letters. Each of them goes into the safe of the Trident submarine and then hopefully when you stop being Prime Minister they take it out and burn it and no one will have ever opened it.

'It is a very big moment,' admits Cameron. 'It's the oddest in a way. You've seen Prime Ministers drive up to Buckingham Palace. You've seen them walking through the door of No. 10. You can't really believe you're doing it yourself, but that bit in your office, writing out the letters, with the shredder . . . it is such an extraordinary thing to have to do, you can't really imagine it until you do it.'

Unlike his two predecessors, Tony Blair and Gordon Brown, David Cameron has made a number of visits to the Royal Navy's 'Vanguard' submarines. As we have seen, in 2008, as Leader of the Opposition, Cameron made his first visit, to HMS *Victorious*, while it was undergoing its extended refit. He returned in February 2012, this time as Prime Minister, just as *Vigilant* was preparing to come out of refit and start sea trials.

The Prime Minister's third visit to a Trident boat took place on 3 April 2013, when he was flown in a Royal Navy helicopter to meet HMS *Victorious* as it returned up the Firth of Clyde after completing the Royal Navy's 100th Vanguard deterrent patrol. He was lowered onto the boat and taken on a detailed tour by *Victorious*'s CO, Commander John Livesey. The Prime Minister presented two of HMS *Victorious*'s crew with their Royal Navy Submariner Dolphins, and two other submariners with the Deterrent Patrol Pins. 'When someone has done a patrol in an SSBN they qualify to wear the Deterrent Pin, which is a silver colour,' said Livesey. He then handed the Prime Minister another pin. 'These gold ones are for people who have done twenty patrols.' 'Twenty patrols,' interrupted a shocked Cameron. 'Twenty deterrent patrols,' continued Livesey, 'that amounts to about four years underwater.' 'That is amazing,' said Cameron, holding the gold Deterrent Pin in his hand as Livesey continued. 'The whole delivering deterrence piece is what is important. The guys do this for months at a time. They take a lot of pride in it. A lot of it is drawn out. It can be eighty days of not a lot happening and then a couple of days of intense activity when it's all-important that everybody does their job. The pride in having the Dolphins is one thing, delivering deterrence is another thing again.' The Prime Minister then presented Warrant Officer Joe Brody with the gold Deterrent Pin in recognition of his twenty patrols. 'Many congratulations. That is a huge achievement,' said Cameron. 'Thank you very much indeed.'

As the Prime Minister continued his tour, an article appeared in the *Daily Telegraph*, in which he explained why, as Prime Minister, 'with ultimate responsibility for the nation's security', he believed the country still required a deterrent.[119] His reasoning was not significantly different from Tony Blair's:

First, we need our nuclear deterrent as much today as we did when a previous British Government embarked on it over six decades ago. Of course, the world has changed dramatically. The Soviet Union no longer exists. But the nuclear threat has not gone away. In terms of uncertainty and potential risk it has, if anything, increased. The significant new factor we have to consider is this: the number of nuclear states has not diminished in recent years – and there is a real risk of new nuclear-armed states emerging. Iran continues to defy the will of the international community in its attempts to develop its nuclear capabilities, while the highly unpredictable and aggressive regime in North Korea recently conducted its third nuclear test and could already have enough fissile material to produce more than a dozen nuclear weapons. Last year North Korea unveiled a long-range ballistic missile which it claims can reach the whole of the United States. If this became a reality it would also affect the whole of Europe, including the UK. Can you be certain how that regime, or indeed any other nuclear-armed regime, will develop? Can we be sure that it won't share more of its technology or even its weapons with other countries? With these questions in mind, does anyone seriously argue that it would be wise for Britain, faced with this evolving threat today, to surrender our deterrent? At the end of the day these issues are matters of judgment. My judgment is that it would be foolish to leave Britain defenceless against a continuing, and growing, nuclear threat.

Second, to those who say we can not afford a nuclear deterrent, I say that the security of our nation is worth the price. Of course, the deterrent is not cheap – no major equipment programme is. But our current nuclear-weapons capability costs on average around 5–6 per cent of the current defence budget. That is less than 1.5 per cent of our annual benefits bill. And the successor submarines are, on average, expected to cost the same once they have entered service. It is a price which I, and all my predecessors since Clement Attlee, have felt is worth paying to keep this country safe.

All governments should, of course, carefully examine all options, but I have seen no evidence that there are cheaper ways of providing a credible alternative to our plans for a successor and I am simply not prepared to settle for something that does not do the job. Furthermore, trying to save money by just relying on the United States to act on our behalf allows potential adversaries to gamble that one day the US might not put itself at risk in order to deter an attack on the UK. Only the retention of our independent deterrent makes clear to any adversary that the devastating cost of an attack on the UK or its allies will always be far greater than anything it might hope to gain.

. . .

Third, there are those who say the only way to protect ourselves properly is to get rid of nuclear weapons entirely. Of course, a world without nuclear weapons is a fine ideal. Britain has taken the lead in pushing for progress towards multilateral disarmament. We operate a minimum, nuclear deterrent – and as part of the Strategic Defence and Security Review we reduced the scale of our deterrent even further. But in the absence of an agreement to disarm multilaterally, those who want us to give up our nuclear weapons entirely must provide evidence that there is no prospect of the UK facing a nuclear threat. I cannot make that argument. I believe that to disarm unilaterally in the hope that others would follow would be an act of naivety, not statesmanship. It would be seen by our adversaries not as wisdom, but weakness.

So as the 101st Vanguard submarine patrol begins, a credible and continuous independent nuclear deterrent remains a crucial component of our national security. It is an insurance policy that the United Kingdom cannot do without. That is why I am determined that we will maintain and renew it for generations to come.[120]

This was a theme upon which Cameron expanded during our interview in October 2013, when asked if he subscribed to the analysis advanced by Sir Frank Cooper, former Permanent Secretary at the Ministry of Defence, that a British Prime Minister would never give up Britain's place as a nuclear-weapons state as long as the memory of 1940, standing alone and the Battle of Britain remained fresh. Then a small group of highly trained young men, operating a small amount of very high-quality equipment, was all that prevented the United Kingdom disappearing as an independent nation. 'I obviously feel the 1940 example strongly in a lot of ways,' said Cameron sitting in his No. 10 office: 'You can't not think that just through those doors there, this amazing decision, correct decision was made to fight on and that is an electrifying thought when you're Prime Minister. But to me the nuclear deterrent argument has always been much more about a dangerous world where others have nuclear weapons, where you might be subject to blackmail. I was brought up somewhere between Aldermaston and Greenham Common. My mother was the local magistrate who had to deal with all the protestors and the whole argument about nuclear deterrence was part of my childhood. So I came down on one side of the argument and I haven't changed my mind since.'[121]

But Cameron does admit that 'it's not unthinkable at some time in the future someone will come to a different decision. I don't think

Britain will give up nuclear deterrence altogether. I think that is out. I'd be very surprised if that happened in my lifetime. What I'm saying is that, at the moment, we have a deterrent that is the real thing, that is the genuine article, it's as good as it could be, it is submarine-based, it's continuously at sea and all the rest of it. I think there are people in politics like the Liberal Democrats arguing for deterrence lite. I think the arguments don't stack up. But could I envisage a future Prime Minister thinking, well, maybe it's worth the risk. I wouldn't take that view, but the argument's out there. I think the idea of a Prime Minister giving up nuclear weapons altogether, I don't think that would happen.'[122]

The implications of doing so would certainly be wide reaching, not only for the United Kingdom, but the United States and the NATO Alliance. Since 1963 the United Kingdom independent nuclear capability has also been part of NATO, providing the Alliance's members with protection under the 'NATO Nuclear Umbrella'. The importance of this contribution is often forgotten in the domestic debate on the UK's continuation as a nuclear-weapons state, but it plainly matters to the NATO Alliance. As David Cameron welcomed HMS *Victorious* back from the 100th deterrent patrol, the NATO Secretary-General, Anders Fogh Rasmussen, wrote to the Defence Secretary, Philip Hammond, to thank the United Kingdom for its nuclear commitment to the NATO Alliance:

Dear Secretary of State,
As our most recent strategic concept highlighted, the greatest responsibility of the alliance is to protect and defend our territory and populations against attack, as set out in Article 5 of the Washington Treaty.

A great example of this has been the UK's independent strategic nuclear forces which, through their constant vigilance and professionalism over these past 45 years, have helped ensure the freedom and security of the allies.

These important UK capabilities will continue to play a crucial role as part of NATO's appropriate mix of nuclear and conventional forces that both deter and defend against threats to our alliance.

With that in mind, I would wish to take this opportunity to congratulate the UK on the successful conclusion of the 100th patrol undertaken by the Vanguard class of submarine under Operation Relentless.

I would be grateful if you could please pass on my thanks to all those Royal Navy personnel who undertake these patrols, their families and those who support this operation for their dedication to this vital mission.

Yours sincerely,

Anders Fogh Rasmussen.[123]

Two political events in 2014/2015 had the potential to cause complications for the 'successor' programme. The first was the September 2014 referendum on Scottish Independence during which the separatist Scottish National Party, which has long favoured unilateral nuclear disarmament, vowed to remove Trident and the 'Vanguard' class submarines from Scotland in the event of a yes vote. UK Government spokesmen repeatedly insisted that 'unilateral disarmament is not an option. We are not planning for Scottish independence and as such it is difficult to estimate the total costs, or how long it would take, to replicate the facilities at Faslane, but it would likely cost taxpayers billions of pounds and take many years.'[124] On 18 September 2014 Scotland voted against becoming an independent country by 55 per cent to 45 per cent. A yes vote would certainly have given the SNP a mandate to remove Trident from Scotland, but exactly how this would have been achieved remains unclear.

The second event was the May 2015 general election. Within days of taking over as Labour's Shadow Defence Secretary in October 2013, Vernon Coaker visited the BAE Shipyard in Barrow to display his personal commitment to Labour retaining an independent deterrent and continuing with the 'Successor' programme. 'In an uncertain and unpredictable world in which other nations possess nuclear weapons and nuclear proliferation remains a deep concern, Labour believes it is right that the United Kingdom retains the minimum credible independent nuclear deterrent.'[125] But given the Labour party's tendency to neuralgia on the nuclear-weapons front, it looked unlikely that a future Labour Government would go straight into 'Main Gate' discussions without a review first. 'We will continue to look at ways in which the Successor programme can be delivered efficiently, through the strategic defence and security and zero based spending reviews we have pledged to conduct under a Labour government,' said Coaker.[126] Yet if such a review (a) lingered long or (b) plumped for an alternative system or (c) went for three Successors instead of four, the 2028 target would be missed and CASD, at the very least, broken for a time, as the Vanguards left service.

Another complication in the months leading up to the election was the increased support for the SNP. Numerous polls projected the SNP securing 54 of Scotland's 59 seats, seriously weakening Labour's ability to secure enough seats to form a majority government. If the projections were accurate and Labour failed to secure the necessary seats outside of Scotland it would have to either govern as a minority government or attempt to form a coalition with other parties. One possible outcome was a Labour and SNP coalition, with the SNP holding the balance of power in any negotiations. The SNP's leader, Nicola Sturgeon, initially said that the non-renewal and removal of Trident from the Clyde was a red-line issue in any post-election negotiations but she later appeared to change position and said that the SNP would simply vote against the 'Main Gate' decision when the vote was held in Parliament.[127]

Despite this apparent shift, the Conservative Party was quick to exploit the possibility of a Labour/SNP coalition. In early April 2015, with less than a month to go until the election, the Conservative Defence Secretary, Michael Fallon, put Trident at the heart of the election debate and claimed that in order to secure support from the SNP the Labour leader, Ed Miliband, would compromise national security and 'stab the United Kingdom in the back' by negotiating away Trident and abandoning any plans to go ahead with the 'Successor' programme. A furious Miliband retorted: 'National security is too important to play politics with. I will never compromise our national security, I will never negotiate away our national security', and the Labour Foreign Secretary, Douglas Alexander, insisted that 'Labour's commitment to continuous at-sea nuclear deterrent was not up for negotiation. The experts say that will require four submarines, we will review that pending any technological advance.'[128]

On 7 May 2015 a Conservative Government was elected with a 12-seat overall majority (331 seats to Labour's 232) on, amongst many other pledges, a mandate to renew Trident and maintain CASD.[129] The Defence Secretary, Michael Fallon, confirmed in Parliament on 8 June 2015, that the Government is 'committed to replacing all four Vanguard submarines with new submarines that will serve this country until at least 2060'.[130]

Given the Conservative majority, 'Successor' will almost certainly go ahead. But the Conservative Government can no longer count on the support of the Labour Party, which on the morning of Saturday 12 September 2015, in the Queen Elizabeth Conference Centre at Westminster, elected the veteran left-wing rebel and Islington North MP, Jeremy Corbyn, as Leader of the Labour Party.

In the hours following his victory, Corbyn, then Vice Chairman of the Campaign for Nuclear Disarmament, made it clear that he intended to take a different approach towards nuclear policy. 'My views on Trident are very well known,' he told Channel 4 News. 'There has to be a discussion about that, and I'm hoping that the party will come together around this issue. We don't need nuclear weapons.'[131] There are deep divisions within the Labour party, with many MPs, including the also newly elected Deputy Leader, Tom Watson, in favour of the 'Successor' programme. 'I think the deterrent has kept the peace in the world for half a century and I hope we can have that debate in the party', said Watson.[132]

So far the debate has been difficult. A review into Labour defence and nuclear policy is underway. The problem is the debate is not just about policy; it is fashioned by long held personal convictions. The Labour leader believes that ridding Britain of nuclear weapons is a moral issue and he is determined to ensure the party adopts a policy of unilateralism. 'Jeremy would not be able to forgive himself if he passed up an opportunity to get Labour to vote against Trident' said one friend of Corbyn.[133] When asked in September if he would ever authorise the use of nuclear weapons, Corbyn said, 'No.'[134] In another interview he said 'I am opposed to nuclear weapons, I am opposed to the holding of and use of nuclear weapons . . . I do not see the use of them as a credible way to do things.'[135] His comments prompted outcry from Labour MPs, with the then Shadow Defence Secretary Maria Eagle, a supporter of the nuclear deterrent, telling the BBC 'I'm surprised he answered the question in the way that he did', saying that it 'undermined to some degree' the policy review process she was responsible for.[136] When the Chief of the Defence Staff, General Sir Nick Houghton, was later asked for a response to Corbyn's vow never to press the nuclear button, the General replied, 'It would worry me if that thought was translated into power, as it were.'[137]

In January 2016, Eagle was removed from the Shadow Defence brief and replaced by the former Shadow Culture Secretary, Emily Thornberry, a unilateralist, who has spoken at Campaign for Nuclear Disarmament (CND) meetings. Hours after Thornberry's appointment, the Shadow Armed Forces Minister, Kevan Jones, resigned from the Labour front bench, citing differences with those now in charge of forming policy. Thornberry maintains that she is approaching the review 'with an open mind', but admits that she is 'extremely sceptical about Trident' and is not 'afraid to ask some very difficult questions.'[138] So far those questions have centered on whether technological developments, such as Cyber attacks and future generations of underwater drones could make Trident ineffective as a nuclear deterrent. 'There are forthcoming generations of drones that can work underseas' said Thornberry. 'At the moment they have two problems; one is communications, and the other is battery life . . . If technology is moving faster than that, then it may well be that Trident will not be able to hide. And if that's right, then if we are to bet everything on mutually assured

90. The Royal Navy's Polaris fleet was retired in 1996 when the four 'Resolution' class SSBNs were replaced by four 'Vanguard' class SSBNs carrying Trident missiles. Here HMS *Resolution* escorts HMS *Vanguard* up the Clyde into Faslane.

91. SSNs continued to exercise traditional Cold War skills and maintain diversions for their crews. In May 1991, HMS *Tireless* surfaced at the North Pole to play cricket against the USS *Pargo*. Note the pitch just to the left of *Tireless*'s fin: to ensure a good bounce, *Tireless* had loaded onboard a 22-yard roll of coir-matting before she sailed.

92. HMS *Artful* on the ship lift at BAE Systems, Barrow Shipyard, waiting to be lowered into the water for the first time, 17 May 2014.

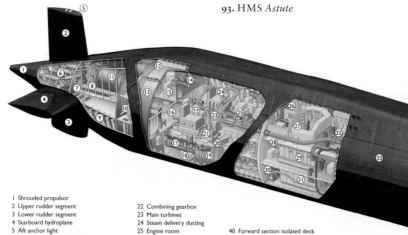

93. HMS *Astute*

1 Shrouded propulsor
2 Upper rudder segment
3 Lower rudder segment
4 Starboard hydroplane
5 Aft anchor light
6 Rudder and hydroplane hydraulic actuators
7 No 4 main ballast tank
8 Propeller shaft
9 High-pressure bottles
10 No 3 main ballast tank
11 Towed array cable drum and winch
12 Main ballast vent system
13 Aft pressure dome
14 Air treatment units
15 Naval stores
16 Propeller shaft thrust block and bearing
17 Circulating water transfer pipes
18 Lubricating oil tank
19 Starboard condenser
20 Main machinery mounting raft
21 Turbo generators, port and starboard

22 Combining gearbox
23 Main turbines
24 Steam delivery ducting
25 Engine room
26 Watertight bulkhead
27 Manoeuvring room
28 Manoeuvring room isolated deck mounting
29 Switchboard room
30 Diesel generator room
31 Static converters
32 Main steam valve
33 Reactor section
34 Part of pressure hull
35 Forward airlock
36 Air handling compartment
37 Waste management equipment
38 Conditioned air ducting
39 Gallery

40 Forward section isolated deck mountings
41 Batteries
42 Junior ratings' mess
43 RESM office
44 Commanding officer's cabin
45 Port side communications office
46 Diesel exhaust mast
47 Snort induction mast
48 SHF/EHF (NEST) mast
49 CESM mast
50 AZL radar mast
51 Satcom mast
52 Integrated comms mast
53 Visual mast – starboard
54 Visual mast – port

55 Navigation mast
56 Bridge fin access
57 Junior ratings' bathroom
58 Senior ratings' bathroom
59 Battery switchroom
60 Control room consoles
61 Sonar operators' consoles
62 Senior ratings bunks
63 Medical berth
64 Weapons stowage and handling compartment
65 Sonar array
66 Maintenance workshop
67 Sonar equipment room
68 Forward hydroplane

94. HMS *Astute* and HMS *Ambush*, part of the Royal Navy's future SSN fleet. Note the Chalfont Special Forces pod on the back of *Ambush*.

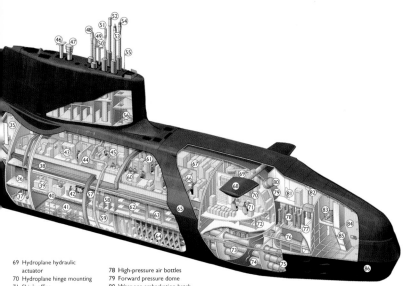

69 Hydroplane hydraulic actuator	78 High-pressure air bottles
70 Hydroplane hinge mounting	79 Forward pressure dome
71 Ship's office	80 Weapons embarkation hatch
72 Junior ratings' berths	81 Gemini craft stowage
73 Torpedo tubes	82 Hinged fairlead
74 Water transfer tank	83 Anchor windlass
75 Torpedo tube bow caps	84 No 1 main ballast tank
76 Air turbine pump	85 Anchor cable locker
77 No 2 main ballast tank	86 Bow sonar

95. (*top left*) HMS *Turbulent* leaves Faslane for a deployment East of Suez, November 2008.

96. (*top right*) HMS *Trenchant* after a dived transit to Singapore, with a number of her anechoic tiles missing.

97–98. In the 1990s, the Service was also equipped with the powerful Tomahawk missile, capable of striking land-based targets hundreds of miles away. Since the end of the Cold War, Royal Navy submarines have fired Tomahawks in a variety of conflicts in the Middle East and Afghanistan.

99. HMS *Vigilant* prepares to dive to carry out a test-firing of a
Trident D5 missile off Cape Canaveral, Florida, on 23 October 2012,
during a DASO. The large telemetery mast is designed to provide
data from the submarine to those monitoring the launch.

100. (*above*) The CO of HMS *Vigilant*,
Commander Mark Lister, completes the
DASO countdown, 23 October 2012.

101. Lieutenant Commander David
O'Connor in the final stages of the DASO
countdown with his finger on the trigger.

102. A Trident D5 missile launched from HMS *Vigilant* reaches into the sky during the October 2012 DASO.

103. Under current plans the British Government is committed to replacing the 'Vanguard' class in the early 2020s. An artist's impression of the 'Vanguard' replacement submarine – currently known simply as 'Successor'.

104–107. The Perisher course. (i) HMS *Talent* off the Isle of Arran; (ii) Commander Hywel Griffith looks on anxiously as one of the Perisher students guides his submarine into an underwater ravine south of Arran; (iii) one of the Perisher students takes the periscope for a quick 'all round look' of the surface; (iv) the attacking frigate taken through the periscope.

108. The successful Perishers: (left to right) Lieutenant Commander Louis Bull, Lieutenant Commander Ian Ferguson, Commander Ryan Ramsey (Teacher), Lieutenant Commander David Burrill and Lieutenant Commander Ben Haskins celebrating in the Ward Room of HMS *Triumph* moments after finding out that they have passed the course. HMS *Triumph*'s Jolly Roger (with Tomahawks) is behind them.

109. How a submarine dies. The laid-up nuclear submarines at Devonport: *Valiant, Warspite, Conqueror, Courageous, Sovereign, Splendid, Spartan, Superb, Trafalgar, Sceptre, Turbulent* and *Tireless*.
They will soon be joined by HMS *Talent*, HMS *Torbay*, HMS *Trenchant* and HMS *Triumph*.

110. *Dreadnought, Churchill* and *Swiftsure* in the basin at Rosyth.

111. *Resolution, Repulse, Renown* and *Revenge* in the basin at Rosyth.

destruction, we have to be assured that it is going to work. And if it can't hide any more, that is a problem.'[139]

Former senior military figures were quick to dismiss such arguments. The former professional head of the Royal Navy, Admiral Lord West of Spithead, described Thornberry's comments as 'nonsense.'[140] Another former First Sea Lord and former Chief of the Defence Staff, the retired submariner Admiral of the Fleet Lord Boyce, pointed out that 'we are more likely to put a man on Mars within the next six months than we are to make the seas transparent within the next 30 years' and during a speech in the House of Lords, spoke of the 'badly informed talk by some people in positions of responsibility. Such statements' he said, 'are totally speculative, they show serious lack of understanding of anti submarine warfare, the science of oceanography, and the science of the impenetrability of water and they are probably being made with the intent of being irresponsibly and wilfully misleading.'[141] Two former Labour Defence Secretaries, Lord Hutton and Lord Robertson of Port Ellen criticised the 'use of spurious arguments and newly created "facts"' and warned that the review process was 'sliding into chaos and incoherence'.[142]

The Labour review continues, as does work on 'Successor'. In November 2015, the Government published a National Security Strategy and Strategic Defence and Security Review: 'A Secure and Prosperous United Kingdom', which modified the 'Sucessor' programme. The estimated in service date of the first 'Successor' class submarine was delayed to the 'early 2030s' and the costs of the now 35-year acquisition programme also increased, to a total of £31 billion (including inflation over the lifetime of the programme), with a contingency of a further £10 billion. The revised cost and schedule estimates reflect the greater understanding of the detailed design of the submarines and their manufacture.[143]

The review also announced plans to implement a number of organizational, managerial and contractual changes, to deliver the programme. A new organization known as the Director General Nuclear, will be established to bring together all the current nuclear teams, including those working for the Chief Scientific Advisor and the Director of Strategic Programmes, to act as the single sponsor for all aspects of the defence nuclear enterprise, from procurement to disposal, with responsibility for submarines, nuclear warheads, skills, related infrastructure and day-to-day nuclear policy. It will occupy a brand new set of secure, bespoke offices on the sixth floor of the Ministry of Defence Main

Building in Whitehall and will be staffed by around 140 people, a significant increase in the 60 to 65 people currently involved in those nuclear teams. It is the only part of the Ministry of Defence that is increasing in size.

The new organization, which one insider described as 'the controlling mind of the entire nuclear enterprise', will be headed by a commercial specialist, with significant experience of delivering large-scale projects on time and on budget. As part of efforts to strengthen arrangements for the procurement and in-service support of nuclear submarines, a new delivery body with the authority and freedom to recruit and retain the best people to manage the submarine enterprise will also be created and efforts to improve performance and investment and develop skills and infrastructure within the necessary industrial suppliers will also be intensified. The review also announced that the Government intended to put 'in place new industrial and commercial arrangements between government and industry, moving away from a traditional single "Main Gate" approach, which is not appropriate for a programme of this scale and complexity, to a staged investment programme.'[144] The Government remains committed to holding a debate in Parliament, but this will be about 'the principle of Continuous At Sea Deterrence and our plans for "Successor"'.[145]

Given the Conservative majority, and the significant number of Labour MPs who support the maintenance of a British nuclear deterrent, work on 'Successor' will almost certainly continue until the 2020 election (due under the Fixed-term Parliaments Act 2011), by which time the first submarine of the 'Successor' class will be taking shape and steelwork for the second will be in full production in the BAE shipyard in Barrow.[146] In addition, a multitude of components and sub-assemblies will be being manufactured for all four boats by hundreds of suppliers. Our hunch is that there will almost certainly be a British bomb with a 'bloody Union Jack on top of it' somewhere in the grey wastelands of the North Atlantic in the 2030s, 2040s and 2050s, carried by one of the submarines currently being laid out on the computer screens in Barrow's Blue Lagoon. When faced with the nuclear question British Prime Ministers, as primary guardians of national security, seem, knowingly or unknowingly, to have been disciples of Cicero, who wrote in *De Legibus*:

Salus populi suprema est lex.
The safety of the people is the chief law.[147]

Epilogue: How a Boat Dies

For all their stealthy and potentially deadly purposes submarines are warm and companionable places to inhabit. Shut the reactor down, get the people out and close the hatches, what do you have left? A cold, inanimate and very large piece of black steel – a huge metallic corpse.

As we have seen throughout this book, the Royal Navy is very good at rites of different kinds (the Remembrance Day service at the beginning). Sailors know how to give a good send-off. On a shining June morning on the parade ground of HMS *Drake* in Devonport it's the turn of HMS *Tireless*, which returned to base two weeks earlier from one last mission when, in August 2013, it was at five minutes' readiness to fire cruise missiles into Syria from somewhere in the Mediterranean, Operation 'Spinney', and in the spring of 2014 was at its maximum diving depth in the southern Indian Ocean searching for the missing Malaysian airliner, MH 370.

A rumour spread through the boat when berthed in Fremantle that yet another final task would fall to *Tireless* – a long sail across the Pacific to show the White Ensign off the Falklands. It was not to be. After a starring part in Anzac Day, *Tireless* set off for her home port, covering 7000 miles in three weeks.

Like all COs, Commander 'Griff' Griffiths loves his boat and it showed as *Tireless* tied up in Devonport. As guests gather on 19 June in HMS *Drake*'s Wardroom before the decommissioning parade, Griffiths says: 'I'm fine now. But bringing the boat alongside for the last time was very emotional.' Hennessy/Jinks ask: 'Did you have a little cry?' Griffiths: 'No. Nearly. I couldn't get through my final pipe.'

The previous night, Griffiths entertained ten of the COs who had commanded *Tireless* since she left Barrow in 1984 for a final dinner on board. 'It was a sombre affair but extremely enjoyable,' he says.

At 0930 on the dot, the ceremony begins as the Band of the Royal Marines, Plymouth, swing on to the parade ground with 'Hearts of Oak', their white helmets dazzling, leading a guard of honour made up of part of *Tireless*'s crew. The others are already on parade divided into junior rates, senior rates and officers with proud wives, children, sweethearts and parents looking on. It's an occasion of quiet dignity punctuated by the occasional shout of 'Daddy' and the cry of seagulls.

At 0946, Rear Admiral Matt Parr and Commander Griffiths are driven to the parade ground. As they step out of the car, the band, with a Pythonesque touch, provides a little burst of Gilbert and Sullivan's *Iolanthe* before launching into a medley of 'We are Sailing' and 'Mr Blue Skies' as they inspect the ranks.

The service begins with 'I Vow to Thee My Country' and 'For Those in Peril on the Sea' before the Chaplain leads the readings and prayers:

'Lord God, we give thanks for the unique and enduring contribution of the Submarine Service to our Nation and the Fleet: We pray for ourselves and all submariners past and present asking for your continued Blessing and protection upon us.

'And now *Tireless* finishes her operational life we give you thanks for the contribution the boat has made to the security of our Nation and maintenance of peace over the years. We recall all those who have served in her and remember the many friendships that have been forged between members of her company in the past and present, we give you thanks that those friendships continue today and pray that they will stand the test of time as we move on to the new appointments and challenges. We ask this through Jesus Christ, our Lord.'

HMS *Tireless*, the old lady of the Cold War, is out of sight of the parade ground hidden behind dockyard buildings and vast cranes. As the service proceeds, she mutates from HMS *Tireless* to just *Tireless* as her White Ensign and Union Flag are taken down.

Griff comes to the microphone, plainly rippling with emotion but still under control. He sings a last song to his beloved boat and his special crew. 'The happiest and most enjoyable years of my career have been in this great boat . . .' 'We will always be proud of our silent accomplishments.'

Of their submarine he tells the crew: 'we understand her traits and her little flaws . . . The personality we ascribe to the boat is inbred in the ship's company. Your professionalism is the essence of that spirit . . . Be it Operation "Spinney" or the search for MH 370, *Tireless* has written herself into the history books . . . You will always carry the *Tireless* spirit with you.'

He recalls the two crew members *Tireless* lost under the ice during the fire of 2007. These are not routine or ritualized words.

Admiral Parr tells the crew they have taken *Tireless* out 'on a high'.

He refers to Operation 'Spinney' and the search for the airliner, and to the clandestine operations *Tireless* has conducted in 'little bits of the colder seas'. 'In the winning of the Cold War,' he says, 'the "T" class was one of those weapons that proved the supremacy of the West in technology and people.' He offers thanks to the families 'for what they have to put up with'.

The band erupts into another burst of Gilbert and Sullivan. A baby cries. The emotion eases. The guard of honour marches off to 'Hearts of Oak' once more and then the band of the Royal Marines departs to 'A Life on the Ocean Wave'.

The party that follows in the Senior Rates' Mess brims with jollity, cheerful families, sausages and patties and a huge cake with a black *Tireless* on a sea of blue icing cut with a sword by the wife of one of the former COs.

Neil Masson, the tactics and sonar officer, takes us down for one last visit to his boat. *Tireless* is flagless, showing patches of rust and clusters of missing anechoic tiles, and is covered with green seaweed like a grass skirt accumulated in the warm tropical and southern seas for which she was never designed. We toast her demise with a cup of tea in the wardroom as the families of the crew pass through the narrow passages. Tomorrow Griffiths will hand *Tireless* over to Devonport dockyard, her passage almost complete. She will then join her older sisters, eighteen Royal Navy nuclear submarines stored at Devonport and Rosyth dockyards, their hulls maintained in a similar way to operational submarines to ensure that they remain safely afloat. *Tireless* is destined for number three basin in Devonport dockyard, where she will sit alongside *Valiant*, *Warspite*, *Courageous*, *Conqueror*, *Splendid*, *Sceptre*, *Superb*, *Spartan*, *Sovereign*, *Trafalgar* and *Turbulent*. Seven others – *Dreadnought*, *Churchill*, *Swiftsure*, and the four Polaris submarines, *Resolution*, *Repulse*, *Renown* and *Revenge* – are berthed in Rosyth dockyard.

As these submarines age the cost of maintaining them is rising. Capacity is also running out as more and more leave service. The Ministry of Defence is looking for an environmentally safe and cost-effective means of disposing of the fleet via a Submarine Dismantling Project. The current plan involves removing radioactive material from each submarine and storing the waste, the vast majority of which is metal from the reactor compartment (less than 1 per cent of the total weight of each submarine). After the radioactive materials have been removed the hull of each submarine will then be broken up and recycled in a similar way to other warships. Until that happens the huge black metallic corpses of these eighteen submarines lie silently at rest.

SUBMARINE PROFILES

Note: Only the main classes of submarines that feature in the book are listed in this section.

Key:

- SLBM – Submarine Launched Ballistic Missile
- SLCM – Submarine Launched Cruise Missile
- SSK – Conventional Submarine
- SSN – Nuclear Powered Submarine
- SSBN – Nuclear Powered Ballistic Missile Submarine
- SSGN – Nuclear Powered Cruise Missile Submarine

S Class SSK (typical example of Group III)

Units: 62 **Displacement:** 1939 orders: 865 tons surfaced / 990 tons submerged; 1940–41 orders: 890 tons surfaced / 990 tons submerged; 1942–43 orders: 854 tons surfaced / 990 tons submerged **Dimensions:** 217 ft x 28 ft, 7 in **Armament:** 7 x 21-in torpedo tubes or 6 x 21-in torpedo tubes (1942–43) / 1 x 3-in gun / 1 x 20mm Oerlikon and variety of configurations **Machinery:** Diesel engines **Speed:** 14.5 kn surfaced / 9 kn submerged **Range:** 6,000 nm at 10 kn / 120 nm at 3 kn **Diving Depth:** 300–350 ft **Complement:** 48 **In Service:** 1939–1950s (either lost, scrapped, sold or modernized)

Converted S Class high-speed target submarine

Units: HMS *Satyr*, HMS *Sceptre*, HMS *Scotsman* (various configurations, including one without a bridge fin), HMS *Selene*, HMS *Seraph*, HMS *Sleuth*, HMS *Solent*, HMS *Statesman*. **Displacement:** 794 tons surfaced / 875 tons submerged **Dimensions:** 217 ft x 15 ft x 23 ft, 11.5-in **Armament:** 6 x 21-in bow torpedo tubes (deactivated) **Machinery:** Paxman high speed diesel generating set **Speed:** 12.5 kn submerged **Diving Depth:** 300–350 ft **Complement:** Varied **In Service:** 1948 – early 1960s

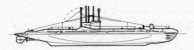

S Class SSK (post-war appearance with snort added)

Units: HMS *Sanguine*, HMS *Scorcher*, HMS *Scythian*, HMS *Sea Devil*, HMS *Sea Scout*, HMS *Seneschal*, HMS *Sentinel*, HMS *Sidon*, HMS *Springer*, HMS *Sturdy*, HMS *Subtle*, HMS *Sirdar*, HMS *Spiteful*, HMS *Sportsman* **Displacement:** 854 tons surfaced / 990 tons submerged **Dimensions:** 217 ft x 28 ft, 7 in **Armament:** 6 x 21-in bow torpedo tubes **Machinery:** Diesel engines **Speed:** 14.5 kn surfaced / 9 kn submerged **Range:** 6,000 nm at 10 kn **Diving Depth:** 300–350 ft **Complement:** Varied **In Service:** 1949–1966

A Class SSK (1951 with snort added)

Units: HMS *Amphion*, HMS *Astute*, HMS *Auriga*, HMS *Aurochs*, HMS *Alderney*, HMS *Alliance*, HMS *Ambush*, HMS *Anchorite*, HMS *Andrew*, HMS *Affray*, HMS *Aneas*, HMS *Alaric*, HMS *Artemis*, HMS *Artful*, HMS *Acheron*, HMS *Alcide* **Displacement:** 1,120 tons standard / 1,375 tons normal / 1,443 tons full load / 1,610 tons submerged **Dimensions:** 279ft, 3 in x 22ft, 3 in x 17 ft, 1 in **Armament:** 6 x 21-in bow torpedo tubes / 4 x 21-in stern torpedo tubes 1 x 4-in/33 QF Mk 23; 1 x 20mm HA (Oerlikon) except *Aeneas*, *Alliance* 2 x 20mm HA (Oerlikon, 1 x 2); 3 x 0.303-in Vickers MG **Machinery:** Diesel engines, electric motors **Range:** 10,500 nm at 11 kn (surface) / 90 nm at 3 kn or 16 nm at 8 kn (submerged) **Diving Depth:** 500 ft **Test Depth:** 600 ft **Complement:** 61 **In Service:** 1945–1970s (modernized in 1950s)

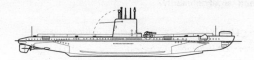

Modified A Class SSK (1958 after streamlining)

Units: HMS *Amphion*, HMS *Astute*, HMS *Auriga*, HMS *Alderney*, HMS *Alliance*, HMS *Ambush*, HMS *Anchorite*, HMS *Andrew*, HMS *Aneas*, HMS *Alaric*, HMS *Artemis*, HMS *Artful*, HMS *Acheron* **Displacement:** 1,120 tons standard / 1,375 tons normal / 1,443 tons full load / 1,610 tons submerged **Dimensions:** 279 ft, 9 in x 22 ft, 3 in x 18 ft **Armament:** 66 x 21-in bow torpedo tubes / 4 x 21-in stern torpedo tubes / 1 x 4-in / 33 QF Mk 23 could be fitted if required **Machinery:** Diesel engines, electric motors **Range:** 11,000 nm at 11 kn (surface) / 114 nm at 3 kn (submerged) / 22 nm at 10 kn (submerged) **Complement:** 64 **In Service:** 1955–1972

T Class SSK (typical example of Group III)

Units (1945): HMS *Talent*, HMS *Tally-Ho*, HMS, *Tantalus*, HMS *Templar*, HMS *Tireless*, HMS *Truculent*, HMS *Truncheon*, HMS *Taciturn*, HMS *Tantivy*, HMS *Telemachus*, HMS *Tradewind*, HMS *Trespasser*, HMS *Truant*, HMS *Thrasher*, HMS *Torbay*, HMS *Trident*, HMS *Taciturn*, HMS *Tapir*, HMS *Taurus*, HMS *Thorough*, HMS *Thule*, HMS *Tiptoe*, HMS *Totem*, HMS *Trenchant*, HMS *Trump*, HMS *Turpin*, HMS *Terrapin*, HMS *Tudor*, HMS *Tuna*, HMS *Taku*, HMS *Tribune*, HMS *Trusty*, HMS *Thermopylae,* HMS *Token*, HMS *Tabard*, HMS *Teredo* **Displacement:** 1,090 tons standard / 1,320 tons normal / 1,424 tons full load **Dimensions:** 273 ft, 6 in x 26 ft, 7 in x 15ft, 10 in **Armament:** 6 x 21-in bow torpedo tubes / 1 x stern torpedo tube **Machinery:** Diesel engines, electric motors **Range:** 11,000 nm at 10 kn **Diving Depth:** 350 ft **Complement:** 50 **In Service:** 1945–1950 (the last unconverted T class left service in July 1963)

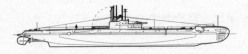

T Class SSK (late 1940s with snort added)

Units (1950): HMS *Talent*, HMS *Tally-Ho*, HMS, *Tantalus*, HMS *Templar*, HMS *Tireless*, HMS *Truculent*, HMS *Truncheon*, HMS *Taciturn*, HMS *Tantivy*, HMS *Telemachus*, HMS *Tradewind*, HMS *Trespasser*, HMS *Thorough*, HMS *Thule*, HMS *Tiptoe*, HMS *Totem*, HMS *Trenchant*, HMS *Trump*, HMS *Turpin*, HMS *Tudor*, HMS *Thermopylae*, HMS *Token*, HMS *Tabard*, HMS *Teredo*, **Displacement:** 1,090 tons standard / 1,320 tons normal / 1,424 tons full load **Dimensions:** 273 ft, 6 in x 26 ft, 7 in x 15 ft, 10 in **Armament:** 6 x 21-in bow torpedo tubes / 1 x stern torpedo tube **Machinery:** Diesel engines, electric motors **Range:** 11,000 nm at 10 kn **Diving Depth:** 350 ft **Complement:** 50 **In Service:** 1948–1950s

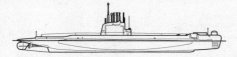

Streamlined T Class SSK

Units: HMS *Tireless*, HMS *Token*, HMS *Tapir*, HMS *Teredo*, HMS *Talent* **Displacement:** 1573 tons submerged **Dimensions:** 273 ft, 6 in x 26 ft, 7 in x 15ft, 10 in (max) **Armament:** 6 x 21-in bow torpedo tubes **Machinery:** Diesel engines, electric motors **Speed:** 15.4 kn submerged **Diving Depth:** 350 ft **Complement:** 68 **In Service:** 1951–1963

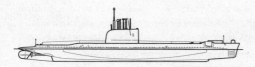

Converted T Class SSK

Units: HMS *Taciturn*, HMS *Turpin*, HMS *Thermopylae*, HMS *Totem*, HMS *Tabard*, HMS *Tiptoe*, HMS *Trump*, HMS *Truncheon* **Displacement:** 1680 tons (dependent on submarine) **Dimensions:** 293 ft (dependent on submarine): HMS *Taciturn* lengthened by 14 ft; HMS *Turpin*, HMS *Thermopylae* and HMS *Totem* lengthened by 12 ft; HMS *Tabard*, HMS *Tiptoe*, HMS *Trump* and HMS *Truncheon* lengthened by 17 feet, 5 in **Armament:** 6 x 21-in bow torpedo tubes **Machinery:** Diesel engines, electric motors **Speed:** 15.4 kn submerged (some exceeded this) **Diving Depth:** 350 ft **Complement:** 68 **In Service:** 1951–1974

HMS *Meteorite*

Units: HMS *Meteorite* **Displacement:** 312 tons surfaced / 337 tons submerged **Dimensions:** 136 ft-2 in x 10 ft-10 in x 14 ft-1 in **Armament:** 2 x bow torpedo tubes **Machinery:** 1 x Walter HTP drive **Speed:** 8.5 kn surfaced / 21.5 kn submerged **Range:** 3,000 nm surfaced at 8 kn / 150 nmi submerged at 20 kn **Diving Depth:** 395 ft **Complement:** 19 **In Service:** 1945–1949

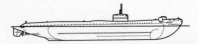

Explorer Class Experimental Submarine

Units: HMS *Explorer*, HMS *Excalibur* **Displacement:** 1,086 tons surfaced / 1,203 tons submerged **Dimensions:** 225 ft, 6 in x 15 ft, 8 in x 14 ft, 5 in **Armament:** None **Machinery:** 2 x hydrogen peroxide experimental drive / diesel electric motors **Speed:** 26 kn surfaced / 26 kn submerged **Diving Depth:** 500 ft **Complement:** Approximately 45 **In Service:** 1956–1964

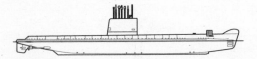

Porpoise/Oberon Class SSK

Units: Porpoise class – HMS *Porpoise*, HMS *Rorqual*, HMS *Narwhal*, HMS *Grampus*, HMS *Finwhale*, HMS *Cachalot*, HMS *Sealion*, HMS *Walrus* / Oberon class – HMS *Oberon*, HMS *Orpheus*, HMS *Odin*, HMS *Olympus*, HMS, *Osiris*, HMS *Onslaught*, HMS *Otter*, HMS *Oracle*, HMS *Ocelot*, HMS *Otus*, HMS *Opossum*, HMS *Onyx* and HMS *Opportune* **Displacement:** 2450 tons **Dimensions:** 295 ft, 2 in x 26 ft, 5 in x 18 ft **Armament:** 6 x 21-in bow torpedo tubes **Machinery:** Diesel engines, electric motors; 2 screws **Fuel:** Diesel oil **Diving Depth:** Over 150 ft **Complement:** 71 **In Service:** 1956–1994

Dreadnought Class SSN

Units: HMS *Dreadnought* **Displacement:** 3,500 tons surfaced / 4,000 tons submerged **Dimensions:** 265 ft, 9 in x 32 ft, 3 in x 26 ft **Armament:** 6 x 21-in bow torpedo tubes **Reactors:** 1 x Westinghouse S5W pressurized water reactor **Range:** Dependent on reactor core life and crew supplies **Speed:** 20 kn surfaced / 28 kn submerged **Diving Depth:** 700 ft **Complement:** 88 **In Service:** 1959–1980

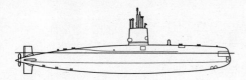

Valiant Class SSN

Units: HMS *Valiant,* HMS *Warspite* / repeat valiants: HMS *Conqueror,* HMS *Courageous,* HMS *Churchill* **Displacement:** 4,400 tons surfaced / 4,900 tons submerged **Dimensions:** 285 ft x 33 ft, 3 in x 27 ft **Armament:** 6 x 21-in bow torpedo tubes **Reactors:** 1 x Rolls Royce pressurized water reactor 1 **Range:** Dependent on reactor core life and crew supplies **Speed:** 20 kn surfaced / 29 kn submerged **Diving Depth:** 750 ft **Complement:** 103 **In Service:** 1966–1994

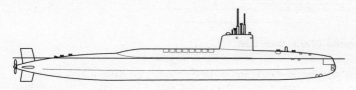

Resolution Class SSBN

Units: HMS *Resolution,* HMS *Repulse,* HMS *Renown,* HMS *Revenge* **Displacement:** 7,500 tons surfaced / 8,500 tons submerged **Dimensions:** 425 ft x 33 ft x 30 ft **Armament:** 6 x 21-in bow torpedo tubes / 16 x submarine launched ballistic missile tubes **Reactors:** 1 x Rolls Royce pressurized water reactor 1 **Range:** Dependent on reactor core life and crew supplies **Speed:** 15 kn surfaced **Diving Depth:** 750 ft **Complement:** 143 **In Service:** 1968–1996

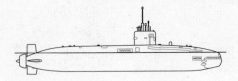

Swiftsure Class SSN

Units: HMS *Swiftsure*, HMS *Sovereign*, HMS *Superb*, HMS *Sceptre*, HMS *Spartan*, HMS *Splendid* **Displacement:** 4,400 tons surfaced / 4,900 tons submerged **Dimensions:** 272 ft x 32 ft, 4 in x 27 ft **Armament:** 5 x 21-inch bow torpedo tubes **Reactors:** 1 x Rolls Royce pressurized water reactor 1 **Range:** Dependent on reactor core life and crew supplies **Speed:** Classified **Diving Depth:** 1,250 ft **Complement:** 116 **In Service:** 1973–2010

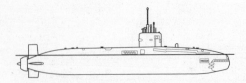

Trafalgar Class SSN

Units: HMS *Trafalgar*, HMS *Turbulent*, HMS *Tireless*, HMS *Torbay*, HMS *Trenchant*, HMS *Talent*, HMS *Triumph* **Displacement:** 4,740 tons surfaced / 5,210 tons submerged **Dimensions:** 280 ft x 32 ft x 31 ft **Armament:** 5 x 21-in bow torpedo tubes **Reactors:** 1 x Rolls Royce pressurized water reactor 1 **Range:** Dependent on reactor core life and crew supplies **Speed:** Classified **Diving Depth:** 1,250 ft **Complement:** 130 **In Service:** 1982 – present

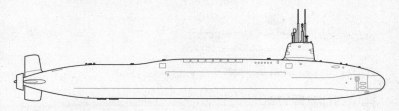

Vanguard Class SSBN

Units: HMS *Vanguard*, HMS *Victorious*, HMS *Vigilant*, HMS *Vengeance*
Displacement: 15,900 tons (submerged) **Dimensions:** 491 ft, 10 in x 42 ft x 39 ft **Armament:** 4
x 21-in bow torpedo tubes **Reactors:** 1 x Rolls Royce pressurized water reactor 2
Range: Dependent on reactor core life and crew supplies **Speed:** Classified
Diving Depth: Classified **Complement:** 135 **In Service:** 1993 – present

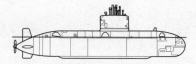

Upholder Class SSK

Units: HMS *Upholder* **Displacement:** 2,205 tons surfaced / 2,465 tons submerged
Dimensions: 230 ft x 24 ft x 25 ft **Armament:** 6 x 21-in torpedo tubes **Machinery:**
Supercharged Paxman Ventura diesel engines / electric motors **Fuel:** Diesel Oil
Speed: 20 kn **Diving Depth:** 820 ft **Complement:** 46 **In Service:** 1990–1994

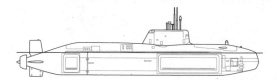

Astute Class SSN

Units: HMS *Astute*, HMS *Ambush*, HMS *Artful*, HMS *Audacious*, HMS *Anson*, HMS
Agamemnon **Displacement:** 7,000 tons surfaced / 7,400 tons submerged **Dimensions:** 318 ft
x 37 ft x 33 ft **Armament:** 6 x 21-inch torpedo tubes; Tomahawk land attack missiles **Reactors:**
1 x Rolls Royce pressurized water reactor 2 **Range:** Dependent on reactor core life and crew
supplies **Speed:** Classified **Diving Depth:** Classified **Complement:** 89 **In Service:** 2010 – present

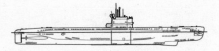

Whiskey Class medium range SSK

Units: 236 **Displacement:** 1,055 tons surfaced / 1350 tons submerged **Dimensions:** 249 ft, 2 in x 20 ft, 8 in x 15 ft, 1 in **Armament:** 6 x 21-in bow torpedo tubes (4 bow, 2 stern) **Machinery:** 2 diesel engines; 4,000 bhp / 2 x electric motors; 2,700 shp / 2 x electric creeping motors; 100 shp / 2 shafts **Speed:** 18.25 kn surfaced / 7 kn surfaced snorkeling / 13 kn submerged **Range:** 22,000 nm surfaced at 9 kn / 443 nm submerged at 2 kn **Diving Depth:** 655 ft **Complement:** 52 **In Service:** 1951 – mid 1990s

Zulu Class long range SSK

Units: 25 (20 x SSK / 5 x SSB) **Displacement:** 1,830 tons surfaced / 2,600 tons submerged **Dimensions:** 296 ft, 10 in x 24 ft, 7 in x 16 ft, 5 in **Armament:** 6 x 21-in bow torpedo tubes / 4 x stern torpedo tubes / 5 units converted to SSB, fitted with enlarged sail for 2 x missiles **Machinery:** 3 x 37D diesel engines; 6,000 bhp / 3 x electric motors; 5,300 shp / 3 shafts (4-bladed propellers) **Speed:** 17 kn surfaced / 15 kn submerged **Range:** 20,000 nm surfaced at economical speed / 9,500 nm on snorkel at 8 kn **Diving Depth:** 655 ft **Complement:** 72 **In Service:** 1954 – mid 1990s (Zulu V last in service)

Juliett SSG

Units: 16 **Displacement:** 3,140 tons surfaced / 4,240 tons submerged **Dimensions:** 281 ft, 9 in x 31 ft, 10 in x 22 ft, 8 in **Armament:** 6 x 21-in bow torpedo tubes / 4 x stern torpedo tubes / 4 x submarine launched cruise missile (SLCM) tubes **Machinery:** 2 diesel engines, 4,000 bhp; 2 x electric motors, 12,000 bhp / 2 shafts **Speed:** 16 kn surfaced / 18 kn submerged **Range:** 9,000 nm at 7 kn **Diving Depth:** 985 ft **Complement:** 78 **In Service:** 1963 – early 1990s

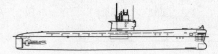

Romeo medium range SSK

Units: 20 **Displacement:** 1,330 tons surfaced / 1,730 tons submerged **Dimensions:** 251 ft, 3 in x 21 ft, 3 in x 15 ft, 1 in **Armament:** 6 x 533mm bow torpedo tubes / 2 x 533mm stern torpedo tubes **Machinery:** 2 x diesel engines **Speed:** 15.5 kn surfaced / 13 kn submerged **Range:** 9,000 nm on snorkel at 9 kn / 350 nm submerged at 2 kn **Diving Depth:** 985 ft **Complement:** 52 **In Service:** 1959 – late 1980s

Foxtrot long range SSK

Units: 40 **Displacement:** 1,950 tons surfaced / 2,400 tons submerged **Dimensions:** 299 ft, 6 in x 24 ft, 7 in x 16 ft, 9 in **Armament:** 6 x bow torpedo tubes / 4 x stern torpedo tubes **Machinery:** 3 x diesel engines **Speed:** 16 kn surfaced / 15.5 kn submerged **Range:** 20,000 nm surfaced at economical speed / 17,900 nm on snorkel at 8 kn / 400 nm submerged at 2 kn **Diving Depth:** 985 ft **Complement:** 78 **In Service:** 1958 – early 1990s

Golf SSB

Units: 22 (9 x Golf 1s / 13 x Golf 11s) **Displacement:** 2,850 tons surfaced / 3,610 tons submerged **Dimensions:** 324 ft, 5 in x 26 ft, 11 in x 26 ft, 7 in **Armament:** 6 x bow torpedo tubes / 4 x stern torpedo tubes / 3 x SLBM tubes (differed in later configurations) **Machinery:** 3 x diesel engines / 3 x electric motors / 3 shafts **Speed:** 14.5 kn surfaced / 12.5 kn submerged **Range:** 9,000 nm on snorkel at 5 kn **Diving Depth:** 985 ft **Complement:** 83 **In Service:** 1959 – early 1990s

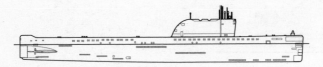

Hotel SSBN (first generation)

Units: 8 **Displacement:** 4,080 tons surfaced / 5,240 tons submerged **Dimensions:** 373 ft, 11 in x 30 ft, 2 in x 25 ft, 3 in **Armament:** 4 x bow torpedo tubes / 2 x stern torpedo tubes **Reactors:** 2 x pressurized water reactors **Speed:** 18 kn surfaced / 26 kn submerged **Range:** Dependent on reactor core life and crew supplies **Diving Dept:** 985 ft **Complement:** 104 **In Service:** 1960 – mid 1990s

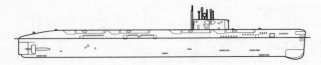

Echo SSGN (first generation); Echo II pictured here

Units: 34 (5 x Echo I / 29 x Echo II) **Displacement:** 4,415 tons surfaced / 5,737 tons submerged **Dimensions:** 378 ft, 6 in x 30 ft, 6 in x 23 ft **Armament:** 6 x bow torpedo tubes / 2 x stern torpedo tubes / 8 SLCM tubes **Reactors:** 2 x pressurized water reactors **Speed:** 14 kn surfaced / 22.7 kn submerged **Range:** Dependent on reactor core life and crew supplies **Diving Dept:** 985 **Complement:** Approximately 90 **In Service:** 1961 – mid 1990s

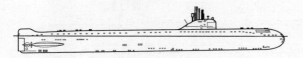

November Class SSN (first generation)

Units: 13 (including K-2, the first Soviet nuclear submarine, which is sometimes referred to as an 'experimental' submarine) **Displacement:** 3,087 tons surfaced / 3,986 tons submerged **Dimensions:** 352 ft, 3 in x 26 ft, 1 in x 21 ft **Armament:** 8 x bow torpedo tubes **Reactors:** 2 x pressurized water reactor / 2 x steam turbines; 35,000 shp / 2 shafts (4 or 6 bladed propellers) **Speed:** 16 kn surfaced / 30 kn submerged **Range:** Dependent on reactor core life and crew supplies **Diving Depth:** 985 ft **Complement:** 110
In Service: 1958 – early 1990s

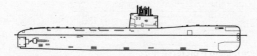

Tango medium range SSK

Units: 18 **Displacement:** 2,640 tons surfaced / 3,560 tons submerged **Dimensions:** 295 ft, 10 in x 28 ft, 3 in x 18 ft, 8 in **Armament:** 6 x bow torpedo tubes / 4 x stern torpedo tubes **Machinery:** 3 x diesel engines; 5,570 bhp / 3 x electric motors; 8,100 shp / 3 shafts (5 bladed propellers) **Speed:** 13 kn surfaced / 15 kn submerged **Range:** 14,000 nm on snorkel at 7 kn / 450 nm submerged at 2.5 kn **Diving Depth:** 985 ft **Complement:** 78
In Service: 1973 – early 2000s

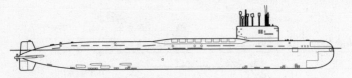

Yankee Class SSBN (second generation)

Units: 34 **Displacement:** 7,760 tons surfaced / 9,600 tons submerged **Dimensions:** 420 ft x 38 ft, 4 in x 26 ft **Armament:** 6 x bow torpedo tubes / 16 x SLBM tubes in Yankee I / 12 x SLBM tubes in Yankee II **Reactors:** 2 pressurized water reactors / 2 steam turbines / 2 x shafts, 5 bladed propellers **Speed:** 15 kn surfaced / 28 kn submerged **Range:** Dependent on reactor core life and crew supplies **Diving Depth:** 1,475 ft **Complement:** 120 **In Service:** 1967 – early 1990s

Victor Class SSN (second generation)

Units: 23 (15 x Victor I / 7 x Victor II) **Displacement:** 3,650 tons surfaced / 4,830 tons submerged **Dimensions:** 305 ft x 34 ft, 9 in x 23 ft, 7 in **Armament:** 4 x bow torpedo tubes / 2 x stern torpedo tubes **Reactors:** 2 x pressurized water reactor / 1 x steam turbine / 1 x shaft **Speed:** 12 kn surfaced / 33 kn submerged **Range:** Dependent on reactor core life and crew supplies **Diving Depth:** 1,300 ft **Complement:** 76 **In Service:** 1967 – early 2000s

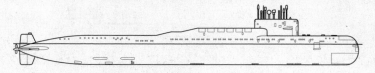

Delta I Class SSBN (second generation)

Units: 22 (18 x Delta 1 / 4 x Delta 11) **Displacement:** 8,900 tons surfaced / 11,000 tons submerged **Dimensions:** 456 ft x 38 ft, 4 in x 27 ft, 6 in **Armament:** 6 x bow torpedo tubes / 12 x SLBM tubes **Reactors:** 2 x pressurized water reactors **Speed:** 15 kn surfaced / 26 kn submerged **Range:** Dependent on reactor core life and crew supplies **Diving Depth:** 1,475 ft **Complement:** 120 **In Service:** 1972 – mid 1990s

Charlie Class SSGN (second generation); Charlie I pictured here

Units: 17 (11 x Charlie 1 / 6 x Charlie 11) **Displacement:** 3,580 tons surfaced / 4,550 tons submerged **Dimensions:** 309 ft, 4 in x 32 ft, 6 in x 24 ft, 7 in **Armament:** 6 x bow torpedo tubes / 8 SLCM tubes (SS-N-7) **Reactors:** 1 x pressurized water reactor / 1 x steam turbine; 18,800 hp / 1 x shaft **Speed:** 26 kn submerged **Range:** Dependent on reactor core life and crew supplies **Diving Depth:** 1,150 ft **Complement:** 90 **In Service:** 1996 – mid 1990s

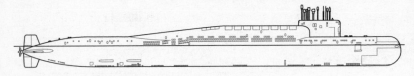

Delta III Class SSBN (second generation)

Units: 14 **Displacement:** 10,600 tons surfaced / 13,000 tons submerged **Dimensions:** 508 ft, 6 in x 38 ft, 4 in x 28 ft, 6 in **Armament:** 6 x bow torpedo tubes / 16 x SLBM tubes
Reactors: 2 x pressurized water reactors **Speed:** 14 kn surfaced / 24 kn submerged
Range: Dependent on reactor core life and crew supplies **Diving Depth:** 1,300 ft
Complement: 130 **In Service:** 1976 – present

Victor III Class SSN (second generation)

Units: 25 **Displacement:** 4,750 tons surfaced / 5,980 tons submerged **Dimensions:** 351 ft, 4 in x 34 ft, 9 in x 26 ft, 3 in **Armament:** 6 x bow torpedo tubes **Reactors:** 2 x pressurized water reactor / 1 x steam turbine / 1 x shaft **Speed:** 18 kn surfaced / 31 kn submerged
Range: Dependent on reactor core life and crew supplies **Diving Depth:** 1,300 ft
Complement: 82 **In Service:** 1977 – present (as of 2015 4 Victor IIIs remain in service with the Russian Navy)

Oscar Class SSGN (third generation); Oscar II pictured here

Units: 13 (2 x Oscar I / 10 x Oscar II) **Displacement:** 12,500 tons surfaced / 22,500 tons submerged **Dimensions:** 472 ft, 4 in x 59 ft, 8 in x 30 ft, 2 in **Armament:** 8 x bow torpedo tubes / 24 x SLCM tubes (24 SS-N-19 missiles) **Reactors:** 2 x OK-560b pressurized water reactors **Speed:** 15 kn surfaced / 30+ kn submerged **Range:** Dependent on reactor core life and crew supplies **Diving Depth:** 600m **Complement:** 107 **In Service:** 1980 – present

Alfa Class SSN (third generation)

Units: 8 (in 1972 the first K-64 suffered a major reactor incident and was taken out of service). **Displacement:** 2,324 surfaced / 3,210 tons submerged **Dimensions:** 261 ft, 2 in x 31 ft, 2 in x 23 ft, 3.5-in **Armament:** 6 x bow torpedo tubes **Reactors:** 1 x lead-bismuth liquid cooled fast reactor / single turbine; 40,000 hp **Speed:** 12 kn surfaced / 41 kn submerged **Range:** Dependent on reactor core life and crew supplies **Diving Depth:** 1,300 ft **Complement:** 29 **In Service:** 1977–1990

Sierra Class SSN (third generation)

Units: 4 (2 x Sierra I / 2 x Sierra II) **Displacement:** 6,300 tons surfaced / 8,300 tons submerged **Dimensions:** 351 ft x 40 ft x 31 ft, 2 in **Armament:** 8 x bow torpedo tubes / **Reactors:** 1 x OK-650a pressurized water reactor **Speed:** 35 kn submerged **Range:** Dependent on reactor core life and crew supplies **Diving Depth:** 1,970 ft **Complement:** 59 **In Service:** 1984 – present

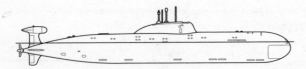

Akula Class SSN (third generation)

Units: 15 (7 x Akula I / 8 x Akula II) **Displacement:** 8,140 tons surfaced / 10,700 tons submerged **Dimensions:** 361 ft, 9 in x 44 ft, 7 in x 31 ft, 9 in **Armament:** 8 x bow torpedo tubes **Reactors:** 1 x OK-650a pressurized water reactor **Speed:** 13 kn surfaced / 33 kn submerged **Range:** Unknown **Diving Depth:** 1,970 ft **Complement:** 73 **In Service:** 1984 – present

Kilo Class SSK

Units: 24 **Displacement:** 2,350 tons surfaced / 3,126 tons submerged **Dimensions:** 242 ft, 1 in x 32 ft, 6 in x 20 ft, 4 in **Armament:** 6 x bow torpedo tubes **Machinery:** 3 diesel engines / 1 x electric motor **Speed:** 11 kn surfaced / 19 kn submerged **Range:** 6,000 nm on snorkel at 7 kn / 400 nm submerged at 3 kn **Diving Depth:** 820 ft **Complement:** 52 **In Service:** 1980 – present

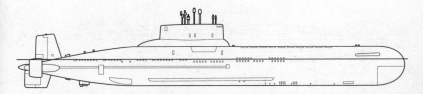

Typhoon Class SSBN (third generation)

Units: 6 **Displacement:** 23,200 tons surfaced / 48,000 tons submerged **Dimensions:** 564 ft, 3 in x 76 ft, 1 in x 36 ft **Armament:** 6 x bow torpedo tubes / 20 x SLBM tubes **Ractors:** 2 x pressurized water reactors / two steam turbines; 50,000 hp each **Speed:** 12–16 kn surfaced / 25–26 kn submerged **Range:** Dependent on reactor core life and crew supplies **Diving Depth:** 1,300 ft **Complement:** 160 **In Service:** 1981 – present (as of 2015 one Typhoon remains in service with the Russian Navy)

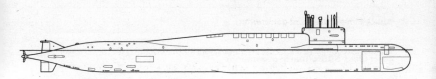

Delta IV Class SSBN (third generation)

Units: 7 **Displacement:** 11,740 tons surfaced / 18,200 tons submerged **Dimensions:** 547 ft, 9 in x 38 ft, 4 in x 28 ft, 10 in **Armament:** 6 x bow torpedo tubes / 16 x SLBM tubes **Reactors:** 2 x pressurized water reactors **Speed:** 14 kn surfaced / 24 kn submerged **Range:** Dependent on reactor core life and crew supplies **Diving Depth:** 1,300 ft **Complement:** 135 **In Service:** 1984 – present

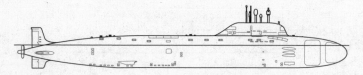

Yasen Class SSN (fourth generation)

Units: 1 (7 planned) **Displacement:** 5,900–9,500 surfaced / 8,600–11,800 submerged
Dimensions: 436.4 ft x 37.7 ft x 27.6 ft **Armament:** 8 x bow torpedo tubes (30 ASW missiles
and/or torpedoes) / 8 x SLCM tubes (24 missiles) **Reactors:** 1 x pressurized water reactor
Speed: 17 kn surfaced / 28–31 submerged **Range:** Dependent on reactor core life and crew
supplies **Diving Depth:** Unknown **Complement:** 80–85 officers **In Service:** 2012 –
present

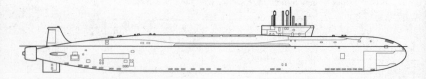

Borei Class SSBN (fourth generation)

Units: 3 (8 planned) **Displacement:** 11,750–14,720 surfaced / 16,750–24,000 submerged
Dimensions: 557 ft, 7 in x 44 ft, 3 in x 32 ft, 8 in **Armament:** 6 x bow torpedo tubes /
12 x SLBM tubes **Reactors:** 2 x pressurized water reactors / 2 diesel generators; 3,400 hp
Speed: 15 kn surfaced / 26 – 29 kn submerged **Range:** Dependent on reactor core
life and crew supplies **Diving Depth:** 1,250–1,475 ft **Complement:** 107–130
In Service: 2013 – present

USS *Albacore*

Units: 1 **Displacement:** 1,517 tons surfaced / 1,810 tons submerged **Dimensions:** 203 ft, 10 in x 29 ft, 4 in x 18 ft, 7 in **Armament:** None **Machinery:** 2 x diesel engines **Speed:** 15 kn surfaced / 27.4 kn submerged **Range:** Unknown **Diving Depth:** 600 ft **Complement:** 37 **In Service:** 1953–1980

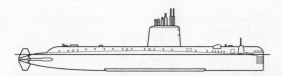

USS *Nautilus*

Units: 1 **Displacement:** 3,180 ton surfaced / 3,500 tons submerged **Dimensions:** 323 ft, 8.5 in x 27 ft, 8 in x 21 ft, 9 in **Armament:** 6 x 21-in bow torpedo tubes **Reactors:** 1 x pressurized water reactor **Speed:** 22 kn surfaced / 23.3 kn submerged **Range:** Dependent on reactor core life and crew supplies **Diving Depth:** 700 ft **Complement:** 104 **In Service:** 1955–1980

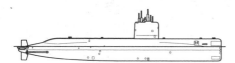

Skate Class SSN

Units: 4 **Displacement:** 2,550 tons surfaced / 2,848 tons submerged **Dimensions:** 267 ft, 8 in x 25 ft x 20 ft, 6 in **Armament:** 6 x 21-in bow torpedo tubes / 2 x 21-in stern torpedo tubes **Reactors:** 1 x S3W pressurized water reactor **Speed:** 15.5 kn surfaced / 18 kn submerged **Range:** Dependent on reactor core life and crew supplies **Diving Depth:** 700 ft **Complement:** 95 **In Service:** 1957–1989

Skipjack Class SSN

Units: 6 **Displacement:** 3,070 tons surfaced / 3,500 tons submerged **Dimensions:** 252 ft x 32 ft x 25 ft **Armament:** 6 x bow torpedo tubes **Reactors:** 1 x pressurized water reactor **Speed:** 15 kn surfaced / 33 kn submerged **Range:** Dependent on reactor core life and crew supplies **Diving Depth:** 700 ft **Complement:** 90 **In Service:** 1959–1990

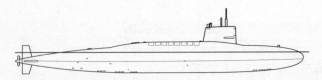

George Washington Class SSBN

Units: 5 **Displacement:** 5,900 tons surfaced / 6,700 tons submerged **Dimensions:** 381 ft, 8 in x 33 ft x 26 ft, 8 in **Armament:** 6 x bow torpedo tubes / 16 x SLBM tubes **Reactors:** 1 x S5W pressurized water reactor **Speed:** 16.5 kn surfaced / 22 kn submerged **Range:** Dependent on reactor core life and crew supplies **Diving Depth:** 700 ft **Complement:** 136 (2 crews) **In Service:** 1960–1985

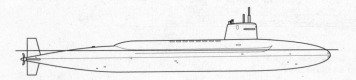

Ethan Allen Class SSBN

Units: 5 **Displacement:** 6,900 tons surfaced / 7,900 tons submerged **Dimensions:** 410 ft, 5 in x 33 ft x 27 ft, 6 in **Armament:** 4 x bow torpedo tubes / 16 x SLBM tubes **Reactors:** 1 x S5W pressurized water reactor **Speed:** 16 kn surfaced / 21 kn submerged **Range:** Dependent on reactor core life and crew supplies **Diving Depth:** 1,300 ft **Complement:** 136 (2 crews) **In Service:** 1961–1992

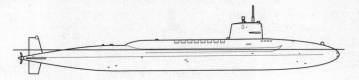

Lafayette Class SSBN

Units: 19 **Displacement:** 7,250 tons surfaced / 8,250 tons submerged **Dimensions:** 425 ft x 33 ft x 27 ft, 9 in **Armament:** 4 x bow torpedo tubes / 16 x SLBM tubes **Reactors:** 1 x S5W pressurized water reactor **Speed:** 16 kn surfaced / 21 kn submerged **Range:** Dependent on reactor core life and crew supplies **Diving Depth:** 1,300 ft **Complement:** 136 (2 crews) **In Service:** 1963–1994

Permit Class SSN

Units: 14 **Displacement:** 3,750 tons surfaced / 3,410 tons submerged **Dimensions:** 278 ft, 6 in x 31 ft, 8 in x 26 ft **Armament:** 4 x bow torpedo tubes **Reactors:** 1 x S5W pressurized water reactor **Speed:** 15 kn surfaced / 27–28 kn submerged **Range:** Dependent on reactor core life and crew supplies **Diving Depth:** 1,300 ft **Complement:** Approximately 88 **In Service:** 1967–1996

Sturgeon Class SSN

Units: 37 **Displacement:** 4,250 tons surfaced / 4,780 tons submerged **Dimensions:** 292 ft x 31 ft, 8 in x 28 ft, 10 in **Armament:** 4 x bow torpedo tubes **Reactors:** 1 x S5W pressurized water reactor **Speed:** 15 kn surfaced / 26–27kn submerged **Range:** Dependent on reactor core life and crew supplies **Diving Depth:** 1,300 ft **Complement:** Approximately 99 **In Service:** 1967–2004

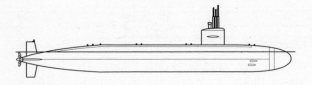

Los Angeles Class SSN

Units: 62 **Displacement:** 6,080 tons surfaced / 6,927 tons submerged **Dimensions:** 362 ft x 33 ft x 32 ft **Armament:** 4 x bow torpedo tubes / 30 units equipped with 12 x vertical SLCM tubes **Reactors:** 1 x S6G pressurized water reactor **Speed:** 33 kn submerged **Range:** Dependent on reactor core life and crew supplies **Diving Depth:** 950 ft **Complement:** 141 **In Service:** 1976 – present

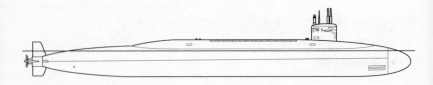

Ohio Class SSBN

Units: 18 (14 x SSBN / 4 x SSGN) **Displacement:** 16,764 tons surfaced / 18,750 tons submerged **Dimensions:** 560 ft x 42 ft x 36 ft, 6 in **Armament:** 4 x 533mm / 24 x SLBM tubes / first 4 of the class converted to SSGNs equipped with 22 x vertical SLCM tubes each carrying up to 7 Tomahawk land attack missiles **Reactors:** 1 x pressurized water reactor **Speed:** 25 kn submerged **Range:** Dependent on reactor core life and crew supplies **Diving Depth:** 985 ft **Complement:** 165 (2 crews) **In Service:** 1981 – present

Seawolf Class SSN

Units: 3 **Displacement:** 9,100 tons submerged **Dimensions:** 353 ft x 40 ft (SSN 23 453 ft x 40 ft) **Armament:** 4 x torpedo tubes / 12 x vertical SLCM tubes **Reactors:** 1 x pressurized water reactor **Speed:** >25 kn submerged **Range:** Dependent on reactor core life and crew supplies **Diving Depth:** >800 ft **Complement:** 140 **In Service:** 1997 – present

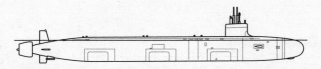

Virginia Class SSN

Units: 12 (as of 2015 – at least 30 planned) **Displacement:** 7,800 tons submerged **Dimensions:** 377 ft x 33 ft **Armament:** 4 x torpedo tubes / 12 x vertical SLCM tubes **Reactors:** 1 x pressurized water reactor **Speed:** >25 kn submerged **Range:** Dependent on reactor core life and crew supplies **Diving Depth:** >800 ft **Complement:** 132 **In Service:** 2004 – present

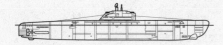

Type XXI U Boat

Units: 170 ordered / over 130 completed **Displacement:** 1,621 tons surfaced / 1,819 tons submerged **Dimensions:** 251 ft, 7 in x 21 ft, 8 in x 20 ft, 8 in **Armament:** 6 x bow torpedo tubes / 4 x 20mm gun **Machinery:** 2 diesel engines **Speed:** 15.6 kn surfaced / 17.2 kn submerged **Range:** 11,150 nm surfaced at 12 kn / 285 nmi submerged at 6 kn **Diving Depth:** 440 ft **Complement:** 57 **In Service:** 1945

Type XVIIB

Units: U-1405, U-1406, U-1407 **Displacement:** 312 tons surfaced / 337 tons submerged **Dimensions:** 136 ft, 2 in x 10 ft, 10 in x 14 ft, 1 in **Armament:** 2 x bow torpedo tubes **Machinery:** 1 x Walter HTP drive **Speed:** 8.5 kn surfaced / 21.5 kn submerged **Range:** 3,000 nm surfaced at 8 kn / 150 nm submerged at 20 kn **Diving Depth:** 395 ft **Complement:** 19 **In Service:** 1945

Notes

INTRODUCTION

1. Conversation with Rear Admiral Simon Lister, Stanton St John, Wiltshire, 20 December 2011 during a session with industrial colleagues on 'Submarine Britain'.

I THE FRANCHISE OF THE DEEP: PERISHER

1. Edward Young, *One of Our Submarines* (Hart-Davis, 1952), p. 115. **2.** Tom Clancy, *The Hunt for Red October* (Naval Institute Press, 1984). Now most readily available in the UK from HarperCollins (1993); Tom Clancy, *Submarine* (Berkley Books, 1993), p. 152. **3.** Conversation with Commander Rémy Thomas, French Navy, 15 April 2012, after spending the weekend on board HMS *Tireless* during the inshore phase of the Perisher course. **4.** Alan Bennett, *Forty Years On and Other Plays* (Faber, 1991), p. 55. **5.** Conversation with Sir Rodric Braithwaite, 30 August 2012. **6.** Commander Hywel ('Griff') Griffiths, CO of HMS *Tireless*, somewhere off the Isle of Arran, 8.15 a.m., Saturday, 14 April 2012.

A Private War Has been Arranged

7. Anthony Preston, *The Royal Navy Submarine Service: A Centennial History* (Conway Maritime Press, 2001), p. 49. **8.** Martin Macpherson, '"Perisher": The Making of a Submarine Commander', in Martin Edmonds (ed.), *100 Years of the Trade: Royal Navy Submarines Past, Present and Future* (Centre for Defence and International Security Studies, 2001), p. 116. **9.** Ibid., p. 119. **10.** Northwood is the location of the Commander Task Force 345, from which the Trident patrols are controlled. **11.** The notion of the 'master-noun' is Tim Blanning's. See his *The Pursuit of Glory: Europe 1648–1815* (Penguin Books, 2008), p. 305. **12.** *The Times*, 'Britain must build on success of Olympics', 9 August 2012. **13.** Ibid. **14.** Edwin Blackburn, 'Scram!', *ORNL Reporter*, No. 19, 2000, p. 6. **15.** Conversation with Commander Hywel Griffiths, 13 July 2012. **16.** Conversation with Captain Andy McKendrick, 22 May 2012. **17.** Conversation with Rear Admiral Ian Corder, 17 September 2012.

2 'THE MOST DANGEROUS OF ALL THE SERVICES':
FROM WORLD WAR TO COLD WAR

1. Hansard, House of Commons Debates, 9 September 1941, Vol. 374, Cols. 67–156.   2. Alfred Roake, 'Cold War Warrior', *Naval Review*, Vol. 82, No. 4, October 1994, pp. 363–72.

Victory

3. Sir Arthur Hezlet, *British and Allied Submarine Operations in World War II* (Royal Navy Submarine Museum, 2002), p. 352.   4. Tim Clayton, *Sea Wolves: The Extraordinary Story of Britain's WW2 Submarines* (Abacus, 2012), p. 389.   5. TNA/ADM/1/19610, Admiral Submarines, 8 August 1945; Michael A. Simpson, *A Life of Admiral of the Fleet Andrew Cunningham: A Twentieth-Century Naval Leader* (Routledge, 2004), p. 91.   6. TNA/ADM/1/19610, Director of Plans, Minute, 2 September 1945.   7. TNA/ADM/1/19610, Memo, 5 April 1946; TNA/ADM/1/19301, meeting held by VCNS to consider the number of submarines that could be kept in service in the post-war Fleet, 28 December 1945.   8. A number of 'U' class submarines were recommissioned in the 1950s. See p. 129.   9. TNA/ADM/1/19610, A. C. Goodall, Minute, 3 August 1945.   10. Edward Young, *One of Our Submarines* (Hart-Davis, 1952), pp. 304–5.   11. Winston Churchill, *The Second World War*, Vol. II: *Their Finest Hour* (Cassell, 1950), p. 598.   12. TNA/ADM/1/18604, 'The Future Development of the Submarine', 22 December 1944.   13. TNA/ADM/1/16396, Director of Anti-U-Boat Division, 29 November, 1944.   14. See Paul Kennedy, *Engineers of Victory: The Problem Solvers Who Turned the Tide in the Second World War* (Allen Lane, 2013).   15. TNA/ADM/1/16396, Admiral Submarine to Secretary of the Admiralty, ' "SCHNORKEL" – Future Policy with Regard to Use in British Submarines', 1 November 1944.   16. TNA/ADM/1/18604, Creasy to Secretary of the Admiralty, 19 January 1945.   17. TNA/ADM/1/16396, Admiral Submarine to Secretary of the Admiralty, ' "SCHNORKEL" – Future Policy with Regard to Use in British Submarines', 1 November 1944.   18. TNA/ADM/1/18604, 'The Future Development of the Submarine', 22 December 1944.   19. Ibid.   20. Ibid.   21. See Harry Hinsley and Edward Thomas, 'The End of the War at Sea', in F. H. Hinsley et al. (eds.), *British Intelligence in the Second World War* (HMSO, 1979–88), Ch. 59, p. 625.   22. Norman Polmar and Jurrien Noot, *Submarines of the Russian and Soviet Navies, 1718–1990* (Naval Institute Press, 1991), p. 137.   23. See David Grier, *Hitler, Dönitz, and the Baltic Sea: The Third Reich's Last Hope, 1944–1945* (Naval Institute Press, 2007), pp. 172–4.   24. Sir Charles S. Lillicrap, Vice President of the Institute of Naval Architects, comments in J. F. Starks, 'German "U"-Boat Design and Production', lecture to the Spring Meeting of the Eighty-Ninth Session of the Institution of Naval Architects, 18 March 1948; for an analysis of how the British evolved anti-submarine

measures to deal with the Type XXI, see Malcolm Llewellyn Jones, 'The Challenge of the Fast Submarine, 1944–1954: Innovation or Evolution?', in Richard Harding (ed.), *The Royal Navy 1930–2000: Innovation and Defence* (Frank Cass, 2005), pp. 135–60.   **25.** Malcolm Llewellyn Jones, *The Royal Navy and Anti-Submarine Warfare, 1917–49* (Routledge, 2006), pp. 81–4.

The Spoils of War

26. TNA/ADM/199/2434, 'Report on the History of the Occupation by the Allies from 5 May to 25 Nov 1945 and the General Activities of the Walterwerke, Kiel, 12 December 1945'.   **27.** Ibid.   **28.** Ibid.   **29.** TNA/ADM/281/143, 'Second Report of Visit to Germany, RE Submarine Construction, July–August, 1945 by Constructor Captain A. J. Sims, D.N.C. Department, Admiralty, Part I'.   **30.** Ibid.   **31.** Norman Polmar and Kenneth Moore, *Cold War Submarines: The Design and Construction of US and Soviet Submarines* (Potomac Books, 2005), p. 34.   **32.** Paul Kemp, *The T-Class: The Classic British Design* (Naval Institute Press, 1990), p. 128; TNA/ADM/1/20045, Minute by Admiral Submarines, 17 July 1945.   **33.** TNA/ADM/1/27774, Engineer-in-Chief, 30 January 1947.   **34.** TNA/ADM/1/27774, Ingolin Underwater Propulsion Project, Minutes of a meeting, 15 February 1946.   **35.** Michael Simpson (ed.), *The Cunningham Papers: Selections from the Private and Official Correspondence of Admiral of the Fleet Viscount Cunningham of Hyndhope*, Vol. 2: *The Triumph of Allied Sea Power 1942–1946* (Ashgate, 2006), p. 267.   **36.** Polmar and Moore, *Cold War Submarines*, p. 39.   **37.** TNA/ADM/167/27, Memorandum by the First Lord of the Admiralty, Development of High Speed Submarines, 29 July 1946.

Modernizing the Wartime Fleet

38. TNA/ADM/1/18604, Director of Naval Construction, Memo, 3 February 1945.   **39.** TNA/ADM/1/18604, Creasy to Secretary of the Admiralty, 'Revised Staff Requirements for the Three Operational Submarines of the 1945 Building Programme', 23 August 1945.   **40.** Ibid.   **41.** TNA/ADM/116/5632, 'Development of Machinery for Fast Underwater Propulsion of Submarines: Naval Staff Answers to Points Raised by E. in C.', 1 September 1949.   **42.** TNA/ADM/1/18604, Creasy to Secretary of the Admiralty, 'Revised Staff Requirements for the Three Operational Submarines of the 1945 Building Programme', 23 August 1945.   **43.** Royal Navy Submarine Museum (RNSM)/ A1944/12, Creasy to Bryant, 12 December 1944.   **44.** TNA/ADM/1/18604, Creasy to Secretary of the Admiralty, 'Revised Staff Requirements for the Three Operational Submarines of the 1945 Building Programme', 23 August 1945.   **45.** Ibid.   **46.** Jack Daniel, 'The Royal Navy and Nuclear Power', lecture to the Spring Meeting of the Eighty-Ninth Session of the Institution of Naval Architects, 18 March 1948.   **47.** TNA/ADM/1/26860, 'Note on the Development of

Nuclear Fuelled Submarines', 5 July 1951. **48.** TNA/ADM/1/27215, 'Submarine Development', 1947. **49.** Ibid. **50.** TNA/ADM/1/18604, Creasy to Secretary of the Admiralty, 'Revised Staff Requirements for the Three Operational Submarines of the 1945 Building Programme', 23 August 1945. **51.** TNA/ADM/1/27774, Engineer-in-Chief, 30 January 1947. **52.** TNA/ADM/1/27774, Ingolin Underwater Propulsion Project, Minutes of a meeting, 15 February 1946. **53.** TNA/ADM/167/27, Memorandum by the First Lord of the Admiralty, 'Development of High Speed Submarines', 29 July 1946. **54.** TNA/ADM/213/1060, CO HMS Meteorite to Captain (S/M), Third Submarine Flotilla, 'Report of Proceedings from 17 March, 1949 to 30 April, 1949'. **55.** TNA/ADM/116/5632, Memorandum on the Development of Machinery for the Fast Underwater Propulsion of Submarines, 1 September 1949. **56.** TNA/ADM/167/134, 'Naval New Construction Programme, 1949/1950', Memorandum by the First Lord of the Admiralty. **57.** TNA/ADM/116/5632, Memorandum on the Development of Machinery for the Fast Underwater Propulsion of Submarines, 1 September 1949. **58.** TNA/DEFE/10/37, D. R.P. (50)123, Submarine Propulsion Development, 31 October 1950. **59.** TNA/ADM/1/23729, The Requirements for an HTP Operational Submarine, 8 September 1952. **60.** John Wise, 'The Royal Navy and the Evolution of the "True Submarine", 1945–1963', in John Jordan (ed.), *Warship 2009* (Conway Maritime Press, 2009). **61.** TNA/ADM/116/5632, Memorandum on the Development of Machinery for the Fast Underwater Propulsion of Submarines, 1 September 1949. **62.** Rodney Carlisle, *Where the Fleet Begins: A History of the David Taylor Research Center, 1898–1998* (University Press of the Pacific, 2003), pp. 250–54; TNA/ADM/116/5632, Memorandum on the Development of Machinery for the Fast Underwater Propulsion of Submarines, 1 September 1949. **63.** Ibid. **64.** TNA/ADM/116/5632, Working Party on Submarine Propulsion, 24 November 1949. **65.** For the best account of the early US nuclear-submarine programme see Richard G. Hewlett and Francis Duncan, *Nuclear Navy 1946–1962* (University of Chicago Press, 1974).

FOSM's Empire

66. This was later known as the 4th Squadron. **67.** TNA/ADM/205/106, Submarine Morale Factors, 1955. **68.** John Coote, *Submariner* (Leo Cooper, 1991), p. 150. **69.** *Daily Telegraph*, Obituary, Vice Admiral Sir Ian McGeoch, 17 August 2007. **70.** Tim Clayton, *Sea Wolves: The Extraordinary Story of Britain's WW2 Submarines* (Abacus, 2012), pp. 230–31. **71.** *Daily Telegraph*, Obituary, Vice Admiral Sir John Roxburgh, 15 April 2004. **72.** John Parker, *The Silent Service* (Headline, 2002), p. 267. **73.** Interview with Rear Admiral John Hervey, 30 January 2013. **74.** Brian Lavery, *Churchill's Navy: The Ships, Men and Organisation, 1939–1945* (Conway, 2006), pp. 212–13. **75.** TNA/ADM/1/19428, Minutes of a Meeting held by VCNS on 3 April 1946, to discuss the size and disposition of the peacetime submarine fleet.

76. TNA/ADM/1/24714, FOSM to Secretary of Admiralty, 3 June 1949. **77.** TNA/ADM/1/2379, Minutes of the 30th Submarine Liaison Meeting, 5 September 1952. **78.** TNA/ADM/1/24714, Director of Naval Training, 16 November 1953. **79.** TNA/ADM/1/24714, Simpson, Minute, 2 February 1954. **80.** John 'Sandy' Woodward, *One Hundred Days: The Memoirs of the Falklands Battle Group Commander* (HarperPress, 2012), p. 51. **81.** Commander J. D. Molyneux, *Submarine Basic Training* (Ministry of Defence Official Publications, 1958). **82.** Woodward, *One Hundred Days*, pp. 51–2. **83.** Sam Fry, *Fruitful Rewarding Years: A Submariner's Story* (The Memoir Club, 2006), p. 41. **84.** Ibid., p. 42. **85.** Interview with Rear Admiral John Hervey, 30 January 2013. **86.** Woodward, *One Hundred Days*, p. 45. **87.** Ibid., p. 45. **88.** Interview with Admiral Sir Peter Herbert, 15 October 2013. **89.** Interview with Rear Admiral John Hervey, 30 January 2013. **90.** Woodward, *One Hundred Days*, p. 55. **91.** Coote, *Submariner*, p. 157. **92.** Vice Admiral Sir Hugh Mackenzie, *The Sword of Damocles* (Periscope Publishing, 2007), p. 161. **93.** Woodward, *One Hundred Days*, p. 56. **94.** IWM/16570, Interview with Geoffrey Jaques, 10 June 1996. **95.** Coote, *Submariner*, p. 156. **96.** IWM/16570, Interview with Geoffrey Jaques, 10 June 1996. **97.** Mackenzie, *Sword of Damocles*, p. 161. **98.** Woodward, *One Hundred Days*, pp. 55–6. **99.** Interview with Rear Admiral John Hervey, 30 January 2013. **100.** 'A Day inside a Clockwork Mouse', *Sydney Morning Herald*, 14 February 1953. **101.** In September and October 1950 Royal Marines from 41 Independent Commando deployed from a US Submarine, the USS *Perch*, and conducted raids on the Korean mainland. **102.** Coote, *Submariner*, p. 150. **103.** Peter Hennessy, *Muddling Through: Power, Politics and the Quality of Government in Postwar Britain* (Victor Gollancz, 1996), pp. 277–83. **104.** TNA/ADM/1/20739, HM Dockyard, Devonport, Operation 'Black-Currant' – January, February & March, 1947. Notes on the provision of D.C. supplies from Submarines. **105.** TNA/ADM/1/20739, Operation Blackcurrant – Technical Report, 6 September 1947. **106.** TNA/ADM/213/595, The Effects of Snorting, from the Royal Navy Physiological Laboratory, April 1947. **107.** 'Five Weeks under Arctic Sea', *Chicago Tribune*, 12 September 1948. **108.** TNA/ADM/213/881, HMS *Ambush*, Report of Snort Patrol, 6 March 1948. **109.** Mackenzie, *Sword of Damocles*, p. 164.

A New Role

110. TNA/ADM/205/53, A Balanced Post War Fleet, 1945. **111.** TNA/ADM/1/24407, N. Abercrombie to Commanders-in-Chief and Flag Officer (Submarines), etc., 8 January 1948. **112.** Quoted in Kemp, *T-Class,* p. 127. **113.** Rear Admiral Martin Wemyss, 'Submarines and Anti-Submarine Operations for the Uninitiated', *RUSI Journal* (September 1981), p. 24. **114.** TNA/ADM/1/21803, FOSM to Secretary of the Admiralty, The Submarine as an Anti-Submarine Weapon, 6 January 1950. **115.** Kemp, *T-Class*, p. 9. **116.** TNA/ADM/1/25252, Memo by Director of Torpedo, Anti Submarine and Mine

Warfare, 6 June 1951. **117.** TNA/ADM/1/25252, FOSM to Secretary of the Admiralty, Submarine Versus Submarine Trials, HMS Alcide and HMS Truncheon, 14 January 1951. **118.** RNSM/A1949, Submarine General Letter, 20 September 1950. **119.** TNA/ADM/189/237, Submarine Weapons and Control Systems, Commander S. A. Hammick, May 1958. **120.** TNA/ADM/189/237, The Future of the Submarine as an A/S Vessel, by Captain A. R. Hezlet, Staff of Flag Officer Submarines. **121.** For more information on sonar development during the early post-war period see Willem Hackman, *Seek and Strike: Sonar, Anti-Submarine Warfare and the Royal Navy, 1914–1954* (HMSO, 1984), pp. 325–54; Norman Friedman, 'Electronics and the Royal Navy', in Harding (ed.), *Royal Navy 1930–2000*, pp. 263–70. **122.** Hackman, *Seek and Strike*, p. 352. **123.** The Type 187 was also used in the 'Porpoise' class which entered service in the late 1950s, see pp. 129–130. **124.** Hackman, *Seek and Strike*, p. 343. **125.** Kemp, *T-Class*, p. 42. **126.** Dan Conley and Richard Woodman, *Cold War Command: The Dramatic Story of a Nuclear Submariner* (Seaforth Press, 2014), pp. 137–8. **127.** Eric Grove, *Vanguard to Trident: British Naval Policy since World War II* (Bodley Head, 1987), p. 227. **128.** Fry, *Fruitful Rewarding Years*, p. 65. **129.** TNA/ADM/189/235, Submarine Progress, 1952. **130.** Ibid.

The Early Cold War

131. Percy Cradock, *Know Your Enemy: How the Joint Intelligence Committee Saw the World* (John Murray, 2002), p. 79. **132.** TNA/ADM/205/83, ACNS Paper, Ships of the Future Navy, 20 April 1949. **133.** Eric Grove and Geoffrey Till, 'Anglo-American Maritime Strategy in the Era of Massive Retaliation, 1945–1960', in John B. Hattendorf and Robert S. Jordan (eds.), *Maritime Strategy and the Balance of Power: Britain and America in the Twentieth Century* (Macmillan, 1989), p. 278. **134.** TNA/CAB/81/132, JIC(46)1(o), Final (Revise), Russia's Strategic Interests and Intentions, 1 March 1946. **135.** TNA/ADM/1/20030, NID/16, Russian Naval Tactics, 10 October 1946. **136.** TNA/ADM/189/237, Discussion to Soviet Scientific Research and Development (Paper read at T.A.S. Conference 1954). **137.** Ibid. **138.** TNA/ADM/1/24407, Rear Admiral Raw to Commander-in-Chief Fleet, Submarine War Plan, Anglo-US Co-Operation. Proposed Further Meeting, 24 July 1951. **139.** Ibid. **140.** Ibid. **141.** TNA/ADM/1/24407, Board Minute, 17 August 1951. **142.** TNA/ADM/1/24407, Rear Admiral Raw to Commander-in-Chief Fleet, Submarine War Plan, Anglo-US Co-Operation. Proposed Further Meeting, 24 July 1951. **143.** TNA/ADM/205/106, Eastern Atlantic Submarine Force, February 1955. **144.** Ibid. **145.** 'Exercise Mainbrace', *Flight International*, 19 September 1952.

'The Admiralty regrets . . .'

146. TNA/ADM/1/22713, Loss of *HMS Truculent*: Admiralty Statement and Report of Board of Enquiry, 1 January 1950–31 December 1950. **147.** One cause of the accident concerned navigation lights. In the 1950s ships over 150 feet in length were equipped with 2 white lights with a lower one forward, to distinguish whether the vessel was on the Port or Starboard bow. They were also equipped with red and green bow lights but these were seldom visible at any distance. Prior to the *Truculent* sinking submarines only carried one white steaming light on top of the periscopes. When *Truculent* went to Port, the submarine was immediately at risk as the *Divina* could only see the single small white light on top of the periscope and was unable to determine if the submarine was turning to Port or Starboard. **148.** Joel Blamey, *A Submariner's Story: The Memoirs of a Submarine Engineer in Peace and in War* (Periscope Publishing, 2002), p. 252. **149.** TNA/ADM/116/5821, Board of Inquiry into the loss of HMS *Affray,* 6 August 1951. **150.** Ibid. **151.** TNA/ADM/116/5821, Shepherd minute, 4 October 1951. **152.** Ibid. **153.** TNA/ADM/116/3821,HeadofNavyLaw,31August1951. **154.**TNA/ADM/205/76, 'A' Class Submarines – Temporary Ban on Operational Use, 27 April 1951. **155.** Hansard, House of Commons Debates, 14 November 1951, Vol. 493, Cols. 980–83. **156.** Alan Gallop, *Subsmash: The Mysterious Disappearance of HM Submarine Affray* (The History Press, 2011), p. 152. **157.** Hansard, House of Commons Debates, 24 January 2012, Vol. 539, Col. 6524. **158.** TNA/ADM/189/235, Captain Bertram Taylor, Submarine Progress – 1952, 11 September 1952.

The Rise of Underwater Intelligence Gathering and Reconnaissance

159. Basil Watson, *Commander-in-Chief: A Celebration of the Life of Admiral of the Fleet The Lord Fieldhouse of Gosport* (Royal Navy Submarine Museum, 2005), p. 42. **160.** John Hervey, 'The Numbers Game', *Naval Review,* Vol. 88, No. 4, October 2000, pp. 402–3. **161.** TNA/CAB/159/34, JIC(60) 49th Meeting, 29 September 1960. **162.** Sherry Sontag and Christopher Drew, *Blind Man's Bluff: The Untold Story of American Submarine Espionage* (Random House, 1999), pp. 1–25. **163.** Jim Ring, *We Come Unseen: The Untold Story of Britain's Cold War Submariners* (John Murray, 2001), p. 83. The Royal Navy's earliest submarine intelligence-gathering operations against the Soviet Union had occurred in December 1939 and the early part of 1940, when four Second World War submarines – HMS *Regulus,* HMS *Rainbow,* HMS *Proteus* and HMS *Perseus* – departed Hong Kong for the waters off Vladivostok to determine whether the German Navy was using Russian facilities to train its U-boat crews. The submarines were also under orders to intercept and capture any German merchant ships attempting to enter Russian ports on the Tartary coast, and to obtain data and experience of operating in frozen Arctic conditions; David Webb, 'Vladivostok'; in Friends of the Royal Navy Submarine Museum, *All Round Look: Year Book 2002/2003* (Royal Navy Submarine Museum, 2003), pp. 14–18. **164.** Young, *One of Our Submarines,* p. 74. **165.** TNA/FO/371/94871, Dinnie to Etherington-Smith, 13 April 1951. **166.** TNA/FO/371/94871, Morgan, Minutes, 19 April 1951. **167.** TNA/FO/371/86803, Memo, 6 May 1950.

168. TNA/FO/371/94871, Morgan, Minutes, 19 April 1951. **169.** TNA/FO/371/100870, Harrison Memo, 8 April 1952. **170.** TNA/FO/371/100870, Hohler to Hanna, 30 April 1952. **171.** TNA/FO/371/100870, Hanna to Hohler, 1 August 1952. **172.** TNA/ADM/1/27784, DNI Memo, 27 February 1953. **173.** TNA/ADM/1/24494, DNI Memo, 19 February 1952. **174.** Ibid. **174.** TNA/ADM/1/27784, Simpson to Secretary of the Admiralty, 28 November 1952. **176.** TNA/ADM/1/27784, DNI Memo, 27 February 1953. **177.** Coote, *Submariner*, p. 184. **178.** TNA/ADM/1/27784, DNI Memo, 27 February 1953.

HMS *Totem* – 'A Slippery Customer'

179. Marine radars: X-band require small antennas, while S-band need larger antennas, are capable of long-range detection and can be used in bad weather. **180.** Quoted in Kemp, *T-Class*, p. 61. **181.** Coote, *Submariner*, p. 185. **182.** Ibid., p. 180. **183.** Quoted in Kemp, *T-Class*, p. 134. **184.** Ibid., p. 180. **185.** Ibid. **186.** Ibid. **187.** TNA/ADM/1/28932, Draft Letter to Commanders-in-Chief, 21 June 1955; references to HMS *Mercury* were removed from the final letter that was circulated around the fleet. **188.** See Richard Aldrich, *GCHQ: The Uncensored Story of Britain's Most Secret Intelligence Agency* (HarperCollins, 2010). **189.** TNA/ADM/1/28932, HMS *Totem*, Report of Operation Cravat, 26 March 1954. **190.** Ibid. **191.** Ibid.; *Indefatigable* and *Implacable* were aircraft carriers being used at the time as training ships. **192.** Coote, *Submariner*, p. 192. **193.** TNA/ADM/1/28932, HMS Totem, Report of Operation Cravat, 26 March 1954. **194.** Ibid. **195.** TNA/ADM/1/28932, Mackay to Commanders-in-Chief, 20 June 1955. **196.** Coote, *Submariner*, p. 192. **197.** '(O) II' probably means the modified 'Ognevoy' class destroyers of the 'Skory' class. **198.** TNA/ADM/1/28292, HMS *Trenchant*, Report of Proceedings – HM Submarines *Sentinel* and *Trenchant*, 12 November 1954. **199.** TNA/ADM/1/28292, Minutes by Director of Naval Intelligence and Director of Undersurface Warfare, 13 August 1954. **200.** TNA/ADM/1/28932, Minute by Head of M Branch, 27 January 1955. **201.** Ibid. **202.** Coote, *Submariner*, p. 193. **203.** Ibid. **204.** LHCMA, Tony Beasley Manuscript, July 2008. There is some confusion as to whether the submarine involved in this incident was HMS *Turpin* or HMS *Totem*. Beaseley's manuscript fails to mention the name of the submarine, but his description of the incident is almost identical to Coote's description of an incident in Coote, *Submariner* pp. 192–3. Richard Aldrich, *GCHQ: The Uncensored Story of Britain's Most Secret Intelligence Agency* (HarperCollins, 2010), p. 138, mentions HMS *Turpin*, but notes that the CO was Coote. Coote was CO of *Totem* at the time, not *Turpin*. **205.** Coote, *Submariner*, p. 193. **206.** LHCMA, Tony Beasley Manuscript, July 2008. **207.** Ibid. **208.** Ibid. **209.** Ibid. **210.** Ibid. **211.** Coote, *Submariner*, p. 193. **212.** See Grove, *Vanguard to Trident*, Ch. 3. **213.** Watson, *Commander-in-Chief*, p. 34. **214.** Aldrich, *GCHQ*, pp. 125–48. **215.** Michael S. Goodman, 'Covering Up Spying in the "Buster" Crabb Affair: A Note', *The International History Review*, 30:4 (2008), pp. 768–84. **216.** TNA/ADM/205/110,

Inglis (DNI) to FOSM, 19 October 1956. **217.** 'Guppy' was an acronym for 'Greater Underwater Propulsive Power'. *Stickleback* was a recently converted 'Guppy IIA' with improved sonar performance. **218.** Coote, *Submariner*, p. 207. **219.** The 'L' class was a Soviet pre-war submarine design still in service although in steadily diminishing numbers in the 1950s. **220.** TNA/ADM/205/110, Coote, USS *Stickleback*, October 1956. **221.** Coote, *Submariner*, p. 208. **222.** TNA/ADM/205/110, Coote, USS *Stickleback*, October 1956.

Operation 'Nightjar'

223. Kemp, *T-Class*, p. 130. **224.** Arthur Bivens, *From Sailboats to Submarines* (Infinity Publishing, 2004), p. 131. **225.** TNA/ADM/1/28939, Admiralty to FOSM, 22 November 1956. **226.** TNA/ADM/1/28939, Head of M Branch Minute, 20 November 1956. **227.** TNA/ADM/1/28939, Admiralty to FOSM, 19 November 1956. **228.** TNA/ADM/1/28939, FOSM, Operation Nightjar, Annex Delta, Instructions for the Conduct of Patrols, 17 November 1956. **229.** TNA/ADM/1/28939, HMS *Tabard*, Operation Nightjar, 19 December 1956. **230.** TNA/ADM/1/28939, HMS *Artful*, Operation Nightjar, 16 December 1956. **231.** TNA/ADM/1/28939, HMS *Tabard*, Operation Nightjar, 19 December 1956. **232.** TNA/ADM/1/28939, Admiralty to FOSM, 30 November 1956. **233.** TNA/ADM/1/28939, Inglis to VCNS, Submarine Patrols, 19 December 1956. **234.** TNA/ADM/1/28939, First Lord's (Lord Hailsham) comments on reports of patrols carried out by HM Submarines *Artful* and *Tabard* – Operation Nightjar, November 1957. **235.** TNA/ADM/1/28939, Elkins to Inglis, 29 November 1956. **236.** *Tirante* was another Guppy IIA submarine. **237.** TNA/ADM/205/110, Elkins to Mountbatten, 31 December 1956. **238.** Ibid. **239.** Ibid. **240.** TNA/ADM/205/169, Inglis to Mountbatten, 4 February 1957. **241.** Ibid. **242.** TNA/ADM/205/169, Mountbatten to Burke, 4 February 1957. **243.** TNA/ADM/205/169, Burke to Mountbatten, 3 March 1957. **244.** TNA/ADM/1/28944, Operation OFFSPRING (HMS/M *Tabard* Submarine patrol), 1957. **245.** Interview with Richard Heaslip, 18 December 2013. **246.** TNA/ADM/1/28933, Operation SANJAK (HMS/M *Turpin* Submarine patrol), 1954–1955; ADM/1/28932, Operation TARTAN (HMS/M *Turpin* Submarine Patrol), 1954–1955; TNA/ADM/1/28944, Operation OFFSPRING (HMS/M *Tabard* Submarine patrol), 1957; TNA/ADM/1/29321, Operation TRIPPER, 1957–9; TNA/ADM/1/29327, Operation ORION, 1958; TNA/ADM/1/29329, Operation ADAMIS, 1958–1959.

Pin-Pricking a Colossus

247. Except where otherwise indicated, the source for the following account of *Taciturn*'s operation is RNSM/A2000/065, M. J. Hurley, 'Early T Boat Patrols in the Cold War' (undated and unpublished manuscript). **248.** Surgeon Captain W. J. Forbes Guild, 'Submarine Living', paper to 'Submarine Medicine and

Submarine Living', the Symposium of Underwater and Aviation Medical Problems at the R. N. Air Medical School, November 1961. **249.** Kemp, *T-Class*, p. 75. **250.** RNSM/A2000/065, Hurley, 'Early T Boat Patrols in the Cold War', (undated and unpublished manuscript). **251.** Watson, *Commander-in-Chief*, p. 43. **252.** Ibid. **253.** Except where otherwise indicated, the source for the following account of *Turpin*'s operation is Alfred Roake, 'Cold War Warrior', *Naval Review*, Vol. 82, No. 4, October 1994, pp. 363-72; the USS *Pueblo* was an American electronic intelligence and signals intelligence ship which was captured by North Korea on 23 January 1968. **254.** Watson, *Commander-in-Chief*, p. 43. **255.** TNA/ADM/205/110, Reid to Mountbatten, Submarines for Jet 56, 8 March 1956. **256.** Roake, 'Cold War Warrior'. **257.** Ibid. **258.** RNSM/A2000/065, M. J. Hurley, 'Early T Boat Patrols in the Cold War' (undated and unpublished manuscript). **259.** Roake, 'Cold War Warrior'. **260.** Ibid.

3 'A NEW EPOCH': TOWARDS THE NUCLEAR AGE

1. TNA/ADM/205/106, Fawkes to Mountbatten, 1955. **2.** Friends of the Royal Navy Submarine Museum, *All Round Look: Year Book 2010/2011* (Royal Navy Submarine Museum, 2010), p. 14. **3.** TNA/ADM/205/163, presentation to Thorneycroft, u/d (May 1962).

The Cold War Fleet

4. TNA/ADM/1/24496, EE, 24 October 1952. **5.** TNA/ADM/1/24897, Minutes of Controller's Meeting with Flag Officer (Submarines) held at Bath, 17 March 1953. **6.** TNA/ADM/1/24897, FOSM to Secretary of the Admiralty, 5 June 1953. **7.** Eric Grove, *Vanguard to Trident: British Naval Policy since World War II* (Bodley Head, 1987), p. 229. **8.** Ibid., pp. 403-9. **9.** TNA/ADM/302/223, Submarine Detection . . . The Present State of the Art and Future Trends, August 1963. **10.** Robert Bud and Philip Gummett, *Cold War, Hot Science: Applied Research in Britain's Defence Laboratories, 1945-1990* (Science Museum, 2002), pp. 166-7. **11.** Norman Friedman, 'Electronics and the Royal Navy', in Harding (ed.), *The Royal Navy 1930-2000: Innovation and Defence* (Frank Cass, 2005), pp. 263-70; Bud and Gummett, *Cold War, Hot Science*, p. 170; Willem Hackman, *Seek and Strike: Sonar, Anti-Submarine Warfare and the Royal Navy, 1914-1954* (HMSO, 1984), pp. 352-3. **12.** TNA/ADM/1/24494, Memorandum on the Characteristics Required by British Submarines, January 1952. **13.** Ibid. **14.** TNA/ADM/1/24494, Director of Plans, Submarines Required for a War with Russia, 1 January 1952. **15.** TNA/ADM/189/235, Submarine Progress, 1952. **16.** TNA/ADM/205/106, FOSM, Present Construction and Modernisation, 1955. **17.** David K. Brown and George Moore, *Rebuilding the Royal Navy: Warship Design since 1945* (Chatham Publishing, 2003), p. 117. **18.** Declan O'Reilly, '*Explorer* and

Excalibur: The Walter Boat, High Test Peroxide and British Submarine Policy 1945-1962. A Study in Technological Failure?', in Martin Edmonds (ed.), *100 Years of the Trade: Royal Navy Submarines, Past, Present & Future* (Centre for Defence and International Security Studies, 2001), pp. 68-74.

The UK Nuclear Programme

19. TNA/PREM/8/1244, Hall to Attlee, 4 May 1950. **20.** TNA/DEFE/7/2055, The Nuclear Powered Submarine, Memorandum by the Admiralty, 14 January 1958. **21.** TNA/ADM/1/26860, Note on the Development of Nuclear Fuelled Submarines, 5 July 1951. **22.** TNA/DEFE/7/2055, DRP(50), Nuclear Fuel Submarines, Memorandum by the Admiralty, 5 June 1950. **23.** Ibid. **24.** Ibid. **25.** Ibid. **26.** TNA/ADM/189/237, Soviet Scientific Research and Development, Paper Read at Torpedo Anti-Submarine Conference, 1954. **27.** Norman Polmar and Kenneth Moore, *Cold War Submarines: The Design and Construction of US and Soviet Submarines* (Potomac Books, 2005), pp. 71-84. **28.** TNA/DEFE/7/2055, DRPS/P(53)5, The Nuclear Submarine, Papers for Naval Review, 26 February 1953. **29.** TNA/ADM/1/26860, ACNS to First Lord/First Sea Lord, 31 January 1950. **30.** TNA/DEFE/7/2055, DRP(50), Nuclear Fuel Submarines, Memorandum by the Admiralty, 5 June 1950. **31.** TNA/ADM/1/26860, Note on the Development of Nuclear Fuelled Submarines, 5 July 1951. **32.** Margaret Gowing (assisted by Lorna Arnold), *Independence and Deterrence: Britain and Atomic Energy, 1945-1952*, Vol. 1: *Policy Making* (Macmillan, 1974), pp. 273-6. **33.** TNA/ADM/1/26860, Note on the Development of Nuclear Fuelled Submarines, 5 July 1951. **34.** TNA/ADM/1/26860, Nuclear Fuelled Submarines, 29 January 1952. **35.** Ibid. **36.** Three submarine designs were considered. The first was a twin-screw design of about 2500 tons with an underwater speed of 25 knots. The second was 3400 tons with a speed of 22 knots. The third was 4500 tons with a speed of 20 knots. The hull diameter had increased from 25 feet in the first design to 31 feet in the third design; Brown and Moore, *Rebuilding the Royal Navy*, p. 120. **37.** TNA/ADM/1/26860, Controller to First Lord, Nuclear Submarine, 29 January 1952. **38.** TNA/DEFE/7/2055, Design Studies on Nuclear Propulsion Units for Naval Vessels, August 1953. **39.** TNA/ADM/1/26860, Nuclear Fuelled Submarines, 29 January 1952. **40.** TNA/PREM/11/2635, The Nuclear Submarine Project, 12 September 1957. **41.** TNA/ADM/1/23729, Minutes of the 30th Submarine Liaison Meeting with the Naval Staff, 19 September 1952. **42.** Ibid. **43.** TNA/ADM/1/23729, Director Tactical & Staff Duties Division, Paper for SDPC Consideration, 12 November 1952. **44.** TNA/1/23729, The Requirements for an HTP Submarine, September 1952. **45.** TNA/DEFE/7/2055, Nuclear Propulsion Systems for the Navy, Note by the Admiralty, 25 June 1953. **46.** Vice Admiral Sir Ted Horlick, 'Nuclear Submarine Propulsion in the RN', The Thomas Lowe Gray Memorial Lecture to the Institution of Mechanical Engineers, January 1982. **47.** The US Navy completed a small experimental midget submarine, X-1, powered by an HTP plant, but it suffered from so many engineering and design problems that it was confined to a museum. **48.** Grove, *Vanguard to Trident*, p. 230.

Securing American Help

49. See Gowing, *Independence and Deterrence*, Vol. 1; Margaret Gowing (assisted by Lorna Arnold), *Independence and Deterrence*, Vol. 2: *Policy Execution* (Macmillan, 1974). **50.** Ian Clark, *Nuclear Diplomacy and the Special Relationship: Britain's Deterrent and America, 1957–1962* (Oxford University Press, 1994), p. 105. **51.** John Baylis, 'Exchanging Nuclear Secrets: Laying the Foundations of the Anglo-American Nuclear Relationship', *Diplomatic History*, 25/1 (2001), p. 36. **52.** TNA/DEFE/7/2055, Exchange of Information with USN on Nuclear Propulsion, Note by Deputy Controller (Research and Development) Admiralty, January 1954. **53.** TNA/DEF/7/2055, Wheeler to Elliot, 5 January 1954. **54.** TNA/ADM/205/106, Director of Plans to First Sea Lord, 19 July 1955. **55.** Admiral Rickover, CBS *60 Minutes* interview with Diane Sawyer, 1984. **56.** Francis Duncan, *Rickover: The Struggle for Excellence* (Naval Institute Press, 2001), p. 81. **57.** TNA/ADM/205/116, Godlet to Harrison Smith, 1 June 1955. **58.** TNA/DEFE/7/2055, Gretton to Roper, Meeting between Rear Admiral Fawkes and Rear Admiral Rickover, 6 September 1955. **59.** TNA/ADM/205/106, Harrison Smith to Engineer-in-Chief, Bath, 7 June 1955. **60.** John Coote, 'How Lord Mountbatten Secured a Reactor from the USN', *United States Naval Institute Proceedings*, 1981. **61.** Quoted in John Coote, *Submariner* (Leo Cooper, 1991), pp. 202–3. **62.** Ignatus Galantin, *Submarine Admiral: From Battlewagons to Ballistic Missiles* (University of Illinois Press, 1995), p. 196, **63.** Coote, 'How Lord Mountbatten . . .' **64.** Bud and Gummett, *Cold War, Hot Science*, p. 161. **65.** Galantin, *Submarine Admiral*, pp. 196–7.

Future Submarine Policy

66. TNA/ADM/205/112, Record of Conversation on 3 November 1955. **67.** Ibid. **68.** TNA/ADM/205/106, Eisenhower to Mountbatten, 23 November 1955. **69.** Philip Ziegler, *Mountbatten: The Official Biography* (HarperCollins, 1985), p. 557. **70.** TNA/ADM/1/26779, Minute by Head of Materiel Branch I, 13 February 1957. **71.** Ibid. **72.** TNA/ADM/205/106, Edwards to Mountbatten, 27 September 1955. **73.** TNA/ADM/205/112, Note of a Conference Held by the First Sea Lord to Discuss Submarine Policy, 2 February 1956. **74.** Ibid. **75.** Norman Polmar, *The Naval Institute Guide to the Soviet Navy* (Naval Institute Press, 5th edn, 1991), pp. 87–93. **76.** TNA/ADM/205/112, Note of a Conference Held by the First Sea Lord to Discuss Submarine Policy, 2 February 1956. **77.** Ibid. **78.** Ibid. **79.** Ibid. **80.** TNA/ADM/205/112, Mountbatten Memo, Conference on Submarine Policy, 26 January 1956. **81.** Ibid.

Collaboration Restored?

82. Hearings before Sub-committee, JCAE, 1958 (85th Cong., 2nd sess., 1958, 516–19).   **83.** Hansard, House of Commons Debates, 20 June 1958, Vol. 544, Col. 1406.   **84.** John Simpson, *The Independent Nuclear State: The United States, Britain and the Military Atom* (Macmillan, 1986), pp. 128–42.   **85.** TNA/ADM/205/112, Elkins to Mountbatten, 19 September 1956.   **86.** TNA/ADM/205/112, Elkins to Admiralty, Exclusive for First Sea Lord, 20 June 1956.   **87.** Ibid.   **88.** John Coote, *Submariner* (Leo Cooper, 1991), p. 203.   **89.** Duncan, *Rickover*, p. 154.   **90.** TNA/ADM/205/112, Meeting with Admiral Rickover at Broadlands on Monday, 20 August 1956.   **91.** TNA/ADM/205/112, Elkins to Mountbatten, 20 June 1956.   **92.** TNA/ADM/205/112, Meeting with Admiral Rickover at Broadlands on Monday 20 August 1956.   **93.** TNA/ADM/205/112, Edwards to Mountbatten, 27 August 1956.   **94.** Ibid.   **95.** TNA/ADM/205/112, Meeting with Admiral Rickover at Broadlands on Monday 20 August 1956.   **96.** Ibid.   **97.** TNA/ADM/205/112, Edwards to Mountbatten, 27 August 1956.   **98.** TNA/ADM/205/112, Meeting with Admiral Rickover at Broadlands on Monday 20 August 1956.   **99.** MB1/I299, Mountbatten to Denning, 4 September 1956, quoted Hill, *Admiral Hyman G. Rickover*   **100.** Ziegler, *Mountbatten*, p. 558.   **101.** TNA/ADM/1/27372, Woods to Secretary of the Admiralty, Staff Requirements for the First Nuclear Submarine, 19 December 1956.   **102.** TNA/ADM/1/27372, Edwards Memo, 22 January 1957.

A Vision of the Future

103. Polmar, *Guide to the Soviet Navy*, p. 38.   **104.** Sergei Khrushchev, *Memoirs of Nikita Khrushchev*, Vol. 2: *Reformer, 1945–1963* (Pennsylvania State University Press, 2006), p. 446.   **105.** Ibid., p. 75.   **106.** Polmar and Moore, *Cold War Submarines*, pp. 93–6.   **107.** TNA/ADM/1/27479, Woods to Secretary of the Admiralty, The Role of the Submarine, 11 August 1956.   **108.** TNA/ADM/205/169, William Hiliman interview with Admiral Arleigh Burke, North American Newspaper Alliance, September 1956.   **109.** TNA/ADM/1/27479, Woods to Secretary of the Admiralty, The Role of the Submarine, 11 August 1956.   **110.** Ibid.   **111.** Ibid.   **112.** Ibid.   **113.** Both these missiles were later cancelled when the Americans decided to concentrate on ballistic systems.   **114.** TNA/ADM/1/27479, The Functions of Our Submarine Forces in Peace and War, 11 August 1956.   **115.** TNA/ADM/1/27479, Minute by Director of Plans, 24 September 1956.   **116.** Ibid.   **117.** Ibid.   **118.** Ibid.   **119.** TNA/ADM/205/112, Note of a Meeting on Submarine Policy Held in First Sea Lord's Room on 8 October, 1956.   **120.** Ibid.   **121.** Ibid.   **122.** Ibid.   **123.** TNA/ADM/1/26286, Fawkes to Secretary of the Admiralty, 1 October 1955.   **124.** TNA/ADM/189/237, Submarine Weapons and Control Systems, Paper by Commander S. A. Hammick, May 1958.   **125.** John Wise, 'The Royal Navy and the Evolution of the "True Submarine", 1945–1963', in John Jordan (ed.), *Warship 2009* (Conway Maritime Press, 2009).   **126.** TNA/ADM/251/114, Investigation of an Incident in HM S/M Explorer, 8 February 1957.   **127.** Polmar and Moore, *Cold War Submarines*, p. 38.   **128.** John Winton, 'The Worst Hangover', *Naval Review*, Vol. 50, No. 1, January 1970.   **129.** RNSM/A1984/52, Commander Chistopher Russell, 'Explorer/Excalibur – What they were like?'.   **130.** Winton, 'The Worst Hangover', p. 245.   **131.** Ibid.   **132.** TNA/ADM/116/5632, Memorandum on the Development of Machinery for the Fast Underwater Propulsion of Submarines, 1 September 1949.

The Special Relationship

133. Hearing before the Sub-committee on Agreements for Cooperation, Joint Committee on Atomic Energy, *Amending the Atomic Energy Act of 1954 – Exchange of Military Information and Material with Allies*, 85th Congress, 2nd Session, p. 162.   **134.** Ibid.   **135.** TNA/DEFE/7/2055, Willis to Pierson, 9 April 1957.   **136.** Sir Alex Smith, *Lock up the Swings on Sundays* (Memoir Club, 1998), pp. 272–3.   **137.** John Jacobsen, 'Mountbatten Asked My Opinion', in Friends of the Royal Navy Submarine Museum, *All Round Look: Year Book 2010/2011* (Royal Navy Submarine Museum, 2010), p. 18.   **138.** Smith, *Lock up the Swings*, pp. 273–4   **139.** Quoted in Vice Admiral Sir Robert Hill, 'Admiral Hyman G. Rickover USN and the UK Nuclear Submarine Propulsion Programme', The Thomas Lowe Gray Memorial Lecture to the Institution of Mechanical Engineers, 19 April 2005.   **140.** Jack Daniel, *The End of an Era: The Memoirs of a Naval Constructor* (Periscope Publishing, 2003), p. 135.   **141.** MB1/I300, First Sea Lord Newsletter, August 1957 quoted in Hill, 'Admiral Hyman G. Rickover'.   **142.** Horlick, 'Nuclear Submarine Propulsion in the RN'; the zero energy reactor used enriched uranium as fuel and ordinary water as the moderator. As it was only capable of producing a few watts of heat it did not require any special cooling.   **143.** Admiral of the Fleet of the Soviet Union S. G. Gorshkov, *Vo Flotskom Stroyu* (military memoirs) (Logos, 1996), p. 175.   **144.** Although the Hotels were initially fitted with the surface-launched R-11 missiles, seven of them, along with fourteen Golfs, were retrofitted with the R-21 underwater-launched missile in the mid-1960s.   **145.** See Owen Cote Jr, *The Third Battle: Innovation in the U.S. Navy's Silent Cold War Struggle with Soviet Submarines* (Naval War College Newport Papers, 2003).   **146.** TNA/DEFE/7/2055, Macmillan to Selkirk, 7 August 1957; TNA/DEFE/7/2055, Selkirk to Macmillan, 12 September 1957.   **147.** TNA/DEFE/7/2055, Selkirk to Macmillan, 12 September 1957.   **148.** Ibid.   **149.** TNA/DEFE/7/2005, Unknown to Way, Nuclear Submarines, 8 October 1957.   **150.** TNA/PREM/11/2635, Thorneycroft to Macmillan, 14 October 1957.   **151.** *Evening News*, 'Atomic Sub Plan May be Scrapped', 23 October 1957; *Daily Telegraph*, 'Check to British A-Ship Plans', 26 October 1957; *Sunday Times*, 'Navy to Lose I-Sub', 1 December 1957.

Operations 'Rum Tub' and 'Strikeback'

152. TNA/ADM/205/169, Mountbatten to Burke, 10 November 1957. **153.** Ziegler, *Mountbatten*, p. 560. **154.** TNA/ADM/189/237, Exercise Rum Tub and the Problem of the Fast Submarine, Paper read by Commander John Coote, Royal Navy, 1957. **155.** TNA/DEFE/7/2055, The Nuclear Powered Submarine, Memorandum by the Admiralty, 14 January 1958. **156.** TNA/ADM/189/237, Exercise Rum Tub and the Problem of the Fast Submarine, Paper read by Commander John Coote, Royal Navy, 1957. **157.** Coote, *Submariner*, p. 213. **158.** Ibid. **159.** TNA/DEFE/7/2055, The Nuclear Powered Submarine, Memorandum by the Admiralty, 14 January 1958. **160.** TNA/ADM/189/237, Exercise Rum Tub and the Problem of the Fast Submarine, Paper read by Commander John Coote, Royal Navy, 1957. **161.** Ibid. **162.** Coote, 'How Lord Mountbatten . . .', 1981. **163.** TNA/ADM/1/27796, Commander-in-Chief Home Fleet to Secretary of the Admiralty, 30 October 1957. **164.** Ibid. **165.** Ibid. **166.** Ibid. **167.** Ibid. **168.** TNA/ADM/189/237, Introduction by Rear Admiral Sir Guy Grantham, Commander-in-Chief Portsmouth, 1957. **169.** TNA/ADM/189/237, Exercise Rum Tub and the Problem of the Fast Submarine, Paper read by Commander John Coote, Royal Navy, 1957. **170.** TNA/ADM/1/27796, Woods to Secretary of the Admiralty, Visit of USS *Nautilus*, 9 November 1957. **171.** Eric Grove, 'Mountbatten as Chief of Naval Staff', Paper presented at 'Aspects of British Defence and Naval Policy in the Mountbatten Era', conference at the University of Southampton, 25–26 September 1990. **172.** TNA/ADM/1/27796, Woods to Secretary of the Admiralty, Visit of USS *Nautilus*, 9 November 1957. **173.** Dwight D. Eisenhower and Harold Macmillan, 'Declaration of Common Purpose between the President of the United States and the Prime Minister of the United Kingdom', Washington DC, 25 October 1957. **174.** Dwight D. Eisenhower, 'Special Message to the Congress Transmitting Agreement with the United Kingdom for Cooperation on Uses of Atomic Energy for Mutual Defense', 3 July 1958.

The Agreement

175. Amending The Atomic Energy Act Of 1954 – Exchange of Military Information and Material with Allies, Congress of the United States, Sub-committee on Agreements for Cooperation Joint Committee on Atomic Energy, 27 February 1958, p. 164. **176.** TNA/DEFE/19/50, Elkins to Brundrett, 9 January 1958. **177.** Hill, 'Admiral Hyman G. Rickover'. **178.** MB1/K208A, Mountbatten to Hamersley, 31 July 1979 quoted in Hill, 'Admiral Hyman G. Rickover'. **179.** Ibid. **180.** Ziegler, *Mountbatten*, pp. 558–9. **181.** MB1/K208A, Mountbatten to Hamersley, 31 July 1979 quoted in Hill, 'Admiral Hyman G. Rickover'. **182.** TNA/DEFE/19/50, Note of a Meeting, 24 January 1958. **183.** TNA/CAB/131/19, D(58) 2nd Meeting, 5 February 1958. **184.** TNA/DEFE/19/50, Note of a Meeting, 24 January 1958. **185.** Hearing before the Sub-committee on Agreements for Cooperation, Joint Committee on Atomic Energy, *Amending the Atomic Energy Act of 1954 – Exchange of Military Information and Material with Allies*, 85th Congress, 2nd Session, 27 February 1958, pp. 501–3. **186.** Ibid. **187.** Ibid. **188.** Ibid.; Baylis, 'Exchanging Nuclear Secrets', p. 48. **189.** TNA/DEFE/7/2055, BJSM to Cabinet Office, 6 March 1957. **190.** TNA/ADM/116/6411, Elkins to Mountbatten, 26 March 1958. **191.** Solly Zuckerman, *Six Men Out of the Ordinary* (Peter Owen, 1992), p. 151. **192.** TNA/ADM/116/6411, Elkins to Mountbatten, 21 April 1958. **193.** TNA/DEFE/13/523, Unknown to Sandys, 30 January 1958. **194.** TNA/ADM/116/6411, Elkins to Mountbatten, 21 May 1958. **195.** Baylis, 'Exchanging Nuclear Secrets', pp. 33–61. **196.** Agreement between Government of the United Kingdom of Great Britain and Northern Ireland and Government of United States of America for Co-operation on the Uses of Atomic Energy for Mutual Defense Purposes, Washington, 3 July 1958.

The *Dreadnought* Programme

197. TNA/DEFE/19/50, Note of a Meeting, 24 January 1958. **198.** TNA/DEFE/72/45, Baker, 9 April 1958. **199.** TNA/DEFE/7/2055, Selkirk to Sandys, 24 March 1958. **200.** TNA/ADM/116/641, Admiralty Board Minutes, 13 February 1958. **201.** TNA/ADM/116/641, Extract from Board Minutes, 13 February 1958. **202.** TNA/DEFE/13/523, Brief about Changes to Agreement, 28 January 1959. **203.** TNA/PREM/11/2635, Macmillan to Selkirk, 9 February 1959. **204.** TNA/DEFE/13/523, Selkirk to Macmillan, 10 February 1959. **205.** TNA/ADM/DEFE/72/45, Le Fanu Memo, 15 March 1961. **206.** TNA/ADM/116/6411, Lang to Makins, 21 February 1958. **207.** Ministry of Defence, 'The United Kingdom's Defence Nuclear Weapons Programme: Plutonium and Aldermaston – An Historical Account', 2000, p. 9. **208.** TNA/DEFE/7/2055, Memo, 18 September 1956. **209.** TNA/DEFE/19/50, First Lord to Sandys, 24 January 1958. **210.** Ibid. **211.** TNA/DEFE/116/6411, Dreadnought Policy, Agenda for Controller's Meeting, 21 March 1958. **212.** TNA/DEFE/69/749, Dreadnought Project, 2 April 1958. **213.** David K. Brown, *Sir Rowland Baker, KB, RCNC: A Personal Appreciation* (MOD, 1996). **214.** TNA/ADM/1/27568, Director of Tactical & Weapons Policy, 15 September 1959. **215.** David K. Brown, *Century of Naval Construction: The History of the Royal Corps of Naval Constructors* (Conway Maritime Press, 1983), p. 235. **216.** Brown and Moore, *Rebuilding the Royal Navy*, p. 121. **217.** TNA/DEFE/69/749, Dreadnought Project, Some Difficulties as of 28th July 1958. **218.** K. Hall and K. C. Mansbridge, 'Materials for Royal Navy Submarines', paper presented at the Symposium on Naval Submarines, London, 17–20 May 1983. **219.** TNA/ADM/1/28241, Alfred J. Sims, HMS *Dreadnought*, paper to the Royal Institute of Naval Architects and Institute of Mechanical Engineers, undated. **220.** Sir Rowland Baker and Professor L. J. Rydill, 'The Building of the Two Dreadnoughts. European

Shipbuilding: One Hundred Years of Change', in *Proceedings of the Third Shipbuilding History Conference at the National Maritime Museum Greenwich, 13–15 April 1983* (National Maritime Museum, 1983). **221.** Anthony Preston, 'The Influence of the Cold War on Submarine Design', in Edmonds (ed.), *100 Years of the Trade*, p. 79. **222.** Bud and Gummett, *Cold War, Hot Science*, pp. 172–3. **223.** Christopher Andrew, *The Defence of the Realm* (Allen Lane, 2009), pp. 483–9. **224.** Bud and Gummett, *Cold War, Hot Science*, pp. 176–7. **225.** Ibid., p. 177. **226.** Preston, 'The Influence of the Cold War on Submarine Design', p. 79. **227.** TNA/DEFE/7/2055, Brief for Sandys, undated. **228.** Daniel, *End of an Era*, p. 139. **229.** TNA/DEFE/7/2055, Brief for Sandys, undated. **230.** Ziegler, *Mountbatten*, p. 558. **231.** TNA/ADM/127198, A Report by Chief Constructor Rowland Baker, Discussions with Admiral Rickover & Mr Mandil, 2–6 February 1959. **232.** Daniel, *End of an Era*, p. 139. **233.** TNA/DEFE/72/45, Le Fanu Memo, 15 March 1961. **234.** Peter Kimm, 'Dinner With Rickover', in *Naval Review*, Vol. 85, No. 1, January 1997, pp. 43–5. **235.** Ibid. **236.** MB/1/I276, Mountbatten to Vice Admiral Sir Charles Lambe, 29 January 1959 quoted in Hill, Admiral Hyman G. Rickover. **237.** John 'Sandy' Woodward, *One Hundred Days: The Memoirs of the Falklands Battle Group Commander* (HarperPress, 2012), p. 57. **238.** Coote, 'How Lord Mountbatten . . .'. **239.** Peter Hammersley, 'The Propulsion System', in John E. Moore, *The Impact of Polaris: The Origins of Britain's Seaborne Nuclear Deterrent* (Richard Netherwood, 1999), p. 155. **240.** Brown, *Century of Naval Construction*, p. 235. **241.** Harry Dickinson, *Wisdom and War: The Royal Naval College Greenwich 1873–1998* (Ashgate, 2012). **242.** Leslie Shore, *Vickers' Master Shipbuilder* (Black Dwarf Publications, 2011), p. 111. **243.** Ibid. **244.** Theodore Rockwell, *The Rickover Effect: How One Man Made a Difference* (Naval Institute Press, 2002), p. 274. **245.** Baker and Rydill, 'The Building of the Two Dreadnoughts'. **246.** Daniel, *End of an Era*, p. 153. **247.** Ibid. **248.** MOD Archive.

The 'Valiant' Class

249. Daniel, *End of an Era*, p. 352. **250.** TNA/ADM/1/26779, Memo from Chairman, Ships' Names Committee, 13 January 1961. **251.** Paul Wrobel, 'U.K. Nuclear Submarines: The Development of the Overall Design,' *Journal of Naval Engineering*, vol. 28, no. 1 (1983), pp. 106–19. **252.** Bud and Gummett, *Cold War, Hot Science*, pp. 168–9. **253.** Brown and Moore, *Rebuilding the Royal Navy*, p. 125. **254.** Horlick, 'Nuclear Submarine Propulsion in the RN'. **255.** Patrick Middleton, *Admiral Clanky Entertains* (Matador, 2010), pp. 147–8. **256.** Hill, 'Admiral Hyman G. Rickover'. **257.** Ibid. **258.** Baker and Rydill, 'The Building of the Two Dreadnoughts'. **259.** Horlick, 'Nuclear Submarine Propulsion in the RN'. **260.** Duncan, *Rickover*, p. 156. **261.** Horlick, 'Nuclear Submarine Propulsion in the RN'. **262.** Professor Jack Edwards, 'Initial Problems of the Submarine Pressurised Water Reactor Design and the

Related Experimental Programme', paper to the Institute of Marine Engineers, January 1962, n.32. **263.** Hammersley, 'The Propulsion System', p. 155. **264.** William Crowe, *The Policy Roots of the Royal Navy 1946–63* (unpublished Ph.D. Thesis, Princeton University, 1965), p. 293. **265.** Hearing before the Sub-committee on Agreements for Cooperation, Joint Committee on Atomic Energy, *Amending the Atomic Energy Act of 1954 – Exchange of Military Information and Material with Allies*, 85th Congress, 2nd Session, 27 February 1958, p. 499. **266.** Solly Zuckerman Archive (SZ)/CSA/122, Zuckerman to Sir Peter Ramsbotham, 30 April 1977. **267.** Peter Hammersley, 'The US/UK Agreement on Nuclear Co-operation', in Friends of the Royal Navy Submarine Museum, *All Round Look: Year Book 2009/2010* (Royal Navy Submarine Museum, 2010), p. 35. **268.** Hearing before the Sub-committee on Agreements for Cooperation, Joint Committee on Atomic Energy, *Amending the Atomic Energy Act of 1954 – Exchange of Military Information and Material with Allies*, 85th Congress, 2nd Session, 27 February 1958, p. 171. **269.** Rockwell, *Rickover Effect*, pp. 274–5. **270.** TNA/DEFE/24/46, Nuclear Submarines Presentation, Opening Remarks by First Lord, 1962. **271.** See Clark, *Nuclear Diplomacy*, p. 338; also Richard Neustadt, *Alliance Politics* (Columbia University Press, 1970).

4 'MOVE DETERRENTS OUT TO SEA':
THE BOMB GOES UNDERWATER

1. Lord Hailsham, *A Sparrow's Flight: Memoirs* (Collins, 1990), p. 283. **2.** TNA/ADM/1/27389, Burke to Mountbatten, 6 February 1959. **3.** Quoted in Ian McGeoch, 'The British Polaris Project: A Study of the British Naval Ballistic Missile System (BNBMS), Its Origins, Procurement and Effect', (M.Phil. Thesis, University of Edinburgh, 1975), p. 89. **4.** Michael Henry, 'A CO's Story', in John E. Moore, *The Impact of Polaris: The Origins of Britain's Seaborne Nuclear Deterrent* (Richard Netherwood, 1999), p. 250.

First Contact

5. Roy Dommett, 'The Blue Streak Weapon', *Prospero: Proceedings from the British Rocket Oral History Programme*, 2 (2005), pp. 24–7. **6.** Ministry of Defence, *Outline of Future Policy*, Cmnd. 124 (Her Majesty's Stationery Office, 1957). **7.** James M. Roherty, *Decisions of Robert S. McNamara: A Study of the Role of the Secretary of Defense* (University of Miami Press, 1970), p. 122. **8.** Rear Admiral I. J. Galantin, 'The Future of Nuclear-Powered Submarines', *US Naval Institute Proceedings*, Vol. 84 (June 1958), pp. 23–35. **9.** Eric Grove, *Vanguard to Trident: British Naval Policy since World War II* (Bodley Head, 1987), p. 256. **10.** Philip Ziegler, *Mountbatten: The Official Biography* (HarperCollins, 1985), p. 593; Sir William Jackson and Lord Bramall, *The Chiefs: The Story of* *the United Kingdom Chiefs of Staff* (Macmillan, 1992), p. 331. **11.** TNA/ADM/205/112, First Sea Lord Conference on Submarine Policy, 2 February 1956. **12.** TNA/ADM/167/149, B.5163, 15 October 1957. **13.** Rear Admiral Peter La Niece, *Not a Nine to Five Job* (Charltons, 1992), p. 131. **14.** TNA/ADM/167/149, B.5163, 15 October 1957. **15.** Peter La Niece, 'First Contact with Polaris by the Royal Navy', in Moore, *Impact of Polaris,* p. 28. **16.** TNA/ADM/1/27375, Selkirk to Hailsham, 1 January 1958. **17.** John Boyes, *Project Emily: Thor IRBM and the RAF* (Tempus, 2008); Ian Clark, *Nuclear Diplomacy and the Special Relationship: Britain's Deterrent and America, 1957–1962* (Oxford University Press, 1994). **18.** TNA/ADM/1/27375, Selkirk to Hailsham, 1 January 1958. **19.** Ken Young, 'The Royal Navy Polaris Lobby, 1955–1962', *Journal of Strategic Studies,* 25/3 (2002), p. 64. **20.** TNA/ADM/205/179, Goodwin to Brockman, 15 April 1957. **21.** TNA/ADM/205/179, Burke to Mountbatten, 14 April 1958. **22.** Richard Moore, *The Royal Navy and Nuclear Weapons* (Frank Cass, 2001), p. 166. **23.** TNA/ADM/205/179, Mountbatten to Burke, 8 May 1958. **24.** LHCMA/McGeoch/GB0099/File7/CDS Data, Garson to McGeoch, 18 March 1994. **25.** Wid Graham, 'Watching Brief 1961–1963', in Moore, *Impact of Polaris,* p. 38. **26.** TNA/ADM/205/179, Burke to Mountbatten, 16 May 1958. **27.** TNA/ADM/205/179, Mountbatten to Burke, 16 September 1958; TNA/ADM/1/27389, Mountbatten minute, 13 February 1959. **28.** TNA/ADM/205/163, Burke to Mountbatten, 28 February 1959. **29.** Ibid. **30.** TNA/DEFE/7/2162, Mountbatten to Sandys, 10 November 1958. As First Sea Lord and later CDS, Mountbatten tended to think he was in charge of everything, down to the smallest detail. He even informed the Ships' Names Committee that 'IF and WHEN' Polaris submarines were acquired by the Royal Navy, they should 'continue the battleship through with an "R" class, starting with SSBN 01 – REVENGE, followed by RENOWN and REPULSE'. All these were the names of traditional capital ships on which Mountbatten had served while a young officer; John Penton, *Solly Zuckerman: A Scientist Out of the Ordinary* (John Murray, 2001), p. 139; TNA/DEFE/19/209, Names for Polaris Submarines, 15 November 1963. **31.** TNA/ADM/1/27389, Lambe to Selkirk, April 1959. **32.** Ziegler, *Mountbatten,* pp. 560–61. **33.** TNA/ADM/1/27389, First Lord to VCNS, DCNS, USS, 25 May 1959. **34.** TNA/CAB/131/23, D(60)1, 24 February 1960; US designed and built standoff air launched ballistic missile with a range of 700 to 1000 nm. **35.** Hansard, House of Commons Debates, 27 April 1960, Vol. 622, Col. 244. **36.** TNA/ADM/1/27389, Polaris, Secretary to VCNS, DCNS, USS, 25 May 1959. **37.** Jackson and Bramall, *Chiefs,* p. 331. **38.** TNA/DEFE/4/124, COS (60) Meetings, 26 and 27 January 1960; TNA/DEFE/7/1328, BND(SG)(59)19(Final), 31 December 1959. **39.** TNA/DEFE/7/1328, BND(SG)(59)19(Final), 31 December 1959; BA/J40, Burke to Mountbatten, 11 April 1960. **40.** Ziegler, *Mountbatten,* p. 594. **41.** Clark, *Nuclear Diplomacy,* pp. 269, 271, 277, 290–96; TNA/PREM/11/2941, Watkinson to Macmillan, 23 September 1960. **42.** TNA/ADM/1/27609, Head of M Branch II, Memo, M.II/679/13/99, Basing Polaris Submarines in the Northern UK,

1959.   **43.** TNA/ADM/1/27389, Admiral BJSM to Admiralty, 9 June 1960.
44. TNA/PREM/11/2941, Washington to Foreign Office, 26 August 1960.
45. TNA/CAB/128/34, CC(60)50th Conclusions, 15 September 1960.   **46.**
TNA/DEFE/7/2162, Record of a meeting between Watkinson and Carrington,
29 September 1960.   **47.** TNA/DEFE7/2162, Harold Watkinson, Polaris Submarines, 19 May 1960.   **48.** TNA/ADM/205/163, Admiralty Organisation for
a 'Polaris' Submarine Programme, Annex A, 12 July 1960; TNA/ADM/205/163,
Admiralty Organisation for a 'Polaris' Submarine Programme, Annex B,
undated (July 1960); TNA/ADM/205/163, Admiralty Organisation for a 'Polaris' Submarine Programme, Annex C, 8 July 1960; TNA/ADM/205/163,
Memorandum by the Secretary, Admiralty Organisation for a 'Polaris' Submarine Programme, Addendum, u/d (July 1960).   **49.** TNA/ADM/205/163,
Extract Board Minutes 5431, 28 July 1960.   **50.** Ibid.   **51.** Richard Baker, *Dry
Ginger: The Biography of Admiral of the Fleet Sir Michael Le Fanu* (W. H.
Allen, 1977), p. 170.   **52.** Ibid.   **53.** Sidney John Palmer, 'Technical Evaluation
1961', in Moore, *Impact of Polaris,* p. 42.   **54.** TNA/ADM/205/163, Items
discussed with CNO by First Sea Lord in Washington, 6 November 1960.   **55.**
TNA/FO/371/159649, Foreign Office to Caccia, 13 January 1961.   **56.**
TNA/ADM/1/29349, Polaris Mission to the United States, Part I – General
Outline of the Visit, p. 4.   **57.** TNA/ADM/281/180, Polaris Mission to
the United States, Appendices, Appendix VIII, The Weapon System,
March 1961.   **58.** Ibid.   **59.** Ibid.   **60.** Ibid.   **61.** TNA/ADM/1/31048, The
Submarine Base of the Future, 31 July, 1961.   **62.** Ibid.   **63.** TNA/
ADM/1/31048, M.I/652/8/61, Minute Sheet No. 4, 2 August 1962.   **64.**
TNA/ADM/205/202, John to Le Fanu, 1 November 1960.   **65.** The records of
the British Nuclear Deterrent Study Group can be found in TNA/DEFE/13/617
and TNA/DEFE/13/618.   **66.** Young, 'Royal Navy Polaris Lobby', pp. 75–6.   **67.**
Stewart Menaul, *Countdown: Britain's Strategic Nuclear Forces* (Hale,
1980), p. 117.   **68.** William Crowe, 'The Policy Roots of the Royal Navy
1946–63' (unpublished Ph.D. Thesis, Princeton University, 1965), p. 284;
TNA/DEFE/13/295, Future of the British Nuclear Deterrent, 8 October 1962;
TNA/ADM 167/158, Board Minutes 5503–5509, 9 November 1961; TNA/ADM/
167/158, Board Minutes 5512, 11 December 1961.   **69.** TNA/ADM167/158,
Board Minutes 5512, 11 December 1961.   **70.** TNA/DEFE/7/2162, Carrington
to Watkinson, 18 December 1961; Carrington gave Watkinson a number of
options: (1) eight sixteen-missile single-purpose Polaris submarines, if the '50%
destruction of 40 cities' criterion was adhered to; (2) four sixteen-missile boats
if the criterion were halved; (3) seven eight-missile hybrid submarines at less cost
than (2) as the hybrids would be substituted for hunter-killers in the building
programme; TNA/DEFE/7/2162, Polaris Submarines, Lawrence-Wilson Minute, 28 December 1961; DEFE/7/2144, Holligan to Price, 26 July 1962;
TNA/DEFE/7/2144, BNDSG, Hybrid Submarines, Note by the Deputy Chief
of Naval Staff, 11 October 1962.   **71.** Fred M. Kaplan, *The Wizards of Armageddon* (Stanford University Press, 1991), pp. 253–6.   **72.** TNA/ADM/1/28839,
Deputy Chief of the Naval Staff to First Sea Lord, 14 November 1962.   **73.**

TNA/ADM/1/28839, Deputy Chief of the Naval Staff, 4 December 1962. **74.** TNA/DEFE/7/2162, Brief by the Admiralty, 14 December 1962. **75.** Clark, *Nuclear Diplomacy*, pp. 355–61. **76.** TNA/ADM/1/28839, Begg on behalf of John to Crawford, 18 December 1962. **77.** *Foreign Relations of the United States*, 1961–1963, Vol. XIII, Memorandum of Conversation, 16 December 1962, pp. 1088–91. **78.** Peter Hennessy, *Muddling Through: Power, Politics and the Quality of Government in Postwar Britain* (Victor Gollancz, 1996), p. 112. **79.** TNA/PREM/11/4147, Record of a Meeting Held at Bali-Hai, The Bahamas, 09.50 am, 19 December 1962. **80.** Hennessy, *Muddling Through*, p. 112. **81.** Harold Macmillan, *At the End of the Day, 1961–1963* (Macmillan, 1973), p. 358. **82.** TNA/T/325/88, 'The Skybolt Story' 1 January 1959 to 31 December 1963. **83.** TNA/PREM/11/4229, Record of a Meeting Held at Bali-Hai, The Bahamas, 9.50 a.m. 19 December 1962. **84.** TNA/PREM/11/4229, Record of a Meeting Held at Bali-Hai, The Bahamas, noon and 12:30pm, 20 December 1962. **85.** TNA/PREM/11/4148, Macmillan to Ormsby-Gore, 26 January 1963 **86.** Cmnd. 1915, *Bahamas Meetings, December 1962: Texts of Joint Communiqués,* (HMSO, 1962). **87.** Harold Macmillan Diary, 23 December 1962. **88.** Gerhard Peters and John T. Woolley, *The American Presidency Project*, 'Partial Transcript of a Background Press Interview with John F Kennedy at Palm Beach', 31 December 1962. **89.** TNA/PREM/11/4147, Paris to the FO, 'Nassau Agreement', 2 January 1963. **90.** Solly Zuckerman, *Monkeys, Men and Missiles: An Autobiography, 1946–1988* (Collins, 1988), p. 399. **91.** TNA/DEFE/13/619, Zuckerman and Scott to Thorneycroft, 21 December 1962; Zuckerman, *Monkeys, Men and Missiles*, p. 261. **92.** TNA/AVIA/65/1840, note of 2 January 1962; TNA/AVIA/65/1840, The Bahamas Conference: Report on Attendance as Part of the Ministry of Defence Party by D/RAE, u/d (December 1962). **93.** Zuckerman, *Monkeys, Men and Missiles*, pp. 262–5. **94.** George Ball, *The Past Has Another Pattern* (Norton, 1982), pp. 264–5. **95.** Donnette Murray, *Kennedy, Macmillan and Nuclear Weapons* (Macmillan, 2000), p. 151; John Baylis, *Anglo-American Defence Relations 1939–1984: The Special Relationship* (Macmillan, 1984), p. 105. **96.** Ball, *The Past Has Another Pattern*, p. 268. **97.** Zuckerman, *Monkeys, Men and Missiles,* p. 265. **98.** TNA/FO/371/16405, Ormsby-Gore, Annual Review of Anglo-American Relations, 1 January 1963. **99.** Alan Dobson, *Anglo-American Relations in the Twentieth Century: Of Friendship, Conflict and the Rise and Decline of Superpowers* (Routledge, 1995), p. 130. **100.** See Oliver Bange, *The EEC Crisis of 1963: Kennedy, Macmillan, de Gaulle and Adenauer in Conflict* (Macmillan, 2000); John Young, *The EEC Crisis of 1963: Kennedy, Macmillan, de Gaulle and Adenauer in Conflict* (Macmillan, 2000). **101.** TNA/ADM/1/28839, Le Fanu, Notes on Bahamas Meeting, u/d (December 1962). **102.** Ibid. **103.** Nigel Ashton, *Kennedy, Macmillan and the Cold War: The Irony of Interdependence* (Palgrave Macmillan), p. 152. **104.** TNA/ADM/1/28839, Controller of the Navy, Notes on Bahamas Meeting, u/d (December 62); Ashton, *Kennedy, Macmillan and the Cold War*, p. 187. **105.** TNA/ADM/1/28839, Deputy Chief of the Naval Staff to First Lord, Polaris, 13 December 1962. **106.** TNA/ADM/1/28839, Brief for the Prime Minister,

Talks with President Kennedy, December 1962. **107.** Hansard, House of Commons Debates, 30 January 1963, Vol. 670, Cols. 955–1074. **108.** TNA/ADM/1/28839, Unaccounted for note by Le Fanu, 22 December 1962. **109.** TNA/ADM/1/28839, Begg on behalf of John to Crawford, 18 December 1962. **110.** Harvey Sapolsky, *The Polaris System Development: Bureaucratic and Programmatic Success in Government* (Harvard University Press, 1972), p. 159. **111.** TNA/ADM/1/27740, Polaris Nuclear Submarine: Admiralty Organisation, May 1960. **112.** Quoted in Moore, *Royal Navy and Nuclear Weapons*, p. 162. **113.** Ibid. **114.** TNA/ADM/1/28839, Letter from John, 31 December 1962. **115.** Rebecca John, *Caspar John* (HarperCollins, 1987), p. 197. **116.** Ibid. **117.** TNA/ADM1/28839, Notes on Bahamas Meeting, u/d (December 1962).

Planning for Polaris

118. Herbert Fitzer, 'The Electrical System and the RN Polaris School', in Moore, *Impact of Polaris*, p. 159. **119.** McGeoch, 'British Polaris Project', p. 60. **120.** TNA/ADM/205/211, Organisation for Polaris, Sitrep by Le Fanu, 16 January 1963. **121.** TNA/ADM/1/28377, Taylor to Osmond, 1 January 1963. **122.** Vice Admiral Sir Hugh Mackenzie, *Sword of Damocles* (Periscope Publishing, 2007), p. 243. **123.** Vice Admiral Sir Hugh Mackenzie, 'Setting Up the Project', in Moore, *Impact of Polaris*, p. 47. **124.** Ibid. **125.** Mackenzie, *Sword of Damocles*, p. 230. **126.** Ken Dunlop, 'Support Planning', in Moore, *Impact of Polaris*, p. 167. **127.** Peter Nailor, *The Nassau Connection: The Organisation and Management of the British Polaris Project* (Her Majesty's Stationery Office, 1988), p. 28. **128.** Nailor, *Nassau Connection*, p. 28; Mackenzie, 'Launching the Project', in Moore, *Impact of Polaris*, p. 55. **129.** TNA/ADM/331/15, The Future Management of the British Naval Ballistic Missile System, Report by Vice Admiral Sir Raymond Hawkins, October 1967. **130.** Sir Rae McKaig, 'Initial Tasks', in Moore, *Impact of Polaris*, p. 78. **131.** IWM/12071, Interview with Vice Admiral Sir Hugh Mackenzie, 3 June 1991. **132.** TNA/ADM/1/28377, Taylor to Hunt, 24 January 1963. **133.** TNA/ADM/1/28362, Synott to Mackenzie, 23 January 1963. **134.** Ibid. **135.** TNA/ADM/1/28362, DGW, Proposed Statement on Polaris Requirements, 8 February 1963. **136.** TNA/ADM/1/28362, What's So Special about the Polaris Programme?, March 1963. **137.** TNA/CAB/131/28, D(63)6, Polaris Submarines: Size and Number, Memorandum by the Minister of Defence, 16 January 1963. At a meeting between Admiralty and Treasury officials, the Treasury referred to the criterion as an 'assumption' and reserved its position; TNA/T/225/2161, Note of a Meeting, 18 January 1963. **138.** Stephen Twigge and Len Scott, *Planning Armageddon: Britain, the United States and the Command of Western Nuclear Forces, 1945–64* (Harwood, 2000), p. 132; TNA/DEFE/13/619, Scott to Thorneycroft, 1 January 1963; the Navy could only build a maximum of seven submarines because of the difficulty of providing crews. **139.** TNA/ADM/1/28839, Deputy Chief of the Naval Staff to First

Lord, 13 December 1962; TNA/DEFE/7/1752, Carrington to Thorneycroft, 31 December 1963. **140.** TNA/DEFE/7/1752, Meeting between Thorneycroft, Carrington and Amery, 3 January 1963. **141.** TNA/ADM/1/28842, Size of the U.K. Polaris Force, (u/d) December 1964. Due to their size as targets, Moscow and Leningrad required 4 and 2 missile hits respectively. 11 missiles on target represented a 7-city deterrent, including Moscow and Leningrad, TNA/DEFE/13/297, Polaris Submarines – Size of UK Force – Appendix to COS/3200, 12 December 1963. **142.** TNA/DEFE/7/1752, D(63)6 – Polaris Submarines, u/d (January 1963). **143.** TNA/DEFE/7/1752, Defence Committee Paper by Thorneycroft, Polaris Submarines, Size and Number, u/d (January 1963). **144.** TNA/ADM/1/28839, Unknown to Carrington, 14 December 1962. **145.** Nailor, *Nassau Connection*, p. 29. **146.** Dunlop, 'Support Planning', in Moore, *Impact of Polaris*, p. 67. **147.** TNA/AVIA/65/1840, Carrington to Thorneycroft, 31 December 1962. **148.** Jack Daniel, *End of an Era: The Memoirs of a Naval Constructor* (Periscope Publishing, 2003), p. 191. **149.** TNA/CAB/131/28, Minutes of a Meeting, 23 January 1963. **150.** Ibid. **151.** TNA/AVIA/65/1840, Follow-up to Nassau, 3 January 1963. **152.** Decision-making was also complicated by different possibilities for re-entry bodies and warheads. The US had developed two re-entry systems for Polaris. The MK I was used on the Polaris A1 and Polaris A2: it carried a large warhead and had no real penetration aid capability. The MK II was to be used on the A3. It used a different warhead and re-entry vehicle and had 'some measure of penetration capability'. The British Skybolt warhead could fit 'without serious modification' into the MK I re-entry body, but this could only be used on the A2, not the A3. If the UK decided to purchase the A3 missile, the Skybolt warhead could be used in different ways. However, all involved 'considerable redesign and flight testing' and there were doubts whether this could 'technically be done by the US within a timescale compatible with UK submarine production'. Alternatively, the UK could use the MK II re-entry system but this would mean developing a completely new warhead. TNA/ADM/1/28839, Appendix III to Report of Polaris Fact Finding Mission to Washington, January 1963. **153.** TNA/ADM/1/28965, U.K. Polaris Mission to U.S., Report by DAS, January 1963. **154.** TNA/ADM/1/28987, POLARIS: Report of Mission to the USA, 16 January 1963. **155.** TNA/DEFE/7/1752, Peck to McMahon, 4 February 1963. **156.** TNA/ADM/1/28839, Carrington to Thorneycroft, 4 March 1963. **157.** Ibid. **158.** TNA/DEFE/13/734, Ormsby-Gore to Macmillan, 28 January 1963. **159.** *Sunday Telegraph*, 'Anglo-US Dispute over Polaris, Britain Asked to Pay for Development Costs, Oversight at Nassau', 27 January 1963. **160.** TNA/ADM/1/28987, Report of Polaris Fact Finding Mission to Washington – January 1963. **161.** Zuckerman, *Monkeys, Men and Missiles,* p. 267. **162.** TNA/PREM/11/4148, Macmillan to Ormsby-Gore, 15 January 1963. **163.** Ibid. **164.** Ibid. **165.** Ibid.; Macmillan, *At the End of the Day,* p. 363. **166.** TNA/PREM/11/4148, Macmillan to Ormsby-Gore, 26 January 1963. **167.** Ibid. **168.** Ibid. **169.** TNA/AVIA/65/1866, Washington to FO, 26 January 1963. **170.** Macmillan, *At the End of the Day*, p. 363.

171. TNA/ADM/1/28978, United States/United Kingdom Polaris Sales Agreement Programme Costs, 1963/4 to 1971/2 Summary of Expenditures by United Kingdom Financial Year.   172. TNA/PREM/11/4148, Macmillan to Maudling, 26 January 1963.   173. TNA/DEFE/13/734, Ormsby-Gore to Macmillan, 28 January 1963.   174. Ibid.   175. TNA/PREM/13/1317, Zuckerman to Wilson, 23 March 1967.   176. TNA/ADM/1/29356 Draft E.P.C. Paper 'Polaris' Submarine Building Programme – Choice of Building Yards, u/d (February 1963).   177. TNA/T/225/2161, Darracott to Dodd, Polaris, 20 February 1963.   178. TNA/DEFE/19/209, Johnson to Director of Naval Contracts, 8 April 1963.   179. Ibid.   180. TNA/ADM/1/28965, Carrington to Mackenzie, 31 January 1963.   181. TNA/ADM/1/28965, Nuclear Submarines – Polaris – Operating Base and Support Facilities – Safety Considerations – Note by Head of M.II, 1 February 1963.   182. TNA/ADM/1/28965, Turner to Mackenzie, 15 February 1963.   183. TNA/ADM/1/28965, Special Military Branch Aquaint, No. 5268, 25 February 1963.   184. For a more detailed discussion of the alternative sites see Malcolm Chalmers and William Walker, *Uncharted Waters: The UK, Nuclear Weapons and the Scottish Question* (Tuckwell Press, 2001), pp. 17–22; TNA/ADM/1/28965, Naval Ballistic Missile Force – UK Operating Base, Report of Working Party, 25 February 1963.   185. TNA/ADM/1/28965, Nuclear Submarines – Polaris Operating Base and Support Facilities – Safety Considerations. Note by Head of M.II, u/d (February 1963).   186. TNA/ADM/1/28965, Polaris – Shore Support, Safety Considerations, 19 February 1963.   187. TNA/ADM/1/28965, Naval Ballistic Missile Force – UK Operating Base, Report of Working Party, 25 February 1963.   188. Although the working party did not let the prospect influence its deliberations, a proposal was under consideration elsewhere in the Admiralty to establish a shore base at Faslane for the Navy's hunter-killer submarines and not, as previously planned, at the submarine depot ship HMS *Maidstone*; TNA/ADM/1/28965, Naval Ballistic Missile System – U.K. Operating Base, 25 April 1963.   189. Ian Mcgeoch, 'Allocated to...', *Naval Review*, Vol. 74, No. 3, July 1986, p. 261.   190. TNA/ADM/1/28965, Requirement for a Naval Ballistic Missile Submarine Operating Base, Appendix A, u/d (April 1963).   191. TNA/DEFE/19/209, RNAD Coulport, Establishment of Depot to Support Ballistic and Other Submarines Operating from the Clyde, 20 June 1963.   192. Sir Simon Cassells, 'The "Bull Pen"', in Moore, *Impact of Polaris,* p. 86.   193. TNA/ADM/157/162, Board Minutes, 5576–5578, 28 March 1963.   194. TNA/DEFE/7/2163, Mackay to Admiralty, 26 February 1963.   195. TNA/CAB/131/28, Nassau Agreement, Record of a Meeting, 25 February 1963; TNA/DEFE/7/2163, Wood to Mackay, 25 January 1963.   196. TNA/DEFE/13/1050, Eighth Joint Annual Report, 1 July 1970.   197. TNA/AVIA/65/1866, Downey to Haviland, 27 February 1963.   198. Alan Pritchard, 'Negotiating the Sales Agreement', in Moore, *Impact of Polaris*, p. 72.   199. TNA/ADM/1/28978, Joint Annual Report (1963) of the Project Officers for the United States and United Kingdom Polaris Program to the Secretary of Defense and to the Minister of Defence.   200. TNA/FO/371/173518, Burden to Rose, 22 March 1963.   201. Daniel, *End of an Era*, p. 160.   202. Francis

Duncan, *Rickover: The Struggle for Excellence* (Naval Institute Press, 2001), pp. 151–3. **203.** IWM/12071, Interview with Vice Admiral Sir Hugh Mackenzie, 3 June 1991. **204.** Harold Macmillan, *Pointing the Way 1959–1961* (Macmillan, 1971), p. 323. **205.** La Niece, *Not a Nine to Five Job*, p. 163. **206.** Peter Nailor, 'Making the Organisation Work', in Moore, *Impact of Polaris*, p. 92.

Constructing Polaris

207. For more information about how the shipbuilders addressed the problem of building the submarines see P. H. Rance, 'Programme and Build Philosophy', paper given to the Nuclear Submarine SSN07 Symposium for equipment, machinery, and systems suppliers, Vickers Limited Shipbuilding Group, Barrow-in-Furness, 28 February 1968 **208.** TNA/DEFE/19/209, P. Sherwell, 28 June 1963. **209.** TNA/DEFE/13/350, Nairne to Healey, 12 May 1965. **210.** John Schank, Jessie Riposo, John Birkler, James Chiesa, *The United Kingdom's Nuclear Submarine Industrial Base,* Vol. 1: *Sustaining Design and Production Resources* (Rand Corporation, 2005), p. 8. **211.** TNA/ADM/1/29270, Staff Requirements of SSBN.01 and Class, 24 February 1963. **212.** For insight into what it takes to design a submarine from scratch see R. J. Daniel, 'Submarine Design', *Naval Forces*, No. 111, 1987. **213.** Jack Daniel, quoted in David K. Brown, *Century of Naval Construction: The History of the Royal Corps of Naval Constructors* (Conway Maritime Press, 1983), p. 239. **214.** TNA/ADM/1/28839, A. J. Sims, Polaris Submarine Programme, 24 January 1963. **215.** TNA/DEFE/19/209, P. T. Williams, Polaris – Second Builder, 24 April 1963; TNA/DEFE/13/953, Nuclear Submarine Programme, Talk for Minister (E), 10 May 1967. **216.** Mackenzie, 'Launching the Project', in Moore, *Impact of Polaris*, p. 66; Daniel, *End of an Era,* p. 157; TNA/ADM/1/167/163, Board Minutes, 5633–5638, 6 February 1964. **217.** Paul Wrobel, 'U.K. Nuclear Submarines: The Development of the Overall Design', *Journal of Naval Engineering*, vol. 28, no. 1 (1983), pp. 106–19. **218.** TNA/ADM/1/29213, Report on the Operation of the Electric Boat Co. Group System, Part 'B' Possible Application of E. B. Group System to Nuclear Naval Production, u/d (September 1963). **219.** Peter La Niece, 'Procuring the Weapon System', in Moore, *Impact of Polaris*, p. 145. **220.** Wirral Archives Service/Cammell Laird Collection/Polaris 1316–1317, Birkenhead Presentation, 8 December 1966. **221.** Mackenzie, 'Launching the Project', in Moore, *Impact of Polaris*, p. 67. **222.** IWM/12071, Interview with Vice Admiral Sir Hugh Mackenzie, 3 June 1991. **223.** Ibid. **224.** Nailor, *Nassau Connection*, p. 73. **225.** 'The Polaris Submarine Programme', *Nuclear Energy*, Nov/Dec 1967, p. 169. **226.** "Resolution": First Polaris Missile Submarine for the Royal Navy', *Shipbuilding and Shipping Record,* 19 October 1967; Nailor, *Nassau Connection*, pp. 77–82. **227.** *The Economist*, 20 April 1968. **228.** Mackenzie, 'Launching the Project', in Moore, *Impact of Polaris*, p. 67. **229.** TNA/DEFE/13/953, BNBMS Building Programme, 23 December 1965. **230.**

Solly Zuckerman, *Six Men Out of the Ordinary* (Peter Owen, 1992); Inconel is a super-alloy made of nickel and chromium, designed to perform in extreme environments. TNA/ADM/1/28982, Le Fanu to Jellicoe, 22 November 1963. **231.** Mackenzie, 'Launching the Project', in Moore, *Impact of Polaris*, p. 65. **232.** TNA/ADM/1/28982, Le Fanu to Jellicoe, 22 November 1963; TNA/ADM/1/28982, Pepper to Hedger, 10 December 1963. **233.** TNA/ADM/1/28842, Mackenzie to Le Fanu, 22 January 1964. **234.** TNA/ADM1/28982, Le Fanu to Zuckerman, u/d (November 1963). **235.** TNA/ADM/1/28842, Le Fanu to Jellicoe, 20 January 1964. **236.** TNA/ADM/1/28842, Mackenzie to Le Fanu, 22 January 1964. **237.** Ibid. **238.** The article criticized nuclear policy and sought to distinguish between nuclear weapons as a deterrent and the futility of using battlefield nuclear weapons. **239.** Zuckerman, *Six Men Out of the Ordinary*, pp. 169–70; Duncan, *Rickover,* pp. 155–6. **240.** TNA/DEFE/7/2165, Zuckerman to Thorneycroft, 12 December 1963; Daniel, *End of an Era*, p. 160; Duncan, *Rickover*, p. 190. **241.** TNA/DEFE/7/2165, Zuckerman to Thorneycroft, 12 December 1963. **242.** TNA/PREM/11/4741, Background Note, u/d (December 1963). **243.** TNA/DEFE/7/2165, Zuckerman to Thorneycroft, 27 November 1963. **244.** TNA/ADM/1/28842, Le Fanu to Jellicoe, 20 January 1964. **245.** TNA/PREM/11/4741, Ormsby-Gore to FO, 24 December 1963. **246.** John Simpson, *The Independent Nuclear State: The United States, Britain and the Military Atom* (Macmillan, 1983), p. 182. **247.** TNA/ADM/1/28842, Le Fanu to Jellicoe, 20 January 1964. **248.** TNA/ADM/1/28842, Mackenzie to Le Fanu, 22 January 1964. **249.** Ibid. **250.** TNA/ADM/1/28842, Le Fanu to Jellicoe, 20 January 1964. **251.** TNA/ADM/1/28842, Le Fanu, record of conversation with Thorneycroft, 24 January 1964. **252.** TNA/ PREM/11/4741, FO to Home, 22 January 1964. **253.** TNA/CAB/130/196, GEN.836/1st mtg., 'Holy Loch', 24 January 1964. **254.** Louis Le Bailly, *The Man around the Engine: Life below the Waterline* (Mason, 1990), pp. 164–5. **255.** TNA/ADM/1/28842, Admiralty to CBNS, 24 January 1964. **256.** TNA/DEFE/13/295, Newell to Thorneycroft, 12 May 1964; Daniel, *End of an Era,* p. 352. **257.** TNA/DEFE/13/295, Newell to Thorneycroft, 12 May 1964. **258.** Ibid.

The Fifth Submarine

259. TNA/ADM/1/28842, Polaris Submarines: Size of Force, u/d (26 September 1963). **260.** TNA/ADM/1/28842, COS 69th Meeting/63, 10 December 1963. **261.** Ibid. **262.** TNA/ADM/1/28842, Director of Plans to First Sea Lord, 13 January 1964; TNA/ADM/1/28842, Jellicoe to Thorneycroft, 2 December 1963. On average the majority of early US Polaris refits overran by about three months, or 27 per cent, for such reasons as strikes and problems connected with the nuclear-propulsion plant; TNA/DEFE/13/953, Memorandum by VCNS, 22 April 1970. **263.** 'The 32 missile load of two submarines constitutes a 20 city deterrent; one submarine, a 7 or 8 city deterrent. These figures allowed for the probabilities of inaccuracies and for the two biggest targets to be hit more than once.' **264.** Mackenzie, 'Launching the Project', in Moore, *Impact of Polaris*, p. 69. **265.** Ibid. **266.** Ibid. **267.** TNA/ADM/167/161, Board Memo, B1470, 25 November 1963; TNA/DEFE/13/296, Hockaday to Mountbatten, 13 December 1963. **268.** TNA/ADM/167/161, Board Memo, B1470, Appendix A, 25 November 1963. **269.** TNA/ADM/167/162, Board Minutes, 5615–5618, 28 November 1963. **270.** TNA/ADM/1/28842, Jellicoe to Thorneycroft, 2 December 1963. **271.** TNA/DEFE/4/160, COS 68th Meeting/63, 3 December 1963. **272.** TNA/ADM/1/28842, Hull to Thorney-croft, u/d (December 1963). **273.** TNA/DEFE/4/160, COS 69th Meeting/63, 10 December 1963. **274.** TNA/ADM/1/28842, Thorneycroft to Douglas-Home, 20 December 1963. **275.** TNA/T/225/2521, Maudling to Douglas-Home, 1 January 1964; throughout 1963, Treasury officials repeatedly asked for assurances from the Admiralty that 'production proposals made no provision for a possible fifth boat' as there was no money in the forward Defence costings. **276.** Lord Home, *The Way the Wind Blows* (Collins, 1976), p. 213. **277.** See *Foreign Relations of the United States, 1964–1968*, Vol. XII, Memorandum of Conversation, 13 February 1964, pp. 17–20. **278.** TNA/225/2521, Fraser to Dodd, 24 February 1964; Hansard, House of Commons Debates, 27 February 1964, Vol. 690, Col. 756. **279.** TNA/CAB/148/1, DO(64) 8th Meeting, 19 February 1964. **280.** TNA/CAB/128/38, CM(14) 14th Conclusions, 25 February 1964. **281.** TNA/CAB/148/1, DO(64) 7th Meeting, 25 February 1964; Michael Quinlan, Oral Evidence taken before the House of Commons Defence Select Committee, 14 March 2006. **282.** TNA/CAB/148/1, DO(64) 9th Meeting, 25 February 1964. **283.** Michael Quinlan, Oral Evidence taken before the House of Commons Defence Select Committee, 14 March 2006. **284.** TNA/CAB/128, CM(14) 14th Conclusions, 25 February 1964. **285.** La Niece, 'Procuring the Weapon System', in Moore, *Impact of Polaris*, p. 145; Sir Hugh Mackenzie, 'The Continuing until It be Thoroughly Finished', in Moore, *Impact of Polaris*, p. 98.

The 1964 General Election

286. Harold Wilson, *The Labour Government 1964–1970: A Personal Record* (Weidenfeld and Nicolson, 1971), p. 40. **287.** Hansard, House of Commons Debates, 12 November 1963, Vol. 684, Cols. 49–51. **288.** The Conservative Party, *The Manifesto of the Conservative and Unionist Party* (1964), p. 6. **289.** The Labour Party, *Let's Go with Labour for the New Britain: The Labour Party's Manifesto for the 1964 General Election* (1964) p. 23. **290.** TNA/DEFE/13/296, Jellicoe to Thorneycroft, 18 December 1963. **291.** Ibid. **292.** TNA/ADM/1/28383, Jellicoe to Le Fanu, 21 November 1963. **293.** TNA/ADM/1/28383, Le Fanu to Jellicoe, 2 December 1963. **294.** Zuckerman, *Monkeys, Men and Missiles*, p. 373. **295.** Nailor, 'Making the Organisation Work', in Moore, *Impact of Polaris*, p. 94. **296.** Nailor, *Nassau Connection*, p. 37. **297.** IWM, 12071, Interview with Vice Admiral Sir Hugh Mackenzie,

3 June 1991.   **298.** Mackenzie, 'Continuing until It be Thoroughly Finished', in Moore, *Impact of Polaris*, p. 99.   **299.** Private correspondence with Jack Daniel, 13 December 2009.   **300.** TNA/DEFE/24/291, Penny to Seaborg, 21 July, 1964.   **301.** Hennessy, *Muddling Through*, p. 113; Nailor, *Nassau Connection*, p. 53.   **302.** Mackenzie, 'Continuing until It be Thoroughly Finished', in Moore, *Impact of Polaris*, p. 99.   **303.** Ibid., p. 98.   **304.** TNA/T/225/2586, Williams to Clifton, 17 July 1964.   **305.** British Areospace Archive BAE/VA/Box10, Polaris Meeting, 3 April 1964.   **306.** *North West Evening Mail*, 15 August 1964.   **307.** Hansard, House of Commons Debates, 23 November 1964, Vol. 702, Col. 995; these are also quoted in Leslie Shore, *Vickers' Master Shipbuilder* (Black Dwarf Publications, 2011), p. 149.   **308.** Ibid.   **309.** Ibid.   **310.** David Gill, 'The Ambiguities of Opposition: Economic Decline, International Cooperation, and Political Rivalry in the Nuclear Policies of the Labour Party, 1963–1964', *Contemporary British History*, 25/2 (2011), p. 263.   **311.** Ziegler, *Mountbatten*, p. 626.   **312.** TNA/DEFE/32/9, COS 58th Mtg/64(2) (Confidential Annex) (SSF), 29 September 1964.   **313.** Ibid.   **314.** Ibid.   **315.** Ibid. In the run-up to, and during, the election campaign, both Wilson and the man who was expected to become Defence Secretary, Denis Healey, had suggested converting the Polaris submarines into hunter-killers, 'a programme of certain and immediate value to the British Navy and to national defence which has been set back five years by the Polaris programme'. They also suggested that they would expand the naval shipbuilding programme by using 'savings made by stopping wasteful expenditure on the politically inspired nuclear programme'; Hansard, House of Commons Debates, 26 February 1964, Vol. 690, Col. 480.   **316.** TNA/DEFE/32/9, COS 58th Mtg/64(2) (Confidential Annex) (SSF), 29 September 1964.   **317.** Zuckerman, *Monkeys, Men and Missiles*, p. 373.   **318.** TNA/ADM/167/162, Board Minutes 5593–5595, 17 July 1963.   **319.** TNA/T/225/2586, Visit To Bath, 22 March 1964.   **320.** Zuckerman also confirms that the Navy had plans for 'the Wilson class of nuclear-powered hunter-killer submarines'; Zuckerman, *Monkeys, Men and Missiles*, p. 374.   **321.** Daniel, *End of an Era*, p. 7.   **322.** TNA/DEFE/24/78, Confidential Annex to COS 59th Meeting/64, 6 October 1964.   **323.** Ibid.   **324.** Ibid.   **325.** TNA/DEFE/24/78, Wright to P.U.S., 16 October 1964.

'Go' or 'No Go': Deciding to Continue

326. Mackenzie, 'Continuing until It be Thoroughly Finished', in Moore, *Impact of Polaris*, p. 101.   **327.** Edmund Dell, *The Chancellors: A History of the Chancellors of the Exchequer, 1945–90* (HarperCollins, 1997), p. 310.   **328.** TNA/DEFE/13/350, CPE to PUS(RN), Appendix 'A', Modification of the Polaris Programme, 19 October 1964.   **329.** Ibid.   **330.** Denis Healey, *Time of My Life* (Michael Joseph, 1989), p. 302.   **331.** Andrew J. Pierre, *Nuclear Politics: The British Experience with an Independent Strategic Force 1939–1970* (Oxford University Press, 1972), pp. 284–5; Lawrence Freedman, *Britain and Nuclear Weapons* (Macmillan, 1980), p. 32; Peter Hennessy, *The Secret State: Preparing* *for the Worst 1945–2010* (Penguin Books, 2010), p. 72; David Gill, 'Strength in Numbers: The Labour Government and the Size of the Polaris Force', *Journal of Strategic Studies*, 33/6 (2010), p. 833. **332.** TNA/DEFE/32/9, COS(I) 5/11/64 (SSF), 5 November 1964. **333.** MB/J61, Mountbatten to Wilson, 19 October 1964; quoted in Ziegler, *Mountbatten*, p. 627. **334.** LHCMA/McGeoch/GB0099/Box6/Mountbatten Research, Interview with Lord Mountbatten, 2 March 1973. **335.** Zuckerman, *Six Men Out of the Ordinary*, p. 147. **336.** Healey, *Time of My Life*, p. 307. **337.** Hennessy, *Muddling Through*, p. 115. **338.** TNA/DEFE/10/510, Defence Council, Minutes, 29 October 1964. **339.** Hansard, House of Commons Debates, 3 March 1966, Vol. 725, Col. 1483; Wilson maintained this in his memoirs, Wilson, *Labour Government 1964–1970*, p. 40. **340.** Hennessy, *Muddling Through*, p. 115. **341.** Ibid. **342.** Ibid., p. 116. **343.** TNA/DEFE/10/510, Defence Council, Minutes of the 9th Meeting, 29 October 1964. **344.** Quoted in Philip Ziegler, *Wilson: The Authorised Life* (Weidenfeld and Nicolson, 1993), p. 208. **345.** TNA/CAB/129/39, CC(64)11th Conclusions, 26 November 1964. **346.** Ibid. **347.** Anthony Shrimsley, *The First Hundred Days of Harold Wilson* (Weidenfeld and Nicolson, 1965), pp. 112–14. **348.** Healey, *Time of My Life*, p. 302. **349.** Richard Crossman, *The Diaries of a Cabinet Minister*, Vol. 1: *Minister of Housing, 1964–1966* (Book Club Associates, 1975), p. 73. **350.** Ibid. **351.** TNA/CAB/130/212, MISC 16, Atlantic Nuclear Force, 11 November 1964. **352.** TNA/DEFE/13/350, Luce to Healey, 6 November 1964. **353.** TNA/DEFE/69/449, Pollock to VCNS, 18 November 1964. **354.** Ibid. **355.** Ibid. **356.** Ibid. **357.** TNA/DEFE/19/129, The Purpose of the U.K.'s Deterrent, 12 October 1971. **358.** TNA/CAB/130/213, MISC, 17/4, Defence Policy, 22 November 1964. **359.** Ibid. **360.** Hennessy, *Muddling Through*, p. 115. **361.** TNA/CAB/130/213, Minutes of a Meeting held at Chequers on Sunday, 22 November 1964. **362.** TNA/DEFE/10/510, Defence Council, Minutes of the 9th Meeting, 29 October 1964; TNA/DEFE/32/9, Minutes of the Chiefs of Staff (Informal) Meeting, 5 November 1964. Healey later acknowledged that there was always 'the hypothetical possibility that the alliance might break up or at least be changed in a radical way; and that, in the event, the (so to speak) bargaining position of this country would be very weak without the strategic nuclear deterrent force'; TNA/DEFE/11/437, Healey to Hull, 20 June 1967. **363.** TNA/CAB/130/213, Minutes of a Meeting held at Chequers on Sunday, 22 November 1964. **364.** TNA/DEFE/13/350, Mayhew to Healey, 17 December 1964. **365.** TNA/DEFE/13/350, Healey to Wilson, 21 December 1964. **366.** TNA/CAB/148/18, OPDC(65), 5th Meeting, 29 January 1965. **367.** Freedman, *Britain and Nuclear Weapons*, p. 34. **368.** Mackenzie, 'Continuing until It be Thoroughly Finished', in Moore, *Impact of Polaris*, p. 104. **369.** Richard Hill, *Lewin of Greenwich: The Authorised Biography of Admiral of the Fleet Lord Lewin* (Cassell, 2000), pp. 326–7. Most literature on Polaris states that the Navy intended to call the submarine HMS *Ramillies*. This is incorrect. When the Ships' Names Committee reconvened in June 1964, to decide on a suitable name, *Redoubtable*, not *Ramillies*, was considered to be the most appropriate 'from the point of view of harmonising with the four previous selections'. However, the Committee eventually recommended breaking from convention and using a two-word name, *Royal Sovereign*, which was approved by the Admiralty Board on 26 June, and then by the Queen on 30 June 1964.

Completing the Programme

370. La Niece, 'Procuring the Weapon System', in Moore, *Impact of Polaris*, p. 152.   **371.** TNA/DEFE/13/350, Third Joint Annual Report (1965); TNA/DEFE/13/700, Fourth Joint Annual Report (1966); TNA/DEFE/13/953, Mackenzie to Mason, 16 May 1967.   **372.** Nailor, *Nassau Connection*, pp. 83–4; TNA/DEFE/13/953, BNBMS, Building Programme, 23 December 1965.   **373.** Comments of Captain A. P. Northey in H. J. Tabb and S. A. T. Warren, 'Quality Control Applied to Nuclear Submarine Construction', *Royal Institute of Naval Architects Quarterly Transactions*, 108 (1966).   **374.** '"Resolution", First Polaris Missile Submarine for the Royal Navy', *Shipbuilding and Shipping Record*, 19 October 1967.   **375.** TNA/DEFE/13/953, Mackenzie to Mason, 16 May 1967; Henry, 'A CO's Story', in Moore, *Impact of Polaris*, p. 242.   **376.** Henry, 'A CO's Story', in Moore, *Impact of Polaris*, p. 238.   **377.** 'Mr Starboard and Mr Port Take Over First Polaris Sub', *Daily Mirror*, 6 October 1965.   **378.** John Wingate, *The Fighting Tenth: The Tenth Submarine Flotilla and the Siege of Malta* (Pen & Sword, 1991), p. 1.   **379.** Henry, 'A CO's Story', in Moore, *Impact of Polaris*, p. 243.   **380.** Ibid., p. 248.   **381.** Ibid.   **382.** Ibid.   **383.** Mackenzie, *Sword of Damocles*, p. 236.   **384.** Ibid.   **385.** TNA/DEFE/13/548, Macdonald to Healey, 5 February 1968.   **386.** Healey, handwritten note, TNA/DEFE/13/547, Begg to Healey, RESOLUTION – Publicity on Commissioning, 15 September, 1967.   **387.** *Daily Express*, 6 November 1967.   **388.** National Meteorological Library and Archive, Great Glasgow Storm, Monday, 15 January 1968.   **389.** Nick Howlett, 'The Faslane Base (2)', in Moore, *Impact of Polaris*, p. 183.   **390.** TNA/T/225/3280, Waller to Adams, 25 November 1968.   **391.** Henry, 'A CO's Story', in Moore, *Impact of Polaris*, p. 247.   **392.** Commander Kenneth Frewer, 'HMS *Resolution*: First Polaris Firing by the First SSBN', *Naval Review*, Vol. 86, No. 4, October 1998, pp. 399–400; a not entirely accurate account can also be found in Louis Le Bailly, 'HMS *Resolution*: First Polaris Test Firing by the First SSBN', *Naval Review*, Vol. 85, No. 1, January 1997, p. 64.   **393.** Hansard, House of Commons Debates, 12 February 1968, Vol. 758, Cols. 286–7W.   **394.** TNA/DEFE/13/548, Begg to Healey, HMS RESOLUTION – DASO Firings, 15 February 1968.   **395.** TNA/DEFE/13/953, Polaris Progress at Birkenhead – Meeting between Minister (E), Mr Philip Hunter and C.P.E. on Wednesday, 17 May, 1967.   **396.** TNA/DEFE/13/953, Statement by Minister (E) at Cammell Lairds on Monday, 30 January 1966; Cammell Laird Collection, Polaris Publicity, 003/0006 1316–1317, Mr Mason's Comments this morning to the Times, u/d.   **397.** TNA/DEFE/13/953, Mason to Healey, 19 December 1967.   **398.** TNA/DEFE/13/547, Mason to Horace Law, 6 December 1967.   **399.** TNA/DEFE/13/953, Private Office (E) Note,

Polaris Programme – Delays at Cammell Lairds – Meeting between Minister (E) and P.U.S.(R. N.) on Tuesday, 10 October. **400.** TNA/DEFE/13/953, Mackenzie to Mason, 6 October 1967. **401.** Mackenzie, *Sword of Damocles*, p. 238. **402.** TNA/DEFE/13/953, Private Office (E) Note, Polaris Programme – Delays at Cammell Lairds – Note of Meeting Held by Minister (E) on 13 December 1968. **403.** Mackenzie, 'Launching the Project', in Moore, *Impact of Polaris*, p. 67. **404.** '. . . a "tinkling" noise was reported from the port loop and subsequent "search and identify" operations established that the inner thermal shield in the outlet nozzle of the emergency cooler had fractured and that 11 of the 12 "skirt" sections were at large in the primary circuit. The location and removal of these pieces of metal proved a difficult and time-consuming operation, necessitating entrance being established and special tools designed and manufactured.' TNA/DEFE/69/700, HMS Renown, Emergency Cooler Thermal Sleeve Failure, 23 December 1968. **405.** TNA/DEFE/13/953, Horace Law to Mason, 2 December 1968. **406.** TNA/DEFE/13/953, Morris to Horace Law, 4 December 1968; TNA/DEFE/13/953, Townsend, HMS Renown, 28 November 1968. **407.** TNA/DEFE/69/700, HMS Renown, Emergency Cooler Thermal Sleeve Failure, 23 December 1968. **408.** TNA/DEFE/69/700, Telegram, u/d (February 1969). **409.** TNA/DEFE/13/953, Defect in HMS Renown, 3 September 1969. **410.** Hansard, House of Commons Debates, 3 December 1969, Vol. 792, Cols. 1485–6. **411.** Basil Watson, *Commander-in-Chief: A Celebration of the Life of Admiral of the Fleet The Lord Fieldhouse of Gosport* (Royal Navy Submarine Museum, 2005), p. 73. **412.** TNA/DEFE/69/700, Renown Bow Damage, 4 July 1974; Jim Ring, *We Come Unseen: The Untold Story of Britain's Cold War Submariners* (John Murray, 2001), p. 152. **413.** In 1966 the US announced that the Soviet Union had started to deploy the 'Galosh' ABM system around a number of its major cities; John Baylis and Kristan Stoddart, 'Britain and the Chevaline Project: The Hidden Nuclear Programme, 1967–1982', *Journal of Strategic Studies*, 26/4 (2003), pp. 124–55; Kristan Stoddart, 'The Wilson Government and the British Responses to Anti-Ballistic Missiles, 1964-1970', *Contemporary British History*, 23/1 (2009), pp. 1–33; Thomas Robb, 'Antelope, Poseidon or a Hybrid: The Upgrading of the British Strategic Nuclear Deterrent, 1970–1974', *Journal of Strategic Studies*, 33/6 (2010), pp. 797–817; for accounts of British assessments of the Russian ABM see Catherine Haddon, 'Union Jacks and Red Stars on Them: UK Intelligence, the Soviet Nuclear Threat and British Nuclear Weapons Policy, 1945–1970' (Unpublished Ph.D. Thesis, Queen Mary University of London, 2008), pp. 253–93; Catherine Haddon, 'British Intelligence, Soviet Missile Defence and the British Nuclear Deterrent', in Matthew Grant (ed.), *The British Way in Cold Warfare: Intelligence, Diplomacy and the Bomb 1945– 1975* (Continuum, 2009); Kristan Stoddart, *Losing an Empire and Finding a Role: Britain, the USA, NATO and Nuclear Weapons, 1964–70* (Palgrave Macmillan, 2012), pp. 37–55, 118–53. **414.** TNA/DEFE/11/437, Wilson to Brown, 24 July 1967; Stoddart, *Losing an Empire*, pp. 118–53; TNA/PREM/13/1316, Healey to Brown, 26 September 1966. **415.** TNA/DEFE/11/437, Note of a

Meeting, 8 June 1967. **416.** TNA/CAB/148/55, OPDO(DR)(67) 54th Meeting, 24 November 1967; TNA/CAB/134/3120, PN(67) 4th Meeting, 5 December 1967. **417.** TNA/CAB/165/600, Trend to Wilson, 4 January 1968. **418.** Wilson, *Labour Government 1964-1970*, p. 479. **419.** Sean Straw and John Young, 'The Wilson Government and the Demise of the TSR 2', *Journal of Strategic Studies*, 20/4 (1997), pp. 18-44. **420.** TNA/CAB/128/43, CC(68) 6th Conclusions, 12 January 1968.

Command and Control Arrangements

421. Andrew Priest, 'The President, the "Theologians" and the Europeans: The Johnson Administration and NATO Nuclear Sharing', *The International History Review*, 33:2, pp. 257-75. **422.** TNA/CAB/134/3120, PN(67) 4th Meeting, Minutes of a Meeting of the PN Committee, 5 December 1967. **423.** TNA/DEFE/24/295, Polaris - Assignment to NATO, 21 October 1966. **424.** TNA/DEFE/13/547, Control and Operation of the UK Polaris Force, 23 October 1967. **425.** TNA/DEFE/13/296, Jellicoe to Thorneycroft, 30 October 1963; the command was redesignated C-in-C Fleet (CINCFLEET) in 1971. **426.** However, as the requirements of the maintenance period had priority CINCWF was authorized to extend the notice and recue the number of missiles carried when essential for maintenance; TNA/DEFE/13/547, Control and Operation of the UK Polaris Force, 23 October 1967. **427.** Hennessy, *Secret State,* pp. 153-310. **428.** TNA/DEFE/13/700, Healey to Wilson, 21 March 1967; many of these are still restricted; TNA/DEFE/223, Nuclear retaliation procedure: correspondence and briefs, 1964-1967; TNA/DEFE/224, Nuclear retaliation procedure: correspondence and briefs, 1967; TNA/DEFE/223, Nuclear retaliation procedure: correspondence and briefs, 1967-1968. **429.** TNA/DEFE/13/547, Control and Operation of the UK Polaris Force, 23 October 1967; TNA/DEFE/11/437, A. Brooke-Turner to J. H. Gibbon, 27 February 1967. **430.** TNA/DEFE/13/700, Wilson to Healey, 10 April 1967. **431.** TNA/DEFE/13/547, Control and Operation of the UK Polaris Force, 23 October 1967. **432.** Ibid. **433.** TNA/PREM/13/2571, Halls to Hodges, 6 September 1967. **434.** TNA/PREM/13/2571, Note for the Record, 29 June 1968. **435.** TNA/PREM/13/2571, Procedure for routine daily tests from No. 10 Downing Street of the closed circuit television link with Polaris HQ, June 1968. **436.** TNA/DEFE/13/547, Draft Note from the Chairman Nuclear Retaliation Procedure Committee to the Secretary of State of the Cabinet, Committee on Nuclear Retaliation Procedures Political Control over the Release of British Nuclear Weapons, u/d. **437.** TNA/PREM/13/2571, Burrough to Halls, 2 May 1968; Stoddart, *Losing an Empire*, p. 127. **438.** TNA/CAB/175/19, Government War Book (69) I, Appendix Z, 2 July 1969; GWB (72), Appendix Z, 24 July 1972; for more information on the Nuclear Deputies see Hennessy, *Secret State*, pp. 258-309. When the Polaris force went to sea Harold Wilson's Nuclear Deputies were the Secretary of State for Economic Affairs and from

March 1968 the Foreign Secretary, Michael Stewart, and the Defence Secretary, Denis Healey. From June 1970 onwards Edward Heath appointed three Nuclear Deputies, the Home Secretary, Reginald Maudling, the Foreign Secretary, Sir Alec Douglas-Home, and the Secretary of State for Defence, Lord Carrington. **439.** TNA/CAB/175/19, Government War Book (69) I, Appendix Z, 2 July 1969; GWB (72), Appendix Z, 24 July 1972. **440.** Ibid. **441.** TNA/DEFE/13/547, Control and Operation of the UK Polaris Force, 23 October 1967. **442.** TNA/DEFE/11/437, Handwritten note, 10 April 1967. **443.** The July 1969 Appendix Z clearly shows that if the AOC-in-C, at RAF Strike Command, was unable to contact the PM or the Nuclear Deputies, he was first to confer with the US Commander 'to ascertain what instructions he has received and co-ordinate the release of air delivered nuclear weapons under joint control'. If this proved impossible then the War Book was clear, 'in the last resort' the AOC-in-C was able to authorize 'on his own responsibility, retaliation by all means at his disposal'. Why this was remained a mystery until the declassification of a brief on UK Nuclear Release Procedures for Michael Foot, in the event of a Labour victory in 1983, which explained that 'The reason for this exception is that the survival of the tactical bomber force can only be achieved by ordering it into the air, after which its ability to retaliate is limited in time by its relatively short endurance. The invulnerability of Polaris submarines makes it unnecessary to delegate equivalent authority to the Commander-in-Chief, Fleet.' TNA/CAB/196/124, Armstrong to Prime Minister, Nuclear Release Procedures and Related Matters, June 1983. **444.** Hennessy, *Secret State*, pp. 310–60. **445.** TNA/CAB/196/124, Armstrong to a possible incoming Labour Prime Minister, Nuclear Release Procedures and Related Matters, June 1983. **446.** TNA/CAB/21/6048, Atlantic Nuclear Force, 1965, The Mechanism of Command and Control: Permissive Action Link, Memorandum by the Ministry of Defence, August 1965. **447.** Private information. **448.** IWM/20513, Interview with Rear Admiral Whetstone, 13 June 2000. **449.** ITN Archive, Britain's Polaris Deterrent, 1967 http://www.channel4.com/news/trident-nuclear-deterrent-review-polaris-itn-archive. **450.** TNA/ADM/1/30970, I. B. C. Macleod, Personnel Working or Associated with Nuclear Weapons – Stability Surveillance, 12 July 1968. **451.** Kenneth Young, 'A Most Special Relationship: The Origins of Anglo-American Nuclear Strike Planning',' *Journal of Cold War Studies*, 9/2 (2007), pp. 5–31. **452.** TNA/DEFE/13/953, Polaris Submarines – Operating Cycles (Memorandum by VCNS), 22 April 1970. **453.** TNA/DEFE/13/350, Annex A to COS 75/66, NATO Targeting of the Polaris Force, 22 June 1966. **454.** TNA/DEFE/24/295, ACNS to CT345, Polaris Targeting Committee, 8 November 1967; a special Polaris Targeting Committee was set up in November 1967 presumably to identify the targets. **455.** TNA/DEFE/19/190, UK Strategic Nuclear Force – Short Term Working Party, 3 June 1971. **456.** TNA/DEFE/13/296, Polaris Submarines –Size of UK Force – Appendix to COS.3200/12/12/63. **457.** Ibid. **458.** IWM/20513, Interview with Rear Admiral Whetstone, 13 June 2000.

On Board a 'Resolution' Class Submarine

459. Henry, 'A CO's Story', in Moore, *Impact of Polaris*, p. 249. **460.** Ibid. **461.** Henry Ellis, 'Providing the People', in Moore, *Impact of Polaris*, p. 226. **462.** IWM/20513, Interview with Rear Admiral Whetstone, 13 June 2000. **463.** Ibid. **464.** *Daily Telegraph Magazine*, no. 165, 1 December 1967. **465.** TNA/DEFE/13/548, Begg to Healey, Publicity for the Polaris Force, 15 May 1968. **466.** J. J. Tall and Paul Kemp, *HM Submarines in Camera, 1901–1996* (Blitz Editions, 1998), p. 220. **467.** Watson, *Commander-in-Chief*, p. 70. **468.** Henry, 'A CO's Story', in Moore, *Impact of Polaris*, p. 242. **469.** John Winton, 'Have Polaris Will Travel', *Naval Review*, Vol. 60, No. 3, July 1972, pp. 233–36. **470.** HMS Renown Commissioning Book. **471.** *Daily Telegraph Magazine*, no. 165, 1 December 1967. **472.** Henry, 'A CO's Story', in Moore, *Impact of Polaris*, p. 250. **473.** IWM/20513, Interview with Rear Admiral Whetstone, 13 June 2000. **474.** IWM/16570, Interview with Geoffrey Jaques, 10 June 1996. **475.** Peter Hennessy and Richard Knight, 'HMS Apocalypse: Deep in the Atlantic, a submarine waits on alert with nuclear missiles that would end the world', *Daily Mail*, 30 November 2008. **476.** Winton, 'Have Polaris Will Travel', pp. 233–34. **477.** Interview with Captain Richard Husk, 15 November 2011. **478.** IWM/16570, Interview with Geoffrey Jaques, 10 June 1996. **479.** Winton, 'Have Polaris Will Travel', p. 234. **480.** Arthur Escreet, 'A Crew Member's Story', in Moore, *Impact of Polaris*, pp. 252–3. **481.** IWM/16570, Interview with Geoffrey Jaques, 10 June 1996. **482.** TNA/DEFE/13/548, Sixth Joint Annual Report (1968). **483.** Captain R. W. Garson, 'It Looks Different from Where You Sit', *Naval Review*, Vol. 85, No. 1, January 1997, pp. 46–7. According to Daniel, fifty seconds after the first missile was launched, shore control at Cape Kennedy reported that it had had to destroy the missile because it was heading towards New York. There is no evidence in the files to support his recollection; Daniel, *End of an Era*, p. 194. **484.** Garson, 'It Looks Different from Where You Sit', pp. 46–7. **485.** Ibid. **486.** See Robb, 'Antelope, Poseidon or a Hybrid', pp. 797–817.

5 MIXING IT WITH THE OPPOSITION: THE COLD WAR IN THE 1960S

1. Interview with Richard Sharpe, 19 November 2013. **2.** TNA/DEFE/25/46, VCNS to PUS MOD, 19 November 1964. **3.** Statement on the Defence Estimates, 1967, Cmnd 3203 (Her Majesty's Stationery Office, 1967). **4.** TNA/DEFE/69/726, The SSN Threat, 1970.

The Cold War at Sea

5. TNA/CAB/158/39, JIC(60)6(Final), The Employment of the Soviet Navy and Soviet Air Forces in the Maritime Role at the Outbreak of Global War – 1960–64, 24 February, 1960. **6.** TNA/ADM/1/27680, The Need to Extend Surveillance Operations, March 1960. **7.** TNA/CAB/158/39, JIC(60)6(Final), 24 February 1960. **8.** TNA/CAB/158/39, JIC(60)25(Final), 6 February 1961. **9.** TNA/CAB/131/25, D(61)8th Meeting, 31 May 1961. **10.** Ibid. **11.** Ibid. **12.** TNA/DEFE/24/46, Nuclear Submarines Presentation, Operational Remarks by V.C.N.S., 1962. **13.** TNA/ADM/1/28093, Hezlet to Begg, 7 August 1962. **14.** TNA/ADM/1/28093, Begg to Hezlet, 15 August 1962. **15.** Ibid.

The Cuban Missile Crisis

16. See Owen Cote Jr, *The Third Battle: Innovation in the U.S. Navy's Silent Cold War Struggle with Soviet Submarines* (Naval War College Newport Papers, 2003); Joseph F. Bouchard, *Command in Crisis: Four Case Studies* (Columbia University Press, 1991), p. 117. **17.** Dino Brugioni, *Eyeball to Eyeball: The Inside Story of the Cuban Missile Crisis* (Random House, 1991), p. 363. **18.** Peter Haydon, *The 1962 Cuban Missile Crisis: Canadian Involvement Reconsidered* (Canadian Institute of Strategic Studies, 1993), p. 137. **19.** Ibid. **20.** Interview with Richard Sharpe, 19 November 2013.

Surveillance

21. TNA/DEFE/13/255, Thorneycroft to Carrington, Special Intelligence Operations, 21 February 1963. **22.** Ibid. **23.** TNA/DEFE/13/255, De Zulueta to Vause, 28 March 1963. **24.** TNA/ADM/1/29149, CO HMS *Onslaught* to FOSNI, Operation Bargold – Patrol Report, 9 July 1963. **25.** TNA/ADM/1/29149, Memo by D.N.I., 30 December 1963. **26.** TNA/ADM/1/29149, Memo by Director of Undersea Warfare, 8 January 1964. **27.** TNA/ADM/1/29149, FOSNI to Secretary of the Admiralty, 28 September 1963. **28.** TNA/ADM/1/29149, Memo by D.N.I., 30 December 1963. **29.** TNA/PREM/11/4721, Trend to Douglas-Home, 17 March 1964. **30.** Ibid. **31.** TNA/PREM/11/4721, First Lord to Douglas-Home, 31 March 1964. **32.** TNA/PREM/11/4723, Jellicoe to Butler, 9 October 1964. **33.** Sam Fry, *Fruitful Rewarding Years: A Submariner's Story* (The Memoir Club, 2006), p. 98. **34.** TNA/PREM/11/4723, Wright to Hockaday, 7 September 1964. **35.** Ibid. **36.** TNA/DEFE/13/255, Hockaday to de Zulueta, 15 February 1963. **37.** IWM/30290, Interview with John Coward, 20 November 2007. **38.** TNA/DEFE/13/405, Thomson to Healey, 6 January 1965. **39.** MOD Archive. **40.** Ibid. **41.** TNA/DEFE/24/20, Peduzie to Wright, 6 December 1965. **42.** TNA/DEFE/13/406, Unknown to Nairne, 26 November 1965.   **43.** TNA/DEFE/13/406, Thomson to Wilson, 26 November 1965.   **44.** TNA/DEFE/13/499, Note on Recent Soviet Submarine Incidents, 7 February 1967.   **45.** MOD Archive.   **46.** Ibid.   **47.** TNA/DEFE/13/499, Nairne to Palliser, 6 February 1967; TNA/PREM/13/1382, DEFENCE. Report of suspected Soviet submarine off Londonderry.   **48.** Ibid.   **49.** TNA/DEFE/24/13, Tait, 21 April 1969.   **50.** TNA/DEFE/24/13, Tait, Annex A, Operation 'Alfa', 21 April 1969.   **51.** Ibid.

Indonesian Confrontation

52. TNA/DEFE/13/405, Thomson to Wilson, 29 January 1965.   **53.** TNA/DEFE/13/405, Thomson to Wilson, 5 May 1965.   **54.** Richard Channon, 'A Close Call', in Friends of the Submarine Museum, *All Round Look: Year Book 2014/2015* (Royal Navy Submarine Museum, 2014), p. 25.   **55.** Patrick Middleton, *Admiral Clanky Entertains* (Matador, 2010), p. 121.   **56.** Channon, 'A Close Call', p. 25.   **57.** Ibid.   **58.** Ibid.   **59.** *Daily Telegraph*, Obituary, Captain John Moore, 23 August 2010.   **60.** Denis Healey, *The Time of My Life* (Michael Joseph, 1989), p. 289.   **61.** Fry, *Fruitful Rewarding Years*, p. 105.

Transformation

62. Anthony Gorst, 'CVA-01', in Richard Harding (ed.), *The Royal Navy 1930–2000: Innovation and Defence* (Frank Cass, 2005), pp. 170–92.   **63.** TNA/DEFE/69/481, Mallalieu to Mayhew, 14 September 1965.   **64.** David Owen, *Time to Declare* (Penguin, 1992), p. 147.   **65.** Middleton, *Admiral Clanky Entertains*, p. 132.   **66.** Interview with Richard Heaslip, 18 December 2013.   **67.** Middleton, *Admiral Clanky Entertains*, p. 149.   **68.** R.G.H., 'The Nuclear Attack Submarine', *Naval Review*, Vol. LV, No. 1, January 1967, pp. 22–6.   **69.** Duncan Redford, *The Submarine: A Cultural History from the Great War to Nuclear Combat* (IB Tauris, 2013), p. 169.   **70.** R.G.H., 'The Nuclear Attack Submarine'.   **71.** Ibid.   **72.** TNA/DEFE/24/46, Healey to PUS MOD, 10 November 1964.   **73.** TNA/DEFE/69/535, DS4 Memo, 8 March 1965.   **74.** P. L. Vosper and A. J. Brown, 'Pumpjet Propulsion – A Splendid British Achievement', *Journal of Naval Engineering*, 36(2) (1996).   **75.** Robert Bud and Philip Gummett, *Cold War, Hot Science: Applied Research in Britain's Defence Laboratories, 1945–1990* (Science Museum, 2002), p. 164.   **76.** David K. Brown and George Moore, *Rebuilding the Royal Navy: Warship Design since 1945* (Chatham Publishing, 2003), pp. 127–8.   **77.** TNA/DEFE/24/238, Future Fleet Working Party, Report, Vol. I, Para. 15, 167; Vol. II, Annex 4, Para. 7; TNA/DEFE/24/149, Options for Meeting the Concept of Operations – the Fleet in 1975', 22 July 1966.   **78.** TNA/DEFE/13/949, Begg to Healey, 13 October 1967.   **79.** Ibid.   **80.** TNA/DEFE/13/949, Begg to Healey, The Roles and Capabilities of the Fleet Submarine, 13 October

1967. **81.** Ibid. **82.** Ibid. **83.** Ibid. **84.** Exocet surface-to-surface missiles were deployed during the 1970s in some ships not fitted with Sea Dart. **85.** TNA/DEFE/13/949, Begg to Healey, The Roles and Capabilities of the Fleet Submarine, 13 October 1967. **86.** Ibid. **87.** Ibid. **88.** TNA/DEFE/13/949, Healey to CA Studies (A. H. Cottrell), SSN Programme, 20 October 1967. **89.** TNA/DEFE/13/949, Begg to Healey, The Roles and Capabilities of the Fleet Submarine, 13 October 1967; TNA/DEFE/13/949, Cottrell to Healey, SSN Programme, 31 October 1967. **90.** TNA/DEFE/13/949, SSN Working Party, Interim Report, December 1967. **91.** Ibid. **92.** Ibid. **93.** TNA/DEFE/24/236, SSNs. The Case for the Present Construction Programme, 1 October 1968. **94.** Jack Daniel, *The End of an Era: The Memoirs of a Naval Constructor* (Periscope Publishing, 2003), p. 191; TNA/PREM/15/584, Dame Irene Ward, MP, to Ted Heath, 6 August 1970. **95.** *North West Evening Mail*, 21 February 1969. **96.** Kenneth Warren, *Steel, Ships and Men: Cammell Laird, 1824–1993* (Liverpool University Press, 1998), p. 291. **97.** Ibid., p. 290. **98.** Toru Takamatsu and Ken Warren, 'A Comparison of Cammell Laird and Hitachi Zosen as Shipbuilders', in Takeshi Abe, Douglas Farnie, David Jeremy et al. (eds.), *Region and Strategy in Britain and Japan: Business in Lancashire and Kansai 1890–1990* (Routledge, 1999), p. 220. **99.** Warren, *Steel, Ships and Men*, p. 291. **100.** For a more general account of the problems with the British shipbuilding industry in the post-war years see Geoffrey Owen, *From Empire to Europe: The Decline and Revival of British Industry since the Second World War* (HarperCollins, 1999), pp. 90–115. **101.** Warren, *Steel, Ships and Men*, p. 291. **102.** TNA/DEFE/13/953, Talk for Minister (E), 'Nuclear Submarine Programme', 10 May 1967. **103.** Ibid. **104.** Ibid. **105.** Ibid. **106.** Sherry Sontag and Christopher Drew, *Blind Man's Buff: The Untold Story of American Submarine Espionage* (Random House, 1999), p. 44. **107.** TNA/DEFE/13/949, HMS Dreadnought – First Commission 1963–1968. **108.** TNA/DEFE/13/949, Wilson to Healey, 17 October 1966. **109.** TNA/PREM/13/798, Healey to Wilson, 25 October 1968. **110.** TNA/DEFE/13/949, HMS Dreadnought – First Commission 1963–1968. **111.** Hansard, House of Commons Debates, 16 November 1966, Vol. 736, Col. 105W. **112.** TNA/DEFE/13/953, Recent Problems in HMS Dreadnought and Their Repercussions on the Refits of Other SSNs and SSBNs', 18 February 1970. **113.** TNA/PREM/13/2936, Zuckerman to Heath, 12 December 1969. **114.** 'Service Experience in "Valiant" and "Warspite"', Nuclear Submarine SSN-07, a Symposium, Wednesday, 28 February 1968, at Vickers Limited Shipbuilding Group, Barrow-in-Furness. **115.** *Daily Telegraph*, 'Atom-sub's torpedo "missed sitting target"', 17 July 1973. **116.** TNA/ADM/256/153, Flag Officer Submarines Paper on Submarine Weapons in the 1970s, 5 February 1969; Michael Pitkeathly and David Wixon, *Submarine Courageous, Cold War Warrior: The Life and Times of a Nuclear Submarine* (The HMS *Courageous* Society, 2010), p. 39. **117.** Fry, *Fruitful Rewarding Years*, pp. 104–5. **118.** TNA/ADM/256/153, Flag Officer Submarines Paper on Submarine Weapons in the 1970s, 5 February 1969. **119.** Ibid. **120.** TNA/ADM/189/237, Long Range Submarine and A/S Torpedoes, Captain G. O. Symonds, 1958. **121.** Ibid. **122.** TNA/ADM/256/153,

Pollock to Controller of the Navy, 5 February 1969. **123.** Ibid. **124.** TNA/ADM/256/153, Flag Office Submarines, Paper on Submarine Weapons in the 1970s, 5 February 1969. **125.** TNA/DEFE/24/398, Flag Officer Submarines' 'Haul Down' Report, 10 November 1969. **126.** Ibid. **127.** TNA/DEFE/24/389, E42, undated.

Up North

128. Interview with Rear Admiral John Hervey, 30 January 2013. **129.** Fry, *Fruitful Rewarding Years*, p. 124. **130.** Ibid. **131.** Patrick Middleton, *Admiral Clanky Entertains* (Matador, 2010), pp. 156–7. **132.** Ibid., p. 157. **133.** Ibid., p. 156. **134.** Interview with Peter Herbert, 15 October 2013. **135.** Ibid. **136.** Ibid. **137.** Ibid. **138.** Nikita Khrushchev, *Khrushchev Remembers: The Last Testament* (Little Brown, 1974), p. 31. **139.** Norman Polmar and Kenneth Moore, *Cold War Submarines: The Design and Construction of US and Soviet Submarines* (Potomac Books, 2005), pp. 97–9. **140.** Iain Ballantyne, *Hunter Killers: The Dramatic Untold Story of the Royal Navy's Most Secret Service* (Orion, 2013), chapters 22–5. **141.** Frank Turvey, *The Silent War*, BBC documentary, episode 1: 'Know Your Enemy', broadcast 5 December 2013. **142.** Interview with John Hervey, 6 January 2014. **143.** Ibid. **144.** Sherry Sontag and Christopher Drew, *Blind Man's Bluff: The Untold Story of American Submarine Espionage* (Random House, 1999), p. 281; Soviet officials told the authors' Russian researcher, Alexander Mozgovoy, that this was the first collision involving a NATO surveillance submarine and Soviet nuclear boat in northern waters. **145.** Interview with John Hervey, 6 January 2014. **146.** *The Times*, 19 October 1968. **147.** The most detailed account of the collision, based in part on interviews with *Warspite's* then XO, Tim Hale, can be found in Ballantyne, *Hunter Killers*, chapters 22–5; see also *North West Evening Mail*, 'Did the Russians Hit Barrow Sub?', 8 July 2006; *North West Evening Mail*, 'Retired Submariner Graham Salmon Says Echo II Doing a "Crazy Ivan" Hit Warspite', 8 July 2006; also, Middleton, *Admiral Clanky Entertains*; Sontag and Drew, *Blind Man's Bluff*, p. 281; Dan Conley and Richard Woodman, *Cold War Command: The Dramatic Story of a Nuclear Submariner* (Seaforth Press, 2014), pp. 88, 101–2; Pitkeathly and Wixon, *Submarine Courageous*, p. 129. **148.** Lord Owen, *The Silent War*, episode 1: 'Know Your Enemy', broadcast 5 December 2013. **149.** Middleton, *Admiral Clanky Entertains*, p. 150. **150.** Correspondence with John Hervey, 14 January 2014. **151.** Ibid. **152.** Interview with Sandy Woodward, 2 April 2012. **153.** TNA/DEFE/13/949, Background Information on the Soviet Navy, 1 December 1969. **154.** Pitkeathly and Wixon, *Submarine Courageous*, p. 241. **155.** Interview with Sandy Woodward, 2 April 2012. **156.** Middleton, *Admiral Clanky Entertains*, p. 158 **157.** Interview with Sandy Woodward, 2 April 2012. **158.** Ibid. **159.** The delay was partly due to Khrushchev's priority for the land-based Strategic Missile Forces that lapsed after his removal, coupled with the difficulty of developing solid-fuel missiles. In the event a compact storable liquid-fuelled missile originally designed for strikes on carrier task forces was reconfigured as a land attack missile; Pavel Podvig, *Russian*

Strategic Nuclear Forces (MIT Press, 2004), pp. 319–20; Pitkeathly and Wixon, *Submarine Courageous*, p. 181. **160.** Podvig, *Russian Strategic Nuclear Forces*, p. 7. **161.** Interview with Sandy Woodward, 2 April 2012. **162.** Ibid. **163.** Ibid. **164.** Ibid. **165.** Ibid. **166.** Middleton, *Admiral Clanky Entertains*, p. 159. **167.** Interview with Sandy Woodward, 2 April 2012. **168.** Ibid. **169.** Ibid. **170.** Ibid. **171.** Ibid. **172.** Ibid. **173.** Ibid.

6 'NO REFUGE IN THE DEPTHS': THE COLD WAR IN THE 1970S

1. Michael Pitkeathly and David Wixon, *Submarine Courageous, Cold War Warrior: The Life and Times of a Nuclear Submarine* (The HMS *Courageous* Society, 2010), p. 129. **2.** TNA/DEFE/69/727, Captain Peter Herbert comments accompanying Submarine ASW Philosophy, Richard Sharpe, 1973.

The Decade of the Passive

3. Norman Polmar, *The Naval Institute Guide to the Soviet Navy* (Naval Institute Press, 5th edn, 1991), p. 40. **4.** Ibid., p. 41. **5.** Henry S. Lowenhaupt, 'How We Identified the Technical Problems of Early Russian Nuclear Submarines', *CIA Studies in Intelligence*, Vol. 18, Fall 1974, pp. 1–9.

SOSUS

6. For more on sonar see John Hervey, *Submarines (Brassey's Sea Power: Naval Vessels, Weapons Systems and Technology, Vol. 7)* (Brassey's, 1994), pp. 91–124. **7.** Rear Admiral Martin Wemyss, 'Submarines and Anti-Submarine Operations for the Uninitiated', *RUSI Journal* (September 1981), p. 24. **8.** Guy Warner, 'The Tactical Challenges of Submarine Operations: A Historical Perspective', unpublished article. **9.** http://www.navy.mil/navydata/cno/n87/usw/issue_25/sosus2.htm. **10.** TNA/AIR/2/18731, draft paper for the ORC, UK Participation in SOSUS, 9 October 1970. **11.** The Americans were also anxious to find a site in the UK at which they could reinstate the Norwegian array, should the political circumstances in Iceland produce a left-wing, anti-American government. **12.** TNA/ADM/189/239, Discussion on Some Operational Aspects of A/S, paper presented to the Fifteenth TAS Conference, May 1958. **13.** TNA/AIR/2/18731, Loose Minute, SOSUS, 11 October 1968. **14.** TNA/AIR/2/18731, Annex to CDRE(INT)404/470, 20 October 1970. **15.** TNA/AIR/2/18731, I. L. Emmett, 10 November 1970. **16.** TNA/AIR/2/18731, D. M. Clause, Minute, 13 October 1970. **17.** TNA/AIR/2/18731, Annex to CDRE(INT)404/470, 20 October 1970. **18.** TNA/AIR/2/18731, J. W. Ring, SOSUS – UK Participation, 22 October. **19.** Charlie Whitham, 'Bargaining over Brawdy: Negotiating the American Military Presence in Wales,

1971', in Luis Nunos Rodriguez and Sergiy Glebov (eds.), *Military Bases: Historical Perspectives, Contemporary Challenges* (IOS Press, 2009), pp. 40–55. **20.** TNA/FCO/82/77, R. J. Murray to C. Lush, 24 May 1971. **21.** TNA/FCO/82/77, N. H. Young to Chamier, Project Backscratch, 15 June 1971. **22.** http://www.navy.mil/navydata/cno/n87/usw/issue_25/sosus2.htm. **23.** Hervey, *Submarines*, p. 93. **24.** MOD Archive. **25.** TNA/DEFE/13/1357, Sea Air Warfare Committee – Operational Concept for Anti-Submarine Warfare, 25 May 1978. **26.** TNA/AIR/2/18731, J. W. Ring, SOSUS – UK Participation, 22 October 1970. **27.** Ibid. **28.** TNA/DEFE/69/726, DIS Brief, 'Your Problem': Soviet New Construction Submarines, 1970. **29.** Ibid. **30.** TNA/CAB/186/9, JIC(A)(71)21, Soviet Maritime Policy, 30 July 1971. **31.** Ibid. **32.** Norman Polmar and Kenneth Moore, *Cold War Submarines: The Design and Construction of US and Soviet Submarines* (Potomac Books, 2005), pp. 158–9. **33.** Pavel Podvig, *Russian Strategic Nuclear Forces* (MIT Press, 2004), p. 625. **34.** TNA/CAB/186/9, JIC(A)(71)21, Soviet Maritime Policy, Report by Joint Intelligence Committee (A), 30 July 1971. **35.** Podvig, *Russian Strategic Nuclear Forces*, pp. 11–12. **36.** Tom Le Marchand, 'Under Ice Operations', *U.S. Naval War College Review*, May–June 1985, pp. 19–27. **37.** Ibid.

Under the Ice

38. William R. Anderson, *Nautilus 90 North* (Tab Books, May 1989). **39.** Polmar and Moore, *Cold War Submarines*, p. 335, fn.28. The available evidence indicates that *Thresher* suffered an electrical bus failure, which shut down the submarine's main coolant pumps and led to a reactor scram. The crew were unable to restart the reactor to regain propulsion and unable to blow ballast to surface. *Thresher* slowly sank towards the ocean floor and imploded at a depth of about 2400 feet (more than 400 feet below her predicted collapse depth). **40.** Pier Horensma, *The Soviet Arctic* (Routledge, 1991), pp. 107–8. **41.** Gary E. Weir, 'Virtual War in the Ice Jungle: "We don't know how to do this"', *Journal of Strategic Studies*, 28:2 (2005), p. 413. **42.** MOD Archive. **43.** Ibid. **44.** Ibid. SSK(N) is a rather confusing non-standard designation occasionally used for hunter-killer nuclear-powered submarines. Normally by this time SSK was used for conventionally powered boats, a reflection of their primary ASW hunter-killer role and SSN for all non-ballistic-missile nuclear-powered boats. **45.** TNA/DEFE/24/99, FOSM, HMS Dreadnought – Proposed Arctic Patrol, 15 December 1970. **46.** MOD Archive. **47.** Ibid. **48.** Ibid. **49.** Ibid. **50.** Ibid. **51.** Ibid. **52.** Ibid. **53.** Ibid.

A Specialist Service?

54. RNSM/A2008, John Roxburgh to Tony Troup, Personal Turnover Notes, 22 June 1972. **55.** Ibid. **56.** *Daily Express*, 3 July 1971. **57.** MOD Archive. **58.** *Daily Telegraph*, 21 October 1971. **59.** RNSM/A2008, John

Roxburgh to Tony Troup, Personal Turnover Notes, 22 June 1972. **60.** Ibid. **61.** RNSM/A2000/061, The Origins of Submarine Sea Training/ Work-Ups (undated manuscript).

A Victor Penetrates the Clyde

62. Rowland White, *Vulcan 607: The Epic Story of the Most Remarkable British Air Attack since WWII* (Corgi Books, 2006), pp. 66–7; Jim Ring, *We Come Unseen: The Untold Story of Britain's Cold War Submariners* (John Murray, 2001), pp. 107–8. **63.** *The Times*, 'Navy Shadows Mystery Sub', 30 January 1973. **64.** Interview with Chris Ward, 23 June 2014. **65.** Ibid. **66.** Interview with Roger Lane-Nott, 25 March 2014. **67.** Interview with Chris Ward, 23 June 2014. **68.** Ibid. **69.** Ibid. **70.** Ibid. **71.** Interview with Roger Lane-Nott, 25 March 2014. **72.** Interview with Chris Ward, 23 June 2014. **73.** Interview with Roger Lane-Nott, 25 March 2014. **74.** Interview with Chris Ward, 23 June 2014. **75.** Interview with Roger Lane-Nott, 25 March 2014. **76.** Interview with Chris Ward, 23 June 2014. **77.** John Roberts, *Safeguarding the Nation: The Story of the Modern Royal Navy* (Seaforth Publishing, 2009), p. 99. **78.** Basil Watson, *Commander-in-Chief: A Celebration of the Life of Admiral of the Fleet The Lord Fieldhouse of Gosport* (Royal Navy Submarine Museum, 2005), p. 118. **79.** Ibid., p. 118; Interview with Richard Sharpe, 19 November 2013.

A New Concept of Operations

80. TNA/DEFE/69/727, Commander Llewelyn, Commander Round-Turner, SSN in Support of Surface Forces – State of the Art, 1973. **81.** Ibid. **82.** TNA/DEFE/48/285,The Nuclear Submarine in Defence of Surface Forces and Interdiction of Enemy Surface Movement, November 1974. **83.** TNA/DEFE/ 69/727, Commander R. G. Sharpe, Submarine ASW Philosophy, 1973. **84.** Ibid. **85.** Weir, 'Virtual War in the Ice Jungle', p. 414. **86.** Ibid. **87.** Pitkeathly and Wixon, *Submarine Courageous*, p. 124. **88.** Interview with Martin Macpherson, 24 May 2013. **89.** Ring, *We Come Unseen*, p. 172. **90.** TNA/DEFE/69/727, Commander R. G. Sharpe, Submarine ASW Philosophy, 1973. **91.** *Daily Telegraph*, 'War Games', 20 August 2000. **92.** Captain Donald Mitchell, 'United States Navy Submarine Development Group Two and Submarine Development Squadron Twelve: A Royal Navy Perspective, 1995 to Date (2007)', June 2007. **93.** TNA/ADM/13/985, Loose minute from Head of DS5, 4 December 1973. **94.** TNA/DEFE/69/653, Future Policy for Special Fit, 29 July 1977. **95.** Ibid. **96.** TNA/DEFE/24/1387, Naval Staff Target 7029 (New Patrol Class Submarine) Supporting Paper, June 1978. **97.** Pitkeathly and Wixon, *Submarine Courageous,* p. 55. **98.** Bruce Schick, *Whales Tales: Recollections of a Diesel Submariner* (DBF Press, 2005), p. 78–9. **99.** RNSM/ A2008, John Roxburgh to Tony Troup, Personal Turnover Notes, 22 June

1972. **100.** MOD Archive. **101.** Ibid. **102.** Pitkeathly and Wixon, *Submarine Courageous,* pp. 128–9. **103.** John 'Sandy' Woodward, *One Hundred Days: The Memoirs of the Falklands Battle Group Commander* (HarperPress, 2012), p. 61. **104.** Ibid., p. 62. **105.** Warner, 'Tactical Challenges of Submarine Operations'. **106.** TNA/DEFE/67/777, Choke Point Operations In Exercise Dawn Patrol 1973. **107.** MOD Archive. **108.** Ibid. **109.** IWM/30390, Interview with John Coward, 20 November 2007. **110.** Ibid. **111.** Watson, *Commander-in-Chief,* p. 120. **112.** IWM/30390, Interview with John Coward, 20 November 2007. **113.** Pitkeathly and Wixon, *Submarine Courageous,* pp. 72–3. **114.** MOD Archive. **115.** Ibid. **116.** Ibid. **117.** Ibid. **118.** Pitkeathly and Wixon, *Submarine Courageous,* pp. 246–7.

The Perils of Special Operations

119. TNA/PREM/16/146, Nichols to Butler, 17 May 1974. **120.** TNA/DEFE/69/226, Ministry of Defence (Navy) surveillance operations reports: Operations Awless, Aver and Artellot: submarine patrols off Northern Ireland, 1 January 1974–31 December 1976. **121.** TNA/DEFE/69/226, C. L. Carver, Headquarters Northern Ireland, Submarine Support for Northern Ireland, 12 June 1975. **122.** TNA/DEFE/69/226, Littlejohns to Captain SM, HMS Dolphin, 28 July 1975. **123.** TNA/DEFE/69/226, RMP Pool, Operation Aweless – Operation Report, 25 January 1975. **124.** Major R. C. Clifford, MBE, RM, 'Five Bells', in Friends of the Royal Navy Submarine Museum, *All Round Look: Year Book 2007/2008* (Royal Navy Submarine Museum, 2008), pp. 44–8. **125.** Ibid. **126.** Ibid. **127.** Ibid. **128.** Duncan Falconer, *First Into Action: A Dramatic Personal Account of Life inside the SBS* (Sphere, 2001), pp. 165–6. **129.** Ibid., pp. 151–65; MOD Archive; Don Camsell, *Black Water: By Strength and by Guile* (Virgin Books, 2001). **130.** Falconer, *First into Action,* pp. 160–65. **131.** Ibid.

The 'Swiftsure' Class

132. Watson, *Commander-in-Chief,* p. 116. **133.** Interview with Martin Macpherson, 24 May 2013. **134.** Vice Admiral Sir Ted Horlick, 'Nuclear Submarine Propulsion in the RN', The Thomas Lowe Gray Memorial Lecture to the Institution of Mechanical Engineers, January 1982, pp. 65–79. **135.** 'Design Philosophy', Nuclear Submarine SSN-07, a Symposium, Wednesday, 28 February 1968, at Vickers Limited Shipbuilding Group, Barrow-in-Furness. **136.** Ibid. **137.** Ibid. **138.** Robert Bud and Philip Gummett, *Cold War, Hot Science: Applied Research in Britain's Defence Laboratories, 1945–1990* (Science Museum, 2002), p. 166. **139.** Core B was the first core designed entirely by Rolls-Royce and Associates. Initial criticality was achieved in June 1968 and Core B was eventually retrofitted into all Core A submarines as well as the 'Swiftsure' class. **140.** Oral History with Dr Waldo Lyon, by Gary E. Weir,

US Naval Historical Center, 16 April 1994, US Navy Operational Archive, Washington DC [OA]. **141.** MOD Archive. **142.** Ibid. **143.** Ibid. **144.** Ibid. **145.** Ibid. **146.** Warner, 'Tactical Challenges of Submarine Operations'. **147.** Ibid. **148.** Pitkeathly and Wixon, *Submarine Courageous*, p. 159. **149.** Warner, 'Tactical Challenges of Submarine Operations'. **150.** TNA/ADM/256/153, Pollock to Controller of the Navy, 5 February 1969. **151.** TNA/DEFE/69/726, The Weapons Programme, 1970. **152.** Ibid. **153.** Ibid. **154.** Ibid.; the SUW-N-1 was a nuclear-armed ASW missile carried by the large Soviet helicopter carriers. **155.** TNA/DEFE/69/727, Anti SLAM tactics, 1970. **156.** TNA/DEFE/24/1782, Updated NSR 6333 (Issue 2) – Underwater to Surface Anti-Ship Guided Weapon System (USGW), a paper by the Navy Department, August 1977. **157.** Ibid. **158.** Pitkeathly and Wixon, *Submarine Courageous*, pp. 195–6. **159.** Ibid. **160.** Hervey, *Submarines*, pp. 100–101. **161.** TNA/DEFE/13/1357, Towed Arrays for Submarines – Sonar 2026, 18 July 1978. **162.** Dan Conley and Richard Woodman, *Cold War Command: The Dramatic Story of a Nuclear Submariner* (Seaforth Press, 2014), pp. 156–7. **163.** Interview with Mark Stanhope, 4 April 2014. **164.** MOD Archive. **165.** Ibid. **166.** Ibid. **167.** Polmar and Moore, *Cold War Submarines*, p. 173.

Operation 'Agile Eagle'

168. MOD Archive. **169.** Roger Lane-Nott, 'Submarine Intelligence and the Cold War', in Michael R. Fitzgerald and Allen Packwood (eds.), *Out of the Cold: The Cold War and Its Legacy* (Bloomsbury Academic, 2013), p. 115. **170.** MOD Archive. **171.** Pitkeathly and Wixon, *Submarine Courageous*, p. 56. **172.** TNA/DEFE/24/1387, Naval Staff Target 7029 (New Patrol Class Submarine) Supporting Paper, 7 June 1978. **173.** MOD Archive. **174.** Ibid. **175.** Ibid. **176.** Interview with Roger Lane-Nott, 25 March 2014. **177.** Rear Admiral J. R. Hill, *Antisubmarine Warfare* (Naval Institute Press, 1984), p. 69.

Assessments of the Soviet Navy

178. TNA/DEFE/13/1357, Sea/Air Warfare Committee – Operational Concept for Anti-Submarine Warfare, 25 May 1978. **179.** Ibid. **180.** TNA/CAB/186/11, JIC(A)(72)3, 25 January 1972. **181.** Gerhardt Thamm, 'Unraveling a Cold War Mystery: The Alfa SSN: Challenging Paradigms, Finding New Truths, 1969–79', *Cold War Studies in Intelligence*, Vol. 52, No. 3, **182.** Polmar and Moore, *Cold War Submarines*, p. 142. **183.** Ibid., p. 177. **184.** Ibid., p. 178. **185.** TNA/DEFE/24/1387, Naval Staff Target 7029 (New Patrol Class Submarine) Supporting Paper, 7 June 1978. **186.** Ibid. **187.** John Lehman Jr, *Command of the Seas: Building the 600 Ship Navy* (Charles Scribner's Sons, 1988), p. 133. **188.** TNA/DEFE/24/1387, Naval Staff Target 7029 (New Patrol Class Submarine) Supporting Paper, 7 June 1978. **189.** Ibid. **190.** Ibid.

7. HOT WAR: THE FALKLANDS CONFLICT

1. IWM/17274, Interview with Roger Lane-Nott, 23 January 1997. **2.** RNSM/HMS *Conqueror* Ship's Book, Woodward to *Conqueror*, 16 June 1982.

Operation 'Journeyman'

3. See Lawrence Freedman, *The Official History of the Falklands Campaign*, Vol. 1: *The Origins of the Falklands War* (Routledge, 2007), pp. 65–8. **4.** Ibid., pp. 69–75. **5.** TNA/DEFE/24/1245, Ure to Moss, Falkland Islands: Rules of Engagement for Naval Forces, 24 November 1977. **6.** TNA/PREM/16/1504, Nigel Brind to Bryan Cartledge, The Falkland Islands, 15 December 1977. **7.** Hansard, House of Commons Debates, 25 January 1983, Vol. 35, Col. 812. **8.** Nigel West, *The Secret War for the Falklands: The SAS, MI6, and the War Whitehall Nearly Lost* (Warner Books, 1998), p. 221. **9.** Freedman, *Official History of the Falklands Campaign*, Vol. 1, pp. 74–5. **10.** Interview with Martin Macpherson, 24 May 2013. **11.** Ibid. **12.** Charles Moore, *Margaret Thatcher: The Authorized Biography*, Vol. 1: *Not for Turning* (Allen Lane, 2013), p. 664. **13.** TNA/CAB/292/37, Falkland Islands Review Committee, Note of an Oral Evidence Session with The Rt Hon. James Callaghan, 18 October 1982. **14.** James Callaghan, *Time and Chance* (HarperCollins, 1987), p. 375. **15.** David Owen, *Time to Declare* (Michael Joseph, 1991), p. 350. **16.** Freedman, *Official History of the Falklands Campaign*, Vol. 1, p. 73. **17.** Ibid. **18.** TNA/FCO/73/283, Owen to Sec. State for Defence, Falkland Islands Task Force, u/d (December 1977). **19.** Ibid. **20.** Paul Hind, 'Dreadnought's Covert Falklands Mission', in Friends of the Submarine Museum, *All Round Look: Year Book 2003/2004* (Royal Navy Submarine Museum, 2004), p. 17.

Operation 'Corporate'

21. For an Argentinean account of Submarine operations in the South Atlantic during the 1982 Falklands Conflict see Mariano Sciaroni, *Malvinas. Tras los submarinos ingleses* (Instituto de Publicaciones Navales 2010). **22.** TNA/DEFE/13/949, SSN, Working Party, Interim Report, Outline Scenarios Not Adopted for Development by the SSN Working Group, 19 December 1967. **23.** According to Nigel West the Falklands story was leaked to the press by 'a submariner at MOD': West, *Secret War for the Falklands*, p. 36. **24.** TNA/CAB/292/57, Falkland Islands Review Committee, Note of an Oral Evidence session held in Room 1/99 Old Admiralty Building, London SW, Monday, 8 November 1982. **25.** Freedman, *Official History of the Falklands Campaign*, Vol. 1, p. 176. **26.** MOD Archive. **27.** Ibid. **28.** TNA/PREM/19/657, Nott to Thatcher, Options for deployment of naval vessels to the Falklands, 29 March

1982. **29.** Jim Ring, *We Come Unseen: The Untold Story of Britain's Cold War Submariners* (John Murray, 2001), p. 182. **30.** Interview with James Taylor, 10 April 2014. **31.** TNA/PREM/19/657, FOC minute to Thatcher, 30 March 1982. **32.** Ibid. (Thatcher annotation), 30 March 1982. **33.** Margaret Thatcher, *The Downing Street Years* (HarperPress, 2012), p. 178. **34.** TNA/PREM/19/657, FOC minute to Thatcher, 30 March 1982. **35.** Freedman, *Official History of the Falklands Campaign*, Vol. 1, p. 200. **36.** Ibid., p. 175. **37.** CAC, Margaret Thatcher Archives (hereafter MTA), ALW/040325/9, Minutes of a Meeting of Defence Operations Executive (Falkland Islands), 30 March 1982. **38.** Freedman, *Official History of the Falklands Campaign*, Vol. 1, p. 175. **39.** TNA/CAB/292/47, Falkland Islands Review Committee, Note of an Oral Evidence Session with the Prime Minister, Margaret Thatcher, 25 October 1982. **40.** Thatcher, *Downing Street Years*, pp. 173–85. **41.** Moore, *Not for Turning*, p. 667. **42.** IWM/31103, Interview with Liam Bradley, April 2008. **43.** Quoted in Lawrence Freedman, *The Official History of the Falklands Campaign*, Vol. 2: *War and Diplomacy* (Routledge, 2007), p. 228. **44.** TNA/FCO/7/4472, Confidential Annex to COS/4/82, 4 April 1982. **45.** Interview with Admiral Sir Peter Herbert, 15 October 2013. **46.** 'Jettisoned bombs just missed *Valiant*', http://www.thenewscentre.co.uk/falklands/commande. htm. **47.** Interview with Doug Littlejohns, 20 November 2013. **48.** Beneath him, officially of equal status but with a lower rank, were Commodore, Amphibious Warfare (COMAW) Commodore Michael Clapp, in charge of the Amphibious Group, and Brigadier Julian Thompson, RM, of the land forces, who would be replaced after the initial landings by Major General Jeremy Moore, RM. **49.** Andrew Gordon, *The Rules of the Game* (John Murray, 2000), p. 587. **50.** *Daily Telegraph*, 'Sir "Sandy" Woodward: A Shy But Decisive Fighter', 11 August 2013. **51.** Freedman, *Official History of the Falklands Campaign*, Vol. 1, p. 175. **52.** Ibid. **53.** TNA/PREM/19/615, Armstrong to Thatcher, The Falkland Islands, 6 April 1982. **54.** TNA/FCO/7/4472, Meeting of the Chiefs of Staff, 5 April 1982. **55.** The War Cabinet was serviced by the Cabinet Secretary, Robert Armstrong, and his deputy, Robert Wade-Gery, who was in charge of foreign and defence liaison in the Cabinet Office. It was also attended by the Chief of Defence Staff, Admiral Terence Lewin, and by the Attorney-General, Sir Michael Havers. Others who attended at various times were the Permanent Secretary at the Foreign Office, Antony Acland, the MOD Permanent Secretary, Frank Cooper, the former Permanent Secretary at the Foreign Office, Sir Michael Palliser, the Foreign Office Legal Advisor, Sir Ian Sinclair, and the Commander-in-Chief Fleet, Admiral Sir John Fieldhouse. **56.** Interview with James Taylor, 10 April 2014. **57.** Hansard, House of Commons Debates, 7 April 1982, Vol. 21, Col. 1045. **58.** TNA/FCO/7/4472, Wright, Chiefs of Staff Meeting, 6 April 1982. **59.** TNA/CAB/148/211, OD(SA)(82)5, Armstrong, Wade-Geary, Colvin note circulated to OD(SA) Committee, Rules of Engagement, 8 April 1982. **60.** TNA/CAB/148/211, OD(SA)82, 3rd Meeting, 8 April 1982. **61.** 'The Falklands War', eds. Andrew Dorman, Michael D. Kandiah and Gillian Staerck, CCBH Oral History Programme, 2005. **62.** Freedman, *Official History of the Falklands Campaign*,

Vol. II, p. 91. **63.** Air Vice Marshal Ron Dick, 'The View from BDLS Washington', *Royal Air Force Historical Society Journal*, no. 30, 2003, pp. 25–35. **64.** TNA/FCO/7/4472,Wright, Chiefs of Staff Meeting, 6 April 1982. **65.** Freedman, *The Official History of the Falklands Campaign*, Vol. 2, pp. 434–5. **66.** Ibid., p. 146. **67.** IWM/17274, Interview with Roger Lane-Nott, 23 January 1997. **68.** Interview with James Taylor, 10 April 2014. **69.** HMS *Splendid*, Report of Proceedings, 23 June 1982. **70.** IWM/17274, Interview with Roger Lane-Nott, 23 January 1997. **71.** Ibid. **72.** Interview with James Taylor, 10 April 2014. **73.** HMS *Spartan,* Report of Proceedings, 23 June 1982. **74.** Tom Clancy, *Submarine* (Berkley Books, 1993), p. xxii. **75.** Freedman, *Official History of the Falklands Campaign*, Vol. 2, p. 219. **76.** IWM/17274, Interview with Roger Lane-Nott, 23 January 1997. **77.** Ibid. **78.** Ibid. **79.** Ibid. **80.** IWM/AU, Interview with Edward John Hogben, 15 December 2010. **81.** Christopher Wreford-Brown, '*Conqueror*'s War Patrol', in John Winton, *The Submariners: Life in British Submarines 1901–1999* (Constable, 2001), p. 281. **82.** HMS *Conqueror*, Report of Proceedings, 1 July 1982. **83.** Freedman, *Official History of the Falklands Campaign*, Vol. II, pp. 208–16. **84.** Jonathan Powis, 'Falklands Memories', *Submarine Review*, January 2008, pp. 61–2. **85.** Ibid., p. 61. **86.** Wreford-Brown, '*Conqueror's* War Patrol', in Winton, *Submariners*, p. 284. **87.** HMS *Conqueror*, Report of Proceedings, 1 July 1982. **88.** Ibid. **89.** TNA/PREM/19/620, Armstrong to Omand, 21 April 1982. **90.** TNA/PREM/19/620, Acland to Armstrong, 22 April 1982. **91.** TNA/PREM/19/620, Wright to Lewin, 22 April 1982. **92.** IWM/17274, Interview with Roger Lane-Nott, 23 January 1997. **93.** Freedman, *Official History of the Falklands Campaign*, Vol. 2, p. 261. **94.** John 'Sandy' Woodward, *One Hundred Days: The Memoirs of the Falklands Battle Group Commander* (HarperPress, 2012), pp. 171–2. **95.** Ibid., p. 173. **96.** Quoted in Mike Rossiter, *Sink the Belgrano* (Bantam Press, 2007), p. 177. **97.** Ibid., p. 134. **98.** Interview with Admiral Sir Peter Herbert, 15 October 2013. **99.** Quoted in Rossiter, *Sink the Belgrano*, p. 134. **100.** Woodward, *One Hundred Days*, p. 173. **101.** HMS *Valiant*, Report of Proceedings, 26 July 1982. **102.** Interview with James Taylor, 10 April 2014. **103.** Interview with James Taylor, 5 December 2013. **104.** HMS *Spartan,* Report of Proceedings, 23 June 1982. **105.** Interview with James Taylor, 5 December 2013. **106.** HMS *Splendid*, Report of Proceedings, 23 June 1982. **107.** IWM/17274, Interview with Roger Lane-Nott, 23 January 1997. **108.** Ibid. **109.** TNA/CAB/148/211, OD(SA)(82)36, The Argentine Aircraft Carrier, Note by the Secretaries, 30 April 1982. **110.** TNA/CAB/148/211, OD(SA)22, 30 April 1982. **111.** Freedman, *The Official History of the Falklands Campaign*, Vol. 2, p. 219. **112.** IWM/17274, Interview with Roger Lane-Nott, 23 January 1997. **113.** Ibid. **114.** Freedman, *Official History of the Falklands Campaign*, Vol. 2, p. 290. **115.** Ibid. **116.** HMS *Spartan,* Report of Proceedings, 23 June 1982. **117.** Woodward, *One Hundred Days*, p. 177. **118.** HMS *Conqueror*, Report of Proceedings, 1 July 1982. **119.** Freedman, *Official History of the Falklands Campaign*, Vol. 2, p. 292. **120.** Woodward, *One Hundred Days*, p. 212. **121.** Ibid. **122.** Ibid.

Sink the *Belgrano*

123. Rossiter, *Sink the Belgrano*, p.223.   **124.** Woodward, *One Hundred Days*, p. 115.   **125.** Interview with Admiral Sir Peter Herbert, 15 October 2013.   **126.** TNA/FCO7/4474, Minutes of COS meeting, 2 May 1982.   **127.** Thatcher, *Downing Street Years*, pp. 214–16; TNA/PREM/19/623, Wade-Gery to Omand, Falklands: Military Decisions, 2 May 1982.   **128.** 'The Falklands War', eds. Dorman, Kandiah and Staerck.   **129.** Churchill College Cambridge, British Diplomatic Oral History Programme, Interview with Robert Wade-Gery, 2000.   **130.** Wreford-Brown, '*Conqueror's* War Patrol', in Winton, *Submariners*, p. 286.   **131.** Dan Conley and Richard Woodman, *Cold War Command: The Dramatic Story of a Nuclear Submariner* (Seaforth Press, 2014), p. 173.   **132.** HMS *Splendid*, Report of Proceedings, 23 June 1982.   **133.** Powis, 'Falklands Memories', p. 64.   **134.** Interview with Tim McClement, 26 September 2013.   **135.** *Daily Telegraph*, 'Falklands Commander: Sinking the Belgrano was Right Thing to Do', 2 April 2012.   **136.** HMS *Conqueror*, Report of Proceedings, 1 July 1982.   **137.** Ibid.   **138.** Wreford-Brown, '*Conqueror's* War Patrol', in Winton, *Submariners*, p. 287.   **139.** Powis, 'Falklands Memories', p. 65.   **140.** HMS *Conqueror*, Report of Proceedings, 1 July 1982.   **141.** Powis, 'Falklands Memories', p. 65.   **142.** Ibid.   **143.** HMS *Conqueror*, Report of Proceedings, 1 July 1982.   **144.** Wreford-Brown, Interview for *War in the Falklands*, ITN Factual, 2002, Episode 1.   **145.** Powis, 'Falklands Memories', p. 64.   **146.** Wreford-Brown, '*Conqueror's* War Patrol', in Winton, *Submariners*, p. 287.   **147.** Ibid., p. 288.   **148.** Woodward, *One Hundred Days*, p. 223.   **149.** Ibid.   **150.** HMS *Spartan*, Report of Proceedings, 23 June 1982.   **151.** Interview with James Taylor, 10 April 2014.   **152.** IWM/17274, Interview with Roger Lane-Nott, 23 January 1997.   **153.** Ibid.   **154.** Exchange between Diana Gould and Margaret Thatcher, *Nationwide*, BBC1, 24 May 1983.   **155.** Freedman, *Official History of the Falklands Campaign*, Vol. 2, p. 295.   **156.** The direction in which the *Belgrano* was sailing has also fuelled conspiracy theories. One, in particular, concerns the fate of *Conqueror's* Control Room log, which disappeared shortly after *Conqueror* returned from the Falklands. What happened to the Control Room log remains a mystery. One theory is that the log book was destroyed due to *Conqueror's* involvement in other Cold War operations prior to and after the Falklands conflict. See Stuart Prebble, *Secrets of the* Conqueror: *The History of Britain's Most Famous Submarine* (Faber & Faber, 2013). HMS *Conqueror's* Report of Proceedings for Operation 'Corporate', the detailed running narrative recorded by watch leaders and the Commanding Officer, has been in the public domain for some time, albeit with a few minor redactions, which appear to relate to intelligence information.   **157.** Freedman, *Official History of the Falklands Campaign*, Vol. 2, p. 296.   **158.** Ibid.   **159.** Ibid.   **160.** 'Belgrano crew "trigger happy"', *Guardian*, 25 May 2003.   **161.** Quoted in Richard Hill, *Lewin of Greenwich: The Authorised Biography of Admiral of the Fleet Lord Lewin* (Cassell, 2000), p. 368.   **162.** Powis, 'Falklands Memories', p. 67.   **163.** Wreford-Brown,

'*Conqueror's* War Patrol', in Winton, *Submariners*, p. 289.   **164.** *Daily Tele-graph*, 'Falklands Commander: Sinking the Belgrano was Right Thing to Do', 2 April 2012.   **165.** Thatcher, *Downing Street Years*, pp. 214–16.   **166.** HMS *Valiant*, Report of Proceedings, 26 July 1982.   **167.** HMS *Spartan*, Report of Proceedings, 23 June 1982.   **168.** Woodward, *One Hundred Days*, p. 207.

Frustration

169. TNA/FOC/7/4474, Meeting of the Defence Operations Executive, 3 May 1982.   **170.** HMS *Spartan*, Report of Proceedings, 23 June 1982.   **171.** IWM/17274, Interview with Roger Lane-Nott, 23 January 1997.   **172.** HMS *Splendid*, Report of Proceedings, 23 June 1982.   **173.** TNA/CAB/148/211, OD(SA)82, 25th Meeting, 4 May 1982.   **174.** Freedman, *Official History of the Falklands Campaign*, Vol. 2, p. 308.   **175.** HMS *Splendid*, Report of Proceedings, 23 June 1982.   **176.** IWM/17274, Interview with Roger Lane-Nott, 23 January 1997.   **177.** TNA/PREM/19/624, Weston to Wright, Chiefs of Staff Committee, 4 May 1982.   **178.** TNA/CAB/148/211, OD(SA), 26th Meeting, 5 May 1982.   **179.** HMS *Splendid*, Report of Proceedings, 23 June 1982.   **180.** Freedman, *Official History of the Falklands Campaign*, Vol. 2, p. 308.   **181.** Ibid., p. 309.   **182.** TNA/PREM/19/647, Pym to Thatcher, 5 May 1982.   **183.** Freedman, *Official History of the Falklands Campaign*, Vol. 2, p. 309.   **184.** Haig proposed arrangements for Argentine withdrawal and a wind-ing down of the military presence, including the British. The creation of some form of international 'interim administration' for the Islands following Argentine withdrawal would operate while long-term sovereignty was negotiated.   **185.** TNA/PREM/19/647, Pym to Thatcher, 5 May 1982.   **186.** TNA/CAB/148/211, OD(SA)(82), 27th Meeting, 6 May 1982.   **187.** TNA/CAB/148/211, OD(SA), 26th Meeting, 5 May 1982.   **188.** John Nott, *Here Today, Gone Tomorrow: Memoirs of an Errant Politician* (Politico's, 2002), p. 294.   **189.** TNA/CAB/148/211, OD(SA)(82), 27th Meeting, 6 May 1982; Freedman, *Official History of the Falklands Campaign*, Vol. 2, p. 310.   **190.** Freedman, *Official History of the Falklands Campaign*, Vol. 2, p. 310.   **191.** Ibid.   **192.** Ibid., pp. 310–11.   **193.** IWM/17274, Interview with Roger Lane-Nott, 23 January 1997.

Valiant Arrives

194. Tom Le Marchand, 'Jettisoned Bombs Just Missed *Valiant*', undated manu-script.   **195.** Ibid.   **196.** HMS *Valiant*, Report of Proceedings, 26 July 1982.   **197.** Ibid.   **198.** HMS *Splendid*, Report of Proceedings, 23 June 1982.   **199.** HMS *Conqueror*, Report of Proceedings, 1 July 1982.   **200.** Ibid.   **201.** Ibid.   **202.** Chris Wreford-Brown, 'Floating Wire Aerials', in Friends of the Sub-marine Museum, *All Round Look: Year Book 2007/2008* (Royal Navy Submarine Museum, 2007), p. 54.   **203.** Rossiter, *Sink the Belgrano*, p. 273.   **204.** Ibid., p. 275.   **205.** HMS *Conqueror*, Report of Proceedings, 1 July 1982.   **206.** The first casualty of the conflict, HMS *Sheffield*, sank on 10 May 1982, six days after being hit by an Exocet missile.   **207.** IWM/17274, Interview with Roger Lane-Nott, 23 January 1997.   **208.** Interview with James Taylor, 10 April 2014.   **209.** HMS *Valiant*, Report of Proceedings, 26 July 1982.   **210.** Ibid.   **211.** TNA/CAB/148/212, OD(SA)(82)58, Argentine Territorial Waters, Note by the Secretaries, 27 May 1982.   **212.** TNA/CAB/148/212, Argentine 12 Nautical Mile Limit (Note by Ministry of Defence Officials), 27 May 1982.   **213.** TNA/CAB/148/211, OD(SA)(82), 46th Meeting, 28 May 1982.   **214.** Ibid.   **215.** Ibid.   **216.** Freedman, *Official History of the Falklands Campaign*, Vol. 2, p. 316.   **217.** CAC/MTA, Margaret Thatcher, *Memoir of the Falklands War*, written at Chequers over Easter 1983, p. 100.   **218.** HMS *Valiant*, Report of Proceedings, 26 July 1982.   **219.** Ibid.   **220.** Ibid.   **221.** Powis, 'Falklands Memories'.   **222.** HMS *Conqueror*, Report of Proceedings, 1 July 1982.   **223.** Ibid.   **224.** John Moore and Richard Compton-Hall, *Submarine Warfare: Today and Tomorrow* (Adler & Adler, 1987), p. 79.   **225.** HMS *Spartan*, Report of Proceedings, 23 June 1982.   **226.** *Daily Telegraph*, 'Fresh Fruit Prize for Unseen Players in Grim Game', 25 June 1982.   **227.** RNSM, HMS *Conqueror* Ship's Book, FOSM to HMS *Conqueror*, 6 June 1982.   **228.** RNSM, HMS *Conqueror* Ship's Book, Conqueror Temporary Memorandum Number 29/82, Enforced Dieting, 2 June 1982.   **229.** HMS *Spartan*, Report of Proceedings, 23 June 1982.

Reinforcements

230. Michael Pitkeathly and David Wixon, *Submarine Courageous, Cold War Warrior: The Life and Times of a Nuclear Submarine* (The HMS *Courageous* Society, 2010), p. 214.   **231.** Ibid., p. 217.   **232.** Andrew Johnson, 'The 116-Day War Patrol', in Winton, *Submariners*, p. 283.   **233.** Ibid., p. 296.   **234.** Michael Clapp and Ewen Southby-Tailyour, *Amphibious Assault Falklands: The Battle of San Carlos Water* (Pen and Sword, 2007), p. 130.   **235.** Ewen Southby-Tailyour, *Exocet Falklands: The Untold Story of Special Forces Operations* (Pen and Sword, 2014), p. 248.   **236.** Ibid., p. 256.   **237.** Ibid., p. 254.   **238.** Johnson, 'The 116-Day War Patrol', in Winton, *Submariners*, p. 296.   **239.** Southby-Tailyour, *Exocet Falklands*, p. 256.   **240.** Ibid., pp. 256–7.

Picket Duty

241. Powis, 'Falklands Memories', p. 67.   **242.** Le Marchand, 'Jettisoned Bombs'.   **243.** HMS *Valiant*, Report of Proceedings, 26 July 1982.   **244.** RNSM, HMS *Valiant*, Navigator Log Book, 1982.   **245.** HMS *Valiant*,

Report of Proceedings, 26 July 1982.   **246.** Nigel West, *Historical Dictionary of Naval Intelligence* (Scarecrow Press, 2010), pp. 39, 62.   **247.** Ibid., p. 39.   **248.** Ibid.   **249.** HMS *Valiant*, Report of Proceedings, 26 July 1982.   **250.** Powis, 'Falklands Memories', p. 67.   **251.** HMS *Valiant*, Report of Proceedings, 26 July 1982.   **252.** Chris Craig, *Call for Fire: Sea Combat in the Falklands and the Gulf War* (John Murray, 1995), p. 76.

Departure

253. Interview with Roger Lane-Nott, 25 March 2014.   **254.** HMS *Splendid*, Report of Proceedings, 23 June 1982.   **255.** The British Naval tradition is to fly a Jolly Roger when returning to port when the sub made a kill at sea. The symbology used was not standard: normally a red bar indicated a warship sunk, but *Conqueror* used a silhouette of a warship in white bunting. Crossed torpedoes were used instead of crossbones under the skull. The dagger for a special operation was not strictly correct as the landing of Special Forces was done into Grytviken after the fall of South Georgia rather than as part of its recapture.   **256.** Interview with Admiral Sir Peter Herbert, 15 October 2013.   **257.** HMS *Valiant*, Report of Proceedings, 26 July 1982.   **258.** Le Marchand, 'Jettisoned Bombs'.   **259.** HMS *Valiant*, Report of Proceedings, 26 July 1982.   **260.** Southby-Tailyour, *Exocet Falklands*, p. 258.

Aftermath

261. *Daily Mail*, 16 March 1983; *Daily Telegraph*, 16 March 1983.   **262.** *Daily Telegraph*, 16 March 1983.   **263.** Pitkeathly and Wixon, *Submarine Courageous*, p. 227.   **264.** Ibid., p. 253.   **265.** MOD Archive.   **266.** Pitkeathly and Wixon, *Submarine Courageous*, p. 225.

The Deterrent

267. Duncan Campbell and John Rentoul, 'All Out War', *New Statesman*, 24 August 1984.   **268.** Freedman, *Official History of the Falklands Campaign*, Vol. 1, p. 49.   **269.** Interview with Toby Elliott, 14 January 2014.   **270.** Ibid.   **271.** Ibid.   **272.** Interview with Admiral Sir Peter Herbert, 15 October 2013.   **273.** *Sunday Times*, 27 November 2005.   **274.** *Guardian,* 22 November 2005; *Sunday Times*, 20 November 2005.   **275.** Sir Lawrence Freedman, *Political Studies Review*, Vol. 5 (2007), pp. 39–44.   **276.** BBC Radio 4, Obituary, 'Margaret Thatcher: Potency and Paradox', first broadcast 8 April 2013.   **277.** Frank Grenier, 'The Falklands War, 1982 – R. N. Submarine involvement', in Friends of the Submarine Museum, *All Round Look: Year Book 2002/2003* (Royal Navy Submarine Museum, 2003), p. 38.   **278.** 'The Falklands War', eds. Dorman, Kandiah and Staerck. **279.** Interview with Admiral Sir Peter Herbert, 15 October 2013. **280.** Thatcher, *Downing Street Years*, p. 174. **281.** John F. Lehman Jr, 'Reflections on the Special Relationship', *Naval History Magazine*, Vol. 26, No. 5 (October 2012). **282.** Powis, 'Falklands Memories', p. 65.

8 MAINTAINING THE DETERRENT: FROM POLARIS TO TRIDENT

1. Peter Hennessy, *Muddling Through: Power, Politics and the Quality of Government in Postwar Britain* (Victor Gollancz, 1996), p. 126. Jim Callaghan originally described his grass-hut meeting with Jimmy Carter for *A Bloody Union Jack on Top of It*, first broadcast on BBC Radio 4 in May 1988. **2.** *Financial Times*, 17 November 1986. **3.** Geoffrey Howe, *Conflict of Loyalty* (Macmillan, 1994), p. 240.

Improving Polaris

4. *Resolution* (Port crew, under the command of Toby Elliott) was on patrol during March and April 1982. Commander Paul Branscombe (Starboard crew) took over command of *Resolution* on 30 April 1982 and departed for patrol in early June 1982. **5.** Interview with Paul Branscombe, 1 April 2014. **6.** CAC/THCR, 3/2/98 f3, Thatcher to Fieldhouse, 10 August 1982. **7.** TNA/CAB/130/1128, MISC 237(69)5, Prospects for nuclear collaboration within NATO including prospects for an Anglo-French nuclear force, 26 March 1969. **8.** TNA/DEFE/24/189, DCSO(R) to AUS, Damage capability of the Polaris force, 26 October 1967. **9.** TNA/DEFE/13/350, Mackenzie to Minister RN, CNS, Controller of the Navy, 2nd PUS (RN), 19 January 1965; TNA/PREM/13/228, Healey to Wilson, 22 January 1965. **10.** Denis Healey, *The Time of My Life* (Michael Joseph, 1989), p. 313. **11.** TNA/CAB/134/3120, PN(67), 4th Meeting, 5 December 1967; for the Kings Norton Working Party Report into the future of AWRE see TNA/CAB/134/3121, Report to the Minister of Technology and the Chairman of the Atomic Energy Authority by the Working Party on Atomic Weapons Establishments, July 1968; for Lord Rothschild's Minority Report see TNA/CAB/134/3121, Part 2, Lord Rothschild's 'Minority Report' dissenting from the findings of the Kings Norton Working Party on Atomic Weapons Establishments, 18 July 1968. **12.** Thomas Robb, 'Antelope, Poseidon or a Hybrid: The Upgrading of the British Strategic Nuclear Deterrent, 1970–1974', *Journal of Strategic Studies*, 33:6, p. 802. **13.** See John Baylis and Kristan Stoddart, 'Britain and the Chevaline Project: The Hidden Nuclear Programme, 1967–1982', *Journal of Strategic Studies*, 26:4 (2003), pp. 128–31; Robb, 'Antelope, Poseidon or a Hybrid', pp. 803–5. **14.** See Robb, 'Antelope, Poseidon or a Hybrid', pp. 805–13.

Chevaline

15. TNA/DEFE/13/1039, A Report on the Progress of the Chevaline Project: The Main Report, 1 April 1976. **16.** TNA/DEFE/13/1039, Meeting British National Criteria for Strategic Deterrence, 10 November 1975. **17.** TNA/ PREM/1564, Hunt to Callaghan, Nuclear Meeting, 28 October, 9.45 a.m., 25 October 1977. **18.** TNA/DEFE/13/1039, Mayne to Secretary of State, 18 November 1975. **19.** TNA/DEFE/13/1039, Carver to Mason, Soviet ABM Cover, 31 March 1976; this file is retained. However, it was temporarily released and is quoted in Kristan Stoddart and John Baylis, 'The British Nuclear Experience: The Role of Ideas, Beliefs and Culture (Part Two)', *Diplomacy and Statecraft*, Issue 23, No. 3 (September 2012), see fn. 38; also, David Owen, *Nuclear Papers* (Liverpool University Press, 2009), p. 10. **20.** CAC/GBR/0014/ DKNS, Rear Admiral David Scott, Polaris, Chevaline and an Encounter with the Russians, 1973–1980, unpublished manuscript. **21.** TNA/DEFE/13/1039, A Report on the Progress of the Chevaline Project: The Main Report, 1 April 1976. **22.** Ibid. **23.** TNA/PREM/16/1564, R. T. Jackling to E. A. J. Fergusson, Chevaline, 26 October 1977. **24.** TNA/PREM/1564, Hunt to Callaghan, Military Nuclear Issues, 25 October 1977. **25.** TNA/PREM/1564, R. T. Jackling to E. A. J. Fergusson, Chevaline, 26 October 1977. **26.** House of Commons Public Accounts Committee, *Ministry of Defence: Chevaline Improvement to the Polaris Missile System, Ninth Report from the Committee of Public Accounts, Session 1981–1982* HC 269 (Her Majesty's Stationery Office, 1982). **27.** Ibid. **28.** Hennessy, *Muddling Through*, p. 121. **29.** TNA/ PREM/16/1978, Mulley to Callaghan, Future of the British Deterrent, 20 December 1978. **30.** House of Commons Defence Select Committee, Sixth Report of Session 1984–1985, 'Trident Programme', HC 479, 10 July 1985, Minutes of Evidence, 27 February 1985, Q 1910. **31.** Most of the files surrounding this re-motoring programme are still classified. 'TNA/DEFE/ 72/224, Joint Motor Life Study Co-ordination Group: status of the UK Polaris rocket motors, January 1981. **32.** TNA/PREM/16/1564, Hunt to Callaghan, 25 October 1977.

Towards Trident

33. Hennessy, *Muddling Through*, pp. 99–130. **34.** TNA/PREM/16/1564, 'Nuclear Studies', Hunt to Callaghan, 13 December 1977. **35.** The 'Neutron Bomb' was intended to kill the enemy on the battlefield by releasing a very high dose of lethal radiation while minimizing the amount of physical damage caused by blast, heat and fallout. **36.** TNA/PREM/16/1564, 'The British Strategic Nuclear Deterrent', Part 2, Conclusions of a Ministerial Meeting Held at No. 10 Downing Street on Friday, 28 October 1977 at 0945. **37.** Ibid. **38.** Ibid. **39.** Ibid.; TNA/PREM/16/1564, Cabinet Nuclear Defence Policy. Note of a Meeting Held at 10 Downing Street on Thursday 1 December 1977 at

10:00 a.m.   **40.** TNA/DEFE/25/335, Mason to Hunt, 12 December 1978; the fullest version of Duff–Mason to reach the National Archives was the Chief of Defence Staff's copy (later taken back by Whitehall because, among other things, targeting details were included). But the file did not contain Part III as it had yet 'to be issued'. Mason circulated the report of his working party to Sir John Hunt's Steering Group on Nuclear Matters on 12 December 1978. This chapter draws on the CDS's file in particular.   **41.** PM/78/138, Letter from David Owen to Jim Callaghan, 11 December 1978 in Owen, *Nuclear Papers*, p. 149.   **42.** See Owen, *Nuclear Papers*.   **43.** TNA/CAB/134/940, HDC(55)3, 'The Defence Implications of Fallout from a Hydrogen Bomb', report by a group of officials, 8 March 1955; see Hennessy, *The Secret State*, pp. 167-78; , J. Hughes, 'The Strath Report: Britain Confronts the H-Bomb, 1954-5'. *History and Technology*, 19:3 (2003), pp. 257–75.   **44.** The name of the Soviet ABM system which was deployed around Moscow.   **45.** TNA/DEFE/25/335, Annex C. Other Criteria. An Assured Second Strike Capability.   **46.** Ibid.   **47.** TNA/PREM/16/1978, Future of the British Deterrent, Hunt to Callaghan, 20 December 1978.   **48.** TNA/PREM/16/1978, Future of the British Deterrent, Owen to Callaghan, 19 December 1978.   **49.** TNA/PREM/16/1978, Future of the British Deterrent, Hunt to Callaghan, 20 December 1978.   **50.** TNA/DEFE/25/433, Quinlan to PS Secretary of State, Nuclear Matters, 18 December 1978.   **51.** TNA/PREM/16/1978, Mason to Callaghan, Future of the British Deterrent, 20 December 1978.   **52.** Ibid.   **53.** TNA/PREM/19/1978, Cabinet Nuclear Defence Policy. Note of a Meeting Held at 10 Downing Street on Thursday, 21 December 1978, at 9.45 a.m.   **54.** Ibid.   **55.** TNA/DEFE/25/335, Factors Relating to Further Consideration of the Future of the United Kingdom Nuclear Deterrent, Part I, paragraph 16.   **56.** TNA/PREM/16/1978, Cabinet Nuclear Defence Policy. Note of a Meeting Held at 10 Downing Street on Thursday, 21 December 1978, at 9.45 a.m.   **57.** TNA/PREM/16/1978, Cabinet Nuclear Defence Policy. Note of a Meeting held at 10 Downing Street on Tuesday, 2 January 1979, at 11.00 a.m.   **58.** Ibid.   **59.** TNA/PREM/16/1978, Prime Minister's Conversation with President Carter: 3:30 p.m., 5 January, at Guadeloupe.   **60.** Ibid.   **61.** Ibid.   **62.** TNA/PREM/16/1978, Hunt to Callaghan, Nuclear Matters. Next Steps, 7 January 1979.   **63.** TNA/PREM/16/1978, Cabinet Nuclear Defence Policy. Note of a Meeting Held at 10 Downing Street on Friday, 19 January 1979, at 10:00 a.m.   **64.** Kenneth O. Morgan, *Callaghan: A Life* (OUP, 1997), pp. 661-2.   **65.** TNA/PREM/16/1978, Cabinet Nuclear Defence Policy. Note of a Meeting Held at 10 Downing Street on Friday, 19 January 1979, at 10:00 a.m.   **66.** TNA/PREM/16/1978, Healey to Callaghan, 8 February 1979.   **67.** TNA/PREM/16/1978, Callaghan to Carter, 27 March 1979.   **68.** TNA/PREM/19/1978, Hunt to Cartledge, 27 March 1979; TNA/PREM/19/1978, Cartledge to Callaghan, 28 March 1979; TNA/PREM/19/1978, Callaghan to Cartledge, 4 May 1979; in 1988, Hennessy asked Jim Callaghan why he had left instructions for Mrs Thatcher to be briefed on Polaris replacement. 'Because,' he replied, 'it was a matter of national importance. I think it is very important that succeeding ministers and succeeding governments and administrations should not know about the political decisions of their predecessors – that is a principle I adhere to. But if the administration, or the Prime Minister, wishes to leave a note for his successor about a matter of greatest national importance, then I think he is entitled to do so': Hennessy, *Muddling Through*, p. 127. **69.** TNA/PREM/16/1978, Cartledge to Hunt, 6 April 1979. **70.** Iain Dale (ed.), *Labour Party General Election Manifestos, 1900–1997* (Routledge/ Politico's, 2000), p. 281. **71.** *1979 Conservative Party General Election Manifesto* (Conservative Party, 1979). **72.** David Butler, *British General Elections since 1945* (Blackwell/Institute of Contemporary British History, 1989), p. 35.

Purchasing Trident

73. Hansard, House of Commons Debates, 24 January 1980, Vol. 977, Col. 681. **74.** Hansard, House of Commons Debates, 15 February 1980, Vol. 147, Col. 384. **75.** CAC/MTA, Sir Robin Day, TV interview with Margaret Thatcher, BBC1, *Panorama*, 8 June 1987. **76.** Hansard, House of Commons Debates, 28 June 1983, Vol. 44, Col. 496. **77.** Hansard, House of Commons Debates, 28 June 1983, Vol. 44, Col. 495. **78.** Rodric Braithwaite, *Across the Moscow River: The World Turned Upside Down* (Yale, 2002), p. 52. **79.** Interview with Toby Elliott, 14 January 2014. **80.** Ibid. **81.** Ibid. **82.** Interview with Paul Branscombe, 1 April 2014. **83.** TNA/PREM/19/159, Note of a Meeting in the Oval Office, The White House, Washington DC, on Monday, 17 December 1979. Present were President Carter, Mrs Thatcher, Cyrus Vance (US Secretary of State), Lord Carrington (Foreign and Commonwealth Secretary), Zbigniew Brzezinski (Carter's National Security Advisor) and Robert Armstrong. **84.** Peter Hennessy, *Cabinets and the Bomb* (OUP, 2007), pp. 87–125. **85.** TNA/CAB/130/1109, MISC 7 (79), 1st Meeting, Cabinet Nuclear Defence Policy, 24 May 1979. **86.** TNA/CAB/130/1109, MISC 7 (79), 1st Meeting, Cabinet Nuclear Defence Policy, 24 May 1979. **87.** Ibid. **88.** TNA/PREM/19/159, Armstrong to Thatcher, Future of the Strategic Deterrent, 4 December 1979. **89.** TNA/PREM/19/416, OD(81)29, The Defence Programme, Note by the Secretary of State for Defence, 2 June 1981. **90.** TNA/PREM/19/159, Armstrong to Thatcher, Future of the Strategic Deterrent, 4 December 1979; the four pillars upon which defence policy was founded were: an independent element of strategic and theatre nuclear forces committed to the Atlantic Alliance; the direct defence of the UK mainland; a major land and air contribution on the European mainland; and a major maritime effort in the Eastern Atlantic and Channel. These pillars were later reiterated in the 1981 Defence Review. **91.** TNA/PREM/19/159, Armstrong to Thatcher, Future of the Strategic Deterrent, 4 December 1979. **92.** TNA/DEFE/69/406, Duff–Mason Report, Part 3 (revised version), 12 October 1979. **93.** TNA/DEFE/25/335, Factors Relating to Further Consideration of the Future of the United Kingdom

Nuclear Deterrent, Part II, Criteria for Deterrence, Annex A: Unacceptable Damage, 30 November 1978. **94.** TNA/PREM/19/159, Armstrong to Thatcher, Future of the Strategic Deterrent, 4 December 1979. **95.** Ibid. **96.** Charles Moore, *Margaret Thatcher: The Authorized Biography*, Volume 1: *Not for Turning* (Allen Lane, 2013), p. 572. **97.** TNA/PREM/19/159, Armstrong to Thatcher, Future of the Strategic Deterrent, 4 December 1979. **98.** Hennessy, *Muddling Through*, p. 128. **99.** TNA/PREM/19/159, Armstrong to Thatcher, Future of the Strategic Deterrent, 4 December 1979. **100.** Ibid. **101.** TNA/PREM/19/159, Armstrong to Thatcher, Future of the Strategic Deterrent, 4 December 1979; TNA/CAB/128/66/25, CC(79), 25th Conclusions, 13 December 1979. **102.** TNA/DEFE/24/2124, Howe to Thatcher, Polaris Successor, 15 June 1980. **103.** TNA/DEFE/25/325, Chiefs of Staff Committee, The Case for Five SSBN, 10 June 1980. **104.** Richard Hill, *Lewin of Greenwich: The Authorised Biography of Admiral of the Fleet Lord Lewin* (Cassell, 2000), p. 327. **105.** TNA/CAB/130/1182, MISC 7(82)1, Anglo-American Negotiations on D5 (Note by the Cabinet Office), March 1982. **106.** TNA/AIR/8/2846, Note by the Directors of Defence Policy, US Strategic Nuclear Force – Chiefs of Staff advice, 2 October 1981. **107.** House of Commons Defence Select Committee, Sixth Report of Session 1984–1985, 'Trident Programme', HC 479, 10 July 1985, Minutes of Evidence, 27 February 1985, Q1890. **108.** House of Commons Defence Select Committee, Sixth Report of Session 1984–1985, 'Trident Programme', HC 479, 10 July 1985, Minutes of Evidence, 30 January 1985. **109.** TNA/CAB/130/1222, MISC7(81)1, Memorandum by the Secretary of State for Defence, Cabinet Nuclear Defence Policy, United Kingdom Strategic Deterrent, 17 November 1981. **110.** Ibid. **111.** TNA/PREM/19/417, Minute of a meeting between the Prime Minister, Defence Minister and Foreign Secretary, 10 February 1981. **112.** TNA/DEFE/24/2123, MISC 7, Frank Cooper, 20 November 1981. **113.** Hill, *Lewin of Greenwich*, p. 324. **114.** John Nott, *Here Today, Gone Tomorrow: Memoirs of an Errant Politician* (Politico's, 2002), p. 220. **115.** TNA/CAB/130/1222, MISC7(81) 1st Meeting, 24 November 1981; TNA/CAB/130/1182, MISC7(82) 1st Meeting, 'Most Confidential Record', 12 January 1982. **116.** TNA/AIR/8/2846, Draft Chiefs of Staff advice to the Secretary of State on UK strategic nuclear forces, 13 October 1981. **117.** Howe, *Conflict of Loyalty*, p. 145. **118.** Nott, *Here Today, Gone Tomorrow*, p. 217. **119.** MISC 7(82) 1st Meeting, 2 March 1982. **120.** TNA/CAB/130/1182, Limited Circulation Annex, MISC 7(82) 2nd Meeting, 4 March 1982. **121.** Ibid. **122.** TNA/CAB/128/75, CC(82) 8th Conclusions, 'Most Confidential Record', 4 March 1982. **123.** Ibid. **124.** Ibid. **125.** TNA/CAB/130/1182, United Kingdom Strategic Deterrent: Missile Processing, 29 July 1982. **126.** TNA/CAB/130/1182, MISC 7(82)4, United Kingdom Strategic Deterrent: Missile Processing, Memorandum by the Secretary of State for Defence, 19 July 1982. **127.** TNA/CAB/130/1182, Impact of Missile Processing on the Independence of the UK Deterrent, Annex A **128.** TNA/CAB/130/1182, MISC 7(82)4, Missile Processing, 29 July 1982 **129.** TNA/CAB/128/75, CC(82) 8th Conclusions, Most Confidential Record, 4 March 1982. **130.** TNA/CAB/

130/1182, United Kingdom Strategic Deterrent: Missile Processing, 29 July 1982. **131.** TNA/DEFE/24/2125, Quinlan memorandum, Polaris successor memorandum, 9 June 1980. **132.** Ibid. **133.** TNA/PREM19/417, Defence Open Government Document 80/23, 'The Future United Kingdom Strategic Nuclear Deterrent Force', 15 July 1980. **134.** Sir Michael Quinlan, 'The British Experience', in Henry D. Sokolski (ed.), *Getting MAD: Nuclear Mutual Assured Destruction, Its Origins and Practice* (Strategic Studies Institute, US Army War College, 2004), p. 273. **135.** During the Cabinet discussion on the purchase of Trident on 4 March 1982, the ethical factor does flicker in the shape of thoughts on the Campaign for Nuclear Disarmament which 'gained from being at least nominally a non-party organisation, and from the support it enjoyed among the young and in some church circles. It was perhaps a pity that the CND's many opponents were not also organised on a non-party basis. The CND rightly stressed the terrible nature of nuclear weapons but failed to recognise that Britain's possession of a strategic nuclear deterrent lessened rather than increased the danger of nuclear war.' TNA/CAB/128/75, CC(82) 8th Conclusions, Most Confidential Record, 4 March 1982. **136.** *Sunday Times*, 14 July 1991. **137.** *Evening Standard*, 10 July 1991. **138.** TNA/CAB/128/75, CC(82) 8th Conclusions, Most Confidential Record, 4 March 1982. **139.** CAC/MTA, Sir Robin Day, TV interview with Margaret Thatcher, BBC1, *Panorama*, 8 June 1987. **140.** Ibid.

The Trident Programme

141. House of Commons Public Accounts Committee, Session 1983–1984, *The UK Trident Programme*, Minutes of Evidence, 26 March 1984, Q2488. **142.** Vice Admiral Sir Ted Horlick, 'Submarine Propulsion in the Royal Navy', The Thomas Lowe Gray Memorial Lecture to the Institution of Mechanical Engineers, January 1981, pp. 950–51. **143.** Ibid. **144.** Ibid. **145.** Norman Friedman, *The Naval Institute Guide to World Naval Weapons Systems, 1997–1998* (Naval Institute Press, 1997), pp. 150–51. **146.** Jack Hool and Keith Nutter, *Damned Un-English Machines: A History of Barrow-Built Submarines* (The History Press, 2003), pp. 296–301. **147.** CAC/MTA, interview from BBC1, *Panorama*, 'The Peace Penalty: Whatever Happened to the Peace Dividend?', 22 March 1993. **148.** CAC/MTA, Speech at the keel laying of HMS *Vanguard*, 3 September 1986. **149.** *Independent*, 6 July 1995. **150.** National Audit Office, Report by the Comptroller and Auditor General, *Ministry of Defence and Property Services Agency: Control and Management of the Trident programme*, HC 27, 1987/88. **151.** House of Commons Defence Committee, *The Progress of the Trident Programme*, HC 337 (Her Majesty's Stationery Office, 1992), p. 6. **152.** John Major, 'Building New Britain', speech, 1 April 1992.

9 THE SILENT VICTORY

1. Admiral Sir William Staveley, RN, Transcript of 'Overview of British Defence Policy and the Relevance of the Northern Flank', at the Conference on Britain and the Security of NATO's Northern Flank, 7–8 May 1986. **2.** Interview with Roger Lane-Nott, 25 March 2014. **3.** Interview with James Taylor, 10 April 2014.

The Cold War Heats Up

4. TNA/DEFE/24/1387, Naval Staff Target 7029 (New Patrol Class Submarine) Supporting Paper, June 1978. **5.** TNA/DEFE/13/1357, Sea/Air Warfare Committee – Operational Concept for Anti-Submarine Warfare, 25 May 1978. **6.** Ibid. **7.** Ibid. **8.** Ibid. **9.** MOD Archive. **10.** Ibid. **11.** A Type II Soviet nuclear submarine was either a 'Charlie' or a 'Victor' class. **12.** MOD Archive. **13.** Ibid. **14.** Dan Conley and Richard Woodman, *Cold War Command: The Dramatic Story of a Nuclear Submariner* (Seaforth Press, 2014), pp. 157–8. **15.** Ibid., pp. 157–9. **16.** Ibid. **17.** Ibid. **18.** Ibid. **19.** Ibid. **20.** MOD Archive. **21.** Ibid. **22.** Ibid. **23.** Michael Pitkeathly and David Wixon, *Submarine Courageous, Cold War Warrior: The Life and Times of a Nuclear Submarine* (The HMS *Courageous* Society, 2010), p. 129; *Independent on Sunday*, 'Report on HMS *Sceptre*'s Collision with an Iceberg', 14 June 1981. **24.** Quoted in Michael Smith, *The Spying Game* (Politico's, 2003), p. 307. **25.** *Yorkshire Post*, 'Day we rammed a Cold War Russian sub', 6 April 2013. **26.** Ibid. **27.** Interview with Doug Littlejohns, 20 November 2013. **28.** Iain Ballantyne, 'Submarining in the Eighties: Angles & Dangles', *Warships International Fleet Review Magazine*, September 2010. **29.** HMS *Sceptre* decommissioned in December 2010. **30.** MOD Archive.

The 1981 Defence Review

31. John Nott, *Here Today, Gone Tomorrow: Memoirs of an Errant Politician* (Politico's, 2002), p. 215. **32.** Eric Grove, *Vanguard to Trident: British Naval Policy since World War II* (Bodley Head, 1987), p. 350. **33.** Ibid. **34.** Christopher A. Ford and David A. Rosenberg, 'The Naval Intelligence Underpinnings of Reagan's Maritime Strategy', *Journal of Strategic Studies*, 28:2 (2005), p. 385. **35.** John B. Hattendorf, *The Evolution of the U.S. Navy's Maritime Strategy 1977–1986* (Department of the Navy, 2006), p. 23. **36.** Ford and Rosenberg, 'Naval Intelligence Underpinnings', p. 381. **37.** Sherry Sontag and Christopher Drew, *Blind Man's Bluff: The Untold Story of American Submarine Espionage* (Random House, 1999), see Chs. 8–12. **38.** Rear Adm. Tom Brooks, USN (retd.) and Capt. Bill Manthorpe, USN (retd.), 'Setting the Record Straight',

Naval Intelligence Professional Quarterly, XII/2 (April 1996), p. 1, quoted in Ford and Rosenberg, 'Naval Intelligence Underpinnings', p. 382. **39.** Eric Grove, 'The Convoy Debate', *Naval Forces*, No. 3 (1985), p. 47. **40.** CIA Historical Review Program, NIE/11/15/28, National Intelligence Estimate, Soviet Naval Strategy and Programs through the 1990s, March 1983, p. 22. Available at http://www.foia.cia.gov/sites/default/files/document_conversions/89801/DOC_0000268225.pdf. **41.** Ibid. **42.** Nott, *Here Today, Gone Tomorrow*, p. 211. **43.** Ibid., p. 231. **44.** Ibid. **45.** Ibid. **46.** Ibid., p. 212. **47.** Ibid., p. 228. **48.** Richard Hill, *Lewin of Greenwich: The Authorised Biography of Admiral of the Fleet Lord Lewin* (Cassell, 2000), p. 332. **49.** TNA/PREM/19/416, Appendix A. **50.** TNA/PREM/19/416, Leach to Thatcher, 18 May 1981. **51.** Nott, *Here Today, Gone Tomorrow*, p. 212. **52.** Cmnd 8288, *The United Kingdom Defence Programme: The Way Forward*, June 1981. **53.** Ibid. **54.** Cmnd 8529-1, *Statement on the Defence Estimates*, 1982. **55.** Ibid. **56.** TNA/DEFE/24/1391, FRC(77)P10, The SSK – Replacement of the Oberon Class.

The 'Upholder' Class

57. RNSM/A2008, Roxburgh to McGeoch, 8 December 1971. **58.** Ibid. **59.** RNSM/A2008, Roxburgh to Blackman, 6 March 1972. **60.** TNA/DEFE/24/1387, Naval Staff Target 7029 (New Patrol Class Submarine) Supporting Paper, 27 November 1977, 7 June 1978. **61.** TNA/DEFE/24/1392, Brief for ACNS(OR) on NST 7029 and Supporting Paper – Submission to the NPC. **62.** TNA/DEFE/24/1391, The New SSK – The Implications of Achieving Commonality of Design between the RN Preferred Option and an Export Version (Paper by the Navy Department). **63.** TNA/DEFE/24/1391, Draft for ACNS(OR)'s Introduction of the SSK NST to the NPC. **64.** HC 455, Defence Committee, Ninth Report. Procurement of Upholder Class Submarines, together with the proceedings of the committee relating to the report, minutes of evidence and memoranda, 1990/91. **65.** Jonathan Powis, 'UK's Upholder Class Boats Go to Canada', *Naval Institute Proceedings* (October 2002). **66.** MOD Archive.

The 'Trafalgar' Class

67. Harry Lambert, *Rolls-Royce, the Nuclear Power Connection: A History of Nuclear Engineering In One of the World's Most Famous Companies* (Rolls-Royce plc, 2009), pp. 64–5. **68.** TNA/DEFE/13/1357, Green to Minister of State, Royal Navy, Sonar Type 2020 (Sonar 2001 Improvements), 8 December 1977. **69.** TNA/DEFE/13/1357, Towed Arrays for Submarines – Sonar 2026, 18 July 1978. **70.** TNA/DEFE/13/1357, Note of a meeting held on 24 August to discuss Sonar 2026. **71.** MOD Archive **72.** Interview with

Martin Macpherson, 24 May 2013. **73.** Lambert, *Rolls-Royce, the Nuclear Power Connection*, p. 77. **74.** Interview with Martin Macpherson, 24 May 2013.

SSNoZ and the Follow-On SSN

75. John F. Schank, Frank W. Lacroix, Robert E Murphy, et al., *Learning From Experience*, Vol. III: *Lessons from the United Kingdom's Astute Submarine Program* (Rand Corporation, 2011). **76.** TNA/DEFE/24/1389, FRC(M)2, Fleet Requirements Committee, Minutes of a Meeting, 10 September 1982. **77.** Ibid. **78.** TNA/DEFE/24/1389, Whetstone to Marsh, A Follow-On SSN, 14 September 1982. **79.** TNA/DEFE/24/1389, A Follow-On SSN: A Paper by the Naval Staff, June 1980. **80.** TNA/DEFE/24/1389, FRC(M)2, Fleet Requirements Committee, Minutes of a Meeting, 10 September 1982. **81.** Interview with Paul Branscombe, 1 April 2014.

The Walker Spy Ring

82. TNA/DEFE/13/1357, Sea/Air Warfare Committee – Operational Concept for Anti-Submarine Warfare, 25 May 1978. **83.** Ibid. **84.** Ibid. **85.** Pitkeathly and Wixon, *Submarine Courageous*, p. 247. **86.** Interview with Roger Lane-Nott, 25 March 2014. **87.** MOD Archive. **88.** Ibid. **89.** A Type III Soviet nuclear submarine is a 'Yankee' or a 'Delta' class. **90.** MOD archive. **91.** Allegedly the USS *Drum*, a 'Sturgeon' class submarine, collided with a Victor III, K-324, in Peter the Great Bay while attempting to obtain photographs of the distinctive pod. An account of the USS *Drum*'s patrol can be found in W. Craig Reed, *Red November: Inside The Secret U. S.–Soviet Submarine War* (William Morrow, 2011), p. 316; Craig maintains that 'As was the case with many Cold War SpecOps, the boat's logs were altered. No record remained, save memories, to validate that the USS *Drum* was ever in the vicinity of a Soviet *Victor III* submarine in Peter the Great Bay.' **92.** Interview with James Perowne, 18 March 2014. **93.** John Prados, 'The Navy's Biggest Betrayal', *Naval History Magazine*, Vol 24, No. 3 (June 2010). **94.** Ibid. **95.** Ibid. **96.** Ibid. **97.** Thomas B. Allen and Norman Polmar, *Merchants of Treason* (Delacorte Press, 1988), pp. 262–3. **98.** Prados, 'The Navy's Biggest Betrayal'. **99.** Norman Polmar and Kenneth Moore, *Cold War Submarines: The Design and Construction of US and Soviet Submarines* (Potomac Books, 2005), p. 377; *New York Times*, 'Weinberger Says The Walkers Gave Soviet Much Key Data', 17 April 1987. **100.** http://www.csmonitor.com/1986/1008/aseni.html/(page)/3 **101.** Interview with Martin Macpherson, 24 May 2013. **102.** Polmar and Moore, *Cold War Submarines*, p. 295. **103.** Ibid., p. 159. **104.** *New York Times*, 'Submarines by Japan and Norway', 22 June 1987; Jere W. Morehead, 'Controlling Diversion: How Can We Convert the Toshiba–

Kongsberg Controversy into a Victory for the West', *New Journal of International Law and Business*, 277 (1988–1989); Polmar and Moore, *Cold War Submarines*, p. 286. **105.** John F. Lehman Jr, *Command of the Seas* (Naval Institute Press, 2001; 2nd edn), p. 133. **106.** Soviet Intentions and Capabilities for Interdicting Sea Lines of Communication in a War with NATO, Interagency Intelligence Memorandum, CIA, http://www.foia.cia.gov/sites/default/files/document_conversions/89801/DOC_0000261312.pdf, 1981. **107.** For more information on this torpedo see John Downing, 'How Shkval Ensured Soviet SSBN Survivability', *Jane's Intelligence Review* (1999). **108.** Ford and Rosenberg, 'Naval Intelligence Underpinnings', pp. 408–9.

'Bearding the Bear in its lair' – US Maritime Strategy

109. Hansard, House of Commons Debates, 3 March 1988, Vol. 128, Col. 1187. **110.** 'Forward Maritime Strategy Options', *The Adelphi Papers*, Vol. 29, Issue 241, 1989, p. 1. **111.** Hattendorf, *Evolution of the U.S. Navy's Maritime Strategy*, p. 36. **112.** Transcript of talk by Admiral Henry C. Mustin, 'The Maritime Strategy', 29 May 1986. **113.** *Daily Telegraph*, 'NATO navies allowed nuclear first strike', 23 March 1984. **114.** Admiral James D. Watkins, 'The Maritime Strategy', in *The Maritime Strategy* (US Naval Institute, 1986), pp. 2–17. **115.** Quoted in Barry Posen, *Inadvertent Escalation: Conventional War and Nuclear Risks* (Cornell University Press, 1991), p. 138. **116.** Ronald Reagan, 'National Security Strategy of the United States' (Jan. 1987), pp. 29–30, quoted in Hattendorf, *Evolution of the U.S. Navy's Maritime Strategy*, p. 25, n.7. **117.** Sir William Staveley, 'British Defence Policy in the North', in Geoffrey Till (ed.), *The Future of British Sea Power* (Macmillan, 1984), p. 70. **118.** Admiral Sir William Staveley, 'Power Factor – Submarine Operations in NATO', *NATO's Sixteen Nations*, Vol. 31, No. 1 (Feb–March 1986). **119.** Vice Admiral Sir Peter Stanford, 'The Current Position of the Royal Navy', in Till (ed.), *Future of British Sea Power*, p. 36. **120.** Staveley, 'British Defence Policy in the North', in Till (ed.), *Future of British Sea Power*, p. 68. **121.** Ibid. **122.** Geoffrey Till, 'A British View', in Till (ed.), *Future of British Sea Power*, p. 118. **123.** Staveley, 'British Defence Policy in the North', in Till (ed.), *Future of British Sea Power*, p. 68. **124.** *Independent*, 'New Navy Plan to attack Soviet subs near bases', 14 April 1987. **125.** Secretary of State for Defence (UK), *Statement on the Defence Estimates 1986* (Her Majesty's Stationery Office, 1986), p. 29. **126.** Cmnd 101-1, '*Why not an Alternative?*': *Statement on the Defence Estimates 1987* (HMSO, 1987). **127.** Interview with Paul Branscombe, 1 April 2014. **128.** Polmar and Moore, *Cold War Submarines*, p. 181. **129.** Interview with Toby Frere, 15 January 2013. **130.** Interview with Martin Macpherson, 24 May 2013. **131.** Interview with Paul Branscombe, 1 April 2014. **132.** Interview with Mark Stanhope, 4 April 2014. **133.** Interview with Toby Frere, 15 January 2013. **134.** Ibid. **135.** Interview with Richard Heaslip, 18 December 2013. **136.** Till, 'A British View', in Till (ed.), *Future of British*

Sea Power p. 121.   **137.** Watkins, 'The Maritime Strategy'.   **138.** Till, 'A British View', in Till (ed.), *Future of British Sea Power* p. 121.   **139.** House of Commons Defence Select Committee, Sixth Report of Session 1987/1988, 'The Future Size and Role of the Royal Navy's Surface Fleet', HC 309, 1988, p. xi.   **140.** Interview with Toby Frere, 15 January 2013.

Arctic Operations

141. Lehman Jr, *Command of the Seas*, p. 138.   **142.** TNA/CAB/186/39, JIC(85)7, Soviet Naval Policy, July 1985.   **143.** The Soviets did not withdraw completely to the Arctic. In 1984 'Delta' class nuclear-powered submarines were deployed into the mid-Atlantic as an 'analogous response' to NATO's deployment of Pershing II and other ground-launched cruise missiles.   **144.** Tom Le Marchand, 'Under-Ice Operations', *U.S. Naval War College Review*, May–June 1985, pp. 21–2.   **145.** Ibid. p. 27.   **146.** Ibid.   **147.** Polmar and Moore, *Cold War Submarines*, pp. 194–7.   **148.** Ibid.   **149.** Ibid.   **150.** Robert C. Stern, *The Hunter Hunted: Submarine versus Submarine Encounters from World War I to the Present* (U.S. Naval Institute Press, 2007), p. 184; Jim Ring, *We Come Unseen: The Untold Story of Britain's Cold War Submariners* (John Murray, 2001), p. 236; *Daily Express*, 24 December 1986.   **151.** Hansard, House of Commons Debates, 2 February 1987, Vol. 109, Cols. 701–83.   **152.** Hansard, House of Commons Debates, 2 February 1987, Vol. 109, Col. 710.   **153.** Ibid.   **154.** Ibid.   **155.** Ibid., Col. 732.   **156.** Ibid.   **157.** Ibid.   **158.** Ibid., Col. 733.   **159.** *Daily Express,* 30 April 1992.

The Prince of Darkness

160. Norman Polmar and Kenneth J. Moore, 'Cold War Strategic ASW', *Undersea Warfare*, Vol. 7, Spring 2005.   **161.** Interview with Toby Frere, 15 January 2013.   **162.** Interview with Martin Macpherson, 24 May 2013.   **163.** Private information.   **164.** MOD Archive.   **165.** Ibid.   **166.** Ibid.   **167.** Ibid.   **168.** Ibid.   **169.** Ibid.   **170.** Ibid.   **171.** Ibid.   **172.** Ibid.   **173.** Ibid.   **174.** Ibid.

The Final Act

175. Stuart Prebble, *Secrets of the Conqueror: The Untold Story of Britain's Most Famous Submarine* (Faber & Faber, 2012).   **176.** *Daily Telegraph*, 20 August 2000.   **177.** The Prime Minister, Margaret Thatcher, took a keen interest in the Royal Navy's submarine operations. Early on in her Premiership, when HMS *Swiftsure* was due to commission into the Royal Navy, the submarine's First Lieutenant, Charles Tibbits, sent a series of Christmas cards to high-profile figures such as the Queen and the Duke of Edinburgh in order to raise the profile of what was then the Royal Navy's newest submarine. The card he sent to the Prime Minister read: 'From one Iron Lady to another'. *Splendid*'s CO, Roger Lane-Nott, was initially worried – 'I thought, I'm not going to last in this Command very long,' he says. But shortly before Christmas a parcel from No. 10 Downing Street arrived addressed to the Captain of HMS *Splendid*. Inside was a card and signed photograph from the Prime Minister. Interview with Roger Lane-Nott, 25 March 2014. **178.** Interview with Mark Stanhope, 4 April 2014. **179.** Interview with Martin Macpherson, 24 May 2013. **180.** MOD Archive. **181.** TNA/ADM/341/32, Anechoic acoustic materials: background of materials; research and development leading to the HMS Churchill anechoic fit and some considerations for future work, 1980 Jan 01–1980 Dec 31 (Retained by Ministry of Defence). **182.** Owen Cote, Jr, *The Third Battle: Innovation in the U.S. Navy's Silent Cold War Struggle with Soviet Submarines* (Naval War College Newport Papers, 2003). **183.** Conley and Woodman, *Cold War Command*, p. 222. **184.** MOD Archive **185.** Ibid. **186.** Ibid. **187.** Ibid.

10 AFTER THE COLD WAR: 1990–TODAY

1. Royal Navy Documentary, *Show of Strength: The Modern State of the Navy* (Fastforward, 1996). **2.** Commander Nick Harrap, 'The Submarine Contribution to Joint Operations: The Role of the SSN in Modern UK Defence Policy', in Martin Edmonds (ed.), *100 Years of the Trade: Royal Navy Submarines Past, Present and Future* (Centre for Defence and International Security Studies, 2001), p. 88.

Uncertainty and Decline

3. House of Commons Defence Select Committee, Sixth Report of Session 1990/1991, 'Royal Navy Submarines' HC 369, 12 June 1991, p. viii. **4.** Alan Clark, *Diaries: In Power 1983–1992* (Phoenix, 2003), 31 January 1990. **5.** House of Commons Defence Select Committee, Sixth Report of Session 1990/1991, 'Royal Navy Submarines' HC 369, 12 June 1991; Michael Pitkeathly and David Wixon, *Submarine Courageous, Cold War Warrior: The Life and Times of a Nuclear Submarine* (The HMS *Courageous* Society, 2010), p. 319. **6.** Pitkeathly and Wixon, *Submarine Courageous*, p. 319. **7.** 'Former submarine commander Captain Stephen Upright is running York's Merchant Adventurers' Hall', *The York Press*, 5 October 2011. **8.** MOD Archive. **9.** Marine Accident Investigation Branch, Report of the Chief Inspector of Marine Accidents into the Collision between the Fishing Vessel Antares and HMS Trenchant with the Loss of Four Lives on 22 November 1990 (HMSO, 1992), pp. 32–4. **10.** House of Commons Defence Select Committee, Sixth Report of Session 1990/1991, 'Royal Navy Submarines' HC 369, 12 June 1991, pp. xiii–xiv. **11.** Ibid., p. xiii. **12.** House of Commons Defence Select Committee, Third Report of Session 1990/1991, 'Options For Change: Royal Navy', HC 226, 27 February 1991, p. xvi. **13.** House of Commons Defence Select

Committee, Sixth Report of Session 1990/1991, 'Royal Navy Submarines', HC 369, 12 June 1991, p. xvi. **14.** Jonathan Powis, 'UK's *Upholder* Class Submarines Go to Canada', *Naval Institute Proceedings* (October 2002). **15.** Ibid. **16.** House of Commons Defence Select Committee, Third Report of Session 1990/1991, 'Options For Change: Royal Navy', HC 226, 27 February 1991, p. xvi. **17.** Ibid. **18.** Ibid. **19.** House of Commons Defence Select Committee, Sixth Report of Session 1990/1991, 'Royal Navy Submarines', HC 369, 12 June 1991. **20.** Ibid., pp. xiv–xv. **21.** Pitkeathly and Wixon, *Submarine Courageous*, p. 318; Dan Conley and Richard Woodman, *Cold War Command: The Dramatic Story of a Nuclear Submariner* (Seaforth Press, 2014), pp. 188–215. **22.** House of Commons Defence Select Committee, Sixth Report of Session 1990/1991, 'Royal Navy Submarines' HC 369, 12 June 1991, pp. xiv–xv. **23.** Powis, 'UK's *Upholder* Class Submarines Go to Canada'. **24.** Interview with Roger Lane-Nott, 25 March 2014. **25.** David Peer, 'Some History of the Upholder-Class Submarines', *Canadian Naval Review* (May 2012).

From Polaris to Trident

26. Interview with Toby Frere, 14 January 2014. **27.** Patrick Middleton, *Admiral Clanky Entertains* (Matador, 2010), p. 168. **28.** House of Commons Defence Select Committee, Sixth Report of Session 1990/1991, 'Royal Navy Submarines', HC 337 (HMSO, 1992), p. 38. **29.** Hansard, House of Commons Debates, 1 February 1996, Vol. 270, Col. 1147. **30.** House of Commons Defence Select Committee, Sixth Report of Session 1990/1991, 'Royal Navy Submarines', HC 369, 12 June 1991, pp. 60–62. **31.** Interview with James Taylor, 10 April 2014. **32.** Interview with Roger Lane-Nott, 25 March 2014; interview with Lord Boyce, 8 April 2014. **33.** Hansard, House of Commons Debates, 20 April 1995, Vol. 258, Col. 224. **34.** John Major's speech at ceremony marking the end of Polaris, 28 August 1996. **35.** BBC Radio 4, *Reflections*, John Major, 13 August 2014. **36.** Interview with Doug Littlejohns, 20 November 2013. **37.** Roger Lane-Nott, 'Submarine Intelligence and the Cold War', in Michael R. Fitzgerald and Allen Packwood (eds.), *Out of the Cold: The Cold War and Its Legacy* (Bloomsbury Academic, 2013), pp. 113–16. **38.** Interview with James Perowne, 18 March 2014. **39.** Interview with Mark Stanhope, 4 April 2014. **40.** Interview with James Taylor, 10 April 2014. **41.** Correspondence with Richard Sharpe, 22 November 2013. **42.** Interview with Mark Stanhope, 4 April 2014. **43.** Hansard, House of Commons Debates, 14 January 1992, Vol. 201, Col. 818. **44.** *Independent*, 'Sub Goes Down in Russian History: Hard-Hit Navies Compare Notes', 3 August 1993. **45.** Interview with Paul Branscombe, 1 April 2014. **46.** Ibid. **47.** Interview with Roger Lane-Nott, 25 March 2014 **48.** Interview with Paul Branscombe, 1 April 2014. **49.** Interview with Roger Lane-Nott, 25 March 2014. **50.** *Independent*, 'Sub Goes Down in Russian History: Hard-Hit Navies Compare Notes', 3 August 1993. **51.** Interview with Paul Branscombe, 1 April 2014. **52.** Interview with Roger Lane-Nott,

25 March 2014. **53.** Interview with Paul Branscombe, 1 April 2014. **54.** 'Trenchant hosts own talk show', *Navy News*, July 1994.

Submarines of the Former Soviet Union

55. 'Russian Sub's Sail Damaged in Collision', *Washington Times*, 27 February 1992; Sherry Sontag and Christopher Drew, *Blind Man's Bluff: The Untold Story of American Submarine Espionage* (Random House, 1999), p. 266. **56.** Commander Submarine Force, US Atlantic Fleet, to Judge Advocate General, Investigation into the Collision between USS Grayling (SSN 646) and a Russian Submarine That Occurred in the Barents Sea on 20 March 1993, 9 June 1993: http://www.jag.navy.mil/library/investigations/uss%20grayling%2020% 20mar%2093.pdf. **57.** 'U.S. and Russian Subs in Collision in Arctic Ocean Near Murmansk', *New York Times*, 23 March 1993. **58.** 'Clinton Pledge May Alter Sub Surveillance', *Chicago Tribune*, 6 April 1993. **59.** MOD Archive. **60.** Ibid. **61.** Ibid. **62.** Ibid. **63.** Ibid. **64.** *Independent*, 'Britain Stops Pointing Its Missiles at Russia', 3 June 1994. **65.** Hans M. Kristensen, 'Russian SSBN Fleet: Modernizing But Not Sailing Much', Federation of American Scientists Strategic Security Blog, 3 May 2013. **66.** Ministry of Defence, *The Strategic Defence Review: Supporting Essays*, Cmnd 3999 (Her Majesty's Stationery Office, 1998), Essay Five, para 13. **67.** Ministry of Defence, *The Strategic Defence Review*, Cmnd 3999 (Her Majesty's Stationery Office, 1998), paras 66–7. **68.** Ibid., para 64. **69.** Ibid., para 70. **70.** Ibid., para 68. **71.** Ministry of Defence, *The Strategic Defence Review: Supporting Essays*, Cmnd 3999 (Her Majesty's Stationery Office, 1998), Essay Five, para 12. **72.** Rear Admiral Rob Stevens, 'Introduction', in Edmonds (ed.), *100 Years of the Trade*, p. vi. **73.** Hansard, House of Commons Debates, 15 January 2001, Vol. 361, Col. 48. **74.** Hansard, House of Commons Debates, 24 October 2000, Vol. 355, Col. 127. **75.** Hansard, House of Commons Debates, 1 November 2000, Vol. 355, Col. 724. **76.** Hansard, House of Commons Debates, 5 February 2001, Vol. 362, Col. 358–9.

Power Projection

77. Harrap, 'The Submarine Contribution to Joint Operations'. **78.** Ibid., p. 87. **79.** Charles Brooking, Ann Hodgetts and Paul Hogan, *Safety Management of the Introduction of the Tomahawk Land Attack Missile into Royal Naval Service* (Frazer–Nash Consultancy, 2010). **80.** Harrap, 'The Submarine Contribution to Joint Operations', p. 87. **81.** House of Commons Defence Select Committee, Fourteenth Report of Session 1999–2000, 'Lessons of Kosovo', HC 347, 24 October 2000, col. 154; *Guardian*, 'Briefing: General Sir Charles Guthrie, Chief of the Defence Staff', 25 March 1999. **82.** House of Commons Defence Select Committee, Fourteenth Report of Session 1999–2000, 'Lessons of Kosovo', HC 347, 24 October 2000, col. 155. **83.** Harrap,

'The Submarine Contribution to Joint Operations', p. 87.   **84.** House of Commons Defence Select Committee, Fourteenth Report of Session 1999–2000, 'Lessons of Kosovo', HC 347, 24 October 2000.   **85.** Harrap, 'The Submarine Contribution to Joint Operations', p. 87.   **86.** Peter Almond, 'Son of a Sub Gives SBS Frogmen New Weapon', *Sunday Times*, January 2004.

Seven Deadly Virtues

87. Stevens, 'Introduction', in Edmonds (ed.), *100 Years of the Trade*, p. xiii.   **88.** Harrap, 'The Submarine Contribution to Joint Operations', pp. 84–6.   **89.** Ibid., p. 86.   **90.** See Stephen Bridgman, *My Bloody Efforts: Life as a Rating in the Modern Royal Navy* (Authorhouse, 2013), p. 533.   **91.** Stevens, 'Introduction', in Edmonds (ed.), *100 Years of the Trade*, p. xiii.   **92.** Ibid.   **93.** Memorandum from the Ministry of Defence (May 2003), House of Commons Defence Select Committee, Eighth Report of Session 2002–03, 'Defence Procurement', HC 694, 23 July 2003.   **94.** Ibid.   **95.** Ibid.

East of Suez

96. 'Former Submarine Commander Captain Stephen Upright is Running York's Merchant Adventurers' Hall', *The York Press*, 5 October 2011.   **97.** 'Nuclear sub shows the flag in Gulf: Royal Navy counters a threat from Iran', *Independent*, 1 April 1993.   **98.** 'Navy joins major Asian exercise', *Independent*, 15 April 1997.   **99.** 'All change for submariners', *Navy News*, August 1998.   **100.** '"Absolute focus" as we launched missiles at Iraq', *Plymouth Herald*, 28 March 2013.   **101.** Ibid.   **102.** *Daily Mirror*, 'Gaddafi Dead: The Incredible Story of the British Nuclear Submarine That Secretly Played a Part in Tyrant's Downfall', *Daily Mirror*, 24 October 2011.   **103.** MOD Press Release, 'Submariners' "Pride" in Libya Operations Medal', 13 December 2012.   **104.** MOD Press Release, 'Royal Navy submarine home from Libyan operations', 4 April 2011.   **105.** MOD Press Release, 'Royal Navy fires cruise missiles at key Libyan targets', 20 March 2011.   **106.** MOD Press Release, 'Submariners' "Pride" in Libya Operations Medal', 13 December 2012.   **107.** MOD Press Release, '*Triumph* home from striking another blow', 20 June 2011.   **108.** 'Horror on board Plymouth nuclear submarine as crew battles to survive', *Plymouth Herald*, 4 June 2014.   **109.** Ibid.   **110.** Ibid.   **111.** Ibid.

Overstretch?

112. Hansard, House of Commons Debates, 21 July 1998, Vol. 317, Col. 201W.   **113.** Ministry of Defence, *Delivering Security in a Changing World: Future Capabilities*, Cmnd 6269 (Her Majesty's Stationery Office, 2004), paras. 2, 5, 7.   **114.** Private information.   **115.** 'HMS *Triumph*: Life on board a

Royal Navy submarine', BBC News Online, 30 July 2012. **116.** Defence
Nuclear Safety Regulator, *Annual Report 2012/2013* (Ministry of Defence,
2013); 'Ageing nuclear submarines could put sailors and public at risk, report
warns', *Guardian*, 4 August 2013. **117.** Ibid. **118.** 'No British submarines to
patrol Falkland Islands', *Sunday Express*, 10 March 2013. **119.** 'England v
Argentina . . . we bring on the sub, Navy sends deadly show of strength', *Sun*,
20 May 2012. **120.** 'Submariner tells of battle to save HMS *Tireless* crewmates engulfed in flames', *Guardian*, 12 February 2009. **121.** Ministry of
Defence, 'Board of Inquiry report into the loss of two Royal Navy Submariners
on board HMS *Tireless* in 2007', 12 June 2008, para 38 and footnote 65. **122.**
'Submariner tells of battle to save HMS *Tireless* crewmates engulfed in flames',
Guardian, 12 February 2009. **123.** Ibid. **124.** 'Parallel Parking in the Arctic
Circle', *The New York Times*, 29 March 2014. **125.** Private information.

11 AND THE RUSSIANS CAME TOO: TODAY AND THE FUTURE

1. Sir Kevin Tebbit speaking at a Royal United Services Institute Seminar on
'Cabinets and the Bomb', 2 September 2009.

The 'Astute' Class

2. (Sea) (ST(S) 7027). **3.** House of Commons Defence Select Committee, Sixth
Report of Session 1990/1991, 'Royal Navy Submarines', HC 369, 12 June 1991, p.
xviii. **4.** John F. Schank, Frank W. Lacroix, Robert E Murphy, et al., *Learning
From Experience*, Vol. III: *Lessons from the United Kingdom's Astute Submarine
Program* (Rand Corporation, 2011), p. 20. **5.** Ibid., pp. 13–14. **6.** Ibid.,
pp. 35–7. **7.** Ibid., p. 40. **8.** 'We're learning from Astute submarine flaws, admiral
promises', *Guardian*, 26 December 2012. **9.** Schank, Lacroix, Murphy, et al.,
Learning From Experience, Vol. III, pp. 43–4. **10.** Ibid., p. 45. **11.** Ibid., p.
46. **12.** Ibid., pp. 45–7. **13.** 'We're learning from Astute submarine flaws, admiral
promises', *Guardian*, 26 December 2012. **14.** Ibid. **15.** Ministry of Defence,
'Report of the Service Inquiry into the Grounding of HMS *Astute* on 22 October
2010', 1 January 2011. **16.** Ibid. **17.** 'Navy sailor gets life sentence for deadly gun
rampage', *Guardian*, 19 September 2011. **18.** 'Trail-blazing U.K. Attack Sub
Proves Itself in the U.S.', *Defence News*, 5 December 2011. **19.** 'Slow, leaky, rusty:
Britain's £10bn submarine beset by design flaws', *Guardian*, 16 November
2012. **20.** 'Britain's nuclear hunter-killer submarines were doomed from the start',
Guardian, 15 November 2012. **21.** Rear Admiral Simon Lister, Letter to the
Guardian, 16 November 2012. **22.** MOD Defence in the Media Brief, 16 Novem-
ber 2012. **23.** Conversation with Rear Admiral Simon Lister on board HMS
Astute, 15 August 2011. **24.** *Sunday Times*, 'Sub's Commando Pod to Attack
Pirates', 2 December 2012. **25.** 'Trail-blazing U.K. Attack Sub Proves Itself in
the U.S.', *Defence News*, 5 December 2011. **26.** Conversation with Ian Brecken-

ridge on board HMS *Astute*, 15 August 2011. **27.** Conversation with Lieutenant Owen Rimmer on board HMS *Astute,* 15 August 2011. **28.** Conversation with Rear Admiral Simon Lister on board HMS *Astute,* 15 August 2011. **29.** Conversation with Lieutenant Commander Rob Tantam on board HMS *Astute,* 15 August 2011. **30.** Conversation with Rear Admiral Simon Lister, on board HMS *Astute,* 15 August 2011. **31.** Ibid. **32.** 'Trail-blazing U.K. Attack Sub Proves Itself in the U.S.', *Defence News,* 5 December 2011. **33.** Conversations with the crew of HMS *Astute,* 15 August 2011. **34.** Ibid. **35.** Ibid. **36.** Ibid. **37.** Conversation with Rear Admiral Simon Lister on board HMS *Astute,* 15 August 2011. **38.** Ibid. **39.** Ibid. **40.** 'Trail-blazing U.K. Attack Sub Proves Itself in the U.S.', *Defence News,* 5 December 2011. **41.** Conversation with Simon Lister on board HMS *Astute,* 15 August 2011. **42.** 'Astute grapples with America's newest submarine in a test of the "best of the best"', *Navy News,* 1 February 2012. **43.** Ibid. **44.** 'Awesome Astute "surpassed every expectation" on her toughest test yet', *Navy News,* 1 March 2012. **45.** 'Navy "will not have enough submarines to protect UK"', *Daily Telegraph,* 16 November 2011; private information.

Current and Future Threats

46. 'Russian subs stalk Trident in echo of Cold War', *Daily Telegraph,* 27 August 2010. **47.** Ibid. **48.** Private information. **49.** 'Nuclear submariner tried to pass secrets to Russians to "hurt" Royal Navy', *Daily Telegraph,* 12 December 2012. **50.** Air Chief Marshal Sir Stephen Dalton, House of Commons Defence Select Committee, Sixth Report of Session 2010–2012, 'The Strategic Defence and Security Review and the National Security Strategy', HC 761, 3 August 2011, Evidence Taken before the Defence Committee, Q241, 11 May 2011. **51.** Air Chief Marshal Lord Stirrup, ibid., Q270, 18 May 2011. **52.** 'Submarine patrols up 50 percent over last year', *Barents Observer,* 14 April 2015. **53.** 'MOD forced to ask US for help in tracking "Russian Submarine"', *Daily Telegraph,* 9 January 2015. **54.** 'Did Russian submarine nearly drag Scottish fishing trawler to a watery grave?', *Daily Telegraph,* 21 March 2015. **55.** Much of the information in the following pages is taken from recent editions of *Jane's Fighting Ships.* See Stephen Saunders, *IHS Jane's Fighting Ships, 2013/2014* (IHS Jane's, 2013). **56.** Freedom of Information Release, JIC(85)7, Soviet Naval Policy, Report by the Joint Intelligence Committee, 19 July 1985. **57.** Private information. **58.** 'Bulava SLBM launch investigation looks towards design faults', *IHS Jane's Defence Weekly,* 13 October 2013. **59.** 'Nuclear subs construction hits post-Soviet peak', *Barents Observer,* 1 July 2014. **60.** Ibid. **61.** Stephen Saunders, *IHS Jane's Fighting Ships, 2012* (IHS Jane's, 2012), pp. 669–70. **62.** Sherry Sontag and Christopher Drew, *Blind Man's Bluff: The Untold Story of American Submarine Espionage* (Random House, 1999), pp. 231–59. **63.** 'This is what the Internet actually looks like: The undersea cables wiring the Earth', CNN Online, 5 March 2014. **64.** 'Top Secret Nuclear Sub Used to Prove North Pole Claim', *Barents Observer,* 29 October 2012. **65.**

'Another Super-Secret Sub in the Pipeline', *Barents Observer*, 14 December 2012. **66.** Private information. **67.** Private information. **68.** Defence Nuclear Safety Regulator, *Annual Report 2012/2013* (Ministry of Defence, 2013); 'Ageing nuclear submarines could put sailors and public at risk, report warns', *Guardian*, 4 August 2013. **69.** 'Navy "running out of sailors to man submarines"', *Daily Telegraph*, 20 August 2012. **70.** Private information.

And the Russians Came Too

71. Though Hennessy did not see the Trident missile burst from the Atlantic off Florida as the test was delayed by a week after an electric storm thwarted it on the appointed day. **72.** 'Navy Detects Russian Sub off U.S. East Coast', CNN Online, 6 November 2012. **73.** Mitch was the Weapons Engineering Officer on board HMS *Vanguard* when Richard Knight and Hennessy recorded *The Human Button* for BBC Radio 4 in 2008. He was the last voice in the documentary, saying: 'One away' as the second movement of Schubert's 'Death and the Maiden' rose up behind the final stage of the firing chain (in the studio that is; not on *Vanguard*). **74.** Conversation with David Young, 10 November 2012; in December 1962 Harold Macmillan commissioned a study of what the UK could manage alone if the United States pulled the plug on the Nassau Agreement he negotiated with John F. Kennedy to provide the UK with Polaris missiles, after his administration had announced the scrapping of the Skybolt stand-off missiles which the British Government was relying upon to prolong the effectiveness of the Royal Air Force's V-bombers; see especially TNA/PREM/11/4148, Amery to Thorneycroft, 15 January 1963; Thorneycroft to Amery, 28 January 1963, reproduced in Peter Hennessy, *Cabinets and the Bomb* (Oxford University Press, 2007), pp. 163–5. **75.** Conversation with Sir Kevin Tebbit, 11 November 2012.

Successor

76. Downing Street Press Briefing, afternoon of 4 December 2006, 'Press Briefing from the Prime Minister's Official Spokesman on Trident'; 'The Future of the United Kingdom's Nuclear Deterrent', Cmnd 6994 (Her Majesty's Stationery Office, 2006); Hansard, House of Commons Debates, 4 December 2006, Vol. 454, Col. 22. **77.** *The Future of the United Kingdom's Nuclear Deterrent*, Cmnd 6994 (Her Majesty's Stationery Office, 2006), p. 7. **78.** Ibid., p. 5. **79.** The letters are dated 7 December 2006 and reproduced in Hennessy, *Cabinets and the Bomb*, pp. 333–7. **80.** 56 Liberal Democrat, one Independent, 2 Social Democrat, 6 Scottish National Party, one Ulster Unionist Party and 3 Plaid Cymru MPs voted against Trident. **81.** Hansard, House of Commons, Debates, 14 March 2007, Vol. 458, Cols. 298–407. **82.** See the letter of 14 March 2007 from the Foreign Secretary, Margaret Beckett, and Defence Secretary, Des Browne, to Dr Adam Whitehead, MP, reproduced in Hennessy, *Cabinets and the Bomb*, pp. 339–41. **83.** For details of these, and extracts from the Cabinet Committee records, see Hennessy, *Cabinets and the Bomb*, pp. 33–59.   **84.** Tony Blair, *A Journey* (Hutchinson, 2010), pp. 635–6.   **85.** Private information.   **86.** Private information.   **87.** Quoted in Peter Hennessy, *The Secret State: Preparing for the Worst 1945–2010* (Penguin Books, 2010), p. 355.   **88.** Quoted in Matthew Parris, 'House of Lords Library Note, Prospects for Nuclear Disarmament and Strengthening Non-Proliferation' (House of Lords, 15 January 2010). Mr Brown delivered his speech at the United Nations on 24 September 2009.   **89.** Private information.   **90.** Private information.   **91.** See Hennessy, *Cabinets and the Bomb*, p. 35.   **92.** Hansard, House of Commons Written Statement, 19 July 2005, Vol. 53WS, Col. 59WS.   **93.** *Freedom Fairness Responsibility: Our Programme for Government* (Cabinet Office, 2010).   **94.** Private information.   **95.** Private information.   **96.** 'Securing Britain in an Age of Uncertainty: The Strategic Defence and Security Review', Cmnd 7948 (Her Majesty's Stationery Office, 2010).   **97.** We have added some private information to that which was published in *Securing Britain in an Age of Uncertainty*.   **98.** Hansard, House of Commons Debates, 18 May 2011, Vol. 528, Col. 355.   **99.** Private information.   **100.** 'PWR-3: A new nuclear plant for the UK's Successor Submarines', *Warship Technology*, January 2013.   **101.** *The United Kingdom's Future Nuclear Deterrent: The Submarine Initial Gate Parliamentary Report* (Ministry of Defence, 2011).   **102.** Ibid., p. 5.   **103.** Discussion at MOD Abbey Wood, Wednesday, 10 August 2011.   **104.** Ronald O'Rourke, 'Navy Ohio Replacement (SSBN[X]) Ballistic Missile Submarine Program: Background and Issues for Congress', *Congressional Research Service*, 25 September 2013.   **105.** Sam LaGrone and Richard Scott, 'Strategic Assets: Deterrent Plans Confront Cost Challenges,' *Jane's Navy International*, December 2011: 17 and 18.   **106.** 'PWR-3: A new nuclear plant for the UK's Successor Submarines', *Warship Technology*, January 2013.   **107.** Defence Nuclear Safety Regulator, *Annual Report 2012/2013* (Ministry of Defence, 2013); 'Ageing nuclear submarines could put sailors and public at risk, report warns', *Guardian*, 4 August 2013.   **108.** 'Hammond challenged over two year delay in revealing nuclear submarine fault', *Daily Telegraph*, 6 March 2014.   **109.** By international agreement underground testing, which in the past has been fundamental to the process used for assuring warhead designs, ended in 1991. In 1998, the British Government ratified the Comprehensive Nuclear Test Ban Treaty, the international agreement prohibiting tests of nuclear weapons. In order to explore how ageing nuclear weapons work, scientists, not only in Britain but worldwide, have been required to develop computer models to simulate nuclear warheads, how they age and how they explode. The United States built the National Ignition Facility (NIF), located in Livermore, California, which uses 192 lasers to create around 4 million joules of energy, while the United Kingdom has built ORION, a smaller relation, 100 times less powerful than NIF.   **110.** Conversation with Tony Johns, 9 June 2011.   **111.** *Defence Budget. Priorities and Choices* (US Department of Defense, January 2012).   **112.** And the contrast does not just consist of outsourcing of the task to private industry. In the summer of 1977, Hennessy, as a *Times* journalist, had undertaken a tour of MOD Research and Development establishments, which included Ships Department Bath and what Hennessy called its

'Deterrent Hut'. This was, like most buildings on the site, a component part of a temporary hospital thrown up rapidly in the late 1930s to take the victims of bombing raids on Bristol lest war came. The security was modern but the building terribly run down. In fact, when it rained there were leaks at one end. The forbearing members of the Royal Corps of Naval Constructors simply moved their blueprints to a dry area when this happened. It was plain they were working on hull designs for a Polaris replacement boat if a future government commissioned one (this was a highly sensitive subject inside the Callaghan Government and the Labour Party). So struck was Hennessy by their zeal that he mentioned it to a nuclear insider. 'Oh yes,' he said, 'their dedication to the deterrent is so keen that if we run out of money, they'll find ways of keeping it going even if they have to fix the Polaris tubes to the Royal Yacht!' **113.** Conversation with Tony Johns, 17 June 2012. **114.** Conversation with Tony Johns, 17 July 2012. **115.** Private information. **116.** Conversation with Tony Johns, 9 June 2011. **117.** 'We're learning from Astute submarine flaws, Admiral promises', *Guardian*, 26 December 2012.

Operation 'Relentless'

118. Private information. **119.** ITN, 'David Cameron touring submarine HMS *Victorious*, 4 April 2013, http://www.itnsource.com/en/shotlist/ITN/2013/04/04/204041310/?s=David%20Cameron%20Trident. **120.** 'We need a nuclear deterrent more than ever', *Telegraph*, 3 April 2013. **121.** Interview with David Cameron, 3 October 2013. **122.** Ibid. **123.** MOD Press Release, 'NATO Praises Royal Navy's Dedication to Delivering Security', 5 March 2013. **124.** 'Yes-vote Scotland would be stuck with Trident "until 2028"', *Daily Telegraph*, 14 August 2014. **125.** 'Vernon Coaker to visit yards building Trident's replacement submarines', *Guardian*, 16 October 2013. **126.** Ibid. **127.** 'Trident row: Cameron defends Fallon's attack on Miliband', *Guardian*, 9 April 2015. **128.** Ibid. **129.** The SNP won 56 seats, Liberal Democrat 8, DUP 8, Others 15. **130.** Hansard, House of Commons Debates, 8 June 2015, Vol. 596, Col. 886. **131.** 'Unions and MPs pour scorn on scrapping Trident', *The Times*, 14 September 2015. **132.** Ibid. **133.** 'Trident: Jeremy Corbyn hopes to alter Labour's stance on nuclear weapons by stripping shadow Cabinet of power', *The Independent*, 9 January 2016. **134.** Jeremy Corbyn Comments on BBC Radio 4 'Today Programme', 30 September 2015 **135.** 'Jeremy Corbyn faces shadow cabinet mutiny over Trident', *The Telegraph*, 30 September 2015 **136.** 'Jeremy Corbyn row after "Id not fire nuclear weapons" comment', *BBC News*, 30 September 2015 **137.** 'Corbyn to complain to MoD about army chief's "political interference"', *The Guardian*, 9 November 2015 **138.** 'Trident Sceptic Thornberry Has "Open Mind"', *Sky News*, 15 January 2016 **139.** Emily Thornberry comments on BBC Radio 4 'Today Programme', 8 February 2016 **140.** Lord West comments to Nick Robinson on BBC Radio 4 'Today Programme', 8 February 2016 **141.** Hansard, House of Lords Debates, 10 February 2016, Vol. Col. 2233. **142.** John Hutton and George Robertson,

'Labour's Trident debate needs to be based on facts', *The Guardian*, 22 February 2016   **143.** 'National Security Strategy and Strategic Defence and Security Review 2015', 20 November 2015   **144.** 'National Security Strategy and Strategic Defence and Security Review 2015', 20 November 2015   **145.** 'National Security Strategy and Strategic Defence and Security Review 2015', 20 November 2015   **146.** Private information   **147.** Marcus Tullius Cicero, *De Legibus*, Book III.

Appendix

Material drawn from a number of sources, including Paul Akermann, *The Encyclopaedia of British Submarines 1901–1955* (Periscope Publishing, 2002); Norman Polmar and Kenneth J. More, *Cold War Submarines: The Design and Construction of U.S. and Soviet Submarines, 1945–2001* (Potomac Books, 2003); Stephen Saunders, *IHS Jane's Fighting Ships, 2012–2013* (IHS Jane's, 2012) and internet sources.

Acknowledgements

The idea of this book was conceived on a perfect early autumn evening in late September 2010 somewhere in the Western Highlands of Scotland during a conversation between Rear Admiral Simon Lister, as he then was, Director Submarines Royal Navy, and Peter Hennessy. They were on the way in a car hired at Inverness Airport to HMS *Vulcan* near Thurso. The plan was to take a scenic route through Lairg. But Hennessy's gifts as a map reader found them instead on the road to Ullapool and somewhere between Dingwall and an exquisite fishing hotel in Scourie, where dinner was just still being served. There a history of the Royal Navy Submarine Service since 1945 commended itself to their joint imaginations. Crucially, later it appealed as well to the mind of the then First Sea Lord, Admiral Sir Mark Stanhope, himself a Submariner. Jinks came on board after completing a PhD on the procurement of the UK Polaris Programme, eager to continue exploring a relatively unexamined area of British history.

A crew-sized collection of gratitude has accumulated since that journey on 30 September 2010. Alongside Vice Admiral Simon Lister, Vice Admiral Ian Corder, the UK Military Representative to NATO and the European Union and Rear Admiral Matt Parr, the Royal Navy's Commander (Operations), NATO's Submarine Force Commander and the head of the Royal Navy's Submarine Service, both opened many doors and offered considerable support, guidance and encouragement. Captain David Pollock, Captain Paul Dunn and Commander Andy Bower, all took time out of their busy day jobs while working in the Naval Staff in London to act as our day-to-day point of contact throughout the project. Special thanks must go to Commander Bower who saw the manuscript through its clearance process with sympathy, firmness and forbearance.

None of this would have been possible without the staff at the Naval Historical Branch, located at the HM Naval Base, Portsmouth. Special thanks to Jock Gardner, the Historian and Deputy Head of the Branch who spent many hours patiently answering questions, facilitating access to material and reading through drafts of the manuscript. Thanks also to Meyrick Young, Clerical Officer, Defence Business Services, Ministry of Defence and George Malcolmson, the archivist at the Royal Navy Submarine Museum in Gosport for providing access to additional source material. We would also like to thank the HMS *Courageous* Association, in particular David Wixon and Michael Pitkeathly for taking us on a tour of the decommissioned *Courageous* in Devonport.

A considerable number of serving and retired Royal Navy and MOD personnel have helped with our research. We would like to thank Admiral of the Fleet The Lord Boyce, Admiral Sir Peter Herbert, Admiral Sir Mark Stanhope, Admiral Sir John Woodward, Vice Admiral Sir Tim McClement, Vice Admiral Sir Toby Frere, Rear Admiral Mark Beverstock, Rear Admiral John Gower, Rear Admiral Steve Lloyd, Rear Admiral Richard Heaslip, Rear Admiral Roger Lane-Nott, Rear Admiral Henry Parker, Rear Admiral James Perowne, Commodore Paul Branscombe, Commodore John Corderoy, Commodore Toby Elliot, Commodore Steve Garrett, Commodore Paul Halton, Commodore David Jarvis, Commodore Doug Littlejohns, Captain Phil Buckley, Captain Paul Burke, Captain Paul Dailey, Captain Mike Davis-Marks, Captain Chris Groves, Captain James Hayes, Captain Paul Jessop, Captain James Taylor, Captain Chris Ward, Captain John Stanley-Whyte, Commander John Aitken, Commander Iain Breckenridge, Commander Dan Clark, Commander Hywel Griffiths, Commander Alan Kennedy, Commander Irvine Lindsay, Commander Mark Lister, Commander John Livsey, Commander Ryan Ramsey, Commander Guy Warner, Lieutenant Commander Andy Johnson, Lieutenant Commander David O'Connor, Andy Mackinder and Heather Hammond.

Three retired submariners in particular, deserve special thanks: Rear Admiral John Hervey, Commodore Martin Macpherson and Captain Richard Sharpe, all helped us with our research and carefully proof read the final manuscript at very short notice. Their contributions have made the book immeasurably better than it would otherwise have been. All errors are ours.

We would also like to thank the crews of HMS *Astute*, HMS *Tireless*, HMS *Trenchant*, HMS *Triumph*, HMS *Talent*, HMS *Vanguard*, HMS *Vigilant*, HMS *Victorious* and HMS *Vengeance* for hosting us on board during our visits and answering our many questions, as well as the Perisher students for allowing us to witness what has to be one of the most demanding training courses in the world.

Many of the companies involved in the UK submarine enterprise have been kind enough to open their doors to help with our research. BAE Systems provided three years sponsorship and financial support for Jinks' PhD Thesis on the UK Polaris programme, much of which appears in this book. We would like to thank Gavin Ireland, who spent a considerable amount of time supporting Jinks' research, Dick Olver, the Chairman of BAE Systems, John Hudson, Managing Director, BAE Systems Maritime and Tony Johns, Managing Director, Maritime – Submarines. At Babcock International Group we would like to thank John Gardner, Group Head of Government Relations, for his considerable assistance over the years, as well as Gavin Leckie, the Submarine Programme Director. At Rolls Royce, we would like to thank Jason Smith and Tony Fletcher. We would also like to thank the scientists and staff at the Atomic Weapons Establishment, Aldermaston.

At Penguin, we would like to thank our editor Stuart Proffitt for his meticulous editing, constant wisdom and friendship, as well as Ben Sinyor, Donald Futers, Richard Duguid, Rebecca Lee, Ilaria Rovera, Rebecca Moldenhauer, Penelope Vogler, Rosie Glaisher, our copyeditor Mark Handsley and Dave Cradduck for the index. Professor Eric Grove also read through a copy of the manuscript and made very helpful suggestions. We would also like to thank our agent David Godwin. Alison Firth at the National Museum of the Royal Navy, Jeremy Postle, the Company Photographer at Babcock International and Michael Vallance, the Senior Photographer at BAE Systems in Barrow-in-Furness all helped with photographs.

Additional thanks must also go to what was the Mile End Group at Queen Mary University, particularly its former Director, Dr Jon Davis, now the Director of Partnerships and The Strand Group at Kings College London's Policy Institute.

Finally, we would like to thank our families for their encouragement and support and for tolerating almost four years of talk about submarines. Hennessy would like to thank his wife Enid, daughters Cecily and Polly, sons-in law Mick and Paul and grandsons Joe and Jack who showed real enthusiasm during the project and relished (or so Hennessy thought) the tales he bought back from the deep. Jinks would like to thank Breege, Richard, Annie and Megan for their considerable support over the years.

Index